Inequalities

Selecta of Elliott H. Lieb

Springer
Berlin
Heidelberg
New York
Barcelona
Hong Kong
London
Milan
Paris
Tokyo

ELLIOTT H. LIEB

Inequalities

Selecta of Elliott H. Lieb

Edited by M. Loss and M.B. Ruskai

Springer

Professor Elliott H. Lieb
Jadwin Hall
Departments of Mathematics and Physics
Princeton University
P.O. Box 708
Princeton, New Jersey 08544-0708, USA

Professor Michael Loss
School of Mathematics
Georgia Tech
Atlanta, GA 30332-0160, USA

Professor Mary Beth Ruskai
Department of Mathematics
University of Massachusetts Lowell
Lowell, MA 01854, USA

Library of Congress Cataloging-in-Publication Data

Lieb, Elliott H. Inequalities : selecta of Elliott H. Lieb /
edited by M. Loss and M.B. Ruskai. p. cm.
Includes bibliographical references.
ISBN 3540430210 (acid-free paper)
1. Inequalities (Mathematics) I. Loss, 1954-II. Ruskai, Mary Beth. III. Title.
QA295 .L54 2002 515'.26--dc21 2002021784
QC173.4.T48 L54 2001 539'.1–dc21 2001041096

ISBN 3-540-43021-0 Springer-Verlag Berlin Heidelberg New York

Springer-Verlag Berlin Heidelberg New York
a member of BertelsmannSpringer Science+Business Media GmbH

http://www.springer.de

Printed in Germany

Printed on acid-free paper SPIN 10854134 55/3141/xo - 5 4 3 2 1 0

Preface

Elliott Lieb made seminal contributions to physics and mathematics. The former are partially collected in the volume "The Stability of Matter: From Atoms to Stars" (Selecta of Elliott Lieb), now in its third edition, which contains some of his papers on the structure of matter, such as its stability and the existence of thermodynamic functions. This new volume is a selection of his contributions to analysis, in particular his work on inequalities.

There are several reasons for publishing this collection of his work. Many of Lieb's results have a substantial impact on analysis, such as his work with Brascamp that determined the sharp constant in Young's inequality. Another example is his work with various collaborators on rearrangement inequalities, which is an area now fully recognized as a very valuable part of analysis.

A second important consideration is the long shelf life of Lieb's work. For example, the paper 'Convex Trace Functions and the Wigner-Yanase-Dyson Conjecture' (No II.3 of the present volume) has been cited in a variety of contexts in the nearly 30 years since its publication, most recently in connection with the theory of quantum computing.

Lieb's work is a fortunate exception to the rule that research papers are usually only accessible to experts. The reader will find that a good background in real analysis and linear algebra is sufficient for understanding and even mastering the content of most of his papers. The vitality of his papers springs from the ideas and not from the technical complexities.

Last but certainly not least is the relevance of Lieb's papers to physics. This is demonstrated by the now famous Lieb-Thirring inequalities. A nice example is also his work with Brascamp about log-concave functions and their application to the one dimensional quantum Wigner crystal. It is satisfying to see how, starting with the simple idea of convexity, a complete understanding of a nontrivial piece of physics emerges.

The papers are grouped around physical and mathematical ideas. We have added commentaries that serve to introduce the papers. Some of these explain aspects of the history of the problem or of the paper, and others point towards further developments. We have chosen to comment mostly on those papers in areas where we have some research expertise. We hope that the reader enjoys this collection of Lieb's papers as much as we do.

Our thanks go to all the publishers who generously supplied the papers free of charge, to the staff of Springer Verlag, and especially to Wolf Beiglböck for his support and his patience.

Atlanta and Lowell, 2002 *Michael Loss and Mary Beth Ruskai*

Contents

Commentaries

Part I. Inequalities Related to Statistical Mechanics and Condensed Matter

Part II. Matrix Inequalities and Combinatorics

Part III. Inequalities Related to the Stability of Matter

Part IV. Coherent States

Part V. Brunn-Minkowski Inequality and Rearrangements

Part VI. General Analysis

Part VII. Inequalities Related to Harmonic Maps

Commentaries

The subject of 'inequalities' was first systematically established by Hardy, Littlewood and Pólya in their book of the same name. The goal, loosely speaking, is to search for an inequality between algebraic or analytic expressions of certain variables that becomes an equality in certain (possibly limiting) cases. The reader may think of Hölder's inequality as an example, but also such things as the dependence upon its shape of the lowest frequency of a drum.

The usefulness and importance of this area can hardly be overstated. Many of the results have entered the 'mathematical subconcious' while others have to be considered as deep contributions to mathematical culture such as the isoperimetric inequality that goes back to ancient times.

What makes this area especially attractive is that it cuts across boundaries of mathematical and scientific disciplines. A case in point is Sobolev's inequality, an indispensable tool in partial differential equations. When looked at from a different angle it is very closely related to the Yamabe problem in differential geometry and, finally, it is the deepest formulation of the uncertainty principle in quantum mechanics.

In physical applications inequalities occur naturally through the principle of least energy. The ultimate goal is to find the lowest value of the energy and to describe the state of the system at that energy, usually as a solution to some partial differential equation. Famous examples are soap bubbles, elasticity theory and the Schrödinger atom. The unifying mathematical theory is known as the calculus of variations, which gave the impetus to much of what is now known as Analysis and Functional Analysis.

The physicist's computational approach to such models is often hampered by the fact that they are not exactly solvable and, as often happens in such cases, they resort to uncontrolled approximations. Inequalities that establish bounds on physical quantities are very useful in such circumstances. One example that illustrates this point nicely is the paper of G. Baym and A.J. Leggett [BL] in which they give rigorous upper bounds to barrier penetration probabilities in many body quantum systems and are able to rule out cold fusion. They use a number of inequalities for Schrödinger operators and well established experimental bounds. Another example that comes to mind is the use of unitarity in estimating scattering cross sections. Often some cross section can be measured and some not. With the help of unitarity one can derive bounds on the latter. One of the most famous inequalities in physics is what is now called Bell's inequality. It is an inequality about classical independent events and delineates in a quantitative way the difference between quantum mechanics and classical mechanics. There have been experiments that violate Bell's inequality [Bel] and give some credence to the opinion that quantum mechanical correlations are not explicable without action at a distance.

Although there is no all encompassing theory of inequalities, many of them have common underlying features. One senses this strongly in Lieb's work which, superficially spans a variety of seemingly unrelated problems. The unity of the whole work is provided by his methods of investigation. As an example, many of Lieb's inequalities that concern physical quantities, such as energy and entropy have a common mathematical source, in that they all can be expressed in one way or another as convexity inequalities. Convexity is at the root of his work on entropy, but it also resurfaces again in the form of the Brunn-Minkowsi inequality, applied to the one-dimensional plasma and rearrangements.

A Selecta gives us the chance to trace some of the historical development of Lieb's work. It is therefore natural to start with his work in statistical mechanics and condensed matter and to see how his interests expanded. It is also interesting to see how Lieb acquired the formidable knowledge of mathematics he is known for; how he summons it all in pursuit of a problem and how it grows with every solved one.

Part I
Inequalities Related to Statistical Mechanics and Condensed Matter

One of the major themes in Lieb's work has been condensed matter physics beginning with one-dimensional models in statistical mechanics, and highlighted by his exact solution of the two-dimensional six-vertex model. His papers in this area could easily form another volume of *Selecta*. In the present volume, we include only those papers in which his analysis led directly to results on inequalities.

I.1, I.2: Both papers are joint work with Mattis with whom he eventually wrote the well-known book *Mathematical Physics in One Dimension* [LM]. They contribute to the understanding of the nature of ferromagnetism. For a many electron system, the total spin value is a good quantum number, provided the interactions are not spin dependent. It is a long-standing problem to understand why the ground state for certain systems has large total spin. Its manifestation is a permanent magnetization of the system as can happen, e.g., in iron and hence the appellation 'ferromagnetism'. There is an old idea of Heisenberg that the sign of a certain integral, called the exchange integral, determines whether a system is ferromagnetic or not. In the first paper it is shown that for one-dimensional quantum systems the ground state energy $E(S)$ with total spin S is monotone increasing as a function of S. Hence, the absolute ground state has spin zero and there is never ferromagnetism in one-dimensional systems even though the exchange integral can have either sign, depending on the potential. Similar ideas were then used by Lieb and Mattis for interacting spin systems in the second paper.

I.9: The first paper, **I.1** has a generalization to positive temperature, that is to an average over all eigenfunctions of the Schrödinger equation, not just the lowest. This was finally achieved 28 years later in a work with Aizenman **I.9** using the Wiener integral representation of the positive temperature state. The use of Wiener integrals to derive rigorous bounds on quantum-mechanical quantities also appears in **III.7**.

I.3–I.6: We describe the history leading up to these papers on quantum mechanical entropy by quoting from the review of Wehrl [We, p. 249],

"Strong subadditivity was known for many years in information theory but not generally called that. For statistical mechanics, at least, it was Robinson and Ruelle (1967) [RR] who coined the word and who first realized that it was important. They proved it in the classical case. Then Lanford and Robinson (1968) [LR] conjectured it in quantum mechanics"

and the notes in Bratteli and Robinson (vol. II, p. 435)[BR],

"Despite attempts by many authors, this conjecture remained unverified for 6 years until Lieb and Ruskai (1973) gave a proof based upon a result of Lieb's" [in **II.3**].

For the statement of the inequality, consider three Hilbert spaces $\mathscr{H}_1$, $\mathscr{H}_2$, $\mathscr{H}_3$ and their tensor products. These spaces correspond to subsystems of a composite system $\mathscr{H}_{123} = \mathscr{H}_1 \otimes \mathscr{H}_2 \otimes \mathscr{H}_3$. Let ϱ_{123} be a density matrix on $\mathscr{H}_{123}$, i.e., a positive semi-definite matrix with trace equal to one. By taking partial traces in the obvious fashion density matrices describing the subsystems are obtained, such as ϱ_{12} on $\mathscr{H}_{12}$. For any of these density matrices, the quantum mechanical entropy can be defined, following von Neumann, as

$$S(\varrho) = -\mathrm{Tr}\varrho \ln \varrho \ .$$

The *strong subadditivity* (SSA) of the quantum entropy can then be stated as

$$S(\varrho_{123}) + S(\varrho_2) \leq S(\varrho_{12}) + S(\varrho_{23}). \tag{1}$$

In **I.3**, which is joint work with Araki, Lieb took a first step and proved the weaker inequality

$$S(\varrho_{123}) \leq S(\varrho_{12}) + S(\varrho_{23}). \tag{2}$$

This sufficed to establish the existence of an infinite volume limit for entropy. In this paper, they also proved the so-called triangle inequality

$$|S(\varrho_1) - S(\varrho_2)| \leq S(\varrho_{12}) \leq S(\varrho_1) + S(\varrho_2)$$

which gives a lower bound and complements the (weak) subadditivity upper bound $S(\varrho_{12}) \leq S(\varrho_1) + S(\varrho_2)$.

Papers **I.4, I.5** with Ruskai contain the proof of SSA. The key was the realization that SSA could be reformulated as a statement about the convexity of the so-called conditional entropy $S(\varrho_{12}) - S(\varrho_1)$ of a two-component system. This reformulation led Lieb to the convexity of a quite different quantity called the Wigner-Yanase-Dyson entropy [WY] and the notion of convex trace function, which will be discussed later (see **II.3**). (Subsequently, it was realized that the convexity of the conditional entropy was equivalent to the more general notion of convexity of relative entropy. Although that term is not used, eq. (3.2) of **I.4** states that the relative entropy is monotone under partial traces.)

Paper **I.6** contains a summary of these results, comparing the differences between classical discrete, classical continuous and quantum entropy. In addition, Lieb considers, and provides counterexamples to, a number of natural generalizations of subadditivity inequalities.

I.7: Simon proved an inequality for the Ising model [Si1] which has been used often to prove exponential decay of correlation functions, among other things. It is a special case of an inequality of Boel and Kasteleyn [BK].

Lieb improved this inequality significantly (it is now often referred to as the Simon-Lieb inequality) with several consequences. These include a more effective method for bounding the phase transition temperature and an algorithm which, in principle, can compute the exact transition temperature to any accuracy (although it is unwieldy in practice). In another direction, Lieb proposed a conjecture which, if true, would extend the results from the Ising model to the plane-rotor model. This conjecture was verified by Rivasseau [Ri].

I.8: Lieb has had a long-standing interest in the Hubbard model, which is an important model in condensed matter physics for describing interacting electrons. With Wu, Lieb showed [LW] that this model is exactly solvable in one-dimension; however, it is much harder in two- and three-dimensions where little is known about its ground state or its behavior at positive temperatures. In the paper under discussion, Lieb proves a result for this model that is analogous to that of **I.1**, namely, that for an even number of electrons and an attractive on-site interaction among them, the ground state has spin zero. This result is independent of the dimension of the system or the structure of the underlying lattice.

If the interaction is repulsive and the underlying lattice is bipartite, then the total spin of the ground state is $S = ||A| - |B||/2$ for the half filled band case. Here $|A|$ and $|B|$ denote the number of vertices in the two components defining the bipartite lattice. One may speak here of itinerant ferromagnetism since the magnetization is created by electrons that are not attached to any specific vertex.

Part II
Inequalities in Algebra and Combinatorics

Lieb's work in statistical mechanics often led to related problems in combinatorics, linear algebra, and maps of operator algebras. Much of this work is of interest in its own right.

It has long been realized that trace-preserving, completely positive maps seem to be the appropriate mathematical structure needed to model noise in quantum communication channels and quantum computers. With recent advances in the development of algorithms that use quantum particles for data encryption, computation and communication, as well as promising experiments toward their implementation, Lieb's work in this area has taken on renewed importance. His work on quantum mechanical entropy, discussed in the previous section has had a particularly strong impact, e.g., Shor [Sh] recently used SSA to prove some important results on channel capacity. In addition, Lieb's later work with Solovej **IV.3** on quantum coherent operators for spin systems and his work with Carlen on Clifford algebras in **II.7, II.9** is likely to have further impact on problems in quantum information theory.

II.1: Schur defined certain functions of $n \times n$ matrices, called "immanants", of which the determinant is a special case. For each irreducible representation λ of the symmetric group S_n, the corresponding immanant of a matrix A is

$$I_\lambda(A) = \frac{1}{m_\lambda} \sum_{P \in S_n} \chi_\lambda(P) A_{1,P1} \cdots A_{n,Pn} ,$$

where m_λ is the degree of the representation and $\chi_\lambda(P)$ is the character of the permutation P. For positive semidefinite matrices Schur proved that the determinant (alternating representation) is the smallest immanant. The permanent corresponds to the identity representation.

In this paper Lieb derives lower bounds for the permanent and conjectures that the permanent is the largest immanant (for positive semidefinite matrices). He goes further and considers generalized immanants in which S_n is replaced by some subgroup of S_n. Schur's proof generalizes to this case, too, i.e., the determinant is the smallest one. Lieb's full conjecture is that the permanent is the largest. In this paper he proves this for a special case of the generalized immanant.

This conjecture has become known as "permanent dominance" and there are many papers devoted to it. For Schur-immanants it has been proved for up to 9×9 matrices and for some special classes of matrices. The permanent is important in several combinatorial and algebraic situations, hence the significance of this conjecture (see [Hai], for example). Only the determinant, however, is invariant under unitary transformations and that is one reason the problem has defied solution up to now.

II.2: Many important coefficients in combinatorics (e.g., the binomial coefficients) are unimodal, meaning that one (or possibly two adjacent) coefficient is the largest and the coefficients drop off monotonically on either side of the maximum.

Harper [Har] showed that the Stirling numbers of the second kind are unimodal. Lieb proved more, namely that these coefficients are log-concave (for Stirling numbers of the first *and* second kind). His technique was to show that the polynomial whose coefficients are Stirling numbers times binomial coefficients times k has only real zeros. Then, by Newton's inequality, the coefficients in the polynomial, when divided by the binomial coefficients, are log concave. This technique has been used many times since. It seems that Lieb was among the the first to employ it. See [St] for a review.

II.3: This paper considers functions that map matrices to complex numbers via the trace of more complex objects involving several matrices. In addition to proving a number of inequalities, it also uncovers surprising relations among them. The heart of the paper is a series of convexity results for functions that are homogeneous and some of them become linear if restricted to commuting matrices. Thus, one could not even have predicted, from examining the special case in which the matrix is a complex number, whether these trace functions would be convex or concave. Proving such results required considerable ingenuity and the use of new techniques. Two of these results merit further discussion.

For positive semidefinite A and B the map

$$A, B \to \mathrm{Tr} A^p K^* B^{1-p} K$$

is jointly concave for any K and for $0 \leq p \leq 1$. Wigner and Yanase [WY] proved this for $p = 1/2$ in connection with a theory about entropy, and briefly discussed the case $p \neq 1/2$, introduced by Dyson (but they did not actually make the explicit concavity conjecture and considered the $p \neq 1/2$ case to be uninteresting for their purposes). The concavity conjecture for $p \neq 1/2$ seems first to have been made by Baumann in his Ph.D. thesis, where he proved it for 2×2 matrices [Bau].

For positive semidefinite A and Hermitian K the map

$$A \to \mathrm{Tr} \exp\{K + \log A\}$$

is concave. This second inequality was an essential ingredient in the original proof of strong subadditivity of entropy. One route to SSA uses the following consequence of this second concavity theorem, which is interesting in its own right,

$$\mathrm{Tr}\, e^{\log R - \log S + \log T} \leq \mathrm{Tr} \int_0^\infty R(S + uI)^{-1} T (S + uI)^{-1} du.$$

This can be regarded as a generalization of the inequality $\mathrm{Tr}\, e^{A+B} \leq \mathrm{Tr}\, e^A e^B$ to three matrices; it is both significant and surprising because the generalization $\mathrm{Tr}\, e^{A+B+C} \leq |\mathrm{Tr}\, e^A e^B e^C|$ is known to be false.

Another proof of SSA, in connection with Minkowski's inequality for matrices, appears in **II.9**.

II.4: The main result of this paper (which is joint with Ruskai) is that for an operator map φ of the type known as "completely positive" the inequality

$$\varphi(A^* A) > \varphi(A^* B)[\varphi(B^* B)]^{-1} \varphi(B^* A)$$

holds. Moreover, it is not necessary that $\varphi(B^* B)$ be invertible, as the proof shows that whenever the left side is finite, the right side can be defined as a suitable limit. In the important special case in which φ is a partial trace, it implies that the map

$$A, K \to K^* A^{-1} K$$

is jointly convex, i.e, the appropriate operator inequality holds.

Completely positive maps play an important role in the theory of operator algebras and arise naturally in the theory of open quantum systems, where the maps are also trace-preserving. Subsequently, Choi [Ch] showed that the requirement of complete positivity can be reduced to two-positivity. Although other proofs have been found, the original argument in **II.4** is extremely simple and the formulation given there seems to have been a breakthrough.

II.5: Lieb introduces complex-valued functions, F, of matrices (which Simon, in his book [Sim] calls "Liebian") satisfying two conditions, namely, that F is positivity preserving. i.e., $B \geq A \geq 0 \Rightarrow F(B) \geq F(A)$, and satisfies the Schwarz-like condition $F(X^*X)F(Y^*Y) \geq |F(X^*Y)|^2$. Examples of such functions are:

immanants, k^{th} elementary symmetric functions of the singular values, and the norm. Additional examples were found by Merris [Me] and others later. Lieb uses the results of **II.4** to show, among other things, that $F(\sum_j X_j^* X_j) F(\sum_j Y_j^* Y_j) \geq |F(\sum_j X_j^* Y_j)|^2$, which is analogous to a log-convexity statement. As a non-obvious special case, he obtains a simple new proof of the inequality (due to Rotfel'd and Seiler-Simon) [Ro][SS]

$$|\det(I + A + B)| \leq \det(I + |A|) \det(I + |B|).$$

II.7: In the early 70's, the same time period as Lieb's work on convex trace functions and entropy, Gross [Gr] developed hypercontractivity estimates and log Sobolev inequalities — tools that proved invaluable in the study of ground states in quantum field theory of bosonic systems. These inequalities are, in fact, special cases of the sharp form of Young's inequality (see **V.3** and **VI.7**). At that time Gross also further developed the non-commutative integration needed for fermionic systems and extended some of his hypercontractivity estimates to this framework. However, these results were not optimal and many open conjectures remained. They were not resolved until 20 years later when Lieb, jointly with Carlen proved optimal hypercontractivity bounds for the fermion oscillator semigroup, and an optimal fermion logarithmic Sobolev inequality. These results, required the following optimal convexity inequality for matrices

$$2^{-2/p} \left(Tr|A + B|^p + Tr|A - B|^p\right)^{2/p} \geq \left(Tr|A|^p\right)^{2/p} + \left(Tr|B|^p\right)^{2/p} .$$

Although this sufficed for the fermionic work, characteristically, Lieb did not stop there but collaborated with Ball and Carlen to prove a host of sharp convexity and smoothness inequalities (see **II.8**). In essence, Ball, Carlen and Lieb extended to the non-commutative C^p ideals almost everything that was known about convexity and smoothness of L^p spaces. Their paper also contains a comprehensive summary of the corresponding L^p inequalities, including new proofs and even a few new inequalities. Some idea of the scope of this work can be seen in the diagram on p. 471 of **II.8**.

II.8: The unit ball in an L^p-space of functions is, of course, convex. One of the important developments in that subject was a quantification of this convexity and also of the smoothness of the balls when $p > 1$, by Clarkson [Cl] and later Hanner [Ha]. The question addressed in this paper is whether these inequalities extend to the trace ideals for matrices, $X \to \{\text{Tr}|X|^p\}^{1/p}$? Hanner's inequality is proved for $1 \leq p \leq 4/3$, for $p = 2$ and for $4 \leq p \leq \infty$. The other cases are still open!

II.9: Minkowski's inequality for functions on the product of two measure spaces states that for $p \geq 1$ and $f \geq 0$, $\left(\int_X \left(\int_Y f(x,y) d\nu\right)^p d\mu\right)^{1/p} \leq \int_Y \left(\int_X f(x,y)^p d\mu\right)^{1/p} d\nu$. The reverse inequality holds for $0 < p \leq 1$. Does this important inequality extend to partial traces of matrices defined on the tensor product of two Hilbert spaces? The answer, proved here jointly with Carlen, is yes.

More interestingly, one can trivially extend Minkowski's inequality to functions of three variables, $f(x, y, z)$ by applying Minkowski's inequality with z fixed, and then integrating over z. The corresponding inequality for matrices defined on the tensor product of three Hilbert spaces is far from trivial. It is proved here for $0 < p \leq 1$ and $p = 2$. Surprisingly, it is shown to be false for $p > 2$. The case $1 < p < 2$ is still open! However, this three Hilbert space theorem for $0 < p \leq 1$ yields one more proof of the strong subadditivity of entropy.

Part III
Inequalities Related to the Stability of Matter

Inequalities related to the stability of matter form an important part of Lieb's work. The constitution of matter, i.e., properties of a system consisting of a large number of electrons interacting with nuclei via electrostatic forces, was one impetus for the development of quantum mechanics. Very early on this theory was successful in explaining the size of the hydrogen atom and of atoms in general. The understanding of its implications for bulk matter is relatively recent and due to Dyson and Lenard [DL]. In the mid-seventies a completely new and much improved version of the Stability of Matter theorem was proved by Lieb and Thirring [LT]. Stability of Matter means that the ground state energy has a lower bound that is proportional to the number of particles involved and, in a nutshell, this is the case because of the Schrödinger equation and the Pauli exclusion principle.

III.1, III.8: The key ingredients of the Lieb-Thirring work on stability of matter are the Lieb-Thirring inequalities. These are generalizations of Sobolev's inequality to orthogonal systems of functions, incorporating in this way the Pauli exclusion principle. These two papers give a detailed account of these results. The key estimate is a bound on the moments of eigenvalues of a Schrödinger operator on $L^2(\mathbb{R}^n)$,

$$-\Delta + V ,$$

where $V(x)$ is a potential and Δ is the Laplace operator. Denote by $\lambda_1 < \lambda_2 \leq \lambda_3 \leq \cdots$ the negative eigenvalues of this operator. The following inequality was proved in **III.1**:

$$\sum_j |\lambda_j|^\gamma \leq C(\gamma) \int V_-(x)^{\gamma+n/2} \mathrm{d}x , \tag{3}$$

where γ is a positive constant whose allowed values, meaning the cases for which $C(\gamma)$ is finite, depends on the dimension n. Here, $V_-(x)$ denotes the negative part of $V(x)$.

The particular case $\gamma = 1$ for $n = 3$ is used in the Lieb-Thirring proof of stability of matter. One of the crucial inequalities is a lower bound on the kinetic energy of N fermions, i.e., the Sobolev-type inequality alluded to before. Consider an N-electron wave function $\psi(x_1, \ldots, x_N)$ normalized by $\int |\psi|^2 \mathrm{d}x_1 \cdots \mathrm{d}x_N = 1$. (We neglect spin here for simplicity.) The Pauli exclusion principle states that any wave function describing fermions must be antisymmetric under interchange of the particle labels. The kinetic energy of such a function is defined to be

$$T(\psi) = \sum_{i=1}^{N} \int_{\mathbb{R}^3} |\nabla_i \psi(x_1, \ldots, x_N)|^2 \mathrm{d}x_1 \cdots \mathrm{d}x_N \ .$$

The one-particle density $\varrho_\psi(x)$ which is a function on $\mathbb{R}^3$ is defined by

$$\varrho_\psi(x) = N \int |\psi(x, x_2 \ldots, x_N)|^2 \mathrm{d}x_2 \cdots \mathrm{d}x_N \ .$$

(Since $|\psi|^2$ is symmetric in its N variables, it does not matter which $N-1$ variables we integrate over.) The following inequality is not hard to deduce from (3) setting $n = 3$ and $\gamma = 1$,

$$T(\psi) \geq c \int_{\mathbb{R}^3} \varrho_\psi(x)^{5/3} \mathrm{d}x \ ,$$

for some explicit constant c. Except for an altered value of the constant c, the right side is precisely the kinetic energy in Thomas-Fermi theory — an approximate theory in which all the physical quantities are expressed in terms of the one particle density ϱ_ψ. Without the antisymmetry (Pauli principle), the best lower bound we would be able to derive is $T(\psi) \geq cN^{-2/3} \int_{\mathbb{R}^3} \varrho_\psi(x)^{5/3} \mathrm{d}x$. An account of this theory and its relation to the stability of matter and thermodynamics can be found in [Li1]. Also mentioned in this paper is the "cheese" theorem which Lieb proved jointly with Lebowitz in [LiLe] and which, according to Mandelbrot [Ma], is a contribution to fractal geometry. The "cheese" theorem is about efficient packing of balls in $\mathbb{R}^n$, whose radii decrease geometrically and which cover $\mathbb{R}^n$ up to a set of measure zero.

III.3: The deepest and, in some sense, most useful form of the Lieb-Thirring inequality is the case $\gamma = 0$ which simply counts the number of eigenvalues of this Schrödinger operator. This bound, which works for all dimension $n \geq 3$ is the celebrated Cwikel-Lieb-Rozenblum bound. It says that the number of negative energy bound states in a Schrödinger operator is measured by the volume of the region in phase space where the energy $p^2 + V(x)$ is negative. This volume is given by $c_n \int V_-(x)^{n/2} \mathrm{d}x$ where c_n is the volume of the n-dimensional unit ball. Three very different proofs were given, one by Rozenblum, one by Cwikel and the one by Lieb, almost at the same time. While Rozenblum's proof [Roz] was first and Cwikel's proof [Cw] is the most general, Lieb's proof furnished the best constant among the three. Further improvements on the constant were given by Lieb in **III.6** and also in [BS]. The sharp constant is not known. All the other relevant Lieb-Thirring bounds (not necessarily with the best constant) can be deduced from this one.

III.4: What makes atomic physics difficult is, besides the kinetic energy, the presence of the Coulomb repulsion among electrons. If we are given a N-electron wave function $\psi(x_1, \ldots, x_N)$ (we neglect the spin), then the Coulomb repulsive energy is given by the expectation value

$$C(\psi) := \sum_{i<j} (\psi, |x_i - x_j|^{-1}\psi) .$$

In the spirit of Thomas-Fermi theory one would like to replace this energy by the repulsive energy of the charge density ϱ_ψ, i.e.,

$$D(\varrho_\psi) := \frac{1}{2}\int \frac{\varrho_\psi(x)\varrho_\psi(y)}{|x-y|}\mathrm{d}x\mathrm{d}y .$$

The difference $C - D$ is called the exchange energy and in this paper, which is jointly with Oxford, an effective bound on the exchange energy was calculated.

In particular it is shown that the inequality

$$C(\psi) - D(\varrho_\psi) \geq -1.68 \int \varrho_\psi(x)^{4/3}\mathrm{d}x \tag{4}$$

holds for all wave functions ψ (irrespective of their symmetry properties). A similar estimate with 1.68 replaced by 8.4 was proved by Lieb earlier [Li2]. Inequality (4) is often used by quantum chemists. It also leads, together with the Lieb-Thirring inequality above, to a lower bound for the true quantum mechanical energy by Thomas-Fermi theory, which, in turn, leads to the celebrated Lieb-Thirring proof of the stability of matter.

Actually, in the original Lieb-Thirring proof of stability of matter [LT], which was much earlier than [Li2], another bound, $C\sqrt{N}\left(\int \varrho^{5/3}\right)^{1/2}$, was used in place of $C\int \varrho^{4/3}$ on the right side of (4).

III.5: This paper deals with general density functional theory rather than a particular theory, such as the Thomas-Fermi theory. Consider a many-body Hamiltonian describing an atom or, more ambitiously a solid, in which the electrons are the only quantum mechanical particles and the nuclei are fixed. Typically, such a Hamiltonian is made up of two types of interactions: single particle interactions such as the interaction of the electrons with the potential V due to the nuclei, and two-body interactions such as the Coulomb repulsion among the electrons. The latter is considered fixed, but the effective one-body potential V will depend upon the particular system under consideration. In a famous 1964 paper, Hohenberg and Kohn [HK] showed that if the one-body potentials for two systems differ by more than a constant, then the single-particle electron densities associated with their respective ground states must also be different. (Since, in general, a given one-particle density can be obtained from many different many-body wave functions, this result was regarded as somewhat surprising.) Thus, knowing the ground state electron density (a function of three space variables) determines the potential V which, in turn, determines the ground state, a function on the high dimensional configuration space.

As a consequence, there must be a universal functional $F(\varrho)$ of the density ϱ that should describe the ground state energy of atoms and maybe their low lying excited states. Despite the obvious difficulty of finding the universal functional $F(\varrho)$, Hohenberg and Kohn's result attracted a good deal of attention and is still being used as the basis for a number of approximation schemes.

In **III.5** Lieb formulated an explicit variational principle for a related functional that does not require knowledge of the mysterious "universal functional" and showed its connection to Hohenberg-Kohn theory and a number of other density functional theories.

Much of this paper may seem rather technical. However, it also establishes the importance of such technicalities for real physical systems, particularly when one wants to use a functional so complicated that one cannot expect every smooth density to arise from a "nice" potential or vice versa.

This important paper attempts to provide a firm mathematical foundation for density functional theory. It proves many results, both fundamental and technical, and raises a number of open question. After first appearing in a collection of papers in honor of Tisza in 1982, it was reprinted (with minor revisions) twice — first in the *Int. J. Quant. Chem.* (the version which appears here) in 1983 and again in a collection of papers on density functional theory in 1985.

III.6: This paper is an unexpected application of Lieb-Thirring inequalities to the computation of the Hausdorff dimension of the attractor for the three dimensional Navier-Stokes equations (modulo the annoying fact that the global solutions of the Navier-Stokes equations are not known to exist). It was realized by Ruelle [Ru] that the Hausdorff dimension can be bounded by sums of eigenvalues of a Schrödinger operator. His estimates used Weyl asymptotics and hence did not give any bounds on the Hausdorff dimension of the attractor for a fluid in a finite volume. It also relied on some unproven conjectures about certain constants. In **III.6**, Lieb implemented Ruelle's idea. Among other results an improvement on the constant of the Cwikel-Lieb-Rozenblum inequality was obtained.

III.7: Consider a Hamiltonian of three nonrelativistic particles interacting with a pair potential of the form $V(x-y)$. The three particles have mass m_1, m_2, m_3 not necessarily the same. Denote the ground state energy by $E(m_1, m_2, m_3)$. This ground state energy can be thought of as the mass of a baryon created by the quarks of mass m_1, m_2, m_3. It was conjectured by particle physicists that the inequality

$$E(m,m,m) + E(m,M,M) \leq 2E(m,m,M) \ ,$$

should hold quite generally for all masses m and M and potentials. Lieb proves this for a certain class of potentials, but also gives some potentials and masses $m > M$ for which this is not true, thus destroying the hope for a potential and mass independent inequality. The proof, characteristically, proceeds via the Trotter formula using the positive definiteness of certain kernels and also works for relativistic forms of the kinetic energies. The condition on the potential V for which the inequality holds appears to be close to optimal.

III.9, III.2: Independent of what has been said in the previous paragraphs, sharp constants in Lieb-Thirring inequalities provide an interesting playing field for analysis. There are a number of conjectures, some of them due to Lieb and Thirring that are open to this day. Already the one-dimensional case ($n = 1$) offers a host of challenging problems. It was relatively recently that Weidl [Wei] was able to establish a Lieb-Thirring inequality in one dimension for $\gamma = 1/2$. Relatively soon after that Lieb, jointly with Hundertmark and Thomas **[III.9]** were able to show that $1/2$ is the sharp constant for this case and that the optimizing potential is a delta function and supports only one bound state. This phenomenon, namely that the optimizing potential should support only one bound state, is believed to hold for all $\gamma < 3/2$ for $n = 1$. It is known that for $\gamma = 3/2$ the classical phase space average also gives the sharp constant. This result, which can be found in **III.1** for the one dimensional case was recently extended by Laptev and Weidl [LaWei] to any dimension $n \geq 1$. In all dimensions and for $\gamma \geq 3/2$ the sharp constant in the Lieb-Thirring inequality is given by the classical constant. Indeed, it is a fairly general argument due to Aizenman and Lieb (see **III.2**) that if the classical constant is sharp for the Lieb-Thirring inequality for some γ it remains the sharp constant for all larger values of γ. Thus the classical constant is sharp for all $\gamma \geq 3/2$.

Part IV
Coherent States

Coherent states, originally discovered by Schrödinger, were fashioned into an incisive tool by Bargmann, Segal and Glauber, which is useful whenever one compares a quantum mechanical model with its classical counterpart.

IV.1: Here Lieb uses the less familiar Bloch coherent states to bound quantum mechanical partition functions Z in terms of classical ones. More precisely, with $f = -\beta^{-1} \ln Z$ denoting the free energy, he proves that

$$f^{\text{classical}}(J, \beta) \geq f^{\text{quantum}}(J, \beta) \geq f^{\text{classical}}(J+1, \beta)$$

where J is the spin of the quantum model, e.g., a Heisenberg ferro- or antiferromagnet consisting of spin J operators, and β denotes the inverse temperature.

The main tool, which is the general idea of coherent states, is to represent (or in some cases approximate) operators as superpositions of projections onto coherent states. For spin systems one can represent any operator on spin space by

$$\int_{\mathbb{S}^2} f(\Omega)\Pi(\Omega)\mathrm{d}\Omega$$

where f is a function on the unit sphere that can be explicitly calculated for a given operator. The projection $\Pi(\Omega)$ is the projection onto the eigenvector of $\Omega \cdot S$ with the largest eigenvalue J. Here Ω is a unit vector in $\mathbb{R}^3$ and $S = (S_1, S_2, S_3)$ are the spin matrices. Upper bounds and lower bounds of this type – obtained with the aid of coherent states – are not restricted to spin systems and are known as Berezin-Lieb inequalities (see also [Ber]).

IV.2: In quantum mechanics the entropy of a state is defined (von Neumann) in terms of the state's nonnegative density matrix ϱ by $S_{\text{quantum}} = -\text{Trace}\varrho \log \varrho = -\sum_j \lambda_j \log \lambda_j$, where the λ_j are the eigenvalues of ϱ and whose sum is one. (The right side of this formula is reminiscent of Shannon's definition of entropy with λ_j being probabilities in his case.) Clearly, $S \geq 0$. The density matrix is $\varrho = Z^{-1} \exp[-\beta H]$, where H is the Hamiltonian, Z^{-1} is a normalization constant, and β is the reciprocal of the temperature. Usually, $H = p^2/2m + V(x)$.

What has been done since the time of Boltzmann is to approximate the quantum entropy by the "classical entropy", $S_{\text{classical}} = -\int \varrho_{\text{class}}(p, x) \log \varrho_{\text{class}}(p, x) dpdx$, where $\varrho_{\text{class}}(p, x) = N' \exp[-(p^2/2m + V(x)/T]$, but this quantity does not have to be positive. Indeed, it tends to $-\infty$ as T tends to zero.

What is a good classical approximation to the quantum entropy that does not have this defect? Wehrl's idea (which also remedies some other defects of the Boltzmann definition) is to take the expectation value of ϱ in a (Schrödinger-Bargmann-Segal-Glauber) coherent state $\psi_{p,x}$, namely, $\varrho_{\text{Wehrl}}(p, x) = (\psi_{p,x}, \varrho\psi_{p,x})$, and then use Boltzmann's formula to define

$$S_{\text{Wehrl}} = -\int \varrho_{\text{Wehrl}}(p, x) \log \varrho_{\text{Wehrl}}(p, x) \mathrm{d}p\mathrm{d}x \,.$$

Wehrl proved that $S_{\text{Wehrl}} > 0$ and conjectured that its minimum value was the number 1, which occurred when ϱ is itself a one-dimensional projector onto a coherent state.

This was proved by Lieb in this paper by combining the knowledge of two sharp constants: in Young's inequality and in the Hausdorff-Young inequality. Moreover, Lieb went on to propose a similar entropy construction for the coherent states appropriate to representations of $SU(2)$ as in **IV.1** and conjectured that one-dimensional projectors would again give the minimum. These conjectures are still open after almost a quarter century, but two special cases have been verified recently by Schupp [Schu].

IV.3: In this paper, which is joint work with Solovej, the ideas of **IV.1** are generalized in that one does not compare the quantum model with spin J with the classical one, but with another quantum spin K model. A consequence is the following inequality between the free energy of the Heisenberg ferromagnet $f_f(J, \beta)$ and the Heisenberg antiferromagnet $f_a(J, \beta)$:

$$f_a\left(J, \frac{J}{J+1}\beta\right) \leq f_f(J, \beta) \leq f_a\left(J, \frac{J+1}{J}\beta\right)$$

and

$$f_f\left(J, \frac{J}{J+1}\beta\right) \leq f_a(J, \beta) \leq f_f\left(J, \frac{J+1}{J}\beta\right) .$$

This is remarkable since on the classical level the two models, the ferromagnet and the antiferromagnet are the same while quantum mechanically they are different. The above estimates delineate the difference. The paper also contains inequalities between f_f and f_f and between f_a and f_a for different J-values.

Part V
Brunn-Minkowski Inequality and Rearrangements

Rearrangement inequalities have a fairly short history starting with the work of Hardy, Littlewood, Polya, Riesz, Faber and Krahn in the thirties. The definition can easily be understood by visualizing a function f of two variables in the plane as a mountain range (without overhangs of course). Cutting this mountain range at a height h displays a level set of the function,

$$S_f(t) := \{x : f(x) > h\} .$$

This will be now 'rearranged' into a disk at the same height with the same Lebesgue measure and centered at the origin. Doing this with all level sets of this function yields level sets of a new 'rearranged' function f^*. This function will be obviously radial and nonincreasing. The value of this operation lies in the fact that it yields various very useful inequalities. A particularly important one is Riesz's inequality, which states that for three integrable nonnegative functions f, g, h

$$\int f(x)g(x-y)h(y)\mathrm{d}x\mathrm{d}y \leq \int f^*(x)g^*(x-y)h^*(y)\mathrm{d}x\mathrm{d}y .$$

Another consequence of rearrangements is the isoperimetric inequality in the form

$$\|\nabla f\|_p \geq \|\nabla f^*\|_p . \tag{5}$$

The case $p = 2$ leads to the famous Faber-Krahn inequality which says that the lowest frequency among drums of equal area is achieved for the disk. The case $p = 1$ can be "understood" using the co-area formula

$$\int |\nabla f(x)|\mathrm{d}x = \int_0^\infty \int_{\{f(x)=t\}} \mathrm{d}\mathscr{H}^{n-1}\mathrm{d}t ,$$

where $\mathscr{H}^{n-1}$ is the $n-1$ dimensional Hausdorff measure. Since the right is given by the 'surface area' of the level set at height t it follows from the isoperimetric inequality that this integral gets smaller when f is replaced by f^*. The proof of this fact is non-trivial and is due to Hilden [Hi]. In general, the level surfaces, of such a function, i.e., the boundary of the level sets, are not surfaces in the usual sense of differential geometry – which is the difficulty with this approach.

V.1: Jointly with Brascamp and Luttinger, Lieb showed that the Riesz rearrangement inequality can be extended to any number of functions. It is interesting to note that the proof of this general rearrangement inequality proceeds via the Brunn-Minkowski inequality. This inequality, in its simplest form, says that the n-th root of the volume of a slice of a convex body in $\mathbb{R}^{n+1}$ is concave as a function of the position of the slice. In particular this implies that the logarithm of this volume is concave. Let us also note that the Brunn-Minkowski inequality provides one of the standard proofs of the isoperimetric inequality.

V.2: This joint paper with Brascamp contains a further application of the Brunn-Minkowski inequality. In this paper they show that the marginals of log concave functions are log concave. Besides the application to rearrangements mentioned above, this innocuous statement has further interesting consequences for Wiener integrals. It is shown in **V.2** that the ground state wave function of a Schrödinger operator in a convex domain with Dirichlet boundary conditions and with a convex potential is itself a log concave function, thereby proving a long-standing conjecture of Payne. Subsequently this result was reproved by [SMYY] using the maximum principle and extended by Borell [Bo] who proved some concavity properties for Wiener integrals jointly in the space and time variables.

A particularly beautiful result related to log concavity is the existence of the quantum mechanical Wigner crystal in one dimension. Consider N particles moving inside an interval which interact among each other via a one-dimensional Coulomb repulsion $-|x|$. The interval also carries a uniform charge distribution of opposite sign. It is not hard to see, classically, that the lowest energy configuration is attained for evenly spaced particles, i.e., they form a periodic array. It was shown in **V.2** that this fact survives in quantum mechanics. Again, it is a consequence of the Brascamp-Lieb argument about log concave functions together with Trotter's formula. The quantum mechanical analog of a periodic array is that the one particle marginal of the ground state of these N particles is a periodic function of period L/N where L is the length of the box. The existence of a Wigner crystal in higher dimensions is still an open problem, even on the classical level.

V.3: This is again an application of rearrangements but with a surprising twist. The idea is to connect rearrangements with the tensor product. Young's inequality is

$$\left|\int f(x)g(x-y)h(y)\mathrm{d}x\mathrm{d}y\right| \le C\|f\|_p\|g\|_q\|h\|_r$$

where

$$\frac{1}{p}+\frac{1}{q}+\frac{1}{r}=2\,.$$

Although Riesz's rearrangement inequality applies here it does not lead to any standard variational problem. A crucial observation is that the inequality is preserved when the variables are doubled, except that C is then replaced by C^2. This leads to a natural iteration scheme. Take radial functions f, g, h, double their variables, i.e., consider $f^{(2)}(s,t) = f(s)f(t)$, etc. and rearrange these functions (which are functions of two variables now). It is clear that Gaussian functions are invariant under this operation but less clear that they are the only ones. The point, however, is that when this operation is repeated indefinitely then a central limit theorem holds, i.e., the sequence converges to 'Gaussian' functions. Loosely speaking, the only functions in infinite dimensions that are invariant under rotations are Gaussian functions. In this fashion it is established in **V.3** that Gaussians are optimizers in Young's inequality. There are many alternative proofs, one by Beckner [Bek] using an idea of Gross [Gr], which is also related to the central limit theorem. Certain special cases of this inequality were considered by Nelson [Ne], Glimm [Gl], Simon [Si2] and Segal [Se] who was the first to point out the fact concerning the doubling of variables for tensor products of positivity preserving kernels from L^p to L^q, p and q arbitrary. If the kernels are not positivity preserving then the doubling of variables works also provided that $q \geq p$. This is known as Beckner's lemma. The 'doubling of variables' idea resurfaces again in **VI.7**.

V.4: Another outcome of these ideas is the generalization of the Brunn-Minkowski inequality to measurable sets which leads among other things to generalizations of inequalities of Prekopa - Leindler. Consider two nonnegative functions f and g on $\mathbb{R}^n$ and pick any number $0 < \lambda < 1$. Consider the function h define by

$$h(z) := \operatorname{esssup}\{f(x)^\lambda g(y)^{1-\lambda} : z = \lambda x + (1-\lambda)y\} .$$

Then

$$\int_{\mathbb{R}^n} h(x)\mathrm{d}x \geq \left(\int_{\mathbb{R}^n} f(x)\mathrm{d}x\right)^\lambda \left(\int_{\mathbb{R}^n} g(x)\mathrm{d}x\right)^{1-\lambda} .$$

This inequality is a generalization of the Brunn-Minkowski inequality. Pick any two *measurable* sets A and B and define their convex addition

$$\lambda A + ((1-\lambda)B := \{z : z = \lambda x + (1-\lambda)y, x \in A, y \in B\} ,$$

then the volumes satisfy the inequalities

$$|\lambda A + (1-\lambda)B| \geq |A|^\lambda |B|^{(1-\lambda)} .$$

When A is a ball and B a single point, then the right side vanishes while the left side is large. Note that the convex addition is all the points z in $\mathbb{R}^n$ such that $(z - \lambda A) \cap ((1-\lambda)B) \neq \emptyset$. In **V.4** the Brunn- Minkowski inequality is proved for the weaker notion of convex addition which is all points z in $\mathbb{R}^n$ such that $(z - \lambda A) \cap ((1-\lambda)B)$ has positive measure.

The Prekopa-Leindler inequality, and hence the Brunn-Minkowski inequality was generalized to Riemannian manifolds in [CMS] by using the Monge transportation problem in a surprising fashion. It seems that the transportation problem plays a role in all of this that has not been fully understood. E.g., the solution of the transportation problem together with the Henstock parametrization, has recently been used by Barthe [Bar] to give an alternative way to determine the sharp constant in Young's inequality.

V.5: In general, rearrangement inequalities are not strict; the L^p norms and Sobolev norms furnish examples. There are however exceptions and the paper **V.5** presents an important example. It says that when the middle function in Riesz's rearrangement inequality is strictly decreasing, then one can have equality in Riesz's rearrangement inequality only if the other two functions are symmetric decreasing about a common point in $\mathbb{R}^n$. This fact is used in **V.5** to prove uniqueness of solutions to Choquard's equation which is, by the way, the same as Pekar's equation that occurs in the study of the polaron model.

This is one of the few general methods to decide the cases of equality in variational problems. It is worth noting that this paper contains a very natural and simple proof of (5) for the case $p = 2$, as a consequence of Riesz's rearrangement inequality in which the middle function g is given by the heat kernel.

V.6, V.7: We have seen before that rearrangements decrease gradient norms. Thus, it is not unreasonable to expect that rearrangements act as a contraction on H^1, the space of square integrable functions whose gradient is also square integrable. It is a fact [CraTa, Chi, CaLo, LiLo] that rearrangement is a contraction on all L^p spaces with $p < \infty$. Simple arguments show that this cannot be true for H^1, however. A reasonable question is whether rearrangement is continuous in H^1. This question was affirmatively answered by Coron [Cor] in one-dimension. In higher dimensions Almgren and Lieb showed that it is wrong in general. The obstruction to this continuity is a peculiar property of functions. Intuitively, functions whose derivative vanish on a set of large measure should have large 'flat' spots. Flat spots reveal themselves in the distribution function as discontinuities. The distribution function is the measure of the level set at height t as a function of t. It is nonincreasing and, by choice, lower semicontinuous. Thus, if the function is constant on a set of positive measure, the distribution function should be discontinuous. Surprisingly, it was discovered in **V.6** that there exist functions in two or more variables whose derivatives vanish on sets of large Lebesgue measure, but that have absolutely continuous distribution functions. Almgren and Lieb construct explicit examples of such functions and call them co-area irregular functions. Returning to the question of the H^1-continuity of rearrangement, Almgren and Lieb pick such a co-area irregular function f and consider the sequence

$$f_j(x) = f(x) + \frac{1}{2j} W_j(x) \sin(jf(x)) \, .$$

Here, the function W_j is a smoothed characteristic function of the set where the derivative of f vanishes. The smoothing can be done in such a way that $\| W_j \|_\infty \leq \sqrt{j}$. It is straightforward to see that ∇f_j converges strongly to ∇f in L^p. Further, some elementary arguments show that the graphs of the rearranged functions f_j^* and f^*, when parametrized by height t intersect when $t = 2m\pi/j$ and the graph of f_j^* sits above the graph of f^* whenever $t = (2m+\sigma)(\pi/j)$, for $0 < \sigma < 1$. As j tends to infinity the points where the two graphs intersect move closer but the 'wiggles' in between persist. In fact it is shown in [**V.6, V.7**] that $\|\nabla f_j^* - \nabla f^*\| \geq c > 0$ for c independent of j.

Almgren and Lieb go far beyond this example by showing that the dichotomy of coarea regularity and irregularity of the limit function precisely characterizes the sequences for which rearrangement is H^1- continuous. All this is explained in great detail in the paper [AlLi], which due to its length, is not part of this Selecta. Instead we chose to reproduce the summary papers **V.6** and **V.7**. In **V.7** the notion of current is used while **V.6** and [AlLi] rely on more 'elementary' methods.

One inequality in the big paper that should be mentioned is what one could call the ultimate version of Riesz's rearrangement inequality. Consider a function $F(u, v)$ with $F(0, 0) = 0$, and assume that the mixed derivative $\partial_u \partial_v F$ is nonnegative. Then

$$\iint F(f(x), g(x))W(ax+by)\mathrm{d}x\mathrm{d}y \leq \iint F(f^*(x), g^*(x))W^*(ax+by)\mathrm{d}x\mathrm{d}y$$

where W is nonnegative and integrable and a and b are nonzero numbers.

Let us remark, finally, that in the case of Steiner symmetrization the situation is completely different. In fact, Steiner symmetrization is continuous in H^1, as was shown by Burchard [Bu].

Part VI
General Analysis

VI.1: This paper deals with a similar issue as in the Lieb-Thirring inequalities. Given an orthonormal family of functions ψ_i, $i = 1, \ldots N$ in $L^2(\mathbb{R}^n)$, define

$$u_i = \left(-\Delta + m^2\right)^{-1/2} \psi_i \quad \text{and} \quad \varrho(x) = \sum_{i=1}^{N} |u_i(x)|^2 .$$

Among other results, Lieb, generalizing previous results by Battle-Federbush [BaFe] and Conlon [Co], proves that

$$\|\varrho\|_p \leq A_n N^{1/p}, \; p = \frac{n}{n-2},$$

where A_n is a universal constant. The remarkable fact about this inequality is that due to the orthogonality of the ψ_i the constant is proportional to $N^{1/p}$. Without the orthogonality condition the best one could have is a factor N.

VI.2: Real analysis is an extremely well developed subject, with many textbooks devoted to it and one is led to believe that all the simple and useful theorems have been discovered. This paper, which is joint work with Brézis adds a further theorem, a simple, but nontrivial fact about pointwise convergence, to this list. It is the physicists way of looking at the problem of pointwise convergence, maybe a reason why it was missed by the masters of the subject. Fatou's lemma states that for any nonnegative sequence of integrable functions f_n that converges pointwise to an integrable function f,

$$\liminf_n \int f_n(x)\mathrm{d}x \geq \int f(x)\mathrm{d}x .$$

In general the inequality is strict. How can sequences of functions fail to converge? There are three intuitive ways: the sequence can oscillate to 'death', it can go 'up the spout' or it can escape to infinity. This is quantified by the correction term to Fatou's lemma, proved in **VI.2**, i.e.,

$$\int f_n(x)\mathrm{d}x = \int f(x)\mathrm{d}x + \int |f(x) - f_n(x)|\mathrm{d}x + o(1) .$$

The reader might feel that such a statement might not mean much but is convinced otherwise when he sees what kind of variational problems can be solved using that fact, examples of which can be found in **VI.2**. Certainly, such a theorem should be standard in any book on real analysis.

VI.3: A very influential paper of Lieb, instigated by Sokal in connection with certain problems of quantum field theory, is his work on the sharp version of the Hardy-Littlewood- Sobolev inequality. This inequality is well known and states that

$$\int_{\mathbb{R}^{2n}} \frac{f(x)g(y)}{|x-y|^{\lambda}} \mathrm{d}x\mathrm{d}y \leq C(n,p,\lambda)\|f\|_p \|g\|_q$$

provided that

$$\frac{1}{p}+\frac{1}{q}+\frac{\lambda}{n}=2\,, \qquad 1<p,q<\infty\,.$$

The problem is to find the sharp constant $C(n, p, \lambda)$ and all the cases of equality. Merely proving the inequality itself with any constant is not easy. The problem was solved in one-dimension by Hardy and Littlewood in the early thirties and shortly thereafter Sobolev generalized it to arbitrary dimensions. One should also mention that Riesz, by an ingenious application of the inequality between the arithmetic and geometric mean derived the one in higher dimension from the case in one-dimension.

The search for the sharp constant is, however, considerably more difficult. This is due to the fact that the functional is invariant under translation and scaling and that makes it very likely that maximizing sequences converge weakly to zero. Note that the translation invariance can be broken by rearrangements but it is unclear, a priori, which mechanism breaks the scale invariance of the prospective optimizer. In **VI.3** Lieb shows quite generally that despite this difficulty, the optimizer always exists. The argument proceeds via rearrangements and using logarithmic variables, which display the scale invariance as translation invariance. He shows that the minimizing sequence cannot tend weakly to zero.

The paper is a real tour de force and contains many more interesting results. Chief among these is the actual evaluation of the sharp constant in the case where $p = q$, including all the cases of of equality. This case is particularly interesting since the functional is conformally invariant. An interesting point is that this large non compact symmetry group, which is generally an obstacle to proving the existence of optimizers, is now used to determine the explicit form of the optimizers themselves. This apparent dichotomy was overcome in [CaLo] to give a simple alternative proof (which uses "competing symmetries") of the sharp HLS inequality for the case $p = q$.

In some sense, the sharp HLS inequality for $p = q$ is the mother of all conformally invariant inequalities. Most of the conformally invariant inequalities, without any hard analysis, can be derived from the HLS inequality. To give one example, by differentiating the HLS inequality at $\lambda = 0$, which can be done provided f, g are non- negative, one obtains the logarithmic HLS inequality. Dual to this is the Onofri-Beckner inequality, which gained great popularity recently [On, Bek2, CaLo2, OPS].

VI.4: It is intuitively clear that if the lowest Dirichlet eigenvalue of a domain is small then the set must be somehow 'fat'. It is surprisingly difficult to give precise notions of this fact, one of which is given in this paper. Given two domains A and B, denote by $\lambda(A)$ the lowest eigenvalue of the Laplacian on this domain with Dirichlet boundary conditions. Then there exists always a translate B_x of B such that

$$\lambda(A \cap B_x) < \lambda(A) + \lambda(B) .$$

In particular if, for some given ball B, $\lambda(A \cap B_x)$ is large for all translations x, then $\lambda(A)$ must be large too. Thus, a set of a granular consistency must have a large Dirichlet eigenvalue. This intuition is also the basis of the following compactness result, which can also be found in **VI.4**. Consider a bounded sequence in $H^1(\mathbb{R}^n)$. In general one can extract a weakly convergent subsequence by the Banach-Alaoglou theorem. In many applications, however, one would like to ensure that the limit is not the zero function. One possibility is that the measure of the level set $\{x : f_n(x) > t\}$ (here t is some positive number) is uniformly bounded below by a constant ε. The reader might immediately object that such a condition is certainly not sufficient, since the sequence might successively split into smaller and smaller islands and, in this way, tend weakly to zero. This is precisely what the result explained above prevents from happening. Such a splitting would increase the $H^1(\mathbb{R}^n)$-norm beyond any bound. Thus the sequence cannot 'pulverize', it has to stay together. It is shown in this paper that there exists a sequence of translations y_n, so that the sequence $f_n(x - y_n)$ has a subsequence that has a nonzero weak limit.

There is a fairly extensive literature on the concentration phenomenon. It has been summarized in the papers [Lio], where it is shown that the limiting behavior of weak sequences can be reduced to studying a finite number of cases. In actual applications it is important to decide which of these cases is at hand and the convergence theorem mentioned above provides one of the tools for deciding that.

VI.5: This paper was the motivation of the above mentioned convergence theorem. The problem is to establish existence of solutions of the following systems of partial differential equations on $\mathbb{R}^n$:

$$-\Delta u_i(x) = g(u_1(x), \ldots, u_N(x)) , \ i = 1, \ldots, N .$$

Here $g_i = \partial_i G$ where G satisfies various assumptions. The solutions of these equations are critical points of the action functional

$$S(u) = \sum_i \int |\nabla u_i(x)|^2 \mathrm{d}x - \int G(u(x)) \mathrm{d}x ;$$

however, this functional is unbounded below. Using the idea of a constrained minimum by Coleman, Glaser and Martin [CGM] i.e., minimizing the first term in the action while keeping the second one fixed, Lieb, together with Brézis, establish existence of solutions of this system of equations and show that they have finite action. Moreover, they show that among all those solutions there is one that has minimal action. This problem is typical for not having obvious compactness properties. Weak sequences tend to leak out to infinity and **VI.4** provides the ideal tool for 'running after them'.

VI.6: In this paper, jointly with Brézis, Lieb improves on a Sobolev inequality with remainder, invented in [BN]. In that paper it was shown that

$$\|\nabla f\|^2 - S_n \|f\|_{2^*} \geq C_p(\Omega)\|f\|_p^2$$

holds for all functions $f \in H_0^1(\Omega)$ and for any $p < n/(n-2)$. The constant $S_n$is the sharp Sobolev constant. The improvement is that the right hand side can be replaced by $\|f\|_{p,w}$, the weak L^p norm, but with $p = n/(n-2)$. This improvement is used again in [DaLi] on the stability of relativistic molecules.

VI.7: Here, Lieb revisits a topic related to Young's inequality. The Fourier transform can be interpreted as a Gaussian integral kernel applied to a function. The Hausdorff-Young inequality says that

$$\|\widehat{f}\|_p \leq C_p \|f\|_q$$

where

$$\frac{1}{p} + \frac{1}{q} = 1 \, , 1 \leq q \leq 2 \, ,$$

and C_p is a universal constant. A famous result of Babenko and Beckner [Bab, Bek] states that the sharp constant in the Hausdorff-Young inequality is attained by Gaussian functions. In this paper Lieb establishes that for $1 < q < 2$ equality can occur only for Gaussian functions. One of the key tools, as in Beckner's lemma mentioned in **V.3**, is the sharp version of Minkowski's inequality. Its application is, however, quite different and allows Lieb to extract information of complex valued functions and not just nonnegative functions as is the case for Young's inequality.

Another high point of this paper is the ultimate version of Young's inequality. In essence it states that any integral of a product of L^p functions with various combination of arguments is maximized by Gaussian functions. This was recently reproved by Frank Barthe [Bar] using the Henstock parametrization and the solution to the Monge- Kantorovich mass transport problem.

VI.8: The following problem is related to **VI.7**. Consider the Woodward 'ambiguity function', which is very closely related to the Wigner transform,

$$A_{f,g}(\tau, \omega) = \int f(t - \tau/2)\overline{g(t + \tau/2)}e^{-2\pi i \omega \tau} dt \, ,$$

where $f \in L^p$ and $g \in L^q$ with $1/p + 1/q = 1$. Lieb calculates sharp bounds on the magnitude of this ambiguity function. These bounds are saturated precisely when f as well as g are Gaussian functions. Depending on the values for p these bounds are upper or lower bounds. Although the paper is much in the spirit of **VI.7**, it is by no means a corollary of that paper.

Part VII
Inequalities Related to Harmonic Maps

In the early eighties Lieb started to develop an interest in problems that have a geometric flavor to them, like liquid crystals or Stability of Matter with magnetic fields. He was also very attracted by the Skyrme model that, although invented in the sixties, got a new lease on life, e.g., through the work of Adkinson, Nappi and Witten [ANW]. Lieb never wrote a research paper (apart from a review paper [Li3]) on that subject but made contributions to the theory of harmonic maps which is a closely related subject.

VII.1: The first paper devoted to this subject, which is joint with Brézis and Coron, is the problem of minimizing the functional

$$\int_D |\nabla\varphi(x)|^2 \mathrm{d}x \tag{6}$$

where D is a domain in $\mathbb{R}^3$ and φ is a map from D into the sphere $\mathbb{S}^2$. Imagine now that the domain has small holes and specify the degree of the map near each hole and on the boundary. The problem is how to minimize the above functional over all those maps that have these specified degrees.

Pair the holes with positive degree with the holes that have negative degree and ignore the holes that have zero degree. Adding up the distances of all these pairs yields a length $L(C)$ that depends on this pairing C. It is shown in **VII.1** that the minimum of (6) equals $8\pi L$ where L is the minimum of $L(C)$ over all pairings. There is no minimizer for this problem. The intuition is that the $|\nabla\varphi(x)|^2$ of an 'approximate minimizer' deteriorates into a measure that is concentrated on these lines connecting the paired holes.

VII.2, VII.4: The following question is now natural. Consider a map from a ball in $\mathbb{R}^3$ into $\mathbb{S}^2$ and fix the map on the boundary, say to be ψ. In many examples the minimizing map, which is called a harmonic map, develops singularities. This is at first surprising since one is dealing with an elliptic equation, but standard regularity results do not apply to elliptic *systems*, however.

Given this boundary data, can one estimate the number of singularities of the harmonic map in the interior? One knows by a result of Schoen and Uhlenbeck that the singularities must be isolated. In this paper with Almgren, a fairly complete solution of that question is given.

What determines the number of singularities is the Dirichlet integral of the boundary data and it does so in a linear fashion. In other words if we denote by $N(\psi)$ the number of singularities then there exists a constant C such that

$$N(\psi) \le C \int_{\partial B} |\nabla\psi(x)|^2 \mathrm{d}S \,. \tag{7}$$

Based on **VII.1** one might guess that the modulus of the Jacobian of ψ integrated over the sphere is also a good candidate for estimating the number of singularities. This is completely false. Almgren and Lieb give an example of a boundary map where this quantity is zero but the corresponding harmonic map has many singularities.

The constant C in (7) is not known, not even remotely. This is due to the fact that the proof relies on compactness arguments. It should be said that there are no assumptions made about the minimizing map concerning its boundary regularity. In fact it is there where the interesting things are happening and where the singularities try to accumulate only to be stopped by the Dirichlet integral of the boundary data. Since the original paper is too long for this volume we reproduce here the summary papers, as we did for **V.6, V.7**

VII.3: How can one compute an area minimizing surface S that spans a closed curve C embedded in $\mathbb{R}^3$? One way, as shown in this paper, is to look at maps from $\mathbb{R}^3$ to the circle $\mathbb{S}^1$ with the property that at each point x of C, and for every little two-dimensional surface containing x and perpendicular to C, the map winds around $\mathbb{S}^1$ exactly once as we wind our way around x. If φ is such a map we can define its energy to be $\mathscr{E}[\varphi] = \int_{\mathbb{R}^3} |\nabla\varphi(x)|^2 dx$. The map that (almost) minimizes this energy (under the 'winding' condition mentioned above) is one that is constant everywhere in $\mathbb{R}^3$ except for a thin layer surrounding S. In this thin layer the map runs around $\mathbb{S}^1$ once as we go from the upper side of the layer to the lower side.

A similar construction can be made for 'minimizing' higher dimensional objects spanning lower dimensional objects in $\mathbb{R}^n$. The basic idea germinated from the paper **VII.1**, where it was shown that the geodesic between two points in $\mathbb{R}^3$? (namely, a straight line segment!) can be generated by minimizing the energy of maps from $\mathbf{R}^3$ to $\mathbb{S}^2$ with prescribed singularities (i.e., winding number 1) at each of the two points.

VII.5: A problem related to harmonic maps is the Ginzburg-Landau problem. In its simplest form it amounts to minimizing a certain energy functional over functions from the unit disc into the complex numbers having given boundary data. There are no general criteria for deciding whether a minimizer has certain symmetry properties of the problem (e.g., rotational symmetry) when the minimizer is a map or a vector field. For functions, there are useful devices like rearrangements but those techniques do not work for maps in general. In this paper it was shown that the minimizer in the radial class is weakly stable. There are better results available due to Mironescu [Mir] who proves that if the minimizer has a zero at the origin, then the minimizer is radial. It is surprising that even for this somewhat rudimentary model, the gap between intuition and proof is still large.

References

[AlLi] Almgren, F. and Lieb, E.H.: Symmetric Decreasing Rearrangement is Sometimes Continuous, Jour. Amer. Math. Soc. **2**, 683–773 (1989).

[ANW] Adkins, G., Nappi, C. and Witten, E.: Static properties of nucleons in the Skyrme model, Nucl. Phys. **B228**, 552–566 (1983).

[Bab] Babenko Izv. Akad. Nauk SSR Ser. Mat. **25**, 531–542, 1961, English. transl. Am. Math. Soc. Transl. (2) **44**, 115–128 (1965).

[Bar] Barthe, F.: Optimal Young's inequality and its converse: a simple proof, Geom. Funct. Anal. **8**, 234–242 (1998).

[Bau] Baumann, F.: Bemerkungen über quantenmechanische Entropie-Ungleichungen, Helv. Phys. Acta **44**, 95–100 (1971).

[BaFe] Battle, G. A. and Federbush, P.: A phase cell cluster expansion for Euclidean field theories, Ann. Physics **142**, 95–139 (1982).

[Bel] Bell, J.: *Speakable and unspeakable in quantum mechanics*, Cambridge University Press, 1987, see especially pp 139–158.

[Bek] Beckner, W.: Inequalities in Fourier Analysis, Ann. Math. **102**, 159–182, (1975).

[Bek2] Beckner, W.: Sharp Sobolev inequalities on the sphere and the Moser-Trudinger inequality, Ann. Math. **138**, 213–242 (1993).

[Ber] Berezin, F.: Convex functions of operators, Mat. Sb. **88**, 268–276 (1972). (Russian)

[BK] Boel, R. J.; Kasteleyn, P. W.: Correlation-function identities and inequalities for Ising models with pair interactions, Commun. Math. Phys. **61**, 191–208 (1978).

[BL] Baym, G. and Leggett, A. J.: Exact upper bound on barrier penetration probabilities in many-body systems: application to "cold fusion". Phys. Rev. Lett. **63**, 191–194 (1989).

[Bo] Borell, C.: Geometric properties of some familiar diffusions in $\mathbb{R}^n$. Ann. Probab. **21**, 482–489 (1993).

[BN] Brézis, H. and Nirenberg, L.: Positive solutions of nonlinear elliptic equations involving critical Sobolev exponents, Comm. Pure Appl. Math. **36**, 437–477 (1983).

[BR] Bratteli, O. and Robinson, D. W.: *Operator Algebras and Quantum Statistical Mechanics* , Vol 2, second edition, Springer Verlag (1996). See Notes and Remarks on Sections 6.2.3, 6.2.4, page 435.

[BS] Blanchard, Ph. and Stubbe, J.: Bound states for Schrödinger Hamiltonians: phase space methods and applications, Rev. Math. Phys. **8**, no. 4, 503–547 (1996).

[Bu] Burchard, A.: Steiner symmetrization is continuous in $W^{1,p}$. Geom. Funct. Anal. **7**, 823–860 (1997).

[CaLo] Carlen, E.A. and Loss, M.: Extremals of functionals with competing symmetries, J. Funct. Anal. **88**, 437–456 (1990).

[CaLo2] Carlen, E.A. and Loss, M.: Competing symmetries, the logarithmic HLS inequality and Onofri's inequality on $\mathbb{S}^n$, GAFA **2**, 90–104 (1992).

[Ch] Choi M.-D.: A Schwarz inequality for positive linear maps on C*-algebras, Ill. J. Math. **18**, 565–574 (1974).

[Chi] Chiti, G.: Rearrangement of functions and convergence in Orlicz spaces, Appl. Anal. **9**, 23–27 (1979).

[Cl] Clarkson, J.A.: Uniformly convex spaces, Trans. Am. Math. Soc. **40**, 396–414, (1936).

[CMS] Cordero-Erausquin, D., McCann, R. J. and Schmuckenschläger, M.: A Riemannian interpolation inequality à la Borell, Brascamp and Lieb. Invent. Math. **146**, 219–257 (2001).

[Co] Conlon, J.: Semi-classical limit theorems for Hartree-Fock theory, Commun. Math.Phys. **88**, 133–150 (1983).

[Cor] Coron, J-M.: The continuity of rearrangement in $W^{1,p}(\mathbb{R})$, Ann. Scuola Norm. Sup. Pisa Cl. Sci (4) **11**, 57–85 (1984).

[CraTa] Crandall, M.G. and Tartar, L.: Some relations between nonexpansive and order preserving mappings, Proc. Amer. Math. Soc. **78**, 385–390 (1980).

[Cw] Cwikel, M.: Weak type estimates for singular values and the number of bound states of Schrödinger operators, Ann. Math. **106**, 93–100 (1977).

[CGM] Coleman, S., Glaser, V. and Martin, A.: Action minima among solutions to a class of Euclidean scalar field equations, Commun. Math. Phys. **58**, 211–221, (1978).

[DaLi] Daubechies, I. and Lieb, E.H.: One Electron Relativistic Molecules with Coulomb Interaction, Commun. Math. Phys. **90**, 497–510 (1983).

[DL] Dyson, F.J. and Lenard, A. Stability of Matter I., J. Math. Phys. **8**, 423–434, (1967) and Stability of Matter II., J. Math. Phys. **9**, 698–711, (1968).

[Gl] Glimm J.: Bose fields with nonlinear self-interaction in two dimensions, Commun. Math. Phys. **8**, 12–25, (1968).

[Gr] Gross, L.: Logarithmic Sobolev inequalities, Amer. J. Math. **97** 1061–1083, (1976).

[Hai] Haiman, M.: Hecke algebra characters and immanant conjectures, J. Amer. Math. Soc. **6**, 569–595 (1993).

[Ha] Hanner, O.: On the uniform convexity of L^p and l^p, Ark. Math. **3**, 239–244, (1956).

[Har] Harper, L.H.: Stirling behavior is asymptotically normal, Ann. Math. Statist. **38**, 410–414, (1967).

[Hi] Hilden K.: Symmetrization of functions in Sobolev spaces and the isoperimetric inequality, Manuscripta Math. **18**, 215–235 (1976).

[HK] Hohenberg, P. and Kohn W.: Inhomogeneous electron gas, Phys. Rev. B **136**, 864–871 (1964)

[LaWei] Laptev, A. and Weidl, T.: Sharp Lieb-Thirring inequalities in high dimensions, Acta Math. **184**, 87–111 (2000).

[Li1] Lieb, E.H.: The Stability of Matter, Rev. Mod. Phys. **48**, 553–569 (1976).

[Li2] Lieb, E.H.: A Lower Bound for Coulomb Energies, Phys. Lett. **70A**, 444–446 (1979).

[Li3] Lieb, E.H.: Remarks on the Skyrme Model, in *Proceedings of the Amer. Math. Soc. Symposia in Pure Math. 54*, part 2, 379–384 (1993). (Proceedings of Summer Research Institute on Differential Geometry at UCLA, July 8–28, 1990.)

[LiLe] Lieb, E.H. and Lebowitz, J.L.: The Constitution of Matter: Existence of Thermodynamics for Systems Composed of Electrons and Nuclei, Adv. in Math. **9**, 316–398 (1972).

[LiLo] Lieb, E.H. and Loss, M.: *Analysis*, Second Edition, Graduate Studies in Mathematics, American Mathematical Society, Providence, Rhode Island, 2000.

[Lio] Lions. P.-L.: The concentration-compactness principle in the calculus of variations. The locally compact case. I., Ann. Inst. H. Poincaré Anal. Non Linéaire **1**, 109–145 (1984). The concentration-compactness principle in the calculus of variations. The locally compact case. II., Ann. Inst. H. Poincaré Anal. Non Linéaire **1**, 223–283 (1984). no. 4,

[LM] Lieb, E. H. and Mattis, D.: *Mathematical Physics in One Dimension*, Academic Press, 1966.

[LR] Lanford III, O. E. and Robinson, D. W.: Mean entropy of states in quantum statistical mechanics, J. Math. Phys. **9**, 1120–1125 (1968).

[LT] Lieb, E.H. and Thirring, W.E.: Bound for the kinetic energy of fermions which proves the Stability of Matter, Phys. Rev. Lett. **35**, 687–689 (1975), Errata **35**, 1116 (1975).

[LW] Lieb, E.H. and Wu, F.Y.: Absence of Mott Transition in an Exact Solution of the Short-Range One-Band Model in One Dimension, Phys. Rev. Lett. *20*, 1445–1448 (1968).

[Ma] Mandelbrot, B.: *The fractal geometry of nature*, W.H. Freeman, revised edition (1983), p. 140.

[Me] Merris, R.: Inequalities for matrix functions, J. Algebra **22**, 451–460 (1972).

[Mir] Mironescu, P.: Les minimiseurs locaux pour l'équation de Ginzburg-Landau sont à symétrie radiale. (French) [Local minimizers for the Ginzburg-Landau equation are radially symmetric] C. R. Acad. Sci. Paris Sér. I Math. **323**, 593–598 (1996).

[Ne] Nelson, E.: The free Markov Field, J. Funct. Anal. **12**, 211–227 (1973).

[On] Onofri, E.: On the positivity of the effective action in a theory of random surfaces, Commun. Math. Phys. **86**, 321–326 (1982).

[OPS] Osgood, B, Phillips, R. and Sarnak, P.: Extremals of determinants of Laplacians, J. Funct. Anal. **80**, 148–211 (1988).

[Ri] Rivasseau, V.: Lieb's correlation inequality for plane rotors, Commun. Math. Phys. **77**, 145–147 (1980)

[Ro] Rotfel'd, S. Yu.: Remarks on the singular numbers of the sum of completely continuous operators, Funct. Anal. Appl. Consultants Bureau trans. **1**, 252–253 (1967).

[Roz] Rozenblum, G.V.: Distribution of the discrete spectrum of singular differential operators, Dokl. Aka. Nauk SSSR **202**, 1012–1015 (1972). The details are given in: Distribution of the discrete spectrum of singular differential operators, Izv. Vyss. Ucebn. Zaved. Matematika **164**, 75–86 (1976) [English transl. Sov. Math. (Iz. VUZ) **20**, 63–71 (1976).]

[RR] Robinson, D. W. and Ruelle, D.: Mean entropy of states in classical statistical mechanics, Commun. Math. Phys. **5**, 288–300 (1967).

[Ru] Ruelle, D.: Large volume limit of the distribution of characteristic exponents in turbulence, Commun. Math. Phys. **87**, 287–302 (1982)

[Schu] Schupp, P.: On Lieb's conjecture for the Wehrl entropy of Bloch coherent states, Comm. Math. Phys. **207**, 481–493 (1999).

[Se] Segal, I.: Construction of nonlinear quantum processes I, Ann. Math. **92**, 462–481, 1970

[Sh] Shor, P. W.: Additivity of the Classical Capacity of Entanglement-Breaking Quantum Channels, arXiv quant-ph/0201149, to appear in Jour. Math. Phys.

[Si1] Simon, B.: Correlation inequalities and the decay of correlations in ferromagnets, Commun. Math. Phys. **77**, 111–126 (1980).

[Si2] Simon, B.: A remark on Nelson's best hypercontractive estimates, Proc. Am. Math. Soc. **55**, 376–378 (1976).

[Sim] Simon, B.: *Trace Ideals and their Applications*, Cambridge Univ. Press 1979.

[SMYY] Singer, I. M., Wong, B., Yau, S.-T. and Yau, Stephen S.-T.: An estimate of the gap of the first two eigenvalues in the Schrödinger operator. Ann. Scuola Norm. Sup. Pisa Cl. Sci. (4) **12**, 319–333 (1985).

[SS] Seiler, E. and Simon, B.: An inequality among determinants, Proc. Nat. Acad. Sci. USA **72**, 3277–3278 (1975).

[St] Stanley, R. P.: Log-concave and unimodal sequences in algebra, combinatorics, and geometry, in *Graph Theory and its Applications: East and West*, Ann. N. Y. Acad. Sci. **576**, 500–535 (1989).

[We] Wehrl, A.: General properties of entropy, Rev. Mod. Phys. **50**, 221–260 (1978).

[Wei] Weidl, T.: On the Lieb-Thirring constants $L_{\gamma,1}$ for $\gamma \geq 1/2$, Commun. Math. Phys. **178**, 135–146 (1996).

[WY] Wigner, E.P. and Yanase M.M.: Information content of distributions, Proc. Acad. Sci. U.S.A. **49**, 910–918 (1963). See also: On the positive semidefinite nature of certain matrix expressions, Canad. J. Math. **16**, 397–406, (1964).

Michael Loss
Mary Beth Ruskai

Part I

Inequalities Related to Statistical Mechanics and Condensed Matter

PHYSICAL REVIEW VOLUME 125, NUMBER 1 JANUARY 1, 1962

Theory of Ferromagnetism and the Ordering of Electronic Energy Levels

ELLIOTT LIEB AND DANIEL MATTIS
Thomas J. Watson Research Center, International Business Machines Corporation, Yorktown Heights, New York

(Received May 25, 1961; revised manuscript received September 11, 1961)

Consider a system of N electrons in one dimension subject to an arbitrary symmetric potential, $V(x_1,\cdots,x_N)$, and let $E(S)$ be the lowest energy belonging to the total spin value S. We have proved the following theorem: $E(S)<E(S')$ if $S<S'$. Hence, the ground state is unmagnetized. The theorem also holds in two or three dimensions (although it is possible to have degeneracies) provided $V(x_1,y_1,z_1;\cdots;x_N,y_N,z_N)$ is separately symmetric in the x_i, y_i, and z_i. The potential need not be separable, however. Our theorem has strong implications in the theory of ferromagnetism because it is generally assumed that for certain repulsive potentials, the ground state is magnetized. If such be the case, it is a very delicate matter, for a plausible theory must not be so general as to give ferromagnetism in one dimension, nor in three dimensions with a separately symmetric potential.

INTRODUCTION

THIS paper consists mainly in the enunciation and proof of a theorem about the ordering of the energy levels of a system of interacting fermions. As such, our primary concern is with mathematics. Without striving to be pedantic, we have endeavored to construct the proof with care and rigor.

We take advantage of the symmetry properties of electron wave functions belonging to various values of total spin-angular-momentum S. In certain cases we are able to order the ground-state energies belonging to the various spins without any explicit numerical calculations. This circumvents the great difficulties of the N-body problem, such as the applicability of perturbation theory, etc.

Our theorem is not without some theoretical consequences. Notably, whenever it is applicable, there can be no ferromagnetism unless one postulates explicitly spin- or velocity-dependent forces. (The theorem does not apply to electrons in a three-dimensional lattice interacting with Coulomb or central forces; but conversely, correct theories of ferromagnetism should not predict ferromagnetism for interactions which *are* covered by the theorem.) However, as the mathematics stand quite independently of such applications, we shall defer further considerations of the physical consequences of this work to the end of the paper (Sec. IV).

In Sec. I we shall state and prove the theorem for a one-dimensional electron system. In order to pass to higher dimensions, it will be necessary to prove further theorems on one-dimensional systems which have no direct relevance to fermions. These will be discussed in Sec. II. Section III will contain the proof of our theorem for certain specialized problems in two or more dimensions. We have added an Appendix on an analogous theorem for certain one-dimensional chains of three-dimensional atoms; the proof uses a different technique from that in Sec. I, insomuch as we switch to the delta function (or lattice gas) representation and use second quantization.

As a preliminary, we should like to recall a well-known theorem on the two-fermion problem, proved many years ago.[1] Consider the general two-particle Hamiltonian[2] (in any number of dimensions)

$$H=\mathbf{p}_1^2+\mathbf{p}_2^2+V(\mathbf{r}_1,\mathbf{r}_2), \tag{1}$$

where V is any symmetric potential. That is to say, the particles may interact with each other and/or with an external potential, the only proviso being that no spin- or velocity-dependent forces are present. The boundary conditions can be anything so long as they are homogeneous. Since the total spin S is a good quantum number, the ground state is either $S=0$ or $S=1$. The theorem states that the ground state always has $S=0$, a statement borne out by the hydrogen molecule, for example.

Since V is real (hence the necessity for excluding velocity-dependent forces) the eigenfunctions of H are real. An eigenstate ${}_M{}^S\Psi$ with $S=0$ must be of the form[3]

$${}_0{}^0\Psi=\Phi_{Sy}(\mathbf{r}_1,\mathbf{r}_2)[(+-)-(-+)], \tag{2}$$

where Φ_{Sy} is a symmetric real function and where we have used an obvious notation for the spin part of the wave function. Alternatively, a state with $S=1$ (and $M=1$, for example) must be of the form

$${}_1{}^1\Psi=\Phi_A(\mathbf{r}_1,\mathbf{r}_2)[(++)], \tag{3}$$

where Φ_A is antisymmetric. In both cases the symmetry property of Φ is determined by the Pauli principle which states that ${}_M{}^S\Psi$ must be antisymmetric.

Now if the ground-state wave function were $S=1$, consider the trial function obtained by taking the absolute value of Φ_A,

$${}_0{}^0\Psi\equiv|\Phi_A|[(+-)-(-+)], \tag{4}$$

which has $S=0$ and satisfies the Pauli principle. Φ_A is the spatial part of the $S=1$ ground-state function. It is readily verified that the variational energy of ${}_0{}^0\Psi$ is the same as that of the supposed ground state, ${}_1{}^1\Psi$ (we shall return to a proof of this in Sec. I). Thus, by

[1] The authors believe this theorem is due to E. P. Wigner.

[2] $\hbar^2/2m=1$.

[3] We shall use the notation ${}_M{}^S\psi$ for a function with a definite S and M value; ${}^S\psi$ if it has only a definite S value; and ${}_M\psi$ if it has only a definite M value.

reductio ad absurdum, it follows that there is always an $S=0$ eigenfunction having an energy at least as low as the lowest $S=1$ function. In fact, as we shall show later, the trial function given above cannot be an eigenfunction unless V is pathologic (e.g., a repulsive core in one dimension). Thus, if $E(S)$ denotes the lowest energy belonging to a given S value, $E(0)<E(1)$ for two particles.

This paper is an extension of the two-particle theorem to an arbitrary number of particles. We are able to do it completely for one dimension, and in certain cases for higher dimensions. The general result is $E(S)\leq E(S+1)$.

I. THE ONE-DIMENSIONAL SYSTEM

We start with the general Hamiltonian[2]

$$H=-\sum_{i=1}^{N}\frac{\partial^2}{\partial x_i^2}+V(x_1,\cdots,x_N), \tag{5}$$

where V is real and symmetric in the N variables $x_1,\cdots,x_N$. Otherwise, it is completely arbitrary. The boundary conditions may be any of the following:

(i) If the particles are in a "box" (i.e., $0\leq x_i\leq L$), then $\Psi=0$ if any $x_i=0$ or L;

(ii) the same as (i) except that $\partial\Psi/\partial x_i=0$ if any $x_i=0$ or L;

(iii) if $-\infty<x_i<\infty$ or $0<x_i<\infty$, then we restrict ourselves to square integrable (bound state) wave functions, assuming these exist. Periodic boundary conditions are excluded, for they require a slight modification of the theorem and a somewhat lengthier analysis.

Although H does not contain the spins explicitly, every eigenfunction belongs to a definite S value, which may take on any of the values $N/2$, $(N/2)-1$, $\cdots$, 0, or $\frac{1}{2}$. If we denote the lowest or ground-state energy belonging to a given S value by $E(S)$, then the theorem to be proved is

Theorem I. If $S>S'$, *then* $E(S)>E(S')$ *unless* V *is pathologic, in which case* $E(S)\geq E(S')$.

The term pathologic potential will be defined in the sequel.

In order to prove this theorem it is first necessary to characterize the spatial part of a wave function of space and spin. To this end, let ${}_M\Psi$ be a wave function satisfying the Pauli principle and having a definite spin azimuthal quantum number M. (That is, $S_z\,{}_M\Psi=M\,{}_M\Psi$, where $S_z=\sum_{i=1}^N S_z^i$.) Then ${}_M\Psi$ may be expanded in the complete set of spin functions having the M value in question. The coefficients of the expansion will be spatial functions. Thus

$${}_M\Psi=\sum_j {}_M\Phi_j(x_1,\cdots,x_N)G_j^M, \tag{6}$$

where G_j^M is a spin function of which a typical one is

$$G_1^M=(\underbrace{--\cdots--}_{p}\underbrace{++\cdots++}_{N-p}),$$

with

$$p=(N/2)-M \tag{8}$$

(i.e., particles $1-p$ have spins down, the rest being up). Since ${}_M\Psi$ is a Pauli function and since the various G_j^M are obtained from each other by a permutation of the spin variables, it follows that the ${}_M\Phi_j$ are related to each other by a permutation of $x_1,\cdots,x_N$. Moreover, the following is easily verified:

(a) If any ${}_M\Phi_j\equiv0$, then all ${}_M\Phi_j\equiv0$.

(b) If Ψ is an eigenfunction of H with energy E, then so is each ${}_M\Phi_j$ separately.

(c) ${}_M\Phi_1$ (henceforth to be denoted simply by ${}_M\Phi$) is nonvanishing and is of the form

$${}_M\Phi=\Phi(x_1,\cdots,x_p|x_{p+1},\cdots,x_N),\quad p=(N/2)-M, \tag{9}$$

where the notation is meant to imply that Φ is antisymmetric in the variables $x_1,\cdots,x_p$ *and* in the variables $x_{p+1},\cdots,x_N$.

(d) Given any spatial function having the symmetry properties of (9) above, it may be used to generate a nonvanishing Pauli function such as ${}_M\Psi\equiv\sum_p(-)^p\times(p\Phi)(pG_1^M)$, where the summation is over all permutations p of the N particles. If Φ is an eigenfunction of H, then so is ${}_M\Psi$.

The next question to consider is what further condition must we impose on ${}_M\Phi$ in order that ${}_M\Psi$ have a definite S value (i.e., $S^2\,{}_M{}^S\Psi=S(S+1)\,{}_M{}^S\Psi$ and $S_z\,{}_M{}^S\Psi=M\,{}_M{}^S\Psi$). For simplicity, let us take $M\geq0$ (i.e., $p\leq N/2$), in which case the S value of Ψ in Eq. (6) could be any of $M, M+1,\cdots,N/2$, or a mixture of these. Suppose we wish $S=M$. Then, a necessary and sufficient condition is $S_+\,{}_M{}^M\Psi=0$ where $S_+=\sum_{i=1}^N S_+^i$. The operator S_+ acts on the G_j^M functions; acting on G_1^M it generates G_1^{M+1} among others. But other G_j^M's also generate G_1^{M+1} and if one sets equal to zero the coefficient of G_1^{M+1} in $S_+\,{}_M{}^M\Psi$, one finds

(e) A necessary and sufficient condition that ${}_M\Psi$ be ${}_M{}^M\Psi$ is that ${}_M{}^M\Phi$ (the coefficient of G_1^M) be of the form (9) and that the bar cannot be moved to the left. By this is meant that ${}_M{}^M\Phi$ cannot be antisymmetrized with respect to the variables $x_p,\cdots,x_N$. In other words,

$$\Big(1-\sum_{j=p+1}^{N}P_{p,j}\Big)\,{}_M{}^M\Phi\equiv0, \tag{10}$$

where $P_{p,j}$ is the simple transposition permutation of x_p and x_j.

(f) If the bar can be moved to the left once, but not twice, then Ψ is, in general, a mixture of ${}^{M+1}{}_M\Psi$ and ${}^M{}_M\Psi$, and so forth.

(g) If $M>0$, we can always lower the M value of Ψ by $S_-\,{}_M\Psi=\text{const}\;{}_{M-1}\Psi\neq0$, and hence the bar can always be moved to the right. In other words, if a function is of the form (9), the bar can always be moved to the right if $p<N/2$, but it cannot always be moved to the left. This is a known result of the theory of the

permutation group which we have proved by recourse to the more generally known theory of angular momentum.

(h) Any function satisfying (9) and (10) may be used to generate a nonvanishing ${}_M{}^S\Psi$ as in (d) above. These remarks tell us that the higher the S value of ${}^S\Phi$, the more antisymmetric it must be. For instance, a totally antisymmetric function always belongs to $S=N/2$, but to any $M\leq N/2$.

Now since S_- and S_+ commute with H, $E(S)$ is degenerate with respect to M value; i.e., $H\,{}_M{}^S\Psi=E(S)\,{}_M{}^S\Psi$ implies there exists ${}_j{}^S\Psi$ with the same eigenvalue, where j can take all $2S+1$ values between S and $-S$. We shall therefore prove Theorem I in the following manner: Let $E(M)$ and ${}_M\Psi$ be the lowest eigenvalue and eigenfunction, respectively, of H belonging to a given M value ≥ 0. We shall show that ${}_M\Psi$ is ${}_M{}^M\Psi$ and hence that $E(M)=E(S)$. We shall further show that $E(M)<E(M+1)$ unless V is pathologic. In other words, we shall show that the lowest eigenfunction of H of the form (9) also satisfies Eq. (10).

Let R be the full domain, all $0\leq x_i\leq L$ [assuming we use boundary condition (i) or (ii) for example], and define $R_M\subset R$ by

$$R_M:\ 0\leq x_1\leq\cdots\leq x_p\leq L$$

and

$$0\leq x_{p+1}\leq\cdots\leq x_N\leq L. \tag{11}$$

Consider the Schrödinger equation in R_M;

$$H\varphi=E\varphi, \tag{12}$$

with boundary conditions:

$$\varphi=0 \text{ on the boundary of } R_M. \tag{13}$$

It is well known that the ground-state function φ_0 of Eqs. (12) and (13) satisfies

$$\varphi_0\geq 0 \quad\text{in}\quad R_M. \tag{14}$$

For suppose (14) were not satisfied and consider $\varphi=|\varphi_0|$. Now $\langle\varphi|\varphi\rangle=\langle\varphi_0|\varphi_0\rangle$ and $H\varphi=E\varphi$ everywhere except where φ_0 vanishes, at which points φ is continuous but has discontinuous derivatives. Thus $H\varphi=E\varphi+\delta$ functions, the latter occurring when φ vanishes. Hence

$$\int_{R_M}\varphi(H\varphi)=E\int_{R_M}\varphi\varphi=E\langle\varphi|\varphi\rangle.$$

Therefore φ satisfies the same boundary conditions as φ_0 and has the same variational energy. This implies that among the ground states of (12) and (13) (assuming the possibility of a degeneracy) there is at least one satisfying (14). The following are also true, although the proof is tedious:

(i) If V is bounded, then in fact $\varphi_0\neq 0$ *inside* R_M.

(j) There can be no degeneracy unless $\varphi_0=0$ *inside* R_M.

We can now define the term "pathologic potential." It is a potential with a sufficiently strong infinity to cause φ_0 to vanish *inside* R_M. An infinite repulsive core is an example. Thus there is, in general, no ground-state degeneracy, but even if there is, one of the eigenfunctions satisfies (14).

Let P be a permutation of the variables $x_1,\cdots,x_p$ and Q a permutation of the variables $x_{p+1},\cdots,x_N$, and define $PQ(R_M)$ as the domain defined by the appropriate permutation of the variables in condition (11). All the domains $PQ(R_M)$ are disjoint, except possibly for the boundaries, and together span R. Let φ be any solution to (12) and (13) and define the function Φ everywhere in R by

$$\Phi=(-)^P(-)^QPQ\varphi \quad\text{in}\quad PQ(R_M). \tag{15}$$

Because of the boundary conditions, (13), it is easily verified that Φ is continuous and has a continuous derivative everywhere in R, and hence satisfies

$$H\Phi=E\Phi \quad\text{in}\quad R. \tag{16}$$

Conversely, any eigenfunction of H in R that satisfies Eq. (9) satisfies (12) and (13) in R_M. Thus if φ_0 is the ground state of (12) and (13), Φ, as defined by (15), is the ground state of H belonging to M.

Now consider

$$\begin{aligned}{}_M{}^M\chi&=\mathrm{Det}\begin{vmatrix}1x_1\cdots x_1{}^{p-1}\\ \vdots\qquad\vdots\\ 1x_p\cdots x_p{}^{p-1}\end{vmatrix}\times\mathrm{Det}\begin{vmatrix}1x_{p+1}\cdots x_{p+1}{}^{N-p-1}\\ \vdots\qquad\vdots\\ 1x_N\ \cdots x_N{}^{N-p-1}\end{vmatrix}\\ &=\prod_{\substack{j,k=1\\ j>k}}^{p}(x_j-x_k)\times\prod_{\substack{j,k=p+1\\ j>k}}^{N}(x_j-x_k).\end{aligned} \tag{17}$$

(Except for a totally symmetric Gaussian factor, this is the solution to the problem of noninteracting one-dimensional electrons in a harmonic oscillator potential.) The function ${}_M{}^M\chi$ clearly satisfies (9), (10), and (14). It is easily verified that if ${}_M{}^Sf$ and ${}_M{}^{S'}g$ are any two functions having different S values but the same M value, then

$$\int_R {}_M{}^Sf\,{}_M{}^{S'}g=P!(N-P)!\int_{R_M}{}_M{}^Sf\,{}_M{}^{S'}g, \qquad (p=\tfrac{1}{2}N-M) \tag{18}$$

and further the right-hand side of (18) vanishes if $S\neq S'$. Since ${}_M{}^M\chi$ and the ground state of H belonging to M, ${}_M\varphi$, are both non-negative in R_M, Eq. (18) implies that ${}_M\varphi$ is not orthogonal to ${}_M{}^M\chi$ in R. If the ground state of (12) and (13) is nondegenerate, then ${}_M\varphi$ can belong to only one S value and this S value must therefore be $S=M$. If one wants $S=M+1$ or higher, it is necessary to go at least to the first-excited state of (12) and (13). In case of degeneracy, at least one of the ground-state functions belongs to $S=M$.

This completes the proof of Theorem I.

II. GENERALIZATION OF THEOREM I, AND THE "POURING PRINCIPLE"

Thus far we have considered eigenfunctions of H which have the symmetry property (9), the only allowed symmetry class for constructing a Pauli function of space and spin. But there are other symmetry classes with their corresponding eigenfunctions—the totally symmetric function, for example. The latter is a Bose function and plays no role for fermions, unless only two particles are involved.

The most widely known classification of symmetry classes is due to Young; but we shall find it convenient to use a slightly modified version of his scheme.[4] It is well known that every function of 2 variables can be written as the sum of an antisymmetric and a symmetric function which *a fortiori* are orthogonal to each other. Thus,

$$\phi(x_1,x_2)=\tfrac{1}{2}(1+p_{12})\phi+\tfrac{1}{2}(1-p_{12})\phi. \tag{19}$$

The operator $1-p_{12}$ is said to antisymmetrize ϕ, while the operator $1+p_{12}$ symmetrizes it. If ϕ is a function of $x_1,\cdots,x_N$, the operator $1-p_{12}$ antisymmetrizes it with respect to the variables x_1, x_2; the operator $(1-p_{13}-p_{23})(1-p_{12})$ antisymmetrizes it with respect to x_1, x_2, x_3, and so forth. It is quite possible that ϕ may be antisymmetrized with respect to x_1 and x_2, but not with respect to x_1, x_2, x_3; $(x_1-x_2)x_3$ is an example of such a function.

Now consider a function ϕ of the variables $x_1, \cdots, x_N$ which is of the form

$$\phi(x_1,\cdots|\cdots x_{N-n_1-n_2}|x_{N-n_1-n_2+1},\cdots,x_{N-n_1}| x_{N-n_1+1},\cdots,x_N), \tag{20}$$

or any of its permutations, by which we mean that it is separately antisymmetric in the n_1 variables $x_{N-n_1+1}, \cdots, x_N$ and in the n_2 variables $x_{N-n_1-n_2+1}, \cdots, x_{N-n_1}$ and so forth, where $n_1 \geq n_2 \geq \cdots$. The bars are to be regarded as unmovable leftwards[5]; e.g., ϕ cannot be antisymmetrized with respect to $x_{N-n_1}, \cdots, x_N$. The function ϕ is said to belong to the symmetry class characterized by the numbers n_1, n_2, etc. For example, a totally symmetric function has N "boxes" with one variable in each, whereas an antisymmetric function has 1 "box" containing all N variables. It is possible to prove the following properties of ϕ:

(a) Any bar may always be moved to the right (i.e., the antisymmetrization procedure which would move the bar to the right gives a nonvanishing result) if the number of variables to the right of the bar is greater than the number to the left.

(b) The largest group of variables with respect to which ϕ may be antisymmetrized is n_1. From among the remainder, the largest group would be n_2, and so forth.

(c) If ϕ and ψ belong to two different symmetry classes, they are orthogonal.

(d) For a function of the form (20), the antisymmetrization operator which moves a bar to the right has as its unique inverse the operator which moves the bar to the left. Therefore, if ψ is obtained from ϕ by moving a bar to the right, it is said to belong to the same class as ϕ.

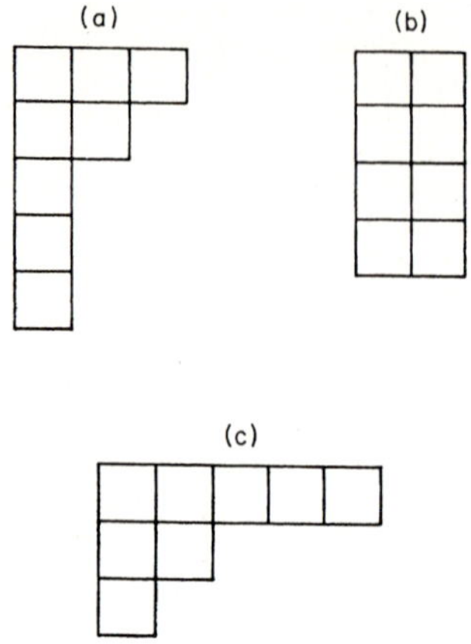

FIG. 1. The tableaux corresponding to several different symmetry classes for eight particles. By the pouring principle $E(a)>E(c)$ and $E(b)>E(c)$, but one cannot say whether (a) or (b) has the lower energy. Note that (c) happens to be the conjugate of (a).

From these remarks we see that by a combination of antisymmetrization and orthogonalization, one can reduce an arbitrary function to a sum of orthogonal functions, each belonging to a different symmetry class. The function in each symmetry class itself is decomposable into a sum of functions of the form (20) and its permutations, although this last decomposition is not unique.

There is a convenient pictorial representation for the symmetry classes called *tableau* illustrated in Fig. 1. One draws a column of n_1 boxes. To the right of it, and starting at the same height, one places a column of n_2 boxes, and so forth. A function of the form (20) is further designated by inserting the appropriate variables in the appropriate column, the order of the variables in any column being immaterial. It will be seen that the lengths of the *rows* decrease from top to bottom; it is therefore possible to define the conjugate to a symmetry class to be the one in which rows are replaced by columns. Thus (a) and (c) in Fig. 1 are conjugate. We shall return to this later, however.

Returning to the problem at hand, it will be appreciated that since H is permutation-invariant, every eigenfunction not only belongs to a definite symmetry class, but is of the form (20). Theorem I, then, tells us about the order of the ground-state energies of functions of the one- and two-column class. Under certain circum-

[4] See, for example, D. E. Rutherford, *Substitutional Analysis* (Edinburgh University Press, Edinburgh, 1948). Young's scheme proceeds by symmetrization, instead of antisymmetrization as used here.

[5] We have departed slightly from the notation in Eq. (9). Formerly the bar was regarded as possibly movable leftwards.

stances it is possible to extend this theorem to more than two-column classes. To do this we need one more concept, which we shall call the "pouring principle."

Definition. If α and β are two different symmetry classes, it is possible to ***pour α into β*** if the bars in the α function can be moved to the right [subject to (*a*) above] so that one gets a function antisymmetric in the same groups of variables as the β function. More formally, if the α tableau has the columns $n_1 \geq n_2 \geq$, and the β tableau has the columns $n_1' \geq n_2' \geq \cdots$, then we require that $n_1 \geq n_1'$; $(n_1 - n_1') + n_2 \geq n_2'$; $(n_1 - n_1' + n_2 - n_2') + n_3 \geq n_3'$; etc., where any missing columns are to be regarded as having $n=0$. If α can be poured into β, we denote this fact by $\alpha \xrightarrow{p} \beta$.

Thus, in Fig. 1, (a) can be poured into (c) and (b) into (c), but neither (a) nor (b) can be poured into the other. Note that if α can be poured into β, then the conjugate of β can be poured into conjugate of α. We can now state the extension of Theorem I in one dimension.

Theorem II. Let α and β be two symmetry classes and let $E(\alpha)$ and $E(\beta)$ be the respective ground-state energies of eigenfunctions of H in the two classes. If α can be poured in β, then $E(\alpha) > E(\beta)$ unless V is pathologic, in which case $E(\alpha) \geq E(\beta)$.

The proof is exactly the same as for Theorem I. One antisymmetrizes the α function until it matches the bars of the β function (a process which does not change the energy of the α function). Next one defines the fundamental region in analogy with (11) above and shows that the ground-state function in this region is positive, and is therefore not orthogonal to a determinantal function which is positive in this region and is known to belong to the β class.

Corollary: Consider[2]

$$H = -\sum_{i=1}^{N} \frac{\partial^2}{\partial x_i^2} - \sum_{i=1}^{M} \frac{\partial^2}{\partial y_i^2} + V(x_1, \cdots, x_N; y_1, \cdots, y_M). \tag{21}$$

If V is symmetric in the x variables, then every eigenfunction belongs to some symmetry class under permutation of these variables. Theorem II obviously is true for this more general problem. If, moreover, V is also separately symmetric in the y variables, although of no particular symmetry under the interchange of an x with a y, then every eigenfunction falls into some symmetry class α in the x variables and some class α' in the y variables. If two functions are characterized by (α,α') and (β,β'), respectively, then $E(\alpha,\alpha') > E(\beta,\beta')$ if α can be poured into β and α' into β'.

III. SEPARATELY SYMMETRIC POTENTIAL IN HIGHER DIMENSIONS

We shall now turn to the proof of the extension of Theorem I to higher dimensions when the potential is separately symmetric. Only two dimensions will be explicitly considered, for the extension to three dimensions is a corollary. The general fermion Hamiltonian is of the form (21) with $M=N$ and V is a symmetric function of the pairs x_1, y_1, etc. A separately symmetric potential is one for which V is, in addition, separately symmetric in the x variables, and in the y variables. The theorem to be proved is:

Theorem III. If V is separately symmetric and if $S > S'$, then $E(S) \geq E(S')$. (Note that Theorem III is not quite so strong as Theorem I, because the equalities can occur without pathologic potentials.)

Before discussing Theorem III, it is necessary to consider the "Kronecker product" of two tableaux. Suppose we have a function of the x's and the y's which is of a definite symmetry class α in the x's and a definite class α' in the y's. Let us consider this function together with all the functions derived from it by permutations, and let us ask if there is some linear combination of these functions which is of a definite symmetry class, β, in the N *pairs* of variables (x_i,y_i). If so, which β classes can be generated and how many independent β functions are there? [By independent β functions we mean functions which cannot be obtained from each other by permutations of the (x_i,y_i) pairs.] This question is analogous to the problem in the theory of angular momentum of combining J_1 and J_2 to give a resultant J. There the answer is given simply by the Clebsch-Gordan theorem: All $J_1+J_2 \geq J \geq |J_1-J_2|$ may be produced once and only once. Unfortunately, there is no such simple rule for tableaux except in two special cases: when β is symmetric, or antisymmetric (i.e., one row, or one column).

Lemma I. To generate a totally antisymmetric function in the pairs x_iy_i, it is required that the x and y tableaux α and α' be conjugates. (To generate a totally symmetric function in the pairs, the x tableau must be the same as the y tableau.)

This is a well-known result from the theory of permutation groups.[4]

As we saw in Sec. I, the spatial part of a Pauli (antisymmetric) function of definite S must be of the two-column type, where $n_1 = (N/2)+S$ and $n_2 = (N/2) - S$. By Lemma I we are led to believe that spin functions of a given S are of the conjugate two-row type. This is indeed the case. There are $\binom{N}{p}$ different G_M functions, and these will be seen to generate all two-row tableaux in which the first row is $(N/2)+M$ or longer. Because the ${}_M\varphi_j$ in Eq. (6) are all derivable from each other, they will not all be linearly independent. The G_j^M's therefore appear only in certain definite linear combinations; if φ is of the two-column type, these linear combinations can be shown to be of the conjugate two-row type.

Any Pauli eigenfunction of H of definite S value is thus seen to be the triple Kronecker product of a function belonging to the α and α' class in the x and y variables, respectively, and of a two-row function of the spin variables. The resultant must be a one-column function in the triplets (x_i,y_i,s_i), where s_i is the spin variable. The

With D.C. Mattis in Phys. Rev. *125*, 164–172 (1962)

problem can be viewed in two ways: a Kronecker product of α and α' must be of the appropriate two-column type; or a Kronecker product of α and the two-row spin function must be conjugate to α'.

To prove Theorem III, we take the latter view. Let α and α' be the tableaux of the spatial function, giving the lowest energy for a given S value and suppose α' is (c) of Fig. 1, so that the (x_i,s_i) class, which we shall denote by β, is (a) of Fig. 1. A typical representative, χ, would have each of the pairs (x_i,s_i) in some box of the (a) tableau and each of the y_i in some box of the (c) tableau. Suppose x has $(x_1,s_1), \cdots, (x_5,s_5)$ in the first column, (x_6,s_6) and (x_7,s_7) in the second, and (x_8,s_8) in the third. It will be seen that χ may be regarded as a five-particle one-dimensional Pauli function in the first five pairs, or as a two-particle Pauli function in the next two pairs.

Now, to prove Theorem III we need the following lemma:

Lemma II. Let χ be an eigenfunction with the following properties: (1) it is of the α and α' symmetry classes in the x and y variables, respectively, and is in fact formed from the lowest eigenfunction of H having these classes; (2) in the (x,s) pairs, it is antisymmetric in the same sets of variables as a function of class β, the conjugate of α', but it is not necessarily itself of the class β; (3) considered as a function of each of the sets of (x,s) pairs in which it is antisymmetric, it has a definite S value (i.e., each column has a definite S value). These S values we shall call $S_1, S_2, \cdots$. Let S be any total S value for all the particles which can be compounded from $S_1, S_2, \cdots$ by the usual Clebsch-Gordan rule. Then $E(S)\leq E(\alpha,\alpha')$.

Proof. By applying the appropriate S_+ and S_- operators of each column the appropriate number of times to χ, we can generate functions having all possible M values in each column and which still have the same energy as χ. These functions are then to be added together with the appropriate coefficients to generate a new function, ${}^S\chi\neq 0$ having the required total S value. ${}^S\chi$ has the same energy as χ. Since the total S^2 operator commutes with all permutations, ${}^S\chi$ may be written as the sum of functions belonging to definite symmetry classes and all having the given S value. One of these tableaux may be β itself, in which case the lemma is proved. If not, then since ${}^S\chi$ is already antisymmetric in the same sets of variables as the β tableau, at least one of the component tableaux, say γ, must be such that $\gamma \xrightarrow{p} \beta$. To make a Pauli function out of this component we should need a function whose y class is γ^+, the conjugate of γ. But γ^+ satisfies $\alpha' \xrightarrow{p} \gamma^+$, and therefore $E(\alpha,\gamma^+)\leq E(\alpha,\alpha')$. Thus, if we carry out the same procedure again with the ground-state function of the (α,γ^+) class, we shall be able to construct a Pauli function having the S value in question and with an energy lower than $E(\alpha,\alpha')$.

Corollary. The ground-state function of a given S value has the properties of χ above, and, moreover, belongs to the symmetry class β in the (x,s) variables.

Proof. If not, then by the procedure of the above proof, we should either be able to lower the energy by changing the y tableau, or else we should end up with the function ${}^S\chi$ having the properties stated in the corollary.

Now we consider the proof of Theorem III. The lowest eigenfunction having a given $S\neq 0$ value is the sum of permutations of a function ${}^S\chi$ having the properties stated in the corollary and therefore has the S values S_1, S_2, etc., in each column of β, the (x,s) tableau. There are three cases to be considered:

(1) If S is not the minimum that can be compounded of S_1, S_2, etc., then we can construct a function of spin $S-1$ and with at least as low an energy.

(2) If S_1, S_2, etc., are already the lowest possible (i.e., either 0 or $\frac{1}{2}$) then (1) above is transparently true.

(3) If S is indeed the minimum of $S_1 \oplus S_2 \oplus$, etc., then lowering some one of these S_i values, say S_1, would permit us to create a function of total S equal to $S-1$. But there does exist a function having a lower energy than ${}^S\chi$ and having the properties of χ listed in Lemma II, except that the first column has the value S_1-1. To see this, we regard the function having the α tableau in the x's as being the sum of functions each belonging to definite symmetry classes in each of the groups of x variables appearing in each column of β. One of these sets of tableaux must be the respective conjugates of S_1, S_2, etc. Looking at χ from this point of view, it is clear from the results of Secs. I and II that we can lower one or more of the S_i at will and at the same time lower the energy.

This completes the proof of Theorem III for two dimensions; the extension to higher dimensions is obvious: We simply treat the s_i, z_i pairs in the same manner as the spins above. Let us remark, however, on the reason for the lack of strict inequality as we had in Theorem I. Suppose the lowest function of $S=1$ had an (x,s) tableau such as (a) in Fig. 1 with $S_1=S_3=\frac{1}{2}$ and $S_2=0$. Since each column already has its lowest possible S value, the only way in which $S=0$ could have a *lower* energy is by having an (x,s) tableau containing only columns of even length, such as (b) of Fig. 1, with $S=0$ in each. But it is not possible to prove that such a function has, in fact, a lower energy, and therefore it is possible for the ground state as well as for excited states to have a degeneracy in more than one dimension. (This degeneracy can be estimated never to exceed $N^{\frac{1}{3}}$, in three dimensions. It is therefore not an extensive property of the system.)

IV. ON THEORIES OF FERROMAGNETISM

Although we could extend them to other particles obeying various statistics, the results of this paper apply most directly to the problem of interacting electrons, and as such have some bearing on the theory of ferromagnetism.

It is well known that ferromagnetism must be a consequence of the electronic interactions, for a noninteracting electron system always obeys the theorem $E(S)\leq E(S+1)$, regardless of the external potential. This is antiferromagnetism, or at most, paramagnetism. Ferromagnetism is assumed to occur when the ground state belongs to a nonvanishing S whose magnitude is proportional to the size of the system. It is also well known that the direct magnetic spin-spin forces are negligibly weak, so that the spatial forces in conjunction with the Pauli principle are held to be responsible for the phenomenon.

The simplest realistic problem which offers some hope of being soluble is the linear chain of three-dimensional atoms. The atomic states are supposed known when the atoms are infinitely far apart, and the problem is to find the new configurations when overlap becomes important. Peierls[6] considers this very problem in the chapter on ferromagnetism in his book. The approximation which he makes is that there is only one orbital state per atom, and he concludes that the electronic interactions can lead to ferromagnetism. However, recent and more realistic calculations[7–9] on such chains have proved the contrary to be true. Also in Sec. I we showed that under no circumstances can a one-dimensional electron system be ferromagnetic with only space-dependent forces; this includes the special case of a chain of one-dimensional "atoms." In the Appendix we also treat a model applicable to an idealized chain of three-dimensional atoms, with similar results. It therefore seems that a linear chain can be magnetic only if the individual atoms have orbital degeneracy, that is, if the single atom displays a magnetic or truly three-dimensional character; but it is not known whether this is a *sufficient* condition for ferromagnetism to occur in a linear system.

Our theorem has no relevance to atomic magnetism *per se* (Hund's rule) because it does not apply to the central force problem in three dimensions. But if we consider ferromagnetism to be an extensive property of a solid, the theorem does have some relevance. For we shall show that it is not merely sufficient to have (i) a band structure, (ii) strong repulsive interactions, and (iii) three dimensions, to produce ferromagnetism. We shall base ourselves on the results of Sec. III.

Suppose, for example, that highly magnetized states of a noninteracting set of electrons lie rather close in energy to the $S=0$ ground state. If one introduces a repulsive interaction potential into the problem, and treats this by lowest-order perturbation theory, certain terms called the "exchange integral" will favor the magnetized states. To lowest order, one may find that the magnetized states have crossed the $S=0$ state, and the interacting system is supposed to become ferromagnetic. But this conclusion would be fallacious if the effect were cancelled by second-order or higher-order terms in the perturbation series (or if the perturbation expansion did not converge). Indeed, we now give an example based on the previous section, which is a case in point. Consider, for example, the unperturbed Hamiltonian to be

$$H_0=\sum p_i^2+\sum_i[V(x_i)+V(y_i)+V(z_i)]. \tag{22}$$

Let $V(x)$ be a periodic potential so that one-electron eigenfunctions are Bloch functions. The potential can be chosen such that the bands display the usual degeneracies and other features of motion in a three-dimensional cubic lattice.

If now we introduce an interaction, say

$$H_I=g^2\sum_{ij}\left(\frac{1}{|x_i-x_j|+d}+\frac{1}{|y_i-y_j|+d}+\frac{1}{|z_i-z_j|+d}\right), \tag{23}$$

or some other repulsive, separately symmetric interaction, the total Hamiltonian H_0+H_I is still subject to our theorem and is *not* ferromagnetic. (The theorem does not exclude paramagnetism, however, for the ground state might be degenerate with states of nonvanishing spin angular momentum.) But what are the conclusions we would reach if we were to apply first-order perturbation theory to H_I? This amounts to calculating the expectation value of H_0+H_I using the Slater determinants appropriate to the unperturbed problem. The unperturbed functions with the most spatial nodes are better, variationally speaking, than those with fewer nodes for *sufficiently large* g^2, and we might be led to conclude that there exist some values of S such that $E(S)<E(0)$, which is erroneous. It is therefore clear that we cannot always trust perturbation theory to properly order the levels, for when it is carried out only to finite order, it might be more accurate for some values of S than for others, depending on the particular features of the problem. The same can be said of variational calculations.

A notorious example of the above is the low-density electron gas with Coulomb interactions which is in a background of compensating positive charge. The expectation value of the Hamiltonian using the unperturbed plane-wave states is lower for the ferromagnetic configuration than for the nonferromagnetic ones, at sufficiently low density. But perturbation theory diverges for this problem,[10] and this ferromagnetism is indeed fictitious. A recent and accurate calculation[11] by

[6] R. E. Peierls, *Quantum Theory of Solids* (Oxford University Press, New York, 1955).
[7] R. K. Nesbet, Phys. Rev. **122**, 1497 (1961).
[8] L. F. Mattheiss, Phys. Rev. **123**, 1209 (1961).
[9] David I. Paul, Phys. Rev. **118**, 92 (1960), and Phys. Rev. **120**, 463 (1960).
[10] M. Gell-Mann and K. A. Brueckner, Phys. Rev. **106**, 364 (1957).
[11] W. J. Carr, Jr., Phys. Rev. **122**, 1437 (1961).

Carr leads that author to conclude that at all but the lowest densities the electronic spins are *antiferromagnetically* aligned.

In concluding, we should recall that our theorem is not valid if there are explicitly spin-dependent forces, or velocity-dependent forces. In the latter case, the eigenfunctions are not real and our method of proof does not apply. Nor does it apply to the Coulomb potential which governs real electrons. But it does serve as a warning that the criterion for ferromagnetism must be rather detailed, and not so broad as to violate the results of this investigation.

We have also found it possible to order many of the energy levels of the Heisenberg Hamiltonian $\sum K_{ij}\mathbf{S}_i\cdot\mathbf{S}_j$ (where the S_i are spins on a lattice in one, two, or three dimensions), by analogy with the calculation in the Appendix. These results will be reported in a subsequent publication.

ACKNOWLEDGMENTS

We should like to thank our colleagues at I.B.M. for many useful discussions, and particularly Dr. D. Jepsen, Dr. T. D. Schultz, and Dr. M. Gutzwiller for their excellent suggestions.

APPENDIX. ONE-DIMENSIONAL LATTICE GAS

We are interested in a theorem analogous to the one in the text, for a chain of *three*-dimensional atoms. This problem cannot be solved in all generality, therefore we are led to consider the following tractable model.

(a) We use a truncated Hamiltonian such that only valence electrons are mobile.

(b) Each atom in the linear chain has only one valence state (capable of double occupancy, however, because of spin degeneracy).

(c) The atoms are at a distance d from their nearest neighbors. This distance is such that only nearest-neighbor overlap is important.

(d) The matrix element for a one-electron hop from site j to $j\pm1$ is K, a constant. Two-electron hops and exchange effects are neglected. This is equivalent to

(e) assuming that aside from the "hopping" matrix elements the Hamiltonian is diagonal, with an energy calculable by specifying which atoms have empty valence states, which have singly occupied valence state, and which have doubly occupied valence states. These assumptions lead directly to Eq. (A4).

Our model reduces, in a certain limit, to the one-dimensional problem of Sec. I. The Appendix provides therefore an alternate proof for Theorem I.

To see this, let us consider one-dimensional space as consisting of discrete points labeled $i=1, 2, \cdots, N$, separated by a distance d. The length of the chain is therefore Nd. Next, introduce the second-quantized Fermi operators c_{is} and $c_{is}^\dagger$, where $s=$"up" or "down" according to the spin coordinate, and

$$[c_{is},c_{js'}]_+=0,\quad [c_{is}^\dagger,c_{js'}]_+=\delta_{ij}\delta_{ss'}. \tag{A1}$$

Now consider the Hamiltonian

$$H=-K\sum_{i,s}(c_{i+1,s}^\dagger c_{is}+c_{i-1,s}^\dagger c_{is}-2c_{is}^\dagger c_{.s}) + \sum_{i,j,s,s'} V(|i-j|)c_{is}^\dagger c_{is}c_{js'}^\dagger c_{js'}. \tag{A2}$$

If we transform to running waves,

$$H\to 2K\sum(1-\cos kd)c_{ks}^\dagger c_{ks} +\sum V_g c_{k+gs}^\dagger c_{ks}c_{k'-gs'}^\dagger c_{k's'}, \tag{A.3}$$

where V_g is the Fourier transform of $V(|i-j|)$. In the limit $K^{-\frac{1}{2}}=d=N^{-1}=0$ this reduces to the problem of a one-dimensional electron gas with two-body forces. The Hamiltonian (A2) is a special case of the problem we shall now consider,

$$H=-K\sum_{i,s}(c_{i+1,s}^\dagger c_{is}+\text{H.c.})+V(\cdots n_j\cdots), \tag{A4}$$

where V is an arbitrary symmetric function of the operators $n_j\equiv(c_{j\uparrow}^\dagger c_{j\uparrow}+c_{j\downarrow}^\dagger c_{j\downarrow})$. This Hamiltonian satisfies (a)–(e), is identical with the general Hamiltonian of Sec. I in the limit $K^{-\frac{1}{2}}=d=0$, and can be shown to commute with the spin operators which, in our new representation, are

$$S_z=\tfrac{1}{2}\sum_i(c_{i\uparrow}^\dagger c_{i\uparrow}-c_{i\downarrow}^\dagger c_{i\downarrow}),\quad S_y=\tfrac{1}{2}i\sum(c_{i\uparrow}^\dagger c_{i\downarrow}-c_{i\downarrow}^\dagger c_{i\uparrow}),$$

and

$$S_x=\tfrac{1}{2}\sum(c_{i\uparrow}^\dagger c_{i\downarrow}+\text{H.c.}). \tag{A5}$$

The problem is soluble because there exists a transformation to Pauli (pseudo-spin) variables, in which the Schrödinger equation can be reduced to a series of algebraic equations. We define the Pauli operators as follows:

$$\begin{aligned} b_{i\downarrow}&\equiv c_{i\downarrow}\exp\{\pi i\sum_{j=1}^{i-1}c_{j\downarrow}^\dagger c_{j\downarrow}\},\\ b_{i\uparrow}&\equiv c_{i\uparrow}\exp\{\pi i[\sum_{j=1}^{N}c_{j\downarrow}^\dagger c_{j\downarrow}+\sum_{j=1}^{i-1}c_{j\uparrow}^\dagger c_{j\uparrow}]\},\end{aligned} \tag{A6}$$

The $b_i^\dagger$'s are given by the Hermitean conjugates of these defining equations. All Pauli operators *commute except* b_{is} and $b_{is}^\dagger$, (for all i and s), which anticommute:

$$b_{is}^\dagger b_{is}+b_{is}b_{is}^\dagger=1. \tag{A7}$$

In terms of these new operators,

$$H=-K\sum(b_{i+1,s}^\dagger b_{is}+\text{H.c.})+V(\cdots n_j\cdots), \tag{A8}$$

where $n_j=b_{j\uparrow}^\dagger b_{j\uparrow}+b_{j\downarrow}^\dagger b_{j\downarrow}$. The Hamiltonian remains simple under this transformation only for this very special case of a linear chain, and nearest-neighbor hops. We now assume $K>0$. Otherwise, K can be made positive by a trivial canonical transformation.

Also, for the moment, let us pretend that the number of electrons is even, and is $2p$. Obviously, p cannot exceed N, the number of sites, as each site can accommodate two electrons at most.

Ground State in $M=0$ Subspace

States of all allowed spin angular momentum can be rotated into the $M=0$ subspace with no change in energy. The ground state here is therefore the ground state of the Hamiltonian. The configurations which form a complete set in this subspace have p electrons with spin up and p with spin down. (A configuration specifies which sites are vacant or occupied); for example, if $p=1$ and $N=2$, the complete set of $M=0$ configurations is

$$\phi_1=b_{1\uparrow}^\dagger b_{1\downarrow}^\dagger|0\rangle,\qquad \phi_2=b_{2\uparrow}^\dagger b_{2\downarrow}^\dagger|0\rangle,$$

$$\phi_3=b_{1\uparrow}^\dagger b_{2\downarrow}^\dagger|0\rangle,\quad\text{and}\quad \phi_4=b_{1\downarrow}^\dagger b_{2\uparrow}^\dagger|0\rangle.$$

The Pauli operators for different particles commute, and therefore, the configurations can all be defined to have the same sign for arbitrary ordering of the operators. The number of distinct configurations is $t\equiv\binom{N}{p}^2$, and we shall label them ϕ_α, where $\alpha=1, 2, \cdots, t$.

Let the ground-state function be

$$\psi_0=\sum f^\alpha\phi_\alpha, \tag{A9}$$

with energy E_0 and real amplitudes f^α. If we denote the eigenvalues of V by V_α,

$$V\phi_\alpha=V_\alpha\phi_\alpha,\quad |V_\alpha|<\infty, \tag{A10}$$

then Schrödinger's equation can be expressed in terms of the amplitudes as

$$-K\sum\nolimits_{\beta(\alpha)} f^{\beta(\alpha)}=(E_0-V_\alpha)f^\alpha. \tag{A11}$$

The index $\beta(\alpha)$ runs over those configurations for which

$$\langle\beta(\alpha)|H-V|\alpha\rangle\neq 0. \tag{A12}$$

We parenthetically observe that a variational function,

$$\psi=\sum g^\alpha\phi_\alpha, \tag{A13}$$

has variational energy W,

$$W=\frac{-K\sum\sum g^\alpha g^{\beta(\alpha)}+\sum V_\alpha(g^\alpha)^2}{\sum(g^\alpha)^2}\geq E_0. \tag{A14}$$

Clearly, all nonzero amplitudes f^α can be chosen positive in the ground state. For if they oscillate in sign, define a trial function by $g^\alpha\equiv|f^\alpha|$. Then by inspection of (A14), $W\leq E_0$. This is a contradiction unless

$$\psi'\equiv\sum|f^\alpha|\phi_\alpha$$

is also a ground-state eigenfunction. Therefore,

$$-K\sum\nolimits_{\beta(\alpha)}|f^{\beta(\alpha)}|=(E_0-V_\alpha)|f^\alpha|. \tag{A15}$$

If any $f^\alpha=0$, then by Eq. (A15) all $f^{\beta(\alpha)}$ also vanish. Since by repeated application of H to any given state, all other states are eventually reached, we can conclude that *all* f^α vanish if some one f^α vanishes. Therefore

$$f^\alpha\neq 0 \text{ for all } \alpha. \tag{A16}$$

(We note that $E_0-V_\alpha<0$ for all α, since the ground-state energy must lie lower than the most favorable potential energy.)

Equations (A11) and (A15) are *incompatible unless* all f^α have the same sign, for if we combine them, we obtain

$$\sum|f^{\beta(\alpha)}|=|\sum f^{\beta(\alpha)}|,\ \text{for all } \alpha. \tag{A17}$$

Hence $\psi'=\pm\psi_0$. That the ground state is nondegenerate follows from the observation that all other eigenfunctions of H must be orthogonal to ψ_0 and therefore must have a change of sign, and therefore cannot obey Eq. (A17).

The spin of ψ_0 is found by noting that ψ_0 is *not* orthogonal to the ground state for $V=0$, because they both contain all configurations of Pauli operators with no changes of sign in the amplitudes. The ground state for $V=0$ can be found by inspection of (A3), and belongs to $S=0$. Therefore, the ground state belongs to $S=0$ in general.

Ground State for $M>0$

By a similar procedure, the ground state in any $M>0$ subspace is found to belong to $S=M$. Since each such subspace contains all states of $S\geq M$, this automatically orders the ground states belonging to the various values of S, whether the number of electrons is even or odd. Denoting by $E(S)$ the lowest energy belonging to total spin S, we have therefore proved the following:

Theorem. $E(S+1)>E(S)$.

Note that the restrictions (a)–(e) preclude "double hops," which is related to so-called "exchange." It is therefore reasonable to assume that if ferromagnetism is possible in a linear chain of the sort we have considered, that these neglected exchange matrix elements would be responsible. However, very recent and accurate calculations[7–9] have shown this exchange mechanism to be rather weak. Nesbet[7] finds that the direct "exchange is small compared with the sum of the various antiferromagnetic effects." Mattheiss[8] finds that the true energy levels of such a chain are accurately approximated by the states of the Heisenberg antiferromagnet with nearest-neighbor interactions. Finally, Paul[9] also concludes that linear chains of atoms in s states are nonferromagnetic. It would be interesting to investigate whether an orbital degeneracy on each atom could lead to ferromagnetism for the linear chain in the same way as it appears to be responsible for the magnetic moment of the O_2 molecule.

JOURNAL OF MATHEMATICAL PHYSICS VOLUME 3, NUMBER 4 JULY-AUGUST 1962

Ordering Energy Levels of Interacting Spin Systems

ELLIOTT LIEB AND DANIEL MATTIS
Thomas J. Watson Research Center, International Business Machines Corporation, Yorktown Heights, New York
(Received October 6, 1961)

The total spin S is a good quantum number in problems of interacting spins. We have shown that for rather general antiferromagnetic or ferrimagnetic Hamiltonians, which need not exhibit translational invariance, the lowest energy eigenvalue for each value of S [denoted $E(S)$] is ordered in a natural way. In antiferromagnetism, $E(S+1) > E(S)$ for $S \geq 0$. In ferrimagnetism, $E(S+1) > E(S)$ for $S \geq \mathcal{S}$, and in addition the ground state belongs to $S \leq \mathcal{S}$. $\mathcal{S}$ is defined as follows: Let the maximum spin of the A sublattice be S_A and of the B sublattice S_B; then $\mathcal{S} \equiv |S_A - S_B|$. Antiferromagnetism is treated as the special case of $\mathcal{S} = 0$. We also briefly discuss the structure of the lowest eigenfunctions in an external magnetic field.

INTRODUCTION

THE general Heisenberg Hamiltonian for interacting spins on a lattice (in any number of dimensions) is

$$H = 2 \sum J_{ij} S_i \cdot S_j. \tag{1}$$

This describes theories of ferromagnetism, ferrimagnetism, and antiferromagnetism, depending on the geometry of the lattice, the structure of the symmetric matrix J_{ij}, and the magnitude of the intrinsic spins (which may vary from site to site). In fact, it is conceivable that these factors be such that the spin system displays a mixture of the three magnetic properties. But we shall restrict the discussion to *ferrimagnetic* arrays, of which a special case is *antiferromagnetism.*

We consider only those arrays for which an A and a B sublattice can be defined. The definition of these two sublattices is circular, and perhaps not unique, for the only requirement in defining them is that there exist a constant $g^2 \geq 0$ such that for all sites $i(A)$ on one sublattice and $i(B)$ on the other,

$$J_{i(A),j(A)} \leq g^2, \qquad J_{i(B),j(B)} \leq g^2, \qquad \text{and} \quad J_{i(A),j(B)} \geq g^2. \tag{2}$$

In general, there might be several ways to decompose the lattice in such a way that (2) is obeyed, or there may be none. In the latter case, the system is not necessarily ferromagnetic, and only explicit solutions will reveal its properties. But if (2) is obeyed, we shall show that one is definitely dealing with ferrimagnetism or antiferromagnetism. Note that the number of sites in each sublattice and the magnitude of the intrinsic spin on each site is irrevalent, so that only the topology of the lattice and the structure of J_{ij} counts. Note also that for $g = 0$, and the A sublattice consisting of the nearest neighbors to the sites on (i.e., intermeshing with) the B sublattice, the requirement (2) gives a tendency for nearest neighbors to align antiparallel and next-nearest neighbors to align parallel, and therefore reduces to the usual definition of ferrimagnetism (when the spins are of unequal magnitude) and of antiferromagnetism (when all spins are equal).

The intrinsic spin of an electron is 1/2, but we may be dealing with various species of magnetized atoms or nuclei, so let the intrinsic spin angular momentum on each site be s_i. The maximum possible spin S_A on the A sublattice is therefore

$$S_A \equiv \sum_{i(A)} s_{i(A)}, \tag{3a}$$

and on the B sublattice

$$S_B \equiv \sum_{i(B)} s_{i(B)}. \tag{3b}$$

Defining

$$\mathcal{S} \equiv |S_A - S_B|, \tag{3c}$$

we shall prove that the ground state of H belongs *at most* to total spin $S = \mathcal{S}$. Moreover, if we denote by $E(S)$ the lowest energy eigenvalue belonging to total spin S, then we shall also prove

$$E(S+1) > E(S) \quad \text{for all} \quad S \geq \mathcal{S},$$
and
$$E(S) > E(\mathcal{S}) \quad \text{for} \quad S < \mathcal{S} \quad \text{and} \quad g^2 = 0. \tag{4}$$

(Antiferromagnetism is when $\mathcal{S} = 0$, and the ground state belongs to total spin zero.) This can be regarded *either* as a theorem in ferri- or antiferromagnetism, *or* as a proof that the conditions in Eq. (2) and above eliminate the possibility of ferromagnetism (insofar as it costs energy to raise the total spin value over and above its ground-state value, and that this ground-state value is far from the maximum permissible value of $S_A + S_B$). It also indicates that a large class of apparently different Hamiltonians (1) have really a similar structure, as summarized in Eq. (4), and in the properties of the corresponding eigenfunctions which we shall find below.

W. Marshall was the first to show[1] that the ground state of an antiferromagnet is a singlet; Elsewhere,[2] we have commented on and strengthened his proof. In the present work, we succeed in removing the requirement of translational invariance, and also apply the method to identify the excited states. The M-subspace arguments presented here were previously found useful in the classification of the states of an electron system, and have been used to disprove the possibility of ferromagnetism in linear chains of atoms in s states.[3]

We shall now restrict the discussion to the special case $g^2 = 0$, until the end of the proof.

M SUBSPACES

With the help of the total spin operator

$$\mathbf{S} \equiv \sum \mathbf{s}_i$$

we can construct two operators which commute with each other and with H, namely, $\mathbf{S}^2$ and $\mathbf{S}_z$, which possess eigenvalues $S(S+1)$ and M, respectively. It is known from the theory of angular momentum that $S \geq |M|$. From the rotational invariance of the Hamiltonian we infer the $(2S+1)$-fold degeneracy of each energy level belonging to S, one degenerate level for each value of M in the range $-S \leq M \leq S$. It therefore follows that every energy eigenvalue has a corresponding eigenfunction (representative) in the $M = 0$ subspace of eigenfunctions; that every energy level except those belonging to $S = 0$ has a representative in the $M = 1$ subspace; similarly for all except $S = 0$ and $S = 1$ in the $M = 2$ subspace, and so forth. The theorem, Eq. (4), will be proved if we can show that the lowest energy in an M subspace belongs to $S = M$, for spin $S + 1$ also has a representative in that subspace and therefore $E(S) < E(S+1)$. If the ground state belongs to $S = S_0$ (we still have to prove that $S_0 \leq \mathcal{S}$), we need only consider the subspaces of $|M| \geq S_0$, for the ground states of the remaining subspaces will always belong to S_0.

The mechanics of the proof are this: The ground state of H in an M subspace is *not* orthogonal to the ground state of a soluble Hamiltonian in the same subspace, and the latter is *known* to belong to $S = M$ for $M \geq \mathcal{S}$; therefore, so does the former. Now let us go into more detail.

PROOF

In an M subspace, choose the basis set to consist of all distinct eigenfunctions of the $\mathbf{s}_i^z$ compatible with eigenvalue M. We denote each configuration in the set by ϕ_a, where a is an index which runs over all members of the set. Shortly, we shall specify a convenient choice of phase for each configuration. But first, perform a canonical transformation on H by letting

$$\mathbf{S}^x_{i(A)} \to -\mathbf{S}^x_{i(A)}, \qquad \mathbf{S}^y_{i(A)} \to -\mathbf{S}^y_{i(A)}, \qquad \mathbf{S}^z_{i(A)} \to +\mathbf{S}^z_{i(A)} \tag{5}$$

but leaving the spins on the B-sublattice invariant. In the *new* language, the Hamiltonian can be written as $H_0 + H_1$, where the diagonal part is

$$H_0 = 2 \sum J_{ij} \mathbf{S}_i^z \mathbf{S}_j^z, \tag{6}$$

and the nondiagonal part is

$$H_1 = -\{ \sum |J_{ij}|\, S_i^+ S_j^- + \text{H.c.}\}. \tag{7}$$

We recall that g^2 of Eq. (2) is zero: the generalization for $g^2 > 0$ comes below.

In a given state ϕ_a, S_i^z has eigenvalue m_i. Choose the phase of ϕ_a in the following manner:

$$\phi_a = C(S_1^+)^{S_1+m_1}(S_2^+)^{S_2+m_2} \cdots (S_N^+)^{S_N+m_N}\chi, \tag{8}$$

where χ is the state in which $m_i = -S_i$, and C is a positive normalization constant. With this definition in mind, it is clear that if we define $K_{\beta a}$ to be

$$K_{\beta a} = \langle \phi_\beta | H_1 | \phi_a \rangle, \tag{9}$$

then

$$K_{\beta a} \leq 0, \quad \text{or equivalently,} \quad K_{\beta a} = -|K_{\beta a}|. \tag{10}$$

The ground state in the M subspace is denoted ψ, belongs to the ground-state energy E_M, and can be expanded in our complete set in terms of the amplitudes f_a,

$$\psi = \sum f_a \phi_a. \tag{11}$$

Since H_0 is diagonal, denote its eigenvalues by e_a,

$$H_0 \phi_a = e_a \phi_a, \tag{12}$$

and therefore the Schrödinger equation reads

$$-\sum_\beta |K_{\beta a}| \cdot f_\beta + e_a f_a = E_M f_a. \tag{13}$$

The variational energy of any trial function exceeds E_M, unless it is also a ground-state eigenfunction.

[1] W. Marshall, Proc. Roy. Soc. (London) **A232**, 48 (1955).
[2] E. Lieb, T. Schultz, D. Mattis, Ann. Phys. **16**, 407 (1961), particularly Appendix B.
[3] E. Lieb and D. Mattis, Phys. Rev. **125**, 164 (1962).

But

$$\psi' = \sum |f_a| \phi_a \tag{14}$$

is a trial function with variational energy E_M, and therefore

$$-\sum |K_{\beta a}| \, |f_\beta| + e_a |f_a| = E_M |f_a|. \tag{15}$$

Moreover,

$$e_a - E_M > 0, \quad \text{for all} \quad a \tag{16}$$

(otherwise, some *one* ϕ_a would be the ground state, which is in general impossible.) Therefore, taking the absolute value of $(e_a - E_M)f_a$ as given be Eq. (11) and combining with Eq. (15), we obtain

$$|(\sum |K_{\beta a}| \, f_\beta)| = \sum |K_{\beta a}| \, |f_\beta|. \tag{17}$$

This is a contradiction unless

$$f_\beta \geq 0 \quad \text{for all} \quad \beta. \tag{18}$$

In general, we have a slightly stronger result,

$$f_\beta > 0, \quad \text{for all} \quad \beta. \tag{19}$$

For, if some f_a vaaished, then Eq. (15) would read:

$$\sum K_{\beta a} |f_\beta| = 0,$$

and by succeeding applications of the Hamiltonian, one could establish that *all* the amplitudes vanished, unless the Hamiltonian splits into sets of noninteracting spins in which case only the weaker result (18) holds. Therefore, in general, all amplitudes are positive and nonvanishing, and hence E_M in nondegenerate. This last statement follows from the impossibility of constructing states orthogonal to ψ without some changes of sign, and consequent violation of the ground-state property (19).

Next consider the special Hamiltonian where $J_{i(A)j(A)} = J_{i(B)j(B)} = 0$ and $J_{i(A)j(B)} = J$, a positive constant. The eigenvalues are readily calculated. The lowest energy belonging to each spin is given by $E(S)$, for $S \geq \mathcal{S}$, and the ground state belongs to $S = \mathcal{S}$.

$$E(S) = J\{S(S+1) - S_A(S_A+1) - S_B(S_B+1)\} \quad \text{for} \quad S \geq \mathcal{S}. \tag{20}$$

By the previous arguments, the ground-state eigenfunctions of this special Hamiltonian in a given M subspace satisfy Eq. (18) or (19) and are therefore not all orthogonal to the corresponding ground state of H. The special Hamiltonian has an $S = M$ ground state in each M subspace, provided $M \geq \mathcal{S}$. Therefore, so does H and this completes the proof for $g^2 = 0$.

When $g^2 > 0$, we have proved the theorem (4) for $H - g^2\mathbf{S}^2$ and it is therefore true *a fortiori* for H. However, the lowest ground state no longer necessarily belongs to $\mathcal{S}$, but belongs to $S \leq \mathcal{S}$.

MAGNETIC FIELD

A magnetic field in the z direction but of arbitrary and variable amplitude B_i modifies H_0 but not H_1, and therefore (18) or (19) are still valid for the ground state in an M subspace. The absolute ground state of the system is no longer necessarily in the $M \leq \mathcal{S}$ subspace nor is S a good quantum number in the presence of such a magnetic field.

ACKNOWLEDGMENT

It is a pleasure to thank Dr. T. D. Schultz for helpful discussion.

Commun. math. Phys. 18, 160—170 (1970)

Entropy Inequalities

Huzihiro Araki
Research Institute for Mathematical Sciences, Kyoto University, Kyoto, Japan

Elliott H. Lieb*
Department of Mathematics
Massachusetts Institute of Technology, Cambridge, Massachusetts 02139, USA

Received March 2, 1970

Abstract. Some inequalities and relations among entropies of reduced quantum mechanical density matrices are discussed and proved. While these are not as strong as those available for classical systems they are nonetheless powerful enough to establish the existence of the limiting mean entropy for translationally invariant states of quantum continuous systems.

I. Introduction

In this note we shall be concerned with inequalities satisfied by the entropies of reduced density matrices. We begin with some definitions and a statement of our main Theorem 1. Section II contains the proof of the main theorem when the dimension is finite. Section III contains some other inequalities that can be derived from Theorem 1 by application of certain transformations. Section IV contains the proof of the main theorem when the dimension is infinite. Section V deals with the application of our theorem to the existence of the mean entropy for translationally invariant states of a quantum continuous system.

Definition 1. *A density matrix, ϱ, on a Hilbert space, H, is a self adjoint non-negative trace class operator on H whose trace is unity.*

Definition 2. *If ϱ is a density matrix,*

$$S(\varrho) = -\operatorname{Tr}\varrho\ln\varrho \tag{1.1}$$

is the entropy associated with ϱ.

Since $0 \leqq \varrho \leqq 1$, we have $-e^{-1} \leqq \varrho \ln\varrho \leqq 0$ and $(\psi_j, (\varrho\ln\varrho)\psi_j) \leqq 0$ for any ψ_j. Hence

$$S = -\sum_j (\psi_j, \varrho\ln\varrho\,\psi_j) \tag{1.2}$$

* Work supported by National Science Foundation Grant GP-9414.

exists for any orthonormal basis $\{\psi_j\}$ and $0 \leqq S \leqq +\infty$. If $S < \infty$ for one basis $\{\psi_j\}$, then the nonnegative operator $-\varrho \ln \varrho$ is in the trace class and hence S is independent of the basis $\{\psi_j\}$ and is finite for all $\{\psi_j\}$. Otherwise S must be $+\infty$. Therefore (1.2) does not depend on the orthonormal basis $\{\psi_j\}$ and defines the right hand side of (1.1).

Definition 3. *If ϱ^{12} is a density matrix on $H^1 \otimes H^2$ then ϱ^1, the reduced density matrix, is a density matrix on H^1 defined by*

$$\varrho^1 = \mathrm{Tr}^2 \varrho^{12}\,. \tag{1.3}$$

Here Tr^2 means the partial trace defined by

$$(x, \varrho^1 y) = \sum_i (x \otimes e_i, \varrho^{12}[y \otimes e_i]),$$

where $\{e_i\}$ is any complete orthonormal basis in H^2 and $x, y \in H^1$.

Notation. If ϱ^{12} is a density matrix on $H^1 \otimes H^2$ then we will denote $S(\varrho^{12})$ by S^{12} and $S(\varrho^1)$ by S^1.

A theorem that is true classically [1] (meaning that all relevant density matrices commute) is the following:

$$S^{123} + S^2 \leqq S^{12} + S^{23}\,. \tag{1.4}$$

We believe that (1.4) is true quantum mechanically as has been conjectured by Lanford and Robinson [2], but have been unable to prove it. We can, however, prove the following which is as good for some applications.

Theorem 1. *Let ϱ^{123} be a density matrix on $H^1 \otimes H^2 \otimes H^3$. Then*

$$S^{123} \leqq S^{12} + S^{23} + \ln \mathrm{Tr}^2 (\varrho^2)^2 \leqq S^{12} + S^{23}\,. \tag{1.5}$$

Furthermore, if $\varrho^2 \otimes I^3$ commutes with ϱ^{23} then

$$S^{123} + S^2 \leqq S^{12} + S^{23}\,. \tag{1.6}$$

II. Proof of Theorem 1 for the Finite Dimensional Case

In this section we prove Theorem 1 when the dimension of $H^1 \otimes H^2 \otimes H^3$ is *finite*. We need two lemmas.

Lemma 1 *(Peierls-Bogolyubov inequality). If R and F are hermitian, $\mathrm{Tr}\, e^R = 1$ and $f \equiv \mathrm{Tr}\, F e^R$, then $\mathrm{Tr}\, e^{R+F} \geqq e^f$.*

Proof. The statement of Peierls' theorem given in Ruelle [3] is not quite the same as the above. To prove Lemma 1 we use Klein's inequality [4]

$$\mathrm{Tr}\{f(A) - f(B) - (A - B)\, f'(B)\} \geqq 0$$

which holds for convex f and hermitian A and B. Take $f(x)=e^x$, $A=R+F$, $B=R+fI$.

Lemma 2 *(Golden-Thompson inequality* [5]*). Let A and B be hermitian. Then*

$$\operatorname{Tr} e^{A+B} \leqq \operatorname{Tr} e^A e^B .$$

Proof of Theorem 1. We first assume that ϱ^{123} is positive definite. Let $\varrho^{123} = \exp R^{123}$, $\varrho^{12} = \exp R^{12}$, etc. We will also denote $R^{12} \otimes I^3$ by R^{12}, etc. Then

$$\Delta = S^{123} - S^{12} - S^{23} = \operatorname{Tr}^{123} \varrho^{123} [-R^{123} + R^{12} + R^{23}] .$$

Using Lemma 1, $e^{\Delta} \leqq \operatorname{Tr}^{123} \exp[R^{123} - R^{123} + R^{12} + R^{23}]$.

Using Lemma 2, $e^{\Delta} \leqq \operatorname{Tr}^{123} \exp(R^{12}) \exp(R^{23}) = \operatorname{Tr}^{123} \varrho^{12} \varrho^{23} = \operatorname{Tr}^2 (\varrho^2)^2$.

Since the eigenvalues of ϱ^2 are in $[0, 1]$, and $\sum_i \lambda_i^2 \leqq (\sum \lambda_i)^2$ for $0 \leqq \lambda_i \leqq 1$, we have $\operatorname{Tr}^2 (\varrho^2)^2 \leqq 1$.

If R^2 commutes with R^{23}, then consider

$$\Delta = S^{123} + S^2 - S^{12} - S^{23} = \operatorname{Tr}^{123} \varrho^{123} [-R^{123} - R^2 + R^{12} + R^{23}] .$$

By Lemma 1, $e^{\Delta} \leqq \operatorname{Tr}^{123} \exp[R^{12} + R^{23} - R^2]$.

By Lemma 2,

$$e^{\Delta} \leqq \operatorname{Tr}^{123} \exp(R^{12}) \exp(R^{23} - R^2) = \operatorname{Tr}^{123} \varrho^{12} \varrho^{23} (\varrho^2)^{-1} = 1 .$$

The case of semidefinite ϱ^{123} follows by the continuity of $\varrho \to S(\varrho)$ for the finite dimensional case. (Note that the statement $-S^2 \leqq \ln \operatorname{Tr}^2 (\varrho^2)^2$ follows trivially from Lemma 1.)

As a corollary we have a well known theorem [6]:

Corollary. *If ϱ^{12} is a density matrix on $H^1 \otimes H^2$ then*

$$S^{12} \leqq S^1 + S^2 .$$

Proof. Interchange 2 and 3 in Theorem 1 and take H^3 to be one dimensional.

III. More Inequalities

The following definition and two lemmas are well known and are repeated here only for the sake of completeness. Matrices here need not be finite dimensional.

Definition 4. *A density matrix ϱ is said to be a pure state if ϱ is a projection operator onto a one-dimensional subspace, i.e. $\varrho x = y(y, x)$ for some fixed y with $|y| = 1$.*

Lemma 3. *Let ϱ^{12} be a pure state density matrix on $H^1 \otimes H^2$. Let $f(\cdot)$ be a real valued function whose domain contains the spectra of ϱ^1 and ϱ^2 and $f(0)=0$. Then*

$$\mathrm{Tr}^1 f(\varrho^1) = \mathrm{Tr}^2 f(\varrho^2).$$

In particular $S^1 = S^2$.

Proof. Let $\varrho^{12}x = (y, x)\, y$, $y = \sum \lambda_i y_{1i} \otimes y_{2i}$, $\lambda_i > 0$, where $\{y_{1i}\}$ and $\{y_{2i}\}$ can be taken to be orthonormal [7]. Let $P(y_{\nu i})$ be the projection on the one dimensional subspace of H^ν containing $y_{\nu i}$. Then $\varrho^\nu = \sum \lambda_i^2 P(y_{\nu i})$. Hence ϱ^1 and ϱ^2 have the same eigenvalues and multiplicities except possibly for the eigenvalue 0. The lemma follows immediately.

Lemma 4. *Let ϱ^1 be a density matrix on H^1. Then there exists a Hilbert space H^2 and a pure state density matrix ϱ^{12} on $H^1 \otimes H^2$ such that*

$$\mathrm{Tr}^2 \varrho^{12} = \varrho^1.$$

Proof. Let $\varrho^1 = \sum \lambda_j P_j$, $\lambda_j > 0$, $P_j x = (y_j, x)\, y_j$ and $\{y_j\}$ be orthonormal (the spectral decomposition). Let $\dim H^2 \geqq \dim H^1$ and let $\{z_j\}$ be an arbitrary orthonormal system in H^2. Let ϱ^{12} be the projection operator on the one dimensional subspace of $H^1 \otimes H^2$ containing the vector $\sum (\lambda_j)^{1/2} y_j \otimes z_j$. Then $\mathrm{Tr}^2 \varrho^{12} = \varrho^1$.

An application of lemmas 3 and 4 is the following:

Theorem 2. *Let ϱ^{123} be a density matrix on $H^1 \otimes H^2 \otimes H^3$. Then*

(a) $S^2 \leqq S^{23} + S^{12} + \ln \mathrm{Tr}^{123} (\varrho^{123})^2 \leqq S^{23} + S^{12}$.

(b) $S^2 \leqq S^{23} + S^{13} + \ln \mathrm{Tr}^1 (\varrho^1)^2 \leqq S^{23} + S^{13}$.

(c) $S^2 \leqq S^1 + S^{12}$.

Proof. We regard ϱ^{123} as a reduction of a pure ϱ^{1234}, whence $S^{123} = S^4$, $S^{12} = S^{34}$, $S^{23} = S^{14}$. Theorem 1 has the alternative forms:

(a′) $S^4 \leqq S^{34} + S^{14} + \ln \mathrm{Tr}^{134} (\varrho^{134})^2$.

(b′) $S^4 \leqq S^{34} + S^{23} + \ln \mathrm{Tr}^2 (\varrho^2)^2$.

(a′) and (b′) are general. In (a′) substitute 2 for 4. In (b′) substitute 2 for 4 and 1 for 2. To derive (c), let H^1 be one dimensional in (b) and then substitute 1 for 3.

Remarks. (1) Any other proof of any one of Theorems 1, 2a and 2b will give an alternative proof of Theorems 1, 2a, 2b and 2c.

(2) If we combine Theorem 2c with the corollary of Theorem 1, we obtain the triangle inequality

$$|S^1 - S^2| \leqq S^{12} \leqq S^1 + S^2. \tag{3.1}$$

[The left hand side should be taken to be 0 if $S^1 = S^2 = +\infty$.]

(3) Another application of Lemmas 3 and 4 is that the conjecture (1.4) is equivalent to

$$S^1 + S^2 \leqq S^{13} + S^{23}. \tag{3.2}$$

(4) We had to appeal to Lemmas 3 and 4 to prove Theorem 2a. A direct proof of Theorem 2a might indicate how to prove (3.2).

IV. Proof of Theorem 1 for the Infinite Dimensional Case

Definition 5. *Let $\{\psi_i^A\}$, $\{\lambda_i^A\}$ and $\{\psi_j^B\}$, $\{\lambda_j^B\}$ be complete orthonormal sets of eigenvectors and corresponding eigenvalues of selfadjoint operators A and B. Assume that $\lambda_i^A \geqq 0$, $\lambda_j^B \geqq 0$ for all i and j. Then we define*

$$\operatorname{Tr} AB \equiv \sum \lambda_i^A \lambda_j^B |(\psi_i^A, \psi_j^B)|^2. \tag{4.1}$$

(The value $+\infty$ is allowed.)

Remark. There exist cases where $\operatorname{Tr} AB < \infty$ and AB is not in the trace class (i.e. $\operatorname{Tr}|AB| = \infty$). If AB is in the trace class, then Definition 5 coincides with the ordinary definition of the trace.

Remark. $\operatorname{Tr} AB = \operatorname{Tr} BA$.

Lemma 5. *The definition of* $\operatorname{Tr} AB$ *is independent of the choice of the complete orthonormal sets of eigenvectors.*

Proof. (a) First we consider the case where A and B are projections. Then $\operatorname{Tr} AB = \sum \lambda_j^B(\psi_j^B, A\psi_j^B) = \operatorname{Tr} BAB$ where the trace of a nonnegative operator BAB is defined as in Definition 2 and is independent of the complete orthonormal sets $\{\psi_i^A\}$ and $\{\psi_j^B\}$.

(b) For general A and B, let $A = \sum_x x P^A(x)$ and $B = \sum_y y P^B(y)$ be the spectral decompositions of A and B. Then $\operatorname{Tr} AB = \sum_{x,y} xy \operatorname{Tr}(P^A(x) P^B(y))$ which is again independent of the complete orthonormal sets. Q.E.D.

In the above, we have used the fact that a sum of positive numbers is independent of the order of the summation irrespective of whether the sum is finite or infinite.

Lemma 6. *Let A^{12} be a bounded nonnegative selfadjoint operator on $H^1 \otimes H^2$ and B^2 be a selfadjoint operator on H^2 with purely discrete nonnegative spectrum. Then* $\operatorname{Tr} A^{12}(I^1 \otimes B^2) = \operatorname{Tr}^2 A^2 B^2$.

Proof. From Definition 5,

$$\operatorname{Tr} AB = \sum \lambda_j^B \|A^{1/2} \psi_j^B\|^2. \tag{4.2}$$

Hence

$$\begin{aligned}\operatorname{Tr} A^{12}(I^1 \otimes B^2) &= \sum_l \lambda_l^B \sum_k (\chi_k \otimes \varphi_l, A^{12} \chi_k \otimes \varphi_l) \\ &= \sum_l \lambda_l^B (\varphi_l, A^2 \varphi_l) \\ &= \operatorname{Tr}^2 A^2 B^2 .\end{aligned}$$

Here $\{\varphi_l\}$ and $\{\lambda_l^B\}$ are a complete orthonormal set of eigenvectors and corresponding eigenvalues of B and $\{\chi_k\}$ is any complete orthonormal set in H^1.

Corollary 6.1. *Let* $A = A^{12} \otimes I^3$, $B = I^1 \otimes B^{23}$ *where* A^{12} *and* B^{23} *are trace class nonnegative operators. Then* $\operatorname{Tr} AB = \operatorname{Tr}^2 A^2 B^2$.

Corollary 6.2. $\operatorname{Tr} \varrho^{ab}(I^a \otimes \ln \varrho^b) = \operatorname{Tr}^b \varrho^b \ln \varrho^b$.

Let $A \geqq 0$, $B = B_1 + B_2$, $B_1 B_2 = B_2 B_1 = 0$, $B_1 \geqq 0$, $B_2 \geqq 0$. Then $\|B^{1/2}\psi\|^2 = \|B_1^{1/2}\psi\|^2 + \|B_2^{1/2}\psi\|^2$ and hence we have, from (4.2),

$$\operatorname{Tr} AB \geqq \operatorname{Tr} AB_1 . \tag{4.3}$$

If ψ is not in the domain of C, we define $\|C\psi\| = \infty$. If $B = \sum x P(x)$ is a spectral decomposition of $1 \geqq B \geqq 0$, then we define $(-\ln B)^{1/2}$ by $\sum_{x \neq 0} (-\ln x)^{1/2} P(x)$ on those vectors ψ which satisfy $P(0)\psi = 0$ and $-\sum (\ln x) \|P(x)\psi\|^2 < \infty$.

Lemma 7. *Let* $\{\psi_i\}$ *be an orthonormal set of vectors,* A *and* B *be nonnegative operators with purely discrete spectrum majorized by* 1. *Then*

$$\sum_i \exp\{-\|(-\ln A)^{1/2}\psi_i\|^2 - \|(-\ln B)^{1/2}\psi_i\|^2 \leqq \operatorname{Tr} AB . \tag{4.4}$$

Proof. (a) First, assume that A and B are of finite rank. Then only a finite number of ψ_i can be in the domain of $(-\ln A)^{1/2}$ and the sum over i reduces to a finite sum. Hence we can discuss the inequality on a finite dimensional space for this case. We may assume that all ψ_i are in the domain of $(-\ln B)^{1/2}$ and $(-\ln A)^{1/2}$. If $A = \sum x P^A(x)$ and $B = \sum x P^B(x)$, we first prove the inequality for $A_\varepsilon = A + \varepsilon P^A(0)$ and $B_\varepsilon = B + \varepsilon P^B(0)$ where $\varepsilon > 0$. From Ref. [3] and the Golden-Thompson inequality, we have

$$\begin{aligned}&\sum_i \exp\{(\psi_i, (\ln A_\varepsilon)\psi_i) + (\psi_i, (\ln B_\varepsilon)\psi_i)\} \\ &\leqq \operatorname{Tr} \exp[\ln A_\varepsilon + \ln B_\varepsilon] \\ &\leqq \operatorname{Tr} A_\varepsilon B_\varepsilon .\end{aligned}$$

By taking the limit $\varepsilon \to 0$, we obtain (4.4) for this case.

(b) Consider the case of a finite number of ψ_i. Let $A = \sum \lambda_n^A P_n^A$ and $B = \sum \lambda_m^B P_m^B$ be spectral decompositions of A and B. (λ_n^A and λ_m^B are distinct.) Let P_{n0}^A and P_{m0}^B be finite subprojections of P_n^A and P_m^B, whose ranges are spanned by $\{P_n^A \psi_i\}$ and $\{P_m^B \psi_i\}$, respectively. Let $A_N = \sum_{<N} \lambda_n^A P_{n0}^A$, $A' = \sum \lambda_n^A P_{n0}^A$, $B_M = \sum_{<M} \lambda_m^B P_{m0}^B$, $B' = \sum \lambda_m^B P_{m0}^B$. From (a), we have (4.4) for A_N and B_M. We also have

$$\|(-\ln A_N)^{1/2}\psi_i\|^2 \uparrow \|(-\ln A')^{1/2}\psi_i\|^2 = \|(-\ln A)^{1/2}\psi_i\|^2 ,$$

$$\|(-\ln B_M)^{1/2}\psi_i\|^2 \uparrow \|(-\ln B')^{1/2}\psi_i|^2 = \|(-\ln B)^{1/2}\psi_i\|^2 .$$

From $B_M \leqq B$, $A_N \leqq A$, we have $\operatorname{Tr} A_N B_M \leqq \operatorname{Tr} A_N B \leqq \operatorname{Tr} AB$ because of (4.3). Hence, by taking the limit $N \to \infty$ and $M \to \infty$, we have (4.4) for A and B.

(c) For the general case, we have (4.4) for any finite subset of ψ_i from (b). Since the sum over i is the supremum of finite sums, we have (4.4) also for the general case. Q.E.D.

Proof of (1.5). Let $\{\psi_i\}$, $\{\lambda_i\}$ be a complete orthonormal set of eigenvectors and corresponding eigenvalues of ϱ^{123}. Let

$$\alpha_i = \exp - \{\|(-\ln \varrho^{12})^{1/2}\psi_i\|^2 + \|(-\ln \varrho^{23})^{1/2}\psi_i\|^2\} .$$

From Lemma 7 and Corollary 6.1, we have

$$\sum_I \alpha_i \leqq \operatorname{Tr} \varrho^{12}\varrho^{23} = \operatorname{Tr}^2(\varrho^2)^2 \leqq 1$$

where the sum is over any subset, I, of indices i. Choose I so that $\sum_I \lambda_i > 0$ whence, from the concavity of the logarithm, we have

$$\Delta_I \equiv \sum_I \lambda_i^I \ln(\alpha_i/\lambda_i^I) \leqq \ln \sum_I \alpha_i$$

where $\lambda_i^I = \lambda_i/(\sum_I \lambda_i)$. For a finite I, we have

$$\Delta_I\left(\sum_I \lambda_i\right) = S_I^{123} + \left(\sum_I \lambda_i\right) \ln\left(\sum_I \lambda_i\right) - S_I^{12} - S_I^{23} ,$$

where

$$S_I^\mu = \sum_I \lambda_i \|(-\ln \varrho^\mu)^{1/2}\psi_i\|^2 , \qquad \mu = 123,\, 12,\, 23 .$$

Hence

$$S_I^{123} + \left(\sum_I \lambda_i\right) \ln\left(\sum_I \lambda_i\right) \leqq S_I^{12} + S_I^{23} + \left(\sum_I \lambda_i\right) \ln \operatorname{Tr}^2(\varrho^2)^2 .$$

From Corollary 6.2, we have $S_I^\mu \uparrow S^\mu$ for $\mu = 12, 23$. We also have $S_I^{123} \uparrow S^{123}$ and $\sum_I \lambda_i \uparrow 1$. Hence we have

$$S^{123} \leqq S^{12} + S^{23} + \ln \operatorname{Tr}^2(\varrho^2)^2 .$$

Q.E.D.

Proof of (1.6). If $S^2 = +\infty$, then $S^{12} + S^{23} = +\infty$ due to Theorem 2a and hence (1.6) holds. [Theorem 2a is a consequence of (1.5), which has been proved above, and Lemma 4.] We now assume that $S^2 < +\infty$.

Let $\{\psi_i\}$ and $\{\lambda_i\}$ be a complete othonormal set of eigenvectors and corresponding eigenvalues of ϱ^{123}. From Corollary 6.2, it follows that ψ_i is in the domain of $(-\ln \varrho^2)^{1/2}$ if $S^2 < \infty$ and $\lambda_i \neq 0$. For a finite set, I, of indices, we have

$$\begin{aligned} &\textstyle\sum_I \exp[\|(-\ln \varrho^2)^{1/2}\psi_i\|^2 - \|(-\ln \varrho^{12})^{1/2}\psi_i\|^2 - \|(-\ln \varrho^{23})\,\psi_i\|^2] \\ &\qquad \leqq \mathrm{Tr}(\varrho^{12}\otimes I^3)(I^1\otimes \{[\varrho^2)^{-1}\otimes I^3]\,\varrho^{23}\}) = \mathrm{Tr}^2\varrho^2 = 1\,. \end{aligned} \tag{4.5}$$

The proof is exactly the same as Lemma 7 except that Lemma 6 should be used instead of Corollary 6.1.

From (4.5), we have

$$S_I^{123} + S_I^2 + (\textstyle\sum_I \lambda_i)\ln(\sum_I \lambda_i) \leqq S_I^{12} + S_I^{23}$$

and hence we have (1.6) by taking the supremum over I.

V. Application to Statistical Mechanics

In this section we prove the existence of the limiting mean entropy for translationally invariant states of quantum continuous systems [8, 9].

We shall restrict our attention to finite closed boxes with a fixed orientation in R^ν and their finite union $\bigcup_j L_j = \Lambda$. If $\Lambda_1 \cap \Lambda_2$ has a lower dimension (or is empty), we say that Λ_1 and Λ_2 are disjoint. For a finite number of mutually disjoint Λ_j, we denote $\cup \Lambda_j$ by $\bigvee_j \Lambda_j$.

Let $H(\Lambda)$ be a Hilbert space for each box Λ such that $H(\bigvee_j \Lambda_j) = \bigotimes_j H(\Lambda_j)$. Let $U_a(\Lambda)$ be a unitary mapping from $H(\Lambda)$ onto $H(\Lambda + a)$ such that $U_a(\Lambda + b)\,U_b(\Lambda) = U_{a+b}(\Lambda)$ and $U_a(\bigvee_j \Lambda_j) = \bigotimes_j U_a(\Lambda_j)$.

A state of a quantum continuous system for our present purpose is a set of density matrices $\varrho(\Lambda)$ for each Λ such that

$$\mathrm{Tr}_{\Lambda_1}\varrho(\Lambda_1 \vee \Lambda_2) = \varrho(\Lambda_2)\,. \tag{5.1}$$

It is translationally invariant if $\varrho(\Lambda + a) = U_a(\Lambda)\,\varrho(\Lambda)\,U_a(\Lambda)^*$ for each Λ.

The entropy $S(\Lambda)$ for each Λ is

$$S(\Lambda) = -\,\mathrm{Tr}_\Lambda \varrho(\Lambda)\ln \varrho(\Lambda)\,. \tag{5.2}$$

If ϱ is translationally invariant, $S(\Lambda + a) = S(\Lambda)$. Let $V(\Lambda)$ be the volume of Λ. Let C_a denote a cube of side length a.

Theorem 3. *If ϱ is a translationally invariant state, then the following limit exists:*

$$S(\varrho) = \lim_{a\to\infty} S(C_a)/V(C_a)\,. \tag{5.3}$$

Proof. Let $L_1 \sim NL_2$ (resp. $L_1 \propto NL_2$) denote the situation where L_1 is equal to (resp. contained in) a disjoint union of N translates of a box L_2.

(a) If $L_1 \sim NL_2$, we have, from the subadditivity (Corollary of Theorem 1),

$$S(L_1) \leqq NS(L_2). \tag{5.4}$$

(b) If $L_1 \propto L_2$ (two boxes are assumed to have the same orientation), then there exists α and β such that $L_1 = (L_2+\alpha) \cap (L_2+\beta)$. By Theorem 2(a), we have

$$2S(L_2) \geqq S(L_1). \tag{5.5}$$

(c) $S(L)=\infty$ for one box L if and only if $S(L')=\infty$ for all boxes L'. This is because, there exists N for any L and L' such that $L \propto NL'$ and hence $S(L') \geqq (2N)^{-1} S(L) = \infty$ by (5.4) and (5.5). Assume that $S(L) \neq \infty$ for all L in the following.

(d) Let Λ be a union of N mutually disjoint boxes L_j such that $V(L_j) \geqq v_0$ and $L_j \propto L$ for a fixed L.

From the subadditivity, we have

$$S(\Lambda) \leqq \sum S(L_j) \leqq 2NS(L).$$

We also have

$$V(\Lambda) = \sum V(L_j) \geqq Nv_0.$$

Hence

$$S(\Lambda)/V(\Lambda) \leqq 2S(L)/v_0. \tag{5.6}$$

(e) Let

$$\alpha_n(a) = S(C_{na})/V(C_{na}), \tag{5.7}$$

$$\alpha_\infty(a) = \inf_n \alpha_n(a). \tag{5.8}$$

From (5.4),

$$\alpha_{mn}(a) \leqq \alpha_n(a). \tag{5.9}$$

(f) Given $\varepsilon > 0$, there exists n such that

$$|\alpha_\infty(a) - \alpha_n(a)| < \varepsilon/3. \tag{5.10}$$

For this n, there exists $l \geqq na$ such that

$$2^{\nu+1} S(C_{na}) \sum_{k=1}^{\nu} \binom{\nu}{k} l^{-k} (na)^{k-\nu} < \varepsilon/3. \tag{5.11}$$

We then have the following estimates for $b > l$:

Let $m \geqq 1$ be an integer such that

$$1 \geqq |(na)^{-1} b - m| \geqq 1/2 .$$

Such an m exists. If $(na)^{-1} b > m$, then C_b is a disjoint union of a translate of C_{mna} and some Λ which, in turn, is a disjoint union of boxes L_j satisfying $L_j \propto C_{na}$ and $V(L_j) \geqq (na/2)^\nu$. If $(na)^{-1} b < m$, then C_{mna} is a disjoint union of C_b and some Λ which is a disjoint union of boxes L_j satisfying the same relations.

By (3.1), we have

$$|S(C_{mna}) - S(C_b)| \leqq S(\Lambda) . \tag{5.12}$$

From (5.6), we have

$$S(\Lambda)\, b^{-\nu} \leqq 2S(C_{na})\,(na/2)^{-\nu}\,(b^{-\nu} V(\Lambda)) .$$

Since

$$b^{-\nu} V(\Lambda) = |(mna/b)^\nu - 1| = \left| \sum_{k=1}^{\nu} \binom{\nu}{k} \left(\frac{mna}{b} - 1 \right)^k \right| \leqq \sum_{k=1}^{\nu} \binom{\nu}{k} (na/b)^k ,$$

we have from (5.12) and (5.11),

$$|S(C_{mna})\, b^{-\nu} - \alpha_1(b)| < \varepsilon/3 . \tag{5.13}$$

Next,

$$\begin{aligned} &|S(C_{mna})\, b^{-\nu} - \alpha_{mn}(a)| \\ &\quad = \alpha_{mn}(a)\, V(\Lambda)\, b^{-\nu} \\ &\quad \leqq \alpha_n(a)\, V(\Lambda)\, b^{-\nu} \\ &\quad < \varepsilon/3 . \end{aligned} \tag{5.14}$$

Finally, from $\alpha_\infty(a) \leqq \alpha_{mn}(a) \leqq \alpha_n(a)$, we have

$$|\alpha_\infty(a) - \alpha_{mn}(a)| < \varepsilon/3 . \tag{5.15}$$

Collecting (5.13), (5.14) and (5.15) together, we have

$$|\alpha_\infty(a) - S(C_b)/V(C_b)| < \varepsilon . \tag{5.16}$$

Q.E.D.

Remark. From the above proof, it is clear that if Λ is restricted to a disjoint union of boxes L_j whose volume is larger than a fixed v_0, then

$$\lim_{\Lambda \to \infty} S(\Lambda)/V(\Lambda) = \alpha_\infty(a)$$

where $\Lambda \to \infty$ in the sense of Van Hove.

Acknowledgements. We should like to thank Dr. D. Ruelle for suggesting this problem and for his constant stimulation and encouragement.

References

1. Ruelle, D.: Statistical mechanics. Proposition (7.2.6). New York: W. A. Benjamin 1969. For related results the reader is referred to: A. Huber in "Mathematical Methods in Solid State and Superfluid Theory", R. C. Clark and G. H. Derrick ed., Oliver and Boyd, Edinburgh, 1969, page 364; H. Falk, "Inequalities of J. W. Gibbs" to appear in Amer. J. Phys., July, 1970.
2. Lanford, O. E., Robinson, D. W.: J. Math. Phys. **9**, 1120—1125 (1968). Partial results have been obtained by F. Baumann and R. Jost, Problems of theoretical physics. Essays dedicated to N. N. Bogoliubov, p. 285. Moscow: Nauka 1969.
3. Ruelle, D.: *op. cit.*, Proposition (2.5.4).
4. — *op. cit.*, Proposition (2.5.2).
5. Golden, S.: Phys. Rev. **137**, B1127 (1965); — C. Thompson, J. Math. Phys. **6**, 1812 (1965).
6. Ruelle, D.: *op. cit.*, Proposition (7.2.10) and Section (7.2.13). See also H. Falk and E. Adler, Phys. Rev. **168**, 185—187 (1968); L. W. Bruch and H. Falk, "Gibbs Inequality in Quantum Statistical Mechanics" to appear in Phys. Rev.
7. Araki, H., Woods, E. J.: Publ. Res. Inst. Math. Sci. Kyoto Univ. Ser. A, **2**, 157—242 (1966). Definition 2.1.
8. Ruelle, D.: *op. cit.*, Section (7.2.13).
9. Miracle-Sole, S., Robinson, D. W.: Commun. Math. Phys. **14**, 235—270 (1969).

Huzihiro Araki
Research Institute
for Mathematical Sciences
Kyoto University
Kyoto/Japan

Elliott H. Lieb
Department of Mathematics
Massachusetts Institute of Technology
Cambridge, Massachusetts 02139, USA

VOLUME 30, NUMBER 10 PHYSICAL REVIEW LETTERS 5 MARCH 1973

A Fundamental Property of Quantum-Mechanical Entropy

Elliott H. Lieb*†
Institut des Hautes Etudes Scientifiques, 91 Bures-sur-Yvette, France

and

Mary Beth Ruskai*†
Department of Mathematics, Massachusetts Institute of Technology, Cambridge, Massachusetts 02139

(Received 26 December 1972)

We have proved the strong subadditivity of quantum-mechanical entropy and the Wigner-Yanase-Dyson conjecture.

There are some properties of entropy, such as concavity and subadditivity, that are known to hold (in classical and in quantum mechanics) irrespective of any assumptions on the detailed dynamics of a system. These properties are consequences of the definition of entropy as

$$S(\rho) = -\operatorname{Tr}\rho\ln\rho \text{ (quantum)}, \tag{1a}$$

$$S(\rho) = -\int\rho\ln\rho \text{ (classical continuous)}, \tag{1b}$$

$$S(\rho) = -\sum\rho_i\ln\rho_i \text{ (classical discrete)}, \tag{1c}$$

VOLUME 30, NUMBER 10 PHYSICAL REVIEW LETTERS 5 MARCH 1973

where Tr means trace, ρ is a density matrix in (1a), and ρ is a distribution function (usually on R^{6N}) in (1b). In (1c) the ρ_i are discrete energy level probabilities.

One such property, strong subadditivity (SSA), was known to hold for classical systems and was only conjectured for quantum systems. The observation that classical entropy has SSA (this, in fact, is a theorem in information theory) and that SSA implies strong results about the thermodynamic limit of entropy per unit volume is due to Robinson and Ruelle.[1] Later, Lanford and Robinson[2] conjectured that SSA holds for quantum systems as well, and Baumann and Jost[3,4] were able to prove this when ρ has a special form. Araki and Lieb[5] proved a weakened form of SSA, but one which held for general ρ and which was sufficient for many of the purposes to which SSA had been put in Ref. 1. The physical significance of SSA is explained below [item (f) of Table I].

Prior to these developments, Wigner and Yanase[6] proposed a different definition of entropy (or negative information) which was generalized by Dyson.[6] The conjecture that this generalized entropy was concave in ρ was also proved by Baumann and Jost[3,7] in special cases, but it was not realized that this concavity problem and the SSA problem were related; in fact they are equivalent.

Here, we wish to announce that both of these problems have been solved affirmatively. The proofs, which are too long for this note, will be given in two papers.[8,9]

A density matrix is a positive semidefinite operator with $\mathrm{Tr}\rho=1$. The Wigner-Yanase-Dyson p-entropy of ρ with respect to a self-adjoint operator (observable) K is

$$S_p(\rho,K)=\tfrac{1}{2}\mathrm{Tr}[\rho^p,K][\rho^{1-p},K], \tag{2}$$

where $[A,B]=AB-BA$ and $0\le p\le 1$ is fixed. We can think of (2) as defined for all $\rho\ge 0$ and ask whether $S_p(\rho,K)$ is concave as a function of ρ. The term $-\frac{1}{2}\mathrm{Tr}\rho K^2$ is obviously concave since it is linear, so the problem reduces to that of the concavity of $\mathrm{Tr}\rho^pK\rho^{1-p}K$. This was proved[6] when $p=\frac{1}{2}$. We have proved the following:

Theorem: For each fixed K (not necessarily self-adjoint), $\mathrm{Tr}\rho^pK^\dagger\rho^rK$ is a concave function of ρ for $\rho\ge 0$ whenever $p\ge 0$, $r\ge 0$, and $p+r\le 1$. This theorem is obviously stronger than necessary.

Returning to the conventional entropy (1a), we suppose that the Hilbert space of the system is a tensor product of three spaces, $H=H^1\otimes H^2\otimes H^3$. Thus, the system has three sets of degrees of freedom; for example, these may be thought of as the degrees of freedom of a gas in three disjoint regions in space (R^3). Given a density matrix ρ^{123} on H, we can define a density matrix ρ^{12} on $H^1\otimes H^2$ by partial trace, i.e., $\rho^{12}=\mathrm{Tr}_3\rho^{123}$. In like manner we can form ρ^{23}, ρ^2, etc., and for each of these we have an entropy given by (1a). Denoting $S(\rho^{123})$ by S^{123}, etc., subadditivity states that

$$S^{12}\le S^1+S^2, \tag{3}$$

while SSA states that

$$S^{123}+S^2\le S^{12}+S^{23}. \tag{4}$$

We first show that S^1-S^{12} is convex in ρ^{12}. This implies SSA because, as was pointed out previously,[5] in the quantum or classical discrete case SSA is equivalent to

$$F\equiv(S^1-S^{12})+(S^3-S^{23})\le 0, \tag{5}$$

but as F is convex in ρ^{123}, it is less than its maximum value on extremal points, which latter are those ρ^{123} that are pure states. For pure states, $F=0$.

We also prove some other related theorems,

TABLE I. Fundamental properties of entropy and their truth (T) or falsity (F) in three kinds of mechanics.

	Classical discrete	Classical continuous	Quantum
(a) $S(\rho)$ is concave in ρ	T	T	T
(b) $S^{12}\le S^1+S^2$	T	T	T
(c) $S(\rho)\ge 0$	T	F	T
(d) $S^{12}\ge S^1$	T	F	F
(e) $S^{12}\ge \lvert S^1-S^2\rvert$	T	F	T
(f) $S^{123}+S^2\le S^{12}+S^{23}$	T	T	T
(g) $S^{12}-S^1$ is concave in ρ^{12}	T	T	T

among which is the following:

Theorem: Let K be self-adjoint on a Hilbert space H and fixed. Then $\mathrm{Tr}[\exp(K+\ln\rho)]$ is a concave function of ρ for $\rho>0$. Closer inspection shows that this theorem is a generalization of the Golden-Thompson inequality[10,11] to three operators.

To conclude, in Table I we append a list of the known fundamental entropy [Eq. (1)] inequalities and their physical significance. The following remarks clarify the physical significance of Table I. The letters (a)–(g) refer to entries in Table I.

(a) states that if two different ensembles are united, the entropy of the resulting ensemble is greater than the average entropy of the component ensembles.[6]

(b) is a statement of subadditivity and is the basic tool for proving that the entropy per unit volume has a thermodynamic limit (which may, however, depend on the particular sequence of domains).

(c) expresses a well-known defect of classical continuous statistical mechanics with respect to the third law of thermodynamics.

(d) expresses an intuitive defect of quantum and classical continuous statistical mechanics. An example occurs when ρ^{12} is a pure state, so that $S^{12}=0$. Thus, the entropy of the universe can remain zero while the entropy of Earth increases without limit.

(e) is a consequence of SSA in the alternative form of inequality (5) (cf. Ref. 5) and is included as partial conpensation for (d).

(f) is the statement of SSA. As a technical tool it allows one to prove that the entropy per unit volume for quantum continuous systems is independent of shape, at least for rectangular parallelepipeds of fixed orientation (cf. Ref. 5). If one is willing to assume that the entropy of every bounded region is finite [which cannot be proved, as (d) is false quantum mechanically], then the limit exists for arbitrary regions in the sense of Van Hove. However, (f) has a more heuristic interpretation. Although the connection between entropy and information is hedged with controversy, we may suppose, along with the Copenhagen school, that when we measure a system its density matrix is reduced to that of a pure state and the entropy is reduced to zero. Thus, entropy measures the information gain in an experiment. $S^{23}-S^2$ can be thought of as the information gained upon measuring a total system (23) when a subsystem (2) is known. In quantum mechanics it may be negative because of (d). $S^{123}-S^{12}\leq S^{23}-S^2$ states that this incremental information is smaller when the initial information [(12) as against (2)] is larger. This, at least, is the interpretation given in information theory.

(g) states that the incremental information, like the entropy itself, increases when two ensembles are united.

We thank D. Ruelle for generating and encouraging our interest in this problem. We are also grateful to R. Jost, O. Lanford, and D. Robinson for their encouragement and for helpful conversations. One author (E.L.) would like to thank the Chemistry Laboratory III, University of Copenhagen, where part of this work was done.

---

*Work partially supported by U. S. National Science Foundation Grant No. GP-31674 X.

†On leave from Department of Mathematics, Massachusetts Institute of Technology, Cambridge, Mass. 02139. Work partially supported by a Guggenheim Memorial Foundation fellowship.

‡Presently, National Research Council Fellow at Department of Physics, University of Alberta, Edmonton 7, Alta., Canada.

[1]D. W. Robinson and D. Ruelle, Commun. Math. Phys. 5, 288 (1967).

[2]O. Lanford, III, and D. W. Robinson, J. Math. Phys. (N. Y.) 9, 1120 (1968).

[3]F. Baumann, Helv. Phys. Acta 44, 95 (1971).

[4]F. Bauman and R. Jost, in *Problems of Theoretical Physics; Essays Dedicated to N. N. Bogoliubov* (Nauka, Moscow, 1969), pp. 285–293.

[5]H. Araki and E. H. Lieb, Commun. Math. Phys. 18, 160 (1970).

[6]E. P. Wigner and M. M. Yanase, Proc. Nat. Acad. Sci. U. S. 49, 910 (1963), and Can. J. Math. 16, 397 (1964).

[7]R. Jost, in *"Quanta"—Essays in Theoretical Physics Dedicated to Gregor Wentzel,* edited by P. G. O. Freund, C. J. Goebel, and Y. Nambu (Univ. of Chicago Press, Chicago, Ill., 1970), pp. 13–19.

[8]E. H. Lieb, "Convex Trace Functions and Proof of the Wigner-Yanase-Dyson Conjecture" (to be published).

[9]E. H. Lieb and M. B. Ruskai, "Proof of the Strong Subadditivity of Quantum Mechanical Entropy" (to be published).

[10]S. Golden, Phys. Rev. B 137, 1127 (1965).

[11]C. J. Thompson, J. Math Phys. (N. Y.) 6, 1812 (1965).

Proof of the strong subadditivity of quantum-mechanical entropy

Elliott H. Lieb[*†]

I.H.E.S., 91 Bures-sur-Yvette, France

Mary Beth Ruskai[*‡]

Department of Mathematics, M.I.T., Cambridge, Massachusetts 02139

(Received 12 April 1973)

We prove several theorems about quantum-mechanical entropy, in particular, that it is strongly subadditive.

1. INTRODUCTION

In this paper we prove several theorems about quantum mechanical entropy, in particular, that it is strongly subadditive (SSA). These theorems were announced in an earlier note,[1] to which we refer the reader for a discussion of the physical significance of SSA and for a review of the historical background. We repeat here a bibliography of relevant papers.[2-9]

The setting for these theorems is as follows:

(a) Given a separable Hilbert space H and a positive, trace-class operator, ρ, on H [i.e., $\rho \geq 0$ means $(\psi,\rho\psi) \geq 0$ for all ψ in H], the entropy of ρ is defined to be

$$S(\rho) \equiv -\operatorname{Tr}\rho \ln\rho = -\sum_{i=1}^{\infty} \lambda_i \ln\lambda_i, \tag{1.1}$$

where Tr means trace, the λ_i are the eigenvalues of ρ, $0 \ln 0 \equiv 0$, and we permit the possibility $S(\rho) = \infty$. In physical applications one also requires that $\operatorname{Tr}\rho = 1$, in which case ρ is called a density matrix.

(b) If $H_{12} = H_1 \otimes H_2$ is the tensor product of two Hilbert spaces and ρ_{12} is a positive, trace-class operator on H_{12}, we can define a positive, trace-class operator, ρ_1, on H_1 by the partial trace, i.e.,

$$\rho_1 \equiv \operatorname{Tr}_2\rho_{12} \tag{1.2}$$

by which we mean

$$(\varphi,\rho_1\psi) = \sum_{i=1}^{\infty} (\varphi \otimes e_i, \rho_{12}[\psi \otimes e_i]) \tag{1.3}$$

for all φ, ψ in H_1 and $\{e_i\}_{i=1}^{\infty}$ any orthonormal basis in H_2. We shall denote $S(\rho_1)$ by S_1, etc. In like manner one can have $H_{123} = H_1 \otimes H_2 \otimes H_3$, and ρ_{123} a positive, trace-class operator on H_{123}, and define ρ_{12} on $H_{12} \equiv H_1 \otimes H_2$, ρ_1 on H_1, etc. by partial traces. When no confusion arises, we shall frequently use the symbol ρ_1 to denote the operator $\rho_1 \otimes 1_2$ on H_{12}.

Our main results are the following two theorems.

Theorem 1: Let $H_{12} = H_1 \otimes H_2$. Then the function

$$\rho_{12} \mapsto S_1 - S_{12} \tag{1.4}$$

is convex on the set of positive, trace-class operators on H_{12}.

Theorem 2 (Strong Subadditivity): Let H_{123} and ρ_{123} be defined as in (b) above. Then

(i) $$S_{123} + S_2 - S_{12} - S_{23} \leq 0 \tag{1.5}$$

and

(ii) $$S_1 + S_3 - S_{12} - S_{23} \leq 0. \tag{1.6}$$

In the next section we prove these theorems in the finite-dimensional case. In Sec. 3 we elucidate the connection between these two theorems and give some related results. Sec. 4 contains the proofs for the infinite-dimensional case and is based on the appendix kindly contributed by B. Simon, to whom we are most grateful.

2. PROOFS OF THEOREMS 1 AND 2 IN THE FINITE-DIMENSIONAL CASE

Proof of Theorem 1: The theorem states that

$$(S_1 - S_{12})(\rho_{12}) \leq \alpha(S_1 - S_{12})(\rho'_{12}) + (1-\alpha)(S_1 - S_{12})(\rho''_{12}) \tag{2.1}$$

where $\rho_{12} = \alpha\rho'_{12} + (1-\alpha)\rho''_{12}$, $0 \leq \alpha \leq 1$, and ρ'_{12} and ρ''_{12} are any positive, trace-class operators on H_{12}. We shall assume that both ρ'_{12} and ρ''_{12} are strictly positive and appeal to continuity of $\rho \mapsto S(\rho)$ in the semidefinite case. Letting

$$\Delta = \alpha \operatorname{Tr}_{12}\rho'_{12}\,(-\ln\rho'_{12} + \ln\rho'_1 + \ln\rho_{12} - \ln\rho_1)$$

and

$$\Gamma = (1-\alpha) \operatorname{Tr}_{12}\rho''_{12}\,(-\ln\rho''_{12} + \ln\rho''_1 + \ln\rho_{12} - \ln\rho_1),$$

one sees that (2.1) is equivalent to $\Delta + \Gamma \leq 0$. We now use Klein's inequality[7,10]:

$$\operatorname{Tr}(-A \ln A + A \ln B) \leq \operatorname{Tr}(B - A). \tag{2.2}$$

(Alternatively, one could use the Peierls–Bogoliubov inequality in a similar way.[2]) We first apply (2.2) to Δ with $A = \rho'_{12}$ and $B = \exp(\ln\rho'_1 + \ln\rho_{12} - \ln\rho_1)$ and then similarly to Γ. Then

$$\begin{aligned}\Delta + \Gamma &\leq \alpha \operatorname{Tr}_{12}[\exp(\ln\rho'_1 + \ln\rho_{12} - \ln\rho_1) - \rho'_{12}] \\ &\quad + (1-\alpha) \operatorname{Tr}_{12}[\exp(\ln\rho''_1 + \ln\rho_{12} - \ln\rho_1) - \rho''_{12}] \\ &\leq \operatorname{Tr}_{12}[\exp(\ln\rho_1 + \ln\rho_{12} - \ln\rho_1) - \rho_{12}] = 0.\end{aligned} \tag{2.3}$$

The second inequality in (2.3) follows from the concavity[11] of $C \mapsto \operatorname{Tr}[\exp(K + \ln C)]$ for positive C applied to $\rho_1 = \alpha\rho'_1 + (1-\alpha)\rho''_1$ with $K = \ln\rho_{12} - \ln\rho_1$. Q.E.D.

Proof of Theorem 2: It has already been pointed out[2] that (1.5) and (1.6) are equivalent; however, we shall prove each statement separately.

(i) Proof of (1.5): We use Klein's inequality, (2.2), with $A = \rho_{123}$ and $B = \exp(-\ln\rho_2 + \ln\rho_{12} + \ln\rho_{23})$. One finds

$$\begin{aligned}F(\rho_{123}) &\equiv S_{123} + S_2 - S_{12} - S_{23} \\ &\leq \operatorname{Tr}_{123}[\exp(\ln\rho_{12} - \ln\rho_2 + \ln\rho_{23}) - \rho_{123}].\end{aligned}$$

We now apply a generalization[11] of the Golden–Thompson inequality, i.e.,

$$\operatorname{Tr}[\exp(\ln B - \ln C + \ln D)] \le \operatorname{Tr} \int_0^\infty B(C + x\mathbf{1})^{-1} D(C + x\mathbf{1})^{-1} dx. \quad (2.4)$$

Thus

$$\begin{aligned} F(\rho_{123}) &\le \operatorname{Tr}_{123}\Big(\int_0^\infty \rho_{12}(\rho_2 + x\mathbf{1})^{-1} \times \rho_{23}(\rho_2 + x\mathbf{1})^{-1} dx - \rho_{123}\Big) \\ &= \operatorname{Tr}_2 \int_0^\infty \rho_2(\rho_2 + x\mathbf{1})^{-1}\rho_2(\rho_2 + x\mathbf{1})^{-1} dx - \operatorname{Tr}_{123}\rho_{123} \\ &= \operatorname{Tr}_2\rho_2 - \operatorname{Tr}_{123}\rho_{123} = 0. \end{aligned}$$

Q.E.D.

(ii) Proof of (1.6): Call the left side of (1.6) $G(\rho_{123})$. Note that $S_1 - S_{12}$ is convex in ρ_{12} by Theorem 1; since ρ_{12} is linear in ρ_{123}, $S_1 - S_{12}$ is convex in ρ_{123}. Thus, $G(\rho_{123})$ is convex in ρ_{123}. In the convex cone of positive matrices, the extremal rays consist of matrices of the form $\rho = \alpha P$ where $\alpha \ge 0$ and P is a one-dimensional projection. If ρ_{123} is extremal, then (see Ref. 2, Lemma 3) $S_1 = S_{23}$ and $S_3 = S_{12}$, so that $G(\rho_{123}) = 0$. Every positive matrix ρ_{123} can be written as a convex combination of extremal matrices; it then follows from the convexity of G that $G(\rho_{123}) \le 0$. Q.E.D.

3. REMARKS AND RELATED RESULTS

We have already noted in the proof of (1.6) that Theorem 1 implies Theorem 2. We now note that the converse is also true and give several alternative proofs of Theorems 1 and 2. We then show that $F(\rho_{123})$ is not convex and give a corollary to Theorem 1.

(A) To show Theorem 2 implies Theorem 1 it suffices to note that [apart from the trivial interchange of the subscripts 1 and 2 in (2.1)] (1.5) is identical to (2.1) for a special choice of ρ_{123}, i.e., $\rho_{123} = \alpha\rho'_{12} \otimes E_3 + (1-\alpha)\rho''_{12} \otimes F_3$ where H_3 is chosen to be two-dimensional and E_3 and F_3 are orthogonal, one-dimensional projections on H_3.

(B) Uhlmann[9] has shown that (1.5) follows from the concavity of $C \mapsto \operatorname{Tr}\exp(K + \ln C)$. This has been shown to be true by Lieb,[11] and an alternate proof was later found by Epstein.[12] Therefore, Uhlmann's remark gives an alternate proof of (1.5).

(C) The proof of (1.6) shows that Theorem 1 implies Theorem 2. However, (1.6) is not equivalent to (1.5) in other contexts.[13] [In fact, (1.6) is false in the classical continuous case.[6]] Therefore, it is instructive to note that one can show that Theorem 1 implies (1.5) directly without using (1.6). Baumann and Jost[3,5] have shown that a special choice of ρ'_{12} and ρ''_{12} in (2.1) implies that $\operatorname{Tr}\int_0^\infty A^*(C + x\mathbf{1})^{-1}A(C + x\mathbf{1})^{-1}dx$ is jointly convex in (A, C) where A and C are matrices with $C > 0$. Lieb has then shown[11] that this implies $C \mapsto \operatorname{Tr}\exp(K + \ln C)$ is concave in C. The last statement was used to prove[11] (2.4) which, as we have already seen, implies (1.5). Alternatively, we have already noted in (B) above that concavity of $C \mapsto \operatorname{Tr}\exp[K + \ln C]$ implies (1.5).

(D) We have already shown that the left side of (1.6), $G(\rho_{123})$, is convex. One might wonder, therefore, if the left side of (1.5), $F(\rho_{123})$, is also convex. In fact, it is not. If it were, one could choose H_2 to be one-dimensional so that

$$F(\rho_{123}) = S_{13} - S_1 - S_3 \equiv E(\rho_{13})$$

would have to be a convex function of ρ_{13}. Take H_1 and H_3 to be two-dimensional and choose ρ'_{13} and ρ''_{13} to be the following orthogonal, one-dimensional projections:

$$\rho'_{13}(i_1, i_3; j_1, j_3) = \tfrac{1}{2}\delta(i_1, i_3)\delta(j_1, j_3)$$

and

$$\rho''_{13}(i_1, i_3; j_1, j_3) = \tfrac{1}{2}[1 - \delta(i_1, i_3)][1 - \delta(j_1, j_3)],$$

where δ is the Krönecker delta. Then $\rho'_1 = \rho''_1 = \frac{1}{2}\mathbf{1}_1$, $\rho'_3 = \rho''_3 = \frac{1}{2}\mathbf{1}_3$, and $E(\rho'_{13}) + E(\rho''_{13}) - 2E(\frac{1}{2}\rho'_{13} + \frac{1}{2}\rho''_{13}) = -2\ln 2 < 0$, which is a contradiction.

(E) It was pointed out in Ref. 11 that if $f(A)$ is a convex function from the set of positive matrices into $\mathbb{R}$, and if it is also homogeneous [i.e., $f(\lambda A) = \lambda f(A)$ for all $\lambda > 0$], then

$$\frac{d}{dx} f(A + xB)\Big|_{x=0} \equiv \lim_{x \downarrow 0} x^{-1}[f(A + xB) - f(A)] \le f(B), \quad (3.1)$$

whenever A, B are positive matrices and the above limit exists. The function $(S_1 - S_{12})(\rho_{12})$ has these properties. To apply (3.1) we compute

$$\begin{aligned} \frac{d}{dx} S(\rho + x\gamma) &= -\frac{d}{dx}\operatorname{Tr}[(\rho + x\gamma)\ln(\rho + x\gamma)] \\ &= -\operatorname{Tr}\gamma\ln(\rho + x\gamma) - \operatorname{Tr}\gamma. \end{aligned}$$

Using this in (3.1) we conclude

Corollary: Let γ_{12} and ρ_{12} be positive, trace-class matrices on H_{12}. Then

$$\operatorname{Tr}_{12}\gamma_{12}\ln\rho_{12} - \operatorname{Tr}_1\gamma_1\ln\rho_1 \le \operatorname{Tr}_{12}\gamma_{12}\ln\gamma_{12} - \operatorname{Tr}_1\gamma_1\ln\gamma_1, \quad (3.2)$$

i.e., for each fixed γ_{12}, the left side of (3.2) achieves its maximum when $\rho_{12} = \gamma_{12}$.

4. EXTENSION TO INFINITE-DIMENSIONS

We can use Theorem A2 to extend Theorems 1 and 2 to infinite dimensions. For simplicity, we confine our discussion to Theorem 1 where $H_{12} = H_1 \otimes H_2$. The extension of Theorem 2 is similar and we point out the necessary changes at the end of this section.

Let E^n_i ($i = 1, 2$ and $n = 1, 2, \cdots$) be sequences of increasing, finite-dimensional projections on H_i, converging strongly to the identity, and define

$$E^n = E^n_1 \otimes E^n_2,$$

$$\rho^n_{12} = E^n\rho_{12}E^n,$$

and

$$\rho^n_1 = \operatorname{Tr}_2\rho^n_{12} = E^n_1(\operatorname{Tr}_2 E^n_2\rho_{12}E^n_2)E^n_1 \quad (4.1)$$

Since the spaces $E^n_i H_i$ are finite dimensional, Theorem 1 is satisfied by ρ^n_{12} on $E^n_1 H_1 \otimes E^n_2 H_2$ for each n. Thus, it suffices to show that the sequences of matrices $\{\rho^n_{12}\}_{n=1}^\infty$ and $\{\rho^n_1\}_{n=1}^\infty$ satisfy the hypotheses of Theorem A2 so that, e.g., $\lim_{n\to\infty} S(\rho^n_{12}) = S(\rho_{12}) = S_{12}$.

To show that $\{\rho^n_{12}\}_{n=1}^\infty$ satisfies Theorem A2, we first note that $E^n \xrightarrow{s} \mathbf{1}_{12}$. If[14] the sequences $A_n \xrightarrow{s} A$ and $B_n \xrightarrow{s} B$, then $A_n B_n \xrightarrow{s} AB$. Consequently, ρ^n_{12} converges to ρ_{12} strongly, and therefore weakly. It follows from the Ritz principle (see Proposition A1) that $\rho^n_{12} = E^n\rho_{12}E^n \prec E^{n+1}\rho_{12}E^{n+1} \prec \rho_{12}$, with $\prec$ as defined in the Appendix. Therefore, the hypotheses of Theorem A2 are satisfied and

$$\lim_{n\to\infty} S(\rho_{12}^n) = S_{12}. \tag{4.2}$$

To show that $\{\rho_1^n\}_{n=1}^{\infty}$ also satisfies Theorem A.2, define $\tilde{\rho}_1^n = \mathrm{Tr}_2 E_2^n \rho_{12} E_2^n$. Then $\rho_1^n = E_1^n \tilde{\rho}_1^n E_1^n$. To show that ρ_1^n converges to ρ_1 weakly, it suffices to show that $\tilde{\rho}_1^n$ converges to ρ_1^n strongly. (In fact, it converges uniformly.) To do this we can assume, without loss of generality, that E_2^n projects on the space spanned by $e_i \cdots e_n$ where $\{e_i : i = 1, 2, \cdots\}$ is an orthonormal basis in H_2. Then

$$(\psi, \tilde{\rho}_1^n \psi) = \sum_{i=1}^{n} (\psi \otimes e_i, \rho_{12} \psi \otimes e_i)$$

for all ψ in H_1, and it follows that

$$\tilde{\rho}_1^n \leq \tilde{\rho}_1^{n+1}, \tag{4.3}$$

and

$$\lim_{n\to\infty} (\psi, (\rho_1 - \tilde{\rho}_1^n)\psi) = \lim_{n\to\infty} \sum_{n+1}^{\infty} (\psi \otimes e_i, \rho_{12}\psi \otimes e_i) = 0 \tag{4.4}$$

Since $\tilde{\rho}_1^n$ is a monotone sequence of positive operators, (4.4) implies that $\tilde{\rho}_1^n \xrightarrow{s} \rho_1$ and therefore $\rho_1^n \xrightarrow{s} \rho_1$. Further, it follows from (4.3), i.e., the monotonicity of $\tilde{\rho}_1^n$, that

$$\begin{aligned} \rho_1^n &\nless E_1^{n+1} \tilde{\rho}_1^n E_1^{n+1} \\ &\leq E_1^{n+1} \tilde{\rho}_1^{n+1} E_1^{n+1} = \rho_1^{n+1} \nless \rho_1. \end{aligned}$$

Thus, Theorem A2 implies

$$\lim_{n\to\infty} S(\rho_1^n) = S(\rho_1) = S_1.$$

The analysis for Theorem 2 is similar. One defines

$$E^n = E_1^n \otimes E_2^n \otimes E_3^n,$$

$$\rho_{123}^n = E^n \rho_{123} E^n,$$

and

$$\rho_{12}^n = \mathrm{Tr}_3 \rho_{123}^n, \text{etc.}$$

ACKNOWLEDGMENTS

We thank D. Ruelle for generating and encouraging our interest in this problem. We are also grateful to R. Jost, O. Lanford, and D. Robinson for their encouragement and for helpful conversations. One author (E.L.) would like to thank the Chemistry Laboratory III, University of Copenhagen, where part of this work was done.

APPENDIX: CONVERGENCE THEOREMS FOR ENTROPY By B. Simon §

We discuss a variety of convergence theorems which are useful in extending entropy inequalities from finite dimensional matrices to infinite dimensional operators on a Hilbert space.

Definition: Let A be a positive compact operator. $\mu_k(A)$ denotes the kth largest eigenvalue of A counting multiplicity.

Definition: Let $s(x)$ be the function on $[0, \infty)$ given by

$$s(x) = \begin{cases} -x \ln x & \text{if } x \geq 0 \\ 0 & \text{if } x = 0. \end{cases}$$

If A is positive and compact, we set

$$S(A) = \sum_{k=1}^{\infty} s(\mu_k(A)),$$

the value infinity being allowed.

Definition: Let A and B be positive, compact operators. We write $A \nless B$ if and only if $\mu_k(A) \leq \mu_k(B)$ for all k.

Definition: Let $\{A_n\}_{n=1}^{\infty}$ and A be positive, compact operators. We write $A_n \xrightarrow{\mu} A$ if and only if $\mu_k(A_n) \longrightarrow \mu_k(A)$ for each fixed k.

Remarks: (1) The topology defined by μ-convergence is, of course, non-Hausdorff. (2) The order $\nless$ is useful because of the following consequence of the Ritz principle:

Proposition A1: Let A be a positive, compact operator and let P be a projection. Then $PAP \nless A$. In particular, if P and Q are projections and $P \leq Q$, then $PAP \nless QAQ$.

The above is false if $\nless$ is replaced by $\leq$.

Theorem A1 (Basic Convergence Theorem): Let B be a positive, compact operator with $S(B) < \infty$. Suppose $\{A_n\}$ and A are given positive, compact operators with

(1) $A_n \xrightarrow{\mu} A$,

(2) $A_n \nless B$ for each n.

Then, $\lim_{n\to\infty} S(A_n) = S(A)$.

Proof: The proof is based on the fact that s is monotone in $[o, e^{-1}]$. Since B is compact, $\mu_k(B) \longrightarrow 0$. Suppose $\mu_N(B) \leq e^{-1}$. By (1) and the continuity of s, $s(\mu_k(A_n)) \longrightarrow s(\mu_k(A))$, each k, and by (2) and the monotonicity of s in $[0, e^{-1}]$, $s(\mu_k(A_n)) \leq s(\mu_k(B))$ for $k \geq N$, each n. Thus by the dominated convergence theorem for sums, $\sum_{k\geq N} s(\mu_k(A_n)) \longrightarrow \sum_{k \geq N} s(\mu_k(A))$. Since $\sum_{k \leq N-1} s(\mu_k(A_n))$ certainly converges, the theorem is proven. Q.E.D.

For applications of Theorem A1, it is convenient to have statements expressed in a more usual form than μ-convergence.

Theorem A2: Let $\{A_n\}$ and A be positive, compact operators. If

(1) $\text{w-}\lim_{n\to\infty} A_n = A$

and

(2) $A_n \nless A$ for all n,

then $\lim_{n\to\infty} S(A_n) = S(A)$.

Proof: We first prove that $A_n \xrightarrow{\mu} A$. Fix k and ϵ. By weak convergence and the min-max principle, it is easy to find a k-dimensional space, V, and an N such that

$$(\psi, A_n \psi) \geq (\mu_k(A) - \epsilon) \|\psi\|^2$$

if $\psi \in V$ and $n \geq N$. But then $\mu_k(A_n) \geq \mu_k(A) - \epsilon$ if $n \geq N$. Since $\mu_k(A) \geq \mu_k(A_n)$ by (2), this means $|\mu_k(A) - \mu_k(A_n)| < \epsilon$ if $n \geq N$ and hence $A_n \xrightarrow{\mu} A$. If $S(A) < \infty$, the theorem then follows from Theorem A1. If $S(A) = \infty$, for any M we can find an L such that

$\sum_{k=1}^{L} s(\mu_k(A)) > M$. However, for L sufficiently large, $S(A_n) \geq \sum_{k=1}^{L} s(\mu_k(A_n))$ and, since $\mu_k(A_n) \longrightarrow \mu_k(A)$, the latter sum can be made arbitrarily close to M. Thus $S(A_n) \longrightarrow \infty$. Q.E.D.

Theorem A3: (Dominated Convergence Theorem for Entropy): Let $\{A_n\}, A$ and B be positive, compact operators and suppose that

(1) $S(B) < \infty$,

(2) $\text{w-lim}_{n\to\infty} A_n = A$,

(3) $A_n \leq B$ (operator inequality !).

Then,

$$\lim_{n\to\infty} S(A_n) = S(A).$$

Proof: Since B is compact, for any $\epsilon > 0$ we can find a finite-dimensional subspace $K \subset H$ such that $(u, Bu) = \|B^{1/2}u\| < \epsilon\|u\|$ for $u \in L$, where L is the orthogonal complement of K. Since $A_n \leq B$, $\|A_n^{1/2}u\| = (u, A_n u) \leq (u, Bu) \leq \epsilon\|u\|$ for all u in L. Since $A_n \xrightarrow{w} A$, $A \leq B$, and $\|A^{1/2}u\| \leq \epsilon\|u\|$ for all u in L also. We now show $A_n \longrightarrow A$ uniformly. Recall that $\|A_n - A\| = \text{supp}\{|(\varphi, (A_n - A)\psi)| : \varphi, \psi \in H, \|\varphi\| = \|\psi\| = 1\}$. Now write $\varphi = f + u$, $\psi = g + v$ where f, g are in K and u, v in L. Then

$$\begin{aligned}(\varphi, (A_n - A)\psi) &= ((f+u), (A_n - A)(g+v)) \\ &\leq (f, (A_n - A)g) + \|A_n^{1/2}f\|^{1/2}\|A_n^{1/2}v\|^{1/2} \\ &\quad + \|A^{1/2}f\|^{1/2}\|A^{1/2}v\|^{1/2} + \|A_n^{1/2}u\|^{1/2}\|A_n^{1/2}g\|^{1/2} \\ &\quad + \|A^{1/2}u\|^{1/2}\|A^{1/2}g\|^{1/2} + \|A_n^{1/2}u\|^{1/2}\|A_n^{1/2}v\|^{1/2} \\ &\quad + \|A^{1/2}u\|^{1/2}\|A^{1/2}v\|^{1/2},\end{aligned}$$

which can be arbitrarily small since $A_n \longrightarrow A$ uniformly on K, $A_n^{1/2}$ and $A^{1/2}$ are bounded on K, $\|A_n^{1/2}u\| < \epsilon$, $\|A^{1/2}u\| < \epsilon$, etc., and $\|f\| \leq \|\varphi\|$, etc. Thus $|(\varphi, (A_n - A)\psi)|$ can be made arbitrarily small independent of φ, ψ (for all φ, ψ with $\|\varphi\| = \|\psi\| = 1$) and thus $\|A_n - A\| \longrightarrow 0$. By the min-max principle, $|\mu_k(A_n) - \mu_k(A)| \leq \|A_n - A\|$. Thus $A_n \xrightarrow{\mu} A$, and (1) implies that Theorem A1 is applicable. Q.E.D.

Example: Let $\{A_n\}, A$ and B be the following operators on H, where $\{\varphi_n\}$ is an orthonormal basis for H:

$$A\varphi_k = 0, \quad \text{each } k,$$

$$A_n \varphi_k = \delta_{nk} e^{-1}\varphi_n,$$

$$B = A_1.$$

Then $A_n \nleq B$, $A_n \longrightarrow A$ strongly, but $S(A_n)$ does not converge to $S(A)$. This example shows that $\leq$ and not $\nleq$ is needed in Theorem A3.

*Work partially supported by U.S. National Science Foundation Grant GP-31674 X.

†On leave from Department of Mathematics, M.I.T., Cambridge, Mass. 02139, U.S.A. Work partially supported by a Guggenheim Memorial Foundation fellowship.

‡Supported in part by the National Research Council of Canada Grant No. NRC-A6595 at the University of Alberta, Edmonton, Canada. Present address: Department of Mathematics, University of Oregon, Eugene. Oregon 97403.

§Princeton University; A. Sloan Fellow.

[1] E. H. Lieb and M. B. Ruskai, Phys. Rev. Letters **30**, 434 (1973).

[2] H. Araki and E. H. Lieb, Commun. Math. Phys. **18**, 160 (1970).

[3] F. Bauman and R. Jost, in *Problems of Theoretical Physics; Essays Dedicated to N. N. Bogoliubov* (Moscow, Nauka, 1969), p. 285.

[4] R. Jost, in *Quanta: Essays in Theoretical Physics Dedicated to Gregor Wentzel*, edited by P. G. O. Freund, C. J. Goebel and Y. Nambu (University of Chicago Press, Chicago, 1970), p. 13.

[5] F Baumann, Helv. Phys. Acta **44**, 95 (1971).

[6] D. W. Robinson and D. Ruelle, Commun. Math. Phys. **5**, 288 (1967).

[7] O. Lanford III and D. W. Robinson, J. Math. Phys. **9**, 1120 (1968).

[8] E. P. Wigner and M. M. Yanase, Proc. Nat. Acad. Sci. **49**, 910 (1963); Can. J. Math. **16**. 397 (1964).

[9] A. Uhlmann. "Endlich Dimensionale Dichtematrizen, II". Wiss. Z. Karl-Marx-University Leipzig, Math-Naturwiss. R. 22, Jg. H. 2, 139 (1973).

[10] D. Ruelle, *Statistical Mechanics: Rigorous Results* (Benjamin, New York, 1969), Theorem 2.5.2.

[11] E. H. Lieb, "Convex Trace Functions and the Wigner-Yanase-Dyson Conjecture", Adv. in Math., to appear Dec. 1973.

[12] H. Epstein, Commun. Math. Phys. **37**, 317 (1973).

[13] M. B. Ruskai, "A Generalization of the Entropy Using Traces on von Neumann Algebras," preprint.

[14] O. Lanford III, in *Statistical Mechanics and Quantum Field Theory* edited by C. De Witt and R. Stora (Gordon and Breach, New York, 1971), p. 174.

BULLETIN OF THE
AMERICAN MATHEMATICAL SOCIETY
Volume 81, Number 1, January 1975

SOME CONVEXITY AND SUBADDITIVITY PROPERTIES OF ENTROPY

BY ELLIOTT H. LIEB[1]

I. **Introduction.** Statistical mechanics is the science of explaining, predicting and understanding the gross, macroscopic attributes of matter (which may be taken to mean mechanical systems with essentially an infinite number of degrees of freedom) in terms of the elementary dynamical laws governing its atomic constituents. The problems that arise are sufficiently complex and intriguing, but at the same time sufficiently well posed, that the subject is nowadays as much a part of mathematics as of physics. The fields of information theory and ergodic theory had their genesis in statistical mechanical modes of thought and are now well established in the mathematics literature; there will be more to come.

Ludwig Boltzmann, who died in 1906, was one of the principal founders of statistical mechanics, and his monument in Vienna contains the following eloquent testimonial to his scientific creativity:

$$S = k \log W. \tag{1}$$

Surely, this hypothesis of Boltzmann [**1**] is one of the most important and daring in statistical mechanics, for it relates S, the macroscopic entropy of a system, to W, the number of microscopic states of the system which have the same, given macroscopic properties. The number k is a universal constant, called Boltzmann's constant, and, for our purposes, we can consider it to be 1.

In these lectures we shall explore some of the abstract properties of entropy, after first giving a precise formulation of it, and will include some recent results (with M. B. Ruskai) which extend formerly known facts about the strong subadditivity of entropy from the domain of classical mechanics to the quantum-mechanical domain. The presentation here will be sketchy and the reader is referred to the original papers [**3**], [**4**], [**5**] for more details.

An expanded version of an invited address delivered before the M.I.T. meeting of the Society on October 27, 1973 by invitation of the Committee to Select Hour Speakers for Eastern Sectional Meetings; received by the editors January 7, 1974.

AMS (MOS) subject classifications (1970). Primary 80A10, 81A81, 82A05, 82A15, 94A15; Secondary 15A45, 28A35, 28A65, 47A99.

Key words and phrases. Entropy, strong subadditivity, convexity, density matrix.

[1] Work supported by National Science Foundation Grant GP 31674 X.

Bull. Amer. Math. Soc. *81*, 1–13 (1975)

II. **Definitions of entropy.** First we shall define abstractly what we mean by entropy in the *classical discrete case*. Let ρ denote a probability measure on an atomic probability space whose points are labelled by $i \in \mathcal{N}$. Hence $\rho(i) \in [0, 1]$ denotes the probability that event i occurs and $\sum_{i=1}^{\infty} \rho(i)=1$. The entropy of ρ is defined by

$$S(\rho) = -\sum_{i=1}^{\infty} \rho(i) \ln \rho(i) \tag{2}$$

with $0 \ln 0 \equiv 0$. As each term in the sum is nonpositive, S is well defined, although it may be $+\infty$. Obviously, S measures the extent to which ρ is "chaotic" or "spread out": If ρ is concentrated on one point (complete certainty) then $S=0$; If $\rho=1/W$ on W points and 0 otherwise, then $S=\ln W$. This last observation establishes the connection between (2) and (1). Clearly there are other functions besides $\rho \rightarrow -\rho \ln \rho$ which have the same qualitative property, but $-\rho \ln \rho$ alone has an important additivity property (additivity of entropy for independent systems) which we shall explain later (cf. equation (16)).

To establish contact with information theory we can define

$$I(\rho) = -S(\rho) \tag{3}$$

to be the information content of ρ (Shannon). The idea behind (3) is the following: Think of the index $i \in \mathcal{N}$ as denoting possible states of a system which is in some definite state j unknown to us. Interpret $\rho(i)$ as an assertion of a priori belief that the system is in the state i. Then, after we measure the system and find it to be in the state j, the new probability function is $\tilde{\rho}(i)=\delta_{i,j}$, $i \in \mathcal{N}$ (Kroenecker delta) and $S(\tilde{\rho})=I(\tilde{\rho})=0$. Thus, our knowledge (information) has increased by $S(\rho)$ and the entropy of the system has decreased by $S(\rho)$. For this reason, it is sometimes said that information is negative entropy. While such an assertion is true by definition (3), it is a matter of dispute whether it has any true physical import.

A generalization of (2) is the *classical continuous case* in which the underlying measure space, Ω, is not atomic and is equipped with a positive measure $d\mu(x)$ (not necessarily finite) and $\rho(x) \geqq 0$ is a probability density. Thus $\int \rho(x)\, d\mu(x)=1$ and

$$S(\rho) = -\int \rho(x) \ln \rho(x)\, d\mu(x). \tag{4}$$

A typical example in statistical mechanics is an N particle system with a Hamiltonian function $H(p, q)$, $p \in \boldsymbol{R}^{3N}$, $q \in \boldsymbol{R}^{3N}$, and $d\mu$ is Lebesgue measure on some subset Ω, of $\boldsymbol{R}^{3N} \times \boldsymbol{R}^{3N}$. Then

$$\rho(p, q) = Z^{-1} \exp[-\beta H(p, q)],$$

where

$$Z = \int_\Omega \exp[-\beta H(p, q)]\, d\mu(p, q), \tag{5}$$

and $\beta = (kT)^{-1}$ with T being the temperature.

Our third definition is the *quantum-mechanical case*. Instead of a measure space, one has a separable Hilbert space $\mathscr{H}$ and ρ is a positive trace-class operator on $\mathscr{H}$ (i.e. ρ is selfadjoint and $(x, \rho x) \geqq 0$, $\forall x \in \mathscr{H}$) with $\operatorname{Tr} \rho = 1$, where Tr is the trace. Such an operator ρ is called a *density matrix*. Then

$$S(\rho) = -\operatorname{Tr} \rho \ln \rho. \tag{6}$$

In a basis in which ρ is diagonal, (6) is seen to be identical to (2); the difference will manifest itself when we try to compare the entropies of two different ρ's which do not commute with each other. In other words, equation (6) is the noncommutative version of equation (2). The typical statistical mechanics example is as in (5), except that H becomes a self-adjoint operator on $\mathscr{H} = L^2(\boldsymbol{R}^{3N})$ and

$$\rho = Z^{-1} e^{-\beta H}, \qquad Z = \operatorname{Tr} e^{-\beta H}. \tag{7}$$

We remark in passing that entropy also plays a role in ergodic theory; given a measure preserving transformation T on a probability space Ω, Kolmogorov and Sinai [**2**] have been able to define the entropy of T by making use of (2) in such a way that the entropy is invariant under isomorphism. There exists an analogous notion of a measure preserving transformation in a Hilbert space setting, but an unsolved problem is to define the analogue of the Kolmogorov-Sinai entropy. In other words, it is not clear how to give an unambiguous definition of the density matrix to use in (6). We shall say no more about ergodic theory in this lecture.

III. **Properties of entropy (one space).** We turn now to a study of some properties of $S(\rho)$ that can be deduced from the definitions (2), (4) and (6). These properties are summarized in Table 1. The proofs we give will not only be sketchy but they will also assume that $\mathscr{H}$ is finite dimensional in the quantum case. The proofs for the infinite dimensional case can be found in [**4**] and [**5**]. When we say that some property is false in some particular case, we mean, of course, that the property does not hold generally and we do not mean that the property never holds.

Property A. $S(\rho) \geqq 0$ (*positivity of entropy*). This property is easily seen to be true in the classical discrete and the quantum cases but is false in the classical continuous case. The difficulty cannot be mitigated by adding a positive constant to the right side of (4) because $S(\rho)$ has no lower bound in the continuous case. Indeed, in the example (5) one sees that if H is nonconstant and has a unique minimum then as $\beta \to \infty$, $S \to -\infty$. Thus,

Bull. Amer. Math. Soc. *81*, 1–13 (1975)

if one believes, as Boltzmann did, that the entropy as we have defined it is the same (apart from an additive constant, possibly) as the physical entropy of a mechanical system, and if the latter is required to be positive, then classical mechanics cannot be valid at very low temperatures. Quantum mechanics must eventually be invoked.

Property B. *$S(\rho)$ is concave in ρ.* By this is meant that if $\rho=\alpha\rho'+(1-\alpha)\rho''$, $0\leqq\alpha\leqq 1$, and $\rho', \rho''\geqq 0$, then

$$S(\rho) \geqq \alpha S(\rho') + (1-\alpha)S(\rho''). \tag{8}$$

The physical interpretation is obvious: if two probability ensembles are mixed (which is not to be confused with the notion of mixing two systems, to be defined later) the entropy increases. This essential property of entropy is true in all three cases. In the classical cases the proof uses Jensen's inequality (i.e. $\int e^{A+B}\,d\mu \geqq \int e^{A}\,d\mu \exp\{\int Be^{A}\,d\mu/\int e^{A}\,d\mu\}$ for A and B real-valued functions). The analogous inequality in the Hilbert space case is the Peierls-Bogoliubov inequality:

$$\operatorname{Tr} e^{A+B} \geqq (\operatorname{Tr} e^{A})\exp\{\operatorname{Tr} Be^{A}/\operatorname{Tr} e^{A}\} \tag{9}$$

for A and B selfadjoint. To apply (9) assume that ρ' and ρ'' are strictly positive and write

$$\begin{aligned} \Delta &\equiv -S(\rho) + \alpha S(\rho') + (1-\alpha)S(\rho'') = \alpha\Delta' + (1-\alpha)\Delta'', \\ \Delta' &\equiv \operatorname{Tr} \rho'\{\ln\rho - \ln\rho'\}, \\ \Delta'' &\equiv \operatorname{Tr} \rho''\{\ln\rho - \ln\rho''\}. \end{aligned} \tag{10}$$

Then, using (9) with $A=\ln\rho'$, $B=\ln\rho-\ln\rho'$

$$e^{\Delta'} \leqq \operatorname{Tr}\exp\{\ln\rho' + \ln\rho - \ln\rho'\} = \operatorname{Tr}\rho = 1.$$

Thus, $\Delta'\leqq 0$. Likewise, $\Delta''\leqq 0$. Q.E.D. The case of ρ', ρ'' semidefinite can be handled by a continuity argument, and the classical cases can be proved in the same way using Jensen's inequality.

IV. **Properties of entropy (two spaces).** The remaining properties refer to the "mixture of different systems". In the classical cases this means that the total probability space, Ω_{12}, is taken to be the Cartesian product of two smaller spaces

$$\Omega_{12} = \Omega_1 \times \Omega_2 \tag{11}$$

and the product measure is assumed. Given a positive function ρ_{12} on Ω_{12} one can define a positive function ρ_1 on Ω_1 by

$$\rho_1(x) = \int \rho_{12}(x, y)\,d\mu_2(y). \tag{12}$$

Similarly, one defines ρ_2 on Ω_2.

In the quantum case, the Hilbert space $\mathscr{H}_{12}$ is taken to be the tensor product of two smaller spaces:

$$\mathscr{H}_{12} = \mathscr{H}_1 \otimes \mathscr{H}_2. \tag{13}$$

Given a positive, trace-class operator ρ_{12} on $\mathscr{H}_{12}$, one defines ρ_1 to be a positive, trace class operator on $\mathscr{H}_1$ by the partial trace operation:

$$\rho_1 \equiv \mathrm{Tr}_2\, \rho_{12}. \tag{14}$$

This means that

$$(x, \rho_1 y) \equiv \sum_{j=1}^{\infty} (x \otimes e_j, \rho_{12}(y \otimes e_j)) \tag{15}$$

for all $x, y \in \mathscr{H}_1$, and where $\{e_j\}_{j=1}^{\infty}$ is an orthonormal basis for $\mathscr{H}_2$. It is easy to prove that the right side of (15) is basis independent. In all cases, note that if ρ_{12} is normalized, then so is ρ_1.

Thus, for a mixed system one has three ρ's to consider, ρ_1, ρ_2 and ρ_{12} on three spaces. Corresponding to these there are three entropies: $S_1 = S(\rho_1)$, $S_2 = S(\rho_2)$ and $S_{12} = S(\rho_{12})$. The two spaces, Ω_1 and Ω_2 (or $\mathscr{H}_1$ and $\mathscr{H}_2$), can have two different physical interpretations (or both at once):

(i) Ω_1 and Ω_2 refer to two different sets of degrees of freedom of one physical system (e.g. the p and q variables in (5), in which case $\Omega_1 \subset \boldsymbol{R}^{3N}$ and $\Omega_2 \subset \boldsymbol{R}^{3N}$);

(ii) Ω_1 and Ω_2 refer to two different physical systems (e.g. the molecules of two different gases) which, under the product, are being thought of as one system.

One says that Ω_1 and Ω_2 (resp. $\mathscr{H}_1$ and $\mathscr{H}_2$) are *independent* if $\rho_{12}(x, y) = \rho_1(x)\rho_2(y)$ (resp. $\rho_{12} = \rho_1 \otimes \rho_2$). In this case it is easy to check the *additivity property*

$$S_{12} = S_1 + S_2. \tag{16}$$

Property C. $S_{12} \geqq S_1$ (*monotonicity of entropy*). This appealing statement is true only in the classical discrete case. It cannot be true generally in the classical continuous case because one could have Ω_1 and Ω_2 independent, so that (16) holds, and at the same time one could have $S_2 < 0$ because of the failure of property A. In the quantum case we note that $S(\rho) = 0$ if ρ is a *pure state*, i.e. ρ is a one-dimensional orthogonal projection on $\mathscr{H}$. (In the classical discrete case, the analogous statement is that ρ is concentrated on one point.) Conversely, $S(\rho) = 0$ implies that ρ is a pure state. Take ρ_{12} to be pure, in which case ρ_1 will not, in general, be pure in the quantum case. Then $S_{12} = 0$ and $S_1 > 0$. The proof of C for the classical discrete case is as follows: Denote the function ρ_{12} on $\mathscr{N} \times \mathscr{N}$ by $\rho(i, j)$ and ρ_1 on $\mathscr{N}$ by $\rho(i)$, $i, j \in \mathscr{N}$. $\rho(i) = \sum_{j=1}^{\infty} \rho(i, j)$. Define

$$\Delta \equiv S_1 - S_{12} = \sum_{i,j=1}^{\infty} \rho(i, j)(\ln \rho(i, j) - \ln \rho(i)).$$

Bull. Amer. Math. Soc. *81*, 1–13 (1975)

Assume that $\rho(i)>0$, $\forall i$. Then, by Jensen's inequality,

$$e^{\Delta} \leqq \sum_{i,j} \frac{\rho(i,j)^2}{\rho(i)} \leqq 1,$$

since $\rho(i) \geqq \rho(i,j)$, $i, j \in \mathcal{N}$. The general case follows by a continuity argument. Q.E.D.

The failure of C to hold presents a serious, but not insoluble problem for physics. It would mean, for example, that, under the second interpretation (ii) of ρ_{12}, the entropy of our planet could increase without limit while the entropy of the universe remains zero. Property E below, which holds in all cases, is partial compensation for the failure of C, but it is not enough. Instead, the resolution of the dilemma comes from further hypotheses about the kinds of ρ_{12}'s that actually occur in physical systems. In particular there are theorems that state that (in all three cases) when systems 1 and 2 are "large enough" then (16) is approximately true. In quantum mechanics $S_2>0$, and so the situation is saved—at least on the macroscopic level. More precisely, for macroscopic systems, S_1, S_2, and S_{12} are proportional to the volumes of the respective systems, whereas the error in (16), $S_{12}-(S_1+S_2)$, is proportional to the area of the surface separating systems 1 and 2.

Property D. $S_{12} \leqq S_1+S_2$ (*subadditivity of entropy*). This is one of the crucial facts about entropy and, fortunately, holds in all three cases. The proof is similar to that of property B: if $\Delta \equiv S_{12}-S_1-S_2$ then, from (9), $e^{\Delta} \leqq \mathrm{Tr}_{12}\, \rho_1 \otimes \rho_2 = 1$.

We now interject two technical lemmas that are relevant to the quantum case.

LEMMA 1. *Let ρ_{12} be a pure state on $\mathcal{H}_{12}$. Then the spectra of ρ_1 and ρ_2 on $\mathcal{H}_1$ and $\mathcal{H}_2$, including multiplicity, are identical except, possibly, for the point zero. In particular, $S_1=S_2$.*

The proof of Lemma 1 is easy and can be found in [**6**], among other places.

LEMMA 2. *Given a positive, trace class operator ρ_1 on a separable Hilbert space $\mathcal{H}_1$, there exist a separable Hilbert space $\mathcal{H}_2$ and a pure state ρ_{12} on $\mathcal{H}_{12}=\mathcal{H}_1 \otimes \mathcal{H}_2$ such that $\rho_1 = \mathrm{Tr}_2\, \rho_{12}$.*

Again, the easy proof can be found in [**6**].

Property E. $S_{12} \geqq |S_1-S_2|$ (*triangle inequality*). We call this the triangle inequality because, when it is combined with D one has

$$S_1 + S_2 \geqq S_{12} \geqq |S_1 - S_2|. \tag{17}$$

However, E is not true in the classical continuous case (because S_{12} can be negative), but it is true in the other cases. To prove E in the quantum case we use Lemma 2 to find a Hilbert space $\mathscr{H}_3$ and a pure state ρ_{123} on $\mathscr{H}_1 \otimes \mathscr{H}_2 \otimes \mathscr{H}_3$ such that $\rho_{12} = \mathrm{Tr}_3\, \rho_{123}$, and we define $\rho_3 = \mathrm{Tr}_{12}\, \rho_{123}$ (resp. $\rho_{23} = \mathrm{Tr}_1\, \rho_{123}$) on $\mathscr{H}_3$ (resp. $\mathscr{H}_2 \otimes \mathscr{H}_3$). But then, by Lemma 1, $S_{12} = S_3 \equiv S(\rho_3)$, $S_1 = S_{23} \equiv S(\rho_{23})$ and, by property D,

$$-S_1 + S_{12} + S_2 = -S_{23} + S_3 + S_2 \geqq 0 \quad \text{Q.E.D.} \tag{18}$$

Finally, while there is no natural analogue of Lemmas 1 and 2 in the classical discrete case, property E is nevertheless true there as well. This is so because $S_{12} \geqq S_1 - S_2$ holds in the special case that ρ_{12} commutes with $\rho_1 \otimes I_2$ and $I_1 \otimes \rho_2$; but this special case is precisely the classical discrete case, provided one thinks of the function ρ_{12} on $\mathscr{N} \times \mathscr{N}$ in the obvious way as a diagonal matrix on $l_1 \otimes l_1$. There exists, in fact, a direct proof of property E for the classical discrete case similar to the proof of property C, but the foregoing detour through the quantum domain is more amusing.

V. **Properties of entropy (three spaces).** Up to this point we have been concerned with the product of two spaces. Now the plot thickens; we consider three spaces and the property of *strong subadditivity* and its variants. Given ρ_{123} on $\Omega_1 \times \Omega_2 \times \Omega_3$ (or $\mathscr{H}_1 \otimes \mathscr{H}_2 \otimes \mathscr{H}_3$) we can, by taking partial traces, define $S_{123} = S(\rho_{123})$, $S_{12} = S(\rho_{12})$, $S_1 = S(\rho_1)$, etc. We list three properties that are closely related (note that G refers to only two spaces).

Property F. $S_{123} + S_2 \leqq S_{12} + S_{23}$ (*strong subadditivity of entropy*).

Property G. $S_{12} - S_1$ *is concave in* ρ_{12}.

Property H. $S_1 + S_2 \leqq S_{13} + S_{23}$.

The significance of properties F, G and H will be discussed after Table 1 is completed. At this point we shall merely indicate the proofs of those three properties. Consider G first. In order to simplify the notation, we shall use Tr to mean $\sum$ or $\int$ in the classical cases. If we use the Jensen or Peierls-Bogoliubov inequality as in (10), we end up trying to prove that

$$f \equiv \alpha\, \mathrm{Tr}_{12} \exp(K + \ln \rho_1') + (1 - \alpha)\mathrm{Tr}_{12} \exp(K + \ln \rho_1'') \leqq 1,$$

with $K = \ln \rho_{12} - \ln \rho_1$ and with an abuse of notation in which ρ_1 stands for $\rho_1 \otimes I_2$. In the two classical cases $f = 1$ since $\rho_1 = \alpha\rho_1' + (1-\alpha)\rho_1''$. For the quantum case we need a lemma [**4**]:

LEMMA 3. *Let K be any fixed selfadjoint $n \times n$ matrix. Then the map from the positive $n \times n$ matrices to the reals defined by $A \to \mathrm{Tr}\exp(K + \ln A)$ is concave.*

The proof [**4**] of Lemma 3 is lengthy, but we can use Lemma 3 to obtain

$$f \leqq \mathrm{Tr}_{12} \exp(K + \ln \rho_1) = 1.$$

Bull. Amer. Math. Soc. *81*, 1–13 (1975)

Next we consider property H. Since ρ_{13} is linear in ρ_{123}, property G implies that $S_{13}-S_1$ is concave in ρ_{123}. Likewise, $S_{23}-S_2$ is concave in ρ_{123}. Thus $\Delta\equiv(S_{13}-S_1)+(S_{23}-S_2)$ is concave in ρ_{123}. In the classical discrete and quantum cases, Δ will have its minimum on extremal states, which are pure states. In that case (see Lemma 1) $S_{13}=S_2$ and $S_{23}=S_1$. In the classical continuous case this argument fails and, in fact, H is false because one could take $\rho_{123}=\rho_{12}\rho_3$ with $S_3<0$.

Finally, we consider property F. By applying the Jensen or Peierls-Bogoliubov inequality, we have to prove that

$$\Delta \equiv \operatorname{Tr}_{123}\exp(\ln\rho_{12}+\ln\rho_{23}-\ln\rho_2)\leqq 1.$$

Clearly, $\Delta=1$ in the two classical cases. In the quantum case an involved argument using Lemma 3 (see [**5**]) shows that $\Delta\leqq 1$. However, there is another way to prove F in the quantum case and this method displays the close connection between properties F and H. As in the proof of property E, introduce a fourth Hilbert space $\mathscr{H}_4$ and ρ_{1234} pure on $\mathscr{H}_{123}\otimes\mathscr{H}_4$ such that $\operatorname{Tr}_4\rho_{1234}=\rho_{123}$. Then $S_{123}+S_2-S_{12}-S_{23}=S_4+S_2-S_{12}-S_{14}$ (by Lemma 1) and this is nonpositive by property H.

REMARK. For the quantum case we have proved G$\Rightarrow$H$\Rightarrow$F. It is also true, however, that F$\Rightarrow$G as may be seen by a special choice of ρ_{123} (see [**5**]).

To complete Table 1 we have to discuss properties I, J, K and L which, as Table 1 shows, are always false and therefore uninteresting. They are mentioned for two reasons: The first is that it occasionally occurs to someone that these properties may be true. Secondly, we shall need to know their falsity in discussing properties F″ and G″ of Table 3.

With the definition $\Delta=S_{123}-S_{12}-S_{13}-S_{23}+S_1+S_2+S_3$, property I states that $\Delta\leqq 0$, while property J states that $\Delta\geqq 0$. We shall prove that these statements are false in the classical discrete case and, *a fortiori*, they are also false in the other two cases. Take $\rho_{123}(i,j,k)=\rho_{12}(i,j)\delta_{j,k}$, whence $S_{123}=S_{12}$, $S_{13}=S_{12}$ and $S_{23}=S_2=S_3$. Then $\Delta=-S_{12}+S_1+S_2$ and this can be positive since ρ_{12} is arbitrary. To demonstrate the falsity of property J an explicit example is required. Let $\rho_{123}(1,1,2)=\frac{1}{3}$, $\rho_{123}(1,2,1)=\frac{1}{3}$, $\rho_{123}(2,1,1)=\frac{1}{3}$ and $\rho_{123}(i,j,k)=0$ otherwise for $i,j,k\in\mathscr{N}$. Then $S_{123}=S_{12}=S_{13}=S_{23}=\ln 3$, $S_1=S_2=S_3=\ln 3-\frac{2}{3}\ln 2$ and $\Delta=\ln 3-2\ln 2<0$.

Property K (resp. L) states that $\Delta\equiv S_{12}-S_1-S_2$ is concave (resp. convex). For K consider ρ_{12} of the form $\rho_{12}(i,j)=\rho_1(i)\delta_{i,j}$. Then $\Delta=-S_1=-S(\rho_1)$, but, as ρ_1 is arbitrary, one can have $-S(\rho_1')-S(\rho_1'')>-2S(\frac{1}{2}\rho_1'+\frac{1}{2}\rho_1'')$ (property B). For L, take $\rho_{12}'(i,j)=\frac{1}{2}\delta_{i,1}\delta_{j,1}+\frac{1}{2}\delta_{i,2}\delta_{j,2}$ and $\rho_{12}''(i,j)=\frac{1}{2}\delta_{i,2}\delta_{j,1}+\frac{1}{2}\delta_{i,1}\delta_{j,2}$, so that $\Delta(\rho_{12}')=\Delta(\rho_{12}'')=-\ln 2$ and $\Delta(\frac{1}{2}\rho_{12}'+\frac{1}{2}\rho_{12}'')=0$.

TABLE 1

	Classical discrete	*Classical continuous*	*Quantum*
A. $S(\rho) \geqq 0$	T	F	T
B. $S(\rho)$ concave in ρ	T	T	T
C. $S_{12} \geqq S_1$	T	F	F
D. $S_{12} \leqq S_1 + S_2$	T	T	T
E. $S_{12} \geqq \|S_1 - S_2\|$	T	F	T
F. $S_{123} + S_2 \leqq S_{12} + S_{23}$	T	T	T
G. $S_{12} - S_1$ concave in ρ_{12}	T	T	T
H. $S_1 + S_2 \leqq S_{13} + S_{23}$	T	F	T
I. $S_{123} + S_1 + S_2 + S_3 \leqq S_{12} + S_{13} + S_{23}$	F	F	F
J. $S_{123} + S_1 + S_2 + S_3 \geqq S_{12} + S_{13} + S_{23}$	F	F	F
K. $S_{12} - S_1 - S_2$ concave in ρ_{12}	F	F	F
L. $S_{12} - S_1 - S_2$ convex in ρ_{12}	F	F	F

TABLE 1. Properties of entropy and their truth (T) or falsity (F) in the three cases.

Thus, Table 1 is complete and we now take up the interpretation of strong subadditivity. There are two. The first is simple but the second is involved and will lead us to the construction of two more tables.

VI. **First interpretation of strong subadditivity.** Property F can be viewed as a generalization of property D, i.e. subadditivity. For concreteness we use the measure space language, but the same idea is applicable in the Hilbert space setting. Suppose that to each Lebesgue measurable subset, Λ, of $\boldsymbol{R}^3$ we associate a probability space $\Omega(\Lambda)$, which can be thought of physically as the set of configurations of a "gas of particles" contained in Λ. We assume that when Λ_1 and Λ_2 are disjoint $\Omega(\Lambda_1 \cup \Lambda_2) = \Omega(\Lambda_1) \times \Omega(\Lambda_2)$. Given a probability density ρ_Λ on each $\Omega(\Lambda)$ we then have an entropy $S(\rho_\Lambda) \equiv S(\Lambda)$ for each Λ. Property D states that

$$S(\Lambda_1 \cup \Lambda_2) \leqq S(\Lambda_1) + S(\Lambda_2) \tag{19}$$

when Λ_1 and Λ_2 are disjoint.

What if Λ_1 and Λ_2 are not disjoint? Define $\Lambda_3 = \Lambda_1 \cap \Lambda_2$, $\Lambda_4 = \Lambda_1 - \Lambda_3$ and $\Lambda_5 = \Lambda_2 - \Lambda_3$. Then, property F states that $S_{345} + S_3 \leqq S_{34} + S_{35}$, or

$$S(\Lambda_1 \cup \Lambda_2) + S(\Lambda_1 \cap \Lambda_2) \leqq S(\Lambda_1) + S(\Lambda_2), \tag{20}$$

which is an appropriate generalization of (19).

VII. **Second interpretation of strong subadditivity: relative entropy.** Given the product of two spaces $\mathscr{H}_1 \otimes \mathscr{H}_2$ (or $\Omega_1 \times \Omega_2$) and a density matrix ρ_{12} on $\mathscr{H}_{12}$ one can *define*

$$S(2 \mid 1) \equiv S_{12} - S_1$$

to be "*the conditional entropy of* 2 *relative to* 1". The term is slightly misleading since $S(2|1)$ depends upon ρ_{12} and not merely on ρ_2, but this is a minor flaw. The real question is whether or not the word entropy is justified in describing $S(2|1)$, which means that we have to check the whole list of properties A to H for $S(2|1)$. In order to do so, strong subadditivity will play a crucial role. It might appear at first glance that since property G involves three spaces and $S(2|1)$ involves two spaces, we shall need a new theorem about four spaces and this will lead us into an endless hierarchy of more and more complex inequalities. This is not so. It turns out, remarkably enough, that we already have all the information we need.

There is, however, one complication. Properties C to H refer to two or more spaces and hence we can let either $\mathscr{H}_2$ or $\mathscr{H}_1$ be such a product. Thus, considering both possibilities we shall end up with two tables (Table 2 and Table 3 respectively). It will be seen that $S(2|1)$ merits the appellation entropy in terms of expanding $\mathscr{H}_2$ (Table 2) fairly well, but that in terms of expanding $\mathscr{H}_1$ it fares poorly. In fact, in Table 3 there are two properties, C″ and H″, that are not only false but their opposites are

TABLE 2

	Classical discrete	Classical continuous	Quantum
A′. $S(2 \mid 1) \geqq 0$	T	F	F
B′. $S(2 \mid 1)$ concave in ρ_{12}	T	T	T
C′. $S(23 \mid 1) \geqq S(2 \mid 1)$	T	F	F
D′. $S(23 \mid 1) \leqq S(2 \mid 1) + S(3 \mid 1)$	T	T	T
E′. $S(23 \mid 1) \geqq \lvert S(2 \mid 1) - S(3 \mid 1) \rvert$	T	F	F
F′. $S(234 \mid 1) + S(3 \mid 1) \leqq S(23 \mid 1) + S(34 \mid 1)$	T	T	T
G′. $S(23 \mid 1) - S(2 \mid 1)$ concave in ρ_{123}	T	T	T
H′. $S(2 \mid 1) + S(3 \mid 1) \leqq S(24 \mid 1) + S(34 \mid 1)$	T	F	F

TABLE 2. Properties of relative entropy, $S(2|1) = S_{12} - S_1$, with respect to the second space and their truth (T) or falsity (F).

true. These are indicated by ~C″ and ~H″. The most important of these is ~C″, namely for three spaces

$$S(2 \mid 13) \leqq S(2 \mid 1). \tag{21}$$

The interpretation of inequality (21), and indeed of $S(2|1)$ itself is obvious. $S(2|1)$ is the incremental information gained by measuring the total system (12) as against merely measuring the subsystem (1). Inequality (21) states that this incremental information gain is less if one knows more to start with (namely (13) as against (1)).

TABLE 3

	Classical discrete	Classical continuous	Quantum
~C″. $S(2 \mid 13) \leqq S(2 \mid 1)$	T	T	T
D″. $S(2 \mid 13) \leqq S(2 \mid 1) + S(2 \mid 3)$	T	F	T
E″. $S(2 \mid 13) \geqq \lvert S(2 \mid 1) - S(2 \mid 3)\rvert$	F	F	F
F″. $S(2 \mid 134) + S(2 \mid 1) \leqq S(2 \mid 13) + S(2 \mid 14)$	F	F	F
~F″. $S(2 \mid 134) + S(2 \mid 1) \geqq S(2 \mid 13) + S(2 \mid 14)$	F	F	F
G″. $S(2 \mid 13) - S(2 \mid 1)$ concave in ρ_{123}	F	F	F
~G″. $S(2 \mid 13) - S(2 \mid 1)$ convex in ρ_{123}	F	F	F
~H″. $S(2 \mid 1) + S(2 \mid 3) \geqq S(2 \mid 14) + S(2 \mid 34)$	T	T	T

TABLE 3. Properties of relative entropy, $S(2|1) \equiv S_{12} - S_1$, with respect to the first space and their truth (T) or falsity (F). The designation ~ indicates that a property is opposite to that in Table 1.

Having set forth Tables 2 and 3 we shall conclude with brief indications of how the entries in the two tables can be checked.

Table 2.

Properties A′ and C′ follow from property C.

Property B′ follows from property G.

Property D′ follows from property F.

Property E′ is true in the classical discrete case because if we define $\Delta \equiv -S_{123} - S_{13} + S_{12} + S_1$ then

$$e^{\Delta} \leqq \sum_i \sum_j \sum_k \frac{\rho_{123}(i,j,k)^2 \rho_{13}(i,k)}{\rho_1(i)\rho_{12}(i,j)}.$$

Bull. Amer. Math. Soc. *81*, 1–13 (1975)

But $\rho_{13}(i, k) \leqq \rho_1(i)$ and $\rho_{123}(i, j, k) \leqq \rho_{12}(i, j)$, so $e^{\Delta} \leqq 1$. In the quantum case, let ρ_{123} be a pure state so that $S_{12} = S_3$ and $\Delta = -S_{13} + S_1 + S_3$. This can be positive (property D) since ρ_{13} is arbitrary. In the classical continuous case take $\rho_{123} = \rho_1\rho_2\rho_3$ and $S_3 < 0$.

Property F′ follows by applying property F to the three spaces $\mathscr{H}_2$, $\mathscr{H}_1 \otimes \mathscr{H}_3$ and $\mathscr{H}_4$ in place of $\mathscr{H}_1$, $\mathscr{H}_2$ and $\mathscr{H}_3$.

Property G′ follows from property G.

Property H′ is true in the classical discrete case since $S_{124} \geqq S_{12}$ and $S_{134} \geqq S_{13}$. It is false in the classical continuous case because one can take $\rho_{1234} = \rho_1\rho_2\rho_3\rho_4$ and $S_4 < 0$. In the quantum case, take $\rho_{1234} = \rho_{14} \otimes \rho_2 \otimes \rho_3$ so that $S_{12} + S_{13} - S_{124} - S_{134} = 2(S_1 - S_{14})$ and this can be positive (property C) since ρ_{14} is arbitrary.

Table 3.

Properties A″ and B″ (not shown) are respectively identical to *A*′ and B′.

Property ~C″ follows from property F.

Property D″ is that $\Delta = (S_{123} - S_{13} - S_{12} + S_1) - (S_{23} - S_3) \leqq 0$. This is true in the classical discrete case by properties F and C. In the classical continuous case it is false because one could take $\rho_{123} = \rho_1\rho_2\rho_3$ with $S_2 < 0$. In the quantum case, use Lemmas 1 and 2: With ρ_{1234} pure on $\mathscr{H}_{123} \otimes \mathscr{H}_4$,

$$\Delta = \tfrac{1}{2}(S_1 + S_4 - S_{13} - S_{34}) + \tfrac{1}{2}(S_1 + S_3 - S_{14} - S_{34}) + \tfrac{1}{2}(S_3 + S_4 - S_{14} - S_{13}) \leqq 0$$

by property H.

Property E″ is that $\Delta = (S_{123} - S_{13} - S_{12} + S_1) + (S_{23} - S_3) \geqq 0$. In the classical continuous case this is false for the same reason that property D″ is false. In the quantum case it is false because $S(2|13)$ can be negative (see A′). In the classical discrete case take $\rho_{123}(i, j, k) = \rho_{12}(i, j)\delta_{j,k}$. Then $S_{23} = S_3 = S_2$ and $S_{123} = S_{13} = S_{12}$, so $\Delta = -S_{12} + S_1$ and this can be negative since ρ_{12} is arbitrary.

Property F″ and its contrary, ~F″, are both false in all three cases. Take $\rho_{1234} = \rho_1 \otimes \rho_{234}$ so that $\Delta \equiv S(2|134) + S(2|1) - S(2|13) - S(2|14) = S_{234} - S_{34} - S_{23} - S_{24} + S_2 + S_3 + S_4$. As ρ_{234} is arbitrary, Δ can be positive or negative (by properties I and J of Table 1).

Property G″ is that $\Delta \equiv S_{123} - S_{13} - S_{12} + S_1$ is a concave function of ρ_{123}. In the classical discrete case take $\rho_{123}(i, j, k) = \rho_{12}(i, j)\delta_{j,k}$ so that $S_{123} = S_{13}$ and $\Delta = -S_{12} + S_1$. If property G″ were true, $-S_{12} + S_1$ would have to be a concave function of ρ_{12}, which is arbitrary, but the contrary is true (property G). To demonstrate the falsity of ~G″ in the classical discrete case, let $\rho_{123} = \rho_1\rho_{23}$ so that $\Delta = S_{23} - S_2 - S_3$ would have to be convex in ρ_{23}. This is false by property L. Since properties G″ and ~G″

are false in the classical discrete case they are *a fortiori* false in the other two cases as well.

Our final task is to prove property $\sim$H″. This is easy to do since

$$S(2\,|\,1) + S(2\,|\,3) - S(2\,|\,14) - S(2\,|\,34)$$
$$= (-S_{124} + S_{12} + S_{14} - S_1) + (-S_{234} + S_{34} + S_{23} - S_3),$$

and this is positive by property F.

ACKNOWLEDGEMENT. The author thanks Professor H. F. Weinberger for his careful reading of this manuscript and for suggesting several improvements and corrections.

References

1. L. Boltzmann, *Ueber die Beziehung zwischen dem zweiten Hauptsatz der mechanischen Waermetheorie und der Wahrscheinlichkeitsrechnung respektive den Saetzen ueber das Waermegleichgewicht*, Wiener Berichte **76** (1877), 373.

2. A. N. Kolmogorov, *A new metric invariant of transient dynamical systems and automorphisms in Lebesgue spaces*, Dokl. Akad. Nauk SSSR **119** (1958), 861–864. (Russian) MR **21** #2035a; Ja. G. Sinaĭ, *On the concept of entropy for a dynamic system*, Dokl. Akad. Nauk SSSR **124** (1959), 768–771. (Russian) MR **21** #2036a.

3. E. H. Lieb and M. B. Ruskai, *A fundamental property of quantum-mechanical entropy*, Phys. Rev. Lett. **30** (1973), 434–436.

4. E. H. Lieb, *Convex trace functions and the Wigner-Yanase-Dyson conjecture*, Advances in Math. **11** (1973), 267–288; See also H. Epstein, *Remarks on two theorems of E. Lieb*, Comm. Math. Phys. **31** (1973), 317–325.

5. E. H. Lieb and M. B. Ruskai, *Proof of the strong subaddivity of quantum-mechanical entropy*, J. Mathematical Phys. **14** (1973), 1938–1941.

6. H. Araki and E. H. Lieb, *Entropy inequalities*, Comm. Math. Phys. **18** (1970), 160–170. MR **42** #1466.

Current address: Department of Mathematics and Physics, Princeton University, Princeton, New Jersey 08540

Permanent address: Department of Mathematics and Physics, Massachusetts Institute of Technology, Cambridge, Massachusetts 02139

Commun. Math. Phys. 77, 127–135 (1980)

Communications in Mathematical Physics

A Refinement of Simon's Correlation Inequality*

Elliott H. Lieb

Departments of Mathematics and Physics, Princeton University, Princeton, NJ 08544, USA

Abstract. A general formulation is given of Simon's Ising model inequality: $\langle\sigma_\alpha\sigma_\gamma\rangle \leqq \sum_{b\in B}\langle\sigma_\alpha\sigma_b\rangle\langle\sigma_b\sigma_\gamma\rangle$ where B is any set of spins separating α from γ. We show that $\langle\sigma_\alpha\sigma_b\rangle$ can be replaced by $\langle\sigma_\alpha\sigma_b\rangle_A$ where A is the spin system "inside" B containing α. An advantage of this is that a finite algorithm can be given to compute the transition temperature to any desired accuracy. The analogous inequality for plane rotors is shown to hold if a certain conjecture can be proved. This conjecture is indeed verified in the simplest case, and leads to an upper bound on the critical temperature. (The conjecture has been proved in general by Rivasseau. See notes added in proof.)

In an accompanying paper [1] in this volume Simon proves a correlation inequality with important consequences. For a finite range pairwise interacting (generalized) Ising ferromagnet (the spins take on values $2M, 2M-2, \ldots, -2M$), Simon shows that

$$\langle\sigma_\alpha\sigma_\gamma\rangle \leqq \sum_{b\in B}\langle\sigma_\alpha\sigma_b\rangle\langle\sigma_b\sigma_\gamma\rangle, \tag{1}$$

where B is any set of spins separating α from γ (i.e. any path from α to γ must run through B). Aizenman and Simon [2] have proved a related inequality for N-component spins. In this paper we shall generalize (1) in the following way: $\langle\sigma_\alpha\sigma_b\rangle$ can be replaced by $\langle\sigma_\alpha\sigma_b\rangle_A$, where A is the connected component of the lattice containing α and B and $\langle\cdot\rangle_A$ denotes expectation values in the A system alone. The possibility of extending this inequality to plane rotors is also discussed, but the proof is carried to completion only in a special case. (See notes added in proof.)

In [1] Simon discusses the consequences of (1) and our generalization. We shall not repeat them, except to note that the most interesting consequence of the extension is that for the first time one has an algorithm for computing the transition temperature, T_c (in the sense that above, but not below T_c there is

* Work partially supported by U.S. National Science Foundation grant PHY-7825390 A01

exponential decay of the two point function $\langle\sigma_0\sigma_x\rangle$), to arbitrary accuracy. Take $\alpha=0$ and let B be the spins on the boundary of a square of side L centered at 0. By boundary we mean all points within a distance R of the geometric boundary, where R is not less than the range of the interaction. The A system is the inside of the square alone. $\langle\sigma_0\sigma_b\rangle_A$ can be computed explicitly, and if

$$\sum_{b\in B}\langle\sigma_0\sigma_b\rangle<1 \tag{2}$$

for some T, then there is exponential decay for that T. This sets an upper bound to T_c. It is easy to see [1], however, that as $L\to\infty$, T_L [the T for which equality holds in (2)] approaches T_c. While the convergence of T_L to T_c is expected to be extremely slow, the mere existence of the algorithm is an interesting matter of principle. It is not known if T_L is necessarily monotone decreasing in L; this is an open question.

A consequence of our generalization is the continuity of the mass gap as function of the interaction, for nearest neighbor ferromagnetic interactions, proven in [1]. A more general stability of the mass gap, m, under perturbations was pointed out by Aizenman (private communication). It is expressed by the *lower semicontinuity of m, as function of the interaction, in the cone of pairwise ferromagnetic interactions of any fixed finite range.* This is proven in the following way. Suppose the (finite range) Hamiltonian H is given and T is such that for the infinite system

$$\langle\sigma_0\sigma_x\rangle<C_\varepsilon\exp[(-m+\varepsilon)|x|]$$

for all x, all $\varepsilon>0$, but not $\varepsilon<0$. m is then the mass gap and it will be assumed that $m>0$. Given $\varepsilon>0$, it is easy to see that for any R there must be a finite box such that

$$\sum_{b\in B}\langle\sigma_0\sigma_b\rangle\exp[\mu|b|]<1 \tag{3}$$

for $\mu=m-\varepsilon$. Conversely, our generalization of (1) shows that if (3) holds with some μ for some box, then the mass gap is not less than μ.

Since condition (3) (with $\mu=m-\varepsilon$) refers to a finite system, by continuity it continues to hold (with $\mu=m-2\varepsilon$) when the Hamiltonian is changed from H to $H+K$ and $\|K\|<\delta_0$, for some $\delta_0>0$ and independent of K. If we also require that $H+K$ is pairwise ferromagnetic and has range $\leqq R$, then (3) (with $\mu=m-2\varepsilon$) and our generalization of (1) imply that the new mass gap is not less than $m-2\varepsilon$.

Simon's proof of (1) uses a graphical expansion. The analysis presented here will not use this explicitly, but instead will use certain "gaussian correlation inequalities" of Newman [3]. While it is true that Newman's inequalities are themselves proved by graphical means, it is hoped that the decomposition of the problem into the two steps given here will be useful.

Let us begin with some definitions. The system under consideration is viewed as the union of two subsystems of spins A and C.

$$A\cap C=B$$

is the set of spins common to both. To say that the B spins separate A from C means that

$$H_{A+C}=H_A+H_C, \tag{4}$$

where the H's are Hamiltonians. The symbols Z denote partition functions, $\langle\cdot\rangle$ denote expectation values and $(\cdot)=Z\langle\cdot\rangle$ denote unnormalized expectation values – all at reciprocal temperature β. Thus, for example,

$$\begin{aligned}(\sigma_A)_A&=\mathrm{Tr}\,\sigma_A\exp(-\beta H_A)\\ Z_A&=(1)_A=\mathrm{Tr}\exp(-\beta H_A)\\ \langle\sigma_A\rangle_A&=(\sigma_A)_A/Z_A.\end{aligned} \tag{5}$$

Here σ_A is some observable in the A system. It may, of course, depend on the B spins since they are in A.

The spins that are mostly relevant to our analysis are the B spins. The word "spin" is to some extent a misnomer, for the only hypothesis is that at each point $b\in B$ there is an independent a-priori probability measure $d\mu_b$ on some measure space Ω_b. For simplicity we take these to be independent of b. Let $\{\phi^n\}$ be a *complete* orthonormal family of functions in $L^2(d\mu)$. The choice of the $\{\phi^n\}$ is important because the hypotheses made later can be expected to hold, if at all, only for special choices. With $\mathbf{n}=(n_1,n_2,\ldots)$ a multi-index on $B=(b_1,b_2,\ldots)$, we denote the following orthonormal functions on $\prod_{b\in B}\Omega_b$:

$$\phi_B^{\mathbf{n}}=\phi_{b_1}^{n_1}\phi_{b_2}^{n_2}\ldots. \tag{6}$$

Example 2(Spin $\frac{1}{2}$ Ising Model). Here $\Omega=\{-1,1\}$, μ gives weight $\frac{1}{2}$ to each point and $\{\phi^n\}=\{\phi^0,\phi^1\}$ with $\phi^0(\sigma)=1$, $\phi^1(\sigma)=\sigma$.

Example 2 (Plane Rotor). Ω is the unit circle $0\leqq\theta<2\pi$, $d\mu(\theta)=d\theta/2\pi$ is the uniform measure, and $\phi^n(\theta)=\exp(in\theta)$ with $n=0,\pm1,\pm2,\ldots$.

The constitution of the remainder of the A and C systems is irrelevant to the general formalism we present. It can be composed of quarks, for example. σ_A (resp. σ_C)will denote observables in the A (resp. C) systems and they can both depend on the B spins. Note that the functions $\phi_B^{\mathbf{n}}$ can be regarded either as A or as C observables.

A formula connecting A, C and $A+C$ expectations is required. In other words, we have to "glue" the A and C systems together to form the $A+C$ system.

Lemma 1.

$$(\sigma_A\sigma_C)_{A+C}=\sum_{\mathbf{n}}(\sigma_A\phi_B^{\mathbf{n}})_A(\sigma_C\bar{\phi}_B^{\mathbf{n}})_C. \tag{7}$$

In particular,

$$Z_{A+C}=\sum_{\mathbf{n}}(\phi_B^{\mathbf{n}})_A(\bar{\phi}_B^{\mathbf{n}})_C. \tag{8}$$

Proof. In a schematic notation, let $x, y,$ and z respectively stand for the B variables, the A variables other than B, and the C variables other than B. The

Commun. Math. Phys. 77, 127–135 (1980)

Boltzmann factor is $M(x,y)N(x,z)$ where $M(x,y)=\exp[-\beta H_A(x,y)]$ and $N(x,z)=\exp[-\beta H_C(x,z)]$. Let the a-priori measure be $d\mu_\beta(x)\,d\mu_\alpha(y)\,d\mu_\gamma(z)$ and let $F(x)=\int d\mu_\alpha(y)\sigma_A(x,y)N(x,y)$, $G(x)=\int d\mu_\gamma(z)\sigma_C(x,z)M(x,z)$. Then, by Parseval's theorem,

$$(\sigma_A\sigma_C)_{A+C}=\int d\mu_\beta(x)F(x)G(x)=\sum_{\mathbf{n}} D_{\mathbf{n}}E_{\mathbf{n}}$$

with $D_{\mathbf{n}}=\int d\mu_\beta(x)\phi_B^{\mathbf{n}}(x)F(x)$ and $E_{\mathbf{n}}=\int d\mu_\beta(x)\bar{\phi}_B^{\mathbf{n}}(x)G(x)$. But this sum on $\mathbf{n}$ is precisely the right side of (7). □

Henceforth we fix the observables σ_A and σ_C, the Hamiltonians H_A and H_C, and make the following hypotheses (with respect to σ_A and σ_C) about the A and C systems.

H.C1 (Positivity). $\langle\sigma_C\bar{\phi}_B^{\mathbf{n}}\rangle_C\geqq 0$ for all $\mathbf{n}$. (9)

H.A1 (The Gaussian-Type Inequality [3]*).* There exists a function $F(\mathbf{n})$, not necessarily nonnegative, of the multi-index such that

$$\langle\sigma_A\phi_B^{\mathbf{m}}\rangle_A\leqq\sum_{\mathbf{n}}F(\mathbf{n})\langle\sigma_A\phi_B^{\mathbf{n}}\rangle_A\langle\bar{\phi}_B^{\mathbf{n}}\phi_B^{\mathbf{m}}\rangle_A \tag{10}$$

for all $\mathbf{m}$ such that $\langle\sigma_C\bar{\phi}_B^{\mathbf{m}}\rangle_C>0$.

The meaning of H.A1 will become clear later when we consider the Ising and plane rotor models as examples. For now we note that comparatively little is required of system C. The main theorem is the following:

Theorem 1. *Under hypotheses H.A1 and H.C1*

$$\langle\sigma_A\sigma_C\rangle_{A+C}\leqq\sum_{\mathbf{n}}F(\mathbf{n})\langle\sigma_A\phi_B^{\mathbf{n}}\rangle_A\langle\bar{\phi}_B^{\mathbf{n}}\sigma_C\rangle_{A+C}\,. \tag{11}$$

Proof. Multiply (11) by Z_AZ_{A+C} and use Lemma 1. We require that

$$\begin{aligned} &Z_A\sum_{\mathbf{m}}(\sigma_A\phi_B^{\mathbf{m}})_A(\bar{\phi}_B^{\mathbf{m}}\sigma_C)_C\\ &\quad\leqq\sum_{\mathbf{m}}\left\{\sum_{\mathbf{n}}F(\mathbf{n})(\sigma_A\phi_B^{\mathbf{n}})_A(\bar{\phi}_B^{\mathbf{n}}\phi_B^{\mathbf{m}})_A\right\}(\bar{\phi}_B^{\mathbf{m}}\sigma_C)_C\,. \end{aligned}\tag{12}$$

Here, $\phi_B^{\mathbf{n}}$ has been regarded as an A observable. In view of H.C1 it suffices to prove (12) for each $\mathbf{m}$ but, if we divide by Z_A^2, this is seen to be H.A1. □

The analogue of Simon's inequality [1] would have $\langle\ \rangle_{A+C}$ instead of $\langle\ \rangle_A$ on the right side of (11). There are then two natural questions: When does the Simon type of inequality hold and when is it weaker than Theorem 1, as it is for the Ising model? The following hypotheses help to answer this.

H.C2. $\langle\bar{\phi}_B^{\mathbf{n}}\rangle_C\geqq 0$, all $\mathbf{n}$. (13)

H.A2 (inequality of the second Griffiths type).

$$\langle\sigma_A\phi_B^{\mathbf{n}}\rangle_A\langle\phi_B^{\mathbf{m}}\rangle_A\leqq\langle\sigma_A\phi_B^{\mathbf{n}}\phi_B^{\mathbf{m}}\rangle_A,\quad\text{all }\mathbf{n} \tag{14}$$

whenever $\langle\bar{\phi}_B^{\mathbf{m}}\rangle_C>0$.

Theorem 2. *Suppose H.A2 and H.C2 hold. Then*

$$\langle \sigma_A \phi_B^{\mathbf{n}} \rangle_A \leqq \langle \sigma_A \phi_B^{\mathbf{n}} \rangle_{A+C}, \quad \text{all } \mathbf{n}. \tag{15}$$

Proof. (15) is equivalent to

$$\sum_{\mathbf{mn}} (\phi_B^{\mathbf{m}})_A (\bar{\phi}_B^{\mathbf{m}})_C (\sigma_A \phi_B^{\mathbf{n}})_A \leqq Z_A \sum_{\mathbf{m},\mathbf{n}} (\sigma_A \varphi_B^{\mathbf{n}} \phi_B^{\mathbf{m}})_A (\bar{\phi}_B^{\mathbf{m}})_C,$$

but this is implied by (13), (14). □

Corollary 1. *Suppose* (11) *and* (15) *hold and* $F(\mathbf{n}) \geqq 0$. *Then*

$$\langle \sigma_A \sigma_C \rangle_{A+C} \leqq \sum_{\mathbf{n}} F(\mathbf{n}) \langle \sigma_A \phi_B^{\mathbf{n}} \rangle_{A+C} \langle \bar{\phi}_B^{\mathbf{n}} \sigma_C \rangle_{A+C}. \tag{16}$$

Moreover, the right side of (16) is not less than the right side of (11).

If $F(\mathbf{n})$ is not nonnegative, (16) can still be proved under a further hypothesis:

H.A3.

$$\langle \sigma_A \phi_B^{\mathbf{m}} \rangle_A \langle \phi_B^{\mathbf{k}} \rangle_A \leqq \sum_{\mathbf{n}} F(\mathbf{n}) \langle \sigma_A \phi_B^{\mathbf{n}} \phi_B^{\mathbf{k}} \rangle_A \langle \bar{\phi}_B^{\mathbf{n}} \phi_B^{\mathbf{m}} \rangle_A. \tag{17}$$

whenever both $\langle \sigma_C \bar{\phi}_B^{\mathbf{m}} \rangle_C > 0$ and $\langle \bar{\phi}_B^{\mathbf{k}} \rangle_C > 0$.

Theorem 3. (16) *holds under hypothese H.C1, H.C2, and HA3.*

The proof of Theorem 3 is an imitation of the proofs of Theorems 1 and 2. Note that under these hypothese one cannot say that (16) is weaker than (11).

The following is a trivial consequence of the definitions

Lemma 2. *If* $F(\mathbf{n}) \geqq 0$ *then H.A1 and H.A2 imply H.A3.*

The Ising Model as an Example

Spin 1/2 Ising Models

The ϕ^n are given in Example 1. We take σ_A and σ_C each to be products of an odd number of spins. H.C1, H.C2, and H.A2 are Griffiths' inequalities. Newman's inequality [3] states, in particular, that if F is a family of partitions of $K = \{1, \ldots, k\}$ into two disjoint subsets then (with $\sigma_D = \sigma_a \sigma_b \ldots \sigma_d$ when $D = \{a, b, \ldots, d\}$)

$$\langle \sigma_K \rangle \leqq \sum_{f \in F} \langle \sigma_{f_1} \rangle \langle \sigma_{f_2} \rangle, \tag{18}$$

whenever $|K| = 2L$ is even and *every partition of* K into L pairs is a refinement of some $f \in F$. Sylvester [14] also gives a proof of (18).

Let the spins in B be labeled $\sigma_1, \ldots, \sigma_M$. In (10), $\mathbf{m}$ can be thought of as a subset of $\{1, \ldots, M\}$. Clearly, $\langle \sigma_A \phi_B^{\mathbf{m}} \rangle_A > 0$ implies that $|\mathbf{m}|$ is odd.

Assume that σ_A is just one spin, σ_α, and, without loss, that $\alpha \notin B$. Taking $K = \{\alpha\} \cup \mathbf{m}$, and all f_1 of the form $\{\alpha, i\}$ with $i \in \mathbf{m}$, (18) implies (10) with

$$F(\mathbf{n}) = 1 \quad \text{if} \quad |\mathbf{n}| = 1$$
$$= 0 \quad \text{otherwise}. \tag{19}$$

Commun. Math. Phys. 77, 127–135 (1980)

[Note: There are more terms on the right side of (10) than the right side of (18). The excess terms are nonnegative by Griffiths first inequality.] In this case we conclude that

$$\langle\sigma_\alpha\sigma_C\rangle_{A+C} \leqq \sum_{b\in B}\langle\sigma_\alpha\sigma_b\rangle_A\langle\sigma_b\sigma_C\rangle_{A+C} \tag{20}$$

as stated in the introduction. It was not assumed that $|C|=1$.

If σ_A is a product of N(odd) spins then (18) implies (10) with

$$\begin{aligned} F(\mathbf{n}) &= 1 \quad \text{if} \quad |\mathbf{n}| = 1, 3, \ldots, N \\ &= 0 \quad \text{otherwise}. \end{aligned} \tag{21}$$

Then (20) changes to

$$\langle\sigma_A\sigma_C\rangle_{A+C} \leqq \sum_{\substack{b\subset B\\ |b|\leqq|A|}}\langle\sigma_A\sigma_b\rangle_A\langle\sigma_b\sigma_C\rangle_{A+C}\,. \tag{22}$$

Other Ising Models

One generalization is to spin $M>\frac{1}{2}$ with $\sigma=2M, \ldots, -2M$. A way to proceed would be to use an appropriate orthonormal basis $\{\phi^n\}$ of dimensions $2M+1$. We have not pursed this possibility. A second method is to use Griffiths' trick [5] of writing a spin M as M ferromagnetically coupled spin $\frac{1}{2}$ spins. H.C1, H.C2, and H.A2 follow from this, as does (20) and (21) by summing over the "component" spins. Much is lost this way, however.

Another generalization, which we shall not explicate, is to allow multi-spin interactions.

The Plane Rotor Model

We consider pairwise ferromagnetic interactions; the interaction between two spins $\vec{\sigma}_a$ and $\vec{\sigma}_b$ is $-J_{ab}\vec{\sigma}_a\cdot\vec{\sigma}_b = -J_{ab}\cos(\theta_a-\theta_b)$, with $J_{ab}\geqq 0$. The basis $\{\phi^n\}$ is given in Example 2.

There is some reason to believe that the analogue of (20) holds in the following sense:

$$\langle\vec{\sigma}_a\cdot\vec{\sigma}_c\rangle_{A+C} \leqq \sum_{b\in B}\langle\vec{\sigma}_a\cdot\vec{\sigma}_b\rangle_A\langle\vec{\sigma}_b\cdot\vec{\sigma}_c\rangle_{A+C}\,. \tag{23}$$

when $\vec{\sigma}_a$ and $\vec{\sigma}_c$ are single spins. In terms of the ϕ^n we have

$$2\vec{\sigma}_a\cdot\vec{\sigma}_c = \phi_a^1\phi_c^{-1} + \phi_a^{-1}\phi_c^1\,.$$

Therefore we require that (9) and (10) hold when σ_C (resp. σ_A) is ϕ_c^1 or ϕ_c^{-1} (resp. ϕ_a^1 or ϕ_a^{-1}) and F is given by (19). With these choices for σ_A and σ_C, (9), and also (14) hold [6]. The difficulty lies with (10).

We do not have a proof of (10), but believe it to be true. A possibility would be to try to imitate the graphical proof [3,4] that is successful in the Ising case. It would then be necessary to deal with directed graphs (digraphs). The following, if it were true would immediately yield a proof of (10):

Conjecture. Let G be a finite direted graph (possibly with several edges between two vertices) and let the valence at vertex i (the number of arrows in minus the number out of i) be M_i. (Clearly, $\Sigma M_i = 0$.) Suppose $M_1 = +1$, and $M_2, M_3, \ldots, M_k < 0$, and $M_i \geqq 0$ otherwise. Let N be the number of subgraphs (subsets of edges) of G, including the empty graph, having valence 0 at each vertex. Let K be the number of subgraphs of G with the following property: vertex 1 has valence $+1$, some vertex j, with $2 \leqq j \leqq k$, has valence -1, and all other valences are 0. Then $N \leqq K$.

There is one special but important case in which the conjecture, and hence (23) holds. Suppose $\vec{\sigma}_a$ is connected to n nearest neighbors, which we take to be B. It is immaterial whether the B spins are connected together, for any such interaction can be regarded as part of H_C. In a v dimensional cubic lattice $n = 2v$.

The graph G in the conjecture then has the following structure: it has $n+1$ vertices and is star-like with edges only between the central vertex V_1 (which is really the original vertex marked a) and the other n. Suppose $M_2 < 0$. Then it is easy to see that the conjecture is verified if the following is true: Let $\tilde{G}$ be the subgraph of G consisting of vertices 1 and 2 and all the edges between them. Let $\tilde{N}$ be the number of valence 0 subgraphs of $\tilde{G}$, and let $\tilde{K}$ be the number of subgraphs of $\tilde{G}$ with $M_1 = +1, M_2 = -1$. Then $\tilde{N} \leqq \tilde{K}$. The easy proof of this is left to the reader.

This simple case can also be conveniently viewed in terms of (10) directly. We require that

$$\langle \phi_a^1 \phi_B^{\mathbf{m}} \rangle \leqq \sum_{b=1}^{n} \langle \phi_a^1 \phi_b^{-1} \rangle \langle \phi_b^1 \phi_B^{\mathbf{m}} \rangle . \tag{24}$$

Both sides of (24) vanish unless $\sum_1^n m_i = -1$. Let $P(m)$ be the Fourier transform of $\exp[\beta \cos\theta]$, namely

$$P(m) = I_m(\beta) > 0, m \in \mathbb{Z}, \tag{25}$$

where I_m is the modified Bessel function. Then (24) reads

$$\{P(0)^n\} \left\{ \prod_{i=1}^{n} P(m_i) \right\} \leqq \sum_{i=1}^{n} \{P(0)^{n-1} P(1)\} \left\{ P(m_i+1) \prod_{j \neq i} P(m_j) \right\} . \tag{26}$$

Suppose $m_1 < 0$, say. It is sufficient to have

$$P(0)P(m_1) \leqq P(1)P(m_1+1) . \tag{27}$$

If both sides of (27) are expanded in a power series in β, (27) is true term by term. This is just the graphical exercise mentioned above. However, the following stronger result, which implies (27), holds.

Lemma 2. *Fix $\beta \geqq 0$. The function $m \to I_m(\beta)$ is log concave on the integers, i.e.*

$$I_m(\beta)^2 \geqq L_{m+1}(\beta) I_{m-1}(\beta), \qquad m \in \mathbb{Z} . \tag{28}$$

Commun. Math. Phys. 77, 127–135 (1980)

Proof. $I_m(\beta)=I_{-m}(\beta)$. (28) is trivial for $m=0$, so it is sufficient to consider $m\geqq 1$. If both sides of (28) are expanded in a power series in β, we claim (28) holds termwise. Use

$$I_m(\beta)=\sum_{j=0}^{\infty}(\beta/2)^{2j+m}[j!(j+m)!]^{-1}.$$

Thus for the coefficient of β^{2m+2t}, $t\geqq 0$, we require

$$\sum_{j=0}^{t} g_0(j-t/2)g_m(j-t/2)\geqq \sum_{j=0}^{t} g_0(j-t/2)g_m(j+1-t/2), \tag{29}$$

where $g_p(x)=[\Gamma(t/2+1+p+x)\Gamma(t/2+1+p-x)]^{-1}\theta(x)$ and $\theta(x)=1$ if $|x|\leqq t/2+p$ and $=0$ otherwise. Now $\Gamma(x)$ is log convex for $x>0$ and hence $g_p(x)$ is even and log concave. Thus, for j an integer in $[0,t]$, $g_0(j-t/2)$ is a positive sum of functions of the form $\mu_a(j)=1$ if $a\leqq j\leqq t-a$ and $=0$ otherwise, for $a=0,1,\dots[t/2]$. Hence, it suffices to have $\sum_{j=a}^{t-a} g_m(j-t/2)-g_m(j+1-t/2)\geqq 0$. But this is true because $g_m(j-t/2)$ is also a positive sum of the $\mu_b(j)$ functions (with $-m\leqq b\leqq[t/2]$) and $\sum_{j=a}^{t-a}\mu_b(j+r)$ is decreasing for $r\geqq 0$, $r\in\mathbb{Z}$. □

Since (23) holds when $\vec{\sigma}_a$ is a single spin and B are its neighbors, we can conclude using (2) that

Theorem 4. *For the plane rotor model on a ν-dimensional hypercubic lattice there is exponential fall-off of the two-point function if $\langle\vec{\sigma}_a\cdot\vec{\sigma}_b\rangle<1/2\nu$ for the two-spin system consisting of a and b alone. This is equivalent to $I_1(\beta)/I_0(\beta)<1/2\nu$ if $J_{ab}=1$ is assumed for the nearest neighbor coupling constant. In particular*

$$\beta_c\geqq 0.52\quad(\nu=2);\qquad \beta_c\geqq 0.34\quad(\nu=3).$$

For $\nu=2$, Fröhlich [7] has shown that $\beta_c\geqq 0.64$, and Aizenman and Simon [8] have shown that $\beta_c\geqq 0.88$.

Acknowledgements. I should like to thank M. Aizenman, J. Bricmont, and B. Simon for helpful conversations.

References

1. Simon, B.: Commun. Math. Phys. **77**, 111–126 (1980). See also Phys. Rev. Lett. **44**, 547–549 (1980)
2. Aizenman, M., Simon, B.: Commun. Math. Phys. **77**, 137–143 (1980)
3. Newman, C.M.: Wahrsch. **33**, 75–93 (1975)
4. Sylvester, G.S.: Commun. Math. Phys. **42**, 209–220 (1975)
5. Griffiths, R.B.: J. Math. Phys. (N.Y.) **10**, 1559–1565 (1969)
6. Ginibre, J.: Commun. Math. Phys. **16**, 310–328 (1970)
7. Fröhlich, J.: Private communication
8. Aizenman, M., Simon, B.: A comparison of plane rotor and ising models. Phys. Lett. A (submitted)

Communicated by A. Jaffe

Received March 21, 1980

Notes added in proof. (1) Simon's inequality (1) for the Ising model is a special case of a class of inequalities and identities discussed by Boel and Kasteleyn [9, 10]. They found necessary and sufficient conditions for such inequalities to hold; therefore (1) can be proved by their methods.

(2) The conjecture in this paper has been proved by Rivasseau [11]. Thus inequality (23) for rotors holds for all subsystems A, not merely for the case of the star graph proved here.

References

9. Boel, R.J., Kasteleyn, P.W.: Commun. Math. Phys. **61**, 191 (1978); Commun. Math. Phys. **66**, 167 (1979); Physica **93** A, 503 (1978)
10. Kasteleyn, P.W., Boel, R.J.: Phys. Lett. **70** A, 220 (1979)
11. Rivasseau, V.: Lieb's correlation inequality for plane rotors, Commun. Math. Phys. **77**, 145–147 (1980)

VOLUME 62, NUMBER 10 PHYSICAL REVIEW LETTERS 6 MARCH 1989

Two Theorems on the Hubbard Model

Elliott H. Lieb

Departments of Physics and Mathematics, Princeton University, P. O. Box 708, Princeton, New Jersey 08544

(Received 12 December 1988)

In the attractive Hubbard model (and some extended versions of it), the ground state is proved to have spin angular momentum $S=0$ for every (even) electron filling. In the repulsive case, and with a bipartite lattice and a half-filled band, the ground state has $S=\frac{1}{2}\,||B|-|A||$, where $|B|$ ($|A|$) is the number of sites in the B (A) sublattice. In both cases the ground state is unique. The second theorem confirms an old, unproved conjecture in the $|B|=|A|$ case and yields, with $|B|\neq|A|$, the first provable example of itinerant-electron ferromagnetism. The theorems hold in all dimensions without even the necessity of a periodic lattice structure.

PACS numbers: 75.10.Lp, 71.20.Ad, 74.65.+n

The importance of the Hubbard model of itinerant electrons is increasingly being appreciated. Because of the model's subtlety, rigorous results and exact solutions are clearly useful bench marks, but these are rare. Two theorems about the ground states are stated and proved here. Parts of them are resolutions of old conjectures while other parts are new. In particular, the assertion that certain versions of the model show ferromagnetic behavior for the half-filled band is, I believe, new and yields the first provable example of itinerant-electron ferromagnetism with finite forces and without *ad hoc* assumptions.

After some preliminary definitions, the two theorems are stated. Each is followed by some remarks about their significance. Finally, the proofs are given. The proofs utilize a new kind of reflection positivity which does not involve the usual spatial reflections but rather reflections in spin space. In fact, spatial symmetry plays no role whatsoever and therefore the theorems apply in the widest generality to any collection of sites; all dimensions and topologies are included. The Hubbard model is not known to satisfy any kind of spatial reflection positivity or infrared bounds and this unfortunate fact has prevented the application of the usual proof techniques[1-3] for establishing the existence of long-range order in periodic lattices.

The Hubbard model on a finite lattice Λ is defined by the Hamiltonian

$$H=\sum_{\sigma}\sum_{x,y\in\Lambda} t_{xy}c^{\dagger}_{x\sigma}c_{y\sigma}+\sum_{x\in\Lambda}U_x n_{x\uparrow}n_{x\downarrow}\,, \tag{1}$$

with the following notation. The operators $c_{x\uparrow}$ and $c_{x\downarrow}$ and their adjoints $c^{\dagger}_{x\sigma}$ satisfy the usual fermion anticommutation relations $\{c^{\dagger}_{x\sigma},c_{y\tau}\}=\delta_{xy}\delta_{\sigma\tau}$ and $\{c_{x\sigma},c_{y\tau}\}=0$. The hopping matrix elements t_{xy} are required to be real and satisfy $t_{xy}=t_{yx}$, but no other *a priori* assumption is made about them (e.g., the condition $t_{xx}=0$ is not assumed, which means that an x-dependent single-particle potential $\sum_{x\sigma}t_{xx}n_{x\sigma}$ is allowed). The reality of the t_{xy}'s is consistent with their interpretation as overlap matrix elements of real operators in real, localized orbitals. The number operators are $n_{x\sigma}=c^{\dagger}_{x\sigma}c_{x\sigma}$, while U_x is the on-site energy which, for theorem 1, is allowed to depend on the site x. The word "lattice" is certainly a misnomer because no particular topology (i.e., periodicity or dimensionality) is assumed; the generality assumed here is that Λ is merely a collection of sites. The number of these sites is denoted by $|\Lambda|$. There is said to be a *bond* between sites x and y if $t_{xy}\neq 0$, and Λ is said to be *connected* if there is a connected path of bonds between every pair of sites. Obviously it is no loss in generality to assume that Λ is connected, and this will always be done here. Finally, Λ is said to be *bipartite* if the sites of Λ can be divided into two disjoint sets A and B such that $t_{xy}=0$ whenever $x\in A$ and $y\in A$ or $x\in B$ and $y\in B$. The bipartite condition is really an assertion about the t_{xy}'s and not about Λ, but the terminology is conventional; it implies, in particular, that $t_{xx}=0$ for each $x\in\Lambda$. The symbols $|A|$ and $|B|$ denote the number of sites in A and B in this case, whence $|\Lambda|=|A|+|B|$. The number of electrons is denoted by N; necessarily $N\leq 2|\Lambda|$.

The aim here is to study the ground state (or states if there is more than one) of H for a given N. Of central importance is the total spin S which is a conserved quantity. The spin operators are the quadratic operators

$$S^z=\tfrac{1}{2}\sum_{x\in\Lambda}(n_{x\uparrow}-n_{x\downarrow}),\ S^+=(S^-)^\dagger=\sum_{x\in\Lambda}c^{\dagger}_{x\uparrow}c_{x\downarrow}\,, \tag{2}$$

and $(S_{\rm op})^2=(S^z)^2+\frac{1}{2}S^+S^-+\frac{1}{2}S^-S^+$, with eigenvalues $S(S+1)$.

Theorem 1 (attractive case).—Assume $U_x\leq 0$ for every x (but U_x is not necessarily constant) and that N is even. No extra assumption about Λ or the t_{xy}'s is made. Then (a) among the ground states of H there is one with spin $S=0$; (b) if $U_x<0$ for every x, the ground state is unique (and hence has $S=0$).

Remarks.—(1) The theorem is "obvious" if all the U_x's are very large, for then the ground state consists of paired electrons on $N/2$ sites of Λ. In the other extreme that each $U_x=0$, theorem 1 is also obvious because one just fills the lowest $N/2$ levels of the $|\Lambda|\times|\Lambda|$ Hermitian hopping matrix $T=\{t_{xy}\}$.

(2) Theorem 1 can be considerably generalized to what is called an *extended Hubbard model*, and more. We can add any *real* operator M to H provided that M satisfies the following two conditions. (A real operator is polynomial in the $c_{x\sigma}$'s and $c_{x\sigma}^{\dagger}$'s with real coefficients.) (i) M commutes with the spin operators S^z and $S^{\pm}$ and conserves both spin-up and spin-down particle numbers; (ii) M can be written as $M^{\uparrow}+M^{\downarrow}-M^{\uparrow\downarrow}$. Here $M^{\uparrow}$ ($M^{\downarrow}$) is real, Hermitian, involves only spin-up (spin-down) operators, and $M^{\uparrow}$ is identical to $M^{\downarrow}$ when the spins are flipped (i.e., $c_{x\uparrow}$ and $c_{x\downarrow}$ are interchanged). The up-down interaction $M^{\uparrow\downarrow}$ can be written as a sum of terms of the form (in which μ merely denotes a summation index) $M^{\uparrow\downarrow}=\sum_\mu V_\mu^{\uparrow}(V_\mu^{\downarrow})^{\dagger}$ in which each V_μ operator is real (but not necessarily Hermitian) and involves only operators for one kind of electron. Again, $V_\mu^{\uparrow}$ must be the spin reflection of $V_\mu^{\downarrow}$ for each μ. The necessary changes in the proof are straightforward (see Ref. 4). It is also easy to extend the proof to some multiband Hubbard models; the details are left to the reader.

Theorem 2 (repulsive case).—Assume $U_x=U$ = positive constant, independent of x. Assume $|\Lambda|$ is even, Λ is bipartite (so that the t_{xy}'s couple only A and B) and $|B|\geq|A|$. No other assumption about Λ or the t_{xy}'s is made. Let $N=|\Lambda|$ (half-filled band). Then the ground state of H is unique [apart from the trivial $(2S+1)$-fold degeneracy] and has spin $S=\frac{1}{2}\times(|B|-|A|)$.

Remarks.—(3) Theorem 2 is considerably more subtle than theorem 1. The assumptions are more stringent. The theorem has long been assumed to be true in the $|B|=|A|$ case ("the half-filled band has spin zero"), but its proof has been elusive.

(4) The fact that $2S=|B|-|A|$ should be no surprise. In the limit $U=0$ we fill the levels of the matrix $T=\{t_{xy}\}$ and one might hastily conclude that S must be zero in this limit. If so, theorem 2 would be contradicted by a continuity argument with respect to U. However, the rank of T is at most $2|A|$ and so T has at least $|\Lambda|-2|A|=|B|-|A|$ zero eigenvalues. The remaining eigenvalues of T come in plus-minus pairs, so that T has at most $|A|$ negative eigenvalues. To achieve $2S=|B|-|A|$, we fill the negative levels twice with opposite spins and place the remaining electrons in the zero levels with a common spin, say, spin up. Thus the ground state is degenerate when $U=0$, but $S=\frac{1}{2}(|B|-|A|)$ is among them. Therefore, there is no contradiction with the continuity argument mentioned above. If, on the other hand, U is very large we know[5] from second-order perturbation theory that H is effectively an isotropic spin-$\frac{1}{2}$ Heisenberg antiferromagnet with Hamiltonian

$$h=(2/U)\sum_{x,y}t_{xy}^2(\mathbf{S}_x\cdot\mathbf{S}_y-\tfrac{1}{4}).$$

For such models it is also known[6] that the ground state is unique and has $2S=|B|-|A|$.

(5) It is easy to construct many regular, periodic lattices in every dimension greater than one with $|B|\neq|A|$. A classic (high-T_c superconductor) two-dimensional example is to start with a square lattice and then intercalate one site in the middle of each bond. The original vertices of the squares are A sites (copper) and the intercalated sites are B sites (oxygen). The half-filled band has three electrons per unit cell and theorem 2 says that then the total net magnetization is $S=\frac{1}{2}(2-1)=\frac{1}{2}$ times the number of unit cells. This was already observed for this lattice in the large-U limit by Mattis.[7]

Whether or not this example is physically realizable is less important than the fact that theorem 2 applied to a periodic lattice with $|B|>|A|$ yields, for the first time, a natural, provable example of an *itinerant-electron model of ferromagnetism*. I use the word ferromagnetism here only in the sense that the spin is extensive, i.e., it is proportional to the number of particles (or cells). Spatial ordering is not implied. A more accurate appelation might be unsaturated ferromagnetism. Still more accurately, ferrimagnetism might be the right word—but technically that word implies a spatial ordering that I am not prepared to prove. In one dimension $|B|=|A|$ by definition and therefore $S=0$; this conclusion coincides with the known result[8] that S is always zero in one dimension with nearest-neighbor nonpositive hopping $t_{xy}\leq 0$ and for *any many-body potential*. There is also the example of Thouless and Nagaoka[9-11] with $N=|\Lambda|-1$, $U=\infty$, and $2S=|\Lambda|-1$, but infinite potentials are crucial for this example.

(6) Theorem 2 also has some extensions similar to some of those described in remark (2) above; hole-particle symmetry is required.

Proof of theorem 1.—S^2 and S^z are conserved and I work in the $S^z=0$ subspace since all competitors have a representative there. That is to say, each eigenstate with a given S value can (by the well known properties of angular momentum) be rotated in spin space to a state with $S_z=0$ without changing its energy. Then there are $n=\frac{1}{2}N$ electrons of each type, spin up and spin down. Let $\{\psi^\alpha\}$ be any orthonormal basis for *one species* of n spinless fermions; there are

$$m=\binom{|\Lambda|}{n}$$

of these and I require that they be *real* (i.e., each ψ^α is a *real*, homogeneous polynomial of order n in the $c_x^{\dagger}$'s acting on the vacuum). A ground state ψ can then be written as $\psi=\sum_{\alpha,\beta}W_{\alpha\beta}\psi_\uparrow^\alpha\otimes\psi_\downarrow^\beta$ with $W_{\alpha\beta}$ as coefficients. This $W_{\alpha\beta}$ is here viewed as a $m\times m$ matrix. Because all operators and basis vectors are real and because the Hamiltonian is symmetric between the up and the down spins, it is obvious that if $W_{\alpha\beta}$ corresponds to a ground state then so does $W_{\beta\alpha}^*=(W^{\dagger})_{\alpha\beta}$, and hence (by linearity) so does $W+W^{\dagger}$ and $i(W-W^{\dagger})$. Thus, for conveni-

VOLUME 62, NUMBER 10 PHYSICAL REVIEW LETTERS 6 MARCH 1989

ence, we may henceforth assume $W=W^\dagger$ (but $W_{\alpha\beta}$ is not assumed to be real). The norm of ψ squared is $\langle\psi|\psi\rangle=\sum_{\alpha,\beta}|W_{\alpha\beta}|^2=\mathrm{Tr}W^2$ and I assume that this is unity. The hopping energy part of $\langle\psi|H|\psi\rangle$ is easily found to be $2\mathrm{Tr}KW^2$ where

$$K_{\alpha\beta}=\langle\psi^\beta|\sum_{x,y}t_{xy}c_x^\dagger c_y|\psi^\alpha\rangle.$$

Clearly K is real and symmetric since each t_{xy} is real. The on-site energy is given by $-\sum_x U_x\mathrm{Tr}(WL_xWL_x)$ with $(L_x)_{\alpha\beta}=\langle\psi^\beta|n_x|\psi^\alpha\rangle$, which is also real and symmetric. The total energy is then

$$E(W)=\langle\psi|H|\psi\rangle=2\mathrm{Tr}KW^2+\sum_x U_x\mathrm{Tr}(WL_xWL_x), \quad (3)$$

and the equation for W, corresponding to the eigenvalue equation $H\psi=e\psi$, is

$$KW+WK+\sum_x U_xL_xWL_x=eW. \quad (4)$$

Now consider the positive semidefinite matrix $|W|$ defined by $|W|^2=W^2$. Obviously, $\mathrm{Tr}W^2=\mathrm{Tr}|W|^2$. Moreover, in an orthonormal basis (not to be confused with the ψ^α basis for the electrons) in which the Hermitian $m\times m$ matrix W is diagonal, with diagonal elements w_i, the Hermitian matrix $|W|$ is also diagonal with elements $|w_i|$. In this diagonal basis I compute

$$\mathrm{Tr}WL_xWL_x=\sum_{i,j}w_iw_j|(L_x)_{ij}|^2\le\sum_{i,j}|w_i||w_j||(L_x)_{ij}|^2$$
$$=\mathrm{Tr}|W|L_x|W|L_x.$$

Since $U_x\le 0$, I conclude that $E(W)\ge E(|W|)$ and therefore that *among the ground states there is one satisfying* $W=|W|\ge 0$. This is the "spin-space reflection positivity" mentioned in the second paragraph. (This part of the proof is an adaptation of that given in Ref. 4.) I choose this positive (possibly semidefinite) matrix W and now make the choice that the ψ^α's are the natural x-space basis for the electrons, i.e., α denotes n points in Λ and $\psi^\alpha=\prod_{x\in\alpha}c_x^\dagger|0\rangle$. (Some arbitrary convention for the sign of the ψ^α's can be made here.) Since $W\ge 0$, it follows that $W_{\alpha\alpha}>0$ for at least one α, for otherwise W vanishes identically. However, the vector $\phi^\alpha=\psi_\uparrow^\alpha\otimes\psi_\downarrow^\alpha$ satisfies $(S_{\mathrm{op}})^2\phi^\alpha=0$ and therefore this ground state ψ has a nonzero projection onto the eigenspace of $(S_{\mathrm{op}})^2$ in which $S=0$. This would be impossible if conclusion (a) were false and thus conclusion (a) must be true.

To prove conclusion (b) I shall prove that necessarily a Hermitian W satisfies $W=|W|$ or $W=-|W|$ for every ground state when $U_x<0$ for every x. This will prove conclusion (b) for the following reason. If there were two normalized, Hermitian ground state W's, say W^1 and W^2 with $W^1\neq\pm W^2$, then for every real constant d, the Hermitian matrix $W^1+dW^2\equiv W_d$ is not zero and defines (after normalization) a ground state, by virtue of the linearity of the eigenvalue Eq. (4). It is easy to verify that there must be a d for which W_d is neither positive nor negative semidefinite, and this contradicts the assertion that the ground state W_d satisfies $W_d=\pm|W_d|$.

With W given then, consider the Hermitian, positive semidefinite matrix $R\equiv|W|-W$ which is also a multiple of a ground state and satisfies (4). If Q denotes the kernel of R, i.e., $Q=\{$vectors V such that $RV=0\}$, then the assertion $W=\pm|W|$ is implied by the following statement which I shall prove: Q is either just the zero vector or else every vector is in Q. Let V be in Q and take the expectation of (4) in this state V, i.e.,

$$\langle V|KR+RK+\sum_x U_xL_xRL_x|V\rangle=e\langle V|R|V\rangle.$$

Since $RV=0$ and, for all x, $U_x<0$ and $\langle V|L_xRL_x|V\rangle\ge 0$ by the positive semidefiniteness of R, I conclude that $\langle V|L_xRL_x|V\rangle=0$ for all x. Since R is positive semidefinite, I conclude that $RL_xV=0$. Thus each L_x maps Q into Q. Now let the matrices in (4) act on V (without taking expectation values). Since $RL_xV=0$ and $RV=0$, I conclude that $RKV=0$. Thus K also maps Q into Q. As before, let α denote a collection of n points in Λ and define $L^\alpha=\prod_{x\in\alpha}L_x$, which is the projector onto the basis vector μ^α in C^m [with components $(\mu^\alpha)_\gamma=\delta_{\alpha\gamma}$] and which has matrix elements $(L^\alpha)_{\gamma\delta}=\delta_{\alpha\gamma}\delta_{\gamma\delta}$. Note that the L_x's commute with each other and so the ordering of the L_x's is unimportant in the definition of L^α. Each L^α maps Q into Q because each L_x does. Since Λ is connected by T, it is easy to see that the α's are connected by K, i.e., for all α and β there are indices $\gamma_1,\gamma_2,\dots,\gamma_p$ for some integer p such that the ordinary (not matrix) product $G_{\beta\alpha}\equiv K_{\beta\gamma_1}K_{\gamma_1\gamma_2}\cdots K_{\gamma_p\alpha}$ is not zero. If Q is not just the zero vector and $V\neq 0$ is in Q then $L^\alpha V\neq 0$ for some fixed choice of α (because $\sum_\alpha L^\alpha$ is the identity). Then $L^\alpha V=z\mu^\alpha$ for some nonzero constant z and with μ^α being the aforementioned basis vector. But then the vector $F\equiv L^\beta KL^{\gamma_1}K\cdots L^{\gamma_p}KL^\alpha V$ is in Q and, in fact, $F=zG_{\beta\alpha}\mu^\beta$. In short, Q contains a complete set of vectors (i.e., every μ^β) because $G_{\beta\alpha}$ is nonzero for every β by virtue of the connectivity. Thus every vector is in Q since Q is a linear space, which implies that $W=\pm|W|$ and which, in turn, implies uniqueness of the ground state. Q.E.D.

Proof of theorem 2.—First make the conventional unitary hole-particle transformation for the spin-up electrons followed by a sign change on the B sublattice, i.e., $c_{x\uparrow}\to\epsilon(x)c_{x\uparrow}^\dagger$ and $c_{x\uparrow}^\dagger\to\epsilon(x)c_{x\uparrow}$ with $\epsilon(x)=1$ for $x\in A$, $\epsilon(x)=-1$ for $x\in B$. The spin-down electrons are unaltered, i.e., $c_{x\downarrow}\to c_{x\downarrow}$. Then $n_{x\uparrow}\to 1-n_{x\uparrow}$ and the transformed H is $\tilde H+UN_\downarrow$ with

$$\tilde H=\sum_\sigma\sum_{x,y\in\Lambda}t_{xy}c_{x\sigma}^\dagger c_{y\sigma}-U\sum_{x\in\Lambda}n_{x\uparrow}n_{x\downarrow}, \quad (5)$$

and with $N_\sigma=\sum_x n_{x\sigma}$. The original number operators N_σ transform as $N_\downarrow\to\tilde N_\downarrow=N_\downarrow$ and $N_\uparrow\to\tilde N_\uparrow=|\Lambda|-N_\uparrow$. The original condition $N=|\Lambda|$ becomes $N_\uparrow=N_\downarrow$. The

Phys. Rev. Lett. *62*, 1201–1204 (1989). Errata *62*, 1927 (1989)

spin operators (2) become the *pseudo-spin operators*

$$\tilde{S}^z = \tfrac{1}{2}(|\Lambda| - N_\uparrow - N_\downarrow),\quad \tilde{S}^+ = \sum_{x\in\Lambda} \epsilon(x) c_{x\uparrow} c_{x\downarrow}, \tag{6}$$

and $S^2 \to (\tilde{S}^z)^2 + \frac{1}{2}\tilde{S}^+\tilde{S}^- + \frac{1}{2}\tilde{S}^+\tilde{S}^+ \equiv (\tilde{S})^2$. The $\tilde{S}$ operators commute with $\tilde{H}$, but so do the spin operators S given in (2). The S operators are the transforms of the pseudo-spin operators in the original variables and are of no special physical interest. The $\tilde{S}$ operators are the ones of interest as far as $\tilde{H}$ is concerned.

As before, I can work in the $\tilde{N}_\uparrow = \tilde{N}_\downarrow = |\Lambda|/2$ subspace, which implies that $N_\uparrow = N_\downarrow = |\Lambda|/2$. The uniqueness part of theorem 2 is then a consequence of conclusion (b) of theorem 1, which also states that the unique ground state $\tilde{\psi}$ of $\tilde{H}$ has $S=0$. This last fact is of secondary importance. The real problem is to prove that $2\tilde{S} = |B| - |A|$. The shortest proof is to return to H and the $S^z=0$ subspace (in the *original variables).* For each $U>0$ the ground state $\psi(U)$ is nondegenerate as has been shown and I want to prove that $2S = |B| - |A|$. The nondegeneracy of the ground state for *all* $U>0$ implies that the S of this unique ground state must be independent of U, for otherwise continuity in U would imply a degeneracy for some value of $U>0$. However, when U is very large, H, as stated before, is equivalent (to leading order in U) to h, the Heisenberg antiferromagnetic Hamiltonian defined in remark (4). As stated there, h *also* has a unique ground state[6] (for $S^z=0$) and this state has $2S = |B| - |A|$. The uniqueness property of h is crucial for it implies a gap (however small it may be) in the spectrum of h. Thus, for large enough U the S value of the ground state of h is identical to that of H. Q.E.D.

The partial support of the U.S. National Science Foundation Grant NO. PHY85-15288 A02 is gratefully acknowledged.

[1]F. J. Dyson, E. H. Lieb, and B. Simon, J. Stat. Phys. **18**, 335 (1978).

[2]J. Fröhlich, R. Israel, E. H. Lieb, and B. Simon, Commun. Math. Phys. **62**, 1 (1978).

[3]J. Frohlich, R. Israel, E. H. Lieb, and B. Simon, J. Stat. Phys. **22**, 297 (1980). The proof of theorem 5.5 on p. 334 is incorrect.

[4]T. Kennedy, E. H. Lieb, and S. Shastry, J. Stat. Phys. (to be published).

[5]P. W. Anderson, Phys. Rev. **115**, 2 (1959).

[6]E. H. Lieb and D. C. Mattis, J. Math. Phys. **3**, 749 (1962). This paper uses a Perron-Frobenius argument to prove that in a certain basis the ground state ψ for the connected case is a positive vector and it is unique. By comparing this ψ with the ψ for a simple soluble model (which is also positive) in the $2S^z = |B| - |A|$ subspace, we concluded $2S \le |B| - |A|$. However, if the comparison is made in the $S^z=0$ subspace, the same argument shows $2S = |B| - |A|$. This simple remark was overlooked in the 1962 paper, but it is crucial for the present work.

[7]D. C. Mattis, Phys. Rev. B **38**, 7061 (1988).

[8]E. H. Lieb and D. C. Mattis, Phys. Rev. **125**, 164 (1962).

[9]Y. Nagaoka, Phys. Rev. **147**, 392 (1966).

[10]D. J. Thouless, Proc. Phys. Soc. London **86**, 893 (1965).

[11]E. H. Lieb, in *Phase Transitions, Proceedings of the Fourteenth Solvay Conference* (Wiley-Interscience, New York, 1971), pp. 45–63.

VOLUME 65, NUMBER 12 PHYSICAL REVIEW LETTERS 17 SEPTEMBER 1990

Magnetic Properties of Some Itinerant-Electron Systems at $T > 0$

Michael Aizenman[(a)]

Courant Institute of Mathematical Sciences, New York University, 251 Mercer Street, New York, New York 10012

Elliott H. Lieb

Departments of Physics and Mathematics, Princeton University, P. O. Box 708, Princeton, New Jersey 08544

(Received 6 June 1990)

The Lieb-Mattis theorem on the absence of one-dimensional ferromagnetism is extended here from ground states to $T > 0$ by proving, *inter alia*, that $M(\beta,h)$, the magnetization of a quantum system in a field $h > 0$, is always *less* than the pure paramagnetic value $M_0(\beta,h) = \tanh(\beta h)$, with $\beta \equiv 1/kT$. Our proof rests on a new formulation in terms of path integrals that holds in any dimension; another of its applications is that the Nagaoka-Thouless theorem on the Hubbard model also extends to $T > 0$ in the sense that $M(\beta,h)$ *exceeds* $M_0(\beta,h)$.

PACS numbers: 75.10.Lp, 05.30.Fk

To many physicists the study of magnetism begins with a Heisenberg model, or something similar, in which localized spins interact with each other. Unfortunately, nature does not present us with such a starting point. Instead, at a reasonably (though not absolutely) fundamental level, we are presented with the Schrödinger equation for electrons whose interaction is, to a very good approximation, spin independent. It is left to the Pauli principle to induce a spin dependence, and for this it is essential that the electrons are *itinerant*, i.e., *not* localized *a priori*.

Given the subtlety of the phenomenon, it is crucial to elucidate the conditions which encourage or discourage the emergence of ferromagnetism. Some years ago Lieb and Mattis[1] showed that one *never* gets ferromagnetism in the ground states of itinerant electrons on a line. The results presented here include an extension of that earlier theorem (by a different method) to positive temperatures, and an opposite statement for systems in any dimension with a certain *parity constraint* (requiring *hard-core* repulsions)—for which we show that parallel spin alignments *are* favored at all temperatures. Examples of the second kind are one-dimensional systems of odd numbers of electrons on a ring (i.e., *periodic* boundary conditions), for which we extend the $T=0$ result of Herring,[2] and a particular case of the Hubbard model whose ground state was shown to be ferromagnetic by Nagaoka[3] and Thouless.[4,5]

We deal with the following general (spin-independent) Hamiltonian for N identical, itinerant particles of mass m ($\hbar = 1$):

$$H = -(2m)^{-1}\sum_{i=1}^{N}\nabla_i^2 + V(x_1,x_2,\ldots,x_N)\,, \tag{1}$$

where V is some potential-energy function[6] which is symmetric in its N variables. Boundary conditions are stated below. We consider simultaneously (a) the continuum model (as just stated) and (b) the lattice model in which the x_i's have integer components and ∇^2 is the second difference operator.

If the particles are spin-$\frac{1}{2}$ fermions (e.g., electrons), the eigenstates of H can be classified according to the eigenvalues (j,s) of the total spin angular momentum operator J and of S—the z component ($j = N/2$, $N/2-1,\ldots,0$ or $\frac{1}{2}$, and $|s| \le j$). The eigenvalues of H depend on j but (for j specified) not on s. The ground-state energies are denoted by $E_0(j)$. Two simple benchmarks are the following: (i) Immobile, noninteracting, particles—with $m \to \infty$ and $V=0$. In this limit $E_0(j)$ is *independent* of j. The system is a *pure paramagnet*, with spins responding to a magnetic field independently of each other. (Recall that $S = \sum \sigma_i^z/2$.) In a uniform field, $\mathbf{h} = (0,0,h)$, the partition function $Z(\beta,h) = \mathrm{Tr}\exp[-\beta(H - \sum \mathbf{h}\cdot\sigma_i)]$ satisfies

$$Z(\beta,h)/Z(\beta,h=0) \equiv \langle e^{2\beta h S}\rangle_{h=0} = [\cosh(\beta h)]^N$$

in this limit, and the magnetization is

$$M(\beta,h) = d[\ln Z(\beta,h)]/d(\beta h) = N\tanh(\beta h)\,. \tag{2}$$

(ii) Noninteracting itinerant system, with $V=0$ and $0 < m < \infty$. The spectrum can be analyzed in terms of one-particle states, and one finds

$$E_0(j+1) > E_0(j) \tag{3}$$

for all $j < N/2$. Thus, the "noninteracting" itinerant system shows a tendency towards *anti*ferromagnetism (resulting from the Pauli principle, which allows two electrons to occupy the same one-particle eigenstate of ∇^2 only if their spins are antialigned). At $T=0$ and weak field the magnetization is 0 or 1.

The theorem of Lieb and Mattis[1] states that (3) always holds in 1D, even when $V \neq 0$. Although the theorem has some practical value for one-dimensional systems, its main conceptual value lies in the observation that theories of ferromagnetism based naively on Heisenberg's "exchange integral" can be misleading since two-body potentials with either ferromagnetic or antiferromagnetic exchange integrals can easily be constructed in one dimension.[7]

Equation (3) states that increasing the spin costs ground-state energy. The first aim of this paper is to provide similarly precise information about the free energy when the inverse temperature β is finite. We denote the partition function with the trace restricted to subspaces with specified values of j, s, or both by $Z_{JS}(\beta, h;j,s)$, $Z_S(\beta,h;s)$, and $Z_J(\beta,h;j)$. When h is omitted, it is understood to take the value $h=0$. We also denote by $Y_{JS}^{(N)}(j,s)$, $Y_J^{(N)}(j)$, and $Y_S^{(N)}(s)$, the dimensions of these spaces for N noninteracting spins. The Y's [see (9) *et seq.*] serve as natural normalization constants for the Z's [because Eqs. (4) and (5) become equalities in the pure paramagnetic case].

Theorem 1.—Let N spin-$\frac{1}{2}$ fermions have the Hamiltonian (1) (with $m<\infty$) in a bounded region $[-L,L]^N$, with Dirichlet ($\psi=0$), Neumann ($\psi'=0$), or sticky $[\psi'/\psi(\pm L)=\lambda_\pm]$ boundary conditions (either in the continuum or on a lattice). Then, for $\beta<\infty$,

$$Z_S(\beta;s+1)/Y_S^{(N)}(s+1)<Z_S(\beta;s)/Y_S^{(N)}(s)\,, \tag{4}$$

$$Z_J(\beta;j+1)/Y_J^{(N)}(j+1)<Z_J(\beta;j)/Y_J^{(N)}(j)\,, \tag{5}$$

for all j and $s\ge 0$. In a field $h>0$ the magnetization satisfies

$$0<M(\beta,h)<N\tanh(\beta h)\,. \tag{6}$$

Furthermore, there are strictly positive functions C_N, $C_{N-2}, C_{N-4},\ldots,C_0$ or C_1 (which depend on β but not on s) such that, for $|s|\le N/2$,

$$Z_S(\beta;s)=\sum_{k=0}^{N} C_k(\beta)Y_S^{(k)}(s)\,, \tag{7}$$

and

$$Z(\beta,h)=\sum_{k=0}^{N} C_k(\beta)[\cosh(\beta h)]^k\,, \tag{8}$$

where it is understood that $C_k\equiv 0$ if $N-k$ is odd, and $Y_S^{(k)}(s)\equiv 0$ if $k<2|s|$.

Remarks.—(1) Equations (7) and (8) are equivalent. They assert that, as far as the total spin is concerned, the thermal equilibrium state is effectively a (positive) superposition of states corresponding to systems of fewer (k) spins with no interaction.

(2) It is false that the susceptibility $\chi(h)\equiv\beta^{-1}\partial M(\beta, h)/\partial h$ does not exceed the pure paramagnetic value $N\cosh^{-2}(\beta h)$. By an elementary calculation using (8), the opposite is *always* the case for *very large* h (with the threshold possibly dependent on N).

(3) Equation (8) is derived below from (13); before turning to it let us note that (4) and (5), which are *not* derivable from each other, follow from (7) by elementary algebra. To do that, let us first recall that by the theory of angular momentum, for $|s|\le j$,

$$Z_{JS}(\beta;j,s)=Z_{JS}(\beta;j,j)=Z_S(\beta;j)-Z_S(\beta;j+1)$$
$$=Z_J(\beta;j)/(2j+1)\,. \tag{9}$$

Similar identities hold for $Y_{JS}^{(N)}(j,s)$, $Y_J^{(N)}(j)$, and $Y_S^{(N)}(s)$, with $Y_S^{(N)}(s)=N!/(N/2+s)!(N/2-s)!$.

To derive (4) from (7), one needs to establish the monotonicity in $|s|$ of $Y_S^{(k)}(s)/Y_S^{(N)}(s)$, for $N-k>0$ even. To do that, it is convenient to write this ratio as a telescopic product, whose factors are the monotone functions

$$Y_S^{(k)}(s)/Y_S^{(k+2)}(s)=[(k/2+1)^2-s^2]/(k+2)\,.$$

Equation (5) is derived by a similar tactic. Equation (6) is an elementary consequence of (8).

In Theorem 1 it is assumed that the interaction has no hard-core repulsion, or other singularities which are strong enough to induce nodal surfaces in the configuration space. The only role of this assumption is to allow (4)–(6) to be stated as *strict* inequalities. For our second result we let H include such repulsive interactions, which are the *sine qua non* for the *parity restriction* which is essential for Theorem 2. The term "dynamically allowed" refers there to those permutations that can be achieved by a motion of the particles that does not pass through a nodal surface induced by the positive divergence of the potential. For example, in 1D permutations are restricted if the interaction includes a hard-core repulsion which disallows particle encounters.

Theorem 2.—Suppose that in a system of N spin-$\frac{1}{2}$ fermions with Hamiltonian (1), in arbitrary dimension and in the continuum or on a lattice, the dynamically allowed permutations are all *even* and not restricted to the identity. (This parity requirement can be satisfied only if H includes a *hard-core repulsion.*) Then: (i) Among the ground states there is one with $j=N/2$. (ii) In a field $h>0$ the magnetization satisfies

$$M(\beta,h)>N\tanh(\beta h)\,. \tag{10}$$

(iii)

$$Z(\beta,h)=\sum_{\{n\}:\sum n_i=N} D_{\{n\}}(\beta)\prod_i\cosh(\beta h n_i) \tag{11}$$

for some $D_{\{n\}}(\beta)\ge 0$, with $\{n\}=\{n_1,n_2,\ldots\}$ varying over partitions of N. $D_{\{n\}}(\beta)>0$ if and only if there is an allowed permutation whose cycle lengths are $n_1,n_2,\ldots$.

Remarks.—(1) The content of (11) is that as far as the z-component S is concerned, at thermal equilibrium the system is in a superposition of states in which the particles form independent "cliques" of sizes n_i whose contribution to S is $\pm n_i/2$. In such a state the system shows enhanced response to magnetic fields. In particular, Eq. (10) [which should be contrasted with (6)] follows from (11) by writing $M(\beta,h)=E(\sum_j n_j\times\tanh(\beta h n_j))$, where $E(\cdot)$ is a normalized average over partitions with weights proportional to $D_{\{n_i\}}\times\prod_i\cosh(\beta h n_i)$, and applying the strict monotonicity of $\tanh(\cdot)$. By a similar argument, all the moments $\langle S^k\rangle$ exceed their paramagnetic values. (A similar clique structure can be inferred for the ferromagnetic Ising model from the Fortuin-Kasteleyn[8] representation.) As-

VOLUME 65, NUMBER 12 PHYSICAL REVIEW LETTERS 17 SEPTEMBER 1990

sertion (i) follows from the uniform bound $Z_S(\beta,N/2)/Z(\beta) \geq 2^{-N}$, implied by (11).

(2) Two examples to which Theorem 2 applies are the following: (a) The Hubbard model in the strong-coupling limit, with one hole on a bipartite lattice. In this case part (i) [in the stronger sense that $E_0(j)$ is strictly lowest at $j=N/2$] was established by Nagaoka and Thouless.[3-5] Our contribution is to extend their result to positive temperatures, in the sense of (10). (b) Systems of *odd* numbers N of spin-$\frac{1}{2}$ fermions in a ring, i.e., in $[0,L]^N$ with periodic boundary conditions, and a *hard-core repulsion*. The dynamically allowed permutations are powers of the cyclic shift $P \equiv (1,2,\dots,N)$, whose parity equals $(-1)^{N-1}=1$. Part (i) was proved in that case by Herring.[2] Herring's and our method also yield[9] $E_0(N/2-1) > E_0(N/2)$.

(3) Theorem 2 has a very restrictive hypothesis which is not satisfied in cases of direct physical interest. Its value lies in two points: (i) the light it sheds on the mechanism of ferromagnetism; (ii) the construction used in the proof, which introduces ideas not ordinarily encountered in the study of electron correlations, e.g., the notion of cliques and the representation (13) which is always valid regardless of the hypothesis.

We turn now to the proof of (8) and (11), by first analyzing the Feynman-Kac path-integral representation[10] of $\mathrm{Tr}\, e^{-\beta H}$:

$$Z(\beta,h) = \int_{\Omega_\beta} \rho(d\omega) \exp\left[-\int_0^\beta V(x(t))\,dt\right](-1)^{\pi(\omega)} \times \exp\left[\beta h \sum_{i=1}^N \sigma_i\right]. \tag{12}$$

Here, $\omega = \{(x(t),\sigma(t)) \mid 0 \leq t \leq \beta\}$, which we refer to as a "path," represents a trajectory of the N particles, with $x=(x_1,\dots,x_N)$, $\sigma=(\sigma_1,\dots,\sigma_N)$. Ω_β is the collection of paths such that (i) the spins $\sigma_i(t) = \pm 1$ are constant in time, i.e., independent of t; (ii) the particle configuration at time β is a permutation, $\pi(\omega)$, of the configuration at time zero, with $(-1)^{\pi(\omega)}$ the parity of that permutation; and (iii) there are no encounters of particles of equal spin. $\rho(d\omega)$ is a (*positive*-) probability measure on Ω_β, determined by just the kinetic component of H, which we now describe in more detail for the two cases considered here.

In the discrete case, with x_i taking values on some lattice, the particle trajectories are piecewise constant functions of time, and hop to neighboring sites independently and at random times. The jumps form a Poisson process, with the probability ρ that a specified particle does not jump during the time interval $t \in [a,b]$ given by $\exp(-|a-b|m)$.

In the continuum case, $\rho(d\omega)$ is a product of Wiener measures constrained to satisfy the periodicity requirement (ii); i.e., $x(t)$ is a "Brownian bridge" with $x_i(\cdot)$ continuous in time. For the Dirichlet Laplacian the paths are killed at the boundary; for Neumann they are "reflected" (in a sense thoroughly discussed in textbooks).

The third condition is absent *a priori*, but it is a standard observation that its inclusion does not affect the value of the integral. (The proof of this version of the Pauli exclusion principle can be based on the symmetry transformation that exchanges the trajectories leading to the earliest equal-spin encounter.) A useful implication for us is that paths in Ω_β have no triple encounters; this follows from the other properties of Ω_β and the fact that the spins take only two values.

The measure in (12) possesses a hidden symmetry which we shall uncover. We are going first to define for each path ω a collection $\Gamma(\omega)$ of loops γ. Each loop will have a winding number $w(\gamma)$. There will be a symmetry group $\mathcal{G}$ having 2^N elements each of which maps Ω_β in a one-to-one way onto Ω_β preserving $\Gamma(\omega)$, $\rho(d\omega)$, $\int V(x(t))dt$, and $(-1)^{\pi(\omega)}$. However, $\mathcal{G}$ does not preserve $\pi(\omega)$ and the spin values $s=\sum\sigma_i$. The values attained by s within the ensemble of paths related by the action of $\mathcal{G}$ are of the form $\sum_{\gamma\in\Gamma(\omega)} \pm w(\gamma)/2$, with all the configurations of $\pm$ signs occurring with equal weights. Symmetrization of the integrand in (12) then yields the following useful representation:

$$Z(\beta,h) = \int_{\Omega_\beta} \rho(d\omega) \exp\left[-\int_0^\beta dt\, V(x(t))\right](-1)^{\pi(\omega)} \times \prod_{\gamma\in\Gamma(\omega)} \cosh[\beta h w(\gamma)]. \tag{13}$$

The loops are associated with the space-time picture of the collection of particle trajectories of the path ω (see Fig. 1). They are drawn by starting from the $t=0$ location of any of the particles, and tracing the particle's location forward in space-time until its first encounter with another particle; at which point the tracing line switches to the world line of the other particle in the reversed orientation in time. Such an orientation switch is repeated whenever a particle encounter is reached. When the trace line reaches the time $t=0$ or β, it reemerges at the

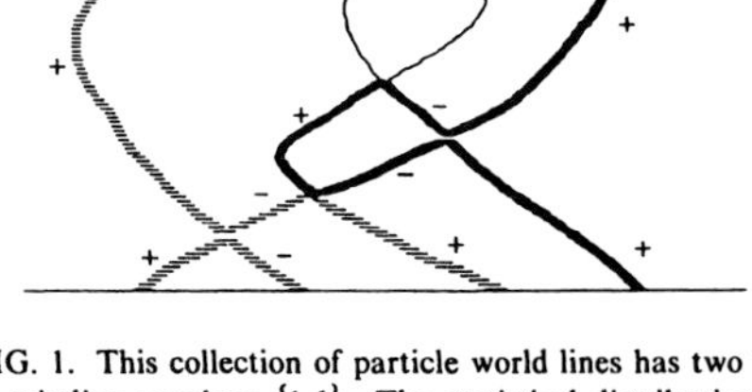

FIG. 1. This collection of particle world lines has two loops with winding numbers $\{1,1\}$. The statistical distribution of s under the symmetry group is that of two noninteracting spin-$\frac{1}{2}$ particles. In dimensions $d>1$, and for 1D periodic boundary conditions, permutations can occur without there being particle encounters, and then the winding numbers can be larger.

VOLUME 65, NUMBER 12 PHYSICAL REVIEW LETTERS 17 SEPTEMBER 1990

same location with time treated as periodic. This is continued until a loop is closed. For each $\omega \in \Omega_\beta$ the loops will all have only finite numbers of orientation reversals, since that number can diverge only if there is a triple encounter. We denote the collection of the distinct loops (γ) by $\Gamma(\omega)$, and define $w(\gamma)$ to be the absolute value of the loop's winding number, i.e., $w(\gamma)=|\int_\gamma dt/\beta|$.

A path $\omega \in \Omega_\beta$ is completely characterized by three things: the initial configuration $x(0)$, the union of the world lines, and the spin assignments along the world lines. It is evident that consistency with the constraints (i)-(iii) is achieved if and only if the spin assignment alternates along each loop, as shown in Fig. 1. By this condition, reading along the $t=0$ line, the admissible values of $2s$ are of the form $\sum_i \sigma_i = \sum_{\gamma\in\Gamma(\omega)} \pm w(\gamma)$.

We now define the group $\mathcal{G}$ to have N commuting generators g_i. In the space-time picture the map g_i flips the spin values $(+\frac{1}{2} \leftrightarrow -\frac{1}{2})$ along the loop γ containing the $t=0$ position of the ith particle. The spin flip is accompanied by the unique rearrangement of particle trajectories needed to maintain the constraints (i)-(iii). The paths ω and $g_i\omega$ differ in the $\pm$sign attached to $w(\gamma)$.

To show that g_i preserves the parity of $\pi(\omega)$, we note that $\pi(g_i\omega)$ is obtained from $\pi(\omega)$ by a product of transpositions associated with the orientation reversals in the loop γ; their number is always even. The invariance of $\rho(d\omega)$ is evident in the lattice case with discrete time steps. The invariance in the continuum case can be deduced by verifying that g_i are continuous maps in the uniform topology of Ω_β and invoking Donsker's theorem[11] on the weak convergence of the random-walk measures to the Wiener measure.

Equation (13)→(8): Evidently, on a line $\pi(\omega)$ can only be the identity, and $w(\gamma)$ can only take the values 0 and 1. Then (13) directly implies (8) with $C_k(\beta)$ given by the integral in (13) restricted to ω's for which the number of loops with $w(\gamma)=1$ is exactly k. The strict positivity is assured by the nonvanishing contributions from paths ω with collisions occurring only within the first $(N-k)/2$ pairs of adjacent particles. Q.E.D.

Equation (13)→(11): The hard-core repulsion precludes encounters, so the loops are dynamically simple but have nontrivial winding numbers, satisfying $\sum w(\gamma) = N$. Q.E.D.

Finally, we remark that Theorem 1 admits a *parastatistical* extension—as did the ground-state theorem of Ref. 1.

This research was supported in part by NSF Grants No. PHY-892067 (M.A.) and No. PHY-8515288-A04 (E.H.L.).

[(a)]Also in the Physics Department. Present address: Departments of Physics and Mathematics, Princeton University, Princeton, NJ 08544.

[1]E. H. Lieb and D. C. Mattis, Phys. Rev. **125**, 164–172 (1962).

[2]C. Herring, Phys. Rev. B **11**, 2056 (1975).

[3]Y. Nagaoka, Phys. Rev. **147**, 392 (1967).

[4]D. Thouless, Proc. Phys. Soc. London **86**, 893 (1965).

[5]E. H. Lieb, in *Phase Transitions, Proceedings of the Fourteenth Chemistry Solvay Conference* (Wiley, New York, 1971), pp. 45–63.

[6]The restrictions on V are fairly lax:

$$\sup_x \int_{|y|\le a} d^{Nd}y \|y\|^{-(Nd-2)} |V(y-x)| \to 0$$

as $a\to 0$. If V consists of k-body interaction, that condition translates to a lower-dimensional condition of similar form, with the weight becoming $\ln\|y\|$ for $kd=2$, and 1 for $kd=1$.

[7]R. E. Peierls, *Quantum Theory of Solids* (Oxford Univ. Press, New York, 1955).

[8]C. M. Fortuin and P. W. Kasteleyn, Physica (Utrecht) **57**, 536 (1972).

[9]At $T=0$ all the powers of P occur with equal weights in Eq. (13). By Eq. (9), $Z_J(\beta,N/2-1)/Z_J(\beta,N/2)\to 0$ as $\beta\to\infty$.

[10]B. Simon, *Functional Integration and Quantum Physics* (Academic, New York, 1979).

[11]M. Donsker, in *Four Papers on Probability*, Memoirs of the American Mathematical Society Vol. 6 (American Mathematical Society, Providence, RI, 1951).

Part II

Matrix Inequalities and Combinatorics

J. of Math. and Mech. *16*, 127–139 (1966)

Proofs of some Conjectures on Permanents

ELLIOTT H. LIEB*

Communicated by S. SHERMAN

I. Introduction. In a recent paper [1] Marcus and Minc have given an admirable summary of the theory of permanents of square matrices, especially with regard to inequalities satisfied by the permanent. Our knowledge of permanents is certainly meager compared to that for determinants, largely for the reason that while the former differs from the latter only in the replacement of minus signs by plus signs, the determinant, and not the permanent, is invariant under unitary transformations.

As far as Hermitian, positive semi-definite (h.p.s.d.) matrices are concerned, however, the few inequalities that are known for the permanent are suspiciously similar to certain special cases of classical inequalities for the determinants of such matrices—the only difference being that the *direction of the inequality is reversed*. In particular, we conjecture that the answer to a question posed in [1], referred to there as problem 2, is given by the following:

Conjecture α. *Let A be an n-square* h.p.s.d. *matrix and let H be a subgroup of S_n, the symmetric group on n objects. Let λ be a representation of H with characters $X^\lambda(P)$ and let $X^\lambda(I) = m$ be the degree of λ. Then*

$$D^\lambda \equiv \frac{1}{m} \sum_{P \in H} X^\lambda(P) \prod_{i=1}^{n} a_{iP(i)} \leqq \operatorname{Per} A, \tag{1}$$

where Per A *is the permanent of* A.

If Per A be replaced by Det A in (1), and if the inequality be reversed, then we would have the classic inequality of Schur [2, 3] which is almost half a century old.

Two immediate corollaries of conjecture α would be two theorems conjectured in [1]:

Conjecture 8 (*M. Marcus and M. Newman*). *If A is* h.p.s.d. *and* $1 \leqq k \leqq n$ *then* Per $A \geqq$ Per α Per β *where α is a principal k-square submatrix and β is the complementary principal $(n - k)$-square submatrix of A. If A is positive definite then equality holds if and only if $A = \alpha \dotplus \beta$.*

* Work supported by the United States Air Force Office of Scientific Research at Yeshiva University, Grant No. AF-AFOSR-713-64.

Journal of Mathematics and Mechanics, Vol. 16, No. 2 (1966).

J. of Math. and Mech. *16*, 127–139 (1966)

***Conjecture* 9** (*M. Marcus*). *Let A be an mk-square* h.p.s.d. *matrix (with $k > 1$) partitioned as follows:*

$$A = \begin{bmatrix} A_{11} \, A_{12} & \cdots & A_{1m} \\ \vdots \quad \vdots & & \vdots \\ A_{m1} A_{m2} & \cdots & A_{mm} \end{bmatrix}$$

in which each A_{ij} is k-square. Let B be the m-square matrix whose (i, j) entry is Per A_{ij} . *Then*

$$\text{Per } A \geqq \text{Per } B.$$

If A is positive definite then equality holds if and only if $A = \sum_{i=1}^{m} A_{ii}$.

If we ignore the "if and only if" statements for equality to hold, then conjecture 8 follows from conjecture α by letting H be the product of the symmetric group on k objects and the symmetric group on the remaining $n - k$ objects and by letting λ be the identity representation. Likewise, conjecture 9 would follow by letting H be the product of the m symmetric groups, each on k objects, together with those permutations that exchange *all* (and not some) of the k objects in one of the m sets with those in another set. Again λ is the identity representation.

Although we cannot prove conjecture α, we can *prove conjecture* 8 *as well as conjecture* 9 *when* $m = 2$. As partial compensation for not being able to prove conjecture 9 generally, we will prove a somewhat deeper theorem for the $m = 2$ case than is contained in conjecture 9. This theorem refers to an arbitrary partitioning of A into four blocks, rather than merely into four square blocks. It also gives the *complete* necessary and sufficient conditions for equality to hold in conjectures 8 and 9. In section III we present an analogous theorem for *determinants*.

II. Statement and Proof of the Theorem on Permanents.

Definition. *An s-square submatrix of a matrix Γ is any matrix formed by the intersection of s rows and columns of Γ. These need not be contiguous.*

Theorem. Let $p > 0$ and $r > 0$ and let A be an $n = (p + r)$-square h.p.s.d. *matrix, partitioned as follows:*

$$A = \begin{bmatrix} \alpha & \gamma \\ \gamma^* & \beta \end{bmatrix} \tag{1}$$

where α is $p \times p$, β is $r \times r$, and γ is $p \times r$. In addition A does not have a zero row. Let $A(\lambda)$ be the n-square matrix obtained from A by replacing α by the matrix $\lambda\alpha$, λ being a complex number. Then all the coefficients in the pth degree polynomial, $P(\lambda) = \text{Per } A(\lambda)$, are real and non-negative. Furthermore, let s be the largest integer for which there exists an s-square submatrix of γ with non-vanishing permanent. Then the coefficients of all terms λ^t in $P(\lambda)$ with $p - s \leqq t \leqq p$, are strictly positive, and the remaining coefficients are zero.

Conjectures 8 and 9 ($m = 2$) follow immediately from this theorem, for Per $A = P(1)$ while Per α Per β is the coefficient of λ^p in $P(\lambda)$. Also, if $p = r$ then Per γ Per γ^* is the constant term in $P(\lambda)$. The conditions for equality are also seen at once to be:

***Addendum to Conjectures* (now theorems) 8 and 9 (m = 2).** *The necessary and sufficient conditions for equality to hold in these two theorems is that either:* (1) *A has an identically zero row or* (2) *$\gamma = 0$, which means that $A = \alpha \dotplus \beta$.*

Proof of theorem. Since A is h.p.s.d. its elements can be represented as $a_{ij} = (f^i, f^j)$, where f^i is an n-dimensional vector with components f_k^i, and $(g, f) \equiv \sum_1^n g_k^* f_k$. Also, since A does not have a zero row, we can assume $a_{ii} = 1$ (all i) without loss of generality. Let $R(k_1, \cdots, k_n)$ be any function defined for all $k_i = 1, \cdots, n$, and define inner product by

$$(R, G) \equiv \sum_{k_1=1}^{n} \cdots \sum_{k_n=1}^{n} R^*(k_1, \cdots, k_n) G(k_1, \cdots, k_n).$$

Define

$$T(i_1, \cdots, i_n \mid k_1, \cdots, k_n) \equiv f_{k_1}^{i_1} \cdots f_{k_n}^{i_n} \equiv T(I, K),$$

where I is the n-tuple $(i_1, \cdots, i_n)$ and similarly for K. Define

$$B(I, J) = a_{i_1 j_1} a_{i_2 j_2}, \cdots, a_{i_n j_n},$$

whence $B(I, J) = (T(I, K), T(J, K))$, where the inner product is taken with respect to the K variables. Let P be the permutation

$$\begin{pmatrix} 1 & \cdots & n \\ \alpha_1 & & \alpha_n \end{pmatrix}$$

and for all functions R, define

$$R(PK) \equiv PR(K) \equiv R(k_{\alpha_1}, \cdots, k_{\alpha_n}).$$

Then

$$\text{Per } A = \sum_{P \varepsilon S_n} B(PI, I),$$

when $I = (1, 2, \cdots, n)$. But since $B(PI, J) = B(I, P^{-1}J)$, we have

$$\text{Per } A = (n!)^{-1} \sum_{P \varepsilon Sn} \sum_{Q \varepsilon S_n} B(PI, QI),$$

when $I = (1, \cdots, n)$. Define $F(K) = T(I, K)$ with $I = (1, \cdots, n)$ and define

$$F^S(K) = \sum_{P \varepsilon S_n} F(PK).$$

Since $T(PI, K) = T(I, P^{-1}K)$ and since $(PF, G) = (F, P^{-1}G)$, we have the important result: Per $A = (n!)^{-1}(F^S, F^S)$.

Now, let S_p be the symmetric group on the indices $1, \cdots, p$ and let S_r be the symmetric group on the indices $p + 1, \cdots, p + r = n$. Define

J. of Math. and Mech. *16*, 127–139 (1966)

$$O_p = \sum_{P \varepsilon S_p} P \quad \text{and} \quad O_r = \sum_{P \varepsilon S_r} P,$$

and let $F'(K) = O_p O_r F(K)$. Since F is a product function, F' will thus be the product of two functions $F'(K) = F_1(K_p)F_2(K_r)$, where K_p denotes the variables $k_1, \cdots, k_p$, and K_r denotes the remaining r variables. Any transposition permutation which is not in S_p or S_r will be called a "cross transposition" and any product of such transpositions will be called a "cross permutation". Thus, every cross permutation has the property that if $1 \leqq q \leqq p$ then either the index q is left undisturbed or else q is replaced by s, where $p + 1 \leqq s \leqq n$. Let X be the set of those cross permutations, P, having the property that all indices q_i $(1 \leqq q_i \leqq p)$ which are "moved" by P remain in serial order after the permutation. Likewise, all indices s_i $(p + 1 \leqq s_i \leqq n)$ which are "moved" by P remain in serial order. The set X has $\binom{n}{p} - 1$ members. Define the operator O as

$$O = I + \sum_{P \varepsilon X} P \tag{2}$$

where I is the identity permutation. Then clearly

$$F^S = OF' \quad \text{and} \quad OF^S = \binom{n}{p} F^S.$$

Hence

$$\text{Per } A = (n!)^{-1}(OF', OF') = (p!\, r!)^{-1}(F', OF') \tag{3}$$

Let us examine the terms on the right hand side of (3). The term $(p!\, r!)^{-1}(F', IF')$ is evidently Per α Per β, which is the coefficient of λ^p in $P(\lambda)$: Define $X_j \subset X$ to be those cross permutations composed of exactly j transpositions, for $1 \leqq j \leqq \min(p, r)$, and $X_0 = I$. Hence $O = \sum_j \bar{O}_j$ where

$$\bar{O}_j = \sum_{P \varepsilon X_j} P.$$

A little reflection will show that the term $(p!\, r!)^{-1}(F', \bar{O}_j F')$ in (3) is the coefficient of λ^{p-j} in $P(\lambda)$.

Consider first the term $(F', F') = p!\, r!$ Per α Per β. It is clearly non-negative, and to show that it is positive it is sufficient to find a function $G(K)$ such that $(G, F') \neq 0$. There always exists a vector V, with components v_k, such that $(V, f^i) \neq 0$ for all $i = 1, \cdots, n$. Let

$$G(K) = \prod_{i=1}^{n} v_{k_i}\, .$$

Plainly $(G, F') \neq 0$.

As for the remaining terms in (3), (F', PF') with $P \varepsilon X_j$ and $j \leqq s$, we claim that each one is real and strictly positive. Consider, for example, the permutation which interchanges the indices 1 and $p + 1$. This term is

$$E = \sum_K F_1^*(k_1, k_2, \cdots, k_p)F_2^*(k_{p+1}, k_{p+2}, \cdots, k_n) \cdot F_1(k_{p+1}, k_2, \cdots, k_p)F_2(k_1, k_{p+2}, \cdots, k_n). \tag{4}$$

Let $\mathbf{x}$ be the variables $(k_2, \cdots, k_p)$ and $\mathbf{y}$ be the variables $(k_{p+2}, \cdots, k_n)$ and define

$$U(\mathbf{x}, \mathbf{y}) = \sum_{k_1} F_1^*(k_1, \mathbf{x})F_2(k_1, \mathbf{y}),$$

so that

$$E = \sum_{\mathbf{x}} \sum_{\mathbf{y}} |U(\mathbf{x}, \mathbf{y})|^2 \geqq 0.$$

To show that E is non-zero, it is sufficient to find a function $H(\mathbf{x}, \mathbf{y})$ such that

$$\sum_{\mathbf{x}} \sum_{\mathbf{y}} H(\mathbf{x}, \mathbf{y})^* \, U(\mathbf{x}, \mathbf{y}) \neq 0.$$

For each of the original vectors f^i, $(i = 1, \cdots, p)$, let $t_i \geqq 1$ be the number of times this vector appears in the set $S_- = (f^1, f^2, \cdots f^p)$. Thus, $p - t_i$ vectors in S_- are distinct from, and not parallel to f^i (recall that all f^i have unit length). Likewise, for $i = p + 1, \cdots, n$, let t_i be the number of times f^i appears in the set $S_+ = (f^{p+1}, \cdots, f^n)$. For each $i = 1, \cdots, n$ we can always find a vector v^i such that $(v^i, f^i) = 0$ and $C_{ij} = (v^i, f^j) \neq 0$ whenever $f^j \neq f^i$. For each $1 \leqq i \leqq p$ and $p + 1 \leqq j \leqq n$ define the function $H^{ij}(\mathbf{x}, \mathbf{y})$ as follows:

$$H^{ij}(\mathbf{x}, \mathbf{y}) = \left(\prod_{\alpha=2}^{t_i} f_{k_\alpha}^i\right)\left(\prod_{\beta=t_i+1}^{p} v_{k_\beta}^i\right)\left(\prod_{\gamma=p+2}^{p+t_j} f_{k_\gamma}^j\right)^*\left(\prod_{\delta=p+t_j+1}^{n} v_{k_\delta}^j\right)^*, \tag{5}$$

where the first and/or the third factors are to be omitted if t_i and/or t_j is equal to one. Clearly

$$\begin{aligned} d_{ij}(k) &= \sum_{\mathbf{x}} \sum_{\mathbf{y}} F_1^*(k, \mathbf{x})F_2(k, \mathbf{y})H^{ij}(\mathbf{x}, \mathbf{y})^* \\ &= f_k^{i*} f_k^j (p - t_i)!\, t_i\,!\, (r - t_j)!\, t_j\,! \left(\prod_{l=t_i+1}^{p} C_{il}^*\right)\left(\prod_{l=p+t_j+1}^{n} C_{jl}\right). \end{aligned} \tag{6}$$

Now, if $s \geqq 1$, there is at least one pair (i, j) with $1 \leqq i \leqq p$ and $p + 1 \leqq j \leqq n$ such that $a_{ij} = (f^i, f^j) \neq 0$. Thus,

$$\sum_{\mathbf{x}} \sum_{\mathbf{y}} H^{ij}(\mathbf{x}, \mathbf{y})^* \, U(\mathbf{x}, \mathbf{y}) = \sum_k d^{ij}(k) \neq 0.$$

By the same method we shall see that all terms (F', PF') are real and strictly positive if P is composed of j cross transpositions and $j \leqq s$. To begin with, we note that if P and Q are any two permutations in the set X_j then $(F', PF') = (F', QF')$. This is a consequence of two facts: firstly that F_1 and F_2 are symmetric functions and hence the order in which their arguments appear is immaterial; secondly that (F', PF') is a sum, as in eqn. (4), over the "dummy" variables, K, which may conveniently be relabeled so that the variables from

J. of Math. and Mech. *16*, 127–139 (1966)

the set K_p which were "moved" by P are labeled $k_1, \cdots, k_j$ and the variables from K_r which were "moved" are labeled $k_{p+1}, \cdots, k_{p+j}$. If we denote by E_j the common value of (F', PF') with $P \in X_j$ then

$$E_j = \sum_k F_1^*(k_1, k_2, \cdots, k_p)F_2(k_{p+1}, k_{p+2}, \cdots, k_n)^* \cdot F_1(k_{p+1}, \cdots, k_{p+j}k_{j+1}, \cdots, k_p)F_2(k_1, \cdots, k_j, k_{p+j+1}, \cdots, k_n). \tag{7}$$

Thus,

$$E_j = \sum_{\mathbf{x}} \sum_{\mathbf{y}} |U(\mathbf{x}, \mathbf{y})|^2 \tag{8}$$

where $\mathbf{x}$ is now $(k_{j+1}, \cdots, k_p)$ and $\mathbf{y}$ is $(k_{p+j+1}, \cdots, k_n)$ and

$$U(\mathbf{x}, \mathbf{y}) = \sum_{k_1} \cdots \sum_{k_j} F_1^*(k_1, \cdots, k_j, \mathbf{x})F_2(k_1, \cdots, k_j, \mathbf{y}). \tag{9}$$

By eqn. (8), E_j is clearly real and non-negative. To prove that E_j is positive for $j \leqq s$, it is only necessary to construct an H function along the lines indicated in eqn. (5), with the property that

$$\sum_{\mathbf{x}} \sum_{\mathbf{y}} H(\mathbf{x}, \mathbf{y})^* \, U(\mathbf{x}, \mathbf{y}) \neq 0.$$

Without loss of generality (but with considerable notational simplification) we may assume that the upper left hand s-square corner of γ has a non-vanishing permanent, *i.e.* the submatrix $a_{\alpha\beta}$ with $\alpha = 1, 2, \cdots, s$ and $\beta = p + 1, p + 2, \cdots, p + s$. This submatrix must necessarily have some j-square submatrix with non-vanishing permanent, and we shall again assume it is in the upper left hand corner of γ, *i.e.*, $a_{\alpha\beta}$ with $\alpha = 1, 2, \cdots, j$ and $\beta = p + 1, p + 2, \cdots, p + j$. Let P_j denote the permanent of this submatrix.

As before, $H(\mathbf{x}, \mathbf{y})$ will be a product function, $H_1(\mathbf{x})H_2(\mathbf{y})$, and H_1 and H_2 will themselves be product functions. To define H_1, divide the set of vectors $S_- = (f^1, \cdots, f^p)$ into two sets $L_- = (f^1, \cdots, f^j)$ and $L_+ = (f^{j+1}, \cdots, f^p)$. Let $h \leqq j$ be the number of distinct vectors in L_- and, without loss of generality, we shall sssume that the vectors in the set $G = (f^1, f^2, \cdots, f^h)$ are distinct. It is always possible to find vectors $g^1, g^2, \cdots, g^h$ such that for all $i = 1, \cdots, h$ and all k, $(g^i, f^k) = 0$ if $f^k \neq f^i$ while $(g^i, f^i) = 1$. It is also always possible to find a vector v such that $(v, f^k) = 0$ if $f^k \in G$, while $C_k = (v, f^k) \neq 0$ whenever $f^k \notin G$.

For all $i = 1, \cdots, h$ let u_i and t_i be the number of times the vector f^i appears in the sets L_+ and S_-, respectively. Define $b_i = \sum_{m=1}^{i} u_m$ for $i > 0$, $b_0 = 0$, and define $a_i = b_{i-1} + 1$ for $i > 0$. Now, let

$$H_1(\mathbf{x}) = \left\{ \prod_{i=1}^{h} \left(\prod_{\alpha=j+a_i}^{j+b_i} g_{k_\alpha}^{u_i} \right) \right\}^* \left(\prod_{\alpha=j+a_h+1}^{p} v_{k_\alpha} \right)^*, \tag{10}$$

where it is understood that any product is to be omitted if its upper index is smaller than its lower. Thus,

(11) $$\sum_{\mathbf{x}} H_1(\mathbf{x})^* F_1^*(k_1, \cdots, k_j, \mathbf{x}) = J^*(k_1, \cdots, k_j) \prod_{i=1}^{h} t_i!\,(p - b_h - j)! \prod_{\alpha=j+a_h+1}^{p} C_\alpha,$$

where J is the symmetric function

(12) $$J(k_1, \cdots, k_j) = O_j[f^1_{k_1} f^2_{k_2} \cdots f^j_{k_j}].$$

Mutatis mutandis we define the function H_2 in the same way, with the result that

(13) $$\sum_{\mathbf{y}} H_2^*(\mathbf{y}) F_2(k_1, \cdots, k_j, k_{p+j+1}, \cdots, k_n) = \varphi J(k_1, \cdots, k_j).$$

Thus,

(14) $$\sum_{\mathbf{x}} \sum_{\mathbf{y}} H(\mathbf{x}, \mathbf{y})^* U(\mathbf{x}, \mathbf{y}) = \psi \sum_{k_1} \cdots \sum_{k_j} |J(k_1, \cdots, k_j)|^2 = \chi P_j.$$

In (13) and (14), φ, ψ, and χ are non-zero constants, thus proving that $E_j \neq 0$.

To complete the proof of the theorem we must show that d_t, the coefficient of λ^t in $P(\lambda)$, is zero whenever $t = p - j$ with $j > s$. This follows from the trivial observation that all contributions to this coefficient involve $(p - j)$ elements of the matrix α, j elements of γ, j elements of γ^* and $(r - j)$ elements of β. Every selection of j elements from γ (such that no row or column is repeated) defines a j-square submatrix of γ. Let M be any such submatrix. Plainly, the only way in which the elements of M enter into d_t is through the combination Per M. Thus, if Per $M = 0$ for all possible j-square submatrices of γ then $d_t = 0$. Q.E.D.

III. Theorem on determinants:

Theorem. *Let $p > 0$ and $r > 0$ and let A be an $n = (p + r)$-square* h.p.s.d. *matrix, partitioned as follows:*

$$A = \begin{bmatrix} \alpha & \gamma \\ \gamma^* & \beta \end{bmatrix}$$

where α is $p \times p$, β is $r \times r$, and γ is $p \times r$. In addition α and β are positive definite. Let s be the largest integer for which there exists an s-square submatrix of γ with non-vanishing determinant. Let $A(\lambda)$ be the matrix obtained from A by replacing α by the matrix $\lambda\alpha$, λ being a complex number. Then the coefficient of λ^{p-t} in the pth degree polynomial $P(\lambda) =$ Det $A(\lambda)$ *is: (a) zero if $t > s$; (b) real and strictly positive if t is even and $t \leqq s$; (c) real and strictly negative if t is odd and $t \leqq s$. Furthermore, if α or β is merely semi-definite, then $P(\lambda)$ vanishes identically.*

The proof of this theorem is almost exactly the same as for the preceding one, except that symmetrizers are replaced by anti-symmetrizers.

J. of Math. and Mech. *16*, 127–139 (1966)

Acknowledgements. This work was supported by the United States Air Force Office of Scientific Research Grant Number AF-AFOSR-713-64. The author also wishes to thank Professor Seymour Sherman for drawing his attention to reference [1].

References

[1] Marvin Marcus & Henry Minc, Permanents, *Amer. Math. Monthly*, **72** (1965) 577–591.
[2] I. Schur, Über endliche Gruppen und Hermitesche Formen, *Math. Z.*, **1** (1918) 184–207.
[3] Marvin Marcus, On two classical results of I. Schur, *Bull. Amer. Math. Soc.*, **70** (1964) 685–688.

Physics Dept.
Northeastern University
Boston, Mass.
Date Communicated: February, 1966

Reprinted from JOURNAL OF COMBINATORIAL THEORY Vol. 5, No. 2, September 1968
 Printed in Belgium

Concavity Properties and a Generating Function for Stirling Numbers

ELLIOTT H. LIEB*†

Physics Department, Northeastern University, Boston, Massachusetts 02115

Communicated by Gian-Carlo Rota

ABSTRACT

The Stirling numbers of the first kind, S_N^k, and of the second kind, σ_N^k, are shown to be strongly logarithmically concave as functions of k for fixed N. This result is stronger than the unimodality conjecture which was heretofore proved only for σ_N^k (Harper). We also introduce a generating function for the σ_N^k which is different from the conventional one but which has a relatively simple closed form expression.

A set of non-negative, real numbers, C_k (defined for $k = 1,..., N$) is said to be *unimodal* if and only if there are two integers a and b (with $1 \leqslant a \leqslant b \leqslant N$) such that: C_k is monotonically non-decreasing for $1 \leqslant k \leqslant a$; C_k is monotonically non-increasing for $b \leqslant k \leqslant N$; $C_k = \text{constant}$ for $a \leqslant k \leqslant b$. In other words, the discrete graph of C_k has no maxima other than the principal one, but the principal maximum may be a plateau extending from a to b.

A stronger property, one which implies unimodality, is *logarithmic concavity* (LC):

$$C_k^2 \geqslant C_{k+1}C_{k-1} \qquad \text{for} \quad k = 2,..., N-1.$$

This means that the function $b_k = \ln(C_k)$ is a concave function of k. *Strong logarithmic concavity* (SLC) means strict inequality in (1) for all k. Obviously, SLC implies not only unimodality but also that the plateau can contain at most two points (i.e., $b - 1 \leqslant a \leqslant b$ in the above definition).

A commonplace example of a SLC set is the binomial coefficients $\binom{N}{k}$ because

$$\binom{N}{k}^2 = \binom{N}{k+1}\binom{N}{k-1}\left(\frac{k+1}{k}\right)\left(\frac{N-k+1}{N-k}\right). \tag{2}$$

* Work supported by National Science Foundation Grant GP-6851.

† Present address: Mathematics Department, M.I.T., Cambridge, Massachusetts, 02139.

J. Combinatorial Theory 5, 203–206 (1968)

A useful strategy for proving SLC is to use *Newton's inequality:*† If the polynomial $Q(x) = \sum_1^N C_k x^k$ has only real roots then, for $k = 2,..., N-1$

$$C_k^2 \geqslant C_{k+1}C_{k-1}\left(\frac{k}{k-1}\right)\left(\frac{N-k+1}{N-k}\right). \tag{3}$$

Two sets that occur in combinatorial analysis are the Stirling numbers of the first kind, $S_N{}^k$, and of the second kind $\sigma_N{}^k$. We shall use Newton's inequality to prove they are SLC as functions of k and hence that they are unimodal.

Stirling numbers of the first kind ($S_N{}^k$): These are defined by $(-1)^{N-k}S_N{}^k$, being the number of permutations of N symbols which have exactly k cycles. It is well known [1] that the generating function is

$$\sum_{k=1}^{N} S_N{}^k x^k = x(x-1)\cdots(x-N+1). \tag{4}$$

Since its roots are real, inequality (3) is true with C_k replaced by $|S_N{}^k|$.

Stirling numbers of the second kind ($\sigma_N{}^k$): These are defined to be the number of ways of partitioning a set of N elements into k non-empty subsets. Harper [2] has shown that the generating function

$$P_H(x) = \sum_{k=1}^{N} \sigma_N{}^k x^k \tag{5}$$

has only non-positive real roots. Using this result (and, incidentally, relying on the fact that the roots are not merely real but non-positive as well) he showed that the $\sigma_N{}^k$ are unimodal and that the plateau can have at most two points. Our point is that the reality of the roots plus Newton's inequality establishes the stronger result (3) for $\sigma_N{}^k$, which is SLC.

Actually, before learning of Harper's result, we proved SLC using a different generating function from (5). Although it gives a slightly weaker result than (3) it seems worth while to mention this function because it appears to be simpler than (5), namely,

$$P(x) = \sum_{k=1}^{N} \sigma_N{}^k k! \binom{N}{k} x^k. \tag{6}$$

† Newton's inequality is extensively discussed in Hardy, Littlewood, and Polya, *Inequalities*, 2nd ed., Cambridge University Press, Cambridge, England, 1959, pp. 104–105 and pp. 51–54. Note that there appears to be a requirement that no two consecutive C_k vanish, but this is irrelevant for our purposes because in that event equation (3) is *a fortiori* true. To obtain our inequality (3), we apply the theorem in the above book to the polynomial $\sum_0^{N-1} C_{k+1}x^k$.

If we can show that the roots of (6) are real we obtain

$$(\sigma_N{}^k)^2 \geqslant \sigma_N^{k+1}\sigma_N^{k-1}\left(\frac{k}{k-1}\right) \tag{7}$$

by setting

$$C_k = \sigma_N{}^k k!\binom{N}{k}$$

in Newton's inequality. To investigate (6) we use the closed form expression [1]:

$$\sigma_N{}^k = \frac{1}{k!}\sum_{p=1}^{k}(-1)^{k-p}\binom{k}{p}p^N. \tag{8}$$

Inserting (8) into (6), changing the order of summation, and setting $k = p + r$:

$$P(x) = \sum_{p=1}^{N}\frac{x^p}{p!}p^N\sum_{r=0}^{N-p}(-1)^r\frac{N!}{(N-p-r)!\,r!}x^r$$

$$= \sum_{p=1}^{N}x^p(1-x)^{N-p}p^N\binom{N}{p}. \tag{9}$$

Equation (9) is one closed form for $P(x)$. Another is obtained by noting that

$$\sum_{p=1}^{N}x^p z^{N-p}p^N\binom{N}{p} = \left(x\frac{\partial}{\partial x}\right)^N(x+z)^N, \tag{10}$$

whence

$$P(x) = \left(x\frac{\partial}{\partial x}\right)^N(x+z)^N\,|_{z=1-x}\,. \tag{11}$$

To show that the roots of $P(x)$ are real consider

$$P_z(x) = \left(x\frac{\partial}{\partial x}\right)^N(x+z)^N$$

for fixed, real z. By Rolle's theorem $P_z(x)$ has N real roots since $(x+z)^N$ has. However, $P_z(x)$ is homogeneous of the N-th degree in z and x, and contains a term x^N. Hence the N roots will be $x = \alpha_i z$ $(i = 1,\ldots, N)$ with

J. Combinatorial Theory 5, 203–206 (1968)

α_i real. The roots of $P(x)$, which we shall call β_i, are given by $\beta_i = \alpha_i(1 - \beta_i)$, and hence are real. (Note that no $\alpha_i = -1$ because then $P(x)$ would have to be of lower degree than N which, as we see from (6), is not the case.)

Acknowledgments

I should like to thank Dr. G. C. Rota and Dr. L. H. Harper for introducing me to the problem and for stimulating discussions.

References

1. M. Abramowitz and I. A. Stegun, eds., *Handbook of Mathematical Functions* (Nat. Bur. Standards Appl. Math. Ser. 55), U. S. Gov't. Printing Office, Washington, D. C., 1964, p. 824.
2. L. H. Harper, Stirling Behavior is Asymptotically Normal, *Ann. Math. Statist.* **38** (1967), 410–414.

Adv. in Math. *11*, 267–288 (1973)

Reprinted from ADVANCES IN MATHEMATICS
Vol. 11, No. 3, December 1973

Printed in Belgium

Convex Trace Functions and the Wigner–Yanase–Dyson Conjecture

ELLIOTT H. LIEB*

Institut des Hautes Etudes Scientifiques, 91440 Bures-sur-Yvette, France

Several convex mappings of linear operators on a Hilbert space into the real numbers are derived, an example being $A \mapsto -\mathrm{Tr}\exp(L + \ln A)$. Some of these have applications to physics, specifically to the Wigner–Yanase–Dyson conjecture which is proved here and to the strong subadditivity of quantum mechanical entropy which will be proved elsewhere.

1. INTRODUCTION

This paper is concerned with certain convex or concave mappings of linear operators on a Hilbert space into the reals. [$f(A)$ is convex if $f(\lambda A + (1-\lambda)B) \leqslant \lambda f(A) + (1-\lambda)f(B)$ for $0 < \lambda < 1$ and $f(A)$ is concave if $-f(A)$ is convex.] These mappings involve the trace operation which plays a central role in quantum statistical mechanics, and it is not surprising, therefore, that the mappings discussed here were motivated by considerations of physics. In particular, Theorem 1 solves affirmatively a conjecture due to Wigner, Yanase, and Dyson [1] about a certain definition of information. In Section 3 we use Theorem 1 to prove other convexity theorems when the Hilbert space is finite dimensional. One of those, Theorem 6, we extend to infinite dimensional spaces in Section 4. Theorem 6 has a physical application; it is the basis for proving that quantum mechanical entropy is strongly subadditive (cf. Refs. [2, 3, 4]). The proof of that fact will be given in a subsequent paper [5].

From the work of Krauss and Bendat and Sherman ([6] and the references quoted therein) it is known that certain convex functions from

* On leave from the Department of Mathematics, M.I.T., Cambridge, Massachusetts 02139. Work partially supported by a Guggenheim Memorial Foundation fellowship and by U.S. National Science Foundation grant GP-31674 X.

Adv. in Math. *11*, 267–288 (1973)

$\mathbb{R}$ to $\mathbb{R}$ extend to operator-valued convex functions. If $f(x)$ is such a function then $A \mapsto \operatorname{Tr} Kf(A)$ (where Tr means trace) is certainly convex when $K > 0$ and fixed. Simple examples are $f(A) = A^{-p}$ and $f(A) = -A^p$ for $A > 0$ and $0 < p \leqslant 1$. However, $A \mapsto \operatorname{Tr} f(A)$ may be convex even when $f(A)$ is not convex as an operator-valued function. Examples of this are $f(A) = e^A$ for A self-adjoint and $f(A) = A^{-p}$ for $p > 1$ and for $A > 0$ (cf. Theorems 8 and 9).

In this paper we shall be concerned with mappings more complicated than those just mentioned. One example, Theorem 6, is

$$A \mapsto -\operatorname{Tr} \exp[L + \ln A]$$

for $A > 0$ and L self-adjoint.

Theorem 1 is our main theorem and Theorems 2, 3, 6, and 7 are derived from it. Theorems 8 and 9 are a side issue and are independent of and simpler than Theorem 1. In Section 5 we remark briefly on the logical connection of Theorems 1, 2, 3, 6, and 7, namely that they can all be derived simply from each other (at least for finite dimensional Hilbert spaces).

2. The Main Theorem and the Wigner–Yanase–Dyson Problem

We begin by proving our main Theorem 1 which constitutes the basis for Theorems 2, 3, 6, and 7 of the next section. Theorem 1 is also the Wigner–Yanase–Dyson (WYD) conjecture [1] (actually, it is a bit stronger) and at the end of this section we shall explain the WYD problem. We also discuss another problem concerning the WYD definition of information [1] and give a partial solution of it.

Theorem 1 will be proved directly for infinite dimensional Hilbert spaces and our notation is the following:

(1) H is a separable Hilbert space with inner product (x, y) which is linear in y and conjugate linear in x.

(2) $\mathscr{B}(H)$ is the set of bounded linear operators from H to H; $\mathscr{B}^s(H) \subset \mathscr{B}(H)$ are the bounded self-adjoint operators; $\mathscr{B}^+(H) \subset \mathscr{B}^s(H)$ are the positive operators ($A \in \mathscr{B}^+(H) \Rightarrow (x, Ax) \geqslant 0, \forall x$); $\mathscr{B}^{++}(H) \subset \mathscr{B}^+(H)$ are the strictly positive operators ($A \in \mathscr{B}^{++}(H) \Rightarrow (x, Ax) > 0, \forall x \neq 0$).

(3) If $A \in \mathscr{B}^+(H)$ and $z \in \mathbb{C}$, we can use the spectral representation of A to define $A^z \in \mathscr{B}(H)$ for $\operatorname{Re}(z) \geqslant 0$. $A^z \in \mathscr{B}^+(H)$ for $z \geqslant 0$.

(4) The $\mathscr{I}_q$ classes: If $A \in \mathscr{B}(H)$ we form $|A| = (A^\dagger A)^{1/2} \in \mathscr{B}^+(H)$. $A \in \mathscr{I}_q(H) \subset \mathscr{B}(H)$ $(q \geqslant 1)$ if $\|A\|_q \equiv (\operatorname{Tr}|A|^q)^{1/q} < \infty$, where Tr means trace. $\mathscr{I}_1(H)$ is the trace class and $\mathscr{I}_2(H)$ is the Hilbert–Schmidt class. $A \in \mathscr{I}_q(H)$ implies that A is compact and that $A \in \mathscr{I}_1(H)$. $\|A\|_q = (\sum_{j=1}^\infty \lambda_j{}^q)^{1/q}$, where the λ_j are the eigenvalues of $|A|$ in decreasing order, including multiplicity. If $A \in \mathscr{B}^+(H)$ but $A \notin \mathscr{I}_1(H)$, it is convenient to define $\operatorname{Tr} A = \infty$.

(5) We recall that if $A \in \mathscr{B}(H)$ and if K is a linear operator (not necessarily bounded) on a dense domain, $D(K)$, in H then AK may have a bounded extension to all of H. If so, it is unique and its adjoint is a bounded extension of $K^\dagger A^\dagger$.

THEOREM 1. *Let K be a linear operator (not necessarily bounded) on H, let $A, B \in \mathscr{B}^+(H)$, and let λ, $0 < \lambda < 1$, be given. Form the convex combination $C = \lambda A + (1-\lambda)B$. Let p and r be given positive real numbers with $p + r \equiv s \leqslant 1$. If $M \equiv C^{p/2}KC^{r/2}$ has an extension to $\mathscr{I}_2(H)$ then*

(1) *$A^{p/2}KA^{r/2}$ and $B^{p/2}KB^{r/2}$ have extensions to $\mathscr{I}_2(H)$ and*

(2) $\lambda \operatorname{Tr} A^{r/2}K^\dagger A^p K A^{r/2} + (1-\lambda) \operatorname{Tr} B^{r/2}K^\dagger B^p K B^{r/2}$

$$\leqslant \operatorname{Tr} C^{r/2}K^\dagger C^p K C^{r/2}, \text{ i.e.,}$$

$$A \in \mathscr{B}^+(H) \mapsto \operatorname{Tr} A^{r/2}K^\dagger A^p K A^{r/2} \text{ is concave.}$$

Proof. (a) We recall the theorem [6] that the map $A \in \mathscr{B}^+(H) \mapsto A^q$ is concave on $\mathscr{B}^+(H)$ when $0 < q \leqslant 1$. Thus, $\lambda A^q \leqslant C^q$ and $\operatorname{Ker}(A^q) = \operatorname{Ker}(A) \supset \operatorname{Ker}(C) = \operatorname{Ker}(C^q)$, and similarly for B. As A, B, and C are bounded, their kernels are closed subspaces and $H = \operatorname{Ker}(C) \oplus \operatorname{Ker}(C)^\perp$. The foregoing inequalities show that for $0 < q \leqslant 1$, $\alpha(q) \equiv A^{q/2}C^{-q/2}$ and $\alpha(q)^\dagger = C^{-q/2}A^{q/2}$ can be extended to bounded operators on $\operatorname{Ker}(C)^\perp$ because $\|A^{q/2}C^{-q/2}\Psi\| \leqslant \lambda^{-1/2}\|\Psi\|$, in the dense set $D_C = \{$vectors with support away from zero in the spectral representation of $C\}$. Similarly, we define $\beta(q) \equiv B^{q/2}C^{-q/2}$. Also, $\alpha(q)$ and $\alpha(q)^\dagger$ can be defined to be zero on $\operatorname{Ker}(C)$ and, thus, are defined on all of H. Clearly, $C^{q/2}\alpha(q)^\dagger = A^{q/2} = \alpha(q)C^{q/2}$. Consequently,

$$A^{p/2}KA^{r/2} = \alpha(p)[C^{p/2}KC^{r/2}]\,\alpha(r)^\dagger = \alpha(p)\,M\alpha(r)^\dagger \in \mathscr{I}_2(H),$$

since $M \in \mathscr{I}_2(H)$. Not only is the first part of the theorem thus proved, but we also see that if $\{\Psi_i\}$ and $\{\varphi_i\}$ are orthonormal bases for $\operatorname{Ker}(C)$ and $\operatorname{Ker}(C)^\perp$, respectively, we can compute traces in the basis $\{\Psi_i\} + \{\varphi_i\}$

Adv. in Math. *11*, 267–288 (1973)

and all terms involving $\{\Psi_i\}$ will vanish. Thus, $\mathrm{Ker}(C)$ is an irrelevant subspace, and we shall, henceforth, assume that $H = \mathrm{Ker}(C)^{\perp}$, i.e., $C > 0$.

(b) With the foregoing definitions, part (2) is equivalent to the following: $\lambda T^A(p, r) + (1 - \lambda)\, T^B(p, r) \leqslant \mathrm{Tr}\, M^\dagger M$ for every $M \in \mathscr{I}_2(H)$, where $T^A(p, r) = \mathrm{Tr}\, \alpha(r)\, M^\dagger \alpha(p)^\dagger \alpha(p)\, M\alpha(r)^\dagger$ and similarly for $T^B(p, r)$.

(c) Let $z = x + iy \in \mathbb{C}$ and consider the operator valued function $\alpha(z) \equiv A^{z/2}C^{-z/2} = A^{iy/2}A^{x/2}C^{-x/2}C^{-iy/2} = A^{iy/2}\alpha(x)\, C^{-iy/2}$. Since $C^{iy/2}$ is unitary and $\| A^{iy/2} \| \leqslant 1$, $\alpha(z)$ is uniformly bounded in $S = \{z \mid 0 \leqslant \mathrm{Re}(z) \leqslant 1\}$. If $\bar{z} = x - iy$, $\alpha(\bar{z})^\dagger = C^{-z/2}A^{z/2}$. For $\Psi \in D_C$, $\alpha(z)\Psi$ is an entire analytic function of z because $C^{-z/2}\Psi$ is entire and $A^{z/2}$ is entire. Hence, by the boundedness of $\alpha(z)$ and a standard density argument, $\alpha(z)\Psi$ is regular on S (continuous on S and analytic in the interior of S) for all $\Psi \in H$. Since weak analyticity implies strong analyticity, we also have that $\alpha(z)$ is strongly continuous on S and is norm analytic in the interior of S. Furthermore, if $A_n \to A$ strongly and if $B \in \mathscr{I}_2(H)$, then $A_nB \to AB$ in the $\mathscr{I}_2(H)$ norm. (This is trivial if B is finite rank, but the finite rank operators are dense in the $\mathscr{I}_2(H)$ norm). Hence, $\alpha(z_1)\, M\alpha(\bar{z}_2)^\dagger$ is $\mathscr{I}_2(H)$ regular on $S \times S$, which means that

$$T^A(z_1, z_2) \equiv \mathrm{Tr}\, \alpha(z_2)\, M^\dagger\alpha(\bar{z}_1)^\dagger\, \alpha(z_1)\, M\alpha(\bar{z}_2)^\dagger$$

is bounded and regular on $S \times S$.

(d) We now set $z_1 = z$, $z_2 = s - z$ and consider $T^A(z) \equiv T^A(z, s - z)$ as a regular function on $\{z \mid 0 \leqslant \mathrm{Re}(z) \leqslant s\}$. By (b) we need to show that $f(p) = \lambda T^A(p) + (1 - \lambda)\, T^B(p) \leqslant \mathrm{Tr}\, M^\dagger M$. By the maximum modulus principle for bounded regular functions on a strip, $|f(p)| \leqslant \max\{\sup_\theta |f(i\theta)|, \sup_\theta |f(s + i\theta)|\}$. We shall consider only the first case, $p = i\theta$, in detail because the second case, $p = s + i\theta$, is parallel. $|f(i\theta)| \leqslant \lambda |\, T^A(i\theta)| + (1 - \lambda)\, |\, T^B(i\theta)|$. Using the facts that for $A \in \mathscr{B}(H)$ and $B \in \mathscr{I}_2(H)$, AB and $BA \in \mathscr{I}_2(H)$, and for $B, C \in \mathscr{I}_2(H)$, $\mathrm{Tr}\, BC = \mathrm{Tr}\, CB$ and $|\, \mathrm{Tr}\, BC \,| \leqslant \frac{1}{2}\mathrm{Tr}\, B^\dagger B + \frac{1}{2}\mathrm{Tr}\, C^\dagger C$, we have that

$$\begin{aligned} 2\, |\, T^A(i\theta)| \leqslant\ & \mathrm{Tr}\, \alpha(s - i\theta)\, M^\dagger\alpha(-i\theta)^\dagger\, \alpha(-i\theta)\, M\alpha(s - i\theta)^\dagger \\ & + \mathrm{Tr}\, \alpha(s + i\theta)\, M^\dagger\alpha(i\theta)^\dagger\, \alpha(i\theta)\, M\alpha(s + i\theta)^\dagger. \end{aligned}$$

However, $\|\, \alpha(-i\theta)^\dagger\alpha(-i\theta)\| \leqslant 1$, so the first term is at most

$$\begin{aligned} \mathrm{Tr}\, \alpha(s - i\theta)\, M^\dagger M\alpha(s - i\theta)^\dagger &= \mathrm{Tr}\, M\alpha(s - i\theta)^\dagger\, \alpha(s - i\theta)\, M^\dagger \\ &= \mathrm{Tr}\, MC^{-i\theta/2}\, \alpha(s)^\dagger\, \alpha(s)\, C^{i\theta/2}M^\dagger \\ &= \mathrm{Tr}\, \alpha(s)^\dagger\, \alpha(s)\, C^{i\theta/2}M^\dagger MC^{-i\theta/2}. \end{aligned}$$

Likewise, the second term is at most $\mathrm{Tr}\,\alpha(s)^\dagger\alpha(s)\,C^{-i\theta/2}M^\dagger MC^{i\theta/2}$. If we add to these the corresponding two terms for $|T^B(i\theta)|$ we obtain

$$\lambda\,|T^A(i\theta)| + (1-\lambda)|T^B(i\theta)| \leqslant \tfrac{1}{2}\,\mathrm{Tr}[\lambda\alpha^\dagger(s)\,\alpha(s) + (1-\lambda)\,\beta^\dagger(s)\,\beta(s)]P, \tag{2.1}$$

where $P = C^{i\theta/2}M^\dagger MC^{-i\theta/2} + C^{-i\theta/2}M^\dagger MC^{i\theta/2} \in \mathscr{B}^+(H)$. As we remarked before, $\lambda A^s + (1-\lambda)\,B^s \leqslant C^s$, whence

$$\lambda\alpha(s)^\dagger\,\alpha(s) + (1-\lambda)\,\beta(s)^\dagger\,\beta(s) = C^{-s/2}[\lambda A^s + (1-\lambda)\,B^s]\,C^{-s/2} \leqslant \mathbb{1}.$$

Substituting this in (2.1) proves the theorem. Q.E.D.

Remark. If $C^{p/2}K$ has an extension to $\mathscr{I}_1(H)$ then so does $A^{p/2}K$ and $B^{p/2}K$ since $\|C^{p/2}K\| \geqslant \lambda^{1/2}\|A^{p/2}K\|$. In this case

$$\mathrm{Tr}\,C^{r/2}K^\dagger C^pKC^{r/2} = \mathrm{Tr}\,C^rK^\dagger C^pK = \mathrm{Tr}\,K^\dagger C^pKC^r$$

and similarly for A and B.

COROLLARY 1.1. *With p and r as in Theorem 1, the function from $\mathscr{B}^+(H)\times\mathscr{B}^+(H)\times\mathscr{B}(H)$ to the nonnegative reals defined by $(A, B, K)\mapsto F(A, B, K) = \mathrm{Tr}\,A^{r/2}K^\dagger B^pKA^{r/2}$*

(1) *is jointly concave in (A, B) and*

(2) *is convex in K.*

Proof. Consider the Hilbert space $H' = H \oplus H$ and define the following operators in $\mathscr{B}(H')$:

$$k\colon (x, y)\mapsto (0, Kx),$$

$$k^\dagger\colon (x, y)\mapsto (K^\dagger y, 0),$$

$$a\colon (x, y)\mapsto (Ax, By), \qquad a\in\mathscr{B}^+(H').$$

Applying Theorem 1 to $\mathrm{Tr}\,a^{r/2}k^\dagger a^pka^{r/2}$ proves the first part. The second part follows from a Schwartz inequality type of argument since $F(A, B, K)$ is nonnegative and quadratic in K. Q.E.D.

COROLLARY 1.2. *With p and r as in Theorem 1, $p + r = s \leqslant 1$, the functions from $\mathscr{B}^+(H)\times\mathscr{B}^+(H)\times\mathscr{B}(H)$ to the nonnegative reals defined by*

$$(A, B, K)\mapsto F_q(A, B, K) = \{\mathrm{Tr}\,A^{r/2}K^\dagger B^pKA^{r/2}\}^q$$

Adv. in Math. *11*, 267–288 (1973)

(1) *are jointly concave in* (A, B) *when* $0 < q \leqslant 1/s$,

(2) *are jointly convex in* (A, B) *when* $q < 0$, *and*

(3) *are convex in* K *when* $q \geqslant \frac{1}{2}$.

Proof. The proof is a standard one for homogeneous concave (or convex) functions [7]. Let $x = (x_1, x_2) \in \mathbb{R}_+{}^2 = \{(x_1, x_2) \mid x_1 \geqslant 0, x_2 \geqslant 0\}$ and define $f(x) = F_1(x_1 A + x_2 A', x_1 B + x_2 B', K)$ for an arbitrary but, henceforth, fixed choice of $A, A', B, B' \in \mathscr{B}^+(H)$. Parts (1) and (2) are equivalent to showing that for all such choices, $f(x)^q$ is concave (or convex). By Corollary 1.1 $f(x)$ is nonnegative, concave and homogeneous of order s, i.e., $f(\lambda x) = \lambda^s f(x)$ for $\lambda \geqslant 0$. For each $\alpha \geqslant 0$, define $G_\alpha = \{x \mid f(x) \geqslant \alpha, x \in \mathbb{R}_+{}^2\}$. It is easily seen from the properties of $f(x)$ that G_α is a convex subset of $\mathbb{R}_+{}^2$ and $G_\alpha = \alpha^{1/s} G_1$ for $\alpha > 0$. Define $k(x) = \sup\{\mu \geqslant 0 \mid x \in G_{\mu^s}\}$ for $x \in \mathbb{R}_+{}^2$. As $x \in G_{f(x)}$, $k(x)$ is verywhere defined. In fact, since $f(x) = \sup\{\alpha \geqslant 0 \mid x \in G_\alpha\}, f(x) = k(x)^s$. Obviously, $k(x)$ is nonnegative and homogeneous of order 1, and, since $k(x) = \sup\{\mu > 0 \mid x \in \mu G_1\}$ when $f(x) \neq 0$, it is easy to check that $k(x)$ is a concave function. For a nonnegative concave function, $k(x)$, $k(x)^p$ is concave when $0 < p \leqslant 1$ and $k(x)^p$ is convex when $p < 0$. This proves parts (1) and (2). For part (3) we define $f(x) = F_1(A, B, x_1 K + x_2 K')$, with $K, K' \in \mathscr{B}(H)$. $f(x)$ is nonnegative, convex, and homogeneous of order 2. We define: $G_\alpha = \{x \mid f(x) \leqslant \alpha, x \in \mathbb{R}_+{}^2\}$ which is convex; $k(x) = \inf\{\mu \geqslant 0 \mid x \in G_{\mu^2}\}$. Then $k(x)$ is nonnegative, convex, and homogeneous of order 1 and $f(x) = k(x)^2$. For any nonnegative convex function, $k(x)$, $k(x)^p$ is convex when $p \geqslant 1$. Q.E.D.

The setting for the next corollary is the following: Let H^1 and H^2 be two separable Hilbert spaces and $H^{12} \equiv H^1 \otimes H^2$ their tensor product. If $A_{12} \in \mathscr{B}^+(H^{12})$ and $A_{12} \in \mathscr{I}_1(H^{12})$ we can define $A_1 \in \mathscr{B}^+(H^1)$ by means of the partial trace, i.e.,

$$A_1 = \mathrm{Tr}^1 A_{12},$$

which means that for $x, y \in H^1$

$$(x, A_1 y) = \sum_i (x \otimes e_i, A_{12}[y \otimes e_i])$$

where $\{e_i\}$ is any orthonormal basis in H^2.

COROLLARY 1.3. *Let* H^2 *be of dimension* $d_2 < \infty$, *let* K *be a linear operator (not necessarily bounded) on* H^1, *and let* $A_{12} \in \mathscr{B}^+(H^{12})$, $A_{12} \in \mathscr{I}_1(H^{12})$ *with* $A_1 = \mathrm{Tr}^1 A_{12}$. *Let* $p > 0$, $r > 0$, $p + r = s \leqslant 1$. *Define* $L = K \otimes 1\!\!1_2$ *on* H^{12}. *If* $A_1^{p/2} K A_1^{r/2}$ *has an extension to* $\mathscr{I}_2(H^1)$ *then* $A_{12}^{p/2} L A_{12}^{r/2}$ *has an extension to* $\mathscr{I}_2(H^{12})$ *and*

$$d_2^{1-s}\, \mathrm{Tr}^1 A_1^{r/2} K^\dagger A_1{}^p K A_1^{r/2} \geqslant \mathrm{Tr}^{12} A_{12}^{r/2} L^\dagger A_{12}^p L A_{12}^{r/2}. \tag{2.2}$$

Proof. If G is the group of unitary transformations on H^2 and if dU is the normalized Haar measure on G then it is easy to see [8] that $B_{12} \equiv \int_G U^\dagger A_{12} U\, dU = d_2^{-1} A_1 \otimes 1\!\!1_2$. Let $F(A_{12})$ be the right side of (2.2). By Theorem 1, $A_{12} \mapsto F(A_{12})$ is concave so

$$F(B_{12}) \geqslant \int_G F(U^\dagger A_{12} U)\, dU.$$

But $F(U^\dagger A_{12} U)$ is independent of U since $(U^\dagger A_{12} U)^p = U^\dagger A_{12}^p U$, $U^\dagger L U = L$, and $\mathrm{Tr}^{12}\, U^\dagger X U = \mathrm{Tr}^{12}\, X$. However, $F(B_{12})$ is the left side of (2.2). Q.E.D.

Remarks. (i) Theorem 1 can be regarded as a special case of Corollary 1.3 as may be seen by taking $H^2 = \mathbb{C}^2$ and

$$A_{12} = \tfrac{1}{2} A \otimes P^a + \tfrac{1}{2} B \otimes P^b,$$

where P^a and P^b are two orthogonal projections on H^2 and $A, B \in \mathscr{B}^+(H^1)$. Then $A_1 = \tfrac{1}{2}A + \tfrac{1}{2}B = C$.

(ii) Similar to Corollary 1.1, we can extend Corollary 1.3 to the following: Let A_{12}, $B_{12} \in \mathscr{B}^+(H^{12})$, A_{12}, $B_{12} \in \mathscr{I}_1(H^{12})$, and $A_1 = \mathrm{Tr}^1 A_{12}$, $B_1 = \mathrm{Tr}^1 B_{12}$, then

$$d_2^{1-s}\, \mathrm{Tr}^1 A_1^{r/2} K^\dagger B_1{}^p K A_1^{r/2} \geqslant \mathrm{Tr}^{12} A_{12}^{r/2} K^\dagger B_{12}^p K A_{12}^{r/2}. \tag{2.3}$$

(iii) When $d_2 = \infty$, Corollary 1.3 makes no sense except when $s = 1$. In that special case, the corollary is true when $d_2 = \infty$. The proof, which we shall not give here, can be constructed in imitation of the proof of Theorem 1 itself. The principal idea is to define $\alpha(q) = A_{12}^{q/2}[A_1^{-q/2} \otimes 1\!\!1_2] \in \mathscr{B}(H^{12})$ and $M = A_1^{p/2} K A_1^{r/2} \in \mathscr{I}_2(H^1)$.

(iv) If we let $A_{12} = 1\!\!1_{12}$ in (2.3) and let K be a projection onto an arbitrary one-dimensional subspace of H^1, we obtain the *operator* inequality (since $A_1 = d_2 1\!\!1_1$)

$$d_2^{1-p} B_1{}^p \geqslant \mathrm{Tr}^2 B_{12}^p$$

for all $B_{12} \in \mathscr{B}^+(H^{12})$ and $0 < p \leqslant 1$.

Adv. in Math. *11*, 267–288 (1973)

The Wigner–Yanase–Dyson Conjecture

In quantum mechanics, a density matrix, ρ, on H satisfies $\rho \in \mathscr{B}^+(H)$, $\rho \in \mathscr{I}_1(H)$ and $\operatorname{Tr} \rho = 1$. The entropy of ρ, as usually defined, is

$$S(\rho) = -\operatorname{Tr} \rho \ln \rho.$$

Wigner and Yanase [1] extended this to the concept of the entropy of ρ relative to a self-adjoint "observable," K, and defined it as $S(\rho, K) \equiv \frac{1}{2}\operatorname{Tr}[\rho^{1/2}, K]^2$, where $[A, B] = AB - BA$. Dyson (cf. Ref. [1]) proposed a generalization of this to

$$S_p(\rho, K) \equiv \tfrac{1}{2} \operatorname{Tr}[\rho^p, K][\rho^{1-p}, K]$$

for $0 < p < 1$. [Actually, Wigner and Yanase defined $I_p(\rho, K) \equiv -S_p(\rho, K)$ which they termed skew information.]

It is well known and easy to prove that $S(\rho)$ is concave in ρ, and the WYD conjecture is that $S_p(\rho, K)$ is concave in ρ for each fixed K. They were able to prove this only when $p = \frac{1}{2}$. In physical applications K may be unbounded, but it is always correct to assume that $\rho^p K$ and $K\rho^p$ have unique extensions to $\mathscr{I}_1(H)$ for all $p > 0$. Thus, (cf. the remark after Theorem 1)

$$S_p(\rho, K) = -\operatorname{Tr} \rho K^2 + \operatorname{Tr} \rho^{1-p} K \rho^p K.$$

The first term is linear and, hence, concave, and the second term is concave by Theorem 1.

Remark. Theorem 1 is stronger than necessary because it allows K to be non-self-adjoint, i.e., $\operatorname{Tr}[\rho^p, K^\dagger][\rho^{1-p}, K]$ is concave in ρ when K is non-self-adjoint. This generalization can be derived from the self-adjoint case when $p = \frac{1}{2}$ by a simple polarization argument, but not when $p \neq \frac{1}{2}$. Baumann and Jost [9, 10] proved the concavity for general p, but for a special class of ρ and H.

Wigner and Yanase properly regarded the concavity of $\rho \mapsto S_p(\rho, K)$ as a necessary requirement in order that $I_p(\rho, K)$ be a sensible definition of information. Another absolute requirement is the subadditivity of $S_p(\rho, K)$. Subadditivity of the ordinary entropy, $S(\rho)$, means (in the terminology preceding Corollary 1.3 and with $\rho_2 \equiv \operatorname{Tr}^2 \rho_{12}$) that

$$S(\rho_{12}) \leqslant S(\rho_1) + S(\rho_2).$$

This inequality is well known.

For $S_p(\rho, K)$, Wigner and Yanase [1] take the following definition: Let K_1 (resp. K_2) be a self-adjoint operator on H^1 (resp. H^2) and define $L = K_1 \otimes \mathbb{1}_2 + \mathbb{1}_1 \otimes K_2$ on H^{12}. The subadditivity condition is that

$$S_p(\rho_{12}, L) \leqslant S_p(\rho_1, K_1) + S_p(\rho_2, K_2). \tag{2.4}$$

It is easy to see that (2.4) is true when $\rho_{12} = \rho_1 \otimes \rho_2$. Wigner and Yanase proved (2.4) when ρ_{12} is a projection onto a one-dimensional subspace of H^{12} and $p = \frac{1}{2}$. In the general case, (2.4) becomes (with $r = 1 - p$)

$$\begin{aligned} &\mathrm{Tr}^1 \rho_1{}^p K_1 \rho_1{}^r K_1 + \mathrm{Tr}^2 \rho_2{}^p K_2 \rho_2{}^r K_2 \\ &\quad \geqslant -2\,\mathrm{Tr}^{12} \rho_{12} [K_1 \otimes K_2] + \mathrm{Tr}^{12} \rho_{12}^p L \rho_{12}^r L. \end{aligned} \tag{2.5}$$

We do not have a proof of this, but when K_1 or K_2 is zero, (2.5) is simply Corollary 1.3. Because (2.5) is true in these three special cases, there is reason to believe it is true generally.

3. Some Finite Dimensional Theorems

In this section we confine our attention to finite dimensional Hilbert spaces over the complex numbers, i.e., $H = \mathbb{C}^n$. Some of the results of this section will be generalized to the infinite dimensional case in the next section by approximation arguments. If $C \in \mathscr{B}^{++}(H)$ then $C > \epsilon \mathbb{1}$ for some $\epsilon > 0$. We remark that for $A, B \in \mathscr{B}^{++}(H)$ and $K \in \mathscr{B}(H)$, $\mathrm{Tr}\, AK^\dagger BK = \mathrm{Tr}(B^{1/2} K A^{1/2})^\dagger (B^{1/2} K A^{1/2}) > 0$ for $K \neq 0$.

THEOREM 2. *The function from $\mathscr{B}^{++}(H) \times \mathscr{B}(H)$ to the nonnegative reals defined by*

$$(A, K) \mapsto F(A, K) = \mathrm{Tr}\, A^{-r} K^\dagger A^{-p} K \tag{3.1}$$

is jointly convex in (A, K) whenever $p \geqslant 0$, $r \geqslant 0$ and $p + r \leqslant 1$, i.e., for all λ, $0 < \lambda < 1$, $\lambda F(A, K) + (1 - \lambda) F(B, L) \geqslant F(C, M)$ when $A, B \in \mathscr{B}^+(H)$; $K, L \in \mathscr{B}(H)$; $C = \lambda A + (1 - \lambda) B$; $M = \lambda K + (1 - \lambda) L$.

Proof. We can think of $\mathscr{B}(H)$ as a complex Hilbert space, $V(H)$, of dimension n^2 with the inner product $K, K' \in \mathscr{B}(H) \mapsto \langle K, K' \rangle = \mathrm{Tr}\, K^\dagger K'$. The linear transformation of $V(H) \to V(H)$ defined by $K \mapsto \alpha K \beta$, with $\alpha, \beta \in \mathscr{B}^{++}(H)$ is Hermitian and positive definite because

Adv. in Math. *11*, 267–288 (1973)

$\langle K, \alpha K\beta\rangle > 0$ when $K \neq 0$. Thus, $K \mapsto F(A, K)$, $K \mapsto F(B, K)$, and $K \mapsto F(C, K)$ are positive definite quadratic forms on $V(H)$. Furthermore, we can define $V_2(H) = V(H) \oplus V(H)$ with inner product $\langle (K, L), (K', L')\rangle = \operatorname{Tr} K^\dagger K' + \operatorname{Tr} L^\dagger L'$. Clearly

$$D \equiv \lambda F(A, K) + (1-\lambda) F(A, L) \qquad \text{and} \qquad N \equiv F(C, \lambda K + (1-\lambda)L)$$

are both positive quadratic forms on $V_2(H)$ and D is definite. With A and B fixed, we form the variational quotient N/D and maximize it with respect to $(K, L) \in V_2(H)$. Using the fact that $\operatorname{Tr} AB = 0$ for all $A \in \mathscr{B}(H)$ implies $B = 0$, we find the eigenvalue equations

$$\begin{aligned} \gamma A^{-p} K A^{-r} &= C^{-p}[\lambda K + (1-\lambda)L]\, C^{-r} \equiv M \in \mathscr{B}(H), \\ \gamma B^{-p} L B^{-r} &= C^{-p}[\lambda K + (1-\lambda)L]\, C^{-r} \equiv M, \end{aligned} \tag{3.2}$$

and the problem is to show that the eigenvalue $\gamma \leqslant 1$. If $\gamma \neq 0$, the equation $\gamma A^{-p} K A^{-r} = M$ has the unique solution $K = \gamma^{-1} A^p M A^r$. Likewise, $L = \gamma^{-1} B^p M B^r$ and $\lambda K + (1-\lambda)L = C^p M C^r$. Thus, finding $\gamma \neq 0$ solutions to (3.2) is equivalent to finding an $M \in \mathscr{B}(H)$ such that $M \neq 0$ and $\lambda A^p M A^r + (1-\lambda)\, B^p M B^r = \gamma C^p M C^r$. However, if we multiply this equation on the left by $M^\dagger$ and take the trace, we see that $\gamma \leqslant 1$ by Theorem 1. Q.E.D.

Corollary 2.1. *Let $p \geqslant 0$, $r \geqslant 0$, $p + r \equiv s \leqslant 1$, and $q \in \mathbb{R}$, $q \neq 0$ be fixed. Then the functions from $\mathscr{B}^{++}(H) \times \mathscr{B}^{++}(H) \times \mathscr{B}(H)$ to the nonnegative reals defined by*

$$(A, B, K) \to F_q(A, B, K) = (\operatorname{Tr} A^{-p} K^\dagger B^{-r} K)^q$$

is jointly convex in (A, B, K) when $q \geqslant (2 - s)^{-1}$.

Proof. The same as for Corollaries 1.1 and 1.2, since $F_1(A, B, K)$ is homogeneous of order $2 - s > 0$. Q.E.D.

We next turn our attention to a function similar to (3.1) from $\mathscr{B}^{++}(H) \times \mathscr{B}(H)$ to the nonnegative reals defined by

$$(A, K) \mapsto Q(A, K) = \operatorname{Tr} \int_0^\infty dx (A + x\mathbb{1})^{-1} K^\dagger (A + x\mathbb{1})^{-1} K.$$

We note that the linear transformation T_A from $V(H)$ (defined in the proof of Theorem 2) to $V(H)$ given by

$$T_A\colon K \mapsto \int_0^\infty dx (A + x\mathbb{1})^{-1} K (A + x\mathbb{1})^{-1} \tag{3.3}$$

is Hermitian and positive definite. In a basis in which A is diagonal it is easy to compute T_A explicitly:

$$T_A\colon \{K_{ij}\} \mapsto \{K_{ij} f(A_i, A_j)\},$$

where the $\{A_i\}$ are the eigenvalues of A and

$$f(x, y) = (x - y)^{-1} \ln(x/y) \quad \text{if} \quad x \neq y, \quad f(x, x) = x^{-1}.$$

The inverse transformation is

$$T_A^{-1}\colon \{K_{ij}\} \mapsto \{K_{ij}/f(A_i, A_j)\},$$

but this is the same as

$$T_A^{-1}\colon K \mapsto \int_0^1 dx\, A^x K A^{1-x} \tag{3.4}$$

as we see by calculating the integral.

THEOREM 3. *The function $Q(A, K)$ from $\mathscr{B}^{++}(H) \times \mathscr{B}(H)$ to the nonnegative reals is jointly convex in (A, K).*

Proof. The proof is exactly the same as for Theorem 2 up to the eigenvalue equation (3.2) which now reads

$$\begin{aligned} \gamma T_A(K) &= T_C(\lambda K + (1 - \lambda) L) \equiv M \in \mathscr{B}(H), \\ \gamma T_B(K) &= T_C(\lambda K + (1 - \lambda) L) = M. \end{aligned} \tag{3.5}$$

If $\gamma \neq 0$ we apply the inversion formula (3.4), whence

$$\int_0^1 dx\, [\lambda A^x M A^{1-x} + (1 - \lambda) B^x M B^{1-x} - \gamma C^x M C^{1-x}] = 0.$$

By multiplying this on the left by $M^\dagger$ and using Theorem 1, we see that $\gamma \leqslant 1$. Q.E.D.

COROLLARY 3.1. *The functions*

$$(A, B, K) \mapsto Q_q(A, B, K) = \left[\mathrm{Tr} \int_0^\infty dx\, (A + x\mathbb{1})^{-1} K^\dagger (B + x\mathbb{1})^{-1} K\right]^q$$

from $\mathscr{B}^{++}(H) \times \mathscr{B}^{++}(H) \times \mathscr{B}(H)$ *to the nonnegative reals are*

(1) *jointly convex in* (A, B, K) *when* $q \geqslant 1$,
(2) *convex in* K *when* $q \geqslant \frac{1}{2}$,
(3) *jointly convex in* (A, B) *when* $q > 0$, *and*
(4) *jointly concave in* (A, B) *when* $-1 \leqslant q < 0$ *and* $K \neq 0$.

Proof. The construction given in Corollary 1.1 allows us to replace A by (A, B). We note that $Q_1(A, B, K)$ is homogeneous of order -1 in (A, B), as may be seen by changing the integration variable to λx, and, hence, $Q_1(A, B, K)$ is homogeneous of order 1 in (A, B, K). The proof of parts (1) and (2) is the same as that given in the second half of Corollary 1.2 for nonnegative, homogeneous, convex functions of positive order. To prove parts (3) and (4) we have to modify the proof given in Corollary 1.2. For $x \in \mathbb{R}_+^2 \backslash \{(0, 0)\}$, we construct $f(x) \equiv Q_1(x_1 A + x_2 A', x_1 B + x_2 B', K)$ with $A, A', B, B' \in \mathscr{B}^{++}(H)$. Assuming that $K \neq 0$ (otherwise there is nothing to prove) we note that $f(x)$ is strictly positive (as remarked before), convex and homogeneous of order $-t = -1$. We shall here give the construction for general $t > 0$ as we shall need it in Corollary 8.1 and Theorem 9. Define the convex sets $G_\alpha = \{x \mid f(x) \leqslant \alpha, x \in \mathbb{R}_+^2 \backslash \{(0, 0)\}\}$, whence $f(x) = \inf\{\alpha > 0 \mid x \in G_\alpha\}$. Define $k(x) \equiv \sup\{\mu > 0 \mid x \in G_{\mu^{-t}}\} = \sup\{\mu > 0 \mid x \in \mu G_1\} = f(x)^{-1/t}$. As before, $k(x)$ is positive, concave, and homogeneous of order 1. The rest follows from the remarks in Corollary 1.2. Q.E.D.

The mapping $T_A : \mathscr{B}(H) \to \mathscr{B}(H)$ defined in (3.3) has a special significance when restricted to the self-adjoint operators, $\mathscr{B}^s(H)$. Let $A \in \mathscr{B}^{++}(H)$ and $K \in \mathscr{B}^s(H)$. Then

$$(d/dx) \ln(A + xK)\,|_{x=0} = T_A(K). \tag{3.6}$$

[To derive this, use the representation

$$\ln A = \int_0^\infty dx\, (1 + x)^{-1} (A - \mathbb{1})(A + x\mathbb{1})^{-1} \tag{3.7}$$

which can be verified in a basis in which A is diagonal; then use

$$-\,(d/dx)(A + xK)^{-1}\,|_{x=0} = A^{-1} K A^{-1} \tag{3.8}$$

for any $K \in \mathscr{B}(H)$, $A \in \mathscr{B}^{++}(H)$.] Using (3.6)–(3.8) we have

$$\begin{aligned}\frac{d^2}{dx^2}\ln(A + xK)\,|_{x=0} &= -2\int_0^\infty dx\,(A + x\mathbb{1})^{-1}K(A + x\mathbb{1})^{-1}K(A + x\mathbb{1})^{-1} \\ &\equiv -R_A(K).\end{aligned} \tag{3.9}$$

We note that $R_A(K) \in \mathscr{B}^{++}(H)$ when $K \neq 0$.

PROPOSITION 4. *For any real number, γ, and $A \in \mathscr{B}^{++}(H)$, $K \in \mathscr{B}^s(H)$, $\gamma^2 R_A(K) + 2\gamma T_A(K) + \mathbb{1} \geqslant 0$.*

Proof.

$$\begin{aligned}0 \leqslant R_A(K + A) &= -(d^2/dx^2)\ln(A + xK + xA)\,|_{x=0} \\ &= -(d^2/dx^2)\{\ln(1 + x)\mathbb{1} + \ln(A + x(1 + x)^{-1}K)\}\,|_{x=0} \\ &= \mathbb{1} + 2T_A(K) + R_A(K).\end{aligned}$$

Now replace K by γK. Q.E.D.

Proposition 4 is not necessary for what follows, but we mention it because it almost, but not quite, implies the following proposition, P: $R_A(K) \geqslant T_A(K)^2$. It does imply that $(\Psi, [R_A(K) - T_A(K)^2]\Psi) \geqslant 0$ when Ψ is any eigenvector of $T_A(K)$. P is false, however, as may be seen by the two-dimensional example: $A = \left(\begin{smallmatrix} a & 0 \\ 0 & b\end{smallmatrix}\right)$ with $a > b > 0$ and $K = \left(\begin{smallmatrix} 0 & 1 \\ 1 & 0\end{smallmatrix}\right)$. Theorem 6 would be trivial to prove if P were true, and, in some sense, it may be taken as a substitute for P.

In order to prove Theorem 6, we need some preliminary remarks and Lemma 5. The remarks are that

$$Q(A, K) = \operatorname{Tr} K^\dagger T_A(K). \tag{3.10}$$

Also, $Q(A + xB, K)$ is differentiable near $x = 0$ for all $B \in \mathscr{B}^s(H)$ and

$$(d/dx)\,Q(A + xB, K)\,|_{x=0} = -\operatorname{Tr} BR_A(K) \tag{3.11}$$

for $K \in \mathscr{B}^s(H)$.

LEMMA 5. *Let $\mathscr{C}$ be a convex cone in a vector space and let $F\colon \mathscr{C} \to \mathbb{R}$ be a convex function which is also right differentiable in the sense that*

$$\lim_{x \downarrow 0} x^{-1}\{F(A + xB) - F(A)\} \equiv G(A, B)$$

Adv. in Math. *11*, 267–288 (1973)

exists for all $A, B \in \mathscr{C}$. *Assume, also, that* F *is homogeneous of order* 1, *i.e.,* $F(\lambda A) = \lambda F(A)$ *for* $\lambda > 0$. *Then* $G(A, B) \leqslant F(B)$. *Conversely, if* F *is two-sided differentiable in the foregoing sense (with equal left and right derivatives), if* $G(A + xB, B)$ *is measurable on* $\{x \mid x \geqslant 0\}$, *if* $G(A, B) \leqslant F(B)$ *and if* F *is homogeneous of order* 1, *then* F *is convex.*

Proof. For all $x > 0$,

$$\begin{aligned} F(A + xB) &= F((1 + x)[(1 + x)^{-1} A + x(1 + x)^{-1} B]) \\ &= (1 + x) F((1 + x)^{-1} A + x(1 + x)^{-1} B) \\ &\leqslant (1 + x)\{(1 + x)^{-1} F(A) + x(1 + x)^{-1} F(B)\} \\ &= F(A) + xF(B). \end{aligned}$$

Subtract $F(A)$ from both sides, divide by x, and take the limit $x \downarrow 0$. For the second part, let $C = \lambda A + (1 - \lambda)B$, and note that

$$\begin{aligned} F(C) - \lambda F(A) &= F(C) - F(\lambda A) \\ &= \int_0^{1-\lambda} dx\, dF(\lambda A + xB)/dx = \int_0^{1-\lambda} dx\, G(A + xB, B) \\ &\leqslant \int_0^{1-\lambda} dx\, F(B) = (1 - \lambda) F(B). \end{aligned}$$

Q.E.D.

We shall return again to this lemma, but for now we note that since $Q(A, K)$ is homogeneous of order 1 on the convex cone $\mathscr{B}^{++}(H) \times \mathscr{B}^{s}(H)$ then, using (3.10) and (3.11),

$$-\operatorname{Tr} BR_A(K) + 2 \operatorname{Tr} MT_A(K) \leqslant \operatorname{Tr} MT_B(M) \tag{3.12}$$

for $A, B \in \mathscr{B}^{++}(H)$ and $K, M \in \mathscr{B}^{s}(H)$.

THEOREM 6. *Let* $L \in \mathscr{B}^{s}(H)$ *be fixed. Then the function from* $\mathscr{B}^{++}(H)$ *to the nonnegative reals, defined by*

$$A \mapsto F_L(A) = \operatorname{Tr} \exp(L + \ln A),$$

is concave on $\mathscr{B}^{++}(H)$ *for all* L.

Proof. Choose $A \in \mathscr{B}^{++}(H)$ and $K \in \mathscr{B}^{s}(H)$ and consider the function $f(x) = \operatorname{Tr} \exp(L + \ln(A + xK))$ which is defined and differentiable for the real variable x in some neighborhood of $\{0\}$. The theorem is equivalent to the statement that $d^2f/dx^2 \leqslant 0$ when $x = 0$ for all choices of A, L, and K. Using the fact (which can be proved by a power series expansion, for example) that

$$\frac{d}{dx} e^{F+xG}|_{x=0} = \int_0^1 dy\, e^{yF} G e^{(1-y)F} = T^{-1}_{\exp(F)}(G)$$

$$\begin{aligned}\frac{d^2f}{dx^2}|_{x=0} &= -\mathrm{Tr}\, BR_A(K) + \mathrm{Tr} \int_0^1 dy\, T_A(K)\, B^y T_A(K)\, B^{(1-y)} \\ &= -\mathrm{Tr}\, BR_A(K) + \mathrm{Tr}\, T_A(K)\, T_B^{-1}[T_A(K)], \qquad (3.13)\end{aligned}$$

where $B = \exp(L + \ln A)$. Now use inequality (3.12) with $M = T_B^{-1}[T_A(K)]$. Q.E.D.

COROLLARY 6.1. *Let k be a positive integer and $p_1, ..., p_k$ positive real numbers with $p_1 + \cdots + p_k = s \leqslant 1$. Let $L \in \mathscr{B}^s(H)$ and $q \in \mathbb{R}$ be fixed. Then the functions from $\mathscr{B}^{++}(H)^k$ to the positive reals defined by*

$$F_q(A_1, ..., A_k) = \left\{\mathrm{Tr}\exp\left[L + \sum_{j=1}^{k} p_j \ln(A_j)\right]\right\}^q$$

(1) *are jointly concave in $(A_1, ..., A_k)$ when $0 < q \leqslant s^{-1}$ and*

(2) *are jointly convex in $(A_1, ..., A_k)$ when $q < 0$.*

Proof. F_1 is homogeneous of order s and we have already explained in Corollaries 1.2 and 2.1 how to treat the cases $q \neq 1$. That F_1 is jointly concave seems like a stronger result than Theorem 6, but, surprisingly, it is not. We have to show that for every choice of $A_1, ..., A_k \in \mathscr{B}^{++}(H)$ and $K_1, ..., K_k \in \mathscr{B}^s(H)$, the k-square second-derivative matrix of $f(x_1, ..., x_k) = F(A_1 + x_1K_1, ..., A_k + x_kK_k)$ is positive semidefinite when $x_1 = \cdots = x_k = 0$. If $\partial^2/(\partial x_i\, \partial x_j) f(x_1, ..., x_k)|_{x_1=\cdots=x_k=0} = m_{ij} = g_{ij} + \delta_{ij}h_i$, we compute

$$g_{ij} = p_i p_j\, \mathrm{Tr}\, T_B^{-1}[T_{A_i}(K_i)]\, T_{A_j}(K_j),$$

$$h_i = -p_i\, \mathrm{Tr}\, BR_{A_i}(K_i) \leqslant 0,$$

where $B = \exp[L + \sum_{j=1}^k p_j \ln(A_j)]$. Clearly, for any $F, G \in \mathscr{B}(H)$, $\mathrm{Tr}\, FT_B^{-1}(G) = \mathrm{Tr}\, GT_B^{-1}(F)$, and $\mathrm{Tr}\, F^\dagger T_B^{-1}(F) \geqslant 0$ so, by a Schwarz inequality argument, $|\mathrm{Tr}\, F^\dagger T_B^{-1}(G)|^2 \leqslant [\mathrm{Tr}\, F^\dagger T_B^{-1}(F)][\mathrm{Tr}\, G^\dagger T_B^{-1}(G)]$.

Adv. in Math. *11*, 267–288 (1973)

Hence, $|g_{ij}|^2 \leqslant g_{ii}g_{jj}$ and $g_{ii} \leqslant -p_i h_i$ by Theorem 6. Thus, for any $\Psi \in \mathbb{C}^k$,

$$\sum_{i=1}^{k}\sum_{j=1}^{k} \overline{\Psi}_i m_{ij} \Psi_j \leqslant \left\{\sum_{i=1}^{k} (-p_i h_i)^{1/2} |\Psi_i|\right\}^2 + \sum_{i=1}^{k} h_i |\Psi_i|^2$$

$$\leqslant (1-s) \sum_{i=1}^{k} h_i |\Psi_i|^2 \leqslant 0$$

since $h_i \leqslant 0$, all i. Q.E.D.

COROLLARY 6.2. *Let $p_1, \ldots, p_k$ be as in Corollary 6.1, and let $\Psi \in H$ with $\|\Psi\| = 1$. Then*

$$(A_1, \ldots, A_k) \mapsto \exp\left[\sum_{j=1}^{k} p_j(\Psi, \ln A_j \Psi)\right]$$

is a concave function on $\mathscr{B}^{++}(H)^k$.

Proof. Let P be the projection onto the one-dimensional subspace of H spanned by Ψ. Then take the limit $e^L \to P$ in Corollary 6.1. Q.E.D.

Remark. If we write $B = e^L$ in Theorem 6, then, by the Trotter product formula, $F_L(A) = \lim_{n\to\infty} F_L^{(n)}(A)$ with $F_L^{(n)}(A) = \mathrm{Tr}(B^{1/n}A^{1/n})^n$. Now, $F_L^{(1)}(A)$ is concave in A since it is linear, $F_L^{(2)}(A)$ is concave in A by Theorem 1 and $F_L(A)$ is concave in A by Theorem 6. Hence, we are led to the following conjecture.

Conjecture. For each fixed $B \in \mathscr{B}^+(H)$ and $n \in \mathbb{Z}^+$, the positive function $\mathrm{Tr}(B^{1/n}A^{1/n})^n$ is concave in $A \in \mathscr{B}^+(H)$.

We remind the reader of the Golden–Thompson inequality (GT) [11, 12] which is that

$$\mathrm{Tr}\, e^{A+B} \leqslant \mathrm{Tr}\, e^A e^B \tag{3.14}$$

for $A, B \in \mathscr{B}^s(H)$. This theorem can be extended to the infinite dimensional case [13]. The obvious generalization of GT to three operators in the form $\mathrm{Tr}\, e^{A+B+C} \leqslant \mathrm{Tr}\, e^A e^B e^C$ is false. We shall now show that Theorem 6 does provide a correct generalization of GT to three operators (and, hence, provides an alternative proof of GT itself).

THEOREM 7. *Let $A, B, C \in \mathscr{B}^s(H)$.*

$$\mathrm{Tr}\, e^C T_{\exp(-A)}(e^B) \geqslant \mathrm{Tr}\, e^{A+B+C}.$$

If A commutes with B then

$$\operatorname{Tr} e^C e^A e^B \geqslant \operatorname{Tr} e^{A+B+C}.$$

Proof. Define $\alpha, \beta \in \mathscr{B}^{++}(H)$ by $\alpha = e^{-A}$ and $\beta = e^B$ and define $L \in \mathscr{B}^s(H)$ by $L = A + C$. Since $\alpha \mapsto -\operatorname{Tr} e^{L+\ln\alpha}$ is homogeneous of order 1 and convex on the cone $\mathscr{B}^{++}(H)$, we can use Lemma 5 to deduce that for $\beta \in \mathscr{B}^{++}(H)$

$$\operatorname{Tr} e^{A+B+C} = \operatorname{Tr}^{\,L+\ln\beta} \leqslant (d/dx) \operatorname{Tr} e^{L+\ln(\alpha+x\beta)} \,|_{x=0} = \operatorname{Tr} e^C T_\alpha(\beta).$$

For the last part, we note that

$$T_\alpha(\beta) = \int_0^\infty dx\, (e^{-A} + x\mathbb{1})\, e^B (e^{-A} + x\mathbb{1})^{-1} = e^A e^B \qquad \text{if} \quad [A, B] = 0.$$

Q.E.D.

Remark. An alternative formulation of Theorem 7 is this: Let $A, C, D \in \mathscr{B}^s(H)$. Then

$$\operatorname{Tr} e^C e^D \geqslant \operatorname{Tr} \exp\left[C + A + \ln \int_0^1 ds\, e^{-As} e^D e^{-A(1-s)}\right]$$

which follows from the definition $T_{\exp(-A)}(e^B) \equiv e^D$ and the inversion formula (3.4).

Additional Theorems

The Theorems proved thus far all rely on Theorem 1 in an essential way. We shall now prove some theorems which appear to be similar to Theorems 2, 3, 6, and 7, but which in reality are less complicated because they can be proved directly by elementary methods.

THEOREM 8. *Let $K \in \mathscr{B}(H)$ and $1 \geqslant p > 0$, $1 \geqslant r > 0$ be fixed. Then the function from $\mathscr{B}^{++}(H)$ to the nonnegative reals defined by*

$$A \mapsto F(A) = \operatorname{Tr} A^{-p} K^\dagger A^{-r} K$$

is convex in A.

Proof. It is sufficient to show that for all $L \in \mathscr{B}^s(H)$, $g(x) \equiv F(A + xL)$ has a positive second derivative at $x = 0$. Let

$$A(x) = A + xL, \qquad B_p = (d/dx)\, A(x)^{-p} \,|_{x=0}, \qquad C_p = (d/dx)\, A(x)^p \,|_{x=0},$$

$$D_p = (d^2/dx^2)\, A(x)^{-p} \,|_{x=0}$$

Adv. in Math. *11*, 267–288 (1973)

and $E_p = (d^2/dx^2)\, A(x)^p \,|_{x=0}$. Using the fact that $A(x)^p A(x)^{-p} = \mathbb{1}$, we compute that $B_p = -A^{-p}C_pA^{-p}$ and $D_p = 2B_pA^pB_p - A^{-p}E_pA^{-p}$. Hence,

$$\begin{aligned}(d^2/dx^2)\, g(x)\,|_{x=0} &= -\mathrm{Tr}\, E_pA^{-p}K^\dagger A^{-r}KA^{-p} - \mathrm{Tr}\, E_rA^{-r}KA^{-p}K^\dagger A^{-r}\\ &\quad + \mathrm{Tr}(\gamma\delta_p + \delta_r\gamma)(\gamma\delta_p + \delta_r\gamma)^\dagger + \mathrm{Tr}(\gamma\delta_p)(\gamma\delta_p)^\dagger\\ &\quad + \mathrm{Tr}(\delta_r\gamma)(\delta_r\gamma)^\dagger,\end{aligned}$$

where $\gamma = A^{-r/2}KA^{-p/2}$ and $\delta_p = A^{-p/2}C_pA^{-p/2}$. The theorem will be proved if we show that $E_p \leqslant 0$, but this fact follows from the integral representation (valid for $1 > p > 0$)

$$A^p = \pi^{-1} \sin \pi p \int_0^\infty x^{p-1}\, dx\, A(A + x\mathbb{1})^{-1}$$

and (3.8). Q.E.D.

COROLLARY 8.1. *Let $1 \geqslant p > 0$ and $1 \geqslant r > 0$, $p + r \equiv s$ and $q \in \mathbb{R}$, $q \neq 0$ be fixed. Then the functions from $\mathscr{B}^{++}(H) \times \mathscr{B}^{++}(H) \times \mathscr{B}(H)$ to the nonnegative reals, defined by*

$$(A, B, K) \mapsto F_q(A, B, K) = (\mathrm{Tr}\, A^{-p}K^\dagger B^{-r}K)^q$$

(1) *are convex in K when $q \geqslant \frac{1}{2}$,*

(2) *are jointly convex in (A, B) when $q > 0$, and*

(3) *are jointly concave in (A, B) when $-1/s \leqslant q < 0$.*

Proof. The same as for Corollary 3.1. We note that the degree of homogeneity in K is 2 while in (A, B) it is $-s$. Q.E.D.

Remark. The map $A \mapsto A^{-p}$ is not convex for $p > 1$, but $A \mapsto \mathrm{Tr}\, A^{-p}$ is convex for $p > 0$. See Theorem 9.

THEOREM 9. *Let k be a positive integer and $p_1, \ldots, p_k$ positive real numbers with $p_1 + \cdots + p_k \equiv s$. Let $L \in \mathscr{B}^s(H)$ and $q \in \mathbb{R}$ be fixed. Then the functions from $\mathscr{B}^{++}(H)^k$ to the positive reals defined by $F_q(A_1, \ldots, A_k) = \{\mathrm{Tr}\exp[L - \sum_{j=1}^k p_j \ln(A_j)]\}^q$*

(1) *are jointly convex in $(A_1, \ldots, A_k)$ when $q \geqslant 0$ and*

(2) *are jointly concave in $(A_1, \ldots, A_k)$ when $-1/s \leqslant q \leqslant 0$.*

Proof. We need only consider the case $q = 1$. The extension to the general case is the same as in Corollary 3.1. When $k = 1$ we define $g(x) = F(A + xK)$, $K \in \mathscr{B}^s(H)$, and compute its second derivative at $x = 0$ to be $p \operatorname{Tr} BR_A(K) + p^2 \operatorname{Tr} T_A(K)\, T_B^{-1}[T_A(K)]$ where $B = \exp(L - p \ln A) > 0$. $R_A(K) \geqslant 0$ and, as remarked in the proof of Corollary 6.1, $\operatorname{Tr} F^+ T_B^{-1}(F) \geqslant 0$ for all $F \in \mathscr{B}(H)$. This last fact is the essence of the proof when $k > 1$. The second derivative matrix is positive definite. Q.E.D.

4. Extension of Theorem 6 to Infinite Dimensions

We fix L, which is assumed to be self-adjoint, and $e^L \in \mathscr{I}_1(H)$, which implies that L has purely discrete spectrum. For $A, B \in \mathscr{B}^+(H)$ and $C \equiv \lambda A + (1 - \lambda)B$, $0 < \lambda < 1$, we want to show that

$$\lambda \operatorname{Tr} e^{L+\ln A} + (1-\lambda) \operatorname{Tr} e^{L+\ln B} \leqslant \operatorname{Tr} e^{L+\ln C}, \tag{4.1}$$

which requires, among other things, giving meaning to these quantities.

Case 1. We first assume that there exist positive numbers ϵ and ω such that $\epsilon\mathbb{1} < A < \omega\mathbb{1}$, $\epsilon\mathbb{1} < B < \omega\mathbb{1}$ so that $\ln A$, etc. can be defined as bounded, self-adjoint operators by means of the spectral representation of A, etc. We define $\alpha = L + \ln A$. Since $\ln A$ is bounded, α is a self-adjoint operator on the domain of L. If we label the eigenvalues of L by $\mu_1(L) \geqslant \mu_2(L) \geqslant \cdots$ we have, by the mini–max principle, that $\mu_k(\alpha) \leqslant \mu_k(L) + \ln \| A \|$ since $\ln A \leqslant \ln \| A \| \, \mathbb{1}$. The convergence of $\sum_1^\infty \exp(\mu_k(L))$ implies that $\mu_k(L) \to -\infty$, which implies that $\mu_k(\alpha) \to -\infty$, which implies that $\exp(\alpha)$ is compact, and since $t(A) \equiv \operatorname{Tr} e^\alpha \leqslant \| A \| \operatorname{Tr} e^L < \infty$, the trace is finite. Now let P_n be the projection onto R_n, the subspace spanned by the first n eigenvectors of L, and define

$$A_n \equiv P_n A P_n + \epsilon(\mathbb{1} - P_n),$$

$$\alpha_n \equiv L + \ln A_n,$$

and likewise for B_n and C_n. Clearly,

$$t(A_n) = \operatorname{Tr} \exp(\alpha_n) = \operatorname{Tr} P_n \exp(L + \ln(A \restriction R_n)) + r_n,$$

where $r_n = (\ln \epsilon) \operatorname{Tr}(\mathbb{1} - P_n)\, e^L$. From this we see that not only is $t(A_n)$ finite, but that the terms involving r_n cancel from both sides of (4.1) leaving an inequality about traces on the finite dimensional space R_n. Thus,

Adv. in Math. *11*, 267–288 (1973)

by Theorem 6, $\lambda t(A_n) + (1-\lambda)\, t(B_n) \leqslant t(C_n)$. All we have to do is prove that $t(A_n) \to t(A)$, etc. Now $A_n < \omega \mathbb{1}$ and $\alpha_n < L + \mathbb{1} \ln \omega$. Since $\mu_k(\alpha_n) \leqslant \mu_k(L) + \ln \omega$, $t(A_n) \to t(A)$ by the dominated convergence theorem if we can show that $\mu_k(\alpha_n) \to \mu_k(\alpha)$ for each k. As L is bounded above, we can find a constant, d, such that $\alpha_n < (d-1)\mathbb{1}$ and $\alpha < (d-1)\mathbb{1}$. Define

$$G_n = (\alpha_n - d\mathbb{1})^{-1} - (\alpha - d\mathbb{1})^{-1} = (\alpha_n - d\mathbb{1})^{-1}(\alpha - \alpha_n)(\alpha - d\mathbb{1})^{-1}$$

and note that $(\alpha_n - d\mathbb{1})^{-1}$ is uniformly bounded, $\alpha_n - \alpha \to 0$ strongly, and $(\alpha - d\mathbb{1})^{-1}$ is compact since e^L is compact. Hence, $\| G_n \| \to 0$. In general, $| \mu_k(A) - \mu_k(B)| \leqslant \| A - B \|$ by the mini–max principle. Thus, $\mu_k(\alpha_n) \to \mu_k(\alpha)$ since $\mu_k((\alpha - d\mathbb{1})^{-1}) = [\mu_k(\alpha) - d]^{-1}$, and the theorem is proved for Case 1.

Case 2. $0 < A < \omega\mathbb{1}$, $0 < B < \omega\mathbb{1}$. If we replace A by $A_\epsilon = A + \epsilon\mathbb{1}$ and B by $B_\epsilon = B + \epsilon\mathbb{1}$, then C is replaced by

$$\lambda(A + \epsilon\mathbb{1}) + (1-\lambda)(B + \epsilon\mathbb{1}) = C + \epsilon\mathbb{1}.$$

Also, A_ϵ, etc. are decreasing in ϵ. Thus, we can *define*

$$\operatorname{Tr} \exp(L + \ln A) \equiv \lim_{\epsilon \downarrow 0} \operatorname{Tr} \exp(L + \ln A_\epsilon) \tag{4.2}$$

because the right trace decreases as ϵ decreases. Theorem 6 is true with this definition because it is true for every $\epsilon > 0$. The usefulness of this definition stems from the following fact (B. Simon, private communication): $\ln A$ is defined as an unbounded self-adjoint operator in a natural way by its spectral decomposition. If $D(\ln A) \cap D(L)$ is dense in H (where $D(\cdot)$ means domain), and if we define $-\alpha$ to be the Friedrichs extension of $-L - \ln A$, then $\operatorname{Tr} e^\alpha = \lim_{\epsilon \downarrow 0} \exp(L + \ln A_\epsilon)$.

Finally, another case to which Theorem 6 can be extended (B. Simon, private communication) is where A and B are positive, not necessarily bounded, self-adjoint operators and $\ln A$ and $\ln B$ are form bounded perturbations of $-L$.

5. On the "Equivalence" of Theorems 1, 2, 3, 6, and 7

Although Theorem 1 is our main theorem and Theorems 2, 3, 6, and 7 appear to be corollaries of it, the fact is that if we could find an independent proof of any of Theorems 2, 3, 6, or 7 the others could be derived from it in a simple way when the Hilbert space is finite dimensional. Thus, in some sense, all five theorems have equal content. We shall indicate very briefly some of the links among these theorems and the reader can easily supply the missing details as well as establish additional connections.

We have previously established the implications $1 \Rightarrow 2$ and $1 \Rightarrow 3 \Rightarrow 6 \Rightarrow 7$.

(a) $2 \Rightarrow 1$. This is easily seen simply by reading the proof of Theorem 2 backwards, i.e., one maximizes the variational quotient N/D with respect to K, where $N = \lambda \operatorname{Tr} A^r K^\dagger A^p K + (1 - \lambda) \operatorname{Tr} B^r K^\dagger B^p K$ and $D = \operatorname{Tr} C^r K^\dagger C^p K$.

(b) $7 \Rightarrow 6$. Theorem 7 was derived from Theorem 6 by using the derivative inequality, Lemma 5. However, we can retrace our steps by using the second part of Lemma 5.

(c) $6 \Rightarrow (3.12)$. If, in (3.12), we fix A, B, K and minimize $\operatorname{Tr} M[T_B(M) - 2T_A(K)]$ with respect to M, we find $M = T_B^{-1}(T_A(K))$. Hence, (3.12) is equivalent to the fact that the right side of (3.13) is negative and this, in term, is equivalent to Theorem 6.

(d) $(3.12) \Rightarrow 3$. The same remark as in (b) suffices.

(e) $3 \Rightarrow 2$. Take 3 in the form of Corollary 3.1 with $q = 1$ and assume that $p + r = s < 1$. For A (resp. B) substitute $\lambda A + \mathbb{1}$ (resp. $\gamma A + \mathbb{1}$), where $\lambda, \gamma \geqslant 0$. Then integrate

$$\int_0^\infty d\lambda\, \lambda^{p-1} \int_0^\infty d\gamma\, \gamma^{r-1} Q_1(\lambda A + \mathbb{1}, \gamma A + \mathbb{1}, K)$$

to obtain a positive constant times $F(A, K)$ of (3.1). When $p + r = 1$ we can appeal to continuity.

Acknowledgments

The author is most grateful to B. Simon for explaining to him how to extend Theorems 1 and 6 to infinite dimensions. He is also indebted to D. Ruelle for stimulating his interest in this problem and to R. Jost, O. Lanford, III, and D. W. Robinson for useful conversations. The hospitality of the Institut des Hautes Études Scientifiques and the Chemistry Laboratory III, University of Copenhagen, where parts of this work were carried out, is grateful acknowledged.

Note added in proof. After this paper was submitted for publication, H. Epstein [14] found alternative proofs of several of these theorems and a proof of the conjecture given after Corollary 6.2.

Adv. in Math. *11*, 267–288 (1973)

References

1. E. P. Wigner and M. M. Yanase, Information content of distributions, *Proc. Nat. Acad. Sci. U.S.A.* **49** (1963), 910–918. See also E. P. Wigner and M. M. Yanase, On the positive semidefinite nature of certain matrix expressions, *Canad. J. Math.* **16** (1964), 397–406.
2. D. W. Robinson and D. Ruelle, Mean entropy of states in classical statistical mechanics, *Commun. Math. Phys.* **5** (1967), 288–300.
3. O. E. Lanford, III, and D. W. Robinson, Mean entropy of states in quantum statistical mechanics. *J. Math. Phys.* **9** (1968), 1120–1125.
4. H. Araki and E. H. Lieb, Entropy inequalities, *Commun. Math. Phys.* **18** (1970), 160–170.
5. E. H. Lieb and M. B. Ruskai, Proof of the strong subadditivity of quantum mechanical entropy, Dec. 1973. *J. Math. Phys.* See also A fundamental property of quantum mechanical entropy, *Phys. Rev. Lett.* **30** (1973), 434–436.
6. J. Bendat and S. Sherman, Monotone and convex operator functions, *Trans. Amer. Math. Soc.* **79** (1955), 58–71. F. Krauss, *Math. Z.* **41** (1936), 18.
7. R. T. Rockafellar, "Convex Analysis," Princeton University, Princeton, NJ, 1970, cf. Theorem 15.3.
8. The idea of representing A_1 in this manner is due to A. Uhlmann, Endlich Dimensionale Dichtermatrizen II, *Wiss. Z. Karl-Marx-Univ. Leipzig Math.-Natur. Reihe* **22** Jg. (1973) H. 2, 139–177.
9. F. Baumann and R. Jost, Remarks on a conjecture of Robinson and Ruelle concerning the quantum mechanical entropy, *in* "Problems of Theoretical Physics; Essays Devoted to N. N. Bogoliubov," pp. 285–293, Nauka, Moscow, 1969.
10. F. Baumann, Bemerkungen Ueber Quantenmechanische Entropie Ungleichungen, *Helv. Phys. Acta* **44** (1971), 95–100.
11. S. Golden, Lower bounds for the Helmhotz function, *Phys. Rev.* **137** (1965), B 1127–1128.
12. C. J. Thompson, Inequality with applications in statistical mechanics, *J. Math. Phys.* **6** (1965), 1812–1813.
13. M. B. Ruskai, Inequalities for traces on Von Neumann algebras, *Commun. Math. Phys.* **26** (1972), 280–289; M. Breitenbecker and H. R. Gruemm, Note on trace inequalities, *Commun. Math. Phys.* **26** (1972), 276–279.
14. H. Epstein, Remarks on two theorems of E. Lieb, *Commun. Math. Phys.*, **31**, (1973) 317–325.

Reprinted from ADVANCES IN MATHEMATICS
Vol. 12, No. 2, February 1974

Printed in Belgium

Some Operator Inequalities of the Schwarz Type

ELLIOTT H. LIEB*

Department of Mathematics, Massachusetts Institute of Technology, Cambridge, Massachusetts 02139

AND

MARY BETH RUSKAI†

Department of Physics, University of Alberta, Edmonton, Alberta, Canada

We prove several operator inequalities which are analogous to the Schwarz inequality.

Our main result is Theorem 2 which gives an inequality for a large class of positive operator maps. Theorem 1, which we present first, is really a special case of Theorem 2. Theorem 1 is, however, Theorem 2 in its simplest and most accessible form, and, for that reason, we prefer to begin with a direct proof of it. After proving Theorem 2, we show how it applies to several well-known maps, e.g., partial traces and projection maps.

Let $\mathscr{B}(H)$ be the set of bounded, linear operators on a Hilbert space, H. The following fact will be useful. If $F, C \in \mathscr{B}(H)$ and $C > 0$ (i.e., $(x, Cx) \geqslant 0, \forall x \in H$), then, for $\epsilon > 0$, $F^*(C + \epsilon I)^{-1}F$ is an increasing net of positive operators in $\mathscr{B}(H)$ as $\epsilon \downarrow 0$. Such a net has a limit in the strong operator topology if and only if it is bounded above [1]. In that case, we define

$$F^*C^{-1}F \equiv s - \lim_{\epsilon \downarrow 0} F^*(C + \epsilon I)^{-1}F \tag{1}$$

and write $F^*C^{-1}F \in \mathscr{B}(H)$.

* Work partially supported by U.S. National Science Foundation Grant GP-31674X and by a Guggenheim Memorial Foundation fellowship.

† Present address: Department of Mathematics, University of Oregon, Eugene, Oregon 97403. Work partially supported by National Research Council of Canada Grant NRC-A6595 and by U.S. National Science Foundation Grant GP-31674X.

THEOREM 1. *If* C, D, F, $G \in \mathscr{B}(H)$, C, $D > 0$ *and* $F^*C^{-1}F$, $G^*D^{-1}G \in \mathscr{B}(H)$, *then* $(F+G)^*(C+D)^{-1}(F+G) \in \mathscr{B}(H)$ *and satisfies*

$$(F+G)^*(C+D)^{-1}(F+G) < F^*C^{-1}F + G^*D^{-1}G. \tag{2}$$

Proof. We first assume that C and D are nonsingular and let

$$X = C^{-1/2}F - C^{1/2}\Lambda,$$
$$Y = D^{-1/2}G - D^{1/2}\Lambda.$$

Then, for all $\Lambda \in \mathscr{B}(H)$,

$$\begin{aligned} 0 < X^*X + Y^*Y = F^*C^{-1}F + G^*D^{-1}G - (F+G)^*(C+D)^{-1}(F+G) \\ + [(C+D)^{1/2}\Lambda - (C+D)^{-1/2}(F+G)]^* \\ \times [(C+D)^{1/2}\Lambda - (C+D)^{-1/2}(F+G)]. \end{aligned}$$

Letting $\Lambda = (C+D)^{-1}(F+G)$, one obtains (2) when C and D are nonsingular. If C and D are arbitrary positive elements of $\mathscr{B}(H)$, note that $C + \epsilon I$ are nonsingular if $\epsilon > 0$. Then

$$\begin{aligned} (F+G)^*(C+D+2\epsilon I)^{-1}(F+G) < F^*(C+\epsilon I)^{-1}F + G^*(D+\epsilon I)^{-1}G \\ < F^*C^{-1}F + G^*D^{-1}G \end{aligned}$$

and the right side is in $\mathscr{B}(H)$ by assumption. Thus $(F+G)^*(C+D+\epsilon I)^{-1}$ $(F+G)$ is bounded above and the strong limit exists and satisfies inequality (2) as desired. Q.E.D.

If, for $0 < \alpha < 1$, we replace F by αF, C by αC, G by $(1-\alpha)G$, and D by $(1-\alpha)D$ in (2), it immediately follows that $\Omega = \{(F, C): C > 0$ and $F, C, F^*C^{-1}F \in \mathscr{B}(H)\}$ is convex. Moreover, the map from Ω into $\mathscr{B}(H)$ given by $(F, C) \to F^*C^{-1}F$ is jointly convex in F and C. When $F = F^*$, $C = [1 - \omega F]$, and $\omega < \| F \|^{-1}$, one obtains the well-known fact [2] that $C \to C^2[1 - \omega C]^{-1}$ is operator convex. Then, when $d\mu(\omega)$ is a positive measure and $a > 0$,

$$f(X) = \int_{-a}^{a} X^2(I - \omega X)^{-1}\, d\mu(\omega)$$

defines a convex operator function on $\{X: \| X \| < 1/a\}$.

THEOREM 2. *Let S be a bounded linear map from a Hilbert space H into a Hilbert space K. Define the positive map $\varphi: \mathscr{B}(K) \to \mathscr{B}(H)$ by*

$$\varphi(X) = S^*XS. \tag{3}$$

Then, for all $A, B \in \mathscr{B}(K)$,

(i) $$\varphi(A^*B)[\varphi(B^*B)]^{-1}\,\varphi(B^*A) \in \mathscr{B}(H)$$

and

(ii) $$\varphi(A^*A) > \varphi(A^*B)[\varphi(B^*B)]^{-1}\,\varphi(B^*A). \tag{4}$$

Proof. Let $\epsilon > 0$, and let $X = A - B\Lambda$, where

$$\Lambda = S[S^*B^*BS + \epsilon I_H]^{-1}\,S^*B^*A \in \mathscr{B}(K).$$

Using $\varphi(X^*X) = S^*X^*XS > 0$ and, e.g., $\varphi(C\Lambda) = \varphi(C)[\varphi(B^*B) + \epsilon I_H]^{-1}\,\varphi(B^*A)$ for all $C \in \mathscr{B}(K)$, one easily finds that for all $\epsilon > 0$,

$$\begin{aligned}\varphi(A^*A) &> \varphi(A^*B)[\varphi(B^*B) + \epsilon I_H]^{-1}\varphi(B^*A)\\ &\quad + \epsilon\varphi(A^*B)[\varphi(B^*B) + \epsilon I_H]^{-2}\,\varphi(B^*A)\\ &> \varphi(A^*B)[\varphi(B^*B) + \epsilon I_H]^{-1}\,\varphi(B^*A).\end{aligned}$$

Therefore, one can take the strong limit as $\epsilon \downarrow 0$, and the theorem is proved. Q.E.D.

We now give some applications of Theorem 2.

Example 1 (*Completely Positive Maps*). Let φ be a normalized, completely positive map (see, e.g., [3, 4]) from one C^* algebra $\mathfrak{A}_1$ into another $\mathfrak{A}_2$. Then [3] there exist $*$-representations π_1 and π_2 of $\mathfrak{A}_1$ and $\mathfrak{A}_2$ as bounded operators on Hilbert spaces K and H, respectively, such that

$$\pi_2(\varphi(X)) = V^*\pi_1(X)V,$$

where V is an isometry from H into K. Therefore, φ satisfies inequality (4). Although $\pi_2[\varphi(A^*B)]\;\pi_2[\varphi(B^*B)]^{-1}\;\pi_2[\varphi(B^*A)] \in \mathscr{B}(H)$, it is not necessarily in $\pi_2(\mathfrak{A}_2)$ unless $\mathfrak{A}_2$ is a von Neumann algebra or $\varphi(B^*B)$ is nonsingular. Letting B be the identity in inequality (4), one obtains Kadison's result [4], $\varphi(A^*A) \geqslant \varphi(A^*)\,\varphi(A)$, for completely positive maps.

Example 2 (*Projection Maps*). Let $\mathcal{O}\!\!\mathcal{l}_1$, $\mathcal{O}\!\!\mathcal{l}_2$ be von Neumann algebras with a unit I such that $I \in \mathcal{O}\!\!\mathcal{l}_2 \subset \mathcal{O}\!\!\mathcal{l}_1$. A projection map (sometimes called conditional expectation) [5–7] τ is a map from $\mathcal{O}\!\!\mathcal{l}_1 \to \mathcal{O}\!\!\mathcal{l}_2$ such that the norm of τ is 1 and $\tau(A_2) = A_2$, $\forall A_2 \in \mathcal{O}\!\!\mathcal{l}_2$. Then τ is completely positive [6], so that, as in Example 1,

$$\tau(A^*A) \geqslant \tau(A^*B)[\tau(B^*B)]^{-1}\,\tau(B^*A). \tag{5}$$

Further, since $\mathcal{O}\!\!\mathcal{l}_2$ was assumed to be a von Neumann algebra, the right side of (5) is in $\mathcal{O}\!\!\mathcal{l}_2$.

Alternatively, by using $\tau(A_1A_2) = \tau(A_1)\,A_2$ $\forall A_1 \in \mathcal{O}\!\!\mathcal{l}_1$, $A_2 \in \mathcal{O}\!\!\mathcal{l}_2$ [7], one can prove (5) directly as in Theorem 3.

Example 3 (*Partial Trace*). Let H_1 and H_2 be Hilbert spaces with H_2 finite-dimensional and consider the map $T_2\colon \mathscr{B}(H_1 \otimes H_2) \to \mathscr{B}(H_1)$ given by

$$(\varphi, T_2(X)\psi) = \sum_{K=1}^{N} (\varphi \otimes e_K, X\psi \otimes e_K), \tag{6}$$

where $\varphi, \psi \in H_1$, $\{e_K\}$ is an arbitrary, orthonormal basis in H_2, and $X \in \mathscr{B}(H_1 \otimes H_2)$. It is easy to prove that the above definition is independent of the basis. If X is of the form $X_1 \otimes X_2$ with $X_1 \in \mathscr{B}(H_1)$ and $X_2 \in \mathscr{B}(H_2)$, then $T_2(X) = \operatorname{tr}(X_2)\,X_1$, where tr is the usual trace on H_2. We call T_2 the partial trace from $\mathscr{B}(H_1 \otimes H_2)$ into $\mathscr{B}(H_1)$.

THEOREM 3. *Let H_1, H_2, and T_2 be as defined above. Then $T_2(A^*B)[T_2(B^*B)]^{-1}T_2(B^*A) \in \mathscr{B}(H_1)$ and*

$$T_2(A^*A) \geqslant T_2(A^*B)[T_2(B^*B)]^{-1}T_2(B^*A) \tag{7}$$

for all $A, B \in \mathscr{B}(H_1 \otimes H_2)$.

Proof. It is easy to see that T_2 is completely positive so that the theorem follows from Theorem 2; it may be instructive, however, to give the following direct proof. Let

$$X = A - B\Lambda,$$

$$\Lambda = [T_2(B^*B) + \epsilon I_1]^{-1}T_2(B^*A) \otimes I_2 .$$

Then the result follows from $T_2(X^*X) > 0$ and $T_2(C[\Lambda_1 \otimes I_2]) = [T_2(C)]\,\Lambda_1$ for all $C \in \mathscr{B}(H_1 \otimes H_2)$ and $\Lambda_1 \in \mathscr{B}(H_1)$, as in Theorem 2.

If H_2 is ∞-dimensional, one can similarly define the partial trace,

T_2, as an unbounded map on an appropriate subset of $\mathscr{B}(H_1 \otimes H_2)$. We shall not, however, do so here. Then T_2 satisfies inequality (7) whenever A^*A and B^*B are in the domain of T_2

Example 4 (*Theorem* 1). Let H_2 be 2-dimensional. Then one can recover Theorem 1 as a special case of Theorem 3 by considering

$$A = [(C + \epsilon I_1) \otimes P + (D + \epsilon I_1) \otimes Q]^{-1/2}[F \otimes P + G \otimes Q],$$
$$B = [(C + \epsilon I_1) \otimes P + (D + \epsilon I_1) \otimes Q]^{1/2},$$

where P and Q are 1-dimensional, orthogonal projections in $\mathscr{B}(H_2)$ and $C, D, F, G \in \mathscr{B}(H_1)$ and $C, D > 0$.

Acknowledgments

Parts of this work were carried out while both authors were at the Institut des Hautes Études Scientifiques, France, the Chemistry Laboratory III, University of Copenhagen, and the Nordic Summer School in Mathematics and the International Summer School in Mathematical Physics of the Mathematics Institute, University of Copenhagen. We are grateful to these three institutes for their generous support and hospitality. One author (E.L.) is also similarly grateful to NORDITA, Copenhagen. We are also indebted to Profs. G. K. Pedersen, E. Størmer, J. Tomiyama, and W. Wils for critical comments and useful discussions.

References

1. P. R. Halmos, "A Hilbert Space Problem Book," Problem 94, D. van Nostrand Co., Princeton, 1967.
2. J. Bendat and S. Sherman, Monotone and convex operator functions, *Trans. Amer. Math Soc.* **79** (1955), 58–71.
3. W. F. Stinespring, Positive functions on C^*-algebras, *Proc. Amer. Math. Soc.* **6** (1955), 211–216.
4. R. V. Kadison, A generalized Schwarz inequality and algebraic invariants for operator algebras, *Ann. of Math.* **56** (1952), 494–503.
5. H. Umegaki, Conditional expectation in operator algebra I, *Tohoku Math. J.* **6** (1954), 177–181.
6. J. Tomiyama, On the product projection of norm one in the direct product of operator algebras, *Tohoku Math. J.* **11** (1959), 303–313.
7. J. Tomiyama, On the projection of norm one in W^*-algebras, *Proc. Japan Acad.* **33** (1957), 608–612.

Reprinted from ADVANCES IN MATHEMATICS
Vol. 20, No. 2, May 1976

Printed in Belgium

Inequalities for Some Operator and Matrix Functions

ELLIOTT H. LIEB*

Departments of Mathematics and Physics, Princeton University, Princeton, New Jersey 08540

In this note we generalize an inequality on determinants (Corollary 3 below) recently proved by Seiler and Simon [1] in connection with some estimates in quantum field theory. Our main result is Lemma 1.

Let H be a Hilbert space and let $\mathscr{B}(H)$ be the bounded linear operators on H. We shall be interested in maps F: $\mathscr{B}(H) \to \mathbf{C}$ that satisfy:

$$B \geqslant A \geqslant 0 \Rightarrow F(B) \geqslant F(A) \geqslant 0 \tag{1}$$

$$F(A^*A)F(B^*B) \geqslant |F(A^*B)|^2. \tag{2}$$

F need not be continuous.

Well known examples of such F's are:

(i) any immanant [2] (of which the determinant and permanent are special cases);

(ii) $F(A) = E_k(A)$, the kth elementary symmetric function of the characteristic roots of A.

That the immanants satisfy (2) is well known [3], and follows from an alternative definition of them as appropriate traces of $\bigotimes^N A$ on the N-fold tensor product of H. Similarly, $E_k(A)$ is a trace of $\bigotimes^k A$ on the k-fold antisymmetric tensor product of H. (1) follows from the fact that $A \geqslant B \geqslant 0 \Rightarrow \bigotimes^k A \geqslant \bigotimes^k B$;

(iii) $F(A) = \| A \|$.

(iv) see Note added in proof.

LEMMA 1. *Let* F: $\mathscr{B}(H) \to \mathbf{C}$ *satisfy* (1) *and* (2). *Let* $\{a_i\}_{i=1}^n$ *and* $\{b_i\}_{i=1}^n$ *be in* $\mathscr{B}(H)$. *Then*

$$F(\Sigma_1^n a_i^* a_i) F(\Sigma_1^n b_i^* b_i) \geqslant |F(\Sigma_1^n a_i^* b_i)|^2.$$

* Work partially supported by National Science Foundation grant MPS71-03375 A03 at M.I.T.

Adv. in Math. *20*, 174–178 (1976)

Proof. Let $C = [\Sigma_1^n a_i^* a_i]^{1/2}$, $D = \Sigma_1^n a_i^* b_i$, $E = \Sigma_1^n b_i^* b_i$. First assume that $|a_i|$ is nonsingular, $\forall\, i$, where $|a| \equiv (a^*a)^{1/2}$; hence C is nonsingular and $E = \Sigma_1^n b_i^* a_i |a_i|^{-2} a_i^* b_i$. By the Schwarz inequality for operators, [4], $E \geqslant |C^{-1}D|^2$. Hence

$$F(C^2)F(E) \geqslant F(C^2)F(|C^{-1}D|^2) \geqslant |F(CC^{-1}D)|^2.$$

If some $|a_i|$ is singular the proof is more involved. Let $\epsilon > 0$.

$$a_i(|a_i|^2 + \epsilon I)^{-1} a_i^* \leqslant I \Rightarrow \Sigma_1^n b_i^* b_i \geqslant s\text{-}\lim_{\epsilon\downarrow 0} \Sigma_1^n b_i^* a_i(|a_i|^2 + \epsilon I)^{-1} a_i^* b_i$$
$$\geqslant s\text{-}\lim_{\epsilon\downarrow 0} D^*(C^2 + \epsilon I)^{-1} D \equiv K$$

by [4]. Ker $C \subset$ Ker a_i and $C \geqslant |a_i| \Rightarrow$ the operator $T_i \in \mathscr{B}(H)$ given by $T_i(C\varphi) = a_i\varphi$ and $T_i : \text{Ker } C \to 0$, is well defined. It is easy to check that $a_i(C^2 + \epsilon I)^{-1/2} \to^s T_i$, and hence $(C^2 + \epsilon I)^{-1/2} a_i^* \to^w T_i^*$. Therefore, if $L \equiv \Sigma_1^n T_i^* b_i$, then $K \geqslant L^*L$ and $F(C^2)F(E) \geqslant F(C^2)F(L^*L) \geqslant |F(CL)|^2$. However, $T_iC = s\text{-}\lim_{\epsilon\downarrow 0} a_i(C^2 + \epsilon I)^{-1/2}(C^2 + \epsilon I)^{1/2} = a_i$. ∎

COROLLARY 2. *With F as in Lemma* 1,

$$F(\Sigma_1^n |c_i^*|)F(\Sigma_1^n |c_i|) \geqslant |F(\Sigma_1^n c_i)|^2.$$

Proof. Write $c_i = U_i |c_i|$, U_i a partial isometry. In Lemma 1 set $a_i = |c_i|^{1/2} U_i^*$, $b_i = |c_i|^{1/2}$. Note that $a_i^* a_i = U_i |c_i| U_i^*$ and $(a_i^* a_i)^2 = U_i c_i^* c_i U_i^* = U_i |c_i| |c_i| U_i^* = c_i c_i^* = |c_i^*|^2$. ∎

Remark. For the determinant function, this says that $[\det \Sigma_1^n |c_i|][\det \Sigma_1^n |c_i^*|] \geqslant |\det \Sigma_1^n c_i|^2$. It is not generally true that $\det(\Sigma_1^n |c_i|) \geqslant |\det \Sigma_1^n c_i|$, Ref. [1]. The example $c_1 = \begin{pmatrix} 1 & -1 \\ -1 & 1 \end{pmatrix}$, $c_2 = \begin{pmatrix} 0 & 1 \\ 0 & 1 \end{pmatrix}$ illustrates this.

COROLLARY 3 (ROTFEL'D, SEILER–SIMON, see note added in proof.)

$$|\det(I + A + B)| \leqslant \det(I + |A|)\det(I + |B|)$$

Proof. Take $c_1 = I$, $c_2 = A$, $c_3 = B$ in Corollary 2. For positive matrices A and B, Corollary 3 is easy to prove, and the general case follows from the fact that $\det(I + |A|) = \det(I + |A^*|)$. ∎

In view of Lemma 1, one might hope that a Hölder type inequality is true, e.g.

$$|\det \Sigma_1^n a_i^* b_i| \leqslant (\det \Sigma_1^n |a_i|^p)^{1/p}(\det \Sigma_1^n |b_i|^q)^{1/q}, \quad p^{-1} + q^{-1} = 1.$$

That this is false in general can be seen by taking $a_i = I$, $\forall i$ and letting $q \to 1$, $p \to 0$. The resulting statement is not true according to the remark following Corollary 2. However, something a bit weaker is true.

DEFINITION 1. With $A = \{a_1, ..., a_n\}$, $p > 0$ and F satisfying (1),

$$F_p(A) \equiv \sup\{F(\Sigma_1^n U_i^* \mid a_i \mid^p U_i)\}^{1/p}$$

where the supremum is over all U_i such that $\| U_i \| \leqslant 1$.

DEFINITION 2. Let $F: \mathscr{B}(H) \to \mathbf{C}$, H separable. We say that F is *analytic* if: (i) Whenever a sequence $A^{(j)}$ converges strongly to A then $F(A^{(j)}) \to F(A)$; (ii) Whenever A_i, B_i and C_i are finite rank operators with A_i, $C_i \geqslant 0$ and $\| B_i \| \leqslant 1$, then $\mathbf{C} \ni z \mapsto F(\Sigma_1^n A_i^z B_i C_i^{1-z})$ is regular in the strip $\Delta = \{z \mid 0 \leqslant \mathrm{Re}(z) \leqslant 1\}$.

LEMMA 4. *Let H be separable and let $F: \mathscr{B}(H) \to \mathbf{C}$ be analytic and satisfy* (1) *and* (2). *Let p, $q > 1$ and $p^{-1} + q^{-1} = 1$. Then*

$$| F(\Sigma_1^n a_i^* b_i)|^2 \leqslant F(\Sigma_1^n \mid a_i \mid^p)^{1/p} F(\Sigma_1^n \mid b_i \mid^q)^{1/q} F_p(A) F_q(B) \leqslant F_p(A)^2 F_q(B)^2.$$

Proof. Owing to the strong sequential continuity of F, we can approximate a_i by a sequence $a_i^{(j)}$, where rank $(a_i^{(j)}) = j$, and $a_i^{(j)} \to^s a_i$ and likewise for b_i. Thus, it is sufficient to prove the lemma for finite rank operators, which we shall henceforth assume. Let $a_i^* = \mid a_i \mid S_i$, $b_i = T_i \mid b_i \mid$, S_i, T_i finite rank partial isometries. Let $\alpha_i = \mid a_i \mid^p$, $\beta_i = \mid b_i \mid^q$. If $F(\Sigma_1^n \beta_i) = 0$ then, by Lemma 1, $0 = F(\Sigma_1^n \beta_i) F(nI) \geqslant F(\Sigma_1^n \beta_i^{1/2})^2$. By induction, $F(\Sigma_1^n \mid b_i \mid^r) = 0$ for some $r \in (0, 1]$. But then $|F(\Sigma_1^n a_i^* b_i)|^2 \leqslant F(\Sigma_1^n \gamma_i^* \gamma_i) F(\Sigma_1^n \mid b_i \mid^r) = 0$, where $\gamma_i = |b_i|^{1-r/2} T_i^* a_i$. Likewise, we can also assume that $F(\Sigma_1^n \alpha_i) \neq 0$ and, a fortiori, $F_p(A)$, $F_q(B) \neq 0$. Now consider the following regular function on Δ:

$$f(z) = F(\Sigma_1^n \alpha_i^z S_i T_i \beta_i^{1-z})^2 \times F(\Sigma_1^n \alpha_i)^{-z} F_p(A)^{-zp} F(\Sigma_1^n \beta_i)^{z-1} F_q(B)^{(z-1)q}.$$

By Lemma 1, the first factor is bounded by $F(\Sigma_1^n \alpha_i^{2\mathrm{Re}(z)}) F(\Sigma_1^n \beta_i^{2-2\mathrm{Re}(z)}) \leqslant F(nM^2 I)^2$, where $M = \max\{\| \alpha_i \|, \| \beta_i \|, 1\}$. Thus, f is bounded in Δ and its maximum modulus occurs on the boundary, $z = i\theta$ or $z = 1 + i\theta$, θ real. In the former case, by Lemma 1

$$| f(i\theta)| \leqslant F(\Sigma_1^n W_i \beta_i W_i^*) F(\Sigma_1^n \beta_i) F(\Sigma_1^n \beta_i)^{-1} F_q(B)^{-q}$$

Adv. in Math. *20*, 174–178 (1976)

with $W_i = \alpha_i^{i\theta} S_i T_i \beta_i^{-i\theta}$. Since $\| W_i \| \leqslant I$, $F(\Sigma_1^n W_i \beta_i W_i^*) \leqslant F_q(B)^q$, and $|f(i\theta)| \leqslant 1$. Likewise $|f(1 + i\theta)| \leqslant 1$, and hence $|f(1/p)| \leqslant 1$. ∎

Remark. When $p = q = 1/2$, Lemma 4 is weaker than Lemma 1. Clearly, some improvement is possible.

We now give a sufficient condition for F to satisfy (1) and (2).

LEMMA 5. *Let* $F: \mathscr{B}(H) \to \mathbf{C}$ *satisfy*

(i) $A \geqslant 0 \Rightarrow F(A) \geqslant 0$,

(ii) $\| U \| \leqslant 1 \Rightarrow |F(UA)| \leqslant F(| A |)$ and $|F(AU)| \leqslant F(| A |)$, $\forall A$.

Then F satisfies (1) *and additionally* $\| U \|, \| W \| \leqslant 1 \Rightarrow |F(UAW)| \leqslant F(| A |)$, $\forall A$. *If, furthermore, H is separable and F is analytic, then F also satisfies* (2).

Proof. If $B \geqslant A \geqslant 0$, let $W \in \mathscr{B}(H)$ be defined by $W(B^{1/2}\varphi) = A^{1/2}\varphi$ and W: $\mathrm{Ker}(B) \to 0$. Since $B \geqslant A \geqslant 0$, W is well defined and $\| W \| \leqslant 1$. Then $A^{1/2} = WB^{1/2}$ and $A = WBW^*$. By (ii), $F(WBW^*) \leqslant F(| WB |)$. If $C^{1/2} = | WB |$ then $C = BW^*WB \leqslant B^2$. As before, $C^{1/2} = U(B^2)^{1/2} = UB$ with $\| U \| \leqslant 1$. By (ii), $F(C^{1/2}) \leqslant F(B)$ and (1) is proved. For the same reason,

$$|F(UAW)| \leqslant F(| UA |) \leqslant F(| A |).$$

As in the proof of Lemma 4, it is sufficient to prove (2) for finite rank operators. Let $\epsilon > 0$ and let $F_\epsilon(A) = F(A) + \epsilon$, whence F_ϵ satisfies (i) and (ii). Let $A, B \in \mathscr{B}(H)$, $A^* = | A | S$, $B = T | B |$ with S and T finite rank partial isometries, and let $\alpha = | A |^2$, $\beta = | B |^2$. Consider the following regular function on Δ:

$$f(z) = F_\epsilon(\alpha^z ST \beta^{1-z}) F_\epsilon(\alpha)^{-z} F_\epsilon(\beta)^{z-1}.$$

If $\gamma_z = \alpha^z ST \beta^{1-z}$ then, for $z \in \Delta$, $\| \gamma_z \| \leqslant M$ for some $M > 0$. Therefore $|F_\epsilon(\gamma_z)| = |F_\epsilon(M^{-1}\gamma_z MI)| \leqslant F_\epsilon(MI)$ by (ii), and $f(z)$ is bounded in Δ. When $z = i\theta$ or $z = 1 + i\theta$ with θ real, $|f(z)| \leqslant 1$ by (ii) and thus (2) is proved for F_ϵ. Since (2) holds for all $\epsilon > 0$, however, with A and B fixed, it holds when $\epsilon = 0$. ∎

To see that the converse of Lemma 5 is false, let C be a fixed positive matrix and let F be the matrix function $F(A) \equiv \det(A + C)$. Then F satisfies (1) and (2) (Lemma 1) and is analytic. However, as the remark after Corollary 2 shows, $|F(UA)| \nleq F(|A|)$ in general when $\|U\| \leqslant 1$.

A possible generalization of Lemma 1 is to have $F: \mathscr{B}(H) \to \mathscr{B}(K)$, where K is another Hilbert space, and to replace (2) either by

$$F(A^*A)^{1/2} F(B^*B) F(A^*A)^{1/2} \geqslant |F(A^*B)|^2$$

or

$$F(B^*B) \geqslant F(A^*B)^* F(A^*A)^{-1} F(A^*B).$$

However, we have not been able to find any example that cannot be treated by the lemmas already stated.

The author thanks E. Seiler and B. Simon for making their result, Corollary 3, available before publication, and for helpful discussions and suggestions.

Notes Added in Proof

(i) The author has learned that Corollary 3, with generalizations, was proved earlier by S. Yu. Rotfel'd [5, 6].

(ii) R. Merris and J. A. Dias da Silva [7, 8] have given other examples of matrix functions satisfying (1) and (2).

References

1. E. Seiler and B. Simon, An inequality among determinants, *Proc. Nat. Acad. Sci. U.S.A.* **72** (1975), 3277–3278.
2. D. E. Littlewood, "The Theory of Group Characters and Matrix Representations of Groups," Oxford University Press, 1940.
3. M. Marcus and H. Minc, "A Survey of Matrix Theory and Matrix Inequalities," Prindle, Weber and Schmidt, Boston, 1964.
4. E. H. Lieb and M. B. Ruskai, Some operator inequalities of the Schwarz type, *Adv. in Math.* **12** (1974), 269–273. Cf. Theorem 1.
5. S. Yu. Rotfel'd, Remarks on the singular numbers of the sum of completely continuous operators, *Funct. Anal. Appl.* Consultants Bureau Trans. **1** (1967), 252–253.
6. S. Yu. Rotfel'd, The singular numbers of the sum of completely continuous operators, "Topics in Math. Phys." (M. Sh. Birman, Ed.), Consultants Bureau Trans. **3** (1969), 73–78.
7. R. Merris, Inequalities for matrix functions, *J. Algebra* **22** (1972), 451–460.
8. R. Merris and J. A. Dias da Silva, Generalized Schur functions, *J. Algebra* **35** (1975), 442–448.

Reprinted from ADVANCES IN MATHEMATICS
Vol. 28, No. 1, April 1978
Printed in Belgium

Positive Linear Maps which are Order Bounded on C^* Subalgebras

MICHAEL AIZENMAN*

Department of Physics, Princeton University, Princeton, New Jersey 08540

E. BRIAN DAVIES†

Department of Mathematics, Princeton University, Princeton, New Jersey 08540

AND

ELLIOTT H. LIEB*

Departments of Mathematics and Physics, Princeton University, Princeton, New Jersey 08540

When studying the third law of thermodynamics for an infinite system, or properties of equilibrium states of such a system (see [3]), one is led to consider the effect of local perturbations.

If $\mathscr{A}$ is the C^* algebra of a system and Λ is a finite region in $\mathbb{Z}^3$ then in an obvious notation

$$\mathscr{A} = \mathscr{A}_\Lambda \otimes \mathscr{A}_{\Lambda^c}.$$

Local perturbations may be defined in various ways, e.g., as positive linear maps $T: \mathscr{A} \to \mathscr{A}$ satisfying $T(1) = 1$ and one of the following two conditions

(i) $T(1 \otimes Y) = 1 \otimes Y$ for all $Y \in \mathscr{A}_{\Lambda^c}$,

(ii) $T(X \otimes Y) = T(X \otimes 1)\, 1 \otimes Y$ for all $X \in \mathscr{A}_\Lambda$ and $Y \in \mathscr{A}_{\Lambda^c}$.

It also makes sense to restrict the definition to completely positive maps [2]. It is therefore of interest to know whether the various definitions are equivalent. That (ii) implies (i) is obvious. It is also clear that, for nontrivial infinite systems, (ii) implies that T is completely positive. It turns out that the slightly surprising implication (i) $\Rightarrow$ (ii) holds, and is a special case of a more general result about positive maps of C^* algebras, which we now state.

* This work was partially supported by NSF Grant MCS 75-21684 A01.

† On leave from the Mathematical Institute, Oxford University. This work was partially supported by NSF Grant MPS 75-11864 A04.

THEOREM. *Let $\mathscr{A}$ be a C^* algebra with identity, $\mathscr{B}$ a C^*-subalgebra of $\mathscr{A}$ with $1 \in \mathscr{B}$, and $\mathscr{C}$ the relative commutant of $\mathscr{B}$ within $\mathscr{A}$. Let T be a positive linear map on $\mathscr{A}$ such that $0 \leqslant B \in \mathscr{B}$ implies*

$$0 \leqslant T(B) \leqslant B. \tag{1}$$

Then $T(\mathscr{C}) \subseteq \mathscr{C}$ and

$$T(BC) = BT(C) \tag{2}$$

for all $B \in \mathscr{B}$ and $C \in \mathscr{C}$.

Proof. We first assume that $\mathscr{B}$ is norm generated by its projections. If $C \in \mathscr{C}$ satisfies $0 \leqslant C \leqslant 1$ and $P \in \mathscr{B}$ satisfies $P = P^2 = P^*$ then $0 \leqslant PC \leqslant P$ so

$$0 \leqslant T(PC) \leqslant T(P) \leqslant P.$$

Therefore

$$0 = (1 - P)\, T(PC) = T(PC)(1 - P).$$

Subtracting the similar equation

$$0 = PT((1 - P)C) = T((1 - P)C)P$$

we obtain

$$0 = T(PC) - PT(C) = T(PC) - T(C)P.$$

It follows that

$$T(PC) = PT(C) = T(C)P. \tag{3}$$

Since the linear span of positive bounded elements of $\mathscr{C}$ is norm dense in $\mathscr{C}$, and the projections span a norm dense subspace of $\mathscr{B}$, Eq. (3) implies Eq. (2) and

$$T(\mathscr{C}) \subset \mathscr{C}.$$

We prove the general case by passing to the universal enveloping algebra. Then $\mathscr{B}^{**}$ is a sub-W^*-algebra of $\mathscr{A}^{**}$ and T^{**} is an ultraweakly continuous positive linear map on $\mathscr{A}^{**}$. Since every nonnegative element B of $\mathscr{B}^{**}$ is an ultraweak limit of nonnegative elements of $\mathscr{B}$ we see that $0 \leqslant B \in \mathscr{B}^{**}$ implies

$$0 \leqslant T^{**}(B) \leqslant B.$$

Applying our special case proved above we deduce that Eq. (2) holds whenever $B \in \mathscr{B} \subseteq \mathscr{B}^{**}$ and $C \in \mathscr{A} \cap (\mathscr{B}^{**})' = \mathscr{A} \cap \mathscr{B}'$, where $'$ refers to the commutant.

Notes. (i) If $\mathscr{A} = \mathscr{B}$ then maps T satisfying Eq. (1) are order bounded in the sense of [1, p. 159]. For this case our result was obtained in a much more general context by Alfsen and Anderson [1, p. 159] and Wils [4].

(ii) If $\mathscr{A} = \mathscr{B} \otimes \mathscr{C}$ where $\mathscr{B}$ contains an $n \times n$ matrix subalgebra, then T is uniquely determined by its restriction to $\mathscr{C}$, which is necessarily n-positive.

Note added in proof. We should like to thank E. Størmer for pointing out that an alternative proof of our theorem may be based on Corollary 1 of [5].

References

1. F. M. Alfsen, "Compact Convex Sets and Boundary Condition," Springer, Berlin, 1971.
2. E. B. Davies, "Quantum Theory of Open Systems," Academic Press, New York, 1976.
3. G. L. Sewell, KMS conditions and local thermodynamical stability of quantum lattice systems, II, *Comm. Math. Phys.* **55** (1977), 53.
4. W. Wils, The ideal center for partially ordered vector spaces, *Acta Math.* **127** (1971), 41.
5. T. B. Andersen, On multipliers and order-bounded operators in C^*-algebras, *Proc. Amer. Math. Soc.* **25** (1970), 896–899.

Commun. Math. Phys. 155, 27–46 (1993)

Communications in Mathematical Physics

Optimal Hypercontractivity for Fermi Fields and Related Non-Commutative Integration Inequalities

Eric A. Carlen[1]* **and Elliott H. Lieb**[2]**

[1] Department of Mathematics, Princeton University, Princeton, New Jersey 08544, USA
[2] Departments of Mathematics and Physics, Princeton University, Princeton, New Jersey 08544–0708, USA

Received April 27, 1992; in revised form July 16, 1992

Dedicated to Huzihiro Araki

Abstract. Optimal hypercontractivity bounds for the fermion oscillator semigroup are obtained. These are the fermion analogs of the optimal hypercontractivity bounds for the boson oscillator semigroup obtained by Nelson. In the process, several results of independent interest in the theory of non-commutative integration are established.

I. Introduction

Observables pertaining to the configuration of a quantum system with n degrees of freedom are operators $Q_1, Q_2, \ldots, Q_n$ which, depending on the system, may or may not commute. Our main concern is with the case in which the configuration variables are amplitudes of certain field modes.

For boson fields, these configuration observables do commute, and the state space $\mathscr{H}$ can be taken as the space of all complex square integrable functions on their joint spectrum. This is the Schrödinger q-space representation, and the fact that in it the state space is a function space, and not just an abstract Hilbert space, is very helpful in the analysis of such systems. As one example, it sometimes turns out that physically interesting operators preserve the cone of positive functions, and this opens the way to the application of the Perron-Frobenius theorem in the study of ground states of such systems.

For fermion fields, the configuration observables do not commute, and this simple q-space representation is not available. However, the non-commutative integration theory of Irving Segal [Se53] permits the construction of a suitable substitute, and in fact it was created with such a purpose in view. This approach to the study of fermion systems has been extensively developed by Gross [Gr72] who, among other things, proved a version of the Perron-Frobenius theorem adapted to

---

* On leave from School of Math., Georgia Institute of Technology, Atlanta, GA 30332, USA
** Work supported by U.S. National Science Foundation grant no. PHY90–19433–A01.

the setting and applied it to prove existence and uniqueness of ground states for certain fermion quantum field models.

The main estimate which enabled Gross to apply his Perron-Frobenius type theorem to fermion fields was a hypercontractivity estimate for the fermion oscillator semigroup. Corresponding hypercontractivity estimates for boson fields had been introduced earlier by Nelson [Ne66] who later obtained the optimal bound for bosons [Ne73].

Our main result, Theorem 4 below, is the corresponding optimal hypercontractivity bound for fermions. Before stating this theorem, we describe its mathematical and physical contexts in some detail because it cannot even be formulated naturally in the conventional Fock space language. (Of course its perturbation theoretic *consequences* can be expressed quite naturally in the usual language.)

The state space for a system of n fermion degrees of freedom is conventionally realized as the Fock space

$$\mathscr{F} = \bigoplus_{j=0}^{n} ((\mathbb{C}^n)^{\wedge j}) . \tag{1.1}$$

The basic "free Hamiltonian" on this space is the fermion number operator

$$\hat{H}_0 = \sum_{j=1}^{n} c_j^* c_j , \tag{1.2}$$

where c_j and c_j^* are the usual fermion annihilation and creation operators acting on $\mathscr{F}$. The fermion oscillator semigroup is the semigroup of operators $\exp(-t\hat{H}_0)$ that it generates.

For n boson degrees of freedom, the state space may be realized *either* as the boson Fock space, or as $L^2(\mathbb{R}^n, (2\pi)^{-n/2} e^{-x^2/2} dx)$. The natural isomorphism between these spaces was pointed out by Segal, and the latter may be regarded as the Schrödinger q-space realization. The boson oscillator semigroup is the semigroup generated by the boson number operator. Though we have just defined it in Fock space terms, it may also be considered as an operator semigroup on $L^2(\mathbb{R}^n, (2\pi)^{-n/2} e^{-x^2/2} dx)$. In this setting, it is **hypercontractive**; i.e., for any finite p greater than 2, there is t_p sufficiently large, for which the semigroup is a contraction from $L^2(\mathbb{R}^n, (2\pi)^{-n/2} e^{-x^2/2} dx)$ to $L^p(\mathbb{R}^n, (2\pi)^{-n/2} e^{-x^2/2} dx)$ for all $t \geqq t_p$. Since t_p is independent of n, this result proved useful in treating perturbation theoretic problems for boson fields. This hypercontractivity inequality cannot be formulated naturally in the Fock space setting because no natural notion of "L^p" can be introduced there. The $L^p(\mathbb{R}^n, (2\pi)^{-n/2} e^{-x^2/2} dx)$ setting is essential.

In order to pass from the Fock space description for a system of n fermion degrees of freedom to a non-commutative analog of the Schrödinger q-space description (in which we have non-commutative analogs of L^p norms), we introduce configuration observables $Q_j = c_j + c_j^*$, and let $\mathscr{C}(\mathscr{Q})$ be the algebra with unit which they generate. This is an operator algebra on the 2^n-dimensional Hilbert space $\mathscr{F}$. As we explain in Sect. II, it is a Clifford algebra naturally associated to $\mathbb{C}^n$ with its usual inner product.

What follows here shall be explained in more detail in Sects. II–IV, but the key fact for our present considerations is that the 2^n distinct monomials in the Q_j are a basis for $\mathscr{C}(\mathscr{Q})$ as a *vector space*. Thus, any A in $\mathscr{C}(\mathscr{Q})$ can be uniquely written as

$$A = \alpha I + \sum_j \alpha_j Q_j + \sum_{j<k} \alpha_{j,k} Q_j Q_k + \cdots + \alpha_{1,\ldots,n} Q_1 \ldots Q_n . \tag{1.3}$$

It is not hard to see that any vector v in $\mathscr{F}$ can be written as $v = A\Omega$ for some unique $A \in \mathscr{C}(\mathscr{Q})$ acting on the vacuum Ω. The operator $\exp(-t\hat{H}_0)$ acts on such vectors v. Thus, $\exp(-t\hat{H}_0)v = B\Omega$ for some unique $B \in \mathscr{C}(\mathscr{Q})$ (which depends on t and A, of course). The map $P_t: A \mapsto B$ is clearly linear, and a simple calculation (elucidated in Sect. IV) shows that

$$P_t A = \alpha I + e^{-t} \sum_j \alpha_j Q_j + e^{-2t} \sum_{j<k} \alpha_{j,k} Q_j Q_k + \cdots + e^{-nt} \alpha_{1,\ldots,n} Q_1 \ldots Q_n \,. \tag{1.4}$$

Instead of considering Fock space, $\mathscr{F}$, and the operator $\exp(-t\hat{H}_0)$, Segal's idea is to consider $\mathscr{C}(\mathscr{Q})$ and the linear operator P_t that acts on $\mathscr{C}(\mathscr{Q})$ as above. Following Segal, we can construct L^p norms on $\mathscr{C}(\mathscr{Q})$, the most important case being $p = 2$, in terms of which one can do quantum mechanics without mentioning Fock space.

To construct the norms, define the linear functional $\tau: \mathscr{C}(\mathscr{Q}) \to \mathbb{C}$ by

$$\tau(A) = \langle \Omega, A\Omega \rangle_{\mathscr{F}} \,. \tag{1.5}$$

It turns out, as we explain in Sect. III that $\tau(A)$ is, for $A \in \mathscr{C}(\mathscr{Q})$, nothing other than the usual normalized trace on $\mathscr{F}$ applied to A.

The basic ingredients of a non-commutative integration theory are an algebra of operators such as $\mathscr{C}(\mathscr{Q})$, and a linear functional on it such as τ. Norms analogous to the L^p norms of commutative integration theory can now be introduced by

$$\|A\|_p = (\tau(A^*A)^{p/2})^{1/p} \quad \text{for} \quad 1 \leqq p < \infty \;;$$

$\|A\|_\infty$ denotes the operator norm of A. Let $\mathscr{C}^p(\mathscr{Q})$ denote $\mathscr{C}(\mathscr{Q})$ equipped with the corresponding norm.

Having expanded our horizons beyond Hilbert space, we can ask for bounds between $\|P_t A\|_q$ and $\|A\|_p$ for different q, p and t. Our main result, Theorem 4, is the optimal **fermion hypercontractivity** inequality; i.e. for all $1 < p \leqq q < \infty$ and all A in $\mathscr{C}^p(\mathscr{Q})$,

$$\|P_t A\|_q \leqq \|A\|_p \quad \text{when} \quad e^{-2t} \leqq \frac{p-1}{q-1}, \tag{1.6}$$

and the t saturating the inequality on the right is the smallest for which the inequality on the left always holds. We prove this for n degrees of freedom with n an arbitrary finite integer. Since the estimate is independent of n, a theorem of Gross [Gr72] implies that it holds as well with infinitely many degrees of freedom.

The result (1.6) was conjectured by Gross [Gr75] who proved it [Gr72] in the special case $p = 2$, $q = 4$. The cases $p = 2$, $q = 2m$, where m is an integer were proved by Lindsay and Meyer [LiMe] following earlier work by Lindsay [Lin] on the case $p = 2$, $q = 2^m$. Since P_t is self adjoint, duality yields a corresponding family of results in which $q = 2$. Until now all other cases (except when $n = 1$ or 2) had remained open. The optimal relation between t, p and q found here for fermion hypercontractivity is the same as that found by Nelson [Ne73] for boson hypercontractivity.

By now there are many proofs of Nelson's inequality. Neveu's elegant proof [Nev], like Nelson's original proof, is based on probabilistic methods. Proofs based on geometric methods have been given by Carlen and Loss [CL90] and by Lieb [Li90] who considers generalizations in which the Mehler kernel (the kernel for the boson oscillator semigroup) is replaced by an arbitrary Gaussian kernel. More references can be found in the bibliography to Gross's article [Gr89, Gr92]. However, none of the existing approaches to the boson problem has been found to solve the fermion problem. See also [Far].

As a corollary to the optimal fermion hypercontractivity inequality, we obtain the optimal **fermion logarithmic Sobolev inequality**:

$$\tau(|A|^2 \ln |A|^2) - (\|A\|_2^2 \ln \|A\|_2^2) \leqq 2\tau(A^* H_0 A)\,, \tag{1.7}$$

where $|A| = (A^*A)^{1/2}$.

This inequality was conjectured by Gross, and proved by him in a weaker form with the constant on the right increased by a factor of ln 3. In studying perturbations of H_0 by multiplication operators V, this inequality plays the same role as does the usual Sobolev inequality in studying perturbations of $-\frac{1}{2}\Delta$ by multiplication operators V on $L^2(\mathbb{R}^n, d^n x)$ [Fe69]. (The multiplication operator associated to a self-adjoint element V of $\mathscr{C}(\mathscr{Q})$ is defined to be the average of left and right multiplication by V, and is denoted here by the same symbol V.) Again, in the standard Fock space setting, there is no natural way to formulate such an a-priori regularity inequality, or even to introduce the notion of a multiplication operator.

This paper is organized as follows: In Sect. II we study the structure of $\mathscr{C}(\mathscr{Q})$ for finite n. It is actually simpler, as well as technically advantageous, to consider it as a subalgebra of the algebra $\mathscr{C}(\mathscr{K})$ generated by the identity, the configuration observables $Q_1, \ldots, Q_n$ and their conjugate momenta $P_1, \ldots, P_n$. Since they both turn out to be certain Clifford algebras, their structure has been worked out long ago with the representation theory of the orthogonal groups. Thus, this section contains no new result but simply introduces notation and prepares the way for what follows. Of particular use are an explicit spin-chain representation of the operator algebra $\mathscr{C}(\mathscr{K})$, and the Jordan-Wigner transform identifying it with the algebra generated by n "hard core bosons."

Section III concerns properties of the spaces $\mathscr{C}^p(\mathscr{K})$ and their norms. The main result here is an optimal uniform convexity inequality for $\mathscr{C}^p(\mathscr{K})$, $1 < p \leqq 2$, which is joint work with Keith Ball. We need only a special case of this inequality here, and a proof is provided in an appendix for the reader's convenience.

Section IV introduces a convenient expression for P_t in terms of the conditional expectation $\pi_{\mathscr{Q}}$ of $\mathscr{C}(\mathscr{K})$ with respect to $\mathscr{C}(\mathscr{Q})$. The main result in this section is an inequality for conditional expectations which enables us to prove that

$$\sup\{\|P_t A\|_q : \|A\|_p = 1\} = \sup\{\|P_t A\|_q : A \geqq 0 \text{ and } \|A\|_p = 1\}\,.$$

Thus, to establish (1.6) in general, we need only consider positive A.

In Sect. V we establish the optimal fermion hypercontractivity bounds and the corresponding optimal fermion logarithmic Sobolev inequality. This is done in several steps. First, using results collected in Sects. II and III we establish that (1.7) holds for $1 < p \leqq 2$ and $q = 2$. At $t = 0$ this is an equality; differentiating it there yields the logarithmic Sobolev inequality (1.7). Gross showed that (1.6) would follow from (1.7), if it were true (as we show here), for all *self adjoint A*. His result rests on a deep inequality he established for positive operators. Since it is *not* in general true that $\|P_t A\|_q \leqq \|P_t |A|\|_q$, as is trivially true in the commutative case, Gross's result does not allow us to conclude (1.6) for *general* (i.e. non-self adjoint) A from (1.7). The results of the Sect. IV do, however, allow us to draw this conclusion.

Finally, in Sect. VI we show that the same hypercontractivity relation (1.6) holds for a mixed system of bosons and fermions.

Non-commutative probability theory has grown into a substantial branch of analysis with a number of physical applications. The mathematical theory is reviewed and developed in [Me85] and [Me86], while other sorts of physical applications, besides those discussed here, are treated in [Da76] and [HuPa] for example.

It is a pleasure to thank Leonard Gross for discussing his results and conjectures with us, and for encouraging us to take up the latter. We are indebted to Keith Ball for his collaboration on the subject of convexity inequalities that led to Theorem 1 [BCL], which is one of the key ingredients in the present work. Thanks are also due to G.-F. Dell'Antonio, A. Jaffe and A. Wightman for useful discussions.

II. Fermions and the Clifford Algebra

We begin by recalling for later use some well known facts about fermions.

The fundamental observables for a system of n fermion degrees of freedom are **configuration operators** $Q_1, Q_2, \ldots, Q_n$ together with their **conjugate momenta operators** $P_1, P_2, \ldots, P_n$ all acting as operators on a complex Hilbert space $\mathscr{H}$ and satisfying the **canonical anticommutation relations**:

$$Q_jQ_k + Q_kQ_j = 2\delta_{jk}\,, \quad P_jP_k + P_kP_j = 2\delta_{jk}\,, \quad P_jQ_k + Q_kP_j = 0\,. \tag{2.1}$$

We denote the complex algebra generated by the identity and the configuration observables by $\mathscr{C}(\mathscr{Q})$, and the complex algebra generated by the identity, the configuration observables and the momentum observables all together by $\mathscr{C}(\mathscr{K})$. $\mathscr{C}(\mathscr{Q})$ is the object of primary interest; but many aspects of its structure are most readily seen within the larger algebra $\mathscr{C}(\mathscr{K})$.

This algebra can be concretely represented as the algebra of observables for a spin$\frac{1}{2}$–chain as follows. We define the matrices

$$I = \begin{bmatrix} 1 & 0 \\ 0 & 1 \end{bmatrix}, \quad U = \begin{bmatrix} 1 & 0 \\ 0 & -1 \end{bmatrix}, \quad Q = \begin{bmatrix} 0 & 1 \\ 1 & 0 \end{bmatrix}, \quad P = \begin{bmatrix} 0 & i \\ -i & 0 \end{bmatrix}.$$

Let $\mathscr{H}$ denote the n-fold tensor product of $\mathbb{C}^2$ with itself:

$$\mathscr{H} = \underbrace{\mathbb{C}^2 \otimes \ldots \otimes \mathbb{C}^2}_{n \text{ times}},$$

and on $\mathscr{H}$ define the operators

$$Q_j = U \otimes \cdots U \otimes Q \otimes I \otimes \cdots \otimes I\,, \quad P_j = U \otimes \cdots U \otimes P \otimes I \otimes \cdots \otimes I\,, \tag{2.2}$$

where the Q and the P occur in the jth places.

The operators $Q_1, \ldots, Q_n, P_1, \ldots, P_n$ just defined are easily seen to satisfy the canonical anticommutation relations. Of great use in studying the algebra $\mathscr{C}(\mathscr{K})$ that they generate is the fact that it is also the algebra generated by n "hard core boson" degrees of freedom. More explicitly, put

$$\hat{Q}_j = I \otimes \cdots I \otimes Q \otimes I \otimes \cdots \otimes I, \quad \hat{P}_j = I \otimes \cdots I \otimes P \otimes I \otimes \cdots \otimes I, \tag{2.3}$$

and call the algebra generated by the operators $\hat{Q}_1, \ldots, \hat{Q}_n, \hat{P}_1, \ldots, \hat{P}_n$ the **hard core boson algebra**, and denote it by $\hat{\mathscr{C}}(\mathscr{K})$. To see that the two algebras coincide, put

$$U_j = I \otimes \cdots I \otimes U \otimes I \otimes \cdots \otimes I \quad \text{and} \quad V_k = \prod_{j=1}^{k-1} U_j \,. \tag{2.4}$$

Then since $P_j Q_j = \hat{P}_j \hat{Q}_j = iU_j$, each V_k belongs to both the hard core boson algebra and $\mathscr{C}(\mathscr{K})$. Moreover

$$\hat{Q}_k = V_k Q_k \quad \text{and} \quad \hat{P}_k = V_k P_k \tag{2.5}$$

with the inverse relation given as well by

$$Q_k = V_k \hat{Q}_k \quad \text{and} \quad P_k = V_k \hat{P}_k \,. \tag{2.6}$$

Thus, Q_k, P_k are in $\hat{\mathscr{C}}(\mathscr{K})$, and $\hat{Q}_k, \hat{P}_k$ are in $\mathscr{C}(\mathscr{K})$.

What we call the hard core boson algebra was initially introduced by Jordan and Klein [JoKl] as a first attempt to implement the Pauli exclusion principle mathematically. The transformation of observables (2.6) was discovered by Jordan and Wigner [JoWi] and used by them to write the algebraic relations characterizing the algebra in the familiar covariant form (2.1); today it is known as the Jordan–Wigner transform. It has been used many times since – for example, in a solution of the two dimensional Ising model by Schultz, Mattis and Lieb, whose paper [SML] can be consulted for references to other applications. It is also the key to Brauer and Weyl's treatment [BrWe] of the Clifford algebra on which the exposition in the rest of this section is largely based. Following them, we now show that $\mathscr{C}(\mathscr{K})$ is the full matrix algebra on $\mathscr{H}$.

First, we introduce a basis in $\mathscr{H}$. Let

$$e_1 = \begin{pmatrix} 1 \\ 0 \end{pmatrix} \quad \text{and} \quad e_{-1} = \begin{pmatrix} 0 \\ 1 \end{pmatrix}$$

be the standard basis of $\mathbb{C}^2$. For each $j = 1, \ldots, n$, let σ_j be either 1 or -1. Then the unit vectors

$$e_{\sigma_1, \ldots, \sigma_n} = e_{\sigma_1} \otimes \cdots \otimes e_{\sigma_n}$$

provide a natural orthonormal basis for $\mathscr{H}$.

Next introduce the fermion creation and annihilation operators $c_j^* = \frac{1}{2}(Q_j - iP_j)$ and $c_j = \frac{1}{2}(Q_j + iP_j)$ and their hard core boson analogs $\hat{c}_j^* = \frac{1}{2}(\hat{Q}_j - i\hat{P}_j)$ and $\hat{c}_j = \frac{1}{2}(\hat{Q}_j + i\hat{P}_j)$. Now put $L(\sigma_1, \ldots, \sigma_n) = \Pi_{j=1}^n B_j$, where $B_j = \hat{c}_j$ if $\sigma_j = 1$ and $B_j = \hat{c}_j \hat{c}_j^*$ if $\sigma_j = -1$. Then $L(\sigma_1, \ldots, \sigma_n)\, e_{\sigma_1, \ldots, \sigma_n} = \Omega$ where

$$\Omega = \underbrace{\begin{pmatrix} 0 \\ 1 \end{pmatrix} \otimes \cdots \otimes \begin{pmatrix} 0 \\ 1 \end{pmatrix}}_{n \text{ times}} = e_{-1, -1, \ldots, -1} \tag{2.7}$$

is a distinguished unit vector in $\mathscr{H}$ called the **ground state**. Also, $L(\sigma_1, \ldots, \sigma_n)$ annihilates all the other basis vectors. Moreover, $L^*(\sigma_1, \ldots, \sigma_n)\,\Omega = e_{\sigma_1, \ldots, \sigma_n}$. Thus,

$$L^*(\tau_1, \ldots, \tau_n)\, L(\sigma_1, \ldots, \sigma_n)\, e_{\sigma_1, \ldots, \sigma_n} = e_{\tau_1, \ldots, \tau_n} \,,$$

and this operator annihilates all other basis vectors. Manifestly this operator belongs to $\hat{\mathscr{C}}(\mathscr{K})$, and hence to $\mathscr{C}(\mathscr{K})$ as well, and the 2^n operators of this kind form a basis for the full matrix algebra on $\mathscr{H}$.

This concrete description of $\mathscr{C}(\mathscr{K})$ in terms of spin–chain observables is the most useful for many purposes. Still, it is also useful to have a characterization of $\mathscr{C}(\mathscr{K})$ which is less dependent on coördinates; i.e. on the choice of fermi configuration observables $Q_1, \ldots, Q_n$.

Toward this end, consider the standard n-dimensional Hilbert space $\mathbb{C}^n$ equipped with its standard inner product $(\cdot, \cdot)$ and complex conjugation. Let $\mathscr{K}$ denote $\mathscr{C}^n$ considered as a real $2n$-dimensional Hilbert space equipped with the inner product

$$\langle x, y\rangle_{\mathscr{K}} = \Re(x, y) .$$

Then complex conjugation on $\mathbb{C}^n$ induces an involutory orthogonal transformation J on $\mathscr{K}$. Let $\mathscr{Q}$ and $\mathscr{P}$ respectively denote the eigenspaces of J corresponding to the eigenvalues $+1$ and -1. The bilinear form on $\mathscr{K}$ given by $\Im(x, y)$ is symplectic so that $\mathscr{K}$ is naturally endowed with the structure possessed by the classical phase space of a system of n linear degrees of freedom. $\mathscr{Q}$ is called the **configuration space**, $\mathscr{P}$ the **momentum space**, and the complex conjugation J is usually identified with time reversal.

The Clifford algebra $\mathscr{C}(\mathscr{K})$ is characterized up to automorphism as the algebra with unit I such that:

(i) There is a linear imbedding $\mathscr{J}: \mathscr{K} \to \mathscr{C}(\mathscr{K})$, and $\mathscr{J}(\mathscr{K})$ generates $\mathscr{C}(\mathscr{K})$.
(ii) For all $x, y \in \mathscr{K}$,

$$\mathscr{J}(x)\mathscr{J}(y) + \mathscr{J}(y)\mathscr{J}(x) = \langle x, y\rangle_{\mathscr{K}} . \tag{2.8}$$

To make contact with our previous concrete description, let $\{q_1, \ldots, q_n\}$ be an orthonormal basis of $\mathbb{C}^n$ consisting of purely real vectors. For each j, let $p_j = iq_j$. Evidently $\mathscr{Q}$ is spanned by $\{q_1, \ldots, q_n\}$ and $\mathscr{P}$ is spanned by $\{p_1, \ldots, p_n\}$. Any $x \in \mathscr{K}$ can then be written as $x = \sum_{j=1}^n \xi_j q_j + \sum_{j=1}^n \eta_j p_j$. Using the notation introduced above, put $\mathscr{J}: \mathscr{K} \to \mathscr{C}(\mathscr{K})$ by

$$\mathscr{J}(x) = \sum_{j=1}^n \xi_j Q_j + \sum_{j=1}^n \eta_j P_j .$$

Let $\{x_1, \ldots, x_{2n}\}$ be any orthonormal basis of $\mathscr{C}(\mathscr{K})$. Then the monomials

$$\mathscr{J}(x_{\alpha_1})\mathscr{J}(x_{\alpha_2}) \cdots \mathscr{J}(x_{\alpha_k})$$

together with I form a basis for the algebra. It is easy to see that the product of any two such monomials is a third. Though the multiplication rule can be simply expressed in terms of certain contraction rules, its precise form is not useful to us here. What is useful to observe is that since the right side of (2.8) is invariant under orthogonal transformations of $\mathscr{K}$, the multiplication law of these monomials does not depend on the choice of the orthonormal basis. For this reason, any orthogonal transformation R of $\mathscr{K}$ induces an automorphism of $\mathscr{C}(\mathscr{K})$ which we shall also denote by R. Indeed, with R: $\mathscr{C}(\mathscr{K}) \to \mathscr{C}(\mathscr{K})$ defined by

$$R(\mathscr{J}(x_{\alpha_1}) \ldots \mathscr{J}(x_{\alpha_k})) = \mathscr{J}(R(x_{\alpha_1})) \ldots \mathscr{J}(R(x_{\alpha_k})) ,$$

R is evidently invertible and by the remarks made just above one easily sees that for all $A, B \in \mathscr{C}(\mathscr{K})$, $R(AB) = R(A)R(B)$, which is to say that R is an automorphism.

Finally, we remark that $\mathscr{C}(\mathscr{K})$ is a $*$-algebra; there is a unique conjugate linear involutory antiautomorphism $A \mapsto A^*$ which is the identity on $\mathscr{J}(\mathscr{K})$. It is given by

$$R(\mathscr{J}(x_{\alpha_1}) \dots \mathscr{J}(x_{\alpha_k}))^* = \mathscr{J}(x_{\alpha_k}) \dots \mathscr{J}(x_{\alpha_1}) .$$

Of course, on $\mathscr{C}(\mathscr{K})$ regarded as a matrix algebra, this is just the usual adjoint. Evidently the automorphism R of $\mathscr{C}(\mathscr{K})$ induced by an orthogonal transformation R of $\mathscr{K}$ is a $*$-automorphism; i.e. $R(A^*) = (R(A))^*$.

The facts that orthogonal transformations of $\mathscr{K}$ induce automorphisms of $\mathscr{C}(\mathscr{K})$, that $\mathscr{C}(\mathscr{K})$ is a full matrix algebra, and that all automorphisms of full matrix algebras are inner; i.e. of the form $A \mapsto SAS^{-1}$ for some nonsingular matrix S, are the basis of Brauer and Weyl's treatment of the spin representations of the orthogonal groups [BrWe]. We will use the fact that all automorphisms of $\mathscr{C}(\mathscr{K})$ are inner several times in what follows.

III. Analysis on the Clifford Algebra

Let $\mathscr{A}$ be a von Neumann algebra of operators on some finite dimensional Hilbert space. By a trace on $\mathscr{A}$ we shall mean a linear functional τ which is positive in the sense that $\tau(A^*A) > 0$ for all non zero A in $\mathscr{A}$, and cyclic in the sense that $\tau(AB) = \tau(BA)$ for all A and B in $\mathscr{A}$. Such a functional is evidently unitarily invariant in the sense that whenever A and U belong to $\mathscr{A}$, and U is unitary, then $\tau(U^*AU) = \tau(A)$. Since $\mathscr{C}(\mathscr{K})$ is a full matrix algebra, it contains all unitaries. Hence any trace on $\mathscr{C}(\mathscr{K})$ must assign the same value to all rank one projections, and thus must be a scalar multiple of the standard trace Tr on the matrix algebra. Henceforth, τ shall denote this trace normalized by the condition that $\tau(I) = 1$ and Tr shall denote the standard unnormalized trace.

In the non-commutative integration theories of Dixmier [Di53] and Segal [Se53], the trace functional τ is the non-commutative analog of the functional that assigns to an integrable function its integral. When the Hilbert space is infinite dimensional, some further regularity properties are required of τ in order to obtain a useful analog. Since all of our estimations will be carried out in the finite dimensional setting, we shall not go into this here, but shall simply refer the reader to these original papers as well as the accounts in [Gr72] and [Ne74].

Norms on $\mathscr{C}(\mathscr{K})$ which are the non-commutative analogs of the L^p norms can now be introduced; namely for $1 \leqq p < \infty$ we put

$$\| A \|_p = (\tau((A^*A)^{p/2}))^{1/p} , \tag{3.1}$$

and denote the operator norm of A by $\| A \|_\infty$.

$\mathscr{C}^p(\mathscr{K})$ shall denote $\mathscr{C}(\mathscr{K})$ equipped with the norm $\|\cdot\|_p$; evidently $\mathscr{C}^2(\mathscr{K})$ is the Hilbert space of $2^n \times 2^n$ matrices equipped with the Hilbert–Schmidt norm.

Consider the monomials

$$E_{[\alpha_1, \dots, \alpha_j; \beta_1, \dots, \beta_k]} = Q_{\alpha_1} \dots Q_{\alpha_j} P_{\beta_1} \dots P_{\beta_k} , \tag{3.2}$$

where $\alpha_1 > \dots > \alpha_j$ and $\beta_1 > \dots > \beta_k$ and $j + k > 0$. Evidently

$$E^*_{[\alpha_1, \dots, \alpha_j; \beta_1, \dots, \beta_k]} E_{[\alpha_1, \dots, \alpha_j; \beta_1, \dots, \beta_k]} = I$$

and thus $\| E_{[\alpha_1, \ldots, \alpha_j; \beta_1, \ldots, \beta_k]} \|_p = 1$ for all p. Moreover

$$\tau(E_{[\alpha_1, \ldots, \alpha_j; \beta_1, \ldots, \beta_k]}) = 0 \,. \tag{3.3}$$

To see this, first consider the case in which $j + k$ is odd. The inversion $x \mapsto -x$ on $\mathscr{K}$ is orthogonal. Hence it induces an automorphism of $\mathscr{C}(\mathscr{K})$, and hence there is an invertible S in $\mathscr{C}(\mathscr{K})$ so that

$$-E_{[\alpha_1, \ldots, \alpha_j; \beta_1, \ldots, \beta_k]} = S E_{[\alpha_1, \ldots, \alpha_j; \beta_1, \ldots, \beta_k]} S^{-1} \,.$$

Then, by using cyclicity of the trace we get the desired result. Next consider the case in which $j + k$ is even and, say, $j > 0$. Then write $E_{[\alpha_1 \ldots, \alpha_j; \beta_1, \ldots, \beta_k]} = Q_{\alpha_1} X$, and note that by (2.1), $Q_{\alpha_1} X = -X Q_{\alpha_1}$. Again the desired conclusion follows from cyclicity of the trace.

It is easy to see from this that

$$\tau(E^*_{[\alpha_1, \ldots, \alpha_j; \beta_1, \ldots, \beta_k]} E_{[\gamma_1, \ldots, \gamma_m; \delta_1, \ldots, \delta_n]}) = 0 \tag{3.4}$$

unless the two monomials coincide. Thus, together with the identity, the assemblage of such monomials forms an orthonormal basis for $\mathscr{C}^2(\mathscr{K})$. Finally observe that since $Qe_- = e_+$,

$$\langle \Omega, E_{[\alpha_1, \ldots, \alpha_j]} \Omega \rangle = 0 \tag{3.5}$$

whenever $j \geqq 1$, and, as indicated, $k = 0$. It now follows that, restricted to $\mathscr{C}(\mathscr{Q})$,

$$\tau(A) = \langle \Omega, A\Omega \rangle \tag{3.6}$$

for all A in $\mathscr{C}(\mathscr{K})$.

Formula (3.6) is very important for us. It permits us to calculate the "physically" relevant quantity $\langle \Omega, A\Omega \rangle$ in terms of the apparently mathematically simpler quantity $\tau(A)$.

Many familiar inequalities for L^p norms hold for the $\mathscr{C}^p$ norms as well [Di53]. This is true in particular of the Hölder inequality

$$\| AB \|_r \leqq \| A \|_p \| B \|_q \qquad \frac{1}{r} = \frac{1}{p} + \frac{1}{q} \,.$$

Certain optimal inequalities expressing the uniform convexity properties of the L^p norms also hold for the $\mathscr{C}^p$ norms, and this fact constitutes one cornerstone of our analysis.

The modulus of convexity δ_p of $\mathscr{C}^p(\mathscr{K})$ is defined by

$$\delta_p(\varepsilon) = \inf \left\{ 1 - \frac{1}{2} \| A + B \|_p : \| A \|_p = \| B \|_p = 1, \; \| A - B \|_p = \varepsilon \right\} \tag{3.7}$$

for $0 < \varepsilon < 2$. For $1 < p < \infty$, δ_p is always positive which means these norms are uniformly convex. Useful geometric information is contained in the rate at which $\delta_p(\varepsilon)$ tends to zero with ε. It is known [TJ74] that for $2 \leqq p < \infty$, $\delta_p(\varepsilon) \sim \varepsilon^p$, but that for $1 < p \leqq 2$, $\delta_p(\varepsilon) \sim \varepsilon^2$.

An optimal expression of this fact is given by the following theorem which was proved jointly with Keith Ball [BCL]:

Theorem 1. (Optimal 2-uniform convexity for matrices). *For all $m \times m$ matrices A and B and all p for $1 \leqq p \leqq 2$,*

$$\left(\frac{\mathrm{Tr}|A+B|^p + \mathrm{Tr}|A-B|^p}{2}\right)^{2/p} \geqq (\mathrm{Tr}|A|^p)^{2/p} + (p-1)(\mathrm{Tr}|B|^p)^{2/p}. \quad (3.8)$$

For $1 < p < 2$, there is equality only when $B = 0$.

This result, which we interpret here as a statement about $\mathscr{C}^p(\mathscr{K})$, is proved in the appendix in the special case that both $A + B$ and $A - B$ are positive; this is the only case in which we shall use it here, and the proof is considerably simpler in this case. The full result is proved in [BCL], in which other geometric inequalities for trace norms are proved as well.

The theorem implies that

$$\delta_p(\varepsilon) \geqq \frac{(p-1)}{2}\left(\frac{\varepsilon}{2}\right)^2 \quad \text{for} \quad 1 < p \leqq 2 \quad (3.9)$$

as one sees by considering $A = (C + D)/2$, $B = (C - D)/2$, $\|C\|_p = \|D\|_p = 1$ and $\|C - D\|_p = \varepsilon$. It is easily seen that the constant $(p-1)/8$ cannot be improved.

We make our main application of this result in Sect. V. There we will also need to know that the norms on $\mathscr{C}^p(\mathscr{K})$ are continuously differentiable away from the origin for $1 < p < \infty$. This is known [Gr75], but a simple proof can be based on inequalities of the form $\delta_p(\varepsilon) \geqq K_p \varepsilon^{r(p)}$ such as we have found above for $1 < p < 2$. This proof, moreover, gives the modulus of continuity of the derivative, and is sketched in the appendix as well. Again, these estimates are independent of the dimension and therefore apply to the case of infinitely many degrees of freedom.

IV. Conditional Expectations and the Fermion Oscillator Semigroup

We are particularly concerned with the subalgebra $\mathscr{C}(\mathscr{Q})$ of $\mathscr{C}(\mathscr{K})$, and the conditional expectation [Di53, Um54] with respect to it shall play a basic role in our investigation. For any A in $\mathscr{C}(\mathscr{K})$, the **conditional expectation** $\pi_{\mathscr{Q}}(A)$ of A with respect to $\mathscr{C}(\mathscr{Q})$ is defined to be the unique element of $\mathscr{C}(\mathscr{Q})$ such that $\tau(B^*\pi_{\mathscr{Q}}(A)) = \tau(B^*A)$ for all B in $\mathscr{C}(\mathscr{Q})$. Otherwise said, $\pi_{\mathscr{Q}}$ is the orthogonal projection from $\mathscr{C}^2(\mathscr{K})$ onto $\mathscr{C}^2(\mathscr{Q})$. It is well known that the conditional expectation is positivity preserving; a familiar argument shows that $\pi_{\mathscr{Q}}(A^*A) \geqq \pi_{\mathscr{Q}}(A)^*\pi_{\mathscr{Q}}(A)$.

We can use the conditional expectation to give a useful expression for the oscillator semigroup for fermion fields.

Let R_θ be the orthogonal transformation of $\mathscr{K}$ given by

$$R_\theta(q_j) = (\cos\theta)q_j + (\sin\theta)p_j \quad (4.1)$$

for each j. Of course R_θ gives the evolution at time θ on phase space $\mathscr{K}$ generated by the classical oscillator Hamiltonian $H(\mathbf{p}, \mathbf{q}) = \sum_{j=1}^{n} \frac{1}{2}(p_j^2 + q_j^2)$. Let R_θ denote the automorphism of $\mathscr{C}(\mathscr{K})$ generated by the orthogonal transformation R_θ as in the first section. For each $t \geqq 0$, define $\theta(t) = \arccos(e^{-t})$ and define the operator P_t on $\mathscr{C}^2(\mathscr{K})$ by

$$P_t A = \pi_{\mathscr{Q}} \circ R_{\theta(t)} \circ \pi_{\mathscr{Q}}^* A. \quad (4.2)$$

Note that $\pi_{\mathscr{Q}}^*$ is the **natural imbedding** of $\mathscr{C}(\mathscr{Q})$ into $\mathscr{C}(\mathscr{K})$, and regarded as such, it is a $*$-automorphism. Formula (4.2) is the analog of the familiar expression for the boson oscillator semigroup on $L^2(\mathscr{Q}, (2\pi)^{-n/2} e^{-q^2/2} d^n q)$, i.e. the Mehler semigroup

$$P_t^{(\text{boson})} A(\mathbf{q}) = \int A(e^{-t}\mathbf{q} + (1 - e^{-2t})^{1/2}\mathbf{p})(2\pi)^{-n/2} e^{-p^2/2} d^n p .$$

Note that since all of the operators on the right in (4.2) are positivity preserving, so is P_t. Also, since the first two operations on the right preserve the $\mathscr{C}^p$-norms, and since the conditional expectation is readily seen to be a contraction from $\mathscr{C}^p(\mathscr{K})$ to $\mathscr{C}^p(\mathscr{Q})$ for each p, it is readily seen that P_t possesses this property as well.

To obtain a more familiar expression for P_t, note that

$$R_{\theta(t)} \circ \pi_{\mathscr{Q}}^*(E_{[\alpha_1, \ldots, \alpha_k]}) = e^{-kt} E_{[\alpha_1, \ldots, \alpha_k]} + (\text{terms annihilated by } \pi_{\mathscr{Q}}) .$$

Hence

$$P_t(E_{[\alpha_1, \ldots, \alpha_k]}) = e^{-kt} E_{[\alpha_1, \ldots, \alpha_k]} . \tag{4.3}$$

Evidently $\{P_t : t \geqq 0\}$ is generated by H_0 where $H_0(E_{[\alpha_1, \ldots, \alpha_k]}) = kE_{[\alpha_1, \ldots, \alpha_k]}$. It is easy to see that under the unitary equivalence between $\mathscr{C}^2(\mathscr{Q})$ and fermion Fock space $\mathscr{F}$ described by Segal [Se56], H_0 is equivalent to the usual number operator, or in other words, the oscillator Hamiltonian on $\mathscr{F}$.

Our primary goal is to prove optimal hypercontractivity bounds for P_t. That is, given $1 < p < q < \infty$ we want to show that for some finite t, P_t is a contraction from $\mathscr{C}^p(\mathscr{Q})$ to $\mathscr{C}^q(\mathscr{Q})$, and to find the smallest such t. Let

$$\| P_t \|_{p \to q} = \sup \{ \| P_t A \|_q : \| A \|_p = 1 \} . \tag{4.4}$$

As a first reduction, we shall show that the supremum on the right in (4.4) can be restricted to the positive operators A with $\| A \|_p = 1$. In the boson case this follows immediately from the fact that, in ordinary probability theory, the absolute value of a conditional expectation is no greater than the conditional expectation of the absolute value.

In general, matters concerning the absolute value in the non-commutative setting are more troublesome than in the commutative setting. An example is provided by the Araki–Yamagami inequality [ArYa] which, specialized to our context, asserts that the map $A \mapsto |A|$ is Lipschitz continuous on $\mathscr{C}^2(\mathscr{K})$ with constant $\sqrt{2}$ instead of the constant 1 which we would have in the commutative setting.

Thus while the conditional expectation in an operator algebra has many properties analogous to those of the conditional expectation in ordinary probability theory [Um54], it is not in general true that $|\pi_{\mathscr{Q}}(A)|$ will be a smaller operator than $\pi_{\mathscr{Q}}(|A|)$. The following theorem expresses a useful property in this direction which does hold, and after proving it we shall show by example that stronger properties do not hold. The theorem and its proof are easily extended to a more general von Neumann algebra setting by the methods in [Ru72].

Theorem 2. (A Schwarz inequality for conditional expectations). *For all A in $\mathscr{C}(\mathscr{K})$ and all p with $1 \leqq p \leqq \infty$,*

$$\| \pi_{\mathscr{Q}}(A) \|_p \leqq \| \pi_{\mathscr{Q}}(|A|) \|_p^{1/2} \, \| \pi_{\mathscr{Q}}(|A^*|) \|_p^{1/2} . \tag{4.5}$$

Remark. If we let $F(A)$ denote $\|\pi_{\mathcal{Q}} A\|_p$, then the same argument which we shall use to prove Theorem 2 also establishes that

$$F(A^*B) \leqq F(A^*A)^{1/2} F(B^*B)^{1/2}. \tag{4.6}$$

In this form, the term "Schwarz inequality," by which we referred to (4.5), is more evidently appropriate. Moreover, inequalities of the type (4.6) are well known in matrix analysis for many familiar functions; for example when $F(A)$ is the determinant of A or the spectral radius of A. Further examples can be found in [MeDS].

In [Li76] it is shown that for a function F that satisfies (4.6), and which is monotone increasing; i.e. satisfies $F(B) \geqq F(A)$ for all $B \geqq A \geqq 0$, the following inequalities hold:

$$F\left(\sum_{j=1}^m A_j^* B_j\right) \leqq F\left(\sum_{j=1}^m A_j^* A_j\right)^{1/2} F\left(\sum_{j=1}^m B_j^* B_j\right)^{1/2}$$

and

$$F\left(\sum_{j=1}^m A_j\right) \leqq F\left(\sum_{j=1}^m |A_j|\right)^{1/2} F\left(\sum_{j=1}^m |A_j^*|\right)^{1/2}.$$

In particular, these inequalities hold for $F(A) = \|\pi_{\mathcal{Q}} A\|_p$.

Specializing the last inequality to the case $m = 1$ then yields (4.5). In our present case however, the proof of the (4.6) is essentially the same as the direct proof of (4.5). Nonetheless, it should not be considered novel that by taking $|A^*|$ into consideration as well as $|A|$, we can obtain a suitable bound on $\|\pi_{\mathcal{Q}} A\|_p$.

Proof. Let $A = U|A|$ be the polar decomposition of A. Then $\|\pi_{\mathcal{Q}}(A)\|_p = \tau(CU|A|)$ for some C in $\mathscr{C}(\mathcal{Q})$ with $\|C\|_{p'} = 1$. Let $C = V|C|$ be the polar decomposition of C. Both V and $|C|$ belong to $\mathscr{C}(\mathcal{Q})$ as well. Thus

$$\begin{aligned}\|\pi_{\mathcal{Q}}(A)\|_p &= \tau(CU|A|^{1/2}|A|^{1/2}) = \tau(|C|^{1/2}U|A|^{1/2}|A|^{1/2}V|C|^{1/2}) \\ &\leqq \tau(|C|^{1/2}U|A|U^*|C|^{1/2})^{1/2}\,\tau(|C|^{1/2}V^*|A|V|C|^{1/2})^{1/2} \\ &= \tau(|C|(U|A|U^*))^{1/2}\,\tau((V|C|V^*)|A|)^{1/2} \leqq \|\pi_{\mathcal{Q}}(U|A|U^*)\|_p^{1/2}\,\|\pi_{\mathcal{Q}}(|A|)\|_p^{1/2}.\end{aligned}$$

Finally, we note that $U|A|U^* = |A^*|$. □

Example. Let A be the matrix $A = \begin{bmatrix} 0 & 1 \\ 0 & 1 \end{bmatrix}$. Then

$$|A| = \sqrt{2}\begin{bmatrix} 0 & 0 \\ 0 & 1 \end{bmatrix} \quad \text{and} \quad |A^*| = \frac{1}{\sqrt{2}}\begin{bmatrix} 1 & 1 \\ 1 & 1 \end{bmatrix}.$$

Note that $|A^*|$ is in $\mathscr{C}(\mathcal{Q})$, but A and $|A|$ are not. One easily finds

$$\pi_{\mathcal{Q}}(A) = \frac{1}{2}\begin{bmatrix} 1 & 1 \\ 1 & 1 \end{bmatrix}, \quad \pi_{\mathcal{Q}}(|A|) = \frac{1}{\sqrt{2}}\begin{bmatrix} 1 & 0 \\ 0 & 1 \end{bmatrix}, \quad \pi_{\mathcal{Q}}(|A^*|) = \frac{1}{\sqrt{2}}\begin{bmatrix} 1 & 1 \\ 1 & 1 \end{bmatrix}.$$

$$\|\pi_{\mathcal{Q}}(A)\|_\infty = 1, \quad \|\pi_{\mathcal{Q}}(|A|)\|_\infty = \frac{1}{\sqrt{2}} \quad \text{and} \quad \|\pi_{\mathcal{Q}}(|A^*|)\|_\infty = \sqrt{2}.$$

Theorem 3. **(P_t has positive maximizers).** *The norm of P_t from $\mathscr{C}^p(\mathscr{Q})$ to $\mathscr{C}^q(\mathscr{Q})$ is achieved on the positive operators*; i.e.

$$\| P_t \|_{p \to q} = \sup \{ \| P_t A \|_q : A \geqq 0 \text{ and } \| A \|_p = 1 \}. \tag{4.7}$$

Proof. Since R_θ and $\pi_{\mathscr{Q}}^*$ are both $*$-automorphisms, $|R_{\theta(t)} \circ \pi_{\mathscr{Q}}^* A| = R_{\theta(t)} \circ \pi_{\mathscr{Q}}^* |A|$ and $|(R_{\theta(t)} \circ \pi_{\mathscr{Q}}^* A)^*| = R_{\theta(t)} \circ \pi_{\mathscr{Q}}^* |A^*|$. Thus $\| P_t A \|_q \leqq \| P_t |A| \|_q^{1/2} \| P_t |A^*| \|_q^{1/2}$, and of course $\| |A^*| \|_p = \| |A| \|_p = \| A \|_p$. □

V. Hypercontractivity for Fermions

Our main result is the following theorem which is established in this section.

Theorem 4. **(Optimal fermion hypercontractivity).** *For all* $1 < p \leqq q < \infty$, $\| P_t \|_{p \to q} = 1$ *exactly when*

$$e^{-2t} \leqq \frac{p-1}{q-1}.$$

The heart of the matter is the following lemma:

Lemma. *For all* $1 < p \leqq 2$, $\| P_t \|_{p \to 2} = 1$ *exactly when* $e^{-2t} \leqq (p-1)$.

Proof. Fix a positive element A of $\mathscr{C}(\mathscr{Q})$. Pick a basis $\{q_1, \ldots, q_n\}$ of $\mathscr{Q}$, and let $\mathscr{C}(\mathscr{Q}_{(n-1)})$ denote the Clifford algebra associated with the span of the first $n-1$ of these basis elements. It is evident from the form of our standard basis of $\mathscr{C}(\mathscr{Q})$ that A can be uniquely decomposed as $A = B + CQ_n$, where B and C belong to $\mathscr{C}(\mathscr{Q}_{(n-1)})$. Then using the Jordan–Wigner transform we can write $A = B + CV_n\hat{Q}_n$. Now write

$$\mathscr{H} = \mathscr{H}_{(n-1)} \otimes \mathbb{C}^2 \tag{5.1}$$

so that B, C and $V_n C$ can be considered as operators on the first factor $\mathscr{H}_{(n-1)}$. Let $u_\pm = (e_+ \pm e_-)/\sqrt{2}$, so that $Qu_\pm = \pm u_\pm$. Then if v is any vector in $\mathscr{H}_{(n-1)}$,

$$\langle (v \otimes u_\pm), A(v \otimes u_\pm) \rangle_{\mathscr{H}} = \langle v, (B \pm CV_n) v \rangle_{\mathscr{H}_{(n-1)}}. \tag{5.2}$$

We see from this that since $A \geqq 0$, so are both $B + CV_n \geqq 0$ and $B - CV_n \geqq 0$. Now let Tr_1 and Tr_2 denote the partial traces over the first and second factors in (5.1), so that with Tr still denoting the full trace, we have $\mathrm{Tr} = \mathrm{Tr}_1 \mathrm{Tr}_2$. Now applying Theorem 1 in the special case which is proved in the appendix

$$\begin{aligned}
\| A \|_p^2 &= \left(\frac{1}{2^n} \mathrm{Tr} |B + CV_n \hat{Q}_n|^p \right)^{2/p} \\
&= \left(\frac{1}{2^{(n-1)}} \right)^{2/p} \left(\frac{\mathrm{Tr}_1 |B + CV_n|^p + \mathrm{Tr}_1 |B - CV_n|^p}{2} \right)^{2/p} \\
&\geqq \left(\frac{1}{2^{(n-1)}} \right)^{2/p} ((\mathrm{Tr}_1 |B|^p)^{2/p} + (p-1)(\mathrm{Tr}_1 |C|^p)^{2/p})
\end{aligned}$$

since V_n is unitary.

Thus $\| A \|_p^2 \geqq \| B \|_p^2 + (p-1) \| C \|_p^2$, where the norms on the right are all norms on $\mathscr{C}(\mathscr{Q}_{(n-1)})$.

Now we make the inductive assumption that the lemma has been established for $\mathscr{C}(\mathscr{Q}_{(n-1)})$. This is clearly the case when $n = 1$. We then have

$$\|A\|_p^2 \geqq \|P_t B\|_2^2 + (p-1)\|P_t C\|_2^2 ,$$

where $e^{-2t} = (p-1)$. But clearly from (4.3) $\|P_t C Q_n\|_2^2 = (p-1)\|P_t C\|_2^2$, where the norms are once again norms on $\mathscr{C}(\mathscr{Q})$. Moreover by (3.4) $P_t B$ and $P_t C Q_n$ are orthogonal. Thus $\|P_t B\|_2^2 + (p-1)\|P_t C\|_2^2 = \|P_t(B + CQ_n)\|_2^2 = \|P_t A\|_2^2$. □

Theorem 5. (Optimal fermion logarithmic Sobolev inequality). *For all $A \in \mathscr{C}(\mathscr{Q})$,*

$$\tau(|A|^2 \ln |A|^2) - (\|A\|_2^2 \ln \|A\|_2^2) \leqq 2\langle A, H_0 A\rangle . \tag{5.3}$$

Proof. By the lemma, $\|P_t A\|_2^2 \leqq \|A\|_{(1+e^{-2t})}^2$ and there is equality at $t = 0$. Both sides are continuously differentiable, and comparing derivatives at $t = 0$ we obtain the result. Indeed,

$$\frac{d}{dp}\|A\|_p = \frac{1}{p}\|A\|_p^{1-p}(\tau(|A|^p \ln |A|) - \|A\|_p^p \ln \|A\|_p) \tag{5.4}$$

and of course

$$\frac{d}{dt}\|P_t A\|_2^2 = \frac{d}{dt}\langle A, P_{2t} A\rangle = -2\langle A, H_0 A\rangle . \quad \square$$

Gross refers to the quadratic form on the right side of (5.3) as the Clifford Dirichlet form since it shares many properties of Dirichlet forms in the ordinary commutative setting. An approach to the development of a theory of Dirichlet forms in the non-commutative setting can be found in [AlHK].

Proof of Theorem 4. By a deep result of Gross, when $A \geqq 0$ and $1 < p < \infty$,

$$\langle A^{p/2}, H_0 A^{p/2}\rangle \leqq \frac{(p/2)^2}{p-1}\langle A, H_0 A^{p-1}\rangle .$$

Replacing A in (5.3) by $A^{p/2}$ and using the inequality just quoted we obtain, following Gross's ideas [Gr75],

$$\tau(A^p \ln A) - \|A\|_p^p \ln \|A\|_p \leqq \frac{p/2}{p-1}\langle A^{p-1}, H_0 A\rangle .$$

By combining this with (5.4) a differential inequality is obtained which implies that $\|P_t A\|_{q(t)}$ is a decreasing function of t when $q(t) = 1 + e^{2t}(p-1)$. This establishes the result for $A \geqq 0$, and by Theorem 3 it is established in general. By (4.3), $P_t I = I$, and therefore $\|P_t\|_{p \to q}$ is always at least 1 for all p and q. That the inequality is best possible follows from a direct computation with one degree of freedom. To be precise, $\|P_t(I + Q_1)\|_q = \|I + e^{-t}Q_1\|_q$ is easily computed and compared with $\|I + Q_1\|_p$ [Gr72]. The first quantity is greater than the second if $e^{-2t} > (p-1)/(q-1)$. □

VI. Hypercontractivity for Bosons and Fermions Together

As a result of the present work and of earlier work on bosons, we know that for t given by $e^{-2t} = (p-1)/(q-1)$, both the fermion and the boson oscillator semigroups are contractive from the appropriate p-spaces to the appropriate q-spaces, and that this value of t is optimal for each case *separately*.

It is natural to expect that the same condition governs hypercontractivity in a situation in which we have bosons and fermions *together*. This is indeed the case, as we now show using Minkowski's inequality in an argument based on Segal's method for showing that the optimal conditions for hypercontractivity with m boson degrees of freedom are the same as for one degree of freedom.

Let $\mu(dx) = (2\pi)^{-m/2} e^{-x^2/2}\, dx$ be the unit Gauss measure on $\mathbb{R}^m$. Then in our mixed setting, with m boson degrees of freedom and n fermion degrees of freedom, the relevant p-space is

$$\mathscr{B}^p = L^p(\mathbb{R}^m, \mu) \otimes \mathscr{C}^p(\mathscr{Q}_{(n)}) ,$$

which may be regarded as consisting of $\mathscr{C}^p(\mathscr{Q}_{(n)})$ valued measurable functions $x \mapsto A(x)$ such that

$$|||A|||_p^p = \int_{\mathbb{R}^m} \|A(x)\|_p^p \mu(dx)$$

is finite. This equation defines the norm on $\mathscr{B}^p$. For $p=2$, $\mathscr{B}^p$ is naturally isomorphic to the tensor product of the symmetric tensor algebra over $\mathbb{C}^m$ and the antisymmetric tensor algebra over $\mathbb{C}^n$ as shown by Segal. On the latter space we have the mixed oscillator semigroup generated by the sum of the boson and fermion number operators

$$\exp\left\{-t\left[\sum_{j=1}^{m} a_j^* a_j + \sum_{j=1}^{n} c_j^* c_j\right]\right\} .$$

Considered as operators on $\mathscr{B}^p$, the operators $\mathscr{P}_t$ which constitute this semigroup are given by

$$\mathscr{P}_t A(x) = \int_{R^m} M_t(x, x') P_t[A(x')]\, \mu(dx') , \tag{6.1}$$

where $M_t(x, x')$ is the Mehler kernel; i.e., the positive integral kernel for the boson oscillator semigroup $P_t^{(\text{boson})}$ discussed in Sect. IV. Of course P_t denotes the fermion oscillator semigroup studied throughout this paper.

Now successively applying Minkowski's inequality, our theorem on optimal fermion hypercontractivity, and Nelson's theorem on optimal boson hypercontractivity, we have for $e^{-2t} \leqq (p-1)/(q-1)$:

$$\begin{aligned} |||\mathscr{P}_t A|||_q^q &= \int_{\mathbb{R}^m} \left\| \int_{\mathbb{R}^m} M_t(x,x') P_t[A(x')]\, \mu(dx') \right\|_q^q \mu(dx) \\ &\leqq \int_{\mathbb{R}^m} \left(\int_{\mathbb{R}^m} M_t(x,x') \|P_t A(x')\|_q \mu(dx') \right)^q \mu(dx) \\ &\leqq \int_{\mathbb{R}^m} \left(\int_{\mathbb{R}^m} M_t(x,x') \|A(x')\|_p \mu(dx') \right)^q \mu(dx) \\ &\leqq \left(\int_{\mathbb{R}^m} \|A(x)\|_p^p \mu(dx) \right)^{q/p} = |||A|||_p^q . \end{aligned}$$

Appendix

Proof of Theorem 1 when $A \pm B \geqq 0$. Let Z and W be the $2m \times 2m$ matrices given by

$$Z = \begin{bmatrix} A & 0 \\ 0 & A \end{bmatrix}, \quad W = \begin{bmatrix} B & 0 \\ 0 & -B \end{bmatrix}.$$

Our goal is to establish that for all r with $0 \leqq r \leqq 1$,

$$\left(\frac{\mathrm{Tr}(A + rB)^p + \mathrm{Tr}(A - rB)^p}{2}\right)^{2/p} \geqq (\mathrm{Tr}(A)^p)^{2/p} + r^2(p-1)(\mathrm{Tr}|B|^p)^{2/p},$$

or what is the same,

$$\mathrm{Tr}(Z + rW)^{2/p} \geqq (\mathrm{Tr}(Z)^p)^{2/p} + r^2(p-1)(\mathrm{Tr}|W|^p)^{2/p}. \tag{A.1}$$

First, note that the null space of $Z + rW$ is exactly the null space of Z for $0 \leqq r < 1$. Thus by carrying out all of the following computations on the orthogonal complement of this fixed null space, we may freely assume that $Z + rW > 0$ for all $0 \leqq r < 1$. Next, both sides of (A.1) agree at $r = 0$, and the first derivatives in r of both sides vanish there as well. We define $\psi(r)$ to be $\mathrm{Tr}(Z + rW)^p$. Then the second derivative in r of the left side of (A.1) satisfies

$$\frac{d^2}{dr^2}(\psi(r))^{2/p} \geqq \frac{2}{p}\psi(r)^{(2-p)/p}\frac{d^2}{dr^2}\psi(r).$$

The second derivative on the right side is just

$$2(p-1)(\mathrm{Tr}|W|^p)^{2/p},$$

and we are left with showing that

$$\frac{1}{p}\psi(r)^{(2-p)/p}\frac{d^2}{dr^2}\psi(r) \geqq (p-1)\,(\mathrm{Tr}|W|^p)^{2/p} \tag{A.2}$$

for all $0 < r < 1$. By redefining Z to be $Z + rW$, it suffices to establish (A.2) at $r = 0$.

Now $\frac{d}{dr}\psi(r) = p(\mathrm{Tr}(Z + rW)^{(p-1)}W)$, since $A \pm B \geqq 0$, $Z + rW \geqq 0$ for small r, and we can use the integral representation

$$(Z + rW)^{(p-1)} = c_p \int_0^\infty t^{(p-1)}\left[\frac{1}{t} - \frac{1}{t + (Z + rW)}\right]dt$$

to conclude that

$$\frac{d^2}{dr^2}\psi(0) = pc_p \int_0^\infty t^{(p-1)}\,\mathrm{Tr}\left[\frac{1}{t+Z}W\frac{1}{t+Z}W\right]dt. \tag{A.3}$$

Consider the right side as a function, $f(Z)$, of Z for fixed W. It is easy to see that f is convex in Z. (Simply replace Z by $Z + tX$, with X self-adjoint, and then differentiate twice with respect to t; the positivity follows from the Schwarz inequality for traces.) Also, $f(UZU^*) = f(Z)$ provided U is unitary and U commutes with W. In a basis in which W is diagonal, we form the set $\mathscr{U}$ consisting of the

2^{2m} distinct diagonal unitary matrices, each with $+1$ or -1 in each diagonal entry. Each of these clearly commutes with W. Then

$$f(Z) = 2^{-2m} \sum_{U \in \mathscr{U}} f(UZU^*) \geqq f\left(2^{-2m} \sum_{U \in \mathscr{U}} UZU^*\right) = f(Z_{\text{diag}}),$$

where Z_{diag} is the matrix that is diagonal in the basis diagonalizing W, and whose diagonal entries are those of Z in this basis.

Replacing Z by Z_{diag} in (A.3), the integration can be carried out, and we obtain

$$\frac{d^2}{dr^2}\psi(0) \geqq p(p-1)\left(\sum_{j=1}^{2m} z_j^{(p-2)} w_j^2\right),$$

where z_j and w_j, respectively, denote the jth diagonal entries of Z and W in a basis diagonalizing W.

Now consider $\psi(0) = \operatorname{Tr}(Z^p)$ as a function of Z. It is clearly convex, and thus by the averaging method just employed, we obtain

$$\psi(0) \geqq \left(\sum_{j=1}^{2m} z_j^p\right).$$

To establish (A.2), we are only left with showing that

$$\left(\sum_{j=1}^{2m} z_j^p\right)^{(2-p)/p} \left(\sum_{j=1}^{2m} z_j^{(p-2)} w_j^2\right) \geqq \left(\sum_{j=1}^{2m} |w_j|^p\right)^{2/p}, \tag{A.4}$$

but this follows immediately from Hölder's inequality.

To complete the proof, observe that equality in (A.1) for $r = 1$ and $1 < p < 2$ implies equality in (A.4) for almost every r in $[0, 1]$. Here, recall that z_j in (A.4) really denotes the jth diagonal element of $Z + rW$; these are the numbers $z_j + rw_j$, where z_j denotes the jth diagonal element of Z.

Let us assume that $w_j \neq 0$ for some j. Then equality in Hölder's inequality (A.4) requires that the vector with positive components $z_j + rw_j$ be proportional to the vector with components $|w_j|$. Thus, for almost every r in $[0, 1]$ we require

$$z_j + rw_j = c(r)|w_j|$$

for some number $c(r)$ that depends on r but not on j. The left side above is a linear function, and thus $c(r) = a + rb$ for some numbers a and b. But then clearly $b = w_j/|w_j|$, and all non-zero eigenvalues of W would necessarily have the same sign. This is impossible since $\operatorname{Tr} W = 0$. □

We now give an application of the uniform convexity implied by this theorem to the differentiability of the $\mathscr{C}^p(\mathscr{H})$ norms. First we recall that for $2 \leqq p < \infty$, the modulus of convexity is given by an analog of an inequality of Clarkson for integrals which Dixmier [Di53] established for traces. Specializing to $\mathscr{C}^p(\mathscr{H})$, this inequality reads

$$\left\|\frac{A+B}{2}\right\|_p^p + \left\|\frac{A-B}{2}\right\|_p^p \leqq \frac{1}{2}[\|A\|_p^p + \|B\|_p^p] \quad 2 \leqq p < \infty,$$

which implies that in this range (cf. (3.7))

$$\delta_p(\varepsilon) \geqq \frac{1}{p}\left(\frac{\varepsilon}{2}\right)^p. \tag{A.5}$$

For any non-zero A in $\mathscr{C}^p(\mathscr{K})$, define $\mathscr{D}(A)$ by

$$\mathscr{D}(A) = \|A\|_p^{(1-p)}|A|^{(p-1)}U^*,$$

where $A = U|A|$ is the polar decomposition of A. Let p' be defined by $1/p + 1/p' = 1$. Then for $1 < p < \infty$, $\|\mathscr{D}(A)\|_{p'} = 1$ and $\tau(\mathscr{D}(A)A) = \|A\|_p$. Moreover, $\mathscr{D}(A)$ is the unique element of $\mathscr{C}^{p'}(\mathscr{K})$ with this property. We call the map $A \mapsto \mathscr{D}(A)$ the **gradient map** on $\mathscr{C}^p(\mathscr{K})$. The next theorem sharpens a result of Gross [Gr75].

Theorem 6. (Hölder continuity of the gradient map on $\mathscr{C}^p(\mathscr{K})$). *For all $1 < p < \infty$, the gradient map is norm continuous. Moreover,*

$$\|\mathscr{D}(A) - \mathscr{D}(B)\|_{p'} \leqq 2\left(p'\frac{\|A-B\|_p}{\|A+B\|_p}\right)^{(p-1)} \quad \textit{for } 1 < p \leqq 2, \tag{A.6}$$

and

$$\|\mathscr{D}(A) - \mathscr{D}(B)\|_{p'} \leqq 4(p-1)\left(\frac{\|A-B\|_p}{\|A+B\|_p}\right) \quad \textit{for } 2 \leqq p < \infty. \tag{A.7}$$

Proof. First observe that for all $1 \leqq p \leqq \infty$,

$$\begin{aligned}\|\mathscr{D}(A) + \mathscr{D}(B)\|_{p'}\|A+B\|_p &\geqq \Re\tau((\mathscr{D}(A) + \mathscr{D}(B))(A+B))\\ &= 2(\|A\|_p + \|B\|_p) - \Re\tau((\mathscr{D}(A) - \mathscr{D}(B))(A-B))\\ &\geqq 2(\|A+B\|_p) - \|\mathscr{D}(A) - \mathscr{D}(B)\|_{p'}\|A-B\|_p.\end{aligned}$$

Thus,

$$1 - \left\|\frac{\mathscr{D}(A) + \mathscr{D}(B)}{2}\right\|_{p'} \leqq \left\|\frac{\mathscr{D}(A) - \mathscr{D}(B)}{2}\right\|_{p'}\left(\frac{\|A-B\|_p}{\|A+B\|_p}\right). \tag{A.8}$$

But, by (A.5) and (3.7), we have

$$1 - \left\|\frac{\mathscr{D}(A) + \mathscr{D}(B)}{2}\right\|_{p'} \geqq \frac{1}{p'}\left\|\frac{\mathscr{D}(A) - \mathscr{D}(B)}{2}\right\|_{p'}^{p'} \tag{A.9}$$

for $1 < p \leqq 2$. By combining (A.9) and (A.8) we obtain (A.6). Similarly, by combining (3.9) with (A.8) we obtain (A.7). □

The continuity of the gradient map for $\mathscr{C}^p(\mathscr{K})$ has been established by Gross [Gr72], but his proof is more involved and does not yield an estimate of the modulus of continuity. It is now easy to establish continuous differentiability of the $\mathscr{C}^p(\mathscr{K})$ norms away from the origin since, with $h(t) = \|A + tB\|_p$, with A different from 0 and with t and s sufficiently small, we have

$$\Re\tau(\mathscr{D}(A+tB)B) \leqq \frac{h(t+s) - h(t)}{s} \leqq \Re\tau(\mathscr{D}(A+(t+s)B)B) \tag{A.10}$$

when s is positive. To see this, observe that $h(t) = \Re\tau(\mathscr{D}(A+tB)(A+tB))$, and that $h(t+s) \geqq \Re\tau(\mathscr{D}(A+tB)(A+(t+s)B))$ by Hölder's inequality. By subtracting the expression for $h(t)$ from the estimate for $h(t+s)$ and dividing by s, we obtain the inequality on the left in (A.10). The inequality on the right is obtained in an analogous manner. When s is negative, the inequalities are clearly reversed. Letting s tend to zero, we obtain

$$\frac{d}{dt} \|A + tB\|_p \bigg|_{t=0} = \Re\tau(\mathscr{D}(A)B) \, .$$

References

[AlHK] Albeverio, S., Høegh-Krohn, R.: Dirichlet forms and Markov semigroups on C^*-algebras. Commun. Math. Phys. **56**, 173–187 (1977)

[ArYa] Araki, H., Yamagami, S.: An inequality for the Hilbert–Schmidt norm, Commun. Math. Phys. **81**, 89–96 (1981)

[BCL] Ball, K., Carlen, E.A., Lieb, E. H.: preprint 1992

[BrWe] Brauer, R., Weyl, H.: Spinors in n dimensions. Am. J. Math. **57**, 425–449 (1935)

[CL91] Carlen, E.A., Loss, M.: Extremals of functionals with competing symmetries. J. Func. Anal. **88**, 437–456 (1991)

[Da76] Davies, E.B.: Quantum Theory of Open Systems. New York: Academic Press, 1976

[Di53] Dixmier, J.: Formes linéaires sur un anneau d'opérateurs, Bull. Soc. Math. France **81**, 222–245 (1953)

[Far] Faris, W.: Product spaces and Nelson's inequality, Helv. Phys. Acta **48**, 721–730 (1975)

[Fe69] Federbush, P.: A partially alternate derivation of a result of Nelson. J. Math. Phys. **10**, 50–52 (1969)

[Gr72] Gross, L.: Existence an uniqueness of physical ground states. J. Funct. Anal. **10**, 52–109 (1972)

[Gr75] Gross, L.: Hypercontractivity and logarithmic Sobolev inequalities for the Clifford–Dirichlet form. Duke Math. J. **43**, 383–396 (1975)

[Gr89] Gross, L.: Logarithmic Sobolev inequalities for the heat kernel on a Lie group and a bibliography on logarithmic Sobolev inequalities and hypercontractivity. In: White Noise Analysis, Mathematics and Applications. Hida et al. (eds.) Singapore: World Scientific, 1990, pp. 108–130

[Gr92] Gross, L.: Logarithmic Sobolev inequalities and contractivity properties of semi-groups. 1992 Varenna summer school lecture notes (preprint)

[HuPa] Hudson, R., Parthasarathy, K.R.: Quantum Itô's formula and stochastic evolutions. Commun. Math. Phys. **93**, 301–323 (1984)

[JoKl] Jordan, P., Klein, O.: Zum Mehrkörperproblem der Quantentheorie. Zeits. für Phys. **45**, 751–765 (1927)

[JoWi] Jordan, P., Wigner, E.P.: Über das Paulische Äquivalenzverbot. Zeits. für Phys. **47**, 631–651 (1928)

[Li76] Lieb, E.H.: Inequalities for some operator and matrix functions. Adv. Math. **20**, 174–178 (1976)

[Li90] Lieb, E.H.: Gaussian kernels have only Gaussian maximizers. Invent. Math. **102**, 179–208 (1990)

[Lin] Lindsay, M.: Gaussian hypercontractivity revisited. J. Funct. Anal. **92**, 313–324 (1990)

[LiMe] Lindsay, M., Meyer, P.A.: preprint, 1991

[MeDS] Merris, R., Dias da Silva, J.A.: Generalized Schur functions. J. Lin. Algebra **35**, 442–448 (1975)

[Me85] Meyer, P.A.: Eléments de probabilités quantiques, exposés I–V. In: Sem. de Prob. XX, Lecture notes in Math. **1204**, New York: Springer, 1985 pp. 186–312

[Me86] Meyer, P.A.: Eléments de probabilités quantiques, exposés VI–VIII. In: Sem. de Prob. XXI, Lecture notes in Math. **1247**, New York: Springer, 1986 pp. 27–80

[Ne66] Nelson, E.: A quartic interaction in two dimensions. In: Mathematical Theory of Elementary Particles, R. Goodman, I. Segal (eds.) Cambridge, MA MIT Press, 1966

[Ne73] Nelson, E.: The free Markov field. J. Funct. Anal. **12**, 211–227 (1973)

[Ne74] Nelson, E.: Notes on non-commutative integration. J. Funct. Anal. **15**, 103–116 (1974)

[Nev] Neveu, J.: Sur l'esperance conditionelle par rapport à un mouvement Brownien. Ann. Inst. H. Poincaré Sect. B. (N.S.) **12**, 105–109 (1976)

[Ru72] Ruskai, M.B.: Inequalities for traces on Von Neumann algebras. Commun. Math. Phys. **26**, 280–289 (1972)

[SML] Schultz, T.D., Mattis, D.C., Lieb, E. H.: Two dimensional Ising model as a soluble problem of many fermions. Rev. Mod. Phys. **36**, 856–871 (1964)

[Se53] Segal, I.E.: A non-commutative extension of abstract integration. Ann. Math. **57**, 401–457 (1953)

[Se56] Segal, I.E.: Tensor algebras over Hilbert spaces II. Ann. Math. **63**, 160–175 (1956)

[Se70] Segal, I.E.: Construction of non-linear local quantum processes: I. Ann. Math. **92**, 462–481 (1970)

[TJ74] Tomczak-Jaegermann, N.: The moduli of smoothness and convexity and Rademacher averages of trace classes $S_p(1 \leqq p < \infty)$. Studia Mathematica **50**, 163–182 (1974)

[Um54] Umegaki, H.: Conditional expectation in operator algebras I. Tohoku Math. J. **6**, 177–181 (1954)

Communicated by A. Jaffe

Invent math. 115, 463–482 (1994)

Inventiones mathematicae

Sharp uniform convexity and smoothness inequalities for trace norms

Keith Ball[1,*], Eric A. Carlen[2,], and Elliott H. Lieb[3,***]**

[1] Department of Mathematics, Texas A&M University, College Station, TX, 77843, USA
[2] School of Mathematics, Georgia Institute of Technology, Atlanta, GA, 30332, USA
[3] Departments of Mathematics and Physics, Princeton University, P.O. Box 708, Princeton, NJ, 08544, USA

Oblatum 7-VII-1993

Summary. We prove several sharp inequalities specifying the uniform convexity and uniform smoothness properties of the Schatten trace ideals C_p, which are the analogs of the Lebesgue spaces L_p in non-commutative integration. The inequalities are all precise analogs of results which had been known in L_p, but were only known in C_p for special values of p. In the course of our treatment of uniform convexity and smoothness inequalities for C_p we obtain new and simple proofs of the known inequalities for L_p.

I Introduction

The concepts of uniform convexity and its dual property, uniform smoothness, play an important role in analysis. After reviewing these concepts in the L_p function spaces, we shall consider their extension to the Schatten trace ideals, C_p, i.e., the setting in which functions are replaced by operators, and integrals are replaced by traces. The emphasis throughout will be on the optimal constants appearing in the various inequalities. These optimal constants are "natural", as will be explained later in the introduction: they are the constants one would obtain from an informed guess using elementary calculus. However, as is often the case in such matters, no ready-made arguments suffice to validate the informed guesses.

A normed space X is said to be **uniformly convex** if, for each $\varepsilon > 0$, there is a $\delta > 0$ such that if x and y are *unit* vectors in X with $\|x - y\| \geqq 2\varepsilon$, then the average $(x + y)/2$ has norm at most $1 - \delta$. A normed space X is said to be **uniformly smooth** if, for all $\varepsilon > 0$, there is a $\tau > 0$ such that if x and y are *unit* vectors in X with $\|x - y\| \leqq 2\tau$, then the average $(x + y)/2$ has norm at least $1 - \varepsilon\tau$.

---

* Work partially supported by US National Science Foundation grant DMS 88-07243
** Work partially supported by US National Science Foundation grant DMS 92-07703
*** Work partially supported by US National Science Foundation grant PHY90-19433 A02

Figuratively speaking, the unit ball of a uniformly convex space is uniformly free of "flat spots", and the unit ball of a uniformly smooth space is uniformly free of "corners". Since the unit ball of X^*, the dual of X, is the polar conjugate of the unit ball of X, it is not difficult to show that X is uniformly convex (and hence reflexive) if and only if X^* is uniformly smooth [D].

Many applications of uniform convexity and smoothness require quantitative versions of these notions. The function δ_X given by

$$\delta_X(\varepsilon) := \inf\{1 - \tfrac{1}{2}\|x + y\| : \|x\| = \|y\| = 1,\ \|x - y\| \geqq 2\varepsilon\} \tag{1.1}$$

is called the **modulus of convexity** of X. (N.B. The function δ_X is frequently defined with ε in place of 2ε. The definition used here simplifies several of the formulae involving δ_X and fits more naturally with the definition of the modulus of smoothness given below.) Clearly, X is uniformly convex if and only if δ_X is strictly positive for every $\varepsilon > 0$.

It might seem natural to define the modulus of smoothness by setting it equal, at τ, to

$$\sup\{1 - \tfrac{1}{2}\|x + y\| : \|x\| = \|y\| = 1,\ \|x - y\| \leqq 2\tau\} \tag{$*$}$$

Clearly, X is uniformly smooth if and only if this supremum is $o(\tau)$ at $\tau = 0$. The definition $(*)$, however, would not be well adapted to the duality between uniform convexity and uniform smoothness. Instead, the function ρ_X given by

$$\rho_X(\tau) := \sup\left\{\frac{\|u + v\| + \|u - v\|}{2} - 1 : \|u\| = 1,\ \|v\| = \tau\right\} \tag{1.2}$$

is called the **modulus of smoothness** of X. This definition arises from $(*)$ if we rewrite the quantity to be maximized there in terms of $u = (x + y)/2$ and $v = (x - y)/2$, and change the constraint from $\|u + v\| = \|u - v\| = 1$ to simply $\|u\| = 1$. For small τ, there is no substantial difference, and it is easy to show (see [Kö]) that X is uniformly smooth if and only if $\lim_{\tau \to 0} \rho_X(\tau)/\tau = 0$.

Lindenstrauss [L] has shown that with these definitions, the modulus of convexity of a normed space X and the modulus of smoothness of its dual X^* are related by

$$\rho_{X^*}(\tau) = \sup\{\tau\varepsilon - \delta_X(\varepsilon) : 0 \leqq \varepsilon \leqq 1\}. \tag{1.3}$$

This is a quantitative versions of Day's duality theorem [D].

Uniform convexity was introduced by Clarkson [C] who proved that every L_p space with $1 < p < \infty$ is both uniformly convex and uniformly smooth. Clarkson proved inequalities which give bounds of the form

$$\delta_{L_p}(\varepsilon) \geqq (\varepsilon/K_{p,r})^r \tag{1.4}$$

where $r = p$ for $2 \leqq p < \infty$, and $r = p/(p - 1)$ for $1 < p \leqq 2$.

Lower bounds of the form (1.4) for the modulus of convexity δ of a normed space are especially useful and, in many applications, particular importance attaches to the best possible value of r. Evidently, such an inequality cannot hold for any $r < 2$. A normed space X is said to be *r-uniformly convex* in case $\delta_X(\varepsilon) \geqq (\varepsilon/C)^r$ for some constant C. (After Eq. (2.6) below, we make an apparently more restrictive definition of r-uniform convexity. The two definitions will be shown to be consistent in Proposition 7, and the present definition is the simplest to use in the introduction.) Clarkson's bounds (1.4) only show that L_p is r-uniformly convex with $r > 2$ for all $p \neq 2$ while, actually, L_p is 2-uniformly convex for $1 < p \leqq 2$.

The 2-uniform convexity of L_p for $1 < p \leqq 2$ follows from a result of Hanner [H], who proved an inequality from which δ_{L_p} can be easily *computed.* Hanner's result is recalled in part (a) of Theorem 2 below. The best constant $K_{p,2}$ in (1.4) seems to have been first determined by Ball and Pisier [BP], who gave a simple direct proof, independent of Hanner's calculation, that L_p is 2-uniformly convex for such p. Their optimal 2-uniform convexity inequality is:

$$\delta_{L_p}(\varepsilon) \geqq \frac{p-1}{2}\varepsilon^2 \quad \text{for} \quad 1 < p \leqq 2. \tag{1.5}$$

Because of the dual nature of the notions of uniform convexity and smoothness, the modulus of smoothness of L_p for $2 \leqq p < \infty$ satisfies an inequality of the form $\rho_{L_p}(\tau) \leqq (\hat{K}_{p,2}\tau)^2$. Again, this is a better estimate than that which follows from Clarkson's inequalities. A more detailed history of these and related inequalities will be presented in Sect. II of our paper.

Less is known about the corresponding inequalities for the trace classes C_p. Clarkson's inequalities were extended to C_p partly by Dixmier [Di], and fully by Klaus [Si], with precisely the same constants and exponents as in the L_p case.

Tomczak-Jaegermann later showed that, as with L_p, C_p is actually 2-uniformly convex for $1 < p \leqq 2$:

$$\delta_{C_p}(\varepsilon) \geqq (\varepsilon/\tilde{K}_{p,2})^2 \quad \text{for} \quad 1 < p \leqq 2. \tag{1.6}$$

Her proof proceeds by establishing the C_p analog of Hanner's inequality when p is an even integer, then deducing the 2-uniform smoothness of C_p for all $p \geqq 2$ from this by interpolation, and then using Lindenstrauss's duality result to obtain the 2-uniform convexity of C_p for $1 < p \leqq 2$. Implicit in her proof is the fact that when $p = 2k/(2k-1)$ for some positive integer k, the sharp constants $\tilde{K}_{p,2}$ for C_p coincide with those of L_p; i.e. $\tilde{K}_{p,2} = (p-1)/2$ for such values of p.

The principal results in our paper are the determination of the best possible constants for all p in Tomczak-Jaegermann's theorem, and the proof that the C_p analog of Hanner's inequality holds for $1 < p \leqq 4/3$, and in the dual range $4 \leqq p < \infty$. Our two main theorems are the following (in which $\|\cdot\|_p$ denotes the L_p or the C_p norm):

Theorem 1 (Optimal 2-uniform convexity) *For* $1 \leqq p \leqq 2$, *the inequality*

$$\left(\frac{\|X+Y\|_p^p + \|X-Y\|_p^p}{2}\right)^{2/p} \geqq \|X\|_p^2 + (p-1)\|Y\|_p^2. \tag{1.7}$$

holds in the following cases:

(a) *X and Y are functions in L_p.*
(b) *X and Y are matrices in C_p.*

If $2 \leqq p \leqq \infty$, *the inequality is reversed.*

The title of this theorem will be explained more fully in Sect. II; the point, of course, is that validity of the inequality (1.7) implies 2-uniform convexity. Part (a) is an unpublished result of Ball and Pisier, and the cases of part (b) for $p = 2k/(2k-1)$ are, as we have said before, implicit in the paper [TJ] of Tomczak-Jaegermann. The rest is new. The constant $p-1$ in (1.7) is clearly seen to be optimal as well as natural from the point of view of elementary calculus: if X and Y are real numbers with $|Y|$ much smaller than $|X|$, then the two sides of (1.7) agree to second order in Y. The fact that (1.7) is true for *all* pairs of *numbers* is Gross's two-point inequality [G].

Theorem 2 (Extension of Hanner's inequality to C_p) *For* $1 \leqq p \leqq 2$, *the inequality*

$$\|X+Y\|_p^p + \|X-Y\|_p^p \geqq (\|X\|_p + \|Y\|_p)^p + |\|X\|_p - \|Y\|_p|^p \tag{1.8}$$

holds in the following cases:

(a) *X and Y are functions in L_p.*
(b) *$p \leqq \frac{4}{3}$ and X and Y are matrices in C_p.*
(c) *X and Y are matrices in C_p such that both $X+Y$ and $X-Y$ are positive semidefinite.*

For $2 \leqq p < \infty$, the inequality is reversed and the restriction in (b) *becomes $p \geqq 4$, and the restriction in* (c) *changes to the restriction that X and Y are positive semidefinite.*

Part (a) is Hanner's inequality, and the cases of part (b) in which $p = 2k$ are due to Tomczak-Jaegermann [TJ]. The rest is new. As we explain in the first proof of Proposition 3 below, the inequality (1.8), whenever it holds, implies the inequality (1.7). Thus, if the conditions under which we establish (1.8) were not more restrictive than those under which we establish (1.7), Theorem 1 would be a corollary of Theorem 2.

The paper is organized as follows: In Sect. II, we review the large number of inequalities bearing on the uniform convexity of L_p spaces. Thus, Sect. II consists largely of known results which are presented because of the light they shed on the problems solved in this paper regarding the uniform convexity of C_p, and those that remain open. To our knowledge, such a systematic compendium of these inequalities has not appeared before, and we hope it will be found useful. There are however some new results and some new, simpler proofs. Finally in Sect. III we prove Theorem 1, and in Sect. IV we prove Theorem 2.

Theorem 1 has been applied by Carlen and Lieb [CL] to prove a conjecture of Gross, which arose in his work on quantum field theory. Other applications of the kinds of the inequalities that we discuss here are given, for example, in Pisier's book [P].

Although all of our theorems are stated and proved in the language of matrices, the proofs go through without any change in the context of linear operators on a Hilbert space. By the results of Ruskai [Ru], they can even be extended to a natural Von Neumann algebra context.

II Uniform convexity and smoothness in L_p

While this section is largely focused on inequalities relating to L^p spaces, we state certain definitions and prove certain results in the general normed space setting so that they are available to us in the next section.

For the rest of the paper, q *denotes the dual index of* p, i.e., $1/p + 1/q = 1$.

The notion of uniform convexity was introduced by Clarkson who proved four inequalities. The two that imply the uniform convexity of L_p spaces are the following, in which x and y are functions in L_p:

$$\left(\left\|\frac{x+y}{2}\right\|_p^p + \left\|\frac{x-y}{2}\right\|_p^p\right)^{1/p} \leqq \left(\frac{\|x\|_p^p + \|y\|_p^p}{2}\right)^{1/p} \quad \text{for} \quad 2 \leqq p \leqq \infty \tag{2.1}$$

and

$$\left(\left\|\frac{x+y}{2}\right\|_p^q + \left\|\frac{x-y}{2}\right\|_p^q\right)^{1/q} \leqq \left(\frac{\|x\|_p^p + \|y\|_p^p}{2}\right)^{1/p} \quad \text{for} \quad 1 \leqq p \leqq 2. \tag{2.2}$$

In the cases $1 \leqq p \leqq 2$ and $2 \leqq p \leqq \infty$, the inequalities in (2.1) and (2.2), respectively, hold in the reversed sense. These are the other two Clarkson inequalities – the ones which imply that the L_p spaces are uniformly smooth. They follow from (2.1) and (2.2) by an elementary duality argument.

The inequality (2.1), which involves *only* p powers and not q powers as well, is simpler to prove, and is known as the "easy" Clarkson inequality. In fact, (2.1) is not only easier to prove, it is actually a consequence of (2.2). This is so because both inequalities can be viewed as statements about the norms of certain linear operators. Viewed as such, (2.1) is weaker than the dual inequality to (2.2). More concretely, for $1 \leqq s,\ t \leqq \infty$, equip $L_s \times L_s$ with the norm $\|\cdot\|_{s,t}$ given by $\|(x, y)\|_{s,t} = ((\|x\|_s^t + \|y\|_s^t)/2)^{1/t}$. Also, define the operator $B: L_{s,t} \to L_{s,t}$ by $B(x, y) = ((x+y)/2, (x-y)/2)$. Then (2.2) is equivalent to the statement that B is a bounded operator from $L_{p,p}$ to $L_{p,q}$ for $1 \leqq p \leqq 2$ with norm $2^{1/p}$. But since B is self-adjoint, it has the same norm as an operator between the dual spaces; i.e. B is bounded from $L_{q,p}$ to $L_{q,q}$ with norm $2^{1/p}$ for $1 \leqq p \leqq 2$. Finally, since $p \leqq q$, $\|(x, y)\|_{q,p}, \leqq \|(x, y)\|_{q,q}$, and hence B has norm $2^{1/p}$ from $L_{q,q}$ to $L_{q,q}$ for $2 \leqq q \leqq \infty$, which is clearly equivalent to (2.1).

Since these bounds on the norm of B (which are equivalent to (2.1) and (2.2)) are log-linear in $1/p$, they can be proved by interpolation between the elementary cases $p = 1$ (Minkowski's inequality for L_1), $p = 2$ (the parallelogram law) and $p = \infty$ (Minkowski's inequality for L_∞), as observed by Boas [Bo]. This same approach was later used by Klaus [Si] to establish the C_p analogs of Clarkson's inequalities.

It is convenient also to have the following inequality, obtained from (2.2) by rearranging some powers of 2. If x and y belong to L_s where either $s = p$ or $s = q$, then

$$\left(\frac{\|x+y\|_s^q + \|x-y\|_s^q}{2}\right)^{1/q} \leqq (\|x\|_s^p + \|y\|_s^p)^{1/p} \quad \text{for} \quad 1 \leqq p \leqq 2. \tag{2.3}$$

Replacing x and y respectively with $x + y$ and $x - y$, and rearranging some powers of 2, we see that the inequality reverses when p and q are interchanged. That is,

$$\left(\frac{\|x+y\|_s^p + \|x-y\|_s^p}{2}\right)^{1/p} \geqq (\|x\|_s^q + \|y\|_s^q)^{1/q} \quad \text{for} \quad 1 \leqq p \leqq 2. \tag{2.4}$$

It follows directly from (2.3) that if x and y are unit vectors in one of the spaces L_p or L_q, and $\|x - y\| = 2\varepsilon$ then

$$\left\|\frac{x+y}{2}\right\| \leqq (1 - \varepsilon^q)^{1/q} \leqq 1 - \frac{\varepsilon^q}{q},$$

so that $\delta_{L_s}(\varepsilon) \geqq \varepsilon^q/q$. Similarly, from (2.4) one sees that $\rho_{L_s}(\tau) \leqq \tau^p/p$.

We next turn to the sort of inequalities that figure in our Theorem 1. We begin with some definitions and general considerations intended to clarify the relations among all the inequalities that we consider.

Let X be a uniformly convex normed space, and suppose that $\delta_X(\varepsilon) \geqq (\varepsilon/C)^r$ for some C and $r > 1$. Then with $\|x\| = \|y\| = 1$ and $\|x - y\| = 2\varepsilon$, we have that

$$\left\|\frac{x+y}{2}\right\|^r \leqq (1-(\varepsilon/C)^r)^r \leqq 1 - \left(\frac{r-1}{r}\right)^{r-1} (\varepsilon/C)^r,$$

and thus, with $K = C(1/t)^{-1/t}$, $1/t + 1/r = 1$,

$$\left\|\frac{x+y}{2}\right\|^r + \left\|\frac{x-y}{2K}\right\|^r \leqq \frac{\|x\|^r + \|y\|^r}{2} \tag{2.5}$$

for all x and y such that $\|x\| = \|y\|$. By replacing x with $x + y$ and y with $x - y$, we find that

$$\frac{\|x+y\|^r + \|x-y\|^r}{2} \geqq \|x\|^r + \|K^{-1}y\|^r \tag{2.6}$$

for all x and y such that $\|x + y\| = \|x - y\|$.

As promised in the introduction (after (1.4)), we now impose a definition of r-uniform convexity that may seem more restrictive than the one we gave before. Proposition 7 below shows that the two definitions are equivalent, *up to constants*. It is the constant in (2.6), figuring in the second definition, that is the main object of our attention.

A normed space X is said to be **r-uniformly convex** for some $r \in [2, \infty)$ if there is a constant K such that (2.6) holds for *all* $x, y \in X$. The best constant K is called the **r-uniform convexity constant** of X. When X is r-uniformly convex, so that (2.6) and hence (2.5) hold, it is immediate from the latter that $\delta_X(\varepsilon) \geqq (\varepsilon/K)^r$. Thus r-uniform convexity implies the validity of a lower bound of the form $\delta_X(\varepsilon) \geqq (\varepsilon/C)^r$ for the modulus of convexity; i.e. the condition under which we called X r-uniformly convex in the introduction.

Similarly, X is said to be **t-uniformly smooth** for some $t \in (1, 2]$ if

$$\frac{\|x+y\|^t + \|x-y\|^t}{2} \leqq \|x\|^t + \|Ky\|^t, \tag{2.7}$$

for some K and *all* $x, y \in X$. The best constant K is called the **t-uniform smoothness constant** of X. We shall show at the end of this section that the t-uniform smoothness constant of a normed space X equals the r-uniform convexity constant of its dual X^* where, as usual, $1/r + 1/t = 1$.

When (2.7) holds, we have that for all x and y with $\|x\| = 1$ and $\|y\| = \tau$

$$\frac{\|x+y\| + \|x-y\|}{2} - 1 \leqq \left(\frac{\|x+y\|^t + \|x-y\|^t}{2}\right)^{1/t} - 1 \leqq (K(1/t)^{1/t}\tau)^t.$$

Hence, by (1.1), t-uniform smoothness implies an estimate of the form $\rho_X(\tau) \leqq (C\tau)^t$. Proposition 7 shows that the reverse implication holds as well.

The parallelogram identity shows that Hilbert space is 2-uniformly convex and 2-uniformly smooth, and it is readily seen that the exponent 2 is the best that can occur for each property. Clarkson's inequality shows that when $1 \leqq p \leqq 2$ then each of L_p and L_q is q-uniformly convex and p-uniformly smooth. As we have remarked, these *exponents* are not in general the best possible, despite the fact that the *constants* in Clarkson's inequalities are always sharp.

The actual situation is the following:

For $1 < p \leqq 2$, L_p *is 2-uniformly convex though no better than p-uniformly smooth while for* $2 \leqq q < \infty$, L_q *is 2-uniformly smooth as well as q-uniformly convex.*

These facts follow from Hanner's inequality (Theorem 2(a) of the introduction) which determines exactly the moduli of convexity and smoothness of all L_p spaces. The optimal 2-uniform convexity inequality is the following:

Proposition 3 (Optimal 2-uniform convexity for L_p) *If* $1 \leqq p \leqq 2$ *and* x *and* $y \in L_p$, *then*

$$\frac{\|x+y\|_p^2 + \|x-y\|_p^2}{2} \geqq \|x\|_p^2 + (p-1)\|y\|_p^2. \tag{2.8}$$

For $2 \leqq p < \infty$, *the inequality is reversed.*

Remark. The inequality (2.8) holds for any normed space for which (1.8) holds, as we will soon show. Inequality (2.8) does not seem to appear in the literature in quite this form but it is probably folk-lore. Ball and Pisier noticed that it follows from Gross's two-point inequality using arguments which (in the context of general Banach lattices) go back to Figiel [F].

The reader will note that (2.8) is not identical to (1.7), but we shall soon see that the validity of (2.8) for *all* L_p spaces implies the validity of the apparently stronger (1.7).

First proof. To deduce (2.8) from Hanner's inequality recall that a special case of Gross's inequality [G] states that if $1 \leqq p \leqq 2$ and a and b are real,

$$\left(\frac{|a+b|^p + |a-b|^p}{2}\right)^{1/p} \geqq (a^2 + (p-1)b^2)^{1/2}.$$

Now if $x, y \in L_p$,

$$\begin{aligned}\left(\frac{\|x+y\|_p^2 + \|x-y\|_p^2}{2}\right)^{1/2} &\geqq \left(\frac{\|x+y\|_p^p + \|x-y\|_q^p}{2}\right)^{1/p} \\ &\geqq \left[\frac{(\|x\|_p + \|y\|_p)^p + |\|x\|_p - \|y\|_p|^p}{2}\right]^{1/p} \\ &\geqq (\|x\|_p^2 + (p-1)\|y\|_p^2)^{1/2},\end{aligned}$$

where we have used, in succession, Hölder's inequality, Hanner's inequality (1.8), and Gross's inequality with $a = \|x\|_p$ and $b = \|y\|_p$. □

Second proof. An alternative proof of Proposition 3 consists simply of showing that for $\|y\|$ small,

$$\frac{\|x+y\|_p^2 + \|x-y\|_p^2}{2} \geqq \|x\|_p^2 + (p-1)\|y\|_p^2 + o(\|y\|_p^2), \tag{2.9}$$

and then observing that this infinitesimal form of (2.8) is equivalent to the full statement. That is, (2.9) is the same as

$$\frac{d^2}{ds^2}\left(\frac{\|x+sy\|_p^p + \|x-sy\|_p^p}{2}\right)\bigg|_{s=0} \geqq p(p-1)\|x\|_p^{p-2}\|y\|_p^2, \tag{2.10}$$

which is easy to establish for L_p functions by elementary calculus. Proposition 3 follows from this by integration with respect to $s \in [0,1]$. □

As will be shown later, inequality (2.10) is also true for matrices in C_p, and this will form the basis of the proof of Theorem 1 given in Sect. III. But there is an important difference between the commutative and non-commutative cases. For *functions* x and y in L_p,

$$\frac{d^2}{ds^2}\left(\frac{\|x+sy\|_p^p + \|x-sy\|_p^p}{2}\right)\bigg|_{s=0} = p(p-1)\int |x|^{p-2}|y|^2, \tag{2.11}$$

and the latter dominates $p(p-1)\|x\|_p^{p-2}\|y\|_p^2$ by Hölder's inequality (since $p<2$). For *matrices*, the analogue of (2.11) is false: one always has

$$\frac{d^2}{ds^2}\left(\frac{\|X+sY\|_p^p + \|X-sY\|_p^p}{2}\right)\bigg|_{s=0} \leqq p(p-1)\,Tr|X|^{p-2}|Y|^2$$

and equality need not hold. The problem is to find a replacement for (2.11) in the non-commutative setting.

Inequality (2.8), as we said, is apparently weaker than (1.7) since

$$\left(\frac{\|x+y\|^p + \|x-y\|^p}{2}\right)^{1/p} \leqq \left(\frac{\|x+y\|^2 + \|x-y\|^2}{2}\right)^{1/2}$$

for $1 \leqq p \leqq 2$. However, a simple doubling argument shows that (1.7) is actually a consequence of the fact that (2.8) holds for all L_p spaces. This argument is also valid for C_p, and we give it in detail in Sect. III, Eq. (3.5)–(3.6). The reader can easily translate the C_p version into the L_p version.

Note also that in the first proof of Proposition 3, which was based on Hanner's inequality, we actually arrived at (1.7) in an intermediate step. In the C_p setting, we do not possess a full analog of Hanner's inequality, and so we shall prove Theorem 1 by adapting the second proof of Proposition 3 to the C_p setting.

The following diagram shows the relationships between the several expressions mentioned above. Connecting lines indicate inequality between the expressions. All of the indicated inequalities hold in both C_p and L_p except for that indicated by the line labeled Hanner, which is only known to hold in C_p in the special cases specified in Theorem 2. In each expression x and y are elements of L_p or of C_p, and q is the index conjugate to p. For $1 \leqq p \leqq 2$, the quantities *increase* as one goes up the page; for $2 \leqq p \leqq \infty$, they *decrease*.

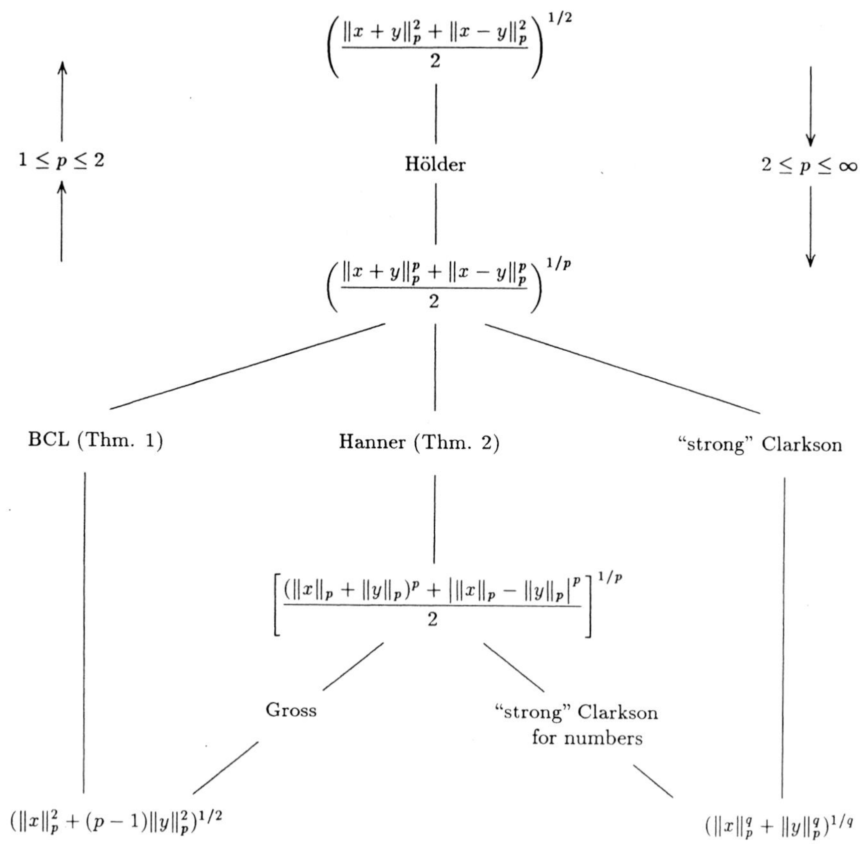

Fig. 1. Relationships among the inequalities

We turn now to a proof of Hanner's inequality for L_p, which yields Clarkson's inequalities along the way, and to a simple duality result which shows that optimal constants obtained for q-uniform convexity of a normed space immediately yield optimal constants for p-uniform smoothness of its dual (and conversely).

Lemma 4 (Variational characterization of sums of pth powers) *For* $1 < p < \infty$ *define* $\alpha = \alpha_p\colon [0, \infty) \to [0, \infty)$ *by*

$$\alpha(r) = (1 + r)^{p-1} + |1 - r|^{p-1} \operatorname{sign}(1 - r).$$

Then for all $x, y \in \mathbb{R}$

$$|x + y|^p + |x - y|^p = \left\{\begin{pmatrix}\sup\\ \inf\end{pmatrix}\right\}\{\alpha(r)|x|^p + \alpha(1/r)|y|^p\colon 0 < r < \infty\}$$

the sup *or* inf *being taken according as* $p \leqq 2$ *or* $p \geqq 2$.

Proof. Assume $1 < p \leqq 2$; the proof for $p \geqq 2$ is similar. Plainly, it may be assumed that $0 < y \leqq 1 = x$. For $r = y$; one readily checks that

$$\alpha(r) + \alpha\left(\frac{1}{r}\right) y^p = (1 + y)^p + (1 - y)^p.$$

To see that $(1 + y)^p + (1 - y)^p \geqq \alpha(r) + \alpha(1/r) y^p$ for all r, it suffices to check that the latter quantity attains its maximum when $r = y$. But

$$\begin{aligned}\frac{d}{dr}(\alpha(r) + \alpha(1/r) y^p) &= \alpha'(r) - \frac{1}{r^2}\alpha'(1/r) y^p \\ &= (p-1)\left[(1 + r)^{p-2} - |1 - r|^{p-2} - \frac{y^p}{r^2}\left(\left(\frac{1}{r} + 1\right)^{p-2} - \left|\frac{1}{r} - 1\right|^{p-2}\right)\right] \\ &= (p-1)\left(1 - \left(\frac{y}{r}\right)^p\right)[(1 + r)^{p-2} - |1 - r|^{p-2}].\end{aligned}$$

Since $p - 2 \leqq 0$ and $1 + r \geqq |1 - r|$, the last factor is non-positive. Thus, the whole is non-negative for $0 < r < y$ and non-positive for $r > y$. □

Proof of Theorem 2(a) (Hanner's inequality) Again assume $1 < p \leqq 2$ and let $x, y \in L_p$. Then

$$\begin{aligned}\|x + y\|^p + \|x - y\|^p &= \int (|x + y|^p + |x - y|^p) = \int \sup_r \{\alpha(r)|x|^p + \alpha(1/r)|y|^p\} \\ &\geqslant \sup_r \int (\alpha(r)|x|^p + \alpha(1/r)|y|^p) = \sup_r \{\alpha(r)\|x\|^p + \alpha(1/r)\|y\|^p\} \\ &= (\|x\| + \|y\|)^p + |\|x\| - \|y\||^p.\end{aligned}$$ □

Proof of (2.2) Let us also show how Lemma 4 can be used to deduce the "hard" Clarkson inequality. This time we shall prove it in the uniform smoothness range, i.e. for L_q with $2 \leqq p < \infty$. Since

$$\|x + y\|_p^p + \|x - y\|_p^p \leqq \alpha(r)\|x\|_p^p + \alpha(1/r)\|y\|_p^p$$

for all r, it is enough to find an r for which the right side equals $2(\|x\|^q + \|y\|^q)^{p/q}$.

Set $u = \|x\|_p^q$, $v = \|y\|_p^q$, $r = v/u$, and assume that $v \leqq u$. Then

$$\begin{aligned}\alpha(r)\|x\|_p^p + \alpha(1/r)\|y\|_p^p &= \alpha(v/u) u^{p-1} + \alpha(u/v) v^{p-1} = 2(u + v)^{p-1} \\ &= 2(\|x\|_p^p + \|y\|_p^p)^{p/q}.\end{aligned}$$ □

We now prove the duality results mentioned earlier. These results hold in general; no reference to L_p or C_p is made.

Lemma 5 (Duality for q-uniform convexity and p-uniform smoothness) *Let X be a normed space with dual X^*. The p-uniform smoothness constant of X (the constant K in* (2.7)) *is equal to the q-uniform convexity constant of X^* (the constant K in* (2.6).

Proof. Suppose that the q-uniform convexity of X^* is K and let $x, y \in X$. We denote norms in X and X^* indiscriminately by $\|\cdot\|$ and trust that the meaning will be clear. There are unit vectors λ and μ in X^* such that $\lambda(x + y) = \|x + y\|$

and $\mu(x-y)=\|x-y\|$. Define $\phi,\psi\in X^*$ by $\phi=Z^{-1/q}\|x+y\|^{p-1}\lambda$ and $\psi=Z^{-1/q}\|x-y\|^{p-1}\mu$ with $Z=(\|x+y\|^p+\|x-y\|^p)/2$. Then, $\|\phi\|^q+\|\psi\|^q=2$ and we have

$$\begin{aligned}\left(\frac{\|x+y\|^p+\|x-y\|^p}{2}\right)^{1/p}&=\frac{\phi(x+y)+\psi(x-y)}{2}=\left(\frac{\phi+\psi}{2}\right)(x)+\left(\frac{\phi-\psi}{2}\right)(y)\\&\leqq\left(\left\|\frac{\phi+\psi}{2}\right\|^q+\left\|\frac{\phi-\psi}{2K}\right\|^q\right)^{1/q}(\|x\|^p+\|Ky\|^p)^{1/p}\\&\leqq\left(\frac{\|\phi\|^q+\|\psi\|^q}{2}\right)^{1/q}(\|x\|^p+\|Ky\|^p)^{1/p}\\&=(\|x\|^p+\|Ky\|^p)^{1/p}.\end{aligned}$$

The first inequality is Hölder's inequality for numbers, and the second is (2.5) with $r=q$. The other implication is similar. □

Lemma 6 (Duality for Hanner's inequality) *Let X be a normed space with dual X^*. Let $1<p\leqq 2$ and $1/p+1/q=1$. Then the validity of*

$$\|\phi+\psi\|^q+\|\phi-\psi\|^q\leqq(\|\phi\|+\|\psi\|)^q+|\,\|\phi\|-\|\psi\|\,|^q \tag{2.12}$$

for all $\phi,\psi\in X^*$ implies the validity of

$$\|y+z\|^p+\|y-z\|^p\geqq(\|y\|+\|z\|)^p+|\,\|y\|-\|z\|\,|^p \tag{2.13}$$

for all $y,z\in X$. Similarly, the validity of (2.13) in X implies the validity of (2.12) in X^*.

Proof. Suppose first that (2.12) holds in X^*. To establish (2.13), we first rewrite it in terms of $u=y+z$ and $v=y-z$, so that what we must show is:

$$2^p(\|u\|^p+\|v\|^p)\geqq(\|u+v\|+\|u-v\|)^p+|\|u+v\|-\|u-v\||^p. \tag{2.14}$$

We may assume without loss of generality that $\|u+v\|=1$, and that $r:=\|u-v\|\leqq 1$. Then the right side of (2.14), which we call R^p, can be rewritten (as in Lemma 4) as

$$R^p=\alpha\|u+v\|^p+\beta\|u-v\|^p=(1+r)^p+(1-r)^p,$$

where $\alpha=(1+r)^{p-1}+(1-r)^{p-1}$ and $\beta=r^{1-p}[(1+r)^{p-1}-(1-r)^{p-1}]$. As in the proof of Lemma 5, we choose unit vectors λ and μ in X^* such that $\lambda(u+v)=\|u+v\|$ and $\mu(u-v)=\|u-v\|$. Then we define

$$\phi=\alpha R^{-p/q}\|u+v\|^{p-1}\lambda\quad\text{and}\quad\psi=\beta R^{-p/q}\|u-v\|^{p-1}\mu.$$

Thus,

$$\begin{aligned}R&=\phi(u+v)+\psi(u-v)=(\phi+\psi)(u)+(\phi-\psi)(v)\\&\leqq\|\phi+\psi\|\,\|u\|+\|\phi+\psi\|\,\|v\|\leqq(\|\phi+\psi\|^q+\|\phi-\psi\|^q)^{1/q}(\|u\|^p+\|v\|^p)^{1/p}.\end{aligned}$$

To complete the demonstration of (2.14), we have to show that $T:=\|\phi+\psi\|^q+\|\phi-\psi\|^q\leqq 2^q$. By (2.12),

$$\begin{aligned}T&\leqq R^{-p}[\alpha\|u+v\|^{p-1}+\beta\|u-v\|^{p-1}]^q+R^{-p}[\alpha\|u+v\|^{p-1}-\beta\|u-v\|^{p-1}]^q\\&=R^{-p}[\alpha+\beta r^{p-1}]^q+R^{-p}[\alpha-\beta r^{p-1}]^q=2^q.\end{aligned}$$

A similar proof works in the other direction to go from (2.13) to (2.12) □

Remark. In our application of Lemma 6 to the proof of Theorem 2, we need the following refinement whose truth is evident from the proof given above. We take $X = C_p$ and $X^* = C_q$. Then the validity of (2.13) with the *extra constraint* that $y + z$ and $y - z$ are *positive semidefinite* matrices implies the validity of (2.12) when ϕ and ψ are *positive semidefinite* matrices.

We close this section with a proposition showing the consistency of the two definitions that we have given for r-uniform convexity; i.e., the ones following (1.4) and (2.6).

Proposition 7 (Equivalence of definitions of r-uniform convexity) *Let X be a normed space. Then* (2.5) *holds for some constant K and all $x, y \in X$ if and only if $\delta_X(\varepsilon) \geqq (\varepsilon/C)^r$ for some constant C. Similarly,* (2.7) *holds for some constant K and all $x, y \in X$ if and only if $\rho_X(\tau) \leqq (C\tau)^r$ for some constant C.*

Proof. We have already seen that (2.5) and (2.7) imply the indicated bounds on δ_X and ρ_X respectively.

Suppose first that $\rho_X(\tau) \leqq (C\tau)^r$ for some constant C. Of course, $1 < r \leqq 2$. Then for all $\|x\| = 1$ and $\|y\| \leqq 1$,

$$\frac{\|x + y\| + \|x - y\|}{2} - 1 \leqq (C\|y\|)^r.$$

Define numbers b and β by

$$b := \frac{\|x + y\| + \|x - y\|}{2} \quad \text{and} \quad \beta := \frac{\|x + y\| - \|x - y\|}{\|x + y\| + \|x - y\|}.$$

Then

$$\left(\frac{\|x + y\|^r + \|x - y\|^r}{2}\right)^{1/r} - \frac{\|x + y\| + \|x - y\|}{2}$$
$$= b\left[\left(\frac{(1 + \beta)^r + (1 - \beta)^r}{2}\right)^{1/r} - 1\right]. \tag{2.15}$$

The function of β on the right side in (2.15) vanishes quadratically at the origin. Thus, a simple estimation using Taylor's theorem shows that it is no greater than $D_r\beta^2$ for some constant D_r depending only on r. Then, since $|\beta| \leqq \|y\|/b \leqq \|y\|$ and $1 \leqq r \leqq 2$, we have from (2.15) and the assumption on ρ_X that

$$\frac{\|x + y\|^r + \|x - y\|^r}{2} \leqq \left(1 + C^r\|y\|^r + D_r\|y\|^2\right)^r \leqq 1 + (K_r\|y\|)^r, \tag{2.16}$$

for all x and y with $\|y\| \leqq \|x\| = 1$, where K_r depends only on C and r. Therefore,

$$\frac{\|x + y\|^r + \|x - y\|^r}{2} \leqq \|x\|^r + K_r^r\|y\|^r \tag{2.17}$$

for all x and y with $\|y\| \leqq \|x\|$. Finally, since we may assume that $K_r \geqq 1$, (2.17) holds for all x and y.

Next, suppose that $\delta_X(\varepsilon) \geqq (\varepsilon/C)^r$ for some constant C. Then by (1.3),

$$\rho_{X^*}(\tau) = \sup\{\tau\varepsilon - \delta_X(\varepsilon) : 0 \leqq \varepsilon \leqq 1\} \leqq \sup\{\tau\varepsilon - (\varepsilon/C)^r : 0 \leqq \varepsilon \leqq \infty\} = (C\tau)^{r'}$$

where $1/r + 1/r' = 1$.Then, by what we have shown above, there is a constant K so that (2.7) is valid in X^*.But then by Lemma 5, (2.5) is valid in X with the same constant K. □

III Optimal 2-uniform convexity inequalities for trace norms

Norms on the space of $n \times n$ matrices which are non commutative analogs of the L_p norms can be defined in terms of the trace by

$$\|X\|_p = (\mathrm{Tr}((X^*X)^{p/2})^{1/p} = (\mathrm{Tr}((XX^*)^{p/2})^{1/p}, \tag{3.1}$$

for $1 \leqq p < \infty$. For $p = \infty$, $\|X\|_p$ denotes the operator norm of X, as usual. The analogy can be made quite close, and it has been developed in a von Neumann algebra context by Segal [Se] and Dixmier [Di] as part of their theories of non-commutative integration.

Many familiar inequalities for L_p norms also hold for the C_p norms. This is true, in particular, of the Hölder inequality

$$\|XY\|_r \leqq \|X\|_p \|Y\|_q, \; 1/r = 1/p + 1/q.$$

There are, however, other inequalities for L_p norms which *do not* hold for the C_p norms. Many examples are connected with the poor behavior of the map

$$X \mapsto |X| = (X^*X)^{1/2}. \tag{3.2}$$

For example, if f and g are complex valued functions in some L_p space, then $\||f| - |g|\|_p \leqq \|f - g\|_p$. This is not true for C_p, and, when $p = 2$, the factor of $\sqrt{2}$ in the Araki-Yamagami inequality [ArY] $\||X| - |Y|\|_2 \leqq \sqrt{2}\|X - Y\|_2$ is optimal whenever $n \geqq 2$.

As we have asserted in Theorems 1 and 2, however, almost all of the optimal inequalities expressing uniform convexity and smoothness properties of L_p spaces have exact analogs which hold for the C_p norms. Most of this section is devoted to the proof of Theorem 1. Before giving the proof we briefly discuss the history of uniform convexity inequalities for C_p as we know it.

The first such inequality was established by Dixmier [Di] who proved the C_p analog of the "easy" Clarkson inequality (2.1) by means of interpolation. As with L_p, this implies that for $2 \leqq p < \infty$, $\delta_{C_p}(\varepsilon) \geqq (1/p)\varepsilon^p$.

Interpolation was later used to establish the analog of the "hard" Clarkson inequality (2.2) which implies the uniform convexity of L_p for $1 < p \leqq 2$. Such a proof has been given by Martin Klaus, and is sketched in [Si]; to some extent it is modeled on Boas' proof [Bo] of (2.2) for L_p. This result implies that for $1 < p \leqq 2$, C_p is at least q-uniformly convex.

Later, Tomczak-Jaegermann showed that for $1 < p \leqq 2$, C_p is actually 2-uniformly convex and C_q is 2-uniformly smooth. Moreover, she showed for $q = 2k$, that the sharp 2-uniform smoothness constants of C_q are the same as those for L_q (so that the corresponding equalities of 2-uniform convexity constants hold by Lemma 5).

Before we prove Theorem 1, note that $t \mapsto t^{p/2}$ is concave for $1 < p \leqq 2$ and therefore (1.7) immediately implies that

$$\frac{\|X+Y\|_p^2 + \|X-Y\|_p^2}{2} \geqq \|X\|_p^2 + (p-1)\|Y\|_p^2, \tag{3.3}$$

which expresses the 2-uniform convexity of C_p in the usual way. It follows that

$$\delta_{C_p}(\varepsilon) \geqq \frac{(p-1)}{2}\varepsilon^2 \quad \text{for} \quad 1 < p \leqq 2, \tag{3.4}$$

and thus the analog of (1.5) holds for C_p.

We now observe that (1.7) is only formally stronger than (3.3). To see that (3.3) implies (1.7), consider the $2n \times 2n$ matrices given in block form by

$$Z = \begin{bmatrix} X & 0 \\ 0 & X \end{bmatrix}, \quad W = \begin{bmatrix} Y & 0 \\ 0 & -Y \end{bmatrix}. \tag{3.5}$$

Then

$$\mathrm{Tr}|Z+W|^p = \mathrm{Tr}|Z-W|^p = (\mathrm{Tr}|X+Y|^p + \mathrm{Tr}|X-Y|^p)$$

and thus,

$$\|Z+W\|_p^2 = \|Z-W\|_p^2 = (\|X+Y\|_p^p + \|X-Y\|_p^p)^{2/p}.$$

Since also $\|Z\|_p^2 = 2^{2/p}\|X\|_p^2$ and $\|W\|_p^2 = 2^{2/p}\|Y\|_p^2$, (3.3) implies

$$\begin{aligned} \|X\|_p^2 + (p-1)\|Y\|_p^2 &= 2^{-2/p}(\|Z\|_p^2 + (p-1)\|W\|_p^2 \\ &\leqq 2^{-2/p}\left(\frac{\|Z+W\|_p^2 + \|Z-W\|_p^2}{2}\right) \\ &= \left(\frac{\|X+Y\|_p^p + \|X-Y\|_p^p}{2}\right)^{2/p}, \end{aligned} \tag{3.6}$$

which is (1.7).

Proof of Theorem 1 First, we reduce to the case in which X and Y are self-adjoint. Consider the $2n \times 2n$ matrices given in block form by

$$C = \begin{bmatrix} 0 & X \\ X^* & 0 \end{bmatrix}, \quad D = \begin{bmatrix} 0 & Y \\ Y^* & 0 \end{bmatrix}.$$

Clearly, if (1.7) holds for the $2n \times 2n$ matrices C and D, it holds for X and Y. Since C and D are self-adjoint, it suffices to prove inequality (3.3) for such matrices. We therefore assume without loss of generality that X and Y are self-adjoint.

Let Z and W be defined in terms of X and Y as in (3.5). Then we can rewrite (1.7) as

$$\mathrm{Tr}(|Z+rW|^p)^{2/p} \geqq (\mathrm{Tr}|Z|^p)^{2/p} + r^2(p-1)(\mathrm{Tr}|W|^p)^{2/p}. \tag{3.7}$$

First, note that without loss of generality we may assume by continuity that the union of the ranges of Z and W span $\mathbb{C}^{2n}$. Then $\det(Z + rW)$ is a polynomial of order exactly $2n$ in r, and it has at most $2n$ zeros for $0 \leqq r \leqq 1$. We will avoid these values of r below in our computations. We define $\psi(r)$ by

$$\psi(r) = \mathrm{Tr}|Z+rW|^p = \mathrm{Tr}((Z^2 + r(ZW+WZ) + r^2W^2)^{p/2}).$$

Then $\psi(r)$ is continuously differentiable and

$$\frac{d}{dr}\psi(r) = (p/2)\mathrm{Tr}((Z^2 + r(ZW + WZ) + r^2 W^2)^{p/2-1})((ZW + WZ) + 2rW^2).$$

With the aid of the integral representation

$$(Z^2 + r(ZW + WZ) + r^2 W^2)^{(p/2-1)}$$

$$= \beta_p \int_0^\infty t^{(p/2-1)} \frac{1}{t + (Z^2 + r(ZW + WZ) + r^2 W^2)} dt, \tag{3.8}$$

we see that $d\psi(r)/dr$ is again continuously differentiable.

Now, both sides of (3.7) agree at $r = 0$, and the first derivatives in r of both sides vanish there as well. Moreover, the second derivative in r of the left hand side of (3.7) satisfies

$$\frac{d^2}{dr^2}(\psi(r))^{2/p} \geqq \frac{2}{p}\psi(r)^{(2-p)/p}\frac{d^2}{dr^2}\psi(r),$$

while the second derivative on the right side of (3.7) is just $2(p-1)(\mathrm{Tr}|W|^p)^{2/p}$. It therefore suffices to show that

$$\frac{1}{p}\psi(r)^{(2-p)/p}\frac{d^2}{dr^2}\psi(r) \geqq (p-1)(\mathrm{Tr}|W|^p)^{2/p} \tag{3.9}$$

for all $0 < r < 1$. By redefining Z to be $Z + rW$, it suffices to establish (3.9) at $r = 0$. Since $Z + rW$ is non-singular, after the redefinition, $|Z|$ will be strictly positive.

We now claim that

$$\left.\frac{d^2}{dr^2}\mathrm{Tr}|Z + rW|^p\right|_{r=0} \geqq \left.\frac{d^2}{dr^2}\mathrm{Tr}||Z| + rW|^p\right|_{r=0}. \tag{3.10}$$

To see this, note that by the integral formula (3.8),

$$\left.\frac{d^2}{dr^2}\mathrm{Tr}|Z + rW|^p\right|_{r=0} = \mathrm{Tr}|Z|^{p-2}W^2 - (p/2)\beta_p \int_0^\infty t^{p/2-1}\mathrm{Tr}$$

$$\times\left(\frac{1}{Z^2 + t}(ZW + WZ)\frac{1}{Z^2 + t}(ZW + WZ)\right)dt. \tag{3.11}$$

The trace under the integral sign consists of four terms which, using the cyclicity of the trace, can be rewritten as

$$\mathrm{Tr}\left(WZ\frac{1}{Z^2 + t}WZ\frac{1}{Z^2 + t}\right) + 3\mathrm{Tr}\left(W\frac{1}{Z^2 + t}WZ^2\frac{1}{Z^2 + t}\right).$$

Since only Z^2 enters the second of these two terms, this term is unchanged when Z is replaced by $|Z|$. Upon writing out the first term in a basis that diagonalizes Z, that term becomes

$$\sum_{i,j=1}^{n}\left(\frac{1}{z_i^2 + t}\right)\left(\frac{1}{z_j^2 + t}\right)|w_{ij}|^2 z_i z_j.$$

Clearly this term, and hence the integral in (3.11), increases when Z is replaced by $|Z|$. The first term in (3.11), being a function of Z^2, is invariant under the substitution, and the assertion (3.10) is established.

Therefore, without loss of generality we may assume that $Z > 0$. Then, of course, $Z + rW > 0$ for all r sufficiently small, and we no longer need to square $Z + rW$ to obtain a positive operator whose powers can be expressed as an integral over its resolvent. Working directly with $Z + rW$, we can use the simpler integral representation

$$(Z + rW)^{(p-1)} = \gamma_p \int_0^\infty t^{(p-1)} \left[\frac{1}{t} - \frac{1}{t + (Z + rW)} \right] dt \tag{3.12}$$

to conclude that

$$\psi''(0) = p\gamma_p \int_0^\infty t^{(p-1)} \mathrm{Tr} \left[\frac{1}{t+Z} W \frac{1}{t+Z} W \right] dt. \tag{3.13}$$

Consider the right side of (3.13) as a function of Z for fixed W. We claim that it is convex in Z. To prove this, it suffices to prove the following inequality for every self-adjoint matrix A:

$$\Delta(A) := \frac{d^2}{ds^2} \mathrm{Tr} \left(\frac{1}{t + (Z + sA)} W \frac{1}{t + (Z + sA)} W \right) \bigg|_{s=0} \geqq 0.$$

There are six terms. If we define $C = (t + Z)^{-1/2} A (t + Z)^{-1/2}$ and $D = (t + Z)^{-1/2} W (t + Z)^{-1/2}$, then the result of the computation is

$$\Delta(A) = 4 \mathrm{Tr}\, C^2 D^2 + 2 \mathrm{Tr}\, CDCD.$$

But by the Schwarz inequality,

$$|\mathrm{Tr}(CDCD)| \leqq \{\mathrm{Tr}(CD^2C)\}^{1/2} \{\mathrm{Tr}(DC^2D)\}^{1/2} = \mathrm{Tr}\, C^2 D^2.$$

Thus, $\Delta(A) \geqq 0$ and the integrand in (3.13) is a convex function of Z.

Now fix W and t, and define

$$F(Z) = \mathrm{Tr} \left[\frac{1}{t+Z} W \frac{1}{t+Z} W \right].$$

Clearly, when U is any unitary matrix that commutes with W, $F(UZU^*) = F(Z)$. Let $\{e_1, \ldots, e_{2n}\}$ be an orthonormal basis of eigenvectors of W. Let $\{U_j\}_{1 \leqq j \leqq 2^{2n}}$ be some enumeration of the 2^{2n} unitary matrices with the property that $U_j e_k = \pm e_k$ for each k. Clearly each of these unitaries commutes with W. Thus, by the convexity of F which we have established in the last paragraph,

$$F(Z) = 2^{-2n} \sum_{j=1}^{2^{2n}} F(U_j Z U_j^*) \geqq F\left(2^{-2n} \sum_{j=1}^{2^{2n}} U_j Z U_j^* \right) = F(Z_{\mathrm{diag}}),$$

where Z_{diag} is the matrix whose diagonal entries, in the basis specified above, are those of Z, and whose off-diagonal entries are all zero. Replacing Z by Z_{diag} in (3.13), the integration can be carried out, and we obtain

$$\psi''(0) \geqq p(p-1) \left(\sum_{j=1}^{2n} z_j^{(p-2)} w_j^2 \right)$$

where z_j and w_j, respectively, denote the jth diagonal entries of Z and W in the W-basis specified above.

Now consider $\psi(0 = \mathrm{Tr}(Z^p)$ as a function of Z. It is clearly convex. Thus,by the averaging method just employed, we obtain

$$\psi(0) \geqq \left(\sum_{j=1}^{2n} z_j^p \right).$$

To establish (3.9), it only remains to check that

$$\left(\sum_{j=1}^{2n} z_j^p \right)^{(2-p)/p} \left(\sum_{j=1}^{2n} z_j^{(p-2)} w_j^2 \right) \geqq \left(\sum_{j=1}^{2n} |w_j|^p \right)^2.$$

but this follows immediately from Hölder's inequality. □

IV Hanner's inequality for matrices

This section is devoted to the proof of parts (b) and (c) of Theorem 2. We begin with the proof of Theorem 2(c), and then show how that implies Theorem 2(b).

Proof of Theorem 2(c) First, let Y be a fixed self-adjoint $n \times n$ matrix, and consider the set M_Y of $n \times n$ self-adjoint matrices given by

$$M_Y := \{X \colon X + Y > 0 \quad \text{and} \quad X - Y > 0\}.$$

Clearly M_Y is convex, and if $X \in M_Y$, then $X > 0$.

We claim that

$$G(X) := \|X + Y\|_p^p + \|X - Y\|_p^p - 2\|X\|_p^p \tag{4.1}$$

is a convex function on M_Y.

By the averaging method employed in the proof of Theorem 1, this convexity would imply that

$$\|X + Y\|_p^p + \|X - Y\|_p^p - \alpha\|X\|_p^p \geqq \|X_{\mathrm{diag}} + Y\|_p^p + \|X_{\mathrm{diag}} - Y\|_p^p - \alpha\|X_{\mathrm{diag}}\|_p^p \tag{4.2}$$

for any $0 \leqq \alpha \leqq 2$, where X_{diag} denotes the diagonal part of X in a basis diagonalizing Y. (Note that if $X \in M_Y$, then $X_{\mathrm{diag}} \in M_Y$.) By Lemma 4 and Hanner's inequality in l_p.

$$\|X_{\mathrm{diag}} + Y\|_p^p + \|X_{\mathrm{diag}} - Y\|_p^p - \alpha(r)\|X_{\mathrm{diag}}\|_p^p \geqq \alpha(1/r)\|Y\|_p^p,$$

for all r, where $\alpha(r)$ is the function defined in Lemma 4. (Here we are making use of the easily checked fact that for $1 \leqq p \leqq 2$, $\alpha(r)$ and $\alpha(1/r)$ never exceed 2.) Combining this with (4.2), we would obtain

$$\|X + Y\|_p^p + \|X - Y\|_p^p \geqq \alpha(r)\|X\|_p^p + \alpha(1/r)\|Y\|_p^p.$$

Then, by another application of Lemma 4, the inequality (1.8) would be established for $1 \leqq p \leqq 2$ for all matrices X and Y such that $X + Y$ and $X - Y$ are positive semidefinite. By Lemma 6, and the remark that follows it, (1.8) would be established for $2 \leqq p < \infty$ and all positive semidefinite matrices X and Y.

It remains to establish the convexity of $G(X)$. We choose a self-adjoint matrix A and define

$$\phi(s) = \|(X + sA) + Y\|_p^p + \|(X + sA) - Y\|_p^p - 2\|(X + sA)\|_p^p. \tag{4.3}$$

Then

$$\frac{d}{ds}\phi(s) = p\mathrm{Tr}[((X + sA) + Y)^{p-1} + ((X + sA) - Y)^{p-1} - 2(X + sA)^{p-1}]A.$$

Using the integral representation formula to compute the next derivative, we have

$$\phi''(0) = p\gamma_p \int_0^\infty t^{(p-1)}\mathrm{Tr}\left(\left[\frac{1}{t+X+Y}A\frac{1}{t+X+Y} + \frac{1}{t+X-Y}A\frac{1}{t+X-Y} - 2\frac{1}{t+X}A\frac{1}{t+X}\right]A\right)dt.$$

This is positive by the convexity of

$$X \mapsto \mathrm{Tr}\frac{1}{t+X}A\frac{1}{t+X}A,$$

which we established in the last section. □

Proof of Theorem 2(b) We use a power doubling argument inspired by the 2-convexification method developed by Figiel and Johnson [FJ]. First, consider the case $p \geqq 4$. As in the proof of Theorem 1, we can assume that X and Y are self-adjoint $n \times n$ matrices. The spectrum of the $2n \times 2n$ matrix

$$\begin{pmatrix} X & Y \\ Y & X \end{pmatrix} \tag{4.4}$$

consists of the union of the spectra of $X + Y$ and of $X - Y$. Thus, the pth power of its C_p norm equals the left side of (1.8). By the same spectral considerations, one sees that the pth power of the C_p norm of the 2×2 matrix

$$\begin{pmatrix} \|X\| & \|Y\| \\ \|Y\| & \|X\| \end{pmatrix} \tag{4.5}$$

equals the right side of (1.8). Thus, our problem is to show that the C_p norm of the 2×2 matrix in (4.5) exceeds the C_p norm of the $2n \times 2n$ matrix in (4.4).

Now

$$\left\|\begin{pmatrix} X & Y \\ Y & X \end{pmatrix}\right\|_p^2 = \left\|\begin{pmatrix} X^2 + Y^2 & XY + YX \\ XY + YX & X^2 + Y^2 \end{pmatrix}\right\|_{p/2}. \tag{4.6}$$

The second matrix is positive semidefinite, and it has the special block form $\left(\begin{smallmatrix} A & B \\ B & A \end{smallmatrix}\right)$. Block matrices of this form are characterized by the fact that they commute with $\left(\begin{smallmatrix} 0 & I \\ I & 0 \end{smallmatrix}\right)$, where I is the $n \times n$ identity matrix. Evidently, all powers of a positive semidefinite block matrix of this special form have the same special form.

Thus, if r is the index conjugate to $p/2$, there is a positive matrix $\left(\begin{smallmatrix} C & D \\ D & C \end{smallmatrix}\right)$ whose C_r-norm is 1 with the property that the norm in (4.6) is realized as

$$\mathrm{Tr}\begin{pmatrix} X^2 + Y^2 & XY + YX \\ XY + YX & X^2 + Y^2 \end{pmatrix}\begin{pmatrix} C & D \\ D & C \end{pmatrix} = 2\mathrm{Tr}(X^2 + Y^2)C + 2\mathrm{Tr}(XY + YX)D$$

$$\leqq 2\|C\|_r(\|X^2\|_{p/2} + \|Y^2\|_{p/2}) + 4\|D\|_r\|X\|_p\|Y\|_p$$

by the Hölder inequality for traces of matrices.

Let us define $\|C\| := \|C\|_r$, $\|D\| := \|D\|_r$, $\|X\| := \|X\|_p$ and $\|Y\| := \|Y\|_p$. The last expression is

$$2\|C\|(\|X\|^2 + \|Y\|^2) + 4\|D\|\,\|X\|\,\|Y\|$$
$$= \mathrm{Tr}\begin{pmatrix} \|C\| & \|D\| \\ \|D\| & \|C\| \end{pmatrix}\begin{pmatrix} \|X\|^2 + \|Y\|^2 & 2\|X\|\,\|Y\| \\ 2\|X\|\,\|Y\| & \|X\|^2 + \|Y\|^2 \end{pmatrix}$$
$$\leqq \left\|\begin{pmatrix} \|C\| & \|D\| \\ \|D\| & \|C\| \end{pmatrix}\right\|_r \left\|\begin{pmatrix} \|X\| & \|Y\| \\ \|Y\| & \|X\| \end{pmatrix}\right\|_p^2 .$$

The positivity of the matrix $\begin{pmatrix} C & D \\ D & C \end{pmatrix}$ guarantees that both $C + D$ and $C - D$ are positive. Since $1 \leqq r \leqq 2$, Theorem 2(c) implies that

$$\left\|\begin{pmatrix} \|C\| & \|D\| \\ \|D\| & \|C\| \end{pmatrix}\right\|_r \leqq \left\|\begin{pmatrix} C & D \\ D & C \end{pmatrix}\right\|_r = 1.$$

Consequently,

$$\left\|\begin{pmatrix} X & Y \\ Y & X \end{pmatrix}\right\|_p^2 \leqq \left\|\begin{pmatrix} \|X\| & \|Y\| \\ \|Y\| & \|X\| \end{pmatrix}\right\|_p^2 ,$$

as required. Finally, by Lemma 6, we obtain the validity of (1.8) for $1 \leqq p \leqq \frac{4}{3}$. □

Remark. For all p, (1.8) holds (with the appropriate direction of inequality) in C_p when $\|Y\| = \|X\|$ since (1.8) is then a special case of the "easy" Clarkson inequality (2.1) which was extended to C_p by Dixmier [Di]. It also holds to leading order for small Y, as one can verify using Theorem 1. We make the natural conjecture that (1.8) holds in C_p for $1 \leqq p \leqq 2$, without the restrictions imposed in part (c) of Theorem 2.

References

[ArYa] Araki, H., Yamagami, S.: An inequality for the Hilbert-Schmidt norm. Commun. Math. Phys. **81**, 89–96 (1981)

[BP] Ball, K., Pisier, G.: Unpublished result; private communication.

[Bo] Boas, R.P.: Some uniformly convex spaces. Bull. Am. Math. Soc. **46**, 304-311 (1940)

[C] Clarkson, J.A.: Uniformly convex spaces. Trans. Am. Math. Soc. **40**, 396-414 (1936)

[CL] Carlen, E., Lieb, E.: Optimal hypercontractivity for fermi fields and related non-commutative integration inequalities. Commun. Math. Phys. "**155**, 27-46 (1993); for a slightly different presentation, see: Optimal two-uniform convexity and fermion hypercontractivity. In: Araki, H., Ito, K.R., Kishimoto, A., Ojima, I. (eds,) Quantum and non-commutative analysis. London New York. Kluwer (in press)

[D] Day, M.: Uniform convexity in factor and conjugate spaces. Ann. Math. **45**, 375–385 (1944)

[Di] Dixmier, J.: Formes linéaires sur un anneau d'opérateurs. Bull. Soc. Math. Fr. **81**, 222–245 (1953)

[F] Figiel, T.: On the moduli of convexity and smoothness. Studia Math. **56**, 121-155 (1976)

[FJ] Figiel, T., Johnson, S.B.: A uniformly convex Banach space which contains no C_p. Compos. Math. **29**, 179-190 (1974)

[Gr] Gross, L.: Logarithmic Sobolev inequalities. Am. J. Math. **97**, 1061–1083 (1975)

[H] Hanner, O.: On the uniform convexity of L^p and l^p. Ark. Math. **3**, 239–244 (1956)

[Kö] Köthe, G.: Topologische lineare Räume, Die Grundlehren der mathematischen Wissen schaften in Einzeldarstellungen, Bd. 107, Springer Berlin Heidelberg New York: 1960

[L] Lindenstrauss, J.: On the modulus of smoothness and divergent series in Banach spaces. Mich. Math. J. **10**, 241–252 (1963)

[P] Pisier, G.: The volume of convex bodies and Banach space geometry. Cambridge: Cambridge University Press, 1989

[Ru] Ruskai, M.B.: Inequalities for traces on Von Neumann algebras. Commun. Math. Phys. **26**, 280–289 (1972)

[Se] Segal, I.E.: A non-commutative extension of abstract integration. Ann. Math. **57**, 401–457 (1953)

[Si] Simon, B.: Trace ideals and their applications. (See p. 22) Cambridge: Cambridge University Press, 1979

[TJ] Tomczak-Jaegermann, N.: The moduli of smoothness and convexity and Rademacher averages of trace classes $S_p(1 \leqq p < \infty)$. Studia Math. **50**, 163–182 (1974)

Amer. Math. Soc. Transl.
(2) Vol. **189**, 1999

A Minkowski Type Trace Inequality and Strong Subadditivity of Quantum Entropy

Eric A. Carlen and Elliott H. Lieb

Dedicated to M. Sh. Birman on his seventieth birthday

ABSTRACT. We consider the following trace function on n-tuples of positive operators:

$$\Phi_p(A_1, A_2, \ldots, A_n) = \mathrm{Tr}\left(\left(\sum_{j=1}^{n} A_j^p\right)^{1/p}\right)$$

and prove that it is jointly concave for $0 < p \leq 1$ and convex for $p = 2$. We then derive from this a Minkowski type inequality for operators on a tensor product of three Hilbert spaces, and show how this implies the strong subadditivity of quantum mechanical entropy. For $p > 2$, Φ_p is neither convex nor concave. We conjecture that Φ_p is convex for $1 < p < 2$, but our methods do not show this.

I. Introduction

Let $\mathcal{P}_{\mathcal{H}}$ denote the set of all positive semidefinite operators on a finite dimensional Hilbert space $\mathcal{H}$ with inner product $\langle \cdot, \cdot \rangle$. Then, for any finite natural number n, any finite $p > 0$, and any finite n-tuple $(A_1, A_2, \ldots, A_n)$ of elements of $\mathcal{P}_{\mathcal{H}}$, define

$$\Phi_p(A_1, A_2, \ldots, A_n) = \mathrm{Tr}\left(\left(\sum_{j=1}^{n} A_j^p\right)^{1/p}\right). \tag{1.1}$$

The main result of this paper is the following

THEOREM 1. *For $0 < p \leq 1$, Φ_p is a jointly concave function of its arguments. For $p = 2$, Φ_p is jointly convex. For $p > 2$, Φ_p is neither convex nor concave.*

We conjecture that Φ_p is jointly convex for $1 < p < 2$. We state all of the theorems in a finite dimensional context, and some of our methods of proof explicitly

---

1991 *Mathematics Subject Classification.* Primary 47A63, 15A90.

The work of the first-named author was supperted by U. S. National Science Foundation grant no. DMS 9500840. The work of the second-named author was supported by U. S. National Science Foundation grant no. PHY95–19433–A01.

involve this finite dimension. Nonetheless, the results themselves do not depend on the dimension, and therefore easily extend to the appropriate trace classes on an infinite dimensional Hilbert space.

We note that the trace in Theorem 1 is essential; the asserted trace inequalities *do not* hold as operators inequalities. If they did, we would have, for example at $p = 2$ that $(A^2 + B^2)^{1/2} \leq A + B$. This is of course not true in general positive operators, as is well known and easily checked.

We shall use Theorem 1 to derive a Minkowski type inequality for traces of operators on a product of three Hilbert spaces. To set this in perspective, recall that the Minkowski inequality says that for non-negative measurable functions f on the Cartesian product of two measure spaces (X, μ) and (Y, ν),

$$\left(\int_X \left(\int_Y f(x,y)d\nu\right)^p d\mu\right)^{1/p} \leq \int_Y \left(\int_X f^p(x,y)d\mu\right)^{1/p} d\nu \tag{1.2}$$

for $p \geq 1$, and that the opposite inequality holds for $0 < p \leq 1$.

A direct analog of (1.2) holds for positive operators A on the tensor product of two Hilbert spaces $\mathcal{H}_1 \otimes \mathcal{H}_2$. To state it, let $\mathrm{Tr}_1 A$ denote the positive operator on $\mathcal{H}_2$ that is given as a quadratic form by

$$\langle v, \mathrm{Tr}_1 Av\rangle = \sum_j \langle u_j \otimes v, A(u_j \otimes v)\rangle,$$

where $v \in \mathcal{H}_2$ and the u_j constitute an orthonormal basis of $\mathcal{H}_1$. As is well known, the quadratic form on the left is independent of the choice of the orthonormal basis on the right. The operator $\mathrm{Tr}_1 A$ so defined is called the *partial trace of A over $\mathcal{H}_1$*. It will be convenient, and generally clearer, in what follows to write Tr_1 also to denote the usual trace on $\mathcal{H}_1$ for operators A on $\mathcal{H}_1$ alone.

The following is the tracial analog of (1.2).

THEOREM 2. *Let A be a positive operator on the tensor product of two Hilbert spaces $\mathcal{H}_1 \otimes \mathcal{H}_2$. Then for all $p \geq 1$,*

$$\left(\mathrm{Tr}_2(\mathrm{Tr}_1 A)^p\right)^{1/p} \leq \mathrm{Tr}_1\left(\left(\mathrm{Tr}_2 A^p\right)^{1/p}\right) \tag{1.3}$$

and inequality (1.3) reverses for $0 < p \leq 1$.

Returning to (1.2), note that it has a trivial extension to functions of three (or more) variables. Though trivial, it has an interesting consequence. If one considers a non-negative measurable function $f(x, y, z)$ on the Cartesian product of three measure spaces (X, μ), (Y, ν) and (Z, ρ), and simply holds z fixed as a parameter, one gets

$$\left(\int_X \left(\int_Y f(x,y,z)d\nu\right)^p d\mu\right)^{1/p} \leq \int_Y \left(\int_X f^p(x,y,z)d\mu\right)^{1/p} d\nu \tag{1.4}$$

pointwise in z for $p \geq 1$. Integrating in z then yields

$$\int_Z \left(\int_X \left(\int_Y f(x,y,z)d\nu\right)^p d\mu\right)^{1/p} d\rho \leq \int_Z \int_Y \left(\int_X f^p(x,y,z)d\mu\right)^{1/p} d\nu d\rho \tag{1.5}$$

for $p \geq 1$, and of course the inequality reverses for $0 < p \leq 1$.

Now, since (1.5) is an equality at $p = 1$, we get another inequality by differentiating (1.5) with respect to p at $p = 1$. This yields an entropy inequality. In fact, using the homogeneity of (1.5), we can normalize f so that it is a probability density. Recall that for any probability density ρ on any measure space (X, μ), the entropy $S(\rho)$ is defined as

$$S(\rho) = -\int_X \rho \ln \rho d\mu \ . \tag{1.6}$$

We denote various marginal densities of f as follows:

$$f_{2,3}(y,z) = \int_X f(x,y,z)d\mu, \qquad f_{1,3}(x,z) = \int_Y f(x,y,z)d\nu,$$
$$f_3(z) = \int_X \int_Y f(x,y,z)d\mu d\nu.$$

Then the derivative of (1.5) at $p = 1$ is

$$S(f_{1,3}) + S(f_{2,3}) \geq S(f_{1,2,3}) + S(f_3), \tag{1.7}$$

which is the strong subadditivity of the classical entropy; see [**L75**].

Now consider operators on the product of three Hilbert spaces, and a density matrix A; i.e., a positive operator on $\mathcal{H}_1 \otimes \mathcal{H}_2 \otimes \mathcal{H}_3$ with $\operatorname{Tr} A = 1$. The entropy $S(A)$ of a density matrix A is defined by

$$S(A) = -\operatorname{Tr}(A \ln A). \tag{1.8}$$

The operator analog of (1.7) is the Lieb–Ruskai [**LR**] strong subadditivity inequality for the quantum mechanical entropy:

$$S(A_{1,3}) + S(A_{2,3}) \geq S(A_{1,2,3}) + S(A_3), \tag{1.9}$$

where, in analogy with our notational conventions for marginal densities, we define

$$A_{1,2,3} = A, \quad A_{2,3} = \operatorname{Tr}_1 A, \quad A_3 = \operatorname{Tr}_1 \operatorname{Tr}_2 A$$

and so forth.

Thus, the differential form of Minkowski type inequality (1.7) is known to hold at $p = 1$ for operators. It is therefore natural to enquire whether there exists an operator analog of the three-variable Minkowski inequality (1.7) for other values of p. Unfortunately, the methods at our disposal suffice to establish this only for $0 < p \leq 1$ and for $p = 2$.

THEOREM 3. *Let A be a positive operator on the tensor product of three Hilbert spaces $\mathcal{H}_1 \otimes \mathcal{H}_2 \otimes \mathcal{H}_3$. Then*

$$\operatorname{Tr}_3 \big(\operatorname{Tr}_2 (\operatorname{Tr}_1 A)^p\big)^{1/p} \leq \operatorname{Tr}_{1,3}\big((\operatorname{Tr}_2 A^p)^{1/p}\big) \tag{1.10}$$

for $p = 2$ and, trivially, $p = 1$, while the reverse inequality holds for $0 < p \leq 1$.

This is, nonetheless, enough to imply the strong subadditivity (1.9): one simply takes the left derivative at $p = 1$.

It is readily seen by considering block-diagonal matrices that the inequality of Theorem 3 implies the convexity of Φ_p for $p = 2$, and the concavity of Φ_p for $0 < p \leq 1$. By the same token, (1.10) cannot hold in general for $p > 2$ since this would imply the convexity of Φ_p for such p, and Theorem 1 precludes this. This is in contrast to Theorem 2, the Minkowski inequality for two spaces, which holds for all $p \geq 1$.

The fact that there is such an easy passage from the Minkowski inequality in two variables to that in three variables may leave one surprised that there should be any difficulty in making the same passage with operators. But difficulty there is. In fact, even the simple version in Theorem 2 seems to require a more intricate proof than does the corresponding statement for integrals—which after all is simply the statement that the unit ball in L^p is convex for $p \geq 1$. In fact, we know of no previous proof of Theorem 2.

We emphasize that there is no operator analog of the pointwise inequality (1.4). That is, if we omit Tr_3 on both sides of (1.10), the result will be two operators on $\mathcal{H}_3$, and these two operators do not satisfy the corresponding operator inequality.

We present a proof of Theorem 2 in Section II. Then in Section III we prove Theorem 1. In Section IV, we recast Theorem 1 into an equivalent form, from which Theorem 3 is readily derived in Section IV. Section V contains a brief comment on a relation between the conjectured convexity for $1 < p < 2$ and a very interesting trace inequality of Birman, Koplienko and Solomyak [**BKS**].

II. Proof of Theorem 2

The following proof of Theorem 2 is given for matrices, but is easily extended to operators as the statement is dimension independent.

Let A be a positive operator on $\mathcal{P}_{\mathcal{H}_1 \otimes \mathcal{H}_2}$, the tensor product of two finite dimensional Hilbert spaces. Suppose first that $p > 1$. We proceed by duality. There is a positive operator B in $\mathcal{P}_{\mathcal{H}_2}$ with $(\mathrm{Tr}_2(B^q)^{1/q}) = 1$ with $1/q + 1/p = 1$ such that

$$\begin{aligned}\left(\mathrm{Tr}_2(\mathrm{Tr}_1 A)^p\right)^{1/p} &= \mathrm{Tr}_2\big(B\,\mathrm{Tr}_1 A\big) = \mathrm{Tr}_{1,2}\big((I \otimes B)A\big) \\ &= \sum_{i,j} \langle u_i \otimes v_j, (I \otimes B)A(u_i \otimes v_j)\rangle = \sum_{i,j} \langle u_i \otimes Bv_j, A(u_i \otimes v_j)\rangle\end{aligned}$$

for any pair of orthonormal bases $\{u_i\}$ and $\{v_j\}$. We now choose the $\{v_j\}$ to be a basis of eigenvectors of B, and let $\{\lambda_j\}$ be the corresponding eigenvalues.

Then the right hand side above becomes

$$\begin{aligned}\sum_{i,j} \lambda_j \langle u_i \otimes v_j, A(u_i \otimes v_j)\rangle &\leq \Big(\sum_j \lambda_j^q\Big)^{1/q} \sum_i \Big(\sum_j \Big(\langle u_i \otimes v_j, A(u_i \otimes v_j)\rangle\Big)^p\Big)^{1/p} \\ &= \sum_i \Big(\sum_j \Big(\langle u_i \otimes v_j, A(u_i \otimes v_j)\rangle\Big)^p\Big)^{1/p}.\end{aligned}$$

Next, by the spectral theorem, for each i and j,

$$\langle u_i \otimes v_j, A(u_i \otimes v_j)\rangle \leq \left(\langle u_i \otimes v_j, A^p(u_i \otimes v_j)\rangle\right)^{1/p}.$$

Using this, one arrives at

$$\begin{aligned}\left(\mathrm{Tr}_2(\mathrm{Tr}_1 A)^p\right)^{1/p} &\leq \sum_i \Big(\sum_j \Big(\langle u_i \otimes v_j, A^p(u_i \otimes v_j)\rangle\Big)\Big)^{1/p} \\ &= \sum_i \left(\langle u_i, \mathrm{Tr}_2\, A^p u_i\rangle\right)^{1/p}.\end{aligned}$$

Now we choose the $\{u_i\}$ to be a basis of eigenvectors of $\mathrm{Tr}_2\, A^p$. Then

$$\sum_i \left(\langle u_i, \mathrm{Tr}_2\, A^p u_i\rangle\right)^{1/p} = \sum_i \langle u_i, \left(\mathrm{Tr}_2\, A^p\right)^{1/p} u_i\rangle = \mathrm{Tr}_1 \left(\mathrm{Tr}_2\, A^p\right)^{1/p}$$

and the desired inequality is proved for $p \geq 1$. Note that this part of the proof works for all $p \geq 1$, not only $1 \leq p \leq 2$.

Now suppose $0 < p \leq 1$, and define $r = 1/p$ and $B = A^p$ so that $A = B^r$. Since $r > 1$, the inequality proved above says $\mathrm{Tr}_1\left(\left(\mathrm{Tr}_2\, B^r\right)^{1/r}\right) \geq \left(\mathrm{Tr}_2\left(\mathrm{Tr}_1\, B\right)^r\right)^{1/r}$. Rewriting this in terms of A and p, and switching the roles of $\mathcal{H}_1$ and $\mathcal{H}_2$, one obtains the desired result for $0 < p \leq 1$. □

III. Proof of Theorem 1

As before we give the proof for matrices. Consider first the case $0 < p < 1$. The proof in this case proceeds by reduction to a theorem of Epstein [**E**] concerning the function

$$A \mapsto \mathrm{Tr}\left((BA^pB)^{1/p}\right)$$

on $\mathcal{P}_{\mathcal{H}}$, where B is any given element of $\mathcal{P}_{\mathcal{H}}$. Epstein's theorem says that this function is concave for $0 < p < 1$.

To apply this, consider first the case $n = 2$ in (1.1), and define

$$\mathcal{A} = \begin{bmatrix} A_1 & 0 \\ 0 & A_2 \end{bmatrix}$$

and

$$\sigma = \begin{bmatrix} 0 & I \\ I & 0 \end{bmatrix}.$$

Then

$$\mathcal{A}^p + \sigma \mathcal{A}^p \sigma = \begin{bmatrix} A_1^p + A_2^p & 0 \\ 0 & A_1^p + A_2^p \end{bmatrix}.$$

But

$$\mathcal{A}^p + \sigma \mathcal{A}^p \sigma = 2\left(\frac{I+\sigma}{2}\right)\mathcal{A}^p\left(\frac{I+\sigma}{2}\right) + 2\left(\frac{I-\sigma}{2}\right)\mathcal{A}^p\left(\frac{I-\sigma}{2}\right).$$

Now define

$$\Pi_{\pm} = \frac{I \pm \sigma}{2}$$

and observe that these are complementary orthogonal projections. Thus,

$$(3.1) \quad 2\,\mathrm{Tr}\left((A_1^p + A_2^p)^{1/p}\right) = 2^{1/p}\,\mathrm{Tr}\left(\left(\Pi_+\mathcal{A}^p\Pi_+\right)^{1/p}\right) + 2^{1/p}\,\mathrm{Tr}\left(\left(\Pi_-\mathcal{A}^p\Pi_-\right)^{1/p}\right).$$

Epstein's theorem, with $A = \mathcal{A}$ and $B = \Pi_{\pm}$, now implies that each term on the right hand side of (3.1) is a concave function of $\mathcal{A}$, which means that the left hand side is a jointly concave function of A_1 and A_2. This concludes the proof for $n = 2$.

One now easily iterates this procedure to obtain the result for all dyadic powers $n = 2^k$, and hence for all n.

To prove the convexity of Φ_2 there are several ways to proceed, but the simplest was pointed out to us by S. Sahi. Namely, let n be given and consider the block matrix $\mathcal{A}$ given by

$$\mathcal{A} = \begin{bmatrix} A_1 & 0 & \dots & 0 \\ A_2 & 0 & \dots & 0 \\ \vdots & \vdots & \dots & \vdots \\ A_n & 0 & \dots & 0 \end{bmatrix}.$$

Then

$$\Phi_2(A_1, A_2, \dots, A_n) = \operatorname{Tr} |\mathcal{A}|,$$

where $|X|$ is the usual operator absolute value; i.e., $\sqrt{X^*X}$. In other words $\Phi_2(A_1, A_2, \dots, A_n)$ is simply the trace norm of $\mathcal{A}$, is therefore clearly jointly convex in $A_1, A_2, \dots, A_n$.

Finally, we show that convexity fails to hold for $p > 2$. To see this, choose any pair A_1,$A_2 \in \mathcal{P}_{\mathcal{H}}$, and any vector v such that

$$\langle v, ((A_1^p + A_2^p)/2)v\rangle < \langle v, ((A_1 + A_2)/2)^p v\rangle \ . \tag{3.2}$$

Note the strict inequality here. It is always possible to find such A_1,A_2 and v for $p > 2$ since, for such p, $X \mapsto X^p$ is not operator convex.

Now let Π_v denote the orthogonal projection onto the span of v, and let $\Pi_v^\perp = I - \Pi_v$ denote its orthogonal complement. Then, for a large number λ to be fixed below, put

$$B = \Pi_v + \lambda \Pi_v^\perp .$$

Then, if Φ_p were convex, we would have

$$\limsup_{t\to 0} pt^{-p}\left(\Phi_p\left(t\frac{A_1 + A_2}{2}, B\right) - \frac{1}{2}\Phi_p(tA_1, B) - \frac{1}{2}\Phi_p(tA_2, B)\right) \le 0 \ . \tag{3.3}$$

However, for small $t > 0$,

$$\begin{aligned}\Phi_p\left(t\frac{A_1 + A_2}{2}, B\right) &= \operatorname{Tr}\left(t^p\left(\frac{A_1 + A_2}{2}\right)^p + B^p\right)^{1/p} \\ &= \operatorname{Tr} B + \frac{t^p}{p}\operatorname{Tr}\left(B^{1-p}\left(\frac{A_1 + A_2}{2}\right)^p\right) + O(t^{2p})\end{aligned}$$

and

$$\frac{1}{2}\Phi_p(tA_1, B) + \frac{1}{2}\Phi_p(tA_2, B) = \operatorname{Tr} B + \frac{t^p}{p}\left(\frac{1}{2}\operatorname{Tr} B^{1-p}A_1^p + \frac{1}{2}\operatorname{Tr} B^{1-p}A_2^p\right) + O(t^{2p}).$$

Thus,

$$\begin{aligned}&\limsup_{t\to 0} pt^{-p}\left(\Phi_p\left(t\frac{A_1 + A_2}{2}, B\right) - \frac{1}{2}\Phi(tA_1, B) - \frac{1}{2}\Phi(tA_2, B)\right) \\ &\quad = \operatorname{Tr}\left(B^{1-p}\left(\frac{A_1 + A_2}{2}\right)^p\right) - \left(\frac{1}{2}\operatorname{Tr} B^{1-p}A_1^p + \frac{1}{2}\operatorname{Tr} B^{1-p}A_2^p\right) \\ &\quad = \left\langle v, \left(\frac{A_1 + A_2}{2}\right)^p v\right\rangle - \left(\frac{\langle v, A_1^p v\rangle}{2} + \frac{\langle v, A_2^p v\rangle}{2}\right) + O(\lambda^{1-p}).\end{aligned}$$

Now taking λ sufficiently large, this last term on the right is strictly positive by (3.2). This contradicts (3.3), and thus convexity does not hold—not even separately. □

IV. Corollary of Theorem 1 and proof of Theorem 3

A corollary of Theorem 1 is obtained by writing the partial trace as an average, and exploiting the convexity and concavity established above. Let A be a positive operator on $\mathcal{H}_1 \otimes \mathcal{H}_2$. Next, suppose the dimension of $\mathcal{H}_2$ is N, and fix some orthonormal basis $\{e_1, e_2, \ldots, e_N\}$. With respect to this basis, define the self-adjoint unitary operators $U_{i,j}$ and $V_{i,j}$ on $\mathcal{H}_2$ by

$$\begin{aligned} U_{i,j} &= I - E_{i,i} - E_{j,j} + E_{i,j} + E_{j,i}, \\ V_i &= I - 2E_{i,i}, \end{aligned}$$

where the i and j are a distinct pair of indices, and $E_{i,j}$ in this basis has the matrix with 1 in the i,jth place, and 0 elsewhere. Let $\mathcal{G}$ be the subgroup of the group of unitary operators on $\mathcal{H}_2$ that is generated by this family together with the identity. Each operator W in this group acts by

$$We_j = (-1)^{s(j)} e_{\pi(j)}$$

for some permutation $\pi(\cdot)$, and some map $s : 1, 2, \ldots, N \mapsto 0, 1$. Thus, the size of the group is $2^N N!$, and the point about it is that any operator on $\mathcal{H}_2$ that commutes with every element of this group is necessarily a multiple of the identity on $\mathcal{H}_2$. Then

$$\frac{1}{2^N N!} \sum_{W \in \mathcal{G}} (I \otimes W^*) A (I \otimes W) = \frac{1}{N} \operatorname{Tr}_2(A) \otimes I_{\mathcal{H}_2}. \tag{4.1}$$

This way of writing partial traces can be traced back to Uhlmann [**U**]. From here one easily arrives at the following result:

THEOREM 4. *For $p > 0$, let the map $\Psi_p(A)$ from positive operators $\mathcal{A}$ on $\mathcal{H}_1 \otimes \mathcal{H}_2$ to $\mathbb{R}_+$ be given by*

$$\Psi_p(A) = \operatorname{Tr}_1\left((\operatorname{Tr}_2 A^p)^{1/p} \right) . \tag{4.2}$$

Then this map is concave for $0 < p \leq 1$, convex for $p = 2$, and neither for $p > 2$.

PROOF. We shall assume that the dimension of $\mathcal{H}_2$ is N so that we may apply the averaging formula introduced above. We then have

$$\begin{aligned} \operatorname{Tr}_1\left((\operatorname{Tr}_2 A^p)^{1/p} \right) &= N^{1/p-1} \operatorname{Tr}_{1,2}\left(\left(\frac{1}{N} \operatorname{Tr}_2 A^p \otimes I_{\mathcal{H}_2} \right)^{1/p} \right) \\ &= N^{1/p-1} \operatorname{Tr}_{1,2}\left(\left(\frac{1}{2^N N!} \sum_{W \in \mathcal{G}} (I \otimes W^*) A^p (I \otimes W) \right)^{1/p} \right) \\ &= N^{1/p-1} \left(\frac{1}{2^N N!} \right)^{1/p} \operatorname{Tr}_{1,2}\left(\left(\sum_{W \in \mathcal{G}} ((I \otimes W^*) A (I \otimes W))^p \right)^{1/p} \right). \end{aligned}$$

The result now follows directly from Theorem 1. □

Notice that the conclusion of Theorem 4 not only follows from Theorem 1, but also implies it. To see this, suppose that the A in Theorem 4 is block diagonal with

$$A = \begin{bmatrix} A_1 & 0 & \dots & 0 \\ 0 & A_2 & \dots & 0 \\ \vdots & \vdots & \dots & \vdots \\ 0 & 0 & \dots & A_n \end{bmatrix}.$$

Then clearly

$$\Psi_p(A) = \Phi_p(A_1, A_2, \dots, A_n).$$

We remark that if our convexity conjecture turns out to be true for $1 < p < 2$, then a proof along the same lines as the proof above of Theorem 4 will prove the conjectured tracial generalization of Minkowski's inequality.

Proof of Theorem 3. Suppose that the dimension of $\mathcal{H}_1$ is N. The left hand side of (1.10) can be written in terms of Ψ_p, namely

$$\mathrm{Tr}_3\big(\mathrm{Tr}_2(\mathrm{Tr}_1 A)^p\big)^{1/p} = \mathrm{Tr}_{1,3}\left(\mathrm{Tr}_2\left(\frac{1}{N}\mathrm{Tr}_1 A\otimes I_{\mathcal{H}_1}\right)^p\right)^{1/p} = \Psi_p\left(\frac{1}{N}\mathrm{Tr}_1 A\otimes I_{\mathcal{H}_1}\right),$$

where the pair of spaces in the definition of Ψ_p is taken to be $\mathcal{H}_2$ and $\mathcal{H}_1\otimes\mathcal{H}_3$.

Then by (4.1) and the convexity of Ψ_p established in Theorem 4,

$$\begin{aligned}\Psi_p\left(\frac{1}{N}\mathrm{Tr}_1 A\otimes I_{\mathcal{H}_1}\right) &= \Psi_p\left(\frac{1}{2^N N!}\sum_{W\in\mathcal{G}}(I\otimes W^*)A(I\otimes W)\right)\\ &\le \frac{1}{2^N N!}\sum_{W\in\mathcal{G}}\Psi_p\Big((I\otimes W^*)A(I\otimes W)\Big).\end{aligned}$$

The last term above is

$$\begin{aligned}&\frac{1}{2^N N!}\sum_{W\in\mathcal{G}}\mathrm{Tr}_{1,3}\Big(\mathrm{Tr}_2\big((I\otimes W^*)A^p(I\otimes W)\big)\Big)^{1/p}\\ &\quad= \frac{1}{2^N N!}\sum_{W\in\mathcal{G}}\mathrm{Tr}_{1,3}\Big((I\otimes W^*)(\mathrm{Tr}_2 A^p)(I\otimes W)\Big)^{1/p}\\ &\quad= \frac{1}{2^N N!}\sum_{W\in\mathcal{G}}\mathrm{Tr}_{1,3}\Big((I\otimes W^*)(\mathrm{Tr}_2 A^p)^{1/p}(I\otimes W)\Big) = \mathrm{Tr}_3\,\mathrm{Tr}_1\big((\mathrm{Tr}_2 A^p)^{1/p}\big),\end{aligned}$$

which is the desired result. □

V. The BKS inequality and the $1 < p < 2$ conjecture

Birman, Koplienko and Solomyak [**BKS**] proved that for $p > 1$, and A and B positive semidefinite operators,

$$\mathrm{Tr}\big(B^p - A^p\big)_+^{1/p} \ge \mathrm{Tr}(B - A)_+, \tag{5.1}$$

where X_+ denotes the positive part of a selfadjoint operator X; i.e., $X_+ = (X + |X|)/2$. In (5.1), neither B nor A needs to be bounded, but it is assumed that $\big(B^p - A^p\big)_+^{1/p}$ is trace class. Though the inequality in (5.1) is only one of several very interesting inqualities proved in [**BKS**], we refer to it here as the BKS inequality.

The proof is in two parts, the first of which is to reduce consideration to the case $B^p \geq A^p$ in which case one has $B = (A^p + C^p)^{1/p}$ with $C > 0$. Then (5.1) becomes

$$\mathrm{Tr}(C + A) \geq \mathrm{Tr}(A^p + C^p)^{1/p} \tag{5.2}$$

for all $A \geq 0$ and $C \geq 0$. It is (5.2) that interests us here.

Clearly (5.2) can be rewritten as

$$\Phi_p(A, C) \leq \Phi_p(A, 0) + \Phi_p(0, C), \tag{5.3}$$

which is a subadditivity property of Φ_p for all $p > 1$. Since Φ_p is homogeneous of degree 1, subadditivity and convexity are the same thing. Thus for $p = 2$, (5.3) is a special case of the convexity of Φ_p proved in Theorem 1, and for $1 < p < 2$, it would be a consequence of the conjectured convexity for these p. However, the BKS inequality holds for all $p > 1$, not only for $1 < p < 2$.

There is a simple proof of (5.2) for matrices. Let

$$M_\pm = \begin{bmatrix} A^{p/2} & 0 \\ \pm C^{p/2} & 0 \end{bmatrix}$$

so that

$$\mathrm{Tr}\big(M_\pm^* M_\pm\big)^{1/p} = \mathrm{Tr}(A^p + C^p)^{1/p}.$$

On the other hand, the spectrum of $M_\pm^* M_\pm$ is the same as the spectrum of $M_\pm M_\pm^*$, so

$$\mathrm{Tr}(A^p + C^p)^{1/p} = \mathrm{Tr}\big(M_\pm M_\pm^*\big)^{1/p}.$$

One computes

$$M_\pm M_\pm^* = \begin{bmatrix} A^p & \pm J \\ \pm J & C^p \end{bmatrix}$$

with $J = A^{p/2}C^{p/2}$. Since $X \mapsto \mathrm{Tr}(X^{1/p})$ is concave for $p > 1$, one has that

$$\mathrm{Tr}(A + C) = \mathrm{Tr}\left(\frac{M_+M_+^* + M_-M_-^*}{2}\right)^{1/p} \geq \mathrm{Tr}\big(A^p + C^p\big)^{1/p}.$$

A recent application of the BKS inequality, and a different proof of (5.1) that holds in the case of unbounded operators, can be found in [**LSS**].

Acknowledgements. We thank T. Ando and F. Hiai for a careful reading of this paper, and for pointing out many misprints in an earlier draft.

References

[AH] T. Ando and F. Hiai, *Hölder type inequalities for matrices*, Preprint, 1997.

[BKS] M. S. Birman, L. S. Koplienko, and M. Z. Solomyak, *Estimates for the spectrum of the difference between fractional powers of two selfadjoint operators*, J. Soviet Math. **19** (1975), no. 3, 1–6.

[E] H. Epstein, *On a concavity theorem of Lieb*, Commun. Math. Phys. **31** (1973), 317–327.

[L75] E. H. Lieb, *Some convexity and subadditivity properties of entropy*, Bull. Amer. Math. Soc. **81** (1975), 1–13.

[LR] E. H. Lieb, and M. B. Ruskai, *Proof of the strong subadditivity of quantum-mechanical entropy*, J. Math. Phys. **14** (1973), 1938–1941.

[LSS] E. H. Lieb, H. Siedentop, and J. P. Solovej, *Stability and instability of relativistic electrons in magnetic fields*, J. Stat. Phys. **89** (1997), 37–59.

[U] A. Uhlmann, *Sätze über Dichtematrizen*, Wiss. Z. Karl-Marx Univ. Leipzig **20** (1971), 633–653.

School of Mathematics, Georgia Institute of Technology, Atlanta, Georgia 30332

Departments of Mathematics and Physics, Princeton University, Princeton, New Jersey 08544–0708

Part III

Inequalities Related to the Stability of Matter

INEQUALITIES FOR THE MOMENTS OF THE EIGENVALUES OF THE SCHRÖDINGER HAMILTONIAN AND THEIR RELATION TO SOBOLEV INEQUALITIES

Elliott H. Lieb*
Walter E. Thirring

1. *Introduction*

Estimates for the number of bound states and their energies, $e_j \leq 0$, are of obvious importance for the investigation of quantum mechanical Hamiltonians. If the latter are of the single particle form $H = -\Delta + V(x)$ in R^n, we shall use available methods to derive the bounds

$$\sum_j |e_j|^\gamma \leq L_{\gamma,n} \int d^n x \, |V(x)|_-^{\gamma+n/2}, \qquad \gamma > \max(0, 1-n/2). \quad (1.1)$$

Here, $|V(x)|_- = -V(x)$ if $V(x) \leq 0$ and is zero otherwise.

Of course, in many-body theory, one is more interested in Hamiltonians of the form $-\sum_i \Delta_i + \sum_{i>j} v(x_i - x_j)$. It turns out, however, that the energy bounds for the single particle Hamiltonian yield a lower bound for the kinetic energy, T, of N fermions in terms of integrals over the single particle density defined by

$$\rho(x) \equiv N \int |\psi(x, x_2, \cdots, x_N)|^2 \, d^n x_2 \cdots d^n x_N \, , \quad (1.2)$$

where ψ is an antisymmetric, normalized function of the N variables $x_i \in R^n$. Our main results, in addition to (1.1), will be of the form

*Work supported by U. S. National Science Foundation Grant MPS 71-03375-A03.

$$T \equiv \sum_{i=1}^{N} \int |\vec{\nabla}_i \psi(x_1 \cdots x_N)|^2 \, d^n x_1 \cdots d^n x_N$$
$$\geq K_{p,n} \left[\int d^n x \, \rho(x)^{p/(p-1)} \right]^{2(p-1)/n} \tag{1.3}$$

when $\max\{n/2, 1\} \leq p \leq 1 + n/2$.

For $N = 1$, $p = n/2$, (1.3) reduces to the well-known Sobolev inequalities. (1.3) is therefore a partial generalization of these inequalities, and we shall expand on this in Section 3.

Our constants $K_{p,n}$ are not always the best possible ones, but nevertheless, they may be useful for many purposes. In particular, in ref. [1], a special case of (1.3) was used to give a simple proof of the stability of matter, with a constant of the right order of magnitude. The result for q species of fermions $(2m = e = \hbar = 1)$ moving in the field of M nuclei with positive charges Z_j is

$$H \geq -1.31\, q^{2/3} N \left[1 + \left(\sum_{j=1}^{M} Z_j^{7/3}/N \right)^{1/2} \right]^2 . \tag{1.4}$$

In particular, if $q = 2$ (spin 1/2 electrons), we have a bound $\sim N$, and if we set $q = N$, we get a bound $\sim N^{5/3}$ if no symmetry requirement is imposed on the wave function; *a fortiori* this is a bound for *bosons*. Our bound implies stability of matter in its intuitive meaning such that the volume occupied by N particles will be $\sim N$ (Bohr radius)^3. To give a formal demonstration of this fact, one might use a method which gives lower bounds for the radii of complex atoms (compare Equation (3.6, 38) of ref. [20]). As a first observation, one calculates the ground state energy of N electrons (with spin) in a harmonic potential. Filling the oscillator levels, one finds

$$\frac{1}{2} \sum_{i=1}^{N} (-\Delta_i + \omega^2 \vec{x}_i^2) \geq \omega N^{4/3} \frac{3^{4/3}}{4} (1 + O(N^{-1/3})) . \tag{1.5}$$

Next, take the expectation value of this operator inequality with the ground state of H, set

$$\omega = \frac{4}{3^{4/3}} \langle -\sum_i \Delta_i \rangle \frac{1}{N^{4/3}} \tag{1.6}$$

and use the virial theorem

$$\langle -\sum_i \Delta_i \rangle = -E_0 \leq 2.08\,N \left[1 + \left(\sum_{j=1}^{M} Z_j^{7/3}/N\right)^{1/2}\right]^2 . \tag{1.7}$$

Altogether we find

$$\langle \sum_{i=1}^{N} \vec{x}_i^2 \rangle \geq \frac{(3N)^{8/3}}{16 \langle -\sum_i \Delta_i \rangle} \geq \frac{3^{8/3} N^{5/3}}{16 \cdot 2.08 \left[1 + \left(\sum_j Z_j^{7/3}/N\right)^{1/2}\right]^2} . \tag{1.8}$$

Thus we have proved that

$$\langle \vec{x}_i^2 \rangle^{1/2} \geq c N^{1/3}, \qquad c = \frac{.75}{\left[1 + \left(\sum_j Z_j^{7/3}/N\right)^{1/2}\right]} . \tag{1.9}$$

Therefore, if the system is not compressed by other forces, so that the virial theorem is valid, it will not collapse, but will adjust its volume to a size proportional to the number of particles. Regarding the Z-dependence, we see that with $Z = Z_j = N/M$ we have (for large Z)

$$\langle \vec{x}_i^2 \rangle^{1/2} \sim M^{1/3} Z^{-1/3} .$$

That is, the mean atomic radius is predicted to be $\geq Z^{-1/3}$. A better result can hardly be expected since for $M = 1$, this is the correct Z-dependence for large Z.

Although we have no results on the best possible constants, $K_{p,n}$, except in a few special cases, experience drawn from computer calculations suggests that there is a critical value $\gamma_{c.n}$ above which the classical value gives a bound:

$$\Big(\sum |e_j|^{\gamma}\Big)_{\text{classical}} = (2\pi)^{-n} \int d^n p\, d^n x\, |p^2 + V(x)|_-^{\gamma}$$

$$\equiv L^C_{\gamma,n} \int |V(x)|_-^{\gamma+n/2}\, d^n x \ ,$$

$$\gamma \geq \gamma_{c,n} \ , \tag{1.10}$$

and where $L^C_{\gamma,n}$, given by the above integral, is

$$L^C_{\gamma,n} = 2^{-n}\pi^{-n/2}\,\Gamma(\gamma+1)/\Gamma(\gamma+1+n/2) \ . \tag{1.11}$$

We conjecture $\gamma_{c,1} = 3/2$, $\gamma_{c,3} \cong .863$ and $\gamma_{c,n} = 0$, all $n \geq 8$. If this conjecture were to be true, the constants in (1.3, 1.4) could be further improved.

In the next section we shall deduce bounds for $\sum_j |e_j|^{\gamma}$ and use them in Section 3 to derive (1.3). In Section 4 we shall discuss our conjectures and support them for $n = 1$ with results from the Korteweg-de Vries equation. Section 5 contains new results added in proof. In Appendix A, generously contributed by J. F. Barnes, further evidence from computer studies is presented. We are extremely grateful to Dr. Barnes for taking an interest in this problem, for without his results we would have been hesitant to put forth our conjectures.

2. *Bounds for Moments of the Eigenvalues*

In this section we shall deduce bounds of the form (1.1), and we shall compare our $L_{\gamma,n}$ with the classical values which one gets by replacing

$$\sum_j |e_j|^{\gamma} \qquad \text{by} \qquad (2\pi)^{-n} \int d^n x\, d^n p\, |p^2 + V(x)|_-^{\gamma} \ .$$

For $n \sim 3$ and $\gamma \sim 1$, the latter are smaller by about an order of magnitude.

Our inequalities are based on the Birman-Schwinger [2, 3] method for estimating N_E, the number of bound states of $H = -\Delta + V(x)$ having an energy $\leq E$. Since

$$\frac{\partial}{\partial E} N_E = \sum_j \delta(E - e_j)$$

we have

$$\sum_j |e_j|^\gamma = \gamma \int_0^\infty d\alpha\, \alpha^{\gamma-1} N_{-\alpha} \,. \tag{2.1}$$

Now, according to Birman-Schwinger [2, 3], for all $\alpha \geq 0$, $m \geq 1$ and $t \in [0, 1]$,

$$N_{-\alpha} \leq \mathrm{Tr}(|V+(1-t)\alpha|_-^{1/2}(-\Delta + t\alpha)^{-1}|V+(1-t)\alpha|_-^{1/2})^m \,. \tag{2.2}$$

REMARKS ABOUT (2.2):

1. We are only interested in potentials such that $V_- \in L^{\gamma+n/2}(R^n)$ for $\gamma \geq \min(0, 1-n/2)$. For such potentials (2.2) is justified, and a complete discussion is given in Simon [4, 5]. Moreover, it is sufficient to consider $V \in C_0^\infty(R^n)$ in (2.2), and in the rest of this paper, and then to use a limiting argument. Such potentials have the advantage that they have only a finite number of bound states [5].

2. Since we are interested in maximizing $\Sigma |e_j|^\gamma / \int |V|_-^{\gamma+n/2}$, we may as well assume that $V(x) \leq 0$, i.e. $V = -|V|_-$. This follows from the max-min principle [4] which asserts that $e_j(V) \geq e_j(-|V|_-)$, all j, including multiplicity.

To evaluate the trace in (2.2), we use the inequality

$$\mathrm{Tr}(B^{1/2} A B^{1/2})^m \leq \mathrm{Tr}\, B^{m/2} A^m B^{m/2} \tag{2.3}$$

when A, B are positive operators and $m \geq 1$. When m is integral and A, B is of our special form, (2.3) is a consequence of Hölder's inequality. For completeness, we shall give a more general derivation of (2.3) in Appendix B.

To calculate

$$\mathrm{Tr}\,|V+(1-t)\alpha|_-^m(-\Delta+t\alpha)^{-m}, \tag{2.4}$$

we shall use an x-representation where $(-\Delta+t\alpha)^{-m}$ is the kernel

$$G_{t\alpha}^{(m)}(x-y) = (2\pi)^{-n}\int d^np\,(p^2+t\alpha)^{-m}\,e^{ip(x-y)} \tag{2.5}$$

if $m > n/2$. Using

$$d^np = \frac{2\pi^{n/2}}{\Gamma(n/2)}\int_0^\infty dp\; p^{n-1}, \tag{2.6}$$

we easily compute

$$G_{t\alpha}^{(m)}(0) = (2\pi)^{-n}\frac{2\pi^{n/2}}{\Gamma(n/2)}(t\alpha)^{-m+n/2}\int_0^\infty p^{n-1}(p^2+1)^{-m}\,dp$$

$$= (4\pi)^{-n/2}\frac{\Gamma(m-n/2)}{\Gamma(m)}(t\alpha)^{-m+n/2} \tag{2.7}$$

if $m > n/2$. Thus,

$$N_{-\alpha} \le (4\pi)^{-n/2}\frac{\Gamma(m-n/2)}{\Gamma(m)}(t\alpha)^{-m+n/2}\int d^nx\,|V(x)+(1-t)\alpha|_-^m. \tag{2.8}$$

Next, we substitute (2.8) into (2.1). If we impose the condition that $t < 1$, it is easy to prove that one can interchange the α and the x integration. Changing variables $\alpha \to (1-t)^{-1}|V(x)|_-\beta$, leads to

$$\sum|e_j|^\gamma \le \gamma(4\pi)^{-n/2}t^{-m+n/2}(1-t)^{m-\gamma-n/2}\frac{\Gamma(\gamma-m+n/2)\Gamma(m-n/2)}{\Gamma(\gamma+1+n/2)}\,m\int d^nx \times |V(x)|_-^{\gamma+n/2} \tag{2.9}$$

provided $n/2 < m < n/2+\gamma$, $m \ge 1$ and $0 < t < 1$. The optimal t is $t = (m-n/2)/\gamma$.

If we put our results together, we obtain the following (see note added in proof, Section 5).

THEOREM 1. *Let* $V_- \in L^{\gamma+n/2}(R^n)$, $\gamma \geq \max(0, 1-n/2)$. *Let* $H = -\Delta + V(x)$, *and let* $e_j \leq 0$ *be the negative energy bound states of* H. *Then*

$$\sum |e_j|^\gamma \leq L_{\gamma,n} \int |V(x)|_-^{\gamma+n/2} \tag{2.10}$$

where

$$L_{\gamma,n} \leq \tilde{L}_{\gamma,n} \equiv \min_m (4\pi)^{-n/2}\gamma^{\gamma+1} \frac{m}{\Gamma\left(\gamma+\frac{n}{2}+1\right)} F\left(m-\frac{n}{2}\right) F\left(\gamma+\frac{n}{2}-m\right), \tag{2.11}$$

and where $F(x) = \Gamma(x)x^{-x}$, $\max\{1, n/2\} \leq m < n/2+\gamma$.

REMARKS:

1. When $\gamma = 0$, $\Sigma|e_j|^0$ means the number of bound states, including zero energy states. For $n \geq 2$, our $\tilde{L}_{0,n} = \infty$. In Section 4, we shall discuss the $\gamma = 0$ case further. See also Section 5.

2. In (2.11), $\tilde{L}_{\gamma,n}$ is the bound we have obtained using the Birman-Schwinger principle. We shall henceforth reserve the symbol $L_{\gamma,n}$ for the quantity

$$L_{\gamma,n} \equiv \sup_V \sum |e_j|^\gamma \Big/ \int |V|_-^{\gamma+n/2} . \tag{2.12}$$

Optimization with respect to m in (2.11) can be done either numerically or analytically in the region where Stirling's formula

$$F(x) \sim e^{-x}\sqrt{2\pi/x} \tag{2.13}$$

can be applied. In [1], for $n = 3$, $\gamma = 1$, we used the value 2 for m. A marginal improvement can be obtained with $m = 1.9$.

If (2.13) were exact, the best m would be

$$\bar{m} = n(\gamma+n/2)/(n+\gamma) . \tag{2.14}$$

Note that as $\gamma \to \infty$, $\bar{m}$ is bounded by n. Using $\bar{m}$, together with (2.13), which is valid when $\gamma n(\gamma+n)^{-1}$ is large,

$$\tilde{L}_{\gamma,n} \sim (4\pi)^{1-n/2} \frac{\gamma^\gamma e^{-\gamma}}{\Gamma(\gamma+n/2)} \left[\frac{n/2}{\gamma+n/2}\right]^{1/2} . \tag{2.15}$$

Finally, we want to compare our bounds with their classical values, $L^C_{\gamma,n}$. From the results of Martin [6] and Tamura [7], one has the following

THEOREM 2. *If* $V(x) \leq 0$ *and* $V \in C_0^\infty(R^n)$, *then*

$$\lim_{\lambda\to\infty} \sum_j |e_j(\lambda V)|^\gamma \Big/ \int |\lambda V|^{\gamma+n/2} = L^C_{\gamma,n} . \tag{2.16}$$

COROLLARY.

$$L_{\gamma,n} \geq L^C_{\gamma,n} . \tag{2.17}$$

Our $\tilde{L}_{\gamma,n}$ satisfies (2.17), in particular in the asymptotic region (2.15), we find

$$\tilde{L}_{\gamma,n}/L^C_{\gamma,n} \approx [4\pi n(\gamma+n/2)]^{1/2} \gamma^{-1/2} . \tag{2.18}$$

We conjecture in Section 4 that for γ sufficiently large, the best possible $L_{\gamma,n}$ should be $L^C_{\gamma,n}$, a result which does not follow from the Birman-Schwinger method employed here. For small γ, we know that $L^C_{\gamma,n}$ is not a bound.

We conclude this section with a theorem about $L_{\gamma,n}$ which will be useful in the discussion of the one-dimensional case in Section 4.

THEOREM 3. *Let* $\gamma \geq 1 + \max(0, 1-n/2)$. *Then*

$$L_{\gamma,n} \leq L_{\gamma-1,n}[\gamma/(\gamma+n/2)] . \tag{2.19}$$

PROOF. Choose $\varepsilon > 0$. We can find a $V \in C_0^\infty(R^n)$, with $V \leq 0$, such that

$$L_{\gamma,n}(V) = \sum_j |e_j(V)|^\gamma \Big/ \int |V|^{\gamma+n/2} \geq L_{\gamma,n} - \varepsilon .$$

Let $g \in C_0^\infty(\mathbf{R}^n)$ be such that $0 \le g(x) \le 1$, $\forall x$, and $V(x) \ne 0$ implies $g(x) = 1$. Let $V_\lambda(x) = V(x) - \lambda g(x)$, $\lambda \le 0$. The functions $|e_j(V_\lambda)|$ are continuous and monotone increasing in λ. Furthermore, there are a finite number of values $-\infty < \lambda_1 \le \lambda_2 \le \cdots \le \lambda_k \le 0$ with λ_j being the value of λ at which $e_j(V_\lambda)$ first appears. λ_1 is finite because V_λ is non-negative for λ sufficiently negative. $e_j(V_\lambda)$ is continuously differentiable on $\Lambda = \{\lambda | 0 \ge \lambda > \lambda_1, \lambda \ne \lambda_i,\ i = 1,\cdots,k\}$ and

$$de_j(V_\lambda)/d\lambda = -\int |\psi_j(x; V_\lambda)|^2 g(x)\, d^n x$$

by the Feynman-Hellman theorem. It is easy to prove that if $f, g \in L^p(\mathbf{R}^n)$, $p > 1$, then

$$h(\lambda) \equiv \int |f(x) - \lambda g(x)|_-^p\, d^n x$$

is differentiable, $\forall\lambda$ and

$$dh/d\lambda|_{\lambda=0} = p \int |f(x)|_-^{p-1} g(x)\, d^n x\ .$$

Thus $L_{\gamma,n}(V_\lambda)$ is piecewise C^1 on Λ and its derivative, $\dot{L}_{\gamma,n}$, is given by

$$\dot{L}_{\gamma,n} = \left[\int |V_\lambda|_-^{\gamma+n/2}\right]^{-1} \Big\{ \gamma \sum_j |e_j(V_\lambda)|^{\gamma-1} \int g(x) |\psi_j(x; V_\lambda)|^2 d^n x - (\gamma+n/2) L_{\gamma,n}(V_\lambda)$$

$$\cdot \int |V_\lambda(x)|_-^{\gamma+n/2-1} g(x)\, d^n x \Big\}\ .$$

By the stated properties of $L_{\gamma,n}$, there exists a $\lambda \in (\lambda_1, 0]$ such that

(i) $\dot{L}_{\gamma,n}(V_\lambda) \ge 0$;

(ii) $L_{\gamma,n}(V_\lambda) \ge L_{\gamma,n} - 2\varepsilon$.

Thus, using the properties of g,

$$0 \le \gamma \sum_j |e_j(V_\lambda)|^{\gamma-1} - L_{\gamma,n}(V_\lambda)(\gamma + n/2) \int |V_\lambda|_-^{\gamma+n/2-1}\ . \tag{2.20}$$

Since ε was arbitrary, (2.20) implies the theorem.

If we use (2.17) together with the fact that $L^C_{\gamma,n} = L^C_{\gamma-1,n}[\gamma/(\gamma+n/2)]$, we have

COROLLARY. *If for some* $\gamma \geq \max(0, 1-n/2)$, $L_{\gamma,n} = L^C_{\gamma,n}$, *then*

$$L_{\gamma+j,n} = L^C_{\gamma+j,n}, \qquad j = 0, 1, 2, 3, \cdots .$$

REMARK. By the same proof

$$L^1_{\gamma,n} \leq L^1_{\gamma-1,n}[\gamma/(\gamma+n/2)] \tag{2.21}$$

(see (3.1) for the definition of $L^1_{\gamma,n}$).

3. *Bounds for the Kinetic Energy*

In this section, we shall use Theorem 1 to derive inequalities of the type (1.3). We recall the definition (2.14) and we further define

$$L^1_{\gamma,n} \equiv \sup_V |e_1|^\gamma \Big/ \int |V|_-^{\gamma+n/2} . \tag{3.1}$$

Clearly,

$$L^1_{\gamma,n} \leq L_{\gamma,n} . \tag{3.2}$$

If $\psi \in \mathcal{H}_{N,n,q}$ = the N-fold antisymmetric tensor product of $L^2(R^n;C^q)$, we can write ψ pointwise as $\psi(x_1,\cdots,x_N;\sigma_1,\cdots,\sigma_N)$ with $x_j \in R^n$, $\sigma_j \in \{1,2,\cdots,q\}$ and $\psi \to -\psi$ if (x_i,σ_i) is permuted with (x_j,σ_j). $q = 2$ for spin 1/2 fermions. We can extend the definition (1.2) to

$$\rho_\sigma(x) \equiv N \sum_{\sigma_2=1}^{q} \cdots \sum_{\sigma_N=1}^{q} \int |\psi(x,x_2,\cdots,x_N;\sigma,\sigma_2,\cdots,\sigma_N)|^2 \, d^nx_2 \cdots d^nx_N . \tag{3.3}$$

We also define

$$T_\psi \equiv \sum_{j=1}^{N} \sum_{\sigma_1=1}^{q} \cdots \sum_{\sigma_N=1}^{q} \int |\nabla_j \psi(\underline{x};\underline{\sigma})|^2 \, d^{nN}\underline{x} \,, \tag{3.4}$$

$$\|\psi\|_2^2 \equiv \sum_{\sigma_1=1}^{q} \cdots \sum_{\sigma_N=1}^{q} \int |\psi(\underline{x};\underline{\sigma})|^2 \, d^{nN}\underline{x} \,. \tag{3.5}$$

Our result is

THEOREM 4. *Let* p *satisfy* $\max\{n/2, 1\} \le p \le 1 + n/2$ *and suppose that* $L_{p-n/2,n} < \infty$. *If* $\|\psi\|_2 = 1$, *then, except for the case* $n = 2$, $p = 1$, *there exists a positive constant* $K_{p,n}$ *such that*

$$T_\psi \ge K_{p,n} \sum_{\sigma=1}^{q} \left[\int \rho_\sigma(x)^{p/(p-1)} \, d^n x \right]^{2(p-1)/n} \tag{3.6}$$

and

$$K_{p,n} \ge \frac{1}{2} n p^{-2p/n} (p-n/2)^{-1+2p/n} (L^1_{p-n/2,n}/L_{p-n/2,n})^{-1+2p/n} L^{1\;-2/n}_{p-n/2,n} . \tag{3.7}$$

Before giving the proof of Theorem 4, we discuss its relation to the well-known Sobolev inequalities [9, 10]:

THEOREM 5 (Sobolev-Talenti-Aubin). *Let* $\nabla\psi \in L^r(R^n)$ *with* $1 < r < n$. *Let* $t = nr/(n-r)$. *Then*

$$\int |\nabla\psi|^r \ge C_{r,n} \left(\int |\psi|^t \right)^{r/t} \tag{3.8}$$

for some $C_{r,n} > 0$.

Talenti [11] and Aubin [21] have given the best possible $C_{r,n}$ (for $n = 3$, $r = 2$, $t = 6$, $C_{2,n}$ is also given in [8] and [12]):

$$C_{r,n} = n\pi^{r/2}\left(\frac{r-1}{n-r}\right)^{1-r}\left\{\frac{\Gamma(1+n-n/r)\Gamma(n/r)}{\Gamma(n)\Gamma(1+n/2)}\right\}^{r/n} . \tag{3.9}$$

Our inequality (3.6) relates only to the $r = 2$ case in (3.8), in which case $t = 2n/(n-2)$. Consider (3.8) with $r = 2$ and $\|\psi\|_2 = 1$. Using Hölder's inequality on the right side of (3.8), one gets

$$\int |\nabla\psi|^2 \geq C_{2,n}\left[\int |\psi|^{2p/(p-1)}\right]^{2(p-1)/n}\left[\int |\psi|^2\right]^{-2(p-n/2)/n} \tag{3.10}$$

whenever $n > 2$ and $p \geq n/2$. However, $C_{2,n}$ is not necessarily the best constant in (3.10) when $p \neq n/2$ ($p = n/2$ corresponds to $r = 2$ in (3.8)). Indeed, Theorem 4 says something about this question.

In the case that $N = 1$ and $q = 1$, Theorem 4 is of the same form as (3.10) (since $\rho = |\psi|^2$ and $\|\psi\|_2 = 1$). We note two things:

1. For $n > 2$ and $p = n/2$, (3.6) agrees with (3.8) except, possibly, for a different constant. We have, therefore, an alternative proof of the usual Sobolev inequality (for the $r = 2$ case). As we shall also show $K_{n/2,n} = C_{2,n}$, so we also have the best possible constant for this case.

2. If $\max\{n/2, 1\} < p \leq 1 + n/2$, Theorem 4 gives an improved version of (3.10), *even if* $n = 1$ or 2 (in which cases $C_{2,n} = 0$, but $K_{p,n} > 0$). For $p > 1 + n/2$, one can always use Hölder's inequality on the $p = 1+n/2$ result to get a nontrivial bound of the form (3.10). However, in Theorem 4, the restriction $p \leq 1 + n/2$ is really necessary. This has to do with the dependence of T_ψ on N rather than on n, as we shall explain shortly.

Next we turn to the case $N > 1$. To illustrate the nature of (3.6), we may as well suppose $q = 1$. To fix ideas, we take a special, but important form for ψ, namely

$$\psi(x_1, \cdots, x_N) = (N!)^{-1/2}\,\mathrm{Det}\,\{\phi^i(x_j)\}_{ij=1}^N \tag{3.11}$$

and where the ϕ^i are orthonormal functions in $L^2(R^n)$. Then, suppressing the subscript σ because $q = 1$,

$$\rho(x) = \sum_{i=1}^{N} \rho^i(x) ,$$

$$\rho^i(x) = |\phi^i(x)|^2 ,$$

$$T_\psi = \sum_{i=1}^{N} t^i ,$$

$$t^i = \int |\nabla\phi^i|^2 . \tag{3.12}$$

Theorem 4 says that

$$\sum_i t^i \geq K_{p,n}\left\{\int \left[\sum_i \rho^i(x)\right]^{p/(p-1)} d^n x\right\}^{2(p-1)/n} . \tag{3.13}$$

If we did not use the orthogonality of the ϕ^i, all we would be able to conclude, using (3.6) with $N = 1$, N times, would be

$$\sum_i t^i \geq K_{p,n} \sum_i \left[\int \rho^i(x)^{p/(p-1)}\, d^n x\right]^{2(p-1)/n} . \tag{3.14}$$

If $p = n/2$, then (3.14) is better than (3.13), by convexity. In the opposite case, $p = 1 + n/2$, (3.13) is superior. For in between cases, (3.13) is decidedly better if N is large and if the ρ^i are close to each other (in the $L^{p/(p-1)}(\mathbf{R}^n)$ sense). Suppose $\rho^i(x) = \rho(x)/N$, $i = 1,\cdots,N$. Then the right side of (3.13) is proportional to $N^{2p/n}$ while the right side of (3.14) grows only as N. This difference is caused by the orthogonality of the ϕ^i, or the Pauli principle.

In fact, the last remark shows why $p \leq 1 + n/2$ is important in Theorem 4. If $\rho^i = \rho/N$, all i, then the best bound, insofar as the N dependence is concerned, occurs when p is as large as possible. It is easy to see by example, however, that the largest growth for T_ψ due to the orthogonality condition can only be $N^{(n+2)/n}$.

PROOF OF THEOREM 4. Let $V(x) \leq 0$ be a potential in $\mathbf{R}^n$ with at least one bound state. If $e_1 = \min\{e_i\}$, then, for $\gamma \in [0,1]$,

$$\sum_j |e_j|^\gamma \geq |e_1|^{\gamma-1} \sum_j |e_j| .$$

Using the definition (2.14) and (3.2), we have that

$$\sum_j |e_j| \leq A_{\gamma,n} \left\{ \int |V|^{\gamma+n/2} \right\}^{1/\gamma} \tag{3.15}$$

$$A_{\gamma,n} = L_{\gamma,n}(L^1_{\gamma,n})^{-1+1/\gamma} \tag{3.16}$$

when $1 \geq \gamma \geq \max(0, 1-n/2)$. (3.15) holds even if V has no bound state. Let π_σ, $\sigma = 1,\cdots,q$, be the projection onto the state σ, i.e. for $\psi \in L^2(\mathbf{R}^n; \mathbf{C}^q)$, $(\pi_\nu \psi)(x,\sigma) = \psi(x,\nu)$ if $\sigma = \nu$ and zero otherwise. Choose $\gamma = p - n/2$. Let $\{\rho_\sigma\}_{\sigma=1}^q$ be given by (3.3) and, for $a_\sigma \geq 0$, $\sigma = 1,\cdots,q$, define

$$h = -\Delta - \sum_{\sigma=1}^{q} a_\sigma \rho_\sigma(x)^{1/(\gamma+n/2-1)} \pi_\sigma \tag{3.17}$$

to be an operator on $L^2(\mathbf{R}^n; \mathbf{C}^q)$ in the usual way. Define

$$H_N = \sum_{i=1}^{N} h_i \tag{3.18}$$

where h_i means h acting on the i-th component of $\mathcal{H}_{N,n,q}$. Finally, let $E = \inf \operatorname{spec} H_N$.

Now, by the Rayleigh-Ritz variational principle

$$E \leq (\psi, H_N \psi) = T_\psi - \sum_{\sigma=1}^{q} a_\sigma \int \rho_\sigma^{p/(p-1)} . \tag{3.19}$$

On the other hand, $E \geq$ the sum of all the negative eigenvalues of h

$$\geq - A_{\gamma,n} \sum_{\sigma=1}^{q} a_\sigma^{p/\gamma} \left\{ \int \rho_\sigma^{p/(p-1)} \right\}^{1/\gamma} \tag{3.20}$$

by (3.15). Combining (3.19) and (3.20) with

$$a_\sigma = \left\{ \int \rho_\sigma^{p/(p-1)} \right\}^{2(\gamma-1)/n} \left\{ \frac{\gamma}{\gamma + n/2} \frac{1}{A_{\gamma,n}} \right\}^{2\gamma/n} ,$$

the theorem is proved.

Note that when $p = 1 + n/2$ (corresponding to $\gamma = 1$ in the proof), $L^1_{1,n}$ does not appear in (3.7). In this case, the right side of (3.7) is the best possible value of $K_{1+n/2,n}$, as we now show.

LEMMA 6. *From* (3.7), *define*

$$L^*_{1,n} \equiv [n/(2K_{1+n/2,n})]^{n/2} (1 + n/2)^{-1-n/2} .$$

Then $L^*_{1,n} = L_{1,n}$.

PROOF. By (3.7), we only have to prove that $L^*_{1,n} \geq L_{1,n}$. Let $V \leq 0$, $V \in C_0^\infty(R^n)$ and let $H = -\Delta + V$. Let $\{\phi_i, e_i\}_{i=1}^N$ be the bound state eigenfunctions and eigenvalues of H. Let ψ and ρ^i be as defined in (3.11), (3.12). Then

$$\sum_i |e_i| = - \int V\rho - T_\psi \leq \|V\|_p \|\rho\|_{p/(p-1)} - T_\psi$$

with $p = 1 + n/2$. Using Theorem 4 for T_ψ, one has that

$$\sum_i |e_i| \le \max_{y>0} \{\|V\|_p y - K_{1+n/2,n} y^{2p/n}\} = L^*_{1,n} \|V\|_p^p .$$

We conclude with an evaluation of $K_{n/2,n}$ for $n > 2$ as promised. By a simple limiting argument

$$K_{n/2,n} \ge \lim_{p\downarrow n/2} \text{(right side of (3.7))} . \qquad (3.21)$$

Our bound (2.11) on $L_{p-n/2,n}$ shows that

$$\lim_{p\downarrow n/2} (L_{p-n/2,n})^{-1+2p/n} = 1 . \qquad (3.22)$$

Hence

$$K_{n/2,n} \ge (L^1_{0,n})^{-2/n} . \qquad (3.23)$$

On the other hand, by the method of Lemma 6 applied to the $N = 1$ case, $K_{n/2,n} \le (L^1_{0,n})^{-2/n}$. The value of $L^1_{0,n}$ is given in (4.24). To be honest, its evaluation requires the solution of the same variational problem as given in [8, 11, 12]. Substitution of (4.24) into (3.23) yields the required result

$$K_{n/2,n} = C_{2,n} = \pi n(n-2)[\Gamma(n/2)/\Gamma(n)]^{2/n} . \qquad (3.24)$$

If we examine (3.23) when $n = 2$, one gets $K_{1,2} \ge 0$ since $L^1_{0,2} = \infty$. This reflects the known fact [5] that an arbitrarily small $V < 0$ always has a bound state in two dimensions. This observation can be used to show that

$$K_{1,2} = 0 . \qquad (3.25)$$

When $n = 1$, the smallest allowed p is $p = 1$. In this case, (3.6) reads

$$T_\psi \ge K_{1,1} \sum_{\sigma=1}^{q} \|\rho_\sigma\|^2_\infty . \qquad (3.26)$$

Using (3.7) and (4.20),

$$K_{1,1} \geq [2L_{1/2,1}]^{-1} . \tag{3.27}$$

If one accepts the conjecture of Section 4 that $L_{1/2,1} = L^1_{1/2,1} = 1/2$, then

$$K_{1,1} = 1 . \tag{3.28}$$

The reason for the equality in (3.28) is that $K_{1,1} = 1$ is well known to be the best possible constant in (3.26) when $q = 1$ and $N = 1$.

4. *Conjecture About* $L_{\gamma,n}$

We have shown that for the bound state energies $\{e_j\}$ of a potential V in n dimensions and with

$$L_{\gamma,n}(V) \equiv \sum_j |e_j|^\gamma / \int |V|^{\gamma+n/2} , \tag{4.1}$$

then

$$L_{\gamma,n} \equiv \sup_{V \in L^{\gamma+n/2}} L_{\gamma,n}(V) \tag{4.2}$$

is finite whenever $\gamma + n/2 > 1$ and $\gamma > 0$. The "boundary points" are

$$\begin{aligned} \gamma &= 1/2 & n &= 1 \\ \gamma &= 0 & n &\geq 2 . \end{aligned} \tag{4.3}$$

We showed that for $n = 1$, $L_{1/2,1} < \infty$. For $\gamma < 1/2$, $n = 1$, there cannot be a bound of this kind, for consider $V_L(x) \equiv -1/L$ for $|x| < L$ and zero otherwise. For $L \to 0$, this converges towards $-2\delta(x)$ and thus has a bound state of finite energy (which is -1 for $-2\delta(x)$). On the other hand,

$$\lim_{L\to 0} \int dx\, |V_L|^{1/2+\gamma} = 0 \quad \text{for} \quad \gamma < 1/2 .$$

For $n = 2$, $\gamma = 0$ is a "double boundary point" and $L_{0,2} = \infty$, i.e. there is no upper bound on the number of bound states in two dimensions. (Cf. [5].)

For $n \geq 3$, $L_{0,n}$ is *conjectured* to be finite (see note added in proof, Section 5); for $n = 3$, this is the well-known $\int |V|_-^{3/2}$ conjecture on the number, $N_0(V)$, of bound states (cf. [5]). The best that is known at present is that

$$N_0(V) \leq c\left[\int |V|_-^{3/2}\right]^{4/3} , \tag{4.4}$$

but for spherically symmetric V, a stronger result is known [8]:

$$N_0(V) \leq I\left(1 + \frac{1}{4}\ln I\right) ,$$

$$I = 4(3\pi^2\, 3^{1/2})^{-1} \int |V|_-^{3/2} . \tag{4.5}$$

In (1.4) and (3.1), we introduced L^C and L^1 and showed that

$$L_{\gamma,n} \geq \max(L^1_{\gamma,n}, L^C_{\gamma,n}) . \tag{4.6}$$

A parallel result is Simon's [22] for $n \geq 3$:

$$N_0(V) \leq D_{n,\varepsilon}(\|V_-\|_{\varepsilon+n/2} + \|V_-\|_{-\varepsilon+n/2})^{n/2}$$

with $D_{n,\varepsilon} \to \infty$ as $\varepsilon \to 0$.

In our previous paper [4], we conjectured that $L_{1,3} = L^C_{1,3}$, and we also pointed out that $L^1_{1,1} > L^C_{1,1}$. A remark of Peter Lax (private communication), which will be explained presently, led us to the following:

CONJECTURE. *For each* n, *there is a critical value of* $\gamma, \gamma_{c,n}$, *such that*

$$L_{\gamma,n} = L^C_{\gamma,n} \qquad \gamma \geq \gamma_c$$

$$L_{\gamma,n} = L^1_{\gamma,n} \qquad \gamma \leq \gamma_c$$

γ_c *is defined to be that* γ *for which* $L^C_{\gamma,n} = L^1_{\gamma,n}$; *the uniqueness of this* γ_c *is part of the conjecture. Furthermore,* $\gamma_{c,1} = 3/2$, $\gamma_{c,2} \sim 1.2$, $\gamma_{c,3} \sim .86$ *and the smallest* n *such that* $\gamma_{c,n} = 0$ *is* $n = 8$.

(A) Remarks on $L^1_{\gamma,n}$

We want to maximize

$$\left|\int [|\psi|^2 V + |\nabla\psi|^2]\, d^n x\right|^{\gamma} \Big/ \int |V|^{\gamma+n/2} \tag{4.7}$$

with respect to V, and where $\int |\psi|^2 = 1$ and $(-\Delta + V)\psi = e_1\psi$. By the variational principle, we can first maximize (4.7) with respect to V, holding ψ fixed. Hölder's inequality immediately yields

$$V(x) = -\sigma |\psi(x)|^{2/(\gamma+n/2-1)}$$

with $\sigma > 0$. The kinetic energy, $\int |\nabla\psi|^2$, is not increased if $\psi(x)$ is replaced by $|\psi(x)|$ and, by the rearrangement inequality [13], this is not increased if $|\psi|$ is replaced by its symmetric decreasing rearrangement. Thus, we may assume that $|V|$ and $|\psi|$ are spherically symmetric, non-increasing functions.

By the methods of [8] or [11], (4.7) can be shown to have a maximum when $\gamma + n/2 > 1$. The variational equation is

$$-\Delta\psi(x) - \sigma\psi(x)^{(\gamma+n/2+1)/(\gamma+n/2-1)} = e_1\psi(x) \tag{4.8}$$

with

$$\sigma = \left\{\frac{\gamma |e_1|^{\gamma-1}}{(\gamma+n/2)\, L^1_{\gamma,n}}\right\}^{1/(\gamma+n/2-1)} . \tag{4.9}$$

Equation (4.8) determines ψ up to a constant and up to a change of scale in x. The former can be used to make $\int \psi^2 = 1$ and the latter leaves (4.7) invariant.

Equation (4.8) can be solved analytically in two cases, to which we shall return later:

(i) $n = 1$, all $\gamma > 1/2$

(ii) $n \geq 3$, $\gamma = 0$.

(B) The One-Dimensional Case

Lax's remark was about a result of Gardner, Greene, Kruskal and Miura [14] to the effect that

$$L_{3/2,1} = L^C_{3/2,1} = 3/16\ . \tag{4.10}$$

To see this, we may assume $V \in C_0^\infty(\mathbf{R})$, and use the theory of the Korteweg-de Vries (KdV) equation [14]:

$$W_t = 6WW_x - W_{xxx}\ . \tag{4.11}$$

There are two remarkable properties of (4.11):

(i) As W evolves in time, t, the eigenvalues of $-d^2/dx^2 + W$ remain invariant.

(ii) $\int W^2\,dx$ is constant in time.

Let $W(x,t)$ be given by (4.11) with the initial data

$$W(x,0) = V(x)\ .$$

Then $L_{3/2,1}(W(\cdot\,,t))$ is independent of t, and may therefore be evaluated by studying its behavior as $t \to \infty$.

There exist traveling wave solutions to (4.11), called solitons, of the form

$$W(x,t) = f(x-ct)\ .$$

Equation (4.11) becomes

$$-c f_x = -f_{xxx} + 6 f f_x\ . \tag{4.12}$$

The solutions to (4.12) which vanish at ∞ are

$$f_a(x) = -2a^2 \cosh^{-2}(ax)$$

$$c = 4a^2\ . \tag{4.13}$$

Any solution (4.13), regarded as a potential in the Schrödinger equation has, as we shall see shortly, exactly one negative energy bound state with energy and wave function

$$e = -a^2$$

$$\psi(x) = \cosh^{-1}(ax) \; . \tag{4.14}$$

Now the theory of the KdV equation says that as $t \to \infty$, W evolves into a sum of solitons (4.13) plus a part that goes to zero in $L^\infty(\mathbf{R})$ norm (but not necessarily in $L^2(\mathbf{R})$ norm). The solitons are well separated since they have different velocities. Because the number of bound states is finite, the non-soliton part of W can be ignored as $t \to \infty$. Hence, for the initial V,

$$\sum |e_j|^{3/2} = \sum_{\text{solitons}} a^3 \tag{4.15}$$

while

$$\int V(x)^2 \, dx \geq \sum_{\text{solitons}} \int f_a(x)^2 \, dx \; . \tag{4.16}$$

Since $4 \int_{-\infty}^{\infty} \cosh^{-4}(x) \, dx = 16/3$, we conclude that

$$L_{3/2,1} = L^C_{3/2,1} = 3/16 \tag{4.17}$$

with equality if and only if $W(x,t)$ is composed purely of solitons as $t \to \infty$. For the same reason,

$$L^1_{3/2,1} = L^C_{3/2,1} \tag{4.18}$$

(cf. (4.21)).

Not only do we have an evaluation of $L_{3/2,1}$, (4.17), but we learn something more. When $\gamma = 3/2$, there is an infinite family of potentials for which $L_{3/2,1}(V) = L_{3/2,1}$, and these may have any number of bound states = number of solitons.

What we believe to be the case is that when $\gamma < 3/2$, the optimizing potential for $L_{\gamma,n}$ has only one bound state, and satisfies (4.8). When

$\gamma > 3/2$, the optimizing potential is, loosely speaking, infinitely deep and has infinitely many bound states; thus $L_{\gamma,n} = L^C_{\gamma,n}$.

An additional indication that the conjecture is correct is furnished by the solution to (4.8). When $\gamma = 3/2$, this agrees with (4.14). In general, one finds that, apart from scaling, the nodeless solution to (4.8) is

$$\psi_\gamma(x) = \Gamma(\gamma)^{1/2}\pi^{-1/4}\Gamma(\gamma-1/2)^{-1/2}\cosh^{-\gamma+1/2}(x)$$

$$V_\gamma(x) = -(\gamma^2-1/4)\cosh^{-2}(x)$$

$$e_1 = -(\gamma-1/2)^2\ . \tag{4.19}$$

Thus,

$$L^1_{\gamma,1} = \pi^{-1/2}\frac{1}{\gamma-1/2}\frac{\Gamma(\gamma+1)}{\Gamma(\gamma+1/2)}\left(\frac{\gamma-1/2}{\gamma+1/2}\right)^{\gamma+1/2}\ . \tag{4.20}$$

When $L^1_{\gamma,1}$ is compared with $L^C_{\gamma,1}$, one finds that

$$L^1_{\gamma,1} \geq L^C_{\gamma,1} \qquad \gamma \leq 3/2$$

$$L^1_{\gamma,1} \leq L^C_{\gamma,1} \qquad \gamma \geq 3/2\ . \tag{4.21}$$

This confirms at least part of the conjecture.

However, more is true. For $\gamma = 3/2$, V_γ has a zero energy single node bound state

$$\phi(x) = \tanh(x)\ .$$

Since V_γ is monotone in γ, it follows that V_γ has only one bound state for $\gamma < 3/2$ and at least two bound states for $\gamma > 3/2$. The (unnormalized) second bound state can be computed to be

$$\phi(x) = \sinh(x)\cosh^{-\gamma+1/2}(x)$$

$$e_2 = -(\gamma-3/2)^2\ . \tag{4.22}$$

In like manner, one can find more bound states as γ increases even further.

Thus we see that the potential that optimizes the ratio $|e_1|^\gamma/\int|V|^{\gamma+1/2}$ automatically has a second bound state when $\gamma > \gamma_c$.

Finally, we remark that Theorem 3, together with (4.10), shows that

THEOREM 7.

$$L_{\gamma,1} = L^C_{\gamma,1} \quad \textit{for} \quad \gamma = 3/2,\ 5/2,\ 7/2,\ \textit{etc.}$$

An application of Theorem 7 to scattering theory will be made in Section 4(D).

(C) Higher Dimensions

We have exhibited the solution to the variational equation (4.8) for $L^1_{\gamma,1}$. When $n > 2$ and $\gamma = 0$, we clearly want to take $e_1 = 0$ in order to maximize $L^1_{0,n}(V)$. (4.8) has the zero energy solution

$$\phi(x) = (1+|x|^2)^{1-n/2}$$

$$V(x) = a\phi(x)^{2/(\gamma+n/2-1)} = n(n-2)(1+|x|^2)^{(2-n)/(n/2-1)} \tag{4.23}$$

(note: $\phi \in L^2(R^n)$ if and only if $n > 4$, but $V \in L^{n/2}(R^n)$ always). This leads to

$$L^1_{0,n} = [\pi n(n-2)]^{-n/2}\,\Gamma(n)/\Gamma(n/2)\ . \tag{4.24}$$

The smallest dimension for which $L^1_{0,n} \leq L^C_{0,n}$ is $n = 8$.

If we suppose that the ratio $L^1_{\gamma,n}/L^C_{\gamma,n}$ is monotone decreasing in γ (as it is when $n = 1$ and as it is when $n = 3$ on the basis of the numerical solution of (4.10) by J. F. Barnes, given in Appendix A), and if our conjecture is correct, then $L_{\gamma,n} = L^C_{\gamma,n}$ for $n \geq 8$. The value of γ_c obtained numerically is

$$\gamma_c = 1.165 \qquad n = 2$$

$$\gamma_c = .863 \qquad n = 3\ . \tag{4.26}$$

The other bit of evidence, apart from the monotonicity of $L^1_{\gamma,n}/L^C_{\gamma,n}$, for the correctness of our conjecture is a numerical study of the energy levels of the potential

$$V_\lambda(x) = \lambda e^{-|x|}, \qquad \lambda > 0 ,$$

in three dimensions. This is given in Appendix A. The energy levels of the square well potential are given in [15, 16]. In both cases, one finds that

$$\lim_{\lambda\to\infty} L_{1,3}(V_\lambda) = L^C_{1,3}$$

and the limit is approached from below. Unfortunately, it is not true, as one might have hoped, that $L_{1,3}(V_\lambda)$ is monotone increasing in λ.

(D) Bounds on One-Dimensional Scattering Cross-Sections

In their study of the KdV equation, (4.11), Zakharov and Fadeev [17] showed how to relate the solution $W(x,t)$ to the scattering reflection coefficient $R(k)$ and the bound state eigenvalues $\{e_j\}$ of the initial potential $V(x)$. There are infinitely many invariants of (4.11) besides $\int W^2$ and these have simple expressions in terms of $R(k)$, $\{e_j\}$.

Thus, for any potential V,

$$\int V^2 = (16/3)\sum |e_j|^{3/2} - 4\int_{-\infty}^{\infty} k^2\, T(k)\,dk \tag{4.27}$$

$$\int V^3 + \tfrac{1}{2} V_x^2 = -(32/5)\sum |e_j|^{5/2} - 8\int_{-\infty}^{\infty} k^4\, T(k)\,dk \tag{4.28}$$

$$\int V^4 + 2VV_x^2 + \tfrac{1}{5} V_{xx}^2 = (256/35)\sum |e_j|^{7/2} - (64/5)\int_{-\infty}^{\infty} k^6\, T(k)\,dk \tag{4.29}$$

where

$$T(k) = \pi^{-1}\ln(1-|R(k)|^2) \le 0 . \tag{4.30}$$

These are only the first three invariants; a recursion relation for the others can be found in [17].

Notice that 3/16, 5/32, 35/256 are, respectively $L^C_{3/2,1}$, $L^C_{5/2,1}$, $L^C_{7/2,1}$. Since $\int V^2 \geq \int |V_-|^2$, (4.27) establishes that $L_{3/2,1} = L^C_{3/2,1}$, as mentioned earlier. For the higher invariants, the signs in (4.28) and (4.29) are not as fortunately disposed and we cannot use these equations to prove Theorem 7. But, given that Theorem 7 has already been proved, we can conclude that

THEOREM 8. *For any nonpositive potential* V(x),

$$\int V_x^2 \geq -16 \int_{-\infty}^{\infty} k^4\, T(k)\, dk\ . \tag{4.31}$$

For any potential V(x),

$$2\int VV_x^2 + (1/5)\int V_{xx}^2 \leq -(64/5)\int_{-\infty}^{\infty} k^6\, T(k)\, dk\ . \tag{4.32}$$

The first inequality, (4.31), is especially transparent: If V(x) is very smooth, it cannot scatter very much.

5. *Note Added in Proof*

After this paper was written, M. Cwikel and Lieb, simultaneously and by completely different methods, showed that the number of bound states, $N_0(V)$ for a potential, V, can be bounded (when $n \geq 3$) by

$$N_0(V) \leq A_n \int |V(x)|_-^{n/2}\, d^n x\ . \tag{5.1}$$

Cwikel exploits the weak trace ideal method of Simon [22]; his method is more general than Lieb's, but for the particular problem at hand, (5.1), his

A_n does not seem to be as good. Lieb's method uses Wiener integrals and the general result is the following:

$$N_{-\alpha}(V) \leq \int d^n x \int_0^\infty dt\, t^{-1} e^{-\alpha t} (4\pi t)^{-n/2} f(t|V(x)|_-) \tag{5.2}$$

for any non-negative, convex function $f:[0,\infty) \to [0,\infty)$ satisfying

$$1 = \int_0^\infty t^{-1} f(t) e^{-t} dt \ . \tag{5.3}$$

For $\alpha = 0$, one can choose $f(t) = c(t-b)$, $t \geq b$, $f(t) = 0$, $t \leq b$. This leads to (5.1), and optimizing with respect to b, one finds that

$$A_3 = 0.116\,, \quad A_4 = 0.0191 \tag{5.4}$$

and, as $n \to \infty$,

$$A_n / L_{0,n}^C = (n\pi)^{1/2} + O(n^{-1/2}) \ . \tag{5.5}$$

Note that $A_3/L_{0,3}^1 = 1.49$, i.e. A_3 exceeds $L_{0,3}$ by at most 49%.

Since $N_{-\alpha}(V) \leq N_0(-|V+\alpha|_-)$, one can use (5.1) and (2.1) to deduce that for $\gamma \geq 0$ and $n \geq 3$,

$$L_{\gamma,n} \leq L_{\gamma,n}^C (A_n / L_{0,n}^C) \ . \tag{5.6}$$

This is better than (2.11), (2.18). In particular, for $n = 3$, $\gamma = 1$, the improvement of (5.6) over (2.11) with $m = 2$ is a factor of 1.83. The factor 1.31 in Equation (1.4) can therefore be replaced by $1.31\ (1.83)^{-2/3} = 0.87$.

APPENDIX A. NUMERICAL STUDIES

John F. Barnes
Theoretical Division
Los Alamos Scientific Laboratory
Los Alamos, New Mexico 87545

I. *Evaluation of* $L^1_{\gamma,n}$, $n = 1, 2, 3$

The figure shows the numerical evaluation of $L^1_{\gamma,n}$ as well as $L^C_{\gamma,n}$. The latter is given in (1.11)

$$L^C_{\gamma,n} = 2^{-n}\pi^{-n/2}\Gamma(\gamma+1)/\Gamma(\gamma+1+n/2) .$$

The former is obtained by solving the differential equation (4.8) in polar coordinates and choosing σ such that $\psi(x) \to 0$ as $|x| \to \infty$. Note that by scaling, one can take $e_1 = -1$, whence

$$(L^1_{\gamma,n})^{-1} = \sigma^{(\gamma+n/2)} \int |\psi(x)|^{(2\gamma+n)/(\gamma-1+n/2)} d^n x .$$

In one dimension, $L^1_{\gamma,1}$ is known analytically and is given in (4.20). Another exact result, (4.24), is

$$L^1_{0,3} = 4\pi^{-2}\, 3^{-3/2} = 0.077997 .$$

The critical values of γ, at which $L^1_{\gamma,n} = L^C_{\gamma,n}$ are:

$$\gamma_{c,1} = 3/2$$

$$\gamma_{c,2} = 1.165$$

$$\gamma_{c,3} = 0.8627 .$$

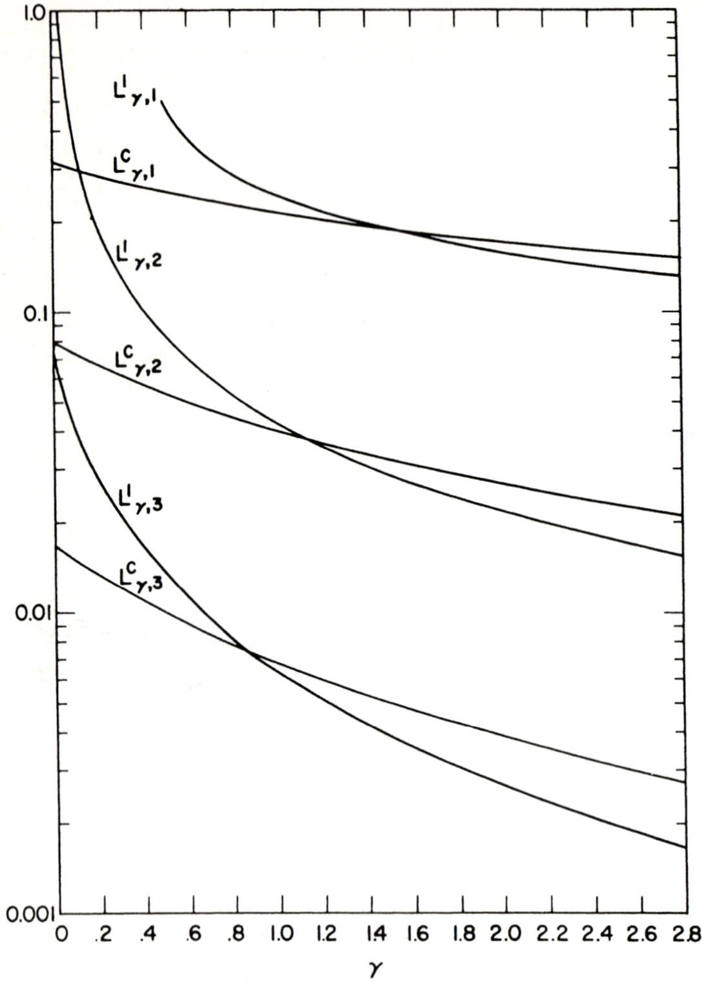

1.0
0.1
0.01
0.001
L
$L^1_{\gamma,1}$
$L^c_{\gamma,1}$
$L^1_{\gamma,2}$
$L^c_{\gamma,2}$
$L^1_{\gamma,3}$
$L^c_{\gamma,3}$
0 .2 .4 .6 .8 1.0 1.2 1.4 1.6 1.8 2.0 2.2 2.4 2.6 2.8
γ

II. *The Exponential Potential*

To test the conjecture that $L_{1,3} = L^{C}_{1,3}$, the eigenvalues of the potential $V_\lambda = -\lambda \exp(-|x|)$ in three dimensions were evaluated for $\lambda = 5, 10, 20, 30, 40, 50$, and 100. These are listed in the table according to angular momentum and radial nodes. These numbers have been corroborated by H. Grosse, and they can be used to calculate $L_{\gamma,3}(V_\lambda)$ for any γ. The final column gives $L_{1,3}(V_\lambda)$, since $\int |V_\lambda|^{5/2} = \lambda^{5/2}(64\pi)/125$. It is to be noted that the classical value $L^{C}_{1,3} = 0.006755$, is approached from below, in agreement with the conjecture, but *not monotonically*.

$$V_\lambda = -\lambda e^{-r}$$

	ℓ	$\lvert e \rvert$	nodes	states	$\sum \lvert e \rvert$	$\dfrac{\sum \lvert e \rvert}{\lambda^{5/2} \frac{64\pi}{125}}$
$\lambda = 5$	0	0.55032	0	$\frac{1}{1}$	$\frac{0.55032}{0.55032}$	0.006120
$\lambda = 10$	0	0.06963	1			
		2.18241	0	2	2.2520	
	1	0.33405	0	$\frac{3}{5}$	$\frac{1.0022}{3.2542}$	0.006398
$\lambda = 20$	0	0.00869	2			
		1.42562	1			
		6.62410	0	3	8.0584	
	1	0.16327	1			
		2.71482	0	6	8.6342	
	2	0.43136	0	$\frac{5}{14}$	$\frac{2.1568}{18.8494}$	0.006551

$V_\lambda = -\lambda e^{-r}$ (continued)

	ℓ	$\|e\|$	nodes	states	$\sum \|e\|$	$\frac{\sum \|e\|}{\lambda^{5/2} \frac{64\pi}{125}}$
$\lambda = 30$	0	0.58894	2			
		3.83072	1			
		11.84999	0	3	16.270	
	1	1.39458	1			
		6.12302	0	6	22.553	
	2	0.00593	1			
		2.36912	0	10	11.875	
	3	0.07595	0	7 26	0.532 51.230	0.006461
$\lambda = 40$	0	0.07676	3			
		1.86961	2			
		6.88198	1			
		17.53345	0	4	26.362	
	1	0.41991	2			
		3.35027	1			
		10.13596	0	9	41.718	
	2	0.93459	1			
		5.03378	0	10	29.842	
	3	1.54738	0	7 30	10.832 108.754	0.006682

$V_\lambda = -\lambda e^{-r}$ (continued)

	ℓ	$\lvert e\rvert$	nodes	states	$\sum \lvert e\rvert$	$\dfrac{\sum \lvert e\rvert}{\lambda^{5/2}\frac{64\pi}{125}}$
$\lambda = 50$	0	0.60190	3			
		3.66447	2			
		10.39110	1			
		23.53215	0	4	38.190	
	1	1.43321	2			
		5.81695	1			
		14.56904	0	9	65.458	
	2	0.07675	2			
		2.45887	1			
		8.19840	0	15	53.670	
	3	0.26483	1			
		3.61626	0	14	27.168	
	4	0.49009	0	9	4.411	
				51	188.897	0.006643
$\lambda = 100$	0	0.39275	5			
		2.91408	4			
		8.29231	3			
		17.44909	2			
		32.07168	1			

$V_\lambda = -\lambda e^{-r}$ (continued)

ℓ	$\lvert e\rvert$	nodes	states	$\sum \lvert e\rvert$	$\dfrac{\sum \lvert e\rvert}{\lambda^{5/2}\frac{64\pi}{125}}$
	56.28824	0	6	117.41	
1	1.10170	4			
	4.76748	3			
	11.62740	2			
	22.79910	1			
	40.45495	0	15	242.25	
2	0.02748	4			
	2.04022	3			
	6.85633	2			
	15.22147	1			
	28.46495	0	25	263.05	
3	0.22692	3			
	3.14743	2			
	9.13429	1			
	19.04073	0	28	220.85	
4	0.52962	2			
	4.37856	1			
	11.56470	0	27	148.26	

$V_\lambda = -\lambda e^{-r}$ (continued)

ℓ	$\lvert e\rvert$	nodes	states	$\sum \lvert e\rvert$	$\dfrac{\sum \lvert e\rvert}{\lambda^{5/2}\frac{64\pi}{125}}$
5	0.88997	1			
	5.69707	0	22	72.46	
6	1.26789	0	$\frac{13}{136}$	$\frac{16.48}{1080.76}$	0.006719

APPENDIX B: PROOF OF (2.3)

THEOREM 9. *Let* $\mathcal{H}$ *be a separable Hilbert space and let* A, B *be positive operators on* $\mathcal{H}$. *Then, for* $m \geq 1$,

$$\operatorname{Tr}(B^{1/2} A B^{1/2})^m \leq \operatorname{Tr} B^{m/2} A^m B^{m/2} . \tag{B.1}$$

REMARK. When $\mathcal{H} = L^2(R^n)$ and A is a kernel $a(x-y)$ and B is a multiplication operator $b(x)$ (as in our usage (2.2)), Seiler and Simon [19] have given a proof of (B.1) using interpolation techniques. Simon (private communication) has extended this method to the general case. Our proof is different and shows a little more than just (B.1).

PROOF. For simplicity, we shall only give the proof when A and B are matrices; for the general case, one can appeal to a limiting argument. For $m = 1$, the theorem is trivial, so assume $m > 1$. Let $C = A^m$ and $f(C) \equiv g(C) - h(C)$, where $g(C) \equiv \operatorname{Tr}(B^{1/2} C^{1/m} B^{1/2})^m$ and $h(C) = \operatorname{Tr} B^{m/2} C B^{m/2}$. Let M^+ be the positive matrices. Clearly $M^+ \ni C \to h(C)$ is linear. Epstein [18] has shown that $M^+ \ni C \to g(C)$ is concave (actually, he showed this for m integral, but his proof is valid generally for $m \geq 1$). Write $C = C^D + C^0$ where C^D is the diagonal part of C in a basis in which B is diagonal. $C_\lambda \equiv C^D + \lambda C^0 = \lambda C + (1-\lambda) C^D$ is in M^+ for $\lambda \in [0,1]$, because $C^D \in M^+$. Then $\lambda \to f(C_\lambda) \equiv R(\lambda)$ is concave on $[0,1]$. Our goal is to show that $R(1) \leq 0$. Since $[C^D, B] = 0$, $R(0) = 0$ and, by concavity, it is sufficient to show that $R(\lambda) \leq 0$ for $\lambda > 0$ and λ small. $h(C_\lambda) = h(C^D)$ for $\lambda \in [0,1]$. Since $f(C)$ is continuous in C, we can assume that C^D is nondegenerate and strictly positive, and that C_λ is positive when $\lambda \geq -\varepsilon$ for some $\varepsilon > 0$. Then $R(\lambda)$ is defined and concave on $[-\varepsilon, 1]$. $\lambda \to C_\lambda^{1/m}$ is differentiable at $\lambda = 0$ and its derivative at $\lambda = 0$ has zero diagonal elements. (To see this, use the representation $C^{1/m} = K \int_0^\infty dx\, x^{-1+1/m} \cdot C(C+xI)^{-1}$.) Likewise, the derivative of $(B^{1/2}(D + \lambda O)B^{1/2})^m$ at $\lambda = 0$ has zero diagonal elements when O has and when D is diagonal. Thus

$$dR(\lambda)/d\lambda|_{\lambda=0} = 0 .$$

Acknowledgment

One of the authors (Walter Thirring) would like to thank the Department of Physics of the University of Princeton for its hospitality.

ELLIOTT H. LIEB
DEPARTMENTS OF MATHEMATICS AND PHYSICS
PRINCETON UNIVERSITY
PRINCETON, NEW JERSEY

WALTER E. THIRRING
INSTITUT FÜR THEORETISCHE PHYSIK
DER UNIVERSITÄT WIEN, AUSTRIA

REFERENCES

[1] E. H. Lieb and W. E. Thirring, Phys. Rev. Lett. *35*, 687 (1975). See Phys. Rev. Lett. *35*, 1116 (1975) for errata.

[2] M. S. Birman, Mat. Sb. *55 (97)*, 125 (1961); Amer. Math. Soc. Translations Ser. 2, *53*, 23 (1966).

[3] J. Schwinger, Proc. Nat. Acad. Sci. *47*, 122 (1961).

[4] B. Simon, "Quantum Mechanics for Hamiltonians Defined as Quadratic Forms," Princeton University Press. 1971.

[5] B. Simon, "On the Number of Bound States of the Two Body Schrödinger Equation – A Review," in this volume.

[6] A. Martin, Helv. Phys. Acta *45*, 140 (1972).

[7] H. Tamura, Proc. Japan Acad. *50*, 19 (1974).

[8] V. Glaser, A. Martin, H. Grosse and W. Thirring, "A Family of Optimal Conditions for the Absence of Bound States in a Potential," in this volume.

[9] S. L. Sobolev, Mat. Sb. *46*, 471 (1938), in Russian.

[10] ———, Applications of Functional Analysis in Mathematical Physics, Leningrad (1950), Amer. Math. Soc. Transl. of Monographs, 7 (1963).

[11] G. Talenti, Best Constant in Sobolev's Inequality, Istituto Matematico, Universitá Degli Studi Di Firenze, preprint (1975).

[12] G. Rosen, SIAM Jour. Appl. Math. *21*, 30 (1971).

[13] H. J. Brascamp, E. H. Lieb and J. M. Luttinger, Jour. Funct. Anal. *17*, 227 (1974).

[14] C. S. Gardner, J. M. Greene, M. D. Kruskal and R. M. Miura, Commun. Pure and Appl. Math. *27*, 97 (1974).

[15] S. A. Moszkowski, Phys. Rev. *89*, 474 (1953).

[16] A. E. Green and K. Lee, Phys. Rev. *99*, 772 (1955).

[17] V. E. Zakharov and L. D. Fadeev, Funkts. Anal. i Ego Pril. *5*, 18 (1971). English translation: Funct. Anal. and its Appl. *5*, 280 (1971).

[18] H. Epstein, Commun. Math. Phys. *31*, 317 (1973).

[19] E. Seiler and B. Simon, "Bounds in the Yukawa Quantum Field Theory," Princeton preprint (1975).

[20] W. Thirring, T7 Quantenmechanik, Lecture Notes, Institut für Theoretische Physik, University of Vienna.

[21] T. Aubin, C. R. Acad. Sc. Paris *280*, 279 (1975). The results are stated here without proof; there appears to be a misprint in the expression for $C_{r,n}$.

[22] B. Simon, "Weak Trace Ideals and the Bound States of Schrödinger Operators," Princeton preprint (1975).

Volume 66A, number 6 PHYSICS LETTERS 26 June 1978

ON SEMI-CLASSICAL BOUNDS FOR EIGENVALUES OF SCHRÖDINGER OPERATORS ☆

Michael AIZENMAN
Department of Physics, Princeton University, Princeton, NJ 08540, USA

and

Elliott H. LIEB
Departments of Mathematics and Physics, Princeton University Princeton, NJ 08540, USA

Received 27 April 1978

Our principal result is that if the semiclassical estimate is a bound for some moment of the negative eigenvalues (as is known in some cases in one-dimension), then the semiclassical estimates are also bounds for all higher moments.

Bounds on the moments of energy levels of Schrödinger operators have been the object of several studies [1, 2, 5–8]. In [1] such bounds were used to obtain a lower bound for the kinetic energy of fermions in terms of their one particle density and thereby prove the stability of matter.

In the notation of [2],

$$\sum_j |e_j(V)|_-^\gamma \leqslant L_{\gamma,n} \int d^n x\, |V(x)|_-^{\gamma+n/2} \tag{1}$$

where $e_j(V)$ are the eigenvalues of $-\Delta + V(x)$ defined in $L^2(\mathbf{R}^n)$ and $|y|_- \equiv \max(-y, 0)$. $L_{\gamma,n}$ denotes the smallest number for which (1) holds independently of V. The case $\gamma = 1$ is the one needed for the kinetic energy bound. It was shown in [2] that $L_{\gamma,n} < \infty$ for $\gamma > \max(0, 1 - n/2)$. Eq. (1) also holds for $\gamma = 0$, $n \geqslant 3$ but the proof is quite different (see refs. [5, 6, 8]).

For $\gamma \geqslant 0$ we use the notation

$$H(x, p) = p^2 + V(x)$$

$$R_n(\gamma, V) = \sum_j |e_j(V)|_-^\gamma \Big/ \int d^n z\, |H(x,p)|_-^\gamma \tag{2}$$

$$R_n(\gamma) = \sup_V \{R_n(\gamma, V)\}\,, \tag{3}$$

and $d^n z = d^n x\, d^n p\, (2\pi)^{-n}$. $R_n(\gamma, V)$ is the ratio of the moments of the binding energies of a quantum mechanical hamiltonian to the moments of its classical analog. The integral in (1) comes from doing the $d^n p$ integration in (2). In the notation of [2]

$$R_n(\gamma) = L_{\gamma,n}/L^c_{\gamma,n}\,. \tag{4}$$

For $V \leqslant 0$, $V \in C_0^\infty(\mathbf{R}^n)$ it is known [3,4,9] that

$$R_n(\gamma, \lambda V) \to 1 \tag{5}$$

as $\lambda \to \infty$, which is the semiclassical limit. Thus

$$R_n(\gamma) \geqslant 1\,. \tag{6}$$

In [2] it was conjectured that $R_n(\gamma) = 1$ for certain γ and n, in particular for $\gamma = 1$, $n = 3$ which is the case of primary physical interest. $R_3(1) = 1$ would imply that the Thomas–Fermi theory of atoms and molecules (together with a modified treatment of the electron-electron repulsion) gives a lower bound to the true Schrödinger ground state energy (see [1]). The only cases where the value of $R_n(\gamma)$ is known are $n = 1$, $\gamma = 3/2, 5/2, 7/2, \ldots$, where $R_1(\gamma) = 1$. Part (a) of the following theorem, together with (6), settles the question for $n = 1$, $\gamma \geqslant 3/2$.

Theorem:

(a) For any n, $R_n(\gamma)$ is a monotone nonincreasing function of γ.

☆ Work partly supported by U.S. National Science Foundation grant MCS 75-21684 A02.

Volume 66A, number 6 PHYSICS LETTERS 26 June 1978

(b) If, for some $\gamma > \max(0, 1 - n/2)$, the supremum in (3) is attained, i.e. $R_n(\gamma) = R_n(\gamma, V)$ for some V with $|V|_- \in L^{\gamma + n/2}$, then $R_n(\cdot)$ is strictly decreasing from the left at γ. In fact

$$\liminf_{\delta \to 0^+} \; [R_n(\gamma - \delta) - R_n(\gamma)]/\delta > 0 .$$

Proof:
(a) Fix V. For $\gamma \geqslant 0, \delta > 0$, let

$$I(\gamma, \delta) = \int_0^1 \mathrm{d}\lambda \, \lambda^{-1+\delta} \, |1 - \lambda|_-^{\gamma} < \infty . \tag{7}$$

By scaling, for any $e \in R$

$$|e|_-^{\gamma+\delta} = I(\gamma, \delta)^{-1} \int_0^\infty \mathrm{d}\lambda \, \lambda^{-1+\delta} \, |e + \lambda|_-^{\gamma} . \tag{8}$$

Thus, for any Schrödinger potential V,

$$\sum_j |e_j(V)|_-^{\gamma+\delta} = I(\gamma, \delta)^{-1} \int_0^\infty \mathrm{d}\lambda \, \lambda^{-1+\delta} \sum_j |e_j(V) + \lambda|_-^{\gamma} . \tag{9}$$

However, $e_j(V) + \lambda$ are the eigenvalues of the potential $V(x) + \lambda$. Therefore, by definition (3),

$$\sum_j |e_j(V) + \lambda|_-^{\gamma} \leqslant R_n(\gamma) \int \mathrm{d}^n z \, |H(x,p) + \lambda|_-^{\lambda} \tag{10}$$

and, by substitution in (9) and using (8),

$$\sum_j |e_j(V)|_-^{\gamma+\delta} \leqslant I(\gamma, \delta)^{-1} \int_0^\infty \mathrm{d}\lambda \, \lambda^{-1+\delta} \, R_n(\gamma) \times \int \mathrm{d}^n z \, |H(x,p) + \lambda|_-^{\gamma} = R_n(\gamma) \int \mathrm{d}^n z \, |H(x,p)|_-^{\gamma+\delta} . \tag{11}$$

Hence

$$R_n(\gamma + \delta, V) \leqslant R_n(\gamma)$$

and therefore

$$R_n(\gamma + \delta) \leqslant R_n(\gamma) . \tag{12}$$

(b) Let $\gamma > \max(0, 1 - n/2)$ and assume that for some V with $|V|_- \in L^{\gamma + n/2}(\mathbf{R}^n)$

$$R_n(\gamma, V) = R_n(\gamma) . \tag{13}$$

In particular (13) implies that

$$0 > e_0 \equiv \inf \operatorname{spec}(-\Delta + V) > \operatorname{ess\,inf} \{V(x)\} . \tag{14}$$

We shall prove that $R_n(\cdot)$ is strictly decreasing from the left at γ by showing that

$$\liminf_{\delta \to 0^+} [R_n(\gamma - \delta) - R_n(\gamma)]/\delta \geqslant R_n(\gamma) \frac{\int \mathrm{d}^n z \, |H(x,p)|_-^{\gamma} \, \mu_\gamma(|e_0/H|)}{\int \mathrm{d}^n z \, |H(x,p)|_-^{\gamma}} > 0 , \tag{15}$$

where

$$\mu_\gamma(t) = \int_t^1 \mathrm{d}\lambda (1 - \lambda)^\gamma / \lambda > (1 - t)^{1+\gamma}/(1 + \gamma) . \tag{16}$$

The key fact which will be used is that the integral in (9) can be cut off from above at $|e_0|$.

For any $\delta > 0$

$$\begin{aligned} \sum_j |e_j(V)|_-^{\gamma} &= I(\gamma - \delta, \delta)^{-1} \times \int_0^{|e_0|} \mathrm{d}\lambda \, \lambda^{-1+\delta} \sum_j |e_j(V) + \lambda|_-^{\gamma-\delta} \\ &\leqslant R_n(\gamma - \delta) \, I(\gamma - \delta, \delta)^{-1} \times \int \mathrm{d}^n z \int_0^{|e_0|} \mathrm{d}\lambda \, \lambda^{-1+\delta} |H(x,p) + \lambda|_-^{\gamma-\delta} \\ &= R_n(\gamma - \delta) \int \mathrm{d}^n z \, |H(x,p)|_-^{\gamma} \\ &\quad - R_n(\gamma - \delta) \, I(\gamma - \delta, \delta)^{-1} \times \int \mathrm{d}^n z \int_{|e_0|}^\infty \mathrm{d}\lambda \, \lambda^{-1+\delta} \, |H(x,p) + \lambda|_-^{\gamma-\delta} . \end{aligned} \tag{17}$$

Therefore, using (2), (13) and $R_n(\gamma - \delta) \geqslant R_n(\gamma)$,

Volume 66A, number 6 PHYSICS LETTERS 26 June 1978

$$[R_n(\gamma-\delta)-R_n(\gamma)]/\delta \geqslant [R_n(\gamma)/\delta I(\gamma-\delta,\delta)] \times \int\limits_{H<e_0} \mathrm{d}^n z\, |H(x,p)|_-^{\gamma} \times \int\limits_{|e_0/H|}^{1} \mathrm{d}\lambda\, \lambda^{-1+\delta}(1-\lambda)^{\gamma-\delta} \Big/ \int \mathrm{d}^n z\, |H(x,p)|_-^{\gamma}\,. \tag{18}$$

Using $\lim_{\delta\to 0^+} \delta I(\gamma-\delta,\delta) = 1$ and Fatou's lemma, we obtain (15). ■

In view of (16) the theorem implies

Corollary 1: If for some $\gamma, R_n(\gamma) = 1$ then $R_n(\bar{\gamma})$ $= 1$ for all $\bar{\gamma} > \gamma$. Moreover, for $\bar{\gamma} > \gamma$ the supremum in (3) is not attained by any potential.

This proves part of a conjecture made in [2] (another part of the conjecture was disproved, for $n \geqslant 7$, in [7]).

In one dimension we can say even more since it is known, [2], that $R_1(3/2) = 1$ ($R_1(\gamma) > 1$ for $\gamma < 3/2$).

Corollary 2: For $\gamma \geqslant 3/2, R_1(\gamma) = 1$.

One may also study bounds like (3) for some restricted classes of potentials, V, as was done in [7] for the spherically symmetric ones (the constants thus obtained are no larger than $R_n(\gamma)$ but it is not known whether any of them are strictly smaller). The theorem and its proof extend to such bounds as long as the class of potentials is closed under the addition of constants.

References

[1] E.H. Lieb and W.E. Thirring, Phys. Rev. Lett. 35 (1975) 687. See Phys. Rev. Lett. 35 (1975) 1116 for errata. Also E.H. Lieb, Rev. Mod. Phys. 48 (1976) 553.

[2] E.H. Lieb and W.E. Thirring, in: Studies in mathematical physics, Essays in honor of V. Bargmann (Princeton Univ. Press, Princeton, N.J., 1976).

[3] A. Martin, Helv. Phys. Acta 45 (1972) 140.

[4] H. Tamura, Proc. Japan Acad. 50 (1974) 19.

[5] M. Cwikel, Ann. of Math. 106 (1977) 93.

[6] E.H. Lieb, Bull. Amer. Math. Soc. 82 (1976) 751.

[7] V. Glaser, H. Grosse and A. Martin, Bounds on the Number of Eigenvalues of the Schrödinger Operator, CERN preprint TH2432 (1977).

[8] G.V. Rosenblum, The distribution of the discrete spectrum for singular differential operators, Isvestia Math. 164 No. 1 (1976) 75.

[9] M.S. Birman and V.V. Borzov, On the asymptotics of the discrete spectrum of some singular differential operators, Topics in Math. Phys. 5 (1972) 19.

Proceedings of the Amer. Math. Soc. Symposia in Pure Math. *36*, 241–252 (1980)

THE NUMBER OF BOUND STATES OF ONE-BODY SCHROEDINGER OPERATORS AND THE WEYL PROBLEM

Elliott H. Lieb[1]

ABSTRACT. If $\tilde{N}(\Omega,\lambda)$ is the number of eigenvalues of $-\Delta$ in a domain Ω in a suitable Riemannian manifold of dimension n, we derive bounds of the form $\tilde{N}(\Omega,\lambda) \leq D_n \lambda^{n/2}|\Omega|$ for *all* Ω, λ, n, Likewise, if $N_\alpha(V)$ is the number of nonpositive eigenvalues of $-\Delta + V(x)$ which are $\leq \alpha \leq 0$, then $N_\alpha(V) \leq L_n \int_M [V - \alpha]_-^{n/2}$ for *all* α and V and $n \geq 3$.

I. INTRODUCTION AND BACKGROUND.

Two closely related problems will concern us here : One is to bound the nonpositive eigenvalues of the one-body Schroedinger operator

$$H = -\Delta + V(x). \tag{1.1}$$

The other is to find an upper bound for

$$\tilde{N}(\Omega,\lambda) = \text{number of eigenvalues of } -\Delta \leq \lambda \tag{1.2}$$

in a domain Ω with Dirichlet boundary conditions.

In both cases the setting is a Riemannian manifold, M, and Δ is the Laplace-Beltrami operator. The only way in which the properties of the manifold will appear in our results will be through the fundamental solution of the heat equation

$$G(x, y; t) = [\exp(t\Delta)](x, y) \tag{1.3}$$

1980 Mathematics Subject Classification 35P15.

[1] Work supported by U.S.National Foundation grants PHYS-7825390 and INT 78-01160.

Proceedings of the Amer. Math. Soc. Symposia in Pure Math. *36*, 241–252 (1980)

evaluated on the diagonal x = y, or else through the Green function

$$(-\Delta + e)^{-1}(x, y) = \int_0^\infty dt\, e^{-et}\, G(x, y; t) \tag{1.4}$$

for $e \geq 0$. (Note: G is defined for the *whole* manifold, not the subdomain Ω in the case of (1.2). One could of course, use G defined for the domain Ω in all the formulas, but then the dependence of the result on Ω will be complicated. It is precisely to avoid this complication that we use the G for the whole of M).

Let us begin with the problem defined by (1.2), which we may term the *Weyl problem.* Weyl [1] proved the asymptotic formula (for suitable domains) :

$$\tilde{N}(\Omega,\lambda) \approx C_n\, \lambda^{n/2}|\Omega| \tag{1.5}$$

where $|\Omega|$ is the Riemannian volume of Ω, n is the dimension of M and

$$\begin{aligned} C_n &= (4\pi)^{-n/2}\, \Gamma(1 + n/2)^{-1} \\ &= (2\pi)^{-n}\, \tau_n , \end{aligned} \tag{1.6}$$

where τ_n is the volume of the unit ball in $\mathbb{R}^n$. The constant C_n is called the *classical constant* for reasons which will become clear later. This result (1.5) is discussed in [2] and [3], §XIII.15, vol.4. The proof uses Dirichlet-Neumann bracketing.

Polya's conjecture is that (1.5) holds in $\mathbb{R}^n$ for all λ and Ω, not just asymptotically. Here we will prove

THEOREM 1. *For all* λ *and* Ω *there exist constants* D_n *and* E_n *(depending on the manifold M) such that*

$$\tilde{N}(\Omega,\lambda) \leq D_n\, \lambda^{n/2}|\Omega| \tag{1.7}$$

if

$$G(x, x; t) \leq A_n\, t^{-n/2}, \ \forall x \in M, \ \forall t > 0 \tag{1.8}$$

for some $A_n < \infty$, *while*

$$\tilde{N}(\Omega,\lambda) \leq (D_n\, \lambda^{n/2} + E_n)|\Omega| \tag{1.9}$$

if

$$G(x, x; t) \leq A_n t^{-n/2} + B_n, \ \forall x \in M, \ \forall t > 0, \tag{1.10}$$

D_n *and* E_n *are proportional to* A_n *and* B_n *respectively.*

In particular (1.8) and (1.7) hold for $\mathbb{R}^n$ (with $A_n = (4\pi)^{-n/2}$) and for many noncompact M, e.g. homogeneous spaces with curvature ≤ 0. (1.10) and (1.9) hold for compact M.

Next we turn to the Schroedinger problem (1.1). Let $E_1(V) \leq E_2(V) \leq \ldots \leq 0$ be the nonpositive eigenvalues of H on $L^2(M)$. If we write

$$V = V_+ - V_- \tag{1.11}$$

with

$$\begin{aligned} V_\pm(x) &= |V(x)| , \quad \text{when } V(x) \gtrless 0 \\ &= 0, \text{ otherwise} \end{aligned} \tag{1.12}$$

then the negative spectrum of H is discrete if $V_- \epsilon L^{n/2}$, for example.

DEFINITION. $N_\alpha(V)$ *is the number of eigenvalues of H which are* $\le \alpha \le 0$.

Our main result is

THEOREM 2. *Suppose* (1.8) *holds and suppose* $n \ge 3$. *Then*

$$N_0(V) \le L_n \int_M V_-(x)^{n/2}\, dx \tag{1.13}$$

for some constant L_n *depending on* M.

There are many remarks to be made about Theorem 2 and its connection with Theorem 1. First, the history of (1.13). Rosenbljum [4] first announced Theorem 2 in 1972. Unaware of this, Simon [5] proved an inequality of the form

$$N_0(V) \le S_{n,\varepsilon}[\,\|V_-\|_{n/2+\varepsilon} + \|V_+\|_{n/2+\varepsilon}]^{n/2}$$

with $S_{n,\varepsilon} \to \infty$ as $\varepsilon \downarrow 0$. Also unaware of [4], Cwikel and myself [6,7] simultaneously found a proof of Theorem 2. Reed and Simon [3] call Theorem 2 the Cwikel - Lieb - Rosenbljum bound. Cwikel's method exploits some ideas in [5]. All three methods are different, Cwikel's and Rosenbljum's are applicable to a wider class of operators than the Schroedinger operator, but my method [7], based on Wiener integrals, which is the one presented here, gives the best constant by far. This result was announced in [7] and the proof was written up in [3], Theorem XIII.12 and in [8]. Because all the technical details can easily be found in [8], the presentation given here will ignore technicalities . Not only am I indebted to B.Simon for his help with the technical details, as just mentioned, but I also wish to acknowledge his role in stimulating my interest in the problem, and for his constant encouragement while the ideas were taking shape.

The connection between the two theorems is

PROPOSITION 3. *Let* $\alpha \le 0$. *Then for all* M

$$\tilde{N}(\Omega,\lambda) \le N_\alpha(\,(\alpha-\lambda)\,\chi_\Omega) \tag{1.14}$$

where χ_Ω *is the characteristic function of* Ω.

PROOF. Let ψ_j be the j th eigenfunction of $-\Delta$ in Ω and let $\tilde{\psi}_j$, defined on all of M, be ψ_j in Ω and zero outside. (1.14) is obtained by using the $\tilde{\psi}_j$ as variational functions for $-\Delta + (\alpha - \lambda)\,\chi_\Omega$ in the Rayleigh - Ritz variational principle. QED.

Similarly, we have

PROPOSITION 4. *if* $0 \le \beta \le -\alpha$, *then for all* M

$$N_\alpha(V) \le N_{\alpha+\beta}(-[V+\beta]_-). \tag{1.15}$$

PROOF. Same as for proposition 3. Alternatively, one can remark that $V(x) \ge -[V+\beta]_-(x) - \beta$, and adding a positive operator can never decrease any eigenvalue. QED.

Proceedings of the Amer. Math. Soc. Symposia in Pure Math. *36*, 241–252 (1980)

In particular, $N_\alpha(V) \le N_0(-[V-\alpha]_-)$ and $N_0(V) \le N_0(-V_-)$, whence we have

COROLLARY 5. *If* $n \ge 3$ *and* (1.13) *holds, then*

$$N_\alpha(V) \le L_n \int_M [V-\alpha]_-(x)^{n/2}\, dx. \tag{1.16}$$

Moreover to prove (1.13) *it is sufficient to consider* V *satisfying* $V(X) \le 0$, $\forall x$.

Proposition 3 will be used to derive Theorem 1 from Theorem 2 (actually from a generalization of Theorem 2, namely Theorem 8 and (4.3), which holds for all n). However, at this point we can, under the assumption that (1.8) holds, deduce (1.7) of Theorem 2 from proposition 3 and Theorem 2. Choosing $\alpha = 0$, we have

$$\begin{aligned}\tilde{N}(\Omega,\lambda) &\le L_n\, \lambda^{n/2} \int_M \chi_\Omega^{\,n/2}\, dx \\ &= L_n\, \lambda^{n/2} |\Omega|\end{aligned} \tag{1.17}$$

for $n \ge 3$.

It is to be emphasized that Theorem 2 is more delicate than Theorem 1. For one thing Theorem 1 holds for all n, whereas Theorem 2 holds only for $n \ge 3$. The analogue of Theorem 2 is definitely false for $n = 1$ and 2. In $\mathbb{R}^n$, at least, an arbitrarily small, nonpositive potential always has a negative eigenvalue when $n < 3$; cf. [3], Theorem XIII.11. For another thing, the best constant in (1.7) is , according to the Polya conjecture, the classical constant C_n (in $\mathbb{R}^n$ at least). However, the best constant L_n is bigger than C_n in general. It is easy to prove [9] (by considering a very "large" V and using Dirichlet-Neumann bracketing) that $L_n \ge C_n$. In [9] it was shown that $L_n > C_n$ for $3 \le n \le 7$ and recently [16] it was shown that $L_n > C_n$ for $n \ge 7$; in both cases explicit examples were constructed.

It is somewhat ironic that although Theorem 2 starts to hold only for $n = 3$, Theorem 1 is easiest to prove for $\mathbb{R}^n$ with $n = 1$. In that case the only domain that need be considered is a finite interval, and there the eigenvalues of $-\Delta$ can be computed explicitly. The Polya conjecture is easily seen to be true.

The intuition behind Theorem 2, and thereby the reason for calling C_n the classical constant is important. In the semiclassical picture of quantum mechanics in $\mathbb{R}^n$, which is similar to WKBJ theory, one has the mystical postulate that "each nice set in phase space $\Gamma = \{(p,x) \mid p \in \mathbb{R}^n,\ x \in \mathbb{R}^n\}$ of volume $(2\pi)^n$ can accomodate one eigenstate of H". This postulate can be made more precise by mean of the Dirichlet-Neumann bracketing method mentioned before. In any event, since the "eigenvalues" of $-\Delta$ are p^2 and the "eigenvalues" of V are V(x), the postulate implies that

$$N_\alpha(V) \approx (2\pi)^{-n} \int_{\mathbb{R}^n\times\mathbb{R}^n} dpdx\ \Theta(\alpha - (p^2+V(x))) \tag{1.18}$$

with $\Theta(a) = 1$, for $a \geq 0$ and $= 0$ for $a < 0$. The p integration in (1.18) for fixed x, is easy to do, namely

$$\int_{p^2\leq \alpha-V(x)} dp = \tau_n\ [V - \alpha]_-(x)^{n/2} \ . \tag{1.19}$$

Thus, (1.18) yields

$$N_\alpha(V) \approx C_n \int [V - \alpha]_-(x)^{n/2}\ dx \ . \tag{1.20}$$

While the chief purpose of this paper is to prove Theorems 1 and 2, quantities of no less interest are the moments of the nonpositive eigenvalues of H.

DEFINITION. *For* $\gamma > 0$

$$\begin{aligned} I_\gamma(V) &= \sum_i |E_i(V)|^\gamma \\ &= \gamma \int_{-\infty}^{0} |\alpha|^{\gamma-1}\ N_\alpha d\alpha \end{aligned} \tag{1.21}$$

$I_0(V)$ *is defined to be* $N_0(V)$.

For $n \geq 3$ we can use Corollary 5 and Fubini's theorem to obtain

$$\begin{aligned} I_\gamma(V) &\leq \gamma\ L_n \int_M dx \int_{-\infty}^{0} |\alpha|^{\gamma-1}[V - \alpha\ ]_-(x)^{n/2}\ d\alpha \\ &= \gamma\ L_n \int_M dx \int_{-V_-(x)}^{0} d\alpha\ |\alpha|^{\gamma-1}(V_-(x) + \alpha)^{n/2} \\ &= L_{\gamma,n} \int_M V_-(x)^{\gamma+n/2}\ dx \end{aligned} \tag{1.22}$$

with

$$L_{\gamma,n} = L_n\ \Gamma(\gamma+1)\ \Gamma(1+n/2)\ \Gamma(1+\gamma+n/2)^{-1} \tag{1.23}$$

There are several things to be said about (1.22). Although it was derived from Corollary 5 under the assumptions (1.13) and $n \geq 3$, it holds much more generally. For example it holds in $\mathbb{R}^n$ for $n = 2$, $\gamma > 0$ and for $n = 1$, $\gamma > 1/2$ provided $\alpha > 0$. This was first given in [9]. In [9] it was stated that it holds for $n = 1$ and $\gamma = 1/2$. That was an error ; it is not known if (1.22) holds for $n = 1$, $\gamma = 1/2$ but it is known [9] that (1.22) does not hold for $n = 1$, $\gamma < 1/2$. In section II we shall briefly mention how to deduce (1.22).

The best constant $L_{\gamma,n}$ in (1.22) is not given by (1.23), as the foregoing remark already indicates. If we use C_n in place of $L_{\gamma,n}$ in (1.23) we have the classical value of $L_{\gamma,n}$ namely,

$$L^C_{\gamma,n} = (4\pi)^{-n/2}\ \Gamma(\gamma+1)\ \Gamma(1+\gamma+n/2)^{-1} \tag{1.24}$$

As in the case of L_n, it is easy to prove that $L_{\gamma,n} \geq L^C_{\gamma,n}$. The classical constant $L^C_{\gamma,n}$ can also be "derived" from the semiclassical assumption as in (1.18), (1.19), namely

Proceedings of the Amer. Math. Soc. Symposia in Pure Math. *36*, 241–252 (1980)

$$\sum_i |E_i|^\gamma \approx (2\pi)^{-n} \int_{\mathbb{R}^n \times \mathbb{R}^n} dp dx \, |p^2 + V(x)|^\gamma \, \Theta(-p^2 - V(x)). \qquad (1.25)$$

If the p integration is done in (1.25), the result is (1.22) with $L^C_{\gamma,n}$.

An important question is :

When is $L_{\gamma,n} = L^C_{\gamma,n}$?

It seems to be true that $L_{\gamma,n} = L^C_{\gamma,n}$ for γ large enough, depending on n. This is known to be true [9, 10] for $n = 1$ and $\gamma \geq 3/2$. In fact [10], if $L_{\gamma_0,n} = L^C_{\gamma_0,n}$ for some γ_0, then equality holds for all $\gamma \geq \gamma_0$.

The case of primary *physical interest* is $\gamma = 1$, $n = 3$, where it is conjectured [9] that equality holds. If this were so, it would have important consequences for physics and it is hoped that someone will be motivated to solve the problem.

We now turn to the proof of Theorems 1 and 2 in the next three sections.

II. THE BIRMAN-SCHWINGER KERNEL

As stated in Corollary 5 we can assume $V(x) \leq 0$. Therefore write $V(x) = -U(x)$, $U \geq 0$.

A useful device for studying the nonpositive eigenvalues of $-\Delta-U$ was discovered by Birman [11] and Schwinger [12]. If $(-\Delta-U)\psi = E\psi$, $E \leq 0$, then

$$\psi = (-\Delta + |E|)^{-1} U\psi \, . \qquad (2.1)$$

Defining $U^{1/2}\psi = \phi$, and multiplying (2.1) by $U^{1/2}$, we have

$$\phi = K_{|E|}(U)\phi \qquad (2.2)$$

where $K_e(U)$, for $e \geq 0$, is the positive *Birman-Schwinger Kernel* given explicitly by

$$K_e(x, y; U) = U(x)^{1/2}(-\Delta + e)^{-1}(x, y)U^{1/2}(y) \, . \qquad (2.3)$$

What (2.2) says is that for every nonpositive eigenvalue, E, of $-\Delta-U$, $K_{|E|}(U)$ has an eigenvalue 1. The converse is also easily seen to hold (see [3,8] for more details). $K_e(U)$ is to be thought of as an operator on L^2 ; we will see that it is compact, when $e > 0$ at least, and U is in a suitable L^p space.

In addition to the advantage that the study of the E's reduces to the study of a compact operator, there is the following important fact: Since $(-\Delta + e)^{-1}$ is operator monotone decreasing as a function of e, so is K_e. Hence (with $V = -U$),

$$N_\alpha(V) = k_{|\alpha|}(U) \equiv \textit{number of eigenvalues of } K_{|\alpha|}(U) \geq 1. \qquad (2.4)$$

(2.4) will be exploited in the following way.

PROPOSITION 6. *Let* $F : \mathbb{R}^+ \to \mathbb{R}^+$ *be any function such that* $F(x) \geq 1$ *for* $x \geq 1$.

Then

$$\mathrm{Tr}\, F(K_e(U)) = \sum_i F(l_e^i(U))$$
$$\geq k_e(U) = N_{-e}(V) \tag{2.5}$$

where Tr *means trace and the* $l_e^i(U)$ *are the eigenvalues of* $K_e(U)$.

For example, consider $\mathbb{R}^n$, $n \leq 3$, and $F(x) = x^2$. Then

$$\mathrm{Tr}\, K_e(U)^2 = \iint U(x)\, U(y)\, [(-\Delta + e)^{-1}(x - y)]^2\, dxdy$$
$$\leq \|U\|_2^2\, \|(-\Delta + e)^{-1}(x)\|_2^2 \tag{2.6}$$

by Young's inequality. The last factor is of the form

$$\|(-\Delta + e)^{-1}(x)\|_2^2 = h_n e^{-2+n/2} \tag{2.7}$$

This shows that $K_e(U)$ is Hilbert-Schmidt when $U \in L^2$ and $e > 0$. When $n > 3$, $h_n = \infty$ but one can show that $K_e(U)$ is compact by considering the trace of a higher power of $K_e(U)$, cf. [9].

At this point we can derive the aforementioned bound, (1.22), on $I_\gamma(V)$. If we use (2.5) and (2.6) and insert the latter in (1.21), the α integration will diverge. The trick [9] is to use Proposition 4 with $2\beta = e = -\alpha$. Thus

$$N_{-e}(V) \leq \mathrm{Tr}\, K_{e/2}([V + e/2]_-)$$
$$\leq \|[U - e/2]_+\|_2^2\, h_n (e/2)^{-2+n/2} \tag{2.8}$$

Inserting (2.8) into (1.21), and doing the α integration first, we obtain

$$I_\gamma(V) \leq 2^{2-n/2}\, \gamma h_n \int dx \int_{-2U(x)}^{0} d\alpha\, |\alpha|^{\gamma-3+n/2} [U(x) + \alpha/2]^2$$
$$= L_{\gamma,n} \int U(x)^{\gamma+n/2} dx \tag{2.9}$$

if $\gamma > 2 - n/2$ and $n \leq 3$.

(2.9) can be extended to other values of n and γ (but with $\gamma > 1/2$ for $n = 1$), cf. [9] .

There is no way, however, to make this method work with $F(x) = x^a$ when $\gamma = 0$ and $n \geq 3$. Quiet reflection shows that if theorem 2 is to be provable by this method then we need $x^{-n/2} F(x) \to 0$ as $x \to 0$ but $x^{-1}F(x)$ remains bounded as $x \to \infty$. The tool we will use to bound $\mathrm{Tr}F(K_e(U))$ for such F's is the Wiener integral. That is the subject of the next section, which is really the main point of this paper.

III. THE WIENER INTEGRAL REPRESENTATION OF $F(K_e(U))$

Let $d\mu_{x,y;t}$ be *conditional* Wiener measure on paths $\omega(t)$ with $\omega(0) = x$ and $\omega(t) = y$. This measure gives a representation for G (cf. (3)) by

$$\int d\mu_{x,y;t}(\omega) = G(x,y;t) = e^{t\Delta}(x, y). \tag{3.1}$$

G itself has the *semigroup property*

$$\int_M G(x, y; t)\, G(y, z; s)\, dy = G(x, z; t+s). \tag{3.2}$$

The well known Feynman-Kac formula [13] is

$$\int d\mu_{x,y;t}(\omega) \exp\left[-\lambda \int_0^t U(\omega(s))ds\right] = e^{-t(-\Delta + \lambda U)}(x, y). \tag{3.3}$$

(Note the signs in (3.3).). Take $e \geq 0$, $\lambda \geq 0$, and $U \geq 0$, multiply (3.3) by $U(x)^{1/2} U(y)^{1/2}$ exp (-et) and integrate with respect to t. The result is

$$\begin{aligned} A &= U(x)^{1/2} U(y)^{1/2} \int_0^\infty dt\, e^{-et} \int d\mu_{x,y;t}(\omega) \exp\left[-\lambda \int_0^t U(\omega(s))ds\right] \\ &= \{U^{1/2}(-\Delta + \lambda U + e)^{-1} U^{1/2}\}(x, y). \end{aligned} \tag{3.4}$$

Now

$$(-\Delta + e)^{-1} = (-\Delta + \lambda U + e)^{-1} + \lambda(-\Delta + \lambda U + e)^{-1} U(-\Delta + e)^{-1} \tag{3.5}$$

Multiplying (3.5) on both sides by $U^{1/2}$ we obtain

$$A = \{K_e(U)\,[1 + \lambda K_e(U)]^{-1}\}(x, y). \tag{3.6}$$

(3.6) can be cast in a more general form. If $g(x) = \exp(-\lambda x)$ and $F(x) = x(1 + \lambda x)^{-1}$ then

$$F(x) = x \int_0^\infty dy\, e^{-y} g(xy) \tag{3.7}$$

and

$$\begin{aligned} U(x)^{1/2} U(y)^{1/2} \int_0^\infty dt\, e^{-et} \int d\mu_{x,y;t}(\omega)\, g\left(\int_0^t U(\omega(s))ds\right) \\ = F(K_e(U))(x, y). \end{aligned} \tag{3.8}$$

Next we want to take the trace of both sides of (3.8). This is obtained by setting $y = x$ and integrating. (To be precise, one must first take $U \in C^\infty$ and then use a limiting argument.) The point to notice is that the x dependence occurs through

$$\int_M dx\, U(x)\, d\mu_{x,x;t}(\omega) = \int_M dx\, d\mu_{x,x;t}(\omega)\, U(\omega(0)). \tag{3.9}$$

By the semigroup property of $\exp[-t(-\Delta + \lambda U)]$, however, the same result is obtained (after the $d\mu$ integration) if $U(\omega(0))$ in (3.9) is replaced by $U(\omega(s))$ for any $0 \leq s \leq t$. Thus,

$$\int_0^\infty dt\, t^{-1} e^{-et} \int_M dx \int d\mu_{x,x;t}(\omega)\, f\left(\int_0^t U(\omega(s))ds\right) = \mathrm{Tr}\, F(K_e(U)). \tag{3.10}$$

with

$$f(x) = xg(x), \tag{3.11}$$

i.e.

$$F(x) = \int_0^\infty dy\, y^{-1} e^{-y} f(xy) \tag{3.12}$$

Now the relations between F, f and g are linear, as are the relations (3.8) and (3.10). The latter therefore extend to a large class of f's. Since we are here not particularly interested in the operator version, (3.8), we will concentrate on the trace version (3.10). The g's of the form $\exp(-\lambda x)$ are norm dense in the continuous functions which vanish at infinity. Furthermore, using the semigroup property of $\exp[-t(-\Delta + U)]$, one can see explicitly that (3.8) and (3.10) hold for f's of the form $x^k \exp(-\lambda x)$, with k a positive integer. By a monotone convergence arguement one arrives at

THEOREM 7. *Let* f *be a nonnegative lower semicontinuous function on* $[0, \infty)$ *satisfying:*

(i) $f(0) = 0$

(ii) $x^p \cdot f(x) \to 0$ *as* $x \to \infty$ *for some* $p < \infty$.

Let $U \geq 0$ *and* $U \in L^p + L^q$ *with* $p = n/2$ $(n \geq 3)$, $p > 1$ $(n = 2)$, $p = 1$ $(n = 1)$ *and* $p < q < \infty$ *Then* (3.10) *holds, with* F *given by* (3.12), *in the sense that both sides may be* $+\infty$

The reader is referred to [8] for details. Obviously the class of f's in Theorem 7 is not the largest possible, but it is more than adequate for our intended application.

The remark that allows us to bound the left side of (3.10) is the following.

THEOREM 8. *Suppose that* f *satisfies the conditions of Theorem 7 and that* f *is also convex. Then*

$$\operatorname{Tr} F(K_e(U)) \leq \int_0^\infty dt\, t^{-1} e^{-et} \int_M dx\, G(x, x; t)\, f(tU(x)). \tag{3.13}$$

PROOF. By Jensen's inequality, for any fixed path $t \to \omega(t)$, $f(\int_0^t U(\omega(s))\, ds) = f(\int_0^t (t^{-1} ds)\, t\, U(\omega(s))) \leq \int_0^t (t^{-1} ds)\, f(tU(\omega(s)))$. By the same remark as that preceding (3.10), $\int_M dx \int d\mu_{x,x;t}(\omega)\, h(\omega(s))$ is independent of s for any fixed function h. Inserting this in (3.10) gives (3.11). QED.

IV. APPLICATIONS OF THEOREM 8

PROOF OF THEOREM 1: We use Proposition 3 and Proposition 6. f is chosen to be of the following form for some $0 < a < b$

$$\begin{aligned} f(x) &= 0, && 0 \leq x \leq a \\ &= b(x - a), && a \leq x. \end{aligned} \tag{4.1}$$

F given by (3.12) is monotone, and the condition that $F(x) = 1$ is that a and b are related by

$$1 = h(a, b) = b\int_a^\infty dy (1 - a/y) e^{-y} = be^{-a} - abE_1(a) \tag{4.2}$$

where E_1 is the exponential integral. Assuming (1.10), (1.9) will be proved; (1.7) follows from the special assumption $B_n = 0$ in what follows.

Proceedings of the Amer. Math. Soc. Symposia in Pure Math. *36*, 241–252 (1980)

In Proposition 3 write $\alpha = -e$ so that $\tilde{N}(\Omega, \lambda) < (3.13)$ with $U = (e+\lambda)\chi_\Omega$. The x integration can be done last in (3.13) in which case $f(tU(x)) = 0$ if $x \notin \Omega$. For $x \in \Omega$ we change variables to $t \to t/(e+\lambda)$. Thus,

$$b|\Omega| \int_a^\infty dt(1 - a/t)(A_n(e+\lambda)^{n/2}t^{-n/2} + B_n)\exp[-et/(e+\lambda)] \geq \tilde{N}(\Omega, \lambda) \ . \tag{4.3}$$

The simplest choice for e, which is arbitrary, is $e = 0$. This will work only when $n \geq 3$ and $B_n = 0$. But for all cases we can choose $e = c\lambda$, $c > 0$ and thereby prove Theorem 1. QED.

PROOF OF THEOREM 2: Proposition 6 is used again and the proof parallels that of Theorem 1. (4.1) and (4.2) are assumed. We change variables in (3.13) to $t \to tU(x)^{-1}$ if $U(x) \neq 0$. If $e = 0$ the result is (1.13) with

$$L_n = A_n \int_0^\infty dt\ t^{-1-n/2} f(t) = 4A_n b a^{1-n/2}[n(n-2)]^{-1} \ . \quad \text{QED.} \tag{4.4}$$

REMARKS: (i) If $\alpha = -e \neq 0$ then the only estimate we have for $N_\alpha(V)$ is contained in Corollary 5, which is valid only for $n \geq 3$. Alternatively, one could try to estimate (3.13) directly with $e \neq 0$, but this is messy. As stated earlier, no inequality of the form (1.16) holds for all α, V when $n = 1$ or 2. But recently Itō [14] has bounded (3.13) when $e \neq 0$ and $n = 2$. He uses the fact that $f(x) \leq bx$ for $x \geq a$ and obtains a complicated upper bound for $N_\alpha(V)$ in terms of $\|V_-\|_2$ and $\|V_-|\ln V_-|^{1/2}\|_2$.

(ii) If $n \geq 3$ and $B_n = 0$ we can choose $e = 0$ in (4.3). This estimate for D_n is, of course, the same as L_n given by (4.4).

As an illustration of how good our bound is let us consider the case of $\mathbb{R}^3$, where $A_3 = (4\pi)^{-3/2}$. We choose $a = 0.25$ in (4.1) and find that $E_1(.25) = 1.0443$ and $b = 1.9315$ according to (4.2). Using (4.4),

$$D_3 = L_3 = 0.1156 \ . \tag{4.5}$$

This value of D_3 can be used in (1.7). When compared with $C_3 = 0.0169$, which is supposed to be the sharp constant, it is not very good. The estimate for D_3 can be improved by using (4.3) with $e = c\lambda$, $c > 0$.

If, however, the same number, L_3, is used in (1.13) the result is quite good. As already stated, the best $L_3 > C_3$. In fact, by an explicit example,

$$L_3 \geq (3\pi)^{-3/2}\Gamma(3)/\Gamma(3/2) = 0.0780, \tag{4.6}$$

cf. [9], eqn. (4.24). It is conjectured that the right side of (4.6) is, in fact, the sharp constant in (1.13) for $\mathbb{R}^3$. In any case our result, (4.5), is off by at most 49%.

As stated in Section 1, a quantity of physical interest is I_1, the sum of the absolute values of the eigenvalues, in $\mathbb{R}^3$. Using the bound (1.22), (1.23), together with (4.5), we have

$$L_{1,3} \leq (2/5)L_3 = .04624 \tag{4.7}$$

This result was announced in [15].

BIBLIOGRAPHY

1. H. Weyl, "Das asymptotische Verteilungsgesetz der Eigenwerte Linearer partieller Differentialgleichungen", Math. Ann. 71 (1911), 441-469.

2. M. Kac, "Can one hear the shape of a drum?", Slaught Memorial Papers, no. 11, Amer. Math. Monthly 73 (1966), no. 4, part II, 1-23.

3. M. Reed and B. Simon, Methods of Modern Mathematical Physics, Acad. Press, N. Y., 1978.

4. G. V. Rosenbljum, "Distribution of the discrete spectrum of singular differential operators", Dokl. Aka. Nauk SSSR, 202 (1972), 1012-1015 (MR 45 #4216). The details are given in "Distribution of the discrete spectrum of singular differential operators", Izv. Vyss. Ucebn. Zaved. Matematika 164 (1976), 75-86. [English trans. Sov. Math. (Iz. VUZ) 20 (1976), 63-71.]

5. B. Simon, "Weak trace ideals and the number of bound states of Schroedinger operators", Trans. Amer. Math. Soc. 224 (1976), 367-380.

6. M. Cwikel, "Weak type estimates for singular values and the number of bound states of Schroedinger operators", Ann. Math. 106 (1977), 93-100.

7. E. Lieb, "Bounds on the eigenvalues of the Laplace and Schroedinger operators", Bull. Amer. Math. Soc. 82 (1976), 751-753.

8. B. Simon, Functional Integration and Quantum Physics, Academic Press, N. Y., to appear 1979.

9. E. Lieb and W. Thirring, "Inequalities for the moments of the eigenvalues of the Schroedinger equation and their relation to Sobolev inequalities", in Studies in Mathematical Physics: Essays in Honor of Valentine Bargmann (E. Lieb, B. Simon and A. Wightman eds.), Princeton Univ. Press, Princeton, N. J., 1976. These ideas were first announced in "Bound for the kinetic energy of fermions which proves the stability of matter", Phys. Rev. Lett. 35 (1975), 687-689, Errata 35 (1975), 1116.

10. M. Aizenman and E. Lieb, "On semi-classical bounds for eigenvalues of Schroedinger operators", Phys. Lett. 66A (1978), 427-429.

11. M. Birman, "The spectrum of singular boundary problems", Math. Sb. 55 (1961), 124-174. (Amer. Math. Soc. Trans. 53 (1966), 23-80).

12. J. Schwinger, "On the bound states of a given potential", Proc. Nat. Acad. Sci. U.S.A. 47 (1961), 122-129.

Proceedings of the Amer. Math. Soc. Symposia in Pure Math. *36*, 241–252 (1980)

13. M. Kac, "On some connections between probability theory and differential and integral equations", Proceedings of the Second Berkeley Symposium on Mathematical Statistics and Probability, Univ. of Calif. Press, Berkeley, 1951, 189-215.

14. K. R. Ito, "Estimation of the functional determinants in quantum field theories", Res. Inst. for Math. Sci., Kyoto Univ. (1979), preprint.

15. E. Lieb, "The stability of matter", Rev. Mod. Phys. 48 (1976), 553-569.

16. V. Glaser, H. Grosse and A. Martin, "Bounds on the number of eigenvalues of the Schroedinger operator", Commun. Math. Phys. 59 (1978), 197-212.

DEPARTMENTS OF MATHEMATICS AND PHYSICS
PRINCETON UNIVERSITY
JADWIN HALL
P.O.BOX 708
PRINCETON, N. J. 08544

With S. Oxford in Int. J. Quant. Chem. *19*, 427–439 (1981)

Improved Lower Bound on the Indirect Coulomb Energy*

ELLIOTT H. LIEB AND STEPHEN OXFORD
Departments of Mathematics and Physics, Princeton University, Princeton, New Jersey 08544, U.S.A.

Abstract

For a Coulomb system of particles of charge e, it has previously been shown that the indirect part of the repulsive Coulomb energy (exchange plus correlation energy) has a lower bound of the form $-Ce^{2/3}\int \rho(x)^{4/3}\,dx$, where ρ is the single particle charge density. Here we lower the constant C from the 8.52 previously given to 1.68. We also show that the best possible C is greater than 1.23.

1. Introduction

In the study of quantum Coulomb systems of charged particles (atoms, molecules, and solids), it is frequently desirable to estimate various energies in terms of the (diagonal) single particle charge density $\rho_\psi(x)$ belonging to a given state ψ of N particles. We will be concerned with the repulsive Coulomb energy

$$I_\psi = \langle\psi|\sum_{i<j}^{N} e_i e_j |x_i - x_j|^{-1}|\psi\rangle, \tag{1}$$

where $e_i > 0$ are the particle charges (by a trivial change we could assume all $e_i < 0$). ψ is *any* normalized state, not necessarily an eigenstate of any given Hamiltonian; it could also be a density matrix. Our goal is to find a lower bound to I_ψ. [See note added in proof.]

We *define*

$$D_\psi = D(\rho_\psi, \rho_\psi) \tag{2}$$

to be the *direct part* of I_ψ, where, in general, we have

$$D(f, g) = \tfrac{1}{2}\iint f(x)g(y)|x-y|^{-1}\,dx\,dy. \tag{3}$$

The density ρ_ψ is the sum $\sum_{i=1}^{N}$ of the N individual single particle charge densities defined by

$$\rho_\psi^i(x) = e_i \sum_{\alpha_1,\ldots,\alpha_N} \int |\psi(x_1, \ldots, x_{i-1}, x, x_{i+1}, \ldots, x_N, \alpha_1, \ldots, \alpha_N)|^2\, d\hat{x}_i. \tag{4}$$

* Work partially supported by U.S. National Science Foundation Grant No. PHY-7825390 A01.

Here the α_j are any quantum numbers (e.g., spin) and $d\hat{x}_i$ means integration over all variables except x_i. The *indirect part* E_ψ of I_ψ is defined by

$$E_\psi + D_\psi = I_\psi. \tag{5}$$

E_ψ includes all "exchange" and "correlation" effects.

It was shown by one of us [1], that for all ψ and $C = 8.52$, we have

$$E_\psi \geq -Ce^{2/3} \int \rho_\psi(x)^{4/3}\, dx. \tag{6}$$

A more complicated bound was given when the e_i are not identical.

The main result of this paper is given in Section 2. We show that

$$E_\psi \geq -C\left[\int \left(\sum_{i=1}^{N} e_i \rho_\psi^i(x)\right)^{4/3} dx\right]^{1/2} \left[\int \rho_\psi(x)^{4/3}\, dx\right]^{1/2}, \tag{7}$$

even in the case that the e_i may be distinct, with $C = 1.68$. With this improved value of C, the bound for E_ψ may have computational value as well as theoretical value.

Inequality (7) reduces to the form of Eq. (6) when all $e_i = e$. Our proof differs from that in Ref. 1 chiefly in avoiding the use of the Hardy–Littlewood maximal function. We also show in Section 3 that the best (i.e., smallest) possible C, in Eq. (6), must be greater than 1.23 by explicitly calculating E for a two-particle wave function ψ.

It should be noted that Eq. (6) is a special case of the lower bound for N-particle states ψ,

$$E_\psi \geq -C_p e^{(5p-6)/(3p-3)} \left(\int \rho_\psi^p\right)^{1/(3p-3)} N^{(3p-4)/(3p-3)}, \tag{8}$$

which is a consequence of Thomas–Fermi theory for $p > \frac{3}{2}$ (see Refs. 1–3). Hölder's inequality applied to Eq. (6) gives Eq. (8) for all $p \geq \frac{4}{3}$ with the same C as in Eq. (6). However, the best C_p in Eq. (8) will in general be smaller than this C and it would be interesting to have an estimate for it. [In Ref. 1 it was incorrectly stated that Thomas–Fermi theory gives Eq. (8) for $p > \frac{4}{3}$. This is wrong because Thomas–Fermi theory is bounded below only for $p > \frac{3}{2}$. Equation (8), for $p = \frac{5}{3}$, was first given in Ref. 3, where it was used to prove the stability of matter.]

The bound, equations (6) or (7), can be examined in a number of ways. One might try to find a constant C in Eq. (6) only for ψ symmetric or antisymmetric, or only for ψ with a particular spin value. However any such restriction on ψ cannot improve the constant in Eq. (6). Let all $e_i = e$. Suppose we take an arbitrary $\psi(x_1, \ldots, x_N; \alpha_1, \ldots, \alpha_N)$, where $\alpha_1, \ldots, \alpha_N$ are arbitrary quantum numbers. Let

$$f_\psi(x_1, \ldots, x_N) = \sum_{\alpha_1 \cdots \alpha_N} |\psi(x_1, \ldots, x_N, \alpha_1, \ldots, \alpha_N)|^2,$$

and let the symmetrized version of f_ψ be given as

$$F(x_1, \ldots, x_N) = \frac{1}{N!} \sum_{P \in S_N} f_\psi(x_{P1}, \ldots, x_{PN}).$$

We can define a symmetric ψ^S such that $E_{\psi^S} = E_\psi$, $I_{\psi^S} = I_\psi$, and $\rho_{\psi^S}(x) = \rho_\psi(x)$. We merely take $\psi^S(x_1, \ldots, x_N) = F(x_1, \ldots, x_N)^{1/2}$. This shows that C cannot be improved by excluding bosonic ψ. In a similar way, we define an antisymmetric ψ^a,

$$\psi^a = F(x_1, \ldots, x_N)^{1/2} \theta(x_1, \ldots, x_N),$$

where θ is any antisymmetric function which takes on the values ± 1 except on a set of measure zero. We see that we cannot improve the constant by excluding ψ with Fermi statistics. A similar construction shows that restrictions on the spin quantum number cannot improve C.

It will be noted that the right-hand side of Eq. (6) is of the form given by Dirac [4] to approximate the exchange energy. There is, however, a difference between the two if spin is taken into account. The Dirac approximation is

$$E_\psi \simeq -Cq^{-1/3} \int \rho_\psi(x)^{4/3}\, dx, \tag{9}$$

with $C = 0.93$ and q is the number of spin states ($q = 2$ for electrons). Dirac computed Eq. (9) using a plane wave determinant for ψ. This determinant depends upon q. In view of Eq. (9) one might infer that a good lower bound to E_ψ should have a $q^{-1/3}$ factor. This, as we have just noted above, is not the case. Another way to say this is the following: a "diagonal" operator, such as the Coulomb repulsion cannot distinguish the spin of a particle. To "see" the spin it is necessary to examine an off-diagonal operator such as the kinetic energy T_ψ. A lower bound for T_ψ does exist [3] and it does have a factor $q^{-2/3}$. In order to have a lower bound for E_ψ that measures q it would be necessary to use an expectation value such as T_ψ that is "off-diagonal." A useful bound of this kind might exist, but a bound such as ours that involves only the "diagonal" quantity ρ_ψ has no q dependence.

It *is* true however, that the constant in Eq. (6) or (7) can be improved if we specify the number of particles. Consider the case of equal charges. Define C_N to be the best constant for Eq. (6) when we consider only N-particle states ψ, i.e.,

$$C_N = \sup_{N \text{ particle } \psi} \left(-\int \rho_\psi(x)^{4/3}\, dx \right)^{-1} E_\psi.$$

In Appendix A we compute $C_1 = 1.092$, following a treatment by Gadre, Bartolotti, and Handy [5]. One notes that C_1 is less than the constant 1.23 computed for the variational two-particle wave function of Section 3, which is itself less than C_2. In Appendix B, we show that this is a general phenomenon, i.e., $C_N \le C_{N+1}$. Presumably, one might take advantage of this if one were only dealing with a fixed low number of particles. However, we have no method available for computing an upper bound for C_N, $N \ge 2$, except 1.68, which holds for all N.

Inequality (7) might be improved upon in another way in the case of unequal charges. One might try to prove that a universal constant C existed such that the following is true (for fixed $0<\alpha\leq 1$):

$$E_\psi \geq -C\left[\int\left(\sum_{i=1}^N e_i^{1/2\alpha}\rho_\psi^i(x)\right)^{4/3} dx\right]^\alpha \left[\int \rho(x)^{4/3}\,dx\right]^{1-\alpha}. \tag{10}$$

It is an open question whether a bound exists when $\alpha=1$ in Eq. (10), called the symmetric form. Hölder's inequality implies that for Eq. (10), (right-hand side, $\alpha=1)\geq$(right-hand side, $\alpha\leq 1$); thus the symmetric form gives the best lower bound. In Ref. 1 a bound with constant 8.52 was proved for $\alpha=\frac{3}{4}$ and in this paper we sacrifice even more "symmetry" in the bound to show that Eq. (10) holds for $\alpha=\frac{1}{2}$ with the improved constant 1.68.

In Ref. 5, the comment was made that no one has yet produced any upper bound for E_ψ. The following simple remark is relevant. There certainly cannot be any upper bound of form $C\int\rho(x)^{4/3}\,dx$. The reason is that $E_\psi/\int\rho(x)^{4/3}\,dx$ can be made arbitrarily positive simply by taking a ψ for two particles of the form

$$\psi = A[k\cdot(x_1-x_2)]\exp(-|x_1|^2-|x_2|^2-\lambda|x_1-x_2|^2,$$

where k is some fixed vector. As λ tends to infinity, E_ψ will tend to $+\infty$ while $\int\rho_\psi^{4/3}\,dx$ will remain bounded. This ψ is antisymmetric.

2. A Lower Bound for E_ψ

We will use an argument similar to that given in Ref. 1 to derive Eq. (7) for $C=1.68$.

We first fix charges $e_1\geq 0,\ldots,e_N\geq 0$. Let $f_\psi(x_1,\ldots,x_N)$ be the particle density associated with an N-particle wave function ψ, as given in Section 1. Let ρ_ψ be the associated single-particle charge density, equation (4), and let $x_1,\ldots,x_N$ be distinct but otherwise arbitrary points in $\mathbf{R}^3$. We take $\mu(y)$ to be some function satisfying the following: (i) μ is non-negative; (ii) μ is spherically symmetric about the origin and $\mu(x)=0$ if $|x|>1$; and (iii) $\int\mu(y)\,dy=1$. Let λ be a positive constant to be determined later.

We now define a function

$$\mu_x(\cdot):\qquad \mu_x(y)\equiv\lambda^3\rho_\psi(x)\mu(\lambda\rho_\psi(x)^{1/3}(y-x)), \tag{11}$$

if $\rho_\psi(x)>0$, and $\mu_x(y)\equiv 0$ if $\rho_\psi(x)=0$. We see that μ_x is a non-negative function which satisfies (i) its integral (with respect to y) is 1 if $\rho_\psi(x)>0$; (ii) it is spherically symmetric about x; (iii) $\mu_x(y)=0$ if $|y-x|>\lambda^{-1}\rho_\psi(x)^{-1/3}$.

We observe that Lemma 1 of Ref. 1 may be applied to this choice of μ. Namely, we prove

Lemma 1:

$$\sum_{1\leq i<j\leq N} e_ie_j|x_i-x_j|^{-1}\geq -D(\rho_\psi,\rho_\psi)+2\sum_{i=1}^N D(\rho_\psi,e_i\mu_{x_i})-\sum_{i=1}^N D(e_i\mu_{x_i},e_i\mu_{x_i}). \tag{12}$$

Proof. It is a well-known fact [1] that the potential ϕ generated by a non-negative spherically symmetric charge distribution of total charge 1 satisfies $\phi(x) \leq 1/|x|$. In particular, taking x_i to be the origin, $\int \mu_{x_i}(y)|x-y|^{-1}\,dy \leq |x-x_i|^{-1}$. Hence

$$\begin{aligned} D(e_i\mu_{x_i}, e_j\mu_{x_j}) &= \tfrac{1}{2}e_ie_j \int \mu_{x_i}(y)|x-y|^{-1}\mu_{x_j}(x)\,dx\,dy \\ &\leq \tfrac{1}{2}e_ie_j \int |x-x_i|\mu_{x_j}(x)\,dx \leq \tfrac{1}{2}e_ie_j|x_i-x_j|^{-1}. \end{aligned} \tag{13}$$

We now observe that

$$D\left(\rho_\psi - \sum_{i=1}^N e_i\mu_{x_i}, \rho_\psi - \sum_{i=1}^N e_i\mu_{x_i}\right) \geq 0 \tag{14}$$

by the positive definiteness of the Coulomb kernel. Expanding the left-hand side of Eq. (14) and rearranging, one has that

$$2\sum_{i<j} D(e_i\mu_{x_i}, e_j\mu_{x_j}) \geq -D(\rho_\psi, \rho_\psi) + 2\sum_{i=1}^N D(\rho_\psi, e_i\mu_{x_i}) - \sum_{i=1}^N D(e_i\mu_{x_i}, e_i\mu_{x_i}). \tag{15}$$

The lemma follows by applying Eq. (13) to the left-hand side of Eq. (15). ■

Now let δ_{x_i} be a point charge distribution of charge one centered at x_i. Adding and subtracting terms on the right-hand side of inequality (12), we get

$$\begin{aligned} \sum_{i<j} e_ie_j|x_i-x_j|^{-1} \geq &-D(\rho_\psi, \rho_\psi) + 2\sum_{i=1}^N D(\rho_\psi, e_i\delta_{x_i}) \\ &-\left(2\sum_{i=1}^N D(\rho_\psi, e_i\delta_{x_i} - e_i\mu_{x_i}) + \sum_{i=1}^N D(e_i\mu_{x_i}, e_i\mu_{x_i})\right). \end{aligned} \tag{16}$$

We now integrate Eq. (16) against $f_\psi(x_1, \ldots, x_N)$, whence

$$\begin{aligned} I_\psi \geq D(\rho_\psi, \rho_\psi) - \Bigg(\sum_{i=1}^N \int 2D(\rho_\psi, \delta_{x_i} - \mu_{x_i})\rho_\psi^i(x_i)\,dx_i \\ + \sum_{i=1}^N \int D(\mu_{x_i}, \mu_{x_i})e_i\rho_\psi^i(x_i)\,dx_i\Bigg). \end{aligned} \tag{17}$$

We wish to find an upper bound of the appropriate kind to the expression in large parentheses in Eq. (17). Let us denote the first sum in large parentheses as (*) and the second as (**). We rewrite (*) as follows:

$$(*) = \int \rho_\psi(x) 2D(\rho_\psi, \delta_x - \mu_x)\,dx = \iint \rho_\psi(y)F_\lambda(\rho_\psi(x), |x-y|)\,dx\,dy, \tag{18}$$

where we define

$$F_\lambda(a, r) = [ar^{-1} - \lambda a^{4/3}\phi(\lambda a^{1/3}r)], \tag{19}$$

and ϕ is the potential generated by our fixed μ. Hereafter we require that μ be bounded. In this case, $F_\lambda(a, r)$, considered as a function of a on $(0, \infty)$, satisfies the following for $r>0$; (i) it is continuously differentiable; (ii) $F_\lambda(0,r)=0$ and $F_\lambda(a,r)=0$ if $\lambda a^{1/3}r>1$. These properties follow from the continuity and differentiability of ϕ and the relation $\phi(x)=1/|x|$ for $|x|>1$. For $a>0$, let $\chi_a(x)=\theta[\rho_\psi(x)-a]$, where $\theta(t)=0$ if $t\leq 0$, and $\theta(t)=1$ if $t>0$. χ_a is the characteristic function of the set $\rho_\psi(x)>a$. By Fubini's theorem and the fundamental theorem of calculus, one has that

$$\int_0^\infty da \int \chi_a(x)\frac{\partial}{\partial a}F_\lambda(a,r)\,dx=\int dx\int_0^{\rho(x)}\frac{\partial}{\partial a}F_\lambda(a,r)\,da=\int dx\,F_\lambda(\rho(x),r), \tag{20}$$

and thus

$$(*)=\int_0^\infty\int_0^\infty da\,db\int dx\,dy\,\chi_a(x)\chi_b(y)\frac{\partial}{\partial a}F_\lambda(a,|x-y|)$$

where we have used the representation $\rho(y)=\int_0^\infty db\,\chi_b(y)$.

We bound $(*)$ as follows: Let $(y)_+=y$ if $y\geq 0$, and $(y)_+=0$ if $y\leq 0$. Then

$$\begin{aligned}(*)&\leq\int_0^\infty\int_0^\infty da\,db\int dx\,dy\,\chi_a(x)\chi_b(y)\Big(\frac{\partial}{\partial a}F_\lambda(a,|x-y|)\Big)_+\\&\leq\int_{a>b}da\,db\int dx\,dy\,\chi_a(x)\Big(\frac{\partial}{\partial a}F_\lambda(a,|x-y|)\Big)_+\\&\quad+\int_{b<a}da\,db\int dx\,dy\,\chi_b(y)\Big(\frac{\partial}{\partial a}F_\lambda(a,|x-y|)\Big)_+.\end{aligned} \tag{21}$$

By scaling properties of $(\partial/\partial a)F_\lambda(a,|x-y|)$, one has that

$$\int\Big(\frac{\partial}{\partial a}F_\lambda(a,|x-y|)\Big)_+dx=\int\Big(\frac{\partial}{\partial a}F_\lambda(a,|x-y|)\Big)_+dy=\lambda^{-2}Ka^{-2/3}, \tag{22}$$

where $K=\int[(\partial/\partial a)F_1(1,|z|)]_+\,dz$ and K only depends on the original choice of μ.

We, therefore, have that

$$\begin{aligned}(*)&\leq\lambda^{-2}K\int dx\Big(\int_0^\infty da\,\chi_a(x)a^{-2/3}\int_0^a db+\int_0^\infty\chi_b(x)\,db\int_0^b a^{-2/3}\,da\Big)\\&=4\lambda^{-2}K\int dx\int_0^\infty\chi_a(x)a^{1/3}\,da=3\lambda^{-2}K\int\rho_\psi^{4/3}(x)\,dx,\end{aligned} \tag{23}$$

where we have used the representation $\rho_\psi^{4/3}(x)=(4/3)\int_0^\infty a^{1/3}\chi_a(x)\,da$.

The second sum $(**)$ in the large parentheses of (17) can be written

$$\begin{aligned}(**)&=\sum_{i=1}^N\int D(\mu_x,\mu_x)e_i\rho_\psi^i(x)\,dx\\&=\lambda D(\mu,\mu)\sum_{i=1}^N\int e_i\rho_\psi^i(x)\rho_\psi(x)^{1/3}\,dx \qquad (24)\\&\leq\lambda D(\mu,\mu)\Big[\int\Big(\sum_{i=1}^N e_i\rho_\psi^i(x)\Big)^{4/3}dx\Big]^{3/4}\Big[\int\rho_\psi(x)^{4/3}\,dx\Big]^{1/4}. \qquad (25)\end{aligned}$$

Equation (24) follows from simple scaling and Eq. (25) is the Hölder inequality.

Optimizing Eqs. (24) and (25) with respect to λ yields

$$(*)+(**)\leq\tfrac{3}{2}[6KD(\mu,\mu)^2]^{1/3}\left[\int\left(\sum_{i=1}^{N}e_i\rho^i(x)\right)^{4/3}dx\right]^{1/2}\left[\int\rho(x)^{4/3}\,dx\right]^{1/2}. \tag{26}$$

A variational argument shows that the optimum choice of μ would be the uniform ball *if* $[(\partial/\partial a)F_\lambda(a,r)]_+$ *were replaced by* $(\partial/\partial a)F_\lambda(a,r)$ [in which case the constant in Eq. (26) would be 1.45]. However, trial and error indicates this choice is also approximately best with the cutoff. We find that $[\partial F_1(1,r)/\partial a]_+=\partial F_1(1,r)/\partial a$ if and only if $r\leq R$ with $R=(5^{1/2}-1)/2$. Then $K=0.6489$ and $D(\mu,\mu)=\frac{3}{5}$. The constant in Eq. (26) is then 1.68.

Thus we reach the conclusion that

$$E_\psi\geq-1.68\left[\int\left(\sum_{i=1}^{N}e_i\rho_\psi^i(x)\right)^{4/3}dx\right]^{1/2}\left[\int\rho_\psi(x)^{4/3}\,dx\right]^{1/2}. \tag{27}$$

3. A Lower Bound for C_2

We now exhibit a lower bound to C_2 [and thus to the best possible C in Eq. (6)] which is greater than C_1.

We choose a singular $\psi(x,y)$, and take $e_1=e_2=1$. Let $t\doteq|x|$ and $s\doteq|y|$ for $x,y\in\mathbf{R}^3$, and let h and e be unit vectors $e=x/|x|$ and $h=y/|y|$. We define

$$|\psi|^2[(t,e),(s,h)]\equiv f\equiv(15/4\pi^2)\delta(1-t-s)\delta(e\cdot h+1)\theta(1-t)\theta(1-s).$$

We check the following:

$$\begin{aligned}\int f[(t,e),(s,h)]s^2\,ds\,dh&=\frac{15}{4\pi^2}\int_0^1\delta(1-t-s)s^2\,dx\\&\quad\times\int_0^{2\pi}\int_0^{\pi}\delta(\cos\psi+1)\sin\psi\,d\psi\,d\phi\,\theta(1-t)\\&=\frac{15}{2\pi}(1-t)^2\theta(1-t)\equiv\rho^1(t).\end{aligned}$$

We have used spherical coordinates to evaluate the above integral with the north pole in the (fixed) "e" direction. Similarly, the s marginal is $\rho^2(s)=\rho^1(s)$. One checks that $\int\rho^1=1$ and hence f is properly normalized. We have that $\rho(t)=2\rho^1(t)$. Trivially, $I_\psi=1$ since the particles are always one unit apart. We have that

$$\int\rho(x)^{4/3}\,dx=4\pi\left(\frac{15}{\pi}\right)^{4/3}\int_0^1(1-t)^{8/3}t^2\,dt=2.084.$$

By Newton's theorem

$$D(\rho,\rho)=(4\pi)^2\int_0^\infty\rho(t)t\int_0^t\rho(s)s^2=3.572.$$

The result is that

$$C \geq C_2 \geq -E_\psi \Big/ \int \rho_\psi(x)^{4/3}\,dx = 1.234.$$

Appendix A: Evaluation of C_1

For one particle $I_\psi = 0$. Therefore C_1 is the maximum of $H(\rho)$ given in Eq. (28). We prove the following theorem and then compute C_1. This was done in Ref. 5. The purpose of this appendix is only to give a rigorous justification of the calculation done in Ref. 5.

Theorem. There exists a symmetric decreasing function ρ which maximizes the functional

$$H(\rho) = \left(\int \rho(x)^{4/3}\,dx\right)^{-1} D(\rho,\rho) \tag{28}$$

over the set $A = \{\rho(x) | \rho(x) \geq 0, \rho \in L^{4/3}(\mathbf{R}^3), \int \rho(x)\,dx = 1\}$.

Proof. For any $\rho \geq 0$ and $\rho \in L^{4/3}(\mathbf{R}^3) \cap L^1(\mathbf{R}^3)$, let $(\rho)_\lambda$ be a scaled version of ρ, i.e.,$(\rho)_\lambda(x) = \lambda^3 \rho(\lambda x)$. It is simple to check that

$$\int (\rho)_\lambda(x)\,dx = \int \rho(x)\,dx,$$

$$\int (\rho)_\lambda(x)^{4/3}\,dx = \lambda \int \rho(x)^{4/3}\,dx,$$

and

$$H[(\rho)_\lambda] = H(\rho). \tag{29}$$

Let $\rho_j \in A$, $j = 1, 2, \ldots$ be such that $\lim_j H(\rho_j) = \sup_{\rho \in A} H(\rho)$. This supremum may *a priori* be infinite, but we will see that it is finite. By scaling ρ_j and using Eq. (29), we may assume henceforth that $\int \rho_j(x)^{4/3}\,dx = 1$. By the Riesz inequality for symmetric decreasing rearrangements [6],† we have $H(\rho^*) \geq H(\rho)$, where ρ^* is the symmetric decreasing rearrangement of ρ. One also has that $\int \rho^*(x)\,dx = \int \rho(x)\,dx$, $\int \rho^*(x)^{4/3}\,dx = \int \rho(x)^{4/3}\,dx$; therefore, by replacing ρ_j by ρ_j^* if necessary, we may also assume that the ρ_j are symmetrically decreasing.

We now use an idea used in Ref. 7. By the symmetric decreasing property of ρ_j, we have [writing $\rho_j(x) = \rho_j(|x|)$]

$$\frac{4\pi}{3} R^3 \rho_j(R) \leq \int_{|x| \leq R} \rho_j(x)\,dx \leq \int \rho_j(x)\,ds = 1.$$

† The 3-dimensional proof can be found in Brascamp, Lieb, and Luttinger, J. Funct. Anal. **17**, 227 (1974).

Hence

$$\rho_j(R)\le k_1/R^3,\ \text{all } j. \tag{30}$$

Similarly,

$$\frac{4\pi}{3}R^3\rho_j^{4/3}(R)\le\int\rho_j(x)^{4/3}\,dx=1,$$

$$\rho_j(R)\le k_2/R^{9/4},\ \text{all } j. \tag{31}$$

We define $f(R)\equiv\min(k_1/R^3, k_2/R^{9/4})$. Since the ρ_j are symmetric decreasing and uniformly bounded by f (which is finite except at 0), by a variant of Helley's theorem [8], some subsequence of the ρ_j (which we continue to denote by ρ_j) converges pointwise almost everywhere to some symmetric decreasing $\rho(x)$ and $\rho(x)\le f(x)$. We will see that $\rho(x)\ne 0$.

We now show that the ρ we have found satisfies the conditions of the theorem. By calculation $D(f,f)<\infty$. We therefore apply the dominated convergence theorem to conclude that

$$\lim_j D(\rho_j,\rho_j)=D(\rho,\rho)<\infty. \tag{32}$$

In particular,

$$0<\sup_A H(\rho)=\lim D(\rho_j,\rho_j)<\infty\ \text{and thus}\ \rho\ne 0. \tag{33}$$

Furthermore, by Fatou's lemma we have that

$$\int\rho(x)^{4/3}\,dx\le\lim_j\int\rho_j(x)^{4/3}\,dx=1, \tag{34}$$

$$\int\rho(x)\,dx\le\lim_j\int\rho_j(x)\,dx=1. \tag{35}$$

Therefore by Eqs. (32)–(34), $H(\rho)\ge\sup_{\rho\in A}H(\rho)$. By Eq. (35) we can multiply ρ by a scalar $\lambda\ge 1$ so that $\int\lambda\rho(x)\,dx=1$. By definition of A, $\lambda\rho\in A$ and

$$H(\lambda\rho)=\lambda^{2/3}H(\rho)\ge H(\rho)\ge\sup_{\rho\in A}H(\rho). \tag{36}$$

It must be that the inequalities in Eq. (36) are actually equalities and thus that $\lambda=1$. We therefore have that ρ belongs to A and maximizes $H(\cdot)$ on that set. ■

We shall now show that the constant C_1 can be calculated. By usual variational arguments [7], one knows that the optimizing ρ satisfies the following variational equations:

$$\phi(x)-\tfrac{4}{3}C_1\rho(x)^{1/3}+\lambda\begin{cases}=0 & \text{if } \rho(x)>0, \quad (37)\\ \le 0 & \text{if } \rho(x)=0, \quad (38)\end{cases}$$

where $\phi(x)$ is the potential generated by ρ,

$$\phi(x)=\int \rho(y)|x-y|^{-1}\,dy, \tag{39}$$

and where λ is a real-valued Lagrange multiplier.

We first note that Eqs. (37) and (38) imply that ρ is compactly supported. If not, then one must have that $\rho(x)>0$ for all x by the symmetric decreasing property of ρ. By letting $|x|\to\infty$ one has that $\lambda=0$, since ρ and ϕ tend to zero in Eq. (37). We then would have by Eq. (37), $\rho(x)=(\text{constant})\phi(x)^3$, where the constant is positive. For sufficiently large $|x|$, we see that $\rho(x)\ge\frac{1}{2}(\text{constant})|x|^{-3}$. This implies that ρ is not integrable, contradicting the fact that $\int\rho(x)\,dx=1$. Let r_0 be the distance at which ρ first vanishes.

We now apply the Laplacian to Eq. (39) and use Eq. (37),

$$-(1/4\pi)\Delta\phi(x)=\rho(x)=\begin{cases}\{\frac{3}{4}C_1^{-1}[\phi(x)+\lambda]\}^3, & |x|\le r_0,\\ 0, & |x|\ge r_0.\end{cases} \tag{40}$$

Let $f(r)=(3\pi/C_1)^{3/2}[\phi(x)+\lambda]/4\pi$. We rewrite Eq. (40) in spherical coordinates

$$r^{-1}\frac{d^2}{dr^2}rf(r)=-f(r)^3,\quad r<r_0. \tag{41}$$

Equations (40) and (41) hold in the distributional sense.

We now argue that $f(r)$ is continuously differentiable and that $f'(0)=0$. This will also imply that Eq. (41) is supplemented by $f(r_0)=0$. In spherical coordinates, one can write

$$\phi(r)=\begin{cases}4\pi r^{-1}\displaystyle\int_0^r\rho(s)s^2\,ds+4\pi\int_r^{r_0}\rho(s)s\,ds, & r\le r_0,\\ r^{-1}, \quad r\ge r_0.\end{cases} \tag{42}$$

We apply Hölder's inequality to the first integral and use the fact that $\rho\in L^{4/3}$,

$$\int_0^r\rho(s)s^2\,ds\le\left(\int_0^r\rho(s)^{4/3}s^2\,ds\right)^{3/4}\left(\int_0^r s^2\,ds\right)^{1/4}\le(\text{constant})r^{3/4}.$$

This and a similar inequality satisfied by the second integral imply that $\phi(r)=O(r^{-1/4})$ near the origin. By Eq. (37) one has that $\rho(r)=O(r^{-3/4})$ near $r=0$. This in turn implies that ϕ is bounded at the origin, and therefore by Eq. (37), ρ is also bounded at the origin. Since ϕ is the potential of a bounded, compactly supported charge distribution, it is also continuous, hence ρ is continuous by Eqs. (37) and (38).

One now can see that ϕ (hence f) is C^1 by examining Eq. (42). Since ρ is continuous and bounded, the first term of Eq. (42) is of the form $r^{-1}g(r)$, where $g(r)$ is continuously differentiable for $r>0$, $g(r)=O(r^3)$ and $g'(r)=O(r^2)$ near the origin. Hence the first term is continuously differentiable for $r\ge0$, and has vanishing derivative at $r=0$. The preceding statement is true of the second term in

Eq. (42) by inspection. Thus $\phi(r)$ is continuously differentiable for $r \geq 0$ and $\phi'(0)=0$. Equation (41) holds in the strong sense because its right-hand side is C^1.

As first noted by Gadre, Bartolotti, and Handy [5], Eq. (41) is the Emden equation of order 3. One may rescale $\rho(x) \to \alpha^3 \rho(\alpha x)$ to ensure that $f(0)=1$. The two conditions $f(0)=1$ and $f'(0)=0$ uniquely determine the solution of the ordinary differential equation (41).

If r_0 is the first zero of the solution, we have that $\rho(r)=0$ if $r \geq r_0$ and $\rho(r)=(3\pi)^{-3/2} C_1^{3/2} f(r)^3$ if $r \leq r_0$. In Ref. 5 it was noted that this equation determines the constant C_1. Namely, we have that

$$\begin{aligned} 1 &= 4\pi \int_0^{r_0} \rho(r) r^2\, dr = 4 \cdot 3^{-3/2} \pi^{-1/2} C_1^{3/2} \int_0^{r_0} r^2 f(r)^3\, dr \\ &= -4 \cdot 3^{-3/2} \pi^{-1/2} C_1^{3/2} \int_0^{r_0} r[rf(r)]''\, dr \\ &= -4 \cdot 3^{-3/2} \pi^{-1/2} C_1^{3/2} r_0^2 f'(r_0). \end{aligned} \tag{43}$$

Emden functions are tabulated [9]. We find that $r_0 = 6.89684$, $f'(r_0) = -0.04243$. Equation (43) then gives $C_1 = 1.092$.

Appendix B: Monotonicity of C_N

We show that $C_N \leq C_{N+1}$, where C_N is defined in Section 1 as the best constant in Eq. (6) for an N-particle state. We consider the case $e_i = e$.

Let $\varepsilon > 0$ be arbitrary but fixed. We let $f_N(x_1, \ldots, x_N)$ be an N-particle density which vanishes for $|x_i| > L$ for $1 \leq i \leq N$, where L is some finite number, and furthermore, let f_N have the property that

$$-\left(e^{2/3} \int \rho_{f_N}(x)^{4/3}\, dx\right)^{-1} E_{f_N} \geq C_N - \varepsilon. \tag{44}$$

A simple approximation argument using dominated convergence shows that L and f_N can be found satisfying Eq. (44).

Let $x_0 \in \mathbf{R}^3$ be chosen such that $|x_0| \geq L + 2R$, where R will be determined later. We define a one-particle density $f_1(x) = (\frac{4}{3}\pi R^3)^{-1} \theta(R - |x - x_0|)$ and we also define the $(N+1)$-particle density $f_{N+1}(x_1, \ldots, x_{N+1}) = f_N(x_1, \ldots, x_N) f_1(x_{N+1})$. One sees that $\rho_{f_{N+1}}(x) = \rho_{f_N}(x) + e f_1(x)$. Since ρ_N and f_1 are never simultaneously nonzero, we have

$$\begin{aligned} e^{2/3} \int \rho_{f_{N+1}}(x)^{4/3}\, dx &= e^{2/3}\left(\int \rho_{f_N}(x)^{4/3}\, dx + e^{4/3} \int f_1(x)^{4/3}\, dx\right) \\ &= e^{2/3} \int \rho_{f_N}(x)^{4/3}\, dx + \frac{e^2 (3/4\pi)^{1/3}}{R}. \end{aligned} \tag{45}$$

We also have that

$$\begin{aligned} I_{f_{N+1}} &= I_{f_N} + e^2 \sum_{i=1}^{N} \int f_N(x_1, \ldots, x_N) f_1(x_{N+1}) |x_i - x_{N+1}|^{-1}\, dx_1, \ldots, dx_{N+1} \\ &\le I_{f_N} + e^2 N/R, \end{aligned} \tag{46}$$

by the definition of f_1. The evident inequality $D_{f_{N+1}} \ge D_{f_N}$ together with Eq. (46) implies that $E_{f_{N+1}} \le E_{f_N} + e^2 N/R$.

This and Eq. (45) imply that

$$\begin{aligned} C_{N+1} &\ge -\left(e^{2/3} \int \rho_{f_{N+1}}(x)^{4/3}\, dx\right)^{-1} E_{f_{N+1}} \\ &\ge -\left(e^{2/3} \int \rho_{f_N}(x)^{4/3}\, dx + \frac{e^2 (3/4\pi)^{1/3}}{R}\right)^{-1} \left(E_{f_N} + \frac{e^2 N}{R}\right). \end{aligned} \tag{47}$$

We now choose R so large that the right-most term in Eq. (47) is greater or equal to

$$-\left(e^{2/3} \int \rho_{f_N}(x)^{4/3}\, dx\right)^{-1} E_{f_N} - \varepsilon.$$

Recalling Eq. (44), we have the result that $C_{N+1} \ge C_N - 2\varepsilon$. Since ε was arbitrary, $C_{N+1} \ge C_N$.

In the case of distinct e_i's, one may define (for some fixed α, $0 < \alpha \le 1$)

$$\begin{aligned} C_N(e_1, \ldots, e_N) = \sup_{f_N(x_1, \ldots, x_N)} & -\left[\int \left(\sum_{i=1}^{N} e_i^{1/2\alpha} \rho^i_{f_N}(x)\right)^{4/3} dx\right]^{-\alpha} \\ & \times \left(\int \rho_{f_N}(x)^{4/3}\, dx\right)^{\alpha - 1} E_{f_N}. \end{aligned}$$

A similar argument shows that these constants also increase, i.e., $C_N(e_1, \ldots, e_N) \le C_{N+1}(e_1, \ldots, e_N, e_{N+1})$, where $e_{N+1} \ge 0$ is arbitrary. Of course Section 2 shows that $C_N(e_1, \ldots, e_N) \le 1.68$ for all N and e_i when $\alpha = \frac{1}{2}$.

Note added in proof:

In the text we proved the inequalities, Eqs. (6) and (7), when ψ is a wave function (pure state), and remarked that the inequalities also hold for a density matrix. To prove this, note that any density matrix, μ, can be written as $\mu = \sum_\beta \psi^\beta >< \psi^\beta$. In the definition, Eq. (4), simply regard β as just one more quantum number to sum over—on the same footing as the α's. The rest of the proof is then the same as in the pure state case.

Bibliography

[1] E. H. Lieb, Phys. Lett. **70A**, 444 (1979).

[2] E. H. Lieb, Rev. Mod. Phys. **48**, 553 (1976).

[3] E. H. Lieb and W. E. Thirring, Phys. Rev. Lett. **35**, 687 (1975); **35**, 1116 (1975) (errata).

[4] P. A. M. Dirac, Proc. Cambridge Philos. Soc. **26**, 376 (1930).

[5] S. R. Gadre, L. J. Bartolotti, and N. C. Handy, J. Chem. Phys. **72**, 1034 (1980).
[6] F. Riesz, J. London Math. Soc. **5**, 162 (1930).
[7] E. H. Lieb, Stud. Appl. Math. **57**, 93 (1977).
[8] W. Feller, *An Introduction to Probability Theory and its Applications*, (*Wiley, New York*, 1966), Vol. 2, p. 261.
[9] *British Association for the Advancement of Science Mathematical Tables*, (Office of the British Assoc., Burlington House, London, 1932) Vol. 2.

Received June 10, 1980.
Accepted for publication September 18, 1980

INTERNATIONAL JOURNAL OF QUANTUM CHEMISTRY, VOL XXIV, 243–277 (1983)

Density Functionals for Coulomb Systems

ELLIOTT H. LIEB

Departments of Mathematics and Physics, Princeton University, P.O.B. 708, Princeton, New Jersey 08544, U.S.A.

Abstract

This paper has three aims: (i) To discuss some of the mathematical connections between N-particle wave functions ψ and their single-particle densities $\rho(x)$. (ii) To establish some of the mathematical underpinnings of "universal density functional" theory for the ground state energy as begun by Hohenberg and Kohn. We show that the HK functional is *not* defined for all ρ and we present several ways around this difficulty. Several less obvious problems remain, however. (iii) Since the functional mentioned above is not computable, we review examples of explicit functionals that have the virtue of yielding rigorous bounds to the energy.

Introduction

It is a pleasure to dedicate this article to Laszlo Tisza on the occasion of his seventy-fifth birthday. As a colleague at MIT he was a source of inspiration and encouragement, especially in drawing our attention to the importance of careful and precise thought in mathematical physics. The subject, if not the content, of this article may therefore not be inappropriate in a book dedicated to Professor Tisza (see the Acknowledgment).

The idea of trying to represent the ground state (and perhaps some of the excited states as well) of atomic, molecular, and solid state systems in terms of the diagonal part of the one-body reduced density matrix $\rho(x)$ is an old one. It goes back at least to the work of Thomas [1] and Fermi [2] in 1927. In 1964 the idea was conceptually extended by Hohenberg and Kohn (HK) [3]. Since then many variations on the theme have been introduced. As the present article is not meant to be a review, I shall not attempt to list the papers in the field. Some recent examples of applications are Refs. 4 and 5. Some recent examples of theoretical papers which will play a role here are Refs. 6–12. A bibliography can be found in the recent review article of Bamzai and Deb [13].

This article has three aims:

(i) To discuss and prove some of the mathematical relations between N-particle functions ψ and their corresponding single-particle densities ρ.

(ii) To discuss the mathematical underpinnings of general density functional theory along the lines initiated by HK. In that theory a universal energy functional $F(\rho)$ is introduced. Despite the hopes of HK, $F(\rho)$ is *not* defined for all ρ because it is not true (see Theorem 3.4) that every ρ (even a "nice" ρ) comes from the ground state of some single-particle potential $v(x)$. This problem can be remedied by replacing the HK functional by the Legendre transform of the energy, as is done here. However, the new theory is also not free of difficulties, and these

 CCC 0020-7608/83/090243-35$04.50

Int. J. Quant. Chem. *24*, 243–277 (1983)

can be traced to the fact that the connection between v and ρ is extremely complicated and poorly understood.

(iii) To present briefly another approach to the ground state energy problem by means of functionals that, while not exact, are explicitly computable and yield upper and lower bounds to the energy.

The analysis in this paper gives rise to many interesting open problems. It is my hope that the incompleteness of the results presented here will be partly compensated if others are encouraged to pursue some of the questions raised by them.

It is not my intention to present a brief for HK theory. However, it deserves to be analyzed for at least two reasons: The HK theory is used by many workers and it gives rise to some deep problems in analysis. While it is my opinion that density functionals are a useful way to approach Coulomb systems, there are other approaches besides the HK approach [e.g., see (iii) above]. Apart from the difficulties mentioned above, the HK approach may be too general because *all* potentials have to be considered. Coulomb potentials are special and do lend themselves to a density functional approach; for example, Thomas–Fermi theory is asymptotically exact as $Z \to \infty$ (see Sect. 5E and Ref. 14). In addition to this question of generality there is also the crucial point that the "universal functional" is very complicated and essentially uncomputable. If one is going to make uncontrolled approximations for this functional, then the general theory is not very helpful.

It is a pleasure to thank Barry Simon for some very helpful conversations and the proofs of Theorems 4.4 and 4.8. I also thank Haïm Brezis for the proof of Theorem 1.3.

1. Single-Particle Densities

The first order of business is to describe the single-particle densities of interest. For simplicity we confine our attention to three dimensions whenever dimensionality is important. $z = (x, \sigma)$ will denote a space-spin variable, that is, $x \in \mathbb{R}^3$ and $\sigma \in \{1, \ldots, q\}$. $q = 2$ for electrons, of course, but one might wish to consider $q = 1$, which would mean that a ferromagnetic state is under consideration. We use the notation

$$\int dz = \sum_{\sigma=1}^{q} \int dx. \tag{1.1}$$

Let $\psi = \psi(z_1, \ldots, z_N)$ be an N-particle function (which may be complex valued). To simplify notation we will not indicate N explicitly except where needed. However, the condition of fixed N is crucial and frequently glossed over. The density functionals that will be introduced later are explicitly N dependent in a highly nontrivial way (see Sect. 4A). ψ is assumed to be *normalized*:

$$\int |\psi|^2 = 1, \tag{1.2}$$

(with $\int = \int dz_1, \ldots, dz_N$) and to have *finite kinetic energy*, that is,

$$T(\psi) = \sum_{i=1}^{N} \int |\nabla_i \psi|^2 < \infty. \tag{1.3}$$

Notes. $f \in L^p$ means that f is a function satisfying $\|f\|_p \equiv \{\int |f|^p\}^{1/p} < \infty$. $f \in H^1$ means that f and each component of ∇f are in L^2. Thus, the above $\psi \in H^1$. If f is differentiable everywhere, then ∇f and $T(f) = \int |\nabla f|^2$ are well defined. Otherwise, the correct definition of $T(f)$ for functions in $L^2(\mathbb{R}^m)$ is

$$T(f) = (2\pi)^{-3} \int k^2 |\hat{f}(k)|^2 \, dk, \tag{1.4}$$

where $\hat{f}$ is the Fourier transform of f. Since $f \in L^2$, $\hat{f}$ exists and $\hat{f} \in L^2$. H^1 is a Hilbert space with inner product $(f, g) = \int f^* g + \int \nabla f^* \cdot \nabla g$.

In most of the following it will be assumed that ψ satisfies the Pauli principle, that is, ψ is antisymmetric. However, some of the theorems are easier for symmetric (i.e., bosonic) ψ with $q = 1$, and occasionally this will be mentioned explicitly. In either case, the symmetry implies that

$$T(\psi) = N \int |\nabla_1 \psi|^2. \tag{1.5}$$

We define the *single-particle density* to be [see Eq. (A.1)]

$$\rho(x) = N \sum_{\sigma} \int |\psi((x, \sigma_1), \ldots, (x_N, \sigma_N))|^2 \, dx_2 \cdots dx_N. \tag{1.6}$$

Notice that $\int \rho(x) \, dx = N$, not 1.

Determinants. If $\phi_1(z), \ldots, \phi_N(z)$ are orthonormal functions, we can form the determinantal

$$\psi(z_1, \ldots, z_N) = (N!)^{-1/2} \det \{\phi_i(z_j)\}, \tag{1.7}$$

which is normalized. Then

$$\rho(x) = \sum_{i=1}^{N} \sum_{\sigma=1}^{q} |\phi_i(x, \sigma)|^2. \tag{1.8}$$

$$T(\psi) = \sum_{i=1}^{N} \sum_{\sigma=1}^{q} \int |\nabla \phi_i(x, \sigma)|^2 \, dx. \tag{1.9}$$

Returning to the general case, the finiteness of $T(\psi)$ implies the following [15].

Theorem 1.1. *$\rho(x)^{1/2} \in L^2(\mathbb{R}^3)$ and $\nabla \rho(x)^{1/2} \in L^2(\mathbb{R}^3)$, that is, $\rho(x)^{1/2} \in H^1(\mathbb{R}^3)$. Moreover, $\int (\nabla \rho^{1/2})^2 \leq T(\psi)$.*

Proof. $\rho^{1/2} \in L^2(\mathbb{R}^3)$ because $\int \rho = N$. Now $\nabla \rho(x) = N \int^\dagger (\nabla_1 \psi)^* \psi + N \int^\dagger \psi^* \nabla_1 \psi$, where $\int^\dagger$ means the integral in (1.6). By the Schwarz inequality,

$$[\nabla \rho(x)]^2 \leq 4N\rho(x) \int^\dagger |\nabla_1 \psi|^2.$$

Thus $\int (\nabla \rho^{1/2})^2 \, dx = \frac{1}{4} \int (\nabla \rho)^2 \rho^{-1} \, dx \leq T(\psi)$. ■

We know $\rho^{1/2} \in H^1(\mathbb{R}^3) = \{f | f \in L^2, \ \nabla f \in L^2\}$. (Here we use the standard convention that $\{A|C\}$ means the set of A such that condition C holds.) To discuss the converse of Theorem 1.1 some definitions are useful.

Definition. $\mathcal{I}_N = \{\rho | \rho(x) \geq 0, \rho^{1/2} \in H^1(\mathbb{R}^3), \int \rho(x)\, dx = N\}$.

Definition. $\mathcal{R}_N = \{\rho | \rho(x) \geq 0, \int \rho(x)\, dx = N, \rho \in L^3(\mathbb{R}^3)\}$.

$\mathcal{R}_N$ contains $\mathcal{I}_N$ by the Sobolev inequality (see Ref. 16) because if $f \in H^1(\mathbb{R}^3)$, then

$$\int |\nabla f(x)|^2\, dx \geq 3(\pi/2)^{4/3} \left[\int |f(x)|^6\, dx\right]^{1/3}. \tag{1.10}$$

Equation (1.10) is true only in three dimensions, but analogous inequalities hold in other dimensions. By Theorem 1.1,

$$T(\psi) \geq 3(\pi/2)^{4/3} \|\rho\|_3.$$

$\mathcal{R}_N$ is clearly a *convex set*; that is, if ρ_1 and $\rho_2 \in \mathcal{R}_N$, then $\rho \equiv \lambda\rho_1 + (1-\lambda)\rho_2 \in \mathcal{R}_N$ for all $0 \leq \lambda \leq 1$. $\mathcal{I}_N$ *is also convex* by the same proof as in Theorem 1.1; that is, by the Schwarz inequality

$$(\nabla\rho)^2 \leq 4\rho[\lambda(\nabla\rho_1^{1/2})^2 + (1-\lambda)(\nabla\rho_2^{1/2})^2].$$

In particular, the *functional* $\int [\nabla\rho^{1/2}]^2$ *is convex*. The convexity of $\mathcal{I}_N$ will be important in Sect. 3.

Definition. A function (or functional) f is convex if

$$f(\lambda x + (1-\lambda)y) \leq \lambda f(x) + (1-\lambda) f(y)$$

for all $0 \leq \lambda \leq 1$ and all x and y in the domain of f.

Theorem 1.2. *Suppose $\rho \in \mathcal{I}_N$. Then for either Bose or Fermi statistics there exists a ψ (which is a determinant in the fermion case) such that* (1.6) *holds and, moreover,*

$$T(\psi) \leq \int [\nabla\rho^{1/2}(x)]^2\, dx \quad (bosons), \tag{1.11}$$

$$T(\psi) \leq (4\pi)^2 N^2 \int [\nabla\rho^{1/2}(x)]^2\, dx \quad (fermions). \tag{1.12}$$

Proof. For bosons the proof is easy; simply take

$$\psi(x_1, \ldots, x_N) = \prod_{i=1}^{N} \left(\frac{\rho(x_i)}{N}\right)^{1/2}.$$

For fermions the construction is much more complicated. Some ideas from Ref. 17 will be used in the following. Write $x = (x^1, x^2, x^3)$ and define

$$f(x^1) = \left(\frac{2\pi}{N}\right) \int_{-\infty}^{x^1} ds \int_{-\infty}^{\infty} dt \int_{-\infty}^{\infty} du\, \rho(s, t, u).$$

Then f is monotone increasing from 0 to 2π. For $k=0,\ldots,N-1$ define

$$\phi^k(x)=[\rho(x)/N]^{1/2}\exp[ikf(x^1)].$$

It is easy to check that the ϕ^k are orthonormal functions in $L^2(\mathbb{R}^3)$. [First do the x^2 and x^3 integrations and then note that the overlap integral is of the form $\int_{-\infty}^{\infty}(df/dx^1)\exp[i(\Delta k)f(x^1)]\,dx^1=\{\exp[i(\Delta k)f(\infty)]-\exp[i(\Delta k)f(-\infty)]\}/i(\Delta k)=0$. Furthermore,

$$N\int|\nabla\phi^k|^2=\int(\nabla\rho^{1/2})^2+\left(\frac{2\pi k}{N}\right)^2\int_{-\infty}^{\infty}g(s)^6\,ds, \tag{1.13}$$

with

$$g(s)^2\equiv\int\int_{-\infty}^{\infty}dt\,du\,\rho(s,t,u).$$

As in Theorem 1.1, we conclude that

$$g\in H^1(\mathbb{R}^1)\qquad\text{and}\qquad\int\left(\frac{dg}{ds}\right)^2ds\leqslant\int(\nabla\rho^{1/2})^2\equiv A.$$

Since

$$g(s)^2=2\int_{-\infty}^{s}g(y)\left(\frac{dg(y)}{dy}\right)dy,$$

we conclude by the Schwarz inequality that $g(s)^4\leqslant 4[\int g^2][\int(dg/dy)^2]$. Thus, the last term in (1.13) is less than $4(2\pi k/N)^2N^2A$. Finally, we take ψ to be a determinant as in (1.7) using the functions $\phi^k(x)\times$(spin up). Equation (1.12) follows by summing on k. ■

Theorem 1.2 is closely related to the results of Gilbert [8] and Harriman [9].

For fermions, the extra factor N^2 in (1.12) is noticeably different from the factor N^0 in Theorem 1.1. Although (1.12) can be improved, it is not easy to do so. In any case, the conclusion is that the map from ψ to $\rho^{1/2}$ given by (1.6) is a map from $H^1(\mathbb{R}^{3N})$ *onto* $H^1(\mathbb{R}^3)$. But the map is clearly not 1:1; different ψ's can give the same ρ.

Question 1. Is this map continuous as a map from $H^1(\mathbb{R}^{3N})$ to $H^1(\mathbb{R}^3)$? That is, if ψ is fixed and ψ_j is a sequence (with corresponding ρ and ρ_j) such that $\int|\psi-\psi_j|^2\to 0$ and $\int|\nabla\psi-\nabla\psi_j|^2\to 0$, does it follow that $\int|\rho^{1/2}-\rho_j^{1/2}|^2\to 0$ and $\int|\nabla\rho^{1/2}-\nabla\rho_j^{1/2}|^2\to 0$?

Question 2. Although the map is not invertible (since it is not 1:1), we can ask the following: Given a sequence $\rho_j^{1/2}$ that converges to $\rho^{1/2}$ in the above $H^1(\mathbb{R}^3)$ since, and given some ψ satisfying (1.6) for ρ, does there exist a sequence ψ_j [related to ρ_j by (1.6)] that converges to ψ in the above $H^1(\mathbb{R}^3)$ sense? [This is equivalent to the statement that the map $\psi\mapsto\rho^{1/2}$ is "open," that is, the map takes open sets in $H^1(\mathbb{R}^{3N})$ into open sets in $H^1(\mathbb{R}^3)$.]

Int. J. Quant. Chem. *24*, 243–277 (1983)

Intuitively, the answer to both questions should be affirmative. The continuity can indeed be proved, but the proof is not entirely elementary. A proof of Theorem 1.3, due to H. Brezis, is given in the appendix.

Theorem 1.3. *The map* $\psi \mapsto \rho^{1/2}$ *given by* (1.6) *is continuous as a map from* $H^1(\mathbb{R}^{3N})$ *to* $H^1(\mathbb{R}^3)$.

I cannot offer any proof of the openness of the map, however. The fact that these questions do not have simple answers should serve as a warning that the connection between ψ and ρ is not as obvious as one might intuitively think.

2. Single-Particle Density Matrices

If ψ is given as before, we can define the *single-particle density matrix*

$$\gamma(x, x') = N \sum_{\sigma} \int \psi((x, \sigma_1), \ldots, (x_N, \sigma_N)) \times \psi((x', \sigma_1), \ldots, (x_N, \sigma_N))^* \, dx_2 \cdots dx_N. \tag{2.1}$$

This definition is different from the usual one because we sum on σ_1 in (2.1). Usually one defines the quantity $\tilde{\gamma}(x, \sigma; x', \sigma')$, so our $\gamma(x, x') = \sum_\sigma \tilde{\gamma}(x, \sigma; x', \sigma)$. Clearly, $\rho(x) = \gamma(x, x)$.

Theorem 2.1. γ *satisfies*
(i) $\operatorname{Tr} \gamma = \int \gamma(x, x)\, dx = N$.
(ii) *As an operator,* $0 \leq \gamma \leq qI$, *for fermions*; *that is,* $0 \leq (f, \gamma f) \leq q(f, f)$. *For bosons,* $0 \leq \gamma \leq NI$.

Proof. (i) is "obvious" but not trivial. The point is that if an operator K is given, then its kernel $K(x, y)$ is defined only almost everywhere. In particular, $K(x, x)$ can be anything. Thus, $\operatorname{Tr} K$ need not be $\int K(x, x)\, dx$. However, (i) can be proved from (2.1). This is left as an exercise. To prove (ii), let $M(x, x') = f(x)f(x')^*$ be a one-particle operator with $(f, f) = 1$. Then $A = \sum_{i=1}^N M(x_i, x_i')$ has as its largest eigenvalue on the antisymmetric space the value q. Moreover, A is clearly positive semidefinite. Thus, $0 \leq (f, \gamma f) = \operatorname{Tr} \gamma M = (\psi, A\psi) \leq q$. ∎

Definition. Let $\gamma(x, y)$ be any kernel. γ is said to be *admissible* if $\operatorname{Tr} \gamma = N$ and $0 \leq \gamma \leq qI$ (fermions) or $0 \leq \gamma \leq N$ (bosons). The set of admissible γ is clearly convex; that is, if γ and δ are admissible, then so is $\alpha\gamma + (1 - \alpha)\delta$ for $0 \leq \alpha \leq 1$.

Now we come to a subtle point. If γ is an admissible operator, we can ask two questions:

Question 3. Does an *N-particle density matrix* Γ always exist, where $\Gamma = \Gamma(z_1, \ldots, z_N; z_1', \ldots, z_N')$, so that γ is given by (2.1) with $\psi\psi^*$ replaced by Γ? (Γ is a density matrix if $0 \leq \Gamma$ and $\operatorname{Tr} \Gamma = 1$. Γ must also satisfy the appropriate symmetry.)

Question 4. Does a ψ always exist so that (2.1) holds; that is, can Γ be chosen to be a *pure state*, namely, $\Gamma = \psi\rangle\langle\psi$?

The answer to question 4 is No! (for fermions). For bosons, the answer is Yes.

The proof of question 3 (which we now call **Theorem 2.2**) has been known for a long time. An explicit construction is given in Ref. 26. An example in which Γ fails to be of the form $\psi\rangle\langle\psi$, for $N=2$ and $q=1$, is the case in which γ has three nonzero eigenvalues $1, \frac{1}{2}, \frac{1}{2}$. To see this, let the normalized eigenvectors of γ be $f(x)$, $g(x)$, and $h(x)$, respectively; that is,

$$\gamma(x, x') = f(x)f(x')^* + \tfrac{1}{2}g(x)g(x')^* + \tfrac{1}{2}h(x)h(x')^*.$$

Let $A = -\gamma(x_1, x_1') - \gamma(x_2, x_2')$ be an operator on the antisymmetric states. Its lowest eigenvalue is $-1-1/2=-3/2$, which is doubly degenerate. If $\Gamma=\psi\rangle\langle\psi$, then ψ must be a ground state since $\operatorname{Tr}\Gamma A = -\operatorname{Tr}\gamma^2 = -1-1/4-1/4=-3/2$. But every ground state is of the form $\psi = 2^{-1/2}\det(f, p)$, where $p = ag + bh$, $|a|^2+|b|^2=1$. But then $\gamma = f\rangle\langle f + p\rangle\langle p$, and this is never of the form $f\rangle\langle f + \frac{1}{2}g\rangle\langle g + \frac{1}{2}h\rangle\langle h$.

The moral of all this is the following: On the one-particle level we can study density matrices $\gamma(x, x')$ or densities, $\rho(x) = \gamma(x, x)$. The former do not always come from pure states $\psi\rangle\langle\psi$. The latter do, as Theorem 1.2 shows. While γ is more complicated than ρ (it has two variables), it has the distinct advantage that the map $\Gamma \mapsto \gamma$ is *linear*! The map $\psi \mapsto \rho$ is *nonlinear*, and this, as will be seen, is the source of some difficulty.

The relation among ψ, Γ, γ, and ρ can be summarized by the following diagram:

$$\psi \mapsto \Gamma \mapsto \gamma \mapsto \rho, \tag{2.2}$$

by which we mean (i) the map $\psi \mapsto \Gamma = \psi\rangle\langle\psi$, (ii) $\Gamma \mapsto \gamma$ by (2.1) with $\psi\psi^*$ replaced by Γ, (iii) $\gamma \mapsto \gamma(x, x) = \rho(x)$. (ii) and (iii) are linear while (i) is nonlinear.

Notation. We shall use the symbol $\psi \mapsto \rho$ (or any other combination such as $\gamma \mapsto \rho$) to indicate that ψ and ρ are related by the above maps.

Technical remarks. Since γ is self-adjoint and trace class, it can always be written in the form

$$\gamma(x, x') = \sum_{j=1}^{\infty} \lambda_j f_j(x) f_j(x')^*, \tag{2.3}$$

where the f_j are orthonormal and $0 \le \lambda_j \le q$ (fermions), $0 \le \lambda_j \le N$ (bosons). If $T \in H^1(\mathbb{R}^{3N})$, by which we mean $\operatorname{Tr} - \sum_{i=1}^N \Delta_i \Gamma < \infty$, then each $f \in H^1(\mathbb{R}^3)$ (see Theorem 1.1). Although $\gamma(x, x)$ is not *a priori* well defined, as stated before, it is well defined almost everywhere (in $\mathbb{R}^3$) by (2.3) and

$$N = \operatorname{Tr}\gamma = \sum_{j=1}^{\infty} \lambda_j. \tag{2.4}$$

3. General Density Functional Theory

The problem that will concern us in calculating the ground state energy for N electrons interacting with each other via a repulsive Coulomb potential

Int. J. Quant. Chem. *24*, 243–277 (1983)

$|x_i - x_j|^{-1}$ and also interacting with a single-particle potential $v(x)$. If $v = 0$, the Hamiltonian is

$$H_0 = K + \sum_{1 \le i < j \le N} |x_i - x_j|^{-1}, \tag{3.1}$$

where K is the kinetic energy operator

$$K = -\sum_{i=1}^{N} \Delta_i \tag{3.2}$$

in units in which $\hbar^2/2m = 1$. Also of interest is the case where $H_0 = K$ alone (see Sec. 4C). Recall that N is fixed and will not be mentioned unless necessary. Also, to simplify matters we shall confine our attention in the following to fermions. However, many of the following results have obvious analogs for bosons.

The total Hamiltonian is

$$H_v = H_0 + V, \tag{3.3}$$

where

$$V = \sum_{i=1}^{N} v(x_i). \tag{3.4}$$

The ground state energy $E(v)$ is *defined* to be

$$E(v) = \inf \{(\psi, H_v \psi) | \psi \in \mathcal{W}_N\}, \tag{3.5}$$

where

$$\mathcal{W}_N = \{\psi | \|\psi\| = 1, T(\psi) < \infty\}. \tag{3.6}$$

Technical remark. Something should be said about the meaning of $(\psi, H_v\psi)$ and about the class of v's under consideration. We shall always interpret $(\psi, H_v\psi)$ in the sense of a quadratic form; in particular, this means that $(\psi, K\psi) \equiv T(\psi)$. It is not assumed that $\Delta\psi \in L^2$. Since $\psi \in H^1$, it is easy to prove that $(\psi, |x_i - x_j|^{-1}\psi)$ is finite for all $i \neq j$. The part containing v is $\int \rho(x) v(x)\, dx$. As $\rho \in L^1$ and $\rho \in L^3$ (since $\nabla \rho^{1/2} \in L^2$), $\rho \in L^p$ for all $1 \le p \le 3$. The integral is then well defined if $v \in L^{3/2} + L^\infty$. This means that *we consider* v*'s that can be written as* $v = v_{3/2} + v_\infty$ *with* $v_{3/2} \in L^{3/2}$ *and with* $|v_\infty|$ *a bounded function*. This choice precludes v's that go to ∞ as $|x| \to \infty$, such as the harmonic oscillator potential. Unbounded potentials can also be handled by the methods given here, but then we have to place additional restrictions on ρ so that $\int v\rho$ makes sense. We restrict ourselves here to $L^{3/2} + L^\infty$ for simplicity of exposition. The class includes Coulomb potentials because $|x|^{-1} = \theta(x)|x|^{-1} + [1 - \theta(x)]|x|^{-1}$ with $\theta(x) = 1$ if $|x| \le 1$, $\theta(x) = 0$, $|x| > 1$. The two terms on the right are in $L^{3/2}$ and in L^∞, respectively.

$L^{3/2} + L^\infty$ is a Banach space with the norm

$$\|v\| = \inf \{\|g\|_{3/2} + \|h\|_\infty | g + h = v\}. \tag{3.7}$$

Technical remark. $\mathcal{R}_N$ is a subset of the Banach space $X = L^3 \cap L^1$. X^*, the dual of X, is $Y = L^{3/2} + L^\infty$. However, the dual of Y is not X because while L^∞ is the dual of L^1, L^1 is not the dual of L^∞. However, $X \subset Y^*$. The duality will be useful.

There may or may not be a minimizing ψ for (3.5), and if there is one it may not be unique (for bosons it is unique because it is a positive function). Any minimizing ψ (called a *ground state*) would satisfy

$$H_v \psi = E(v)\psi \tag{3.8}$$

in the distributional sense. The proof of this assertion is not difficult. For example, a minimizing ψ will not exist if v is an attractive square well and if N is too large; the extra, unbound electrons will simply "leak away" to infinity. In such a case, $E(v)$ would still have physical significance. It would be the ground state energy for *fewer* than N particles.

There are three simple, but important, properties of $E(v)$:

Theorem 3.1. (i) *$E(v)$ is concave in v: that is,*

$$E(v) \geq \alpha E(v_1) + (1-\alpha)E(v_2), \tag{3.9}$$

for all v_1, v_2, $0 \leq \alpha \leq 1$ and $v = \alpha v_1 + (1-\alpha)v_2$.

(ii) *$E(v)$ is monotone decreasing: that is, if $v_1(x) \leq v_2(x)$ for all x, then $E(v_1) \leq E(v_2)$.*

(iii) *$E(v)$ is continuous in the $L^{3/2} + L^\infty$ norm and is, moreover, locally Lipschitz. In particular, $E(v)$ is finite.*

Proof. (i) If $\psi \in \mathcal{W}_N$, then

$$(\psi, H_v\psi) = \alpha(\psi, H_{v_1}\psi) + (1-\alpha)(\psi, H_{v_2}\psi) \geq \alpha E(v_1) + (1-\alpha)E(v_2).$$

(ii) $(\psi, H_{v_2}\psi) \geq (\psi, H_{v_1}\psi) \geq E(v_1)$.

(iii) Fix v_0 and let $\delta = v - v_0$. We want to show that when $\|\delta\| \leq L/3$, $|E(v) - E(v_0)| < C\|\delta\|$ for some C, independent of v. [Here, L is the constant in (1.10).] Since $E(v)$ is concave, it is sufficient to show that for some fixed D, $E(v) - E(v_0) \geq D$ whenever $\|\delta\| = L/3$; because if $0 \leq \gamma \leq 1$,

$$\gamma[E(v_0+\delta) - E(v_0)] \leq E(v_0 + \gamma\delta) - E(v_0) \leq \gamma[E(v_0) - E(v_0 - \delta)].$$

Let $E(v, \frac{1}{2})$ denote (3.5) with K replaced by $K/2$. Then

$$E(v) \geq E(v_0, \tfrac{1}{2}) + \inf\,(\psi, [\tfrac{1}{2}K + \Sigma\delta(x_i)]\psi).$$

The last term is bounded by $-LN/2$ because $\delta = g + h$ with $\|g\|_{3/2} < L/2$ and $\|h\|_\infty < L/2$. Thus,

$$\int \delta\rho > -(L/2)[\|\rho\|_3 + N].$$

But $(\psi, K\psi)/2 \geq (L/2)\|\rho\|_3$ by (1.10) and Theorem 1.1. Finally, note that $E(v_0, \frac{1}{2}) - E(v_0)$ is a constant, D', independent of v. ■

Now we begin the study of density functional theory in the manner of Hohenberg and Kohn. Their work is based on the following theorem [3]:

Theorem 3.2. *Suppose ψ_1 (respectively, ψ_2) is a ground state for v_1 (respectively v_2) and $v_1 \neq v_2 +$ constant. Then $\rho_1 \neq \rho_2$.*

Proof. Suppose $\rho_1 = \rho_2 = \rho$. $\psi_1 \neq \psi_2$ because they satisfy different Schrödinger equations, (3.8). [Note. To prove this we must know that $v_1\psi = v_2\psi$ implies that $v_1 = v_2$. This, in turn, requires that $\psi(x)$ does not vanish on a set of positive measure. This technical point is discussed in remark (ii) preceding Theorem 3.5.] Moreover, ψ_2 (respectively, ψ_1) does not satisfy (3.8) for v_1 (respectively, v_2). Therefore,

$$E(v_1) < (\psi_2, H_{v_1}\psi_2) = E(v_2) + \int (v_1 - v_2)\rho.$$

Likewise, $E(v_2) < E(v_1) + \int (v_2 - v_1)\rho$. This is a contradiction. ∎

Hohenberg and Kohn assume that every ρ comes from some ψ that is a ground state for some v. For such ρ they *define* the functional

$$F_{\mathrm{HK}}(\rho) = E(v) - \int v\rho, \tag{3.10}$$

and we shall retain this definition for $\rho \in \mathscr{A}_N$, where

$$\mathscr{A}_N = \{\rho \mid \rho \text{ comes from a ground state}\}. \tag{3.11}$$

$\mathscr{A}_N \neq \mathscr{I}_N$, as remarked earlier, and it is not convex (see Theorem 3.4)! The definition given by (3.10) requires Theorem 3.2, according to which there is a unique v (up to a constant) associated with ρ. We can also define

$$\mathscr{V}_N = \{v \mid H_v \text{ has a ground state}\}. \tag{3.12}$$

It then follows easily that for $v \in \mathscr{V}_N$

$$E(v) = \min\left\{F_{\mathrm{HK}}(\rho) + \int v\rho \mid \rho \in \mathscr{A}_N\right\}. \tag{3.13}$$

This is *the* HK *variational principle*, but it is important to note that it holds only for $v \in \mathscr{V}_N$, which is unknown, and that the variation is restricted to the unknown set $\mathscr{A}_N$.

We also do not know what F_{HK} is, and that is a very serious problem. But there are also conceptual problems, which will be addressed here.

If F is to be used in a variational principle, it is clearly desirable that F be a convex functional. In particular, it should be defined everywhere on $\mathscr{I}_N$, or at least on some known convex subset of $\mathscr{I}_N$.

The domain of F_{HK} (i.e., $\mathscr{A}_N$) is not all of $\mathscr{I}_N$ and it is not convex. This last fact is closely connected with the following difficulty: One can define a functional for all ρ in $\mathcal{I}_N$ by*

$$\tilde{F}(\rho)=\inf\{(\psi, H_0\psi)|\psi\mapsto\rho, \psi\in\mathcal{W}_N\}. \tag{3.14}$$

It then follows trivially that

$$E(v)=\inf\left\{\tilde{F}(\rho)+\int v\rho|\rho\in\mathcal{I}_N\right\}, \tag{3.15}$$

$$\tilde{F}(\rho)=F_{\mathrm{HK}}(\rho), \qquad \text{if} \quad \rho\in\mathcal{A}_N. \tag{3.16}$$

So far, so good. The difficulty is that $\tilde{F}$ is not convex either. However, $\tilde{F}$ has one important property that is proved in the appendix.

Theorem 3.3. *For each ρ in $\mathcal{I}_N$ there is a $\psi\in\mathcal{W}_N$ such that $\tilde{F}(\rho)=(\psi, H_0\psi)$. In other words, the infimum in* (3.14) *is a minimum.*

The following functional F is one choice for "the density functional" that remedies the difficulties mentioned so far:

$$F(\rho)=\sup\left\{E(v)-\int v\rho|v\in L^{3/2}+L^\infty\right\}. \tag{3.17}$$

We shall explore the properties of F, but it, too will be seen to have subtle difficulties of its own.

Remarks. (i) (3.17) defines $F(\rho)$ for all $\rho\in\mathcal{R}_N$, not just $\mathcal{I}_N$, provided F is interpreted in the extended sense as a function that can have the value $+\infty$. In fact, (2.17) defines F on the much larger set $X=L^3\cap L^1$, without the restrictions $\rho(x)\geq 0$ and $\int\rho=N$. As Theorem 3.5 shows, however, it is only necessary to consider F on the convex subset $\mathcal{I}_N$ of X.

(ii) Recall that F depends explicitly on N through E.

(iii) Since F is the supremum of a family of linear functionals, it is convex.

(iv) Theorem 3.8 shows that $F(\rho)=+\infty$ if $\rho\notin\mathcal{I}_N$. There is an alternative definition of F, namely, F', by which F' is finite on the set

$$\mathcal{I}'_N\equiv\{\rho|\rho(x)\geq 0 \text{ and } \nabla\rho^{1/2}\in L^2\},$$

without requiring $\int\rho=N$. This is

$$F'(\rho)=\left(\int\rho\right)F\left(\rho\Big/\int\rho\right), \qquad \rho\not\equiv 0,$$

$$F'(0)=0. \tag{3.18}$$

It is easy to check that the convexity and lower semicontinuity (a concept to be defined later) of F carry over to F'. This definition has the virtue that F' is finite

* Levy [10] also defined $\tilde{F}(\rho)$ which he called Q, and derived (3.15). He did not prove Theorem 3.3, but assumed the existence of a minimizing ψ. Also, he did not establish the connection between $\tilde{F}$ and the Legendre transform, F (Theorem 3.7). In Ref. 11, Levy proved Theorem 3.4(ii), independently and virtually at the same time as myself, using essentially the same construction. See Ref. 12 for additional remarks about Q.

Int. J. Quant. Chem. *24*, 243–277 (1983)

on a dense subset of the set of nonnegative functions in X. However, this does not change the theory in any important way, so we shall continue to use the definition given by (3.17).

(v) Other characterizations of F, directly in terms of $\tilde{F}$, are given in Theorem 3.7, and in Eqs. (4.5)–(4.7).

There is an obvious relation between F and $\tilde{F}$, namely,

$$F(\rho) \le \tilde{F}(\rho) \qquad \text{for all} \quad \rho \in \mathcal{I}_N, \tag{3.19}$$

since $E(v) \le \tilde{F}(\rho) + \int v\rho$ for all $\rho \in \mathcal{I}_N$. Furthermore, since F is convex and $\tilde{F}$ is not convex (by Theorem 3.4), there are ρ's in $\mathcal{I}_N$ for which $F(\rho) < \tilde{F}(\rho)$.

First we prove that not all ρ's come from ground states. The essential ingredient is the existence of v with a degenerate ground state. (Such v's, incidentally, preclude the existence of a map $v \mapsto \rho$.)

Theorem 3.4. *Let $N > q$ = number of spin states. Then*

(i) *$F(\rho)$) is not convex*

(ii) *There exists a $\rho \in \mathcal{I}_N$ that does not come from a ground state ψ. Moreover this ρ is a convex combination of ρ's that do come from a ground state.*

Proof. Let v be a spherically symmetric potential having a ground state and with the property that its ground state has orbital angular momentum $L \ge 1$. We assume the degeneracy is no greater than necessary, namely $M = 2L+1$. The orthonormal ground states are $\psi_1, \ldots, \psi_M$ and $\psi_i \mapsto \rho_i$. Under simultaneous rotation of all N coordinates, they transform as a basis for the M-dimensional irreducible representation of $O(3)$. The following fact is easy to prove: (a) If $\bar{\rho} = M^{-1}\Sigma\rho_i$, then $\bar{\rho}(x)$ is spherically symmetric: that is, $\bar{\rho}$ depends only on $r = |x|$.

A second fact that will be needed is (b): if ϕ is any ground state (and hence a linear combination of the ψ_i) and $\phi \mapsto \rho$, then ρ is not spherically symmetric. This fact must follow from some group-theoretic agreement, but I have not found one. However, it is not hard to see that (b) is equivalent to (c): There exists a perturbation of v, $v(x) \mapsto v(x) + \lambda w(x)$, with w bounded and of compact support, so that to first order in λ the M-fold degeneracy is broken. Such pairs v and w certainly exist, so we can henceforth assume that (b) holds. [A proof that a v satisfying (b) exists is the following. First, take the case that $H_0 = K$, that is, independent particles. The ground states are determinants. Choose v so that the ground state has $L \ge 1$, in which case (b) obviously holds. Next, consider $H = K + \lambda\Sigma|x_i - x_j|^{-1} + V$. Angular momentum is still conserved and for sufficiently small λ the ground state will have the same L and, by continuity of the ground states, (b) will continue to hold for small λ. We are interested in $\lambda = 1$ but, under the scaling

$$x \to x/\lambda, \qquad v(x) \to \lambda^{-2} v(x/\lambda) = v'(x),$$

the v, λ problem is converted into the v', $\lambda = 1$ problem. Thus, v' has the desired properties. I thank B. Simon for this remark.]

Clearly $\tilde{F}(\rho_i) = \text{constant} = D = E(v) - \int v\rho_i$. We claim $\tilde{F}(\bar{\rho}) > D$, thereby proving lack of convexity. Obviously, $\tilde{F}(\bar{\rho}) \geq D$, for otherwise we could use $\bar{\rho}$ instead of ρ_i in (3.15). (Note: $\int v\bar{\rho} = \int v\rho_i = \text{constant} = C$.) Suppose $\tilde{F}(\bar{\rho}) = D$. Then $\bar{\rho}$ comes from some ϕ that must be a ground state for v. But (a) and (b) show this to be impossible. Thus $\tilde{F}(\bar{\rho}) > D$. Moreover, $\bar{\rho}$ cannot come from any ground state ψ for any other v'. If it did, then

$$E(v') = \tilde{F}(\bar{\rho}) + \int v'\bar{\rho} > M^{-1}\sum\left(\tilde{F}(\rho_i) + \int v'\rho_i\right).$$

This implies that for some $1 \leq i \leq M$, $\tilde{F}(\rho_i) + \int v'\rho_i < E(v')$, which is a contradiction. ∎

Remarks. (i) The foregoing proof holds just as well if $(\psi, H_0\psi)$ is replaced by $T(\psi)$ in the definition of $\tilde{F}$. [This functional will be denoted by $\tilde{T}(\rho)$ and the analog of (3.17) by $T(\rho)$.] In other words, the interelectron Coulomb repulsion plays no role in Theorem 3.4 (see Sec. 4C).

(ii) There are other ρ's that do not come from some v, namely, those $\rho \in \mathcal{I}_N$ that vanish on a nonempty open set. If $v \in L^{3/2} + L^\infty$ and ψ is its ground state, then ψ cannot vanish in an open set by the unique continuation theorem [18]. (Strictly speaking, this theorem is only known to hold for $v \in L^3_{\text{loc}}$, but it is believed to hold for $L^{3/2} + L^\infty$.) Presumably, such ρ's can, in many cases, be obtained as limits in which $v \to \infty$ on the open set. Therefore, if the set of allowed v's can be extended properly to include infinite v's, the existence of such ρ's may not have any particular importance. The question is very delicate, however, as Englisch and Englisch [7] showed recently. Even for one particle there are densities which never vanish but which do not come from any v, even if one allows density matrices (see Sect. 4B) instead of pure states. These densities have regions in which they are "small" so that the obvious v (defined by $v = \rho^{-1/2}\Delta\rho^{1/2}$) has the property that $-\Delta + v$ cannot be defined as a semibounded operator.

Theorem 3.5.

$$E(v) = \inf\left\{F(\rho) + \int \rho v \,|\, \rho \in L^3 \cap L^1\right\}, \tag{3.20}$$

$$E(v) = \inf\left\{F(\rho) + \int \rho v \,|\, \rho \in \mathcal{I}_N\right\}. \tag{3.21}$$

Remark. The right sides of (3.20) and (3.21) are automatically concave functionals, which is a property we already proved for E.

Proof. Let $M^-(v)$ [respectively, $M^+(v)$] be the infimum in (3.20) [respectively, (3.21)]. Obviously, $M^-(v) \leq M^+(v)$. First, pick v_0. Clearly $F(\rho) \geq E(v_0) - \int v_0\rho \equiv F_1(\rho)$. Therefore,

$$M^-(v) \geq \inf\left\{F_1(\rho) + \int \rho v \,|\, \rho \in L^3 \cap L^1\right\}.$$

Int. J. Quant. Chem. *24*, 243–277 (1983)

and hence $M^-(v_0) \geq E(v_0)$. Second, by (3.19), $F(\rho) \leq \tilde{F}(\rho)$, so that $M^+(v) \leq \inf\{\tilde{F}(\rho) + \int \rho v | \rho \in \mathcal{I}_N\} = E(v)$. ■

Let us pause briefly to review the situation. Three functionals have been defined:

$$\begin{aligned} &F_{\mathrm{HK}} && \text{defined on} \quad \mathcal{A}_N \subset \mathcal{I}_N; \\ &\tilde{F} && \text{defined on} \quad \mathcal{I}_N \subset L^3 \cap L^1; \\ &F && \text{defined on} \quad X = L^3 \cap L^1. \end{aligned}$$

Of these, only F is convex and only $\tilde{F}$ and F satisfy the variational principle for *all* v.

The next step is to find out something of the nature of F. It is at this point that the analysis becomes complicated and where difficulties and incompleteness arise. The basic reason is that the connection between v and ρ is anything but simple. We have $X = L^3 \cap L^1$ and its dual $X^* = L^{3/2} + L^\infty$. Although X is not the dual of X^*, it is a subset of X^{**}, the dual of X^*.

Definitions. (i) A sequence $\rho_n \in X$ is said to converge to $\rho \in X$ $(\rho_n \to \rho)$ if and only if $\|\rho_n - \rho\|_3 \to 0$ and $\|\rho_n - \rho\|_1 \to 0$. This is also called *norm convergence*. ρ_n *converges weakly* to ρ $(\rho_n \rightharpoonup \rho)$ if and only if $\int v(\rho_n - \rho) \to 0$ for all $v \in Y = X^*$. Clearly, strong convergence implies weak convergence.

(ii) A functional f on X is *continuous* (or norm continuous) if and only if $\rho_n \to \rho$ implies $f(\rho_n) \to f(\rho)$. *Weak continuity* requires the concept of nets to define but, if f is weakly continuous, then whenever $\rho_n \rightharpoonup \rho$, $f(\rho_n) \to f(\rho)$. Weak continuity implies norm continuity.

(iii) A real functional f on X is *lower semicontinuous* (l.s.c.) if and only if $\rho_n \to \rho$ implies $f(\rho) \leq \liminf f(\rho_n)$. *Weak lower semicontinuity* requires nets to define, but if f is weakly l.s.c. then $\rho_n \rightharpoonup \rho$ implies $f(\rho) \leq \liminf f(\rho_n)$. (Weak) lower semicontinuity is equivalent to the following: $\{\rho | f(\rho) \leq \lambda\}$ is (weakly) closed for all real λ.

Remarks. (i) Weak lower semicontinuity always implies lower semicontinuity, but not conversely. It is a theorem of Mazur [19], however, that if f is *convex* and norm l.s.c., then it is automatically weakly l.s.c.

(ii) The function $\rho(x) \equiv 0$ is *not* in the $L^3 \cap L^1$ weak closure of $\mathcal{I}_N$.

The reader may be puzzled by all these definitions, especially lower semicontinuity, because finite convex functions on $\mathbb{R}^n$ are always continuous. Unfortunately, this is not true in infinite-dimensional spaces such as the space X we are considering. Even l.s.c. cannot be taken for granted.

Theorem 3.6. *$F(\rho)$ is weakly (and hence also norm) lower semicontinuous.*

Proof.

$$K_\lambda \equiv \{\rho | F(\rho) \leq \lambda\} = \left\{\rho | E(v) - \int v\rho \leq \lambda \text{ for all } v \in Y\right\}.$$

Now if $\rho_n \to \rho$ in norm and $\rho_n \in K_\lambda$ then, for each $v \in Y$.

$$E(v) - \int v\rho = \lim \left(E(v) - \int v\rho_n \right) \leq \lambda.$$

Therefore K_λ is norm closed, so that F is norm l.s.c. Weak l.s.c. is a consequence of Mazur's theorem. ■

Next we define the *convex envelope* (CE).

Definition. Let f be a real functional defined on a subset A of X. $f(\rho)$ is allowed to be $+\infty$, but not for all $\rho \in A$. CE f is defined on all of X as follows: CE $f(\rho) = \sup \{g(\rho) | g$ is weakly l.s.c., g is convex on X, and $g(\rho') \leq f(\rho')$ for all $\rho' \in A\}$.

It is easy to check that CE f is convex and weakly l.s.c. and CE $f(\rho) \leq f(\rho)$ for all $\rho \in A$. However, CE $f(\rho)$ may be $+\infty$ for some ρ.

The function of interest is CE $\tilde{F}$ with $A = \mathcal{I}_N$. Note that A is convex and that $\tilde{F}$ (and hence CE $\tilde{F}$) is finite on A by Theorem 1.2. Since CE $\tilde{F} \leq \tilde{F}$ on A, it is obvious from (3.19) and Theorem 3.6 that $F \leq$ CE $\tilde{F}$ on X. On the other hand, suppose we use CE $\tilde{F}$ instead of $\tilde{F}$ in (3.15). This gives a new function, which we call E'. Clearly $E' \leq E$. Then, if E' is used in (3.17), we get a new function F', and $F' \leq F$. However, an infinite-dimensional generalization of Fenchel's theorem [29] (which uses the Hahn–Banach theorem) states that if the original function (in our case, CE $\tilde{F}$) is convex and weakly l.s.c. on X, then its double Legendre transform (in our case F') is equal to the original function. Thus, $F' =$ CE $\tilde{F}$ and we have

Theorem 3.7. *$F(\rho) =$ CE $\tilde{F}(\rho)$ for all $\rho \in L^3 \cap L^1$.*

The reader may wonder what Theorem 3.7 is good for; the following is an example of the usefulness of the foregoing functional analysis (see Theorem 4.3).

Theorem 3.8. *For all $\rho \in L^3 \cap L^1$ let*

$$G(\rho) \equiv \int (\nabla \rho(x)^{1/2})^2 \, dx \qquad \text{if } \rho \in \mathcal{I}_N$$

$$= +\infty \qquad \text{otherwise.}$$

Then $F(\rho) \geq G(\rho)$, for all $\rho \in L^3 \cap L^1$.

Proof. G is obviously convex on X [see the remark after (1.10)]. We claim that G is norm l.s.c. (Note: The norm in question is $L^3 \cap L^1$, not the H^1 norm on $\rho^{1/2}$.) If so, we are done because G is then weakly l.s.c. and, by Theorem 1.1, $G \leq \tilde{F}$ on $\mathcal{I}_N$: but then $G \leq$ CE $\tilde{F} = F$.

To prove norm l.s.c., let ρ_n be any sequence in X with $\rho_n \to \rho$; that is, $\|\rho_n - \rho\|_1 \to 0$ and $\|\rho_n - \rho\|_3 \to 0$. We can assume that $G = \lim G(\rho_n)$ exists and is finite, and we have to show that $G \geq G(\rho)$. We can also assume $\rho(x) \geq 0$ a.e. because if $\rho < 0$ on a set S of positive measure, then, for sufficiently large n,

$\rho_n < 0$ on some set of positive measure; hence $\rho_n \notin \mathcal{I}_N$ and $G(\rho_n) = \infty$. For a similar reason we can assume $\int \rho = N$. Since $G(\rho_n) < \infty$, $\rho_n \in \mathcal{I}_N$. Thus, if we define $g_n = \rho_n^{1/2}$ and $g = \rho^{1/2}$, we have: (a) g_n is bounded in H^1; (b) $g_n^2 \to g^2$ in L^3 and L^1. By the Banach–Alaoglu theorem there is an $f \in H^1$ such that $g_n \rightharpoonup f$ and $\nabla g_n \rightharpoonup \nabla f$ weakly in L^2. Clearly $f(x) \geq 0$. It is not hard to prove that if $g_n \rightharpoonup f$ in L^2 and $g_n^2 \to g^2$ in L^1, then $g = f$. Hence $\nabla g = \nabla f$, and thus $\nabla g_n \rightharpoonup \nabla g$. But since $\int (\nabla g)^2$ is H^1-norm continuous, it is H^1 weakly l.s.c., so that $\lim G(\rho_n) \geq G(\rho)$. ■

Theorem 3.8 is certainly not obvious. Among other things it says that if $\rho \notin \mathcal{I}_N$ (and such ρ's can be quite smooth and innocent looking), then there exists a sequence of potentials $v_n \in L^{3/2} + L^\infty$ such that $E(v_n) - \int v_n \rho \to \infty$. The reader is asked to reflect on this fact. Another interesting fact is that F is convex and finite on $\mathcal{I}_N$, but infinite off $\mathcal{I}_N$. However, the complement of $\mathcal{I}_N$ (in X) is dense (in the X norm) in $\mathcal{I}_N$ and $\mathcal{I}_N$ is dense in the cone of nonnegative functions in X.

The following upper bound complements Theorem 3.8.

Theorem 3.9. *If $\rho \in \mathcal{I}_N$, then*

$$F(\rho) \leq \tilde{F}(\rho) \leq (4\pi)^2 N^2 G(\rho) + \tfrac{1}{2} \iint \rho(x)\rho(y)|x-y|^{-1}\, dx\, dy. \tag{3.22}$$

Proof. Use the definition (3.14). By Theorem 1.2 there is a determinantal ψ, with $\psi \mapsto \rho$, such that (1.12) holds. With this ψ we can calculate the Coulomb repulsion $I = (\psi, \Sigma |x_i - x_j|^{-1} \psi)$. I has a direct term, given in (3.22), plus an exchange term. The latter is negative, as is well known, since $|x-y|^{-1}$ is a positive definite kernel. Thus $\tilde{F}(\rho) \leq$ right side of (3.22). Then use (3.19). ■

Remark. By one of Sobolev's inequalities,

$$D \equiv \iint \rho(x)\rho(y)|x-y|^{-1}\, dx\, dy \leq (\text{const.}) \|\rho\|_{6/5}^2.$$

By Hölder's inequality $\|\rho\|_{6/5}^4 \leq \|\rho\|_1^3 \|\rho\|_3$. Again, by Sobolev's inequality, (1.10), $\|\rho\|_3 \leq (\text{const.}) G(\rho)$. Thus,

$$D \leq (\text{const.}) N^{3/2} G(\rho)^{1/2} \leq (\text{const.})[N + N^2 G(\rho)].$$

To continue the study of F the following concept is needed.

Definition. Let f be a real functional on a subset A of a Banach space B, and let $\rho_0 \in A$. A linear functional l on B is said to be a *tangent functional* (TF) *at* ρ_0 if and only if for all $\rho \in A$

$$f(\rho) \geq f(\rho_0) - l(\rho - \rho_0). \tag{3.23}$$

l may not be unique. If l is continuous, then l is a *continuous tangent functional* at ρ_0.

l is a continuous linear functional on X if and only if it has the form $\int v\rho$ with $v \in X^* = Y$. If f is convex, then at every point ρ_0 at which f is finite, f has at least one TF. This is guaranteed by the Hahn–Banach theorem. However, f may have no *continuous* TF at ρ_0.

The functional of interest is obviously F. In general, $F \le \tilde{F}$, but the following says something about those ρ for which $F(\rho) = \tilde{F}(\rho)$.

Theorem 3.10. *Let $\rho_0 \in \mathcal{I}_N$. The following are equivalent:*
(1) *$F(\rho_0) = \tilde{F}(\rho_0)$ and F has a continuous* TF *at ρ_0.*
(2) *$\rho_0 \in \mathcal{A}_N$.*
(3) *$\tilde{F}$ has a continuous* TF *at ρ_0; that is, $\tilde{F}(\rho) \ge \tilde{F}(\rho_0) - \int v(\rho - \rho_0)$ for $\rho \in \mathcal{I}_N$.*
(4) *(3) and (5) hold with the same v.*
(5) *$E(v) = \tilde{F}(\rho_0) + \int v\rho_0$ for some v.*
(6) *(5) holds and, in addition, $v \in \mathcal{V}_N$ and $v \mapsto \rho_0$.*
(7) *$\tilde{F}$ has a continuous* TF *at ρ_0 and v is unique up to a constant. Moreover, F has the same continuous* TFs *at ρ_0, and no others.*

Proof. (1)$\Rightarrow$(3): For $\rho \in \mathcal{I}_N$,

$$\tilde{F}(\rho) \ge F(\rho) \ge F(\rho_0) - \int v(\rho - \rho_0) = \tilde{F}(\rho_0) - \int v(\rho - \rho_0).$$

(3)$\Rightarrow$(4): Let $F_1(\rho) = \tilde{F}(\rho_0) - \int v(\rho - \rho_0) \le \tilde{F}(\rho)$. Then

$$\tilde{F}(\rho_0) + \int v\rho_0 \ge E(v) \ge \inf\left\{F_1(\rho) + \int \rho v \,|\, \rho \in \mathcal{I}_N\right\} = \tilde{F}(\rho_0) + \int v\rho_0.$$

(4)$\Rightarrow$(5), (7)$\Rightarrow$(3), (6)$\Rightarrow$(5): All trivial.

(5)$\Rightarrow$(1): $\tilde{F}(\rho_0) + \int \rho_0 v = E(v) \le F(\rho_0) + \int \rho_0 v \Rightarrow F(\rho_0) \ge \tilde{F}(\rho_0) \Rightarrow F(\rho_0) = \tilde{F}(\rho_0)$. Then, for all $\rho \in X$, $F(\rho) + \int \rho v \ge E(v) = F(\rho_0) + \int \rho_0 v$.

(2)$\Rightarrow$(5): By (3.16).

(5)$\Rightarrow$(2), (6): By Theorem 3.3, $\tilde{F}(\rho_0) = (\psi, H_0\psi)$ for some ψ with $\psi \mapsto \rho_0$. Then $E(v) = (\psi, H_0\psi) + \int v\rho_0 \Rightarrow \rho_0 \in \mathcal{A}_N$, $v \in \mathcal{V}_N$, and $v \mapsto \rho_0$. Thus (1)–(6) are equivalent and (7)$\Rightarrow$(3). Now we show that (1)–(6)$\Rightarrow$(7). If v is a continuous TF for F, then v is a continuous TF for $\tilde{F}$ [by the proof of (1)$\Rightarrow$(3)]. If v is a continuous TF for $\tilde{F}$, then $F(\rho) \ge E(v) - \int v\rho$, so v is a continuous TF for F. Suppose $\tilde{F}$ has two continuous TFs v and w with $v - w \ne$ constant. Then $E(v) = \tilde{F}(\rho_0) + \int v\rho_0$ and $E(w) = \tilde{F}(\rho_0) + \int w\rho_0$. Since $\rho_0 \in \mathcal{A}_N$, this is impossible by Theorem 3.2. ■

It should be noted that the only place that the HK Theorem 3.2 entered in the analysis of F was in establishing the uniqueness (modulo constants) in (7).

Now we turn to two important questions whose answers we cannot give but that are obviously important for the theory. We replaced F_{HK} by F because F_{HK} was not defined on all of $\mathcal{I}_N$. Theorem 3.10 states that on $\mathcal{A}_N$, where F_{HK} is defined, $F = \tilde{F} = F_{HK}$ and F has an essentially unique *continuous* TF.

Question 5. For which points of $\mathcal{I}_N$ does F have a continuous TF? Where there is one, is it unique (modulo adding a constant to v)?

Question 6. If F has a continuous TF at $\rho_0 \in \mathcal{I}_N$ given by some $v \in L^{3/2}+L^\infty$, is this $v \in \mathcal{V}_N$?

Questions 5 and 6 have alternative formulations, given below.

Theorem 3.11. *Let $\rho_0 \in \mathcal{I}_N$ and $v \in L^{3/2}+L^\infty$. v is not necessarily in $\mathcal{V}_N$. Then, for all ρ,*

$$F(\rho) \geq F(\rho_0) - \int v(\rho - \rho_0) \qquad (\textit{continuous}\ \text{TF}) \tag{3.24}$$

if and only if

$$E(v) = F(\rho_0) + v \int \rho_0 \qquad [\textit{minimum in}\ (3.21)]. \tag{3.25}$$

Proof. Assume (3.24) and let $M_v(\rho)$ be its right side. Then

$$E(v) \geq \inf\left\{M_v(\rho) + \int \rho v\right\} = F(\rho_0) + \int v\rho_0 \geq E(v).$$

For the converse,

$$F(\rho) + \int v\rho \geq E(v) = F(\rho_0) + \int v\rho_0. \quad \blacksquare$$

Question 5 is equivalent to the following: For which $\rho_0 \in \mathcal{I}_N$ is there a v such that (3.25) holds? Is this v unique (up to constants)? Question 6 is the following: If (3.25) holds, is $v \in \mathcal{V}_N$?

Some insight into the continuous TFs of F are provided by the Bishop–Phelps theorem. We refer the reader to Ref. 20 for this as well as other interesting facts about convexity. A definition is needed.

Definition. Let F be a real functional on a real Banach space B with dual B^* (the set of continuous linear functionals on B). $b^* \in B^*$ is said to be *F-bounded* if there is a constant C (depending on b^* but not on b) such that $F(b) \geq b^*(b) + C$ for all $b \in B$.

In our case $B = X$ and F is our density functional.

Theorem 3.12. *Every $v \in X^* = L^{3/2} + L^\infty$ is F bounded.*

Proof. By Theorem 3.8, $F(\rho) = \infty$ if $\rho \notin \mathcal{I}_N$, so we only have to consider $\rho \in \mathcal{I}_N$ and prove that $G(\rho) \geq \int v\rho + C$ for some C. The proof of this is identical to the last part of the proof of Theorem 3.1. $\blacksquare$

The Bishop–Phelps theorem is the following.

Theorem 3.13. *Let F be a l.s.c. convex functional on a real Banach space B. (Note: Norm and weak l.s.c. are identical.) F can take the value $+\infty$, but not everywhere. Then*

(i) *The continuous tangent functionals to F (over all of B) are B^*-norm dense in the set of F-bounded functionals in B^**

(ii) *Suppose* $b_0 \in B$ *and* $b_0^* \in B^*$ *with* $F(b_0) < \infty$. *For every* $\varepsilon > 0$ *there exists* $b_\varepsilon \in B$ *and* $b_\varepsilon^* \in B^*$ *such that* $\|b_\varepsilon^* - b_0^*\|_{B^*} \leq \varepsilon$ *and*

$$\varepsilon \|b_\varepsilon - b_0\|_B \leq F(b_0) + b_0^*(b_0) - \inf\{F(x) + b_0^*(x) | x \in B\}.$$

Moreover, b_ε^* *is tangent to* F *at* b_ε, *namely* $F(b) \geq F(b_\varepsilon) - b_\varepsilon^*(b - b_\varepsilon)$ *for all* b.

The significance of Theorem 3.13(i) is the following. There are certainly many v's in Y that are not in $\mathscr{V}_N$. (Example: Suppose $v \in L^{3/2}$ and $\|v\|_{3/2} < L$, where L is the constant in (1.10). Then $(\psi, H_v\psi) > 0$ for all ψ, but $E(v) = 0$ because we can always take a sequence ψ_n that "leaks away to infinity.") Let $v \notin \mathscr{V}_N$, whence $\tilde{F}(\rho) + \int v\rho$ does not have a minimum. What Theorem 3.13 says is that there always exists a sequence v_n (not necessarily in $\mathscr{V}_N$) such that

(a) $F(\rho) + \int \rho v_n$ has a minimum at some $\rho_n \in \mathscr{I}_N$ and this minimum is $E(v_n)$.
(b) $F(\rho) \geq F(\rho_n) - \int v_n(\rho - \rho_n)$ for all ρ.
(c) $v_n \to v$ in the $L^{3/2} + L^\infty$ norm.

Point (c) means the following: $v_n = v + g_n + h_n$ with $\|g_n\|_{3/2} \to 0$ and $\|h_n\|_\infty \to 0$. In particular, if $v \in L^{3/2}$ with $\|v\|_{3/2} < L$, then $\|v + g_n\| < L$ for large n. Hence $v + g_n \notin \mathscr{V}_N$. If $v_n \in \mathscr{V}_N$, then it can only be because of the (vanishingly small) L^∞ piece h_n.

One consequence of Theorem 3.13(ii) is the following.

Theorem 3.14. *Let* $\rho_0 \in \mathscr{I}_N$. *Then there exists a sequence* $\rho_n \in \mathscr{I}_N$ *such that*

(i) $\rho_n \to \rho_0$ *in* $L^3 \cap L^1$ *norm.*
(ii) F *has a continuous* TF *at* ρ_n.

Proof. Given $n > 0$, by (3.17) there exists v_n such that $E(v_n) - \int \rho_0 v_n > F(\rho_0) - 1/n$. Hence

$$F(\rho) \geq E(v_n) - \int \rho v_n \geq F(\rho_0) - \int v_n(\rho - \rho_0) - 1/n.$$

Take $\varepsilon = 1$ in Theorem 3.13. There exists $w_n \in Y$ such that w_n is a continuous TF at some ρ_n and

$$\|\rho_n - \rho_0\| \leq F(\rho_0) + \int v_n \rho_0 - Z,$$

with

$$Z = \inf\left\{F(\rho) + \int v_n \rho | \rho \in X\right\}.$$

By the above,

$$Z \geq F(\rho_0) + \int \rho_0 v_n - n^{-1}. \quad \blacksquare$$

4. Additional Remarks about Density Functionals

A. The N-Dependence of F

As was stressed earlier, any functional F that satisfies (3.20) or (3.21) must depend explicitly on the particle number N. This fact is unavoidable and

Int. J. Quant. Chem. *24*, 243–277 (1983)

frequently overlooked. Let us denote the N dependence by $F(N,\rho)$. It might be hoped that F is jointly convex in N and ρ in the sense that for $N\geq 2$

$$F(N+1,\rho_1)+F(N-1,\rho_2)\geq 2F(N,\tfrac{1}{2}\rho_1+\tfrac{1}{2}\rho_2). \tag{4.1}$$

This convexity definitely does not hold as a general feature, as will be demonstrated.

The importance of convexity is shown by the following.

Theorem 4.1. *Consider the following two statements about any two functionals, F and E:* (i) *$F(N,\rho)$ is jointly convex in N and ρ in the sense of* (4.1). (ii) *$E(N,v)$ is convex in N for all fixed v; that is, for $N\geqq 2$*

$$E(N+1,v)+E(N-1,v)\geq 2E(N,v). \tag{4.2}$$

(a) *If* (3.17) *holds, then* (ii) *implies* (i).
(b) *If either* (3.20) *or* (3.21) *holds, then* (i) *implies* (ii).

Proof. (a) For each v, $E(N,v)-\int v\rho$ is jointly (N,ρ) convex. By (3.17), $F(N,\rho)$ is the supremum of such convex functions and hence is convex.

(b) Pick $\varepsilon>0$. For $N+1$ there is a ρ_+ such that

$$A\equiv F(N+1,\rho_+)+\int\rho_+v\leq E(N+1,v)+\varepsilon.$$

Likewise,

$$B\equiv F(N-1,\rho_-)+\int\rho_-v\leq E(N-1,v)+\varepsilon.$$

For the N problem, define $2\rho=\rho_++\rho_-$. Then

$$2\left\{F(N,\rho)+\int\rho v\right\}\leq A+B.$$

Since this holds for all $\varepsilon>0$, (ii) is proved. ■

Equation (4.2) has a simple physical meaning. The ionization potential increases as the number of electrons is decreased. This is intuitively expected to be true, but if it is true, it must be because of some special property of the Coulomb repulsion. A non-Coulombic counterexample is given below.

The kinetic energy functional $\tilde{T}(N,\rho)$ is not even convex in ρ (Theorem 3.4), but the Legendre transform $T(N,\rho)$ is jointly convex. This is so because $E(N,v)$ is indeed convex in N for independent particles as the explicit expression for E, as the sum of the first N eigenvalues (counted with an extra multiplicity q) shows.

What about the convexity of F when the Coulomb repulsion is included? While it has been conjectured that $E(N,v)$ is convex in N (for *all* v) in the case of Coulomb repulsion, this has never been proved. It has not even been proved that $E(3,v)+E(1,v)\geq 2E(2,v)$.

Lest the reader think that convexity in N is a general feature, we present a counterexample. Replace $|x|^{-1}$ by the hard-core repulsion $\theta(x)=\infty$ if $|x|<1$ and $\theta(x)=0$ otherwise. Pick four distinct points x_0, y_1, y_2, y_3 in $\mathbb{R}^3$ such that $|y_i-y_j|>1$ for all $i\neq j$ but $|x_0-y_i|<1$ for all i. Let $v(x)=-2\lambda<0$ in small balls about the y_i, $v(x)=-3\lambda$ in a small ball about x_0, and $v(x)=0$, otherwise. If the kinetic energy be neglected, then $E(1,v)=-3\lambda$, $E(2,v)=-4\lambda$, and $E(3,v)=-6\lambda$. Convexity does not hold. This can be turned into a proper example by letting λ be sufficiently large so that the kinetic energy can effectively be neglected; it is also possible to replace the hard core by a soft core.

Remark. The foregoing example is not applicable if θ is replaced by $|x|^{-1}$, thereby keeping alive the hope that convexity holds in the Coulomb case. The reason is the following: Given any four points x_0, y_1, y_2, y_3, let

$$|x_0-y_1|=\max_i\{|x_0-y_i|\}.$$

Then

$$|x_0-y_1|^{-1}\leq|y_1-y_2|^{-1}+|y_1-y_3|^{-1}.$$

The proof of this is left as an exercise, as well as the implication that if the kinetic energy is neglected, then convexity holds in the Coulomb case.

Question 7. For the case of Coulomb repulsion, is $F(N,\rho)$ jointly convex in N and ρ?

B. Density Matrices

Another possible modification of the theory of Sect. 3 is to replace densities $\rho(x)$ by single-particle admissible density matrices $\gamma(x,x')$. (See Questions 3 and 4 in Sec. 2. We do not restrict ourselves to γ's that come from pure states $\psi\rangle\langle\psi$.) This set of γ's is convex, and $\tilde{F}(\gamma)$, defined analogously to (3.14), is convex [see the proof of Theorem 4.1(b)].

Despite the attractive feature just mentioned, there are three drawbacks to the approach:

(i) The problems about continuous tangent functionals remain and may even be more complex than before.

(ii) The original aim of the theory was to express the energy in terms of $\rho(x)$ and not $\gamma(x,x')$.

(iii) While the set of admissible γ's is well defined, it is not easy to identify. Given some γ, it is easy to verify that $\mathrm{Tr}\,\gamma=N$, but it is difficult to verify that $0\leq\gamma\leq qI$.

Still another possible modification is to retain ρ but to consider all N-particle density matrices Γ instead of merely pure states $\psi\rangle\langle\psi$. In other words, consider $\Gamma\mapsto\rho$ instead of $\psi\mapsto\rho$ and define

$$F_{\mathrm{DM}}(\rho)=\inf\{\mathrm{Tr}\,H_0\Gamma|\Gamma\mapsto\rho\} \tag{4.3}$$

on $\mathcal{I}_N$ and $F_{\mathrm{DM}}(\rho)=+\infty$ otherwise. Because $\Gamma\mapsto\rho$ is linear, F_{DM} is convex on $\mathcal{I}_N$. (Note: The example in Theorem 3.4 does not yield nonconvexity of F_{DM}.)

Obviously, the analog of (3.15) holds, namely,

$$E(v)=\inf\left\{F_{\mathrm{DM}}(\rho)+\int \rho v \middle| \rho\in\mathcal{I}_N\right\}. \tag{4.4}$$

Since F_{DM} is convex, (4.4) can be used directly instead of (3.20) or (3.21).

Both F and F_{DM} are convex. The amusing fact is that

$$F(\rho)=F_{\mathrm{DM}}(\rho), \qquad \rho\in\mathcal{I}_N. \tag{4.5}$$

Equation (4.5) is not at all obvious, but it does say that the modification does not change the theory in any way. Equation (4.5) also yields another characterization of F. Equation (4.5) is proved in Theorem 4.3.

First, Γ is admissible if and only if

$$\Gamma(\mathbf{z},\mathbf{z}')=\sum_{i=1}^{\infty}\lambda_i\psi_i(\mathbf{z})\psi_i(\mathbf{z}')^*$$

with $0\le\lambda_i$, $\Sigma\lambda_j=1$, and the ψ_i are orthonormal. If $\psi_i\mapsto\rho_i$, then

$$\operatorname{Tr} H_0\Gamma=\Sigma\lambda_i(\psi_i, H_0\psi_i).$$

Thus we conclude that for all $\rho\in\mathcal{I}_N$

$$F_{\mathrm{DM}}(\rho)=\inf\left\{\sum_{i=1}^{\infty}\lambda_i\tilde{F}(\rho_i)\middle|\sum\lambda_i\rho_i=\rho,\ \rho_i\in\mathcal{I}_N,\ \lambda_i\ge0,\ \sum\lambda_i=1\right\}. \tag{4.6}$$

A simpler expression (which has to be proved) is

$$F_{\mathrm{DM}}(\rho)=\inf\left\{\sum\lambda_i\tilde{F}(\rho_i)\middle|\sum\lambda_i\rho_i=\rho,\ \rho_i\in\mathcal{I}_N,\ \lambda_i\ge0,\ \sum\lambda_i=1\right\}, \tag{4.7}$$

where the sums in (4.7) are restricted to finite sums. In view of (4.5), (4.7) is an alternative characterization of $F(\rho)$ for $\rho\in\mathcal{I}_N$.

Theorem 4.2. *Equation* (4.7) *is true.*

Proof. Pick $\varepsilon>0$. Using (4.6), let $\{\lambda_i,\rho_i\}$ be an infinite sequence satisfying $\Sigma\lambda_i\rho_i=\rho, \rho_i\in\mathcal{I}_N$, and $F_{\mathrm{DM}}(\rho)\ge\Sigma\lambda_i\tilde{F}(\rho_i)-\varepsilon$. Since $\Sigma\lambda_i=1$ and $\Sigma\lambda_i\tilde{F}(\rho_i)<\infty$, there exists K such that $A\equiv\sum_{i=K}^{\infty}\lambda_i\le\varepsilon$ and $B\equiv\sum_{i=K}^{\infty}\lambda_i F(\rho_i)\le\varepsilon$. Assume $A>0$ for otherwise we are done. By Theorem 1.1 and the convexity of $G(\rho)=\int(\nabla\rho^{1/2})^2$,

$$\varepsilon\ge\sum_K^{\infty}\lambda_i\tilde{F}(\rho_i)\ge\sum_K^{\infty}\lambda_i G(\rho_i)\ge AG(\rho^K)$$

with $\rho^K=\sum_K^{\infty}\lambda_i\rho_i/A\in\mathcal{I}_N$. By Theorem 3.9 and the remark following it,

$$\tilde{F}(\rho^K)\le C[N^2G(\rho^K)+N].$$

Therefore the finite sequence $\{\lambda_i,\rho_i\}_{i=1}^{K}$ with $\{\lambda_K,\rho_K\}=\{A,\rho^K\}$ satisfies $\Sigma\lambda_i\tilde{F}(\rho_i)\le F_{\mathrm{DM}}(\rho)+\varepsilon CN(N+1)+\varepsilon$. ■

Theorem 4.3. *Equation* (4.5) *is true.*

Proof. The easy part is that for $\rho \in \mathcal{I}_N$, $F_{\mathrm{DM}}(\rho) \geq F(\rho)$. By (4.4). $E(v) \leq F_{\mathrm{DM}}(\rho) + \int \rho v$ for all v. Hence, by (3.17), $F_{\mathrm{DM}}(\rho) \geq F(\rho)$. The hard part is contained in Corollary 4.5, which will be assumed for now. Then: (i) $F_{\mathrm{DM}}(\rho) \leq \tilde{F}(\rho)$ by (4.6); (ii) F_{DM} is convex and l.s.c. Hence $F_{\mathrm{DM}}(\rho) \leq \mathrm{CE}\, \tilde{F}(\rho) = F(\rho)$ by Theorem 3.7. ∎

Theorem 4.4. *Suppose* $\{\rho_n\}$ *and* $\rho \in \mathcal{I}_N$ *and* $\rho_n \rightharpoonup \rho$ *weakly in* L^1. *Then there exists a density matrix* Γ, *with* $\Gamma \mapsto \rho$, *such that* $\mathrm{Tr}\, H_0 \Gamma \leq \liminf F_{\mathrm{DM}}(\rho_n)$.

The proof of Theorem 4.4, due to Barry Simon, is given in the appendix.

Corollary 4.5. (i) F_{DM} *is* (*norm and weakly*) *l.s.c.*

(ii) *If* $\rho \in \mathcal{I}_N$, *there exists a density matrix* Γ *with* $\Gamma \mapsto \rho$ *such that* $\mathrm{Tr}\, H_0 \Gamma = F_{\mathrm{DM}}(\rho)$ (*see Theorem* 3.3).

Proof. (i) If $\rho_n \to \rho$, $F_{\mathrm{DM}}(\rho) \leq \mathrm{Tr}\, H_0 \Gamma \leq \liminf F_{\mathrm{DM}}(\rho_n)$. Norm l.s.c. implies weak l.s.c.

(ii) Take $\rho_n = \rho$ in Theorem 4.4. ∎

C. The Kinetic Energy Functional

Kohn and Sham (KS) [30] define a kinetic energy functional $T_{\mathrm{KS}}(\rho)$. There are several other possible kinetic energy functionals and we shall explore their interrelations, as well as the fact that T_{KS} does *not* have a property assumed by KS. KS define the *exchange and correlation functional* $E_{\mathrm{xc}}(\rho)$ by

$$F_{\mathrm{HK}}(\rho) \equiv \tfrac{1}{2} \iint \rho(x)\rho(y)|x-y|^{-1}\, dx\, dy + T_{\mathrm{KS}}(\rho) + E_{\mathrm{xc}}(\rho). \tag{4.8}$$

F_{HK} and T_{KS} are defined on different subsets of $\mathcal{I}_N$, so E_{xc} is defined only on a third unknown subset of $\mathcal{I}_N$. This difficulty can be remedied by using $\tilde{F}$ and $\tilde{T}$ in (4.8), but there is another point that should be stressed: There is no reason to believe that E_{xc} is convex on $\mathcal{I}_N$.

First, let us give some definitions. These use K instead of H_0 but otherwise are self-explanatory (with the aid of the equation numbers on the left):

$$\begin{aligned}
&(3.5){:} && E'(v) && \text{on } L^{3/2}+L^{\infty}; \\
&(3.10){:} && T_{\mathrm{KS}}(\rho) && \text{on } \mathcal{A}'_N \quad (3.11); \\
&(3.14){:} && \tilde{T}(\rho) && \text{on } \mathcal{I}_N; \\
&(3.17){:} && T(\rho) && \text{on } L^3 \cap L^1.
\end{aligned} \tag{4.9}$$

($T(\psi) = (\psi, K\psi)$ was defined in (1.3) but it is quite different from $T(\rho)$ above. It is hoped that this notational lapse will not be confusing.) All the previous theorems [except for 3.9, wherein the last term in (3.22) should be omitted] carry over to these quantities. The primes on $E'(v)$ and $\mathcal{A}'_N$ indicate that these are different from before. Since Theorem 3.4 still holds, $\mathcal{A}'_N$ is not $\mathcal{I}_N$. It is left as an exercise to show that $\mathcal{A}_N \neq \mathcal{A}'_N$.

Question 8. What is $\mathcal{A}_N \cap \mathcal{A}'_N$?

There is one more kinetic energy functional that can be defined on $\mathcal{I}_N$, namely,

$$T_{\det}(\rho) = \inf\{(\psi, K\psi) | \psi \mapsto \rho, \psi \in \mathcal{W}_N, \psi \text{ is a determinant}\}. \tag{4.10}$$

Clearly, $T_{\det}(\rho) \geq \tilde{T}(\rho)$. The question to be addressed is whether $T_{\det} = \tilde{T}$. The answer is No!, *not even on all of* $\mathcal{A}'_N$. KS assumed implicitly that $T_{KS}(\rho) = T_{\det}(\rho)$ for $\rho \in \mathcal{A}'_N$; any such ρ minimizes $K + V$, but it is not true that such a $\rho(x)$ can always be written as $\sum_{i=1}^N |\psi_i(x)|^2$ with the ψ_i being orthonormal functions on $\mathbb{R}^3$. (Spin is a complication that is ignored at this point for simplicity.) In other words, not every ground state of $K + V$ is a determinant when degeneracy is present. I thank B. Simon for drawing my attention to this subtlety and for the construction in Theorem 4.8, which is reminiscent of the construction in Theorem 3.4.

Of course, $T_{KS} = \tilde{T}$ on $\mathcal{A}'_N$ by definition. Also $\tilde{T} = T$ on $\mathcal{A}'_N$ by Theorem 3.10. The following shows that there are cases in which $\tilde{T} = T_{\det}$.

Theorem 4.6. *Suppose $\rho \in \mathcal{A}'_N$, so that $K + V$ has a ground state. If this ground state is nondegenerate, then $T_{\det}(\rho) = \tilde{T}(\rho)$.*

Proof. The ψ that minimizes $(\psi, [K + V]\psi)$ is, of course, a determinant. ■

The following analog of Theorem 3.3 will be needed for Theorem 4.8.

Theorem 4.7. *Let $\rho \in \mathcal{I}_N$. Then there exists a determinant that minimizes $(\psi, K\psi)$ under the condition that $\psi \mapsto \rho$, $\psi \in \mathcal{W}_N$, and ψ is a determinant. Thus,* (4.10) *is actually a minimum.*

Proof. Let D_j be a sequence of determinants with

$$D_j \mapsto \rho \qquad \text{and} \qquad \lim\,(D_j, KD_j) = T_{\det}(\rho).$$

The proof of Theorem 3.3 shows that ψ exists such that (i) $\psi \mapsto \rho$; (ii) $(\psi, K\psi) = T_{\det}(\rho)$; (iii) $D_j \to \psi$ strongly in L^2. It suffices to show that ψ is a determinant. Let f_j^i, $i = 1, \ldots, N$, be the orthonormal single-particle functions of D_j. By the Banach–Alaoglu theorem, N functions $f^1, \ldots, f^N$ exist so that (after passing to a subsequence) $f_j^i \to f^i$ weakly. The f^i are not necessarily orthonormal. The function

$$P_j(z_1, \ldots, z_N) = \prod f_j^i(z_i)$$

then converges weakly to $P = \Pi f^i$. This so because any $\psi \in L^2(\mathbb{R}^{3N})$ can be approximated in norm by sums of product functions. Therefore,

$$D_j \rightharpoonup (N!)^{1/2} \det[f^i(z_j)] \equiv D \qquad \text{weakly.}$$

But $D_j \to \psi$, so $D = \psi$. ■

Theorem 4.8. *Let $N = 7$ and $q = 1$. Then there is a $\rho \in \mathcal{A}'_N$ such that $T_{\det}(\rho) > \tilde{T}(\rho)$.*

Proof. Take $v(x) = |x|^{-1}$, the hydrogen potential. The eigenvalues of $-\Delta + v$ are $-1/4$ (onefold), $-1/16$ (fourfold), $-1/36$ (ninefold). All other eigenvalues are greater than $-1/36$. The ground state for $N = 7$ and $q = 1$ is $\binom{9}{2} = 36$-fold degenerate, and a basis for this eigenspace consists of the determinants $(7!)^{-1/2}\det(1S, 2S, 2P_1, 2P_2, 2P_3, f, g)$ where f and g are any orthonormal functions in the nine-dimensional space M spanned by $S, P_1, P_2, P_3, D_1, \ldots, D_5$ (an orthonormal set for the $3S$, $3P$, and $3D$ waves). Let $d(f, g)$ denote the above normalized determinant and let

$$3^{1/2}\psi = d(S, D_1) + d(D_2, D_3) + d(D_4, D_5).$$

Then $\psi \mapsto \rho$ with $\rho = \rho_a + \rho_b$ and

$$\rho_a(x) = |1S(x)|^2 + |2S(x)|^2 + \sum_{i=1}^{3} |2P_i(x)|^2,$$

$$3\rho_b(x) = |S(x)|^2 + \sum_{i=1}^{5} |D_i(x)|^2.$$

Clearly $\rho \in \mathscr{A}'_N$ since ψ is a ground state.

If $T_{\mathrm{det}}(\rho) = \tilde{T}(\rho)$, then there exists a determinant ϕ with $\phi \mapsto \rho$ and such that ϕ *must be* a ground state. Therefore $\phi = d(f, g)$ for some orthonormal $f, g \in M$. Thus,

$$|f(x)|^2 + |g(x)|^2 = \rho_b(x). \tag{4.11}$$

I claim that this is impossible. Write $f = A + D$ and $g = B + d$, with A and B being linear combinations of S and the P_i while D and d are linear combinations of the D_i. Now the S, P, and D waves behave as $|x|^0$, $|x|^1$, $|x|^2$, respectively, near the origin. By examining the behavior of (4.11) near the origin we conclude that

$$|D(x)|^2 + |d(x)|^2 = \tfrac{1}{3}\sum_{1}^{5} |D_i(x)|^2.$$

Since all the D_i waves have the same radial wave functions, this is really an equality about spherical harmonics Y_{2m}. The right side of the last equality is spherically symmetric, so the problem is to find two linear combinations F and G of the Y_{2m} such that

$$|F(\Omega)|^2 + |G(\Omega)|^2 = \text{constant} > 0.$$

This is impossible, and the proof is left as an exercise. (It is easily carried out if the following five basis functions are used: xyr^{-2}, xzr^{-2}, yzr^{-2}, $3x^2r^{-2} - 1$, $3y^2r^{-2} - 1$, with $r^2 = x^2 + y^2 + z^2$.) ■

Remarks. (i) $N = 7$ is not special; it was chosen for convenience in the proof.

(ii) An alternative way of viewing Theorem 4.8 is following. Suppose $K + V$ has a degenerate ground state, so that the ground eigenspace G is more than one-dimensional. $\psi \in G$ is a linear combination of determinants. Consider a perturbation w of v, namely, $v \to v + \lambda w$. In first-order perturbation theory, $V + \lambda W$ picks out a subspace g of G as the new ground eigenspace. If g is one dimensional, then g consists of one determinant since the ground eigenspace of $V + \lambda W$ always contains determinants (see Theorem 4.6). Now we ask, if $\psi_0 \in G$

Int. J. Quant. Chem. *24*, 243–277 (1983)

and $\psi_0 \mapsto \rho_0$, can w be chosen so that g is one dimensional and $g = \{\psi_0\}$? Alternatively, can w be chosen so that $\min\{\int w\rho | \psi \mapsto \rho$ and $\psi \in G\}$ occurs uniquely for $\rho = \rho_0$? If so, ψ_0 is a determinant. Theorem 4.8 says that there can be a ρ_0 such that no w can pick it out uniquely.

Even though $T_{\mathrm{det}}(\rho) > \tilde{T}(\rho)$ for some ρ, T_{det} still satisfies the variational principle for $E(v)$.

Theorem 4.9. *For all* $v \in L^{3/2} + L^{\infty}$

$$E'(v) = \inf\left\{T_{\mathrm{det}}(\rho) + \int \rho v | \rho \in \mathcal{I}_N\right\}. \tag{4.12}$$

Proof. Equation (4.12) is equivalent to the following:

$$\begin{aligned} E'(v) &\equiv \inf\{(\psi, [K+V]\psi) | \psi \in \mathcal{W}_N\} \\ &= \inf\{(\psi, [K+V]\psi) | \psi \in \mathcal{W}_N, \psi \text{ is a determinant}\} \\ &\equiv \tilde{E}(v). \end{aligned}$$

Clearly $E'(v) \le \tilde{E}(v)$. Consider the operator $-\Delta + v(x)$. We *define* its "eigenvalues" $e_1 \le e_2, \ldots$ (here, spin degeneracy is included) by the min-max principle:

$$e_{n+1} = \sup\{\varepsilon_n(\phi_1, \ldots, \phi_n)\},$$

where

$$\begin{aligned} \varepsilon_n(\phi_1, \ldots, \phi_n) = \inf\{(\phi, [-\Delta + v]\phi) | \phi \in H^1, \|\phi\|_2 = 1 \\ \text{and } \phi \text{ is orthogonal to } \phi_1, \ldots, \phi_n\}. \end{aligned}$$

From this definition, it follows by a standard argument that

$$E_N(v) \equiv \sum_{i=1}^{N} e_i = \inf\left\{\sum_{i=1}^{N} (\phi_i, [-\Delta + v]\phi_i) | \phi_1, \ldots, \phi_N \text{ are orthonormal}\right\}. \tag{4.13}$$

But this least infimum equals

$$\begin{aligned} \inf\left\{\sum_{i=1}^{\infty} \lambda_i(\phi_i, [-\Delta + v]\phi_i) | \phi_1, \phi_2, \ldots, \text{ are orthonormal,} \right. \\ \left. 0 \le \lambda_i \le 1 \text{ and } \sum_{i=1}^{\infty} \lambda_i = N\right\}. \end{aligned}$$

This is easy to verify.

Let $\psi \in \mathcal{W}_N$ and let $\gamma = \Sigma \lambda_i f_i \rangle\langle f_i$ be its one-particle density matrix (including spin and with the f^i-orthonormal. $0 \le \lambda_i \le 1$, $\Sigma \lambda_i = N$). Then

$$(\psi, [K+V]\psi) = \Sigma \lambda_i (f_i, [-\Delta + v] f_i).$$

Thus $E'(v) \ge E_N(v)$. But $E_N(v) = \tilde{E}(v)$ by inspection. ■

Remark. This proof gives a formula for $E'(v)$, namely, $E_N(v)$.

The situation is complicated, so let us summarize it. T_{KS} is defined only on $\mathscr{A}'_N$, the set of ρ's that come from ground states for some v. $\mathscr{A}'_N$ has a smaller subset, $\mathscr{A}''_N$, in which ρ comes from a determinantal ground state, $\mathscr{A}''_N$ includes, but is larger than, $\mathscr{A}'''_N$, the set of ρ's that come from nondegenerate ground states. (Note: By Theorem 3.2 any ρ comes from a unique v (up to constants). Thus, if ρ comes from a determinant in a degenerate ground eigenspace, then $\rho \notin \mathscr{A}'''_N$.) On $\mathscr{A}''_N$ we have

$$T_{\mathrm{KS}}(\rho) = T_{\mathrm{det}}(\rho) = \tilde{T}(\rho).$$

Elsewhere on $\mathscr{A}'_N$,

$$T_{\mathrm{det}}(\rho) > T_{\mathrm{KS}}(\rho) = \tilde{T}(\rho).$$

Thus, there are two choices for (4.8): either T_{KS} or T_{det}. On $\mathscr{B}_N$, the complement of $\mathscr{A}'_N$, T_{KS} is not defined (but $\tilde{T}$, T_{det}, and T are defined). The preferred functional here is $T(\rho)$ because it is convex and hence most manageable. On $\mathscr{B}_N$, $T(\rho)$ is probably strictly less than $\tilde{T}(\rho) \leq T_{\mathrm{det}}(\rho)$—at least this is so when T has a continuous tangent functional (Theorem 3.10). Such points are dense (Theorem 3.14). In any case, since T, $\tilde{T}$, and T_{det} can be interchangeably used in (4.12); it makes no difference which is used as far as $E'(v)$ is concerned.

5. Some Density Functionals That Are Bounds

In this section we forego the abstract functional theory of the previous sections and instead expound a different philosophy. Rather than pursuing "the correct density functional," which seems to be uncomputable, we shall content ourselves here with finding upper and lower bounds to the various quantities of interest in terms of $\rho(x)$. This latter program can provide rigorous bounds on ground state energies that, while they may not always be extremely accurate, do have a proper place in our conceptual scheme. Some of these bounds will be briefly displayed here; the interested reader is referred to the original papers for proofs.

It should be remembered that if one has bounds on two quantities (e.g., T and I; see below) and even if these bounds are optimal, then, in general, the sum of the bounds is *not* optimal for the sum of the two quantities (e.g., $T+I$).

A. Kinetic Energy Lower Bound

Lieb and Thirring (LT) [21] (also see Ref. 16) proved (for fermions in three dimensions) that if $\psi \mapsto \rho$, then (for all N)

$$T(\psi) \geq K^{c}(4\pi)^{-2/3} q^{-2/3} \int \rho(x)^{5/3}\, dx, \tag{5.1}$$

where K^{c} is the "classical" value $(3/5)(6\pi^2)^{2/3}$. LT conjectured that (5.1) holds in three dimensions with the $(4\pi)^{-2/3}$ deleted. [Note: Although an analog of (5.1) holds in all dimensions, the corresponding constant is definitely less than K^{c} in one and two dimensions.] In Ref. 22 (also see Ref. 16) $(4\pi)^{-2/3}$ was replaced by $1.496(4\pi)^{-2/3}$.

Int. J. Quant. Chem. 24, 243–277 (1983)

Incidentally, the statement $T(\psi) \geq Kq^{-2/3} \int \rho^{5/3}$ for all ψ, all N, and some K is equivalent to the following [21]. Let v be any nonpositive potential in $L^{5/2}(\mathbb{R}^3)$ and let $e_1 \leq e_2 \leq \cdots$, be the negative eigenvalues (if any) of $-\Delta + v(x)$ counting degeneracy, but not counting the q-fold degeneracy. Then

$$\sum e_i \geq -L \int |v(x)|^{5/2}\, dx$$

with $K = (3/5)(2/5L)^{2/3}$.

B. Kinetic Energy Upper Bound

There is, of course, no upper bound for $T(\psi)$ in terms of ρ. March and Young (MY) [17] proposed that for all $\rho \in \mathcal{I}_N$ there is a *determinantal* ψ, with $\psi \mapsto \rho$, such that

$$T(\psi) \leq q^{-2/n} K^{\mathrm{c}} \int \rho(x)^{(n+2)/n}\, dx + \int [\nabla \rho^{1/2}(x)]^2\, dx, \tag{5.2}$$

where n is the dimension and $K^{\mathrm{c}} = \pi^2/3$ for $n = 1$. (Compare (5.2) with Theorem 1.2.) They proved (5.2) for $n = 1$, but their proof for $n > 1$ has an error. Equation (5.2) for $n > 1$ is still an *open problem*. The MY construction for $n = 1$ motivated the construction in the proof of Theorem 1.2.

C. Lower Bound for the Indirect Part of the Coulomb Repulsion

Let Γ be a density matrix (which may be a pure state, $\Gamma = \psi\rangle\langle\psi)$ with $\Gamma \mapsto \rho$. Let

$$I(\Gamma) = \mathrm{Tr}\left\{\Gamma \sum_{1 \leq i < j \leq N} |x_i - x_j|^{-1}\right\} \tag{5.3}$$

be the Coulomb repulsive energy. The *indirect part* of this energy, $E(\Gamma)$, is *defined* by

$$I(\Gamma) = D(\rho) + E(\Gamma), \tag{5.4}$$

with

$$D(\rho) = \tfrac{1}{2} \iint \rho(x)\rho(y)|x-y|^{-1}\, dx\, dy \tag{5.5}$$

being the *direct part*.

In Ref. 23 it was shown that

$$E(\Gamma) \geq -C \int \rho(x)^{4/3}\, dx \tag{5.6}$$

with $C = 8.5$. In Ref. 24 this was improved to $C = 1.68$. The sharp (i.e., best) C in (5.6) is not known, but it is larger than 1.23.

It is well known that in any pure, *determinantal* state, $E(\Gamma)<0$. For other states, $E(\Gamma)$ can be positive. Indeed, for any fixed ρ there is no upper bound for $E(\Gamma)$ (see Ref. 24).

There is no q-dependence in (5.6) and, indeed, (5.6) holds for all statistics (i.e., C does not depend on statistics). This is explained in Ref. 24. The Dirac approximation has $C_D q^{-1/3}$ in (5.6) with $C_D = 3(6/\pi)^{1/3}/4 = 0.93$, but this q dependence is an artifact of the particular q-dependent determinantal ψ used to evaluate E from (5.4).

It should be noted that the bound

$$I(\Gamma) \geq D(\rho) - C\int \rho^{4/3} \tag{5.7}$$

is *not convex* in ρ. It is not even positive. These two faults lead to absurd conclusions when the right side of (5.7) is used in Thomas–Fermi–Dirac theory (see Ref. 25).

Since $\Gamma \mapsto \rho$ is linear,

$$\tilde{I}(\rho) = \inf \{I(\Gamma) | \Gamma \mapsto \rho\} \tag{5.8}$$

is convex in ρ. In other words, an optimal positive, convex lower bound must exist. Any reader who is devoted to abstract density functional theory, in the spirit of Sec. 3 or (5.8), should try to guess a plausible form for $\tilde{I}(\rho)$. (Proving it is another matter.) It will quickly be seen that $\tilde{I}(\rho)$ must be extremely complicated, and to say that it is "nonlocal" is an understatement. To see this, consider $N=2$ and ρ consisting of two "bumps," ρ_1 and ρ_2, very far apart. As long as $\int \rho_1 = \int \rho_2 = 1$, $\tilde{I}(\rho) - D(\rho) \approx 0$, independently of ρ_1 and ρ_2. But when $\int \rho_1 > 1$, $\int \rho_2 < 1$, then $\tilde{I}(\rho) - D(\rho)$ depends heavily on ρ_1 but not on ρ_2. The reason is that in the former case the two electrons can be far apart in the two bumps; in the latter case the two electrons must partly be close together in the first bump.

A problem that is physically more relevant and that illustrates the hidden complexity of density functional theory is the following problem about induced dipolar (or Van der Waals) forces raised in Ref. 25. When two atoms are a distance R apart, and R is large, there is an attraction $-R^{-6}$ (neglecting retardation effects). This attraction comes from the Coulomb repulsion, but it is not a static effect. The atomic dipole moment is almost zero. (There are, in fact, tiny dipole moments, but these are opposite in sign by symmetry, and hence repulsive. They must exist by the Feynmann–Hellman theorem: dE/dR = electric potential at the nucleus. I thank C. Herring for this remark.) There is almost no static dipole moment because to create one would cost a polarization energy αd^2. The attractive energy is $-d^2R^{-3}$ and, if $R^{-3} < \alpha$, $d=0$ for minimum energy. The cause of the $-R^{-6}$ energy is more subtle, but it has a semiclassical basis: The electrons in each atom *move in phase* while maintaining the spherical symmetry about each atom. The energy cost is then αd^4 and the minimum energy occurs when $2\alpha d^2 = R^{-3}$. Thus, the $-R^{-6}$ attraction comes from the fact that the electron

Int. J. Quant. Chem. *24*, 243–277 (1983)

cloud cannot be thought of as a simple "fluid." This effect is somehow built into $\tilde{I}(\rho)$, but an explicit form of $\tilde{I}(\rho)$ that will produce this effect has yet to be displayed.

D. A Variational Principle

$E(v)$, given by (3.5), satisfies (by definition) the well-known variational principle

$$E(v) \leq (\psi, H_v \psi). \tag{5.9}$$

Can an upper bound for $E(v)$ be given in terms of ρ alone? If (5.2) were true, then, for any $\rho \in \mathcal{I}_N$,

$$E(v) \leq \text{right side of (5.2)} + D(\rho) + \int v\rho. \tag{5.10}$$

[See the remark about $E(\Gamma)$ for determinants in Sec. 5C.]

An upper bound for $E(v)$ can, indeed, be given in terms of the one-particle density *matrices* $\gamma(z, z')$ as follows [26]:

Let γ be any admissible one-particle density matrix ($0 \leq \gamma \leq I$, $\text{Tr}\,\gamma = N$). (Note: γ includes spin. It was called $\bar{\gamma}$ in Sec. 2). Then

$$E(v) \leq \text{Tr}\,\gamma(-\Delta + v(x)) + \tfrac{1}{2} \int K_2(z, z')|x - x'|^{-1}\,dz\,dz', \tag{5.11}$$

where $\int dz = \sum_\sigma \int dx$ and

$$K_2(z, z') = \gamma(z, z)\gamma(z', z') - |\gamma(z, z')|^2. \tag{5.12}$$

The form (5.11) is well known *if* γ came from a pure state $\psi\rangle\langle\psi$ with ψ being a determinant. The point about (5.11) is that it holds for *all* admissible γ. Incidentally, the minimum of (5.11) over all admissible γ occurs when γ comes from a determinantal ψ. In other words, *the best* Hartree–Fock function minimizes (5.11), but (5.11) is interesting precisely because this HF function is unknown.

E. Thomas–Fermi Theory

This theory (see Ref. 25 for an exposition) does not yield bounds and therefore does not properly belong here. However, it illustrates the usefulness of the bounds in Secs. 5A–C.

The TF functional is

$$\mathscr{E}^{\text{TF}}(\rho) = K^c q^{-2/3} \int \rho^{5/3} + D(\rho) + \int v\rho, \tag{5.13}$$

while the TF Weizsaecker functional $\mathscr{E}^{\text{TFW}}(\rho)$ is the right-hand side of (5.10). If $-C \int \rho^{4/3}$ is added to the right-hand side of (5.13), the result is TF Dirac theory.

The TF energy for N particles is defined by

$$E^{\mathrm{TF}} = \inf\left\{\mathscr{E}^{\mathrm{TF}}(\rho) \mid \int \rho = N\right\}, \tag{5.14}$$

and similarly for E^{TFW} and E^{TFD}.

Now suppose that v is an atomic or molecular potential, that is,

$$v(x) = -\sum_{j=1}^{k} z_j |x - R_j|^{-1}, \tag{5.15}$$

with the $z_j > 0$. It is a fact [14] that under the scaling $z_j \to \lambda z_j$ and $N \to \lambda N$, as $\lambda \to \infty$

$$E^{\mathrm{TF}}/E(v) \to 1, \tag{5.16}$$

where $E(v)$ is the true ground state energy. Note that (5.16) also holds if E^{TF} is replaced by E^{TFW} or E^{TFD} (see Ref. 25).

Thus we see that if the conjecture in Sec. 5A holds, then, combining (5.1) with (5.7), TFD theory is a lower bound that is asymptotically exact. Similarly, if (5.2) holds, then, as remarked in Sec. 5C TFW theory is an upper bound that is asymptotically exact.

F. Two-Body Density Matrices

If one is willing to go beyond the one-body density ρ or one-body density matrix γ and consider the two-body reduced density matrix $\gamma^{(2)}$, then $E(v)$ is directly and exactly expressible in terms of $\gamma^{(2)}$, since H_v has only one- and two-body terms. The problem is that it is very difficult to decide when a given $\gamma^{(2)}$ is, in fact, the reduction of an admissible N-body density matrix Γ. This is called the *N-representability problem* and it has not been solved. (This is to be compared with the fact that there is a simple necessary and sufficient condition for a one-body γ to be N-representable; see Sec. 2).

It is possible, however, to find some necessary conditions and some sufficient conditions for $\gamma^{(2)}$ to be N-representable. Using these, bounds on $E(v)$ can be derived. Since this approach is outside the scope of this article, we refer the reader to the excellent review of Percus [27].

Appendix: Proofs of Theorems 1.3, 3.3, and 4.4

The following proof of Theorem 1.3 is due to H. Brezis (pivate communication.)

Proof. For simplicity of presentation we take $N = 2$ and $q = 1$ (no spin). Therefore we have $\psi(x, y)$ and

$$F(x) = \left(\int |\psi(x, y)|^2\, dy\right)^{1/2} = [\rho(x)/2]^{1/2}.$$

Suppose that $\psi_n \to \psi$ in H^1 $(\mathbb{R}^3 \times \mathbb{R}^3)$; that is, $\psi_n \to \psi$ and $\nabla\psi_n \to \nabla\psi$ in L^2. We want to show that $F_n \to F$ in $L^2(\mathbb{R}^3)$ and $\nabla F_n \to \nabla F$ in $L^2(\mathbb{R}^3)$. The former is trivial:

By the Schwarz inequality,

$$\left(\int |\psi_n(x,y)|^2\,dy \int |\psi(x,y)|^2\,dy\right)^{1/2} \geq \tfrac{1}{2}\int \psi_n(x,y)^*\psi(x,y)\,dy + \tfrac{1}{2}\int \psi(x,y)^*\,\psi_n(x,y)\,dy.$$

Therefore

$$\int |F_n(x)-F(x)|^2\,dx \leq \int dx \int |\psi_n(x,y)-\psi(x,y)|^2\,dy.$$

and the right-hand side converges to zero.

The proof that $\nabla F_n \to \nabla F$ is the difficult one. It is sufficient to prove convergence for *some* subsequence n_j, $j=1,2,\ldots$. (If $\nabla F_n \not\to \nabla F$, then there is some subsequence and some $\varepsilon>0$ such that $\|\nabla F_{n_j} - \nabla F\| > \varepsilon$. But then this subsequence clearly does not have a subsequence which converges to ∇F.) Now since $\psi_n \to \psi$ in H^1 there is some subsequence and some function $G \in H^1$ such that

$$|\psi_{n_j}(x,y)| \leq G(x,y) \text{ and } |\nabla\psi_{n_j}(x,y)| \leq G(x,y),$$

a.e. in $\mathbb{R}^6$. (The proof of this fact is the same as the first half of the proof of the Riesz–Fischer lemma that L^1 is complete.) Henceforth, we shall replace n_j by n. We shall also assume, for simplicity, that $F_n(x)$ and $F(x)>0$ for all x (otherwise, an approximation argument can be used).

Now

$$2\nabla F_n(x) = A_n(x)/F_n(x) + \text{c.c.}$$

with

$$A_n(x) = \int B_n(x,y)\,dy \qquad \text{and} \qquad B_n(x,y) = \psi_n(x,y)^*\nabla\psi_n(x,y).$$

As we saw, $F_n \to F$ in L^2, so, by passing to a subsequence, we can assume $F_n(x) \to F(x)$ a.e. Furthermore, a.e.

$$|B_n(x,y)| \leq G(x,y)^2 \in L^1(\mathbb{R}^6).$$

By passing to a subsequence we can assume $\nabla\psi_n \to \nabla\psi$ and $\psi_n \to \psi$ a.e. Thus, by dominated convergence, $B_n \to B$ in $L^1(\mathbb{R}^6)$. For this subsequence $|B_n(x,y) - B(x,y)| \to 0$, a.e. ($\mathbb{R}^6$). Then, for a.e. x, $|B_n(x,y)-B(x,y)| \to 0$ a.e. y. Thus, by dominated convergence $\int dy |B_n(x,y)-B(x,y)| \to 0$, a.e. x. In other words, for some subsequence, $\nabla F_n(x) \to \nabla F(x)$, a.e.

Finally, we note that, by the Schwarz inequality,

$$|\nabla F_n(x)|^2 \leq \int |\nabla\psi_n(x,y)|^2\,dy \leq \int G(x,y)^2\,dy \equiv C(x)^2.$$

Since C is a fixed L^2 function, $\nabla F_n \to \nabla F$ in L^2 by dominated convergence. ■

Proof of Theorem 3.3. Let ψ_j (with $\psi_j \mapsto \rho$) be a minimizing sequence for $\tilde{F}(\rho)$. The ψ_j are obviously bounded in $H^1(\mathbb{R}^{3N})$, so, by the Banach–Alaoglu theorem, there is a $\psi \in H^1(\mathbb{R}^{3N})$ such that $\psi_j \rightharpoonup \psi$ weakly in $H^1(\mathbb{R}^{3N})$. Obviously, ψ has the same symmetry as the ψ_j. It is well known that under weak limits positive quadratic forms decrease. Thus

$$\tilde{F}(\rho) = \lim (\psi_j, H_0\psi_j) \geq (\psi, H_0\psi).$$

If we can show that $\psi \mapsto \rho$, we are done. To do so it is sufficient to prove that $\psi_j \to \psi$ strongly because if $\psi \mapsto \tilde{\rho}$ we have, by the easy part of Theorem 1.3 that $\rho^{1/2} = \rho_j^{1/2} \to \tilde{\rho}^{1/2}$ in L^2, so that $\tilde{\rho} = \rho$.

Strong convergence will be proved by showing that $\int |\psi|^2 = 1$. Let S be the characteristic function of some bounded set in $\mathbb{R}^{3N}$. By the Rellich–Kondrachov theorem [28] there is a subsequence (which can be chosen independent of S) of the ψ_j such that $S\psi_j$ converges strongly (in L^2) to $S\psi$. Pick $\varepsilon > 0$ and let χ be the characteristic function of a bounded set in $\mathbb{R}^3$ such that

$$\varepsilon > \int \rho(1-\chi) \equiv \int |\psi_j|^2 \sum_j [1-\chi(x_i)].$$

But

$$\sum [1-\chi(x_i)] \geq 1 - S,$$

where $S = \Pi\chi(x_i)$. Thus, $\int |\psi_j|^2 S \geq 1-\varepsilon$. Since $|\psi_j|^2 S \to \int |\psi|^2 S$, we have that $\int |\psi|^2 \geq \int |\psi|^2 S \geq 1-\varepsilon$ for all $\varepsilon > 0$. ■

Remark. The symmetry of ψ was not needed in this proof provided one generalizes definition (1.6) to

$$\rho(x) = \sum_\sigma \sum_{i=1}^N \int |\psi(z_1, \ldots, (x, \sigma_i), \ldots, z_N)|^2 \, dx_1 \cdots dx_{i-1} \, dx_{i+1} \cdots dx_N. \quad \text{(A.1)}$$

The following proof of Theorem 4.4 is due to B. Simon (private communication). It is closely related to the proof of Theorem 3.3 just given.

Proof. Without loss, replace H_0 by $h^2 = H_0 + 1$ in the definitions. h^{-1} is a bounded operator. We can assume that $g_n \equiv F_{\mathrm{DM}}(\rho_n) < \infty$, $g \equiv \lim g_n$ exists, and $\operatorname{Tr} h\Gamma_n h = \operatorname{Tr} \Gamma_n h^2 \leq g_n + 1/n$ with $\Gamma_n \mapsto \rho_n$. Thus, $y_n \equiv h\Gamma_n h$ is uniformly bounded in the trace norm. The dual of the compact operators, com, is the trace class operators t, and $\gamma \in t$ takes $A \in \text{com}$ into $\operatorname{Tr} \gamma A$. A sequence $\gamma_n \in t$ converges to $\gamma \in t$, in the weak* topology, if and only if $\operatorname{Tr} \gamma_n A \to \operatorname{Tr} \gamma A$ for all $A \in \text{com}$.

The Banach–Alaoglu theorem states that a norm-closed ball of finite radius in t is compact in the weak* topology. For us this means that there exists y with $\operatorname{Tr} y < \infty$, so that, for a subsequence, $\operatorname{Tr} y_n A \to \operatorname{Tr} yA$ for every compact A. Clearly, $y \geq 0$ and therefore $\underline{\lim} \operatorname{Tr} y_n \geq \operatorname{Tr} y$. Also, y obviously has the correct (Pauli) symmetry. If we can show that $\Gamma \equiv h^{-1} y h^{-1}$ (which is in trace class) satisfies $\Gamma \mapsto \rho$, we are done. To do this we shall show that if $\Gamma \mapsto \rho'$, then $\int (\rho_n - \rho') f \to 0$ for any $f \in L^\infty$. This would mean that $\rho_n \rightharpoonup \rho'$ weakly in L^1. But since $\rho_n' \rightharpoonup \rho$ in L^1, $\rho' = \rho$.

Int. J. Quant. Chem. *24*, 243–277 (1983)

As in the proof of Theorem 3.3, for any $\varepsilon > 0$ there is a χ (=characteristic function of a bounded set in $\mathbb{R}^3$) such that

$$\int \rho(1-\chi) < \varepsilon \qquad \text{and} \qquad \int \rho'(1-\chi) < \varepsilon.$$

Since $\rho_n \rightharpoonup \rho$, $\int \rho_n(1-\chi) < \varepsilon$ for n sufficiently large. If

$$\phi_n(x_1, \ldots, x_N) = \Gamma_n(x_1, \ldots, x_N; x_1, \ldots, x_N)$$

(after summing on spins), and similarly for ϕ, we have (as in Theorem 3.3)

$$\int \phi_n(1-S) < \varepsilon \qquad \text{and} \qquad \int \phi(1-S) < \varepsilon$$

where $S = \Pi\chi(x_i)$. In view of this, it is sufficient to show that

$$\int_n P \to \int \phi P \qquad \text{with} \qquad P = S \sum_i f(x_i).$$

Let $P = P(x_1, \ldots, x_N)$ be any bounded functions of compact support and let M_P be the operator (in L^2) of multiplication by P. It is a fact that $A_P \equiv h^{-1} M_P h^{-1}$ is compact. (This is essentially the same as the Rellich–Kondrachov theorem used in Theorem 3.3.) Therefore

$$\text{Tr}\, \Gamma_n M_P = \text{Tr}\, y_n A_P \to \text{Tr}\, y A_P = \text{Tr}\, \Gamma M_P.$$

■

Acknowledgment

This work was partially supported by U.S. National Science Foundation grant No. PHY-7825390-A02. This paper is a revised version of a paper with the same title that appeared in *Physics as Natural Philosophy*: *Essays in Honor of Laszlo Tisza on his 75th Birthday*, H. Feshbach and A. Shimony, Eds. (M.I.T. Press, Cambridge, 1982), pp. 111–149.

Bibliography

[1] L. H. Thomas, Proc. Camb. Phil. Soc. **23**, 542 (1927).
[2] E. Fermi, Rend. Accad. Naz. Lincei **6**, 602 (1927).
[3] P. Hohenberg and W. Kohn, Phys. Rev. B **136**, 864 (1964).
[4] M. M. Morell, R. G. Parr and M. Levy, J. Chem. Phys. **62**, 549 (1975).
[5] R. G. Parr, S. Gadre and L. J. Bartolotti, Proc. Natl. Acad. Sci. USA **76**, 2522 (1979).
[6] R. A. Donnelly and R. G. Parr, J. Chem. Phys. **69**, 4431 (1978).
[7] H. Englisch and R. Englisch, "Hohenberg–Kohn theorem and non-v-representable densities," Physica A, to be published.
[8] T. L. Gilbert, Phys. Rev. B **6**, 211 (1975).
[9] J. E. Harriman, Phys. Rev. A **6**, 680 (1981).

[10] M. Levy, Proc. Natl. Acad. Sci. USA **76**, 6062 (1979).
[11] M. Levy, Phys. Rev. A **26**, 1200 (1982).
[12] S. M. Valone, J. Chem. Phys. **73**, 1344 (1980); *ibid.* **73**, 4653 (1980).
[13] A. S. Bamzai and B. M. Deb, Rev. Mod. Phys. **53**, 95 (1981). *Erratum*, **53**, 593 (1981).
[14] E. H. Lieb and B. Simon, Adv. Math. **23**, 22 (1977). See also Thomas–Fermi theory revisited, Phys. Rev. Lett. **31**, 681 (1973). See also Refs. 16 and 25.
[15] M. Hoffmann-Ostenhof, and T. Hoffmann-Ostenhof, Phys. Rev. A **16**, 1782 (1977).
[16] E. H. Lieb, Rev. Mod. Phys. **48**, 553 (1976).
[17] N. H. March and W. H. Young, Proc. Phys. Soc. **72**, 182 (1958).
[18] M. Reed and B. Simon, *Methods of Modern Mathematical Physics* (Academic, New York, 1978), Vol. 4.
[19] S. Mazur, Studia Math. **4**, 70 (1933).
[20] R. B. Israel, *Convexity in the Theory of Lattice Gases* (Princeton U.P., Princeton NJ, 1979).
[21] E. H. Lieb and W. E. Thirring, "Inequalities for the moments of the eigenvalues of the Schrödinger hamiltonian and their relation to Sobolev inequalities," in *Studies in Mathematical Physics*, E. H. Lieb, B. Simon, and A. S. Wightman, Eds. (Princeton U.P., Princeton, NJ, 1976). See also Phys. Rev. Lett. 687 (1975); *Errata*, **35**, 1116 (1975).
[22] E. H. Lieb, Am. Math. Soc. Proc. Symp. Pure Math. **36**, 241 (1980).
[23] E. H. Lieb, Phys. Lett. A **70**, 444 (1979).
[24] E. H. Lieb and S. Oxford, Int. J. Quantum Chem. **19**, 427 (1981).
[25] E. H. Lieb, Rev. Mod. Phys. **53**, 603 (1981); *Errata*, **54**, 311 (1982).
[26] E. H. Lieb, Phys. Rev. Lett. **46**, 457 (1981); *Erratum*. **47**, 69 (1981).
[27] J. K. Percus, Int. J. Quantum Chem. **13**, 89 (1978).
[28] R. A. Adams, *Sobolev Spaces* (Academic Press, New York, 1975).
[29] W. Fenchel, Can. J. Math. **1**, 23 (1949).
[30] W. Kohn and L. J. Sham, Phys. Rev. A **140** 1133 (1965).

Received October 19, 1982
Accepted for publication March 11, 1983

Commun. Math. Phys. 92, 473–480 (1984)

Communications in Mathematical Physics

On Characteristic Exponents in Turbulence

Elliott H. Lieb*

Departments of Mathematics and Physics, Princeton University, P.O. Box 708, Princeton, NJ 08544, USA

Abstract. Ruelle has found upper bounds to the magnitude and to the number of non-negative characteristic exponents for the Navier-Stokes flow of an incompressible fluid in a domain Ω. The latter is particularly important because it yields an upper bound to the Hausdorff dimension of attracting sets. However, Ruelle's bound on the number has three deficiences: (i) it relies on some unproved conjectures about certain constants; (ii) it is valid only in dimensions $\geqq 3$ and not 2; (iii) it is valid only in the limit $\Omega \to \infty$. In this paper these deficiences are remedied and, in addition, the final constants in the inequality are improved.

Ruelle [1] has derived upper bounds on the magnitude and number of non-negative characteristic exponents of the Navier-Stokes equation for the flow of an incompressible fluid in a domain $\Omega \in \mathbb{R}^d$. The bound on the number, $\tilde{N}(\mu)$ [defined in (42)], is particularly interesting because it leads to an upper bound on the Hausdorff dimension of a compact attracting set [1, Corollary 2.3]. Unfortunately, the bounds in [1] on $\tilde{N}(\mu)$, unlike those on the magnitude, have certain deficiencies which are

(i) They rely for their validity on some conjectured, but as yet unproved, relations between the sharp constants in two known inequalities.

(ii) They are valid only for $d \geqq 3$.

(iii) Because Weyl's asymptotic formula for the eigenvalues of the Laplacian in Ω is used, the inequalities are not valid for any fixed Ω, but only in the limit $\Omega \to \infty$.

In this paper a different proof of Ruelle's inequality for the number will be given so that the above three deficiencies are remedied. The result is contained in Eqs. (40)–(43).

Let $v : \Omega \to \mathbb{R}^d$ denote a solution to the Navier-Stokes equation, and let $\mu_1 \geqq \mu_2 \geqq \ldots$ be the characteristic exponents corresponding to a probability measure $\varrho(dv)$ on the space of solutions that is ergodic with respect to the Navier-

* Work partially supported by U.S. National Science Foundation grant No. PHY-8116101-A01

Commun. Math. Phys. 92, 473–480 (1984)

Stokes time evolution. Ruelle shows [1] that for all $n \geqq 1$

$$\sum_{i=1}^{nd} \mu_i \leqq -d \sum_{i=1}^{n} \langle e_i \rangle \equiv -d \langle E_n \rangle . \tag{1}$$

The brackets $\langle \cdot \rangle$ denote average with respect to ϱ, $E_n = \sum_1^n e_i$, and the $e_i = e_i(v)$ are ordered such that $e_1 \leqq e_2 \leqq \ldots$ and are the eigenvalues of the Schrödinger operator

$$H = -\nu\Delta - w(x) \tag{2}$$

with Dirichlet boundary conditions on Ω. Here, ν is the kinematic viscosity and $w(x) \geqq 0$ with

$$w(x)^2 = [(d-1)/4d] \sum_{i,j} (\partial v_i/\partial x_j + \partial v_j/\partial x_i)^2 = [(d-1)/2\nu d]\, \varepsilon(x) . \tag{3}$$

The quantity $\varepsilon(x)$ is the rate of energy dissipation per unit mass in the flow v. In (2), (3) and henceforth, explicit dependence of the various quantities on v is understood but not explicitly indicated unless necessary.

One might try to take additional advantage of the fact that $\operatorname{div} v = 0$ but, as in [1], we shall merely assume that w is some given non-negative function. It will, however, be assumed, as in [1], that

$$w \in L^{1+d/2}(\Omega) . \tag{4}$$

Remark. The definition (3) has a factor $(d-1)/d$, which is an improvement over that in [1]. The reason is the following: Ruelle starts with an operator on $L^2(\mathbb{R}^d) \otimes \mathbb{R}^d$ given by $\mathscr{H} = -\nu\Delta\delta_{ij} + W_{ij}(x)$, where $W_{ij}(x)$ is the $d \times d$ symmetric matrix $W_{ij}(x) = (\partial v_i/\partial x_j + \partial v_j/\partial x_i)/2$. Ruelle notes that the eigenvalues of $\mathscr{H}$ will satisfy (1) if $w(x)$ in (2) is the largest eigenvalue of the matrix $W_{ij}(x)$. This he estimates by $(\operatorname{Tr} W^2)^{1/2}$, and this leads to (3) without $(d-1)/d$. Since $\operatorname{div} v = 0$, however, $\operatorname{Tr} W = 0$. If $\lambda_1 \geqq \lambda_2 \geqq$ are the eigenvalues of W, then $\operatorname{Tr} W^2 = \sum \lambda_i^2$ and $\operatorname{Tr} W = \sum \lambda_i$. But $(d-1)\sum_2^d \lambda_i^2 \geqq \left(\sum_2^d \lambda_i\right)^2 = \lambda_1^2$, and hence $(d-1)\operatorname{Tr} W^2 \geqq d\lambda_1^2$. In addition to the condition $\operatorname{div} v = 0$, $\mathscr{H}$ is supposed to be restricted to the space of divergenceless functions. This restriction might improve (1) but, as in [1], it will not be used here.

The domain $\Omega \in \mathbb{R}^d$ is assumed to be an open set of finite volume $|\Omega|$; boundedness is not required. Condition (4) insures that the quadratic form on $H_0^1(\Omega)$, defined by

$$Q(\phi, \phi) = \nu \int |\nabla\phi|^2 - \int w\phi^2 , \tag{5}$$

is bounded below and thus defines H as a self-adjoint operator. (Integrals, here and henceforth, are over Ω.) For our purposes, self-adjointness is not important; the only important consideration is the max – min principle which can be used as a definition of the e_i:

$$E_n = \inf_{\phi} \sum_{i=1}^{n} Q(\phi_i, \phi_i) , \tag{6}$$

where $\phi = \{\phi_1, \ldots, \phi_n\}$ is any L^2 orthonormal set in $H_0^1(\Omega)$. It is, in fact, the right side of (6) that enters in the derivation of the bound (1).

The goal is to find upper bounds on the following two quantities

(a) $$E(\gamma)\equiv \sum_{e_i\leqq 0} |e_i|^\gamma, \qquad \gamma\geqq 0, \tag{7}$$

(b) $$E_n \text{ itself for fixed } n. \tag{8}$$

It is a consequence of (1) that for $\gamma\geqq 1$

$$\sum_{\mu_i\geqq 0} \mu_i^\gamma \leqq d \sum_{\langle e_i\rangle\leqq 0} |\langle e_i\rangle|^\gamma \leqq d\langle E(\gamma)\rangle. \tag{9}$$

This is Karamata's theorem which, more generally, states that when $f:\mathbb{R}\to\mathbb{R}$ is convex and non-decreasing then (1) implies that

$$\sum_{i=1}^{nd} f(\mu_i)\leqq d\sum_{i=1}^{n} f(-\langle e_i\rangle)\leqq d\left\langle \sum_{i=1}^{n} f(-e_i)\right\rangle \tag{10}$$

for all n. If, in addition, $f(t)=0$ for $t\leqq 0$, then

$$\sum_{\mu_i\geqq 0} f(\mu_i)\leqq d\sum_{\langle e_i\rangle\leqq 0} f(-\langle e_i\rangle)\leqq d\left\langle \sum_{e_i\leqq 0} f(-e_i)\right\rangle. \tag{11}$$

[Actually, Karamata's inequality gives the left-hand inequalities in (9)–(11). The right-hand inequalities come from Jensen's inequality $\langle f(a)\rangle\geqq f(\langle a\rangle)$.]

It is (9) that gives information about the magnitude of the μ_i. The bound used in [1] [except for the factor $(d-1)/d$ in (3)] was

$$E(\gamma)\leqq L_{\gamma,d}\nu^{-d/2}\int w(x)^{\gamma+d/2}\,dx. \tag{12}$$

The present knowledge about (12) is the following:

(1) $L_{\gamma,d}<\infty$ for $\gamma>\frac{1}{2}(d=1)$, $\gamma>0\,(d=2)$, $\gamma\geqq 0\,(d\geqq 3)$. No such bound exists for $\gamma<\frac{1}{2}(d=1)$ or $\gamma=0\,(d=2)$. The case $\gamma=\frac{1}{2}$, $d=1$ does not seem to have been settled. (The claim in [2] that $L_{1/2,1}<\infty$ is not justified.) Bounds on $L_{1,3}$ were first given by Lieb and Thirring in [3] and on $L_{\gamma,d}$ for $\gamma>0$ $(d=2,3)$ and $\gamma>\frac{1}{2}$ $(d=1)$ in [2]. Bounds on $L_{0,d}$, $d\geqq 3$, were first given by Cwikel [4], Lieb [5, 6], and Rosenbljum [7]. The best upper bound for $L_{0,3}$ is in [6], namely $0.0780=4\pi^{-2}3^{-3/2}\leqq L_{0,3}\leqq 0.1156$. The lower bound is from [2, Eq. (4.24)]. Recently, by a simpler method, Li and Yau [8] derived upper bounds for $L_{0,d}$, $d\geqq 3$ which they claimed was better than that in [6]; unfortunately a numerical error was made in [8] and their bound for $L_{0,3}$ is three times larger than that in [6].

(2) The sharp constant $L_{\gamma,d}$ in (12) cannot depend on Ω, i.e. $L_{\gamma,d}(\Omega)=L_{\gamma,d}(\mathbb{R}^d)$. To see this, assume that $0\in\Omega$ and, given w on $\mathbb{R}^d$, consider $w_c(x)=c^2w(cx)$ on Ω. Then let $c\to\infty$. This situation is in contrast with the $|\Omega|$ dependent bound for E_n to be derived later.

(3) There is a natural "guess" for $L_{\gamma,d}$ given by the semiclassical formula

$$E(\gamma)\approx(2\pi)^{-d}\iint dp\,dx|\nu p^2-w(x)|_-^\gamma=L_{\gamma,d}^c\nu^{-d/2}\int w(x)^{\gamma+d/2}dx \tag{13}$$

with $|a|_-=\max(0,-a)$. An easy integration gives

$$L_{\gamma,d}^c=2^{-d}\pi^{-d/2}\Gamma(\gamma+1)/\Gamma(\gamma+1+d/2). \tag{14}$$

(4) It is a fact [2] that

$$L_{\gamma,d}\geqq L_{\gamma,d}^c. \tag{15}$$

Commun. Math. Phys. 92, 473–480 (1984)

In [2, 3] it was conjectured that $L_{1,d}=L^c_{1,d}$ for $d\geqq 3$. It is known [2] that for each $d\leqq 7$ there is a $\gamma_d>0$ such that $L_{\gamma,d}>L^c_{\gamma,d}$ when $\gamma<\gamma_d$. When $d=1$ or 2, $\gamma_d>1$. It is also known [9] that $L_{\gamma,1}=L^c_{\gamma,1}$ for $\gamma\geqq 3/2$. In fact [9] the ratio $R_d(\gamma)=L_{\gamma,d}/L^c_{\gamma,d}$ is monotone non-increasing in γ; thus if $R_d(\gamma_0)=1$ for some γ_0, then $R_d(\gamma)=1$ for all $\gamma\geqq\gamma_0$. Glaser et al. [10] have shown that $L_{0,d}>L^c_{0,d}$ for $d\geqq 7$. They also evaluate $L_{0,4}$ exactly (it is a Sobolev constant) *provided* w is restricted to be spherically symmetric. For related results see [11].

(5) Inequality (12) for $\gamma=1$ is equivalent [2] to

$$\sum_{i=1}^{n}\int|\nabla\phi_i(x)|^2\,dx\geqq K_d\int\varrho_\phi(x)^{1+2/d}\,dx\,, \tag{16}$$

where the $\{\phi_i\}$ is any L^2 orthonormal set in $H^1(\mathbb{R}^d)$ [or $H^1_0(\Omega)$] and

$$\varrho_\phi(x)=\sum_{i=1}^{n}|\phi_i(x)|^2\,. \tag{17}$$

The sharp constants in (12) and (16) are related by

$$L_{1,d}=[d/2K_d]^{d/2}(1+d/2)^{-1-d/2}\,. \tag{18}$$

[Note: If n is specified then the sharp constant in (16) may depend on n, i.e. $K_d(n)$. K_d, the sharp constant in (16), (18) is defined to be $\sup_n K_d(n)$.] Corresponding to $L^c_{1,d}$ in (14) there is a classical value K^c_d given by (18):

$$K^c_d=4\pi d\Gamma(1+d/2)^{2/d}/(2+d)\,. \tag{19}$$

By (15), $K_d\leqq K^c_d$.

An inequality related to (16), and which will be used later in the event that $K_d<K^c_d$, was proved by Li and Yau [8]. It depends on the volume $|\Omega|$ and it is the relation obtained by setting $K_d=K^c_d$ and $\varrho(x)=n/|\Omega|$ in (16). For any orthonormal set in $H^1_0(\Omega)$

$$\sum_{i=1}^{n}\int|\nabla\phi_i(x)|^2\,dx>K^c_d n^{1+2/d}|\Omega|^{-2/d}\,. \tag{20}$$

(The strict inequality in (20) is, in fact, implied by the proof in [8].)

Before turning to our estimate for E_n let us make a few additional remarks about (12).

(α) Combining (3), (9), (12) we see that the right side of (12) is suitable for passing to the "infinite volume" limit, i.e. in some vague sense it is proportional to the volume. The upper bound we shall obtain later for the quantity introduced in [1],

$$N(w)=\text{smallest } n \text{ such that } E_n>0\,, \tag{21}$$

will also have this extensivity property. By (1), $dN(w)$ is related to number of non-negative characteristic exponents and an upper bound on $N(w)$ will yield a bound on the number of non-negative characteristic exponents [see (43)].

(β) The bound on $N(w)$ in [1] relied on the fact that $L_{0,d}<\infty$ (which is true if and only if $d\geqq 3$) and on the conjecture that $L_{1,d}<L^c_{0,d}$. While $L^c_{0,d}/L^c_{1,d}=1+d/2>1$, the best bound published so far [6] for $L_{1,3}$ is

$$L_{1,3}\leqq(6.844)L^c_{1,3}=0.04624\,, \tag{22}$$

and this exceeds $L^c_{0,3}=0.01689$. However, the bound can be improved slightly to 0.04030 [see (51) below].

(γ) Inequality (12) can be used to derive a lower bound for each e_n. If $e_n(V)$ is the nth eigenvalue for the potential V in place of $-w$ in (2) then, for any number e, it is clear that $e_n(-w) \geqq e_n(-(w+e)_+)+e$. Take $\gamma=0$ in (12) and set $e=e_n$. Then the number of non-positive eigenvalues for $V=-(w+e_n)_+$ is at least n, and (12) yields

$$n \leqq L_{0,d} \nu^{-d/2} \int (w(x)+e_n)_+^{d/2}\, dx\,. \tag{23}$$

The integral on the right side of (23) is finite if $e_n<0$ or if $|\Omega|$ is finite. It is also monotone in e_n and thus (23) yields a lower bound for e_n.

Now we turn to our main goal which is an upper bound for E_n. Let $\phi_1, \ldots, \phi_n$ be the eigenfunctions corresponding to $e_1 \leqq e_2 \leqq \ldots \leqq e_n$. [By virtue of (6) and a limiting argument, any approximating orthonormal set such that $\sum_1^n Q(\phi_i, \phi_i) \leqq E_n+\varepsilon$ will suffice.] Let $\varrho_\phi(x)=\sum_1^n |\phi_i(x)|^2$.

By (6), (16) and with $p=1+2/d$,

$$E_n \geqq F(\varrho_\phi)\,, \tag{24}$$

with

$$F(\varrho) \equiv \nu K_d \|\varrho\|_p^p - \int w\varrho \tag{25}$$

which in turn is greater or equal to

$$G(\varrho) \equiv \nu K_d \|\varrho\|_p^p - \|w\|_{p'}\, \|\varrho\|_p\,. \tag{26}$$

Thus,

$$E_n \geqq \inf\{F(\varrho) \mid \textstyle\int \varrho = n, \varrho(x) \geqq 0\} \tag{27}$$

$$\geqq \inf\{G(\varrho) \mid \textstyle\int \varrho = n, \varrho(x) \geqq 0\}\,. \tag{28}$$

However, $\|\varrho\|_p |\Omega|^{1/p'} \geqq \int \varrho$, and therefore if we define the function J, (for $X>0$), and $\tilde{E}_n$ by

$$J(X)=\nu K_d X^p - \|w\|_{p'} X\,, \tag{29}$$

$$\tilde{E}_n = \inf\{J(X) \mid X \geqq n|\Omega|^{-1/p'}\}\,, \tag{30}$$

we have that

$$E_n > \tilde{E}_n\,. \tag{31}$$

The strict inequality in (31) is justified by the fact that ϱ_ϕ cannot satisfy the Hölder inequality after (28), i.e. ϱ_ϕ cannot be constant in Ω.

[It is left as an exercise, using the fact that $\|\varrho\|_p / \|\varrho\|_1$ can be made arbitrarily large, that $\tilde{E}_n$ is indeed the infimum in (27).]

The minimum in (30) can be computed to be

$$\begin{aligned} E_n > \tilde{E}_n &= J(n|\Omega|^{-1/p'}), \qquad && n \geqq |\Omega|^{1/p'} X_0\,, \\ &= J(X_0)\,, && n \leqq |\Omega|^{1/p'} X_0\,, \end{aligned} \tag{32}$$

where $J'(X_0)=0$, namely

$$p\nu K_d X_0^{p-1} = \|w\|_{p'}\,. \tag{33}$$

Commun. Math. Phys. *92*, 473–480 (1984)

In particular, if $n|\Omega|^{-1/p'}$ is greater than or equal to the value $X_1>0$ such that $J(X_1)=0$, then $E_n>0$. Therefore, $N(w)$ defined by (21) satisfies

$$N(w)\leqq \mathscr{S}|\Omega|^{1/p'}\{\|w\|_{p'}/\nu K_d\}^{1/(p-1)}$$
$$=\mathscr{S}|\Omega|\,\{\textstyle\int dx\, w(x)^{1+d/2}/|\Omega|\}^{d/(2+d)}(\nu K_d)^{-d/2}\,. \tag{34}$$

The symbol $\mathscr{S}$ denotes "the smallest integer $\geqq$."

If $K_d<K_d^c$, the constant in (34), namely $K_d^{-d/2}$, can be improved as follows. Let $0<\delta<1$ and split the kinetic energy term in two parts using (16) and (20). Thus,

$$E_n\geqq F_\delta(\varrho_\phi)\,, \tag{35}$$

with

$$F_\delta(\varrho)=(1-\delta)\nu K_d^c n^p|\Omega|^{1-p}+\delta\nu K_d\|\varrho\|_p^p-\textstyle\int w\varrho\,. \tag{36}$$

As before,

$$E_n>\tilde{E}_n(\delta)=(1-\delta)\nu K_d^c n^p|\Omega|^{1-p}+\inf\{\delta\nu K_d X^p-\|w\|_{p'}X\,|\,X\geqq n|\Omega|^{-1/p'}\}\,. \tag{37}$$

Previously, in (32), we discussed the inf in (37). Thus $\tilde{E}_n(\delta)\geqq 0$ if n satisfies the following two conditions:

$$n\geqq|\Omega|^{1/p'}X_0\delta^{-1/(p-1)}\,,\qquad [\text{see } (32),(33)]\,, \tag{38}$$

$$n^{p-1}\nu|\Omega|^{1-p}\{(1-\delta)K_d^c+\delta K_d\}\geqq\|w\|_{p'}|\Omega|^{-1/p'}\,. \tag{39}$$

Condition (39) implies that $\tilde{E}_n(\delta)\geqq 0$, provided (38) is satisfied. Choose δ so that (38) and (39) are the same, namely

$$\delta=K_d^c[2K_d/d+K_d^c]^{-1}\,.$$

Inserting this in (37), we have as before

$$N(w)\leqq\mathscr{S}A_d|\Omega|\nu^{-d/2}\{\textstyle\int dx\, w(x)^{1+d/2}/|\Omega|\}^{d/(d+2)}\,, \tag{40}$$

$$(A_d)^{2/d}=[2K_d+dK_d^c]\,[(d+2)K_dK_d^c]^{-1}\,. \tag{41}$$

The inequality (40), (41) is our main result.

We now wish to relate (40), (41) to the turbulence problem, i.e. we want to find an upper bound to

$$\tilde{N}(\mu)\equiv\text{smallest } n \text{ such that } \sum_{i=1}^{n}\mu_i<0\,. \tag{42}$$

By (1),

$$\tilde{N}(\mu)\leqq d\{\text{smallest integer such that } \langle E_n\rangle>0\}$$
$$\leqq d\{\text{smallest integer such that } \langle\hat{E}_n\rangle\geqq 0\}\,,$$

where, for each w,

$$\hat{E}_n\equiv\sup\{\tilde{E}_n(\delta)\,|\,0\leqq\delta\leqq 1\}\,.$$

For each fixed n and δ, $\tilde{E}_n(\delta)$, and $\tilde{E}_n$ are functions of $t\equiv\|w\|_{p'}^{p'}$. Denote them by $\tilde{E}_n(\delta,t)$ and $\tilde{E}_n(t)$. Direct calculation using (32), shows that $\tilde{E}_n(t)$ is a convex function of t (*not* $t^{1/p'}=\|w\|_{p'}$). Since $\tilde{E}_n(\delta,t)$ differs from $\tilde{E}_n(t)$ in a trivial way, $\tilde{E}_n(\delta,t)$ is also a convex function of t. Since $\hat{E}_n(t)$ is the supremum of convex functions, is too is convex in t. By Jensen's inequality $\langle\hat{E}_n\rangle\geqq\hat{E}_n(\langle t\rangle)$.

Thus, by expressing the right side of (40) in terms of $\|w\|_{p'}^{p'}$ and then averaging with respect to $\varrho(dv)$ we obtain the bound sought in [1]:

$$\tilde{N}(\mu) \leqq \mathscr{S} d A_d |\Omega| v^{-d/2} \{ \iint \varrho(dv)\, dx\, w_v(x)^{1+d/2} / |\Omega| \}^{d/(d+2)} . \tag{43}$$

Finally, let us record some available information about the constants in (41). Using (19) we have

$$\begin{aligned} K_1^c &= \pi^2/3 = 3.290 , \\ K_2^c &= 2\pi = 6.283 , \\ K_3^c &= 3(6\pi^2)^{2/3}/5 = 9.116 . \end{aligned} \tag{44}$$

To bound K_d a bound on $L_{1,d}$ is needed.

$d=1$: The bound in [2, Eq. (2.11)] with $m=1$, $n=1$ is

$$L_{1,1} \leqq (4\pi)^{-1/2} \Gamma(5/2)^{-1} \Gamma(1/2)^2 (1/2)^{-1} = 4/3 . \tag{45}$$

$d=2,3$: In this case we use the formula [6]

$$\sum_{e_i \leqq 0} |e_i|^\gamma = \gamma \int_{-\infty}^{0} |e|^{\gamma-1} N_e de , \tag{46}$$

where N_e is the number of eigenvalues of $H \leqq e$. In [6] it is shown (with $v=1$) that

$$N_e \leqq (4\pi)^{-d/2} \int dx \int_0^\infty dt\, t^{-1-d/2} e^{et} f(tw(x)) , \tag{47}$$

with $f(t) = \max\{0, b(t-a)\}$ and

$$1/b = \int_a^\infty (1 - a/y) e^{-y} dy . \tag{48}$$

Inserting (47) in (46), then doing the e integration, then the t integration [after a change of variable to $tw(x)$] and finally the x integration, one finds

$$L_{\gamma,d} \leqq b (4\pi)^{-d/2} \Gamma(\gamma+1) (d/2 - 1 + \gamma)^{-1} (d/2 + \gamma)^{-1} a^{1-d/2-\gamma} . \tag{49}$$

The optimum constant a satisfies

$$a e^a \int_1^\infty e^{-ay} dy/y = (d/2 + \gamma - 1)/(d/2 + \gamma) . \tag{50}$$

When $\gamma = 1$ we take $a=0.61$, $b=3.6807$ for $d=2$ and $a=1.02$, $b=6.9358$ for $d=3$. Inserting this in (49) yields

$$\begin{aligned} L_{1,2} &\leqq 0.24008 , \\ L_{1,3} &\leqq 0.040304 . \end{aligned} \tag{51}$$

Using (18)

$$\begin{aligned} K_1 &\geqq 1/12 = 0.08333 , \\ K_2 &\geqq 1.0413 , \\ K_3 &\geqq 2.7709 , \end{aligned} \tag{52}$$

Commun. Math. Phys. 92, 473–480 (1984)

which, by (41), leads to

$$A_1 = 2.050\,,$$
$$A_2 = 0.5597\,, \tag{53}$$
$$A_3 = 0.1329\,.$$

This value for A_3 can be compared with the value in [1, Footnote 7], which is obtained under the conjectured assumptions $L_{0,3}=0.0780$ and $L_{1,3}=L_{1,3}^c$, namely

$$L_{0,3}[1-(L_{1,3}^c/L_{0,3}^c)^{2/5}]^{-3/2}=0.459\,. \tag{54}$$

If $K_3=K_3^c$, which is conjectured to be true, (41) yields

$$A_3=(K_3^c)^{-3/2}=0.03633\,. \tag{55}$$

In addition to the improvement in (53) over (54), we also note the additional factor $(d-1)/d$ in (3) which yields a factor $[(d-1)/d]^{d/4}$ when the right sides of (40), (43) are expressed in terms of $\varepsilon(x)$. This factor is 0.7378 for $d=3$ and 0.7071 for $d=2$.

Acknowledgement. I should like to thank David Ruelle for stimulating this work and for several helpful conversations.

References

1. Ruelle, D.: Large volume limit of the distribution of characteristic exponents in turbulence. Commun. Math. Phys. **87**, 287–302 (1982)
2. Lieb, E., Thirring, W.: Inequalities for the moments of the eigenvalues of the Schrödinger Hamiltonian and their relation to Sobolev inequalities. In: Studies in mathematical physics: essays in honor of Valentine Bargmann, Lieb, E., Simon, B., Wightman, A. (eds.), pp. 269–303. Princeton, NJ: Princeton University Press 1976
3. Lieb, E., Thirring, W.: Bound for the kinetic energy of fermions which proves the stability of matter. Phys. Rev. Lett. **35**, 687–689 (1975); **35**, 1116 (1975) (Erratum)
4. Cwikel, M.: Weak type estimates for singular values and the number of bound states of Schrödinger operators. Ann. Math. **106**, 93–100 (1977)
5. Lieb, E.: Bounds on the eigenvalues of the Laplace and Schrödinger operators. Bull. Am. Math. Soc. **82**, 751–753 (1976); the details appear in [6]
6. Lieb, E.: The number of bound states of one-body Schrödinger operators and the Weyl problem. Proc. Am. Math. Soc. Symp. in Pure Math., Osserman, R., Weinstein, A. (eds.), Vol. 36, pp. 241–252 (1980). Much of this material is reviewed in Simon, B.: Functional integration and quantum physics, pp. 88–100. New York: Academic Press 1979
7. Rosenbljum, G.: Distribution of the discrete spectrum of singular differential operators. Dokl. Akad. Nauk SSSR **202**, 1012–1015 (1972) (MR45 No. 4216). The details are given in: Distribution of the discrete spectrum of singular differential operators. Izv. Vyss. Ucebn. Zaved. Matem. **164**, 75–86 (1976) [English transl. Sov. Math. (Iz. VUZ) **20**, 63–71 (1976)]
8. Li, P., Yau, S.-T.: On the Schrödinger equation and the eigenvalue problem. Commun. Math. Phys. **88**, 309–318 (1983)
9. Aizenman, M., Lieb, E.: On semiclassical bounds for eigenvalues of Schrödinger operators. Phys. Lett. **66** A, 427–429 (1978)
10. Glaser, V., Grosse, H., Martin, A.: Bounds on the number of eigenvalues of the Schrödinger operator. Commun. Math. Phys. **59**, 197–212 (1978)
11. Grosse, H.: Quasiclassical estimates on moments of the energy levels. Acta Phys. Austr. **52**, 89–105 (1980)

Communicated by A. Jaffe

Received October 21, 1983

Phys. Rev. Lett. *54*, 1987–1990 (1985)

Baryon Mass Inequalities in Quark Models

Elliott H. Lieb
Departments of Mathematics and Physics, Princeton University, Princeton, New Jersey 08544
(Received 28 February 1985)

Recently conjectured three- (and more-) body mass inequalities are investigated for the quark models of baryons in which it is assumed that baryon masses are the ground-state energies of Schrödinger-type operators with pair potentials V. It is proved that these inequalities hold (even with a "relativistic" form for the kinetic energy) if V belongs to a certain class (which includes many potentials commonly used), but that they do not hold for all V (even in the nonrelativistic case). One example of our results is $2M(cqs) \geqslant M(cqq) + M(css)$.

PACS numbers: 12.70.+q, 12.35.Eq, 12.35.Ht, 12.40.Qq

In the nonrelativistic quark model of mesons and baryons, the masses of these composite particles are estimated by the ground-state energies of simple two- (for mesons) or three- (for baryons) body Schrödinger operators with ordinary pair potentials. A question of some recent interest is whether one can derive inequalities among these masses (in this model) that hold for all (or a large class of) pair potentials. Such "potential-independent" results about the masses of mesons and baryons are obviously conceptually interesting.

The two- and three-body Hamiltonians have the form

$$H^{(2)} = T_1(\mathbf{x}_1) + T_2(\mathbf{x}_2) + V(\mathbf{x}_1, \mathbf{x}_2), \tag{1}$$

$$H^{(3)} = T_1(\mathbf{x}_1) + T_2(\mathbf{x}_2) + T_3(\mathbf{x}_3) + V_{12}(\mathbf{x}_1, \mathbf{x}_2) + V_{23}(\mathbf{x}_2, \mathbf{x}_3) + V_{13}(\mathbf{x}_1, \mathbf{x}_3). \tag{2}$$

Here, $\mathbf{x}_1$, $\mathbf{x}_2$, $\mathbf{x}_3$ are the particle (quark) coordinates and the quantities in parentheses [e.g., $(\mathbf{x}_1, \mathbf{x}_2)$] denote the particle coordinates on which the operators act. (The dimensionality of space is unimportant in most of the following.) The subscripts (e.g., 1 or 12) are simply labels to designate possibly different operators (e.g., T_1 and T_2 might be kinetic energy operators with respective masses m_1 and m_2, but no special form of T_j other than that it is a one-body operator is assumed at this point). The potentials, V, are usual multiplication operators; initially, neither translation invariance [$V(\mathbf{x}, \mathbf{y}) = V(\mathbf{x} - \mathbf{y})$] nor symmetry [$V(\mathbf{x}, \mathbf{y}) = V(\mathbf{y}, \mathbf{x})$] is assumed. The ground-state energy of a Hamiltonian is

$$E = \inf\langle \psi, H\psi \rangle, \quad \text{with } \langle \psi | \psi \rangle = 1. \tag{3}$$

The absolute ground state is implied, i.e., no symmetry restriction is imposed on ψ. The reason is that there are enough internal quantum members (color and flavor) associated with the particles so that the Pauli principle can be satisfied by "internal" antisymmetry.

The known two-body inequality[1-3] concerns the following situation: Fix $V(\mathbf{x}_1, \mathbf{x}_2) = V(\mathbf{x}_1 - \mathbf{x}_2)$ in (1) and let T_1 and T_2 be even functions of the momentum only (i.e., they are translation and inversion invariant). Consider three different systems in which the first two terms in (1) are respectively (a) T_1, T_1; (b) T_1, T_2; (c) T_2, T_2. The desired inequality is

$$E(a) + E(c) \leqslant 2E(b). \tag{4}$$

This can be proved by noting that $E(a)$ is also the ground-state energy of $\hat{H}_a = 2T_1(1) + V(\mathbf{x}_1 - \mathbf{x}_2)$ and similarly for $E(c)$. {Proof: Since $H_a = \frac{1}{2}\hat{H}_a + \frac{1}{2}\tilde{H}_a$ with $\tilde{H}_a = 2T_1(2) + V(\mathbf{x}_1 - \mathbf{x}_2)$ and since $\tilde{H}_a$ is, by the evenness of T_1, unitarily equivalent to $\hat{H}_a$, we have $E(a) \geqslant \hat{E}(a)$. Conversely, $\hat{E}(a) = \inf\langle \phi | h | \phi \rangle$, $\langle \phi | \phi \rangle = 1$, with ϕ being a one-particle function and $h = 2T(\mathbf{x}) + V(\mathbf{x})$. As a variational function in (3) for $E(a)$ take

$$\psi(\mathbf{x}_1, \mathbf{x}_2) = C\phi(\mathbf{x}_1 - \mathbf{x}_2)\exp[-\epsilon(\mathbf{x}_1 + \mathbf{x}_2)^2]$$

and let $\epsilon \to 0$. This yields $E(a) \leqslant \hat{E}(a)$.} The operator equality $2H(b) = \hat{H}(a) + \hat{H}(b)$ implies (4). Inequalities similar to (4) have been proved in a field-theoretic context by Weingarten[4] and Witten.[5] Note that (4) need not hold if the T_i are arbitrary single-particle operators [e.g., $T = p^2 + U(x)$]. Counterexample I, at the end of this paper, shows this.

An inequality relating $H^{(3)}$ to $H^{(2)}$ has also been derived.[6-8] Let T_1, T_2, and T_3 be arbitrary single-particle operators. In (2) let V_{12}, V_{23}, and V_{13} be arbitrary two-body operators (not necessarily multiplication operators). Consider three two-body problems in (1): (a) $H_a = T_1 + T_2 + 2V_{12}$; (b) $H_b = T_2 + T_3 + 2V_{23}$; (c) $H_c = T_1 + T_3 + 2V_{13}$. Then $2H^{(3)} = H_a + H_b + H_c$, which implies that

$$2E^{(3)} \geqslant E(a) + E(b) + E(c). \tag{5}$$

If the V's and T's are identical then $2E^{(3)} \geqslant 3E^{(2)}$ or, in terms of baryon and meson masses, $m_B \geqslant 3m_M/2$.

The purpose of this paper is to investigate n-body inequalities with $n \geq 3$. I am grateful to P. Taxil and J. M. Richard for communicating the following problem to me and for their very helpful correspondence on the subject.

In (2), let $V_{12}=V_{13}=V_{23}=V$ with $V(\mathbf{x},\mathbf{y})=V(\mathbf{x}-\mathbf{y})=V(\mathbf{y}-\mathbf{x})$. Let T_i be nonrelativistic kinetic energy operators: $T_i=p^2/2m_i$. Denote the energy by $E(m_1,m_2,m_3)$. Given m and M, is it true that

$$E(m,m,m)+E(m,M,M)\leq 2E(m,m,M)? \quad (6)$$

The physically interesting case is $m<M$, where m is the mass of a u or d quark and M is the mass of a strange quark, in which case (6) is related to the Gell-Mann[9]–Okubo[10] mass formula.

Unfortunately (6) is not true for all m, M, and V, as counterexample II (with $m>>M$) given at the end of this paper shows. Although it was thought[3] that a straightforward "convexity" argument similar to the proof of (4) and (5) would yield (6), a recent critique[11] (and our counterexample) dispels this idea. For $m<<M$, I do not have a counterexample, and the status of (6) could conceivably be different in the two cases $m<M$ and $m>M$.

It will be shown here that for suitable V, (6) is indeed true for all m and M. This class of V is large enough to include many of the potentials actually used in these quark-model calculations. As partial compensation for the restriction on V, a larger class of one-body operators, T, will be allowed [in particular, the "relativistic" expression $T(p)=(p^2c^2+m^2c^4)^{1/2}$]. Furthermore, our result extends to $n>$ three bodies.

First, let us define the one-body operators, T, to be considered. It will always be assumed that the kernel $K_\beta(\mathbf{x},\mathbf{y})\equiv(e^{-\beta T})(\mathbf{x},\mathbf{y})$ is real for all $\mathbf{x},\mathbf{y}$ and all $\beta>0$. This condition is automatically satisified when $T=T(p)+U(x)$ with U and T real and $T(p)=T(-p)$. We also define a special subclass A by saying that $T\in A$ if $K_\beta(\mathbf{x},\mathbf{y})\geq 0$ for all $\mathbf{x}$, $\mathbf{y}$ and all $\beta>0$. Examples of such T's in A are[12]

$$T=p^2/2m+U(x), \quad (7)$$

$$T=(p^2c^2+m^2c^4)^{1/2}+U(x), \quad (8)$$

for any real $U(x)$. [Remark: Hidden in this, and the following, is the tacit assumption that various operators such as (1), (2), (7), and (8) are bounded below and self-adjoint. This restricts the singularities of U and V in well-known ways.]

Next we define a class, B, of two-body potentials, V. We say that $V\in B$ if $V(\mathbf{x},\mathbf{y})=V(\mathbf{y},\mathbf{x})$ and the kernel $L_\beta(\mathbf{x},\mathbf{y})=\exp[-\beta V(\mathbf{x},\mathbf{y})]$ is positive semidefinite for all $\beta>0$. [This means that $\int f(\mathbf{x})f(\mathbf{y})L_\beta(\mathbf{x},\mathbf{y})d\mathbf{x}\,d\mathbf{y}\geq 0$ for all f for which the integral is absolutely convergent.] B is a cone, namely if $V_1\in B$ and $V_2\in B$ then $aV_1+bV_2\in B$ for all a and $b\geq 0$. A subset of B that is physically important is $\hat{B}\subset B$ consisting of translation-invariant potentials $V(\mathbf{x}-\mathbf{y})$. In this case, the ordinary functions $F_\beta(\mathbf{x})=\exp[-\beta V(\mathbf{x})]$ are known as infinitely divisible distributions. $V\in\hat{B}$ if and only if the Fourier transform (FT) of F_β is positive (as a distribution) for all $\beta>0$. The Levy-Khintchine formula[12] provides a necessary and sufficient condition for $V\in\hat{B}$, but it is not particularly transparent.

The physically interesting case is $V(\mathbf{x})=V(r)$, with $r=|\mathbf{x}|$. Call this class $\tilde{B}\subset\hat{B}$. I shall give two different sufficient conditions for $V(r)\in\tilde{B}$. The first is dimension dependent and is due to Askey[13] who proves that $f(r)$ has a positive FT in d dimensions if, for all $r>0$,

$$(-1)^j f^{(j)}(r)\geq 0, \quad 0\leq j\leq 2+d/2, \quad (9)$$

where $f^{(j)}\equiv d^jf/dr^j$, $j=0,1,2,\ldots$. (When $d=1$, the result goes back to W. H. Young and to Polya.) If $V(r)$ satisfies

$$(-1)^j V^{(j)}(r)\leq 0, \quad 1\leq j\leq 2+d/2, \quad (10)$$

then F_β will satisfy (9). Thus, in three dimensions, it suffices that $V(r)$ satisfies $\dot{V}\geq 0$, $\ddot{V}\leq 0$, and $\dddot{V}\geq 0$. In many derivations of V from lattice gauge field theory, one automatically gets $\dot{V}\geq 0$ and $\ddot{V}\leq 0$ for the $q\bar{q}$ potential by reflection positivity. (10) requires only the additional condition $\dddot{V}\geq 0$.

Another sufficient condition for $V\in\tilde{B}$ is this: Write $V(r)=W(r^2)$ and demand that for all $s\geq 0$ and $j\geq 1$,

$$(-1)^j W^{(j)}(s)\geq 0. \quad (11)$$

If (11) holds then $g(s)\equiv e^{-\beta W(s)}$ satisfies $(-1)^j\times g^{(j)}(s)\geq 0$ and, by Bernstein's theorem, is the Laplace transform of a positive measure ν_β on $[0,\infty)$, whence

$$\exp[-\beta V(\mathbf{x})]=\int_0^\infty \exp(-tx^2)\,d\nu_\beta(t),$$

which is obviously positive definite. Some potentials in $\tilde{B}$ of this form are (with a and $b\geq 0$)

$$V(r)=(ar^2+b)^q, \quad 0\leq q\leq 1$$
$$=-(ar^2+b)^q, \quad q\leq 0. \quad (12)$$

In particular, $V(r)=ar^p-br^{-q}$, $a,b,q,p\geq 0$, $p\leq 2$, which is a choice frequently employed,[2,14] is allowed. Again, note that $\tilde{B}$ is a cone which implies that $V(r)\in\tilde{B}$ if it can be decomposed into $V=V_1+V_2$ with V_1 satisfying (10) and V_2 satisfying (11).

Now our theorem can be stated: Let T_1, T_2, T_3 be three one-body operators, with $T_1\in A$. Let V_1, V_2, and V_3 be two-body potentials with $V_1\in B$. (T_2 and T_3 do not have to be in A and V_2 and V_3 do not have to be in B.) Consider the three three-body Hamiltoni-

VOLUME 54, NUMBER 18 PHYSICAL REVIEW LETTERS 6 MAY 1985

ans

$$H_a = T_1(\mathbf{x}_1) + T_2(\mathbf{x}_2) + T_2(\mathbf{x}_3) + V^{22},$$

$$H_b = T_1(\mathbf{x}_1) + T_2(\mathbf{x}_2) + T_3(\mathbf{x}_3) + V^{23}, \tag{13}$$

$$H_c = T_1(\mathbf{x}_1) + T_3(\mathbf{x}_2) + T_3(\mathbf{x}_3) + V^{33},$$

$$V^{ij} = V_i(\mathbf{x}_1, \mathbf{x}_3) + V_j(\mathbf{x}_1, \mathbf{x}_2) + V_1(\mathbf{x}_2, \mathbf{x}_3). \tag{14}$$

Then

$$E(a) + E(c) \leq 2E(b). \tag{15}$$

Note that T_1, T_2, and T_3 are unrelated, as are V_1, V_2, V_3. Thus, the theorem covers the case of three different quarks, e.g., mass(cqq) + mass(css) $\leq 2 \times$ mass(cqs) if one assumes $V_{qq} = V_{ss} = V_{qs} \in B$.

Proof.—Choose some $X \equiv (\mathbf{x}_1, \mathbf{x}_2, \mathbf{x}_3)$ and let $Z_\beta = (e^{-\beta H})(X,X)$. As $\beta \to \infty$, $\beta^{-1} \ln Z_\beta \to -E$, so that it suffices to prove $Z_\beta(a) Z_\beta(c) \geq Z_\beta(b)^2$ for all $\beta > 0$. The Trotter product formula asserts that $Z_\beta = \lim_{N\to\infty} Z_\beta^{(N)}$ with

$$Z_\beta^{(N)} = [(e^{-\beta T/N} e^{-\beta V/N})^N](X,X),$$

and it suffices to prove the inequality for each N. $Z_\beta^{(N)}$ gives us a polygonal path approximation. Let X_1 be a path of x_1 [i.e., $N+1$ points $\mathbf{x}_1(0) = \mathbf{x}_1, \mathbf{x}_1(\beta/N), \ldots, \mathbf{x}_1(N\beta/N) = \mathbf{x}_1$] and similarly for X_2, X_3. The contribution of the single-particle terms to a path in $Z_\beta^{(N)}(a)$ has the form $F_a = F_1(X_1) F_2(X_2) F_2(X_3)$. For H_b and H_c, the terms are

$$F_b = F_1(X_1) F_2(X_2) F_3(X_3),$$

$$F_c = F_1(X_1) F_3(X_2) F_3(X_3).$$

By assumption $F_1(X) \geq 0$ and F_2 and F_3 are real. The contribution of V to a path is of the form $G_2(X_1, X_3) G_2(X_1, X_3) G_1(X_2, X_3)$ for H_a, $G_2 G_3 G_1$ for H_b, and $G_3 G_3 G_1$ for H_c. By assumption, G_2 and G_3 are real functions and G_1 is positive definite {reason: G_1 is of the form

$$\prod_{j=1}^{N} \exp[-(\beta/N) V_1(\mathbf{x}_2(\beta j/N), \mathbf{x}_3(\beta j/N))],$$

and the N-fold tensor product of positive semidefinite operators is positive semidefinite}. Thus $Z_\beta^{(N)}(a)$ has the form

$$\int d^{3(N-1)}X_1\, F_1(X_1) \{\int d^{3(N-1)}X_2\, d^{3(N-1)}X_3\, G_2(X_1, X_2) F_2(X_2) G_2(X_1, X_3) F_2(X_3) G_1(X_2, X_3)\}.$$

A similar form holds for $Z_\beta^{(N)}(b)$ and $Z_\beta^{(N)}(c)$. For any real number λ, the quantity

$$\Delta(\lambda) = Z_\beta^{(N)}(a) + \lambda^2 Z_\beta^{(N)}(c) - 2\lambda Z_\beta^{(N)}(b)$$

can be written

$$\Delta(\lambda) = \int d^{3(N-1)}X_1\, F_1(X_1) \{\int d^{3(N-1)}X_2\, d^{3(N-1)}X_3\, J_\lambda(X_1, X_3) J_\lambda(X_1, X_2) G_1(X_2, X_3)\}$$

with $J_\lambda(X,Y) = G_2(X,Y) F_2(Y) - \lambda G_3(X,Y) F_3(Y)$. Since G_1 is positive semidefinite, the inner integral (in curly braces) is nonnegative. Since $F_1 \geq 0$, $\Delta(\lambda) \geq 0$. Minimizing $\Delta(\lambda)$ with respect to λ yields $Z_\beta^{(N)}(a) Z_\beta^{(N)}(c) \geq [Z_\beta^{(N)}(b)]^2$, Q.E.D.

We remark that the theorem can obviously be extended easily to appropriate two-body operators V (instead of merely potentials), but we shall not indicate this explicitly.

Another extension is to replace V in (14) by a genuine three-body potential $V(\mathbf{x}_1, \mathbf{x}_2, \mathbf{x}_3)$ (the same V for a,b,c) with the property that for every fixed $\mathbf{x}_1$ and $\beta > 0$, $e^{-\beta V}$ is positive semidefinite as a function of $\mathbf{x}_2$ and $\mathbf{x}_3$.

The theorem can also be extended to $n > 3$ bodies as follows. Let $T_1, \ldots, T_n$ be one-body operators with $T_3, \ldots, T_n \in A$. Let W be an arbitrary $(n-2)$-body potential, U an arbitrary $(n-1)$-body potential, and V a two-body potential in B. Let

$$H_{ij} = T_i(\mathbf{x}_1) + T_j(\mathbf{x}_2) + \sum_{k=3}^{n} T_k(\mathbf{x}_k) + W(\mathbf{x}_3, \ldots, \mathbf{x}_n)$$
$$+ U(\mathbf{x}_1, \mathbf{x}_3, \ldots, \mathbf{x}_n) + U(\mathbf{x}_2, \mathbf{x}_3, \ldots, \mathbf{x}_n) + V(\mathbf{x}_1, \mathbf{x}_2). \tag{16}$$

Then $E(H_{11}) + E(H_{22}) \leq 2E(H_{12})$.

Again, obvious generalizations suggest themselves for $n > 3$ as they did for $n = 3$.

Counterexample I.—Inequality (4) is false for arbitrary T_1, T_2. Take $T_1 = p^2 + U(x)$, $T_2 = p^2$, and $V(\mathbf{x}, \mathbf{y}) = V(\mathbf{x} - \mathbf{y})$. Let U be a deep, narrow square well and let V be smooth, but with $V(0) \equiv Q$ a local maximum. In (4) we have (essentially) that $E(a) = e + e + Q$, where e is the ground-state energy of T_1. Similarly, $E(b) = e + E_1$ and $E(c) = E_2$, where E_n is the ground-state energy of $h_n = np^2 + V(\mathbf{x})$. For (4) to hold would require $2E_1 \geq E_2 + Q$, but this is false for Q large enough.

Counterexample II.—Inequality (6) can be false. We shall show that with $m = \infty$ and V an infinite square well [$V(\mathbf{x}) = 0$ for $|\mathbf{x}| \leq 1$, and $V(\mathbf{x}) = \infty$ otherwise] the inequality in (6) is reversed and strict (i.e., $\leq$ is

Phys. Rev. Lett. *54*, 1987–1990 (1985)

replaced by >). By continuity, (6) will continue to fail for m finite, but large, and V bounded and smooth. When $m=\infty$, one simply fixes the coordinates of those particles with mass m, then does the minimization in (3) for the other variables and, finally, minimizes the energy with respect to the assumed fixed coordinates (i.e., the Born-Oppenheimer approximation is exact). For (m,m,m) we obviously fix $\mathbf{x}_1=\mathbf{x}_2=\mathbf{x}_3=0$ (since one cannot do better than $V_{12}=V_{13}=V_{23}=0$), so that $E(m,m,m)=0$. For (m,m,M) we note that if V_{12} is ignored, we should clearly take $\mathbf{x}_1=\mathbf{x}_2$. [This follows from concavity or by noting that given any $\psi(\mathbf{x}_3)$ one should place $\mathbf{x}_1$ and $\mathbf{x}_2$, which are independent, at the point x that minimizes the effective potential $|\psi|^2 * V$.] But $\mathbf{x}_1=\mathbf{x}_2$ also minimizes V_{12}; hence $\mathbf{x}_1=\mathbf{x}_2$ is the best choice. Thus $E(m,m,M)=E(h)$ with $h=p^2/2M+2V(\mathbf{x})$. Likewise $E(m,M,M)=E(\hat{h})$ with

$$\hat{h}=p_2^2/2M+p_3^2/2M+V(\mathbf{x}_2)+V(\mathbf{x}_3)+V(\mathbf{x}_2-\mathbf{x}_3).$$

Since $V_{23}\geq 0$, $E(\hat{h})>2E(\tilde{h})>0$ with $\tilde{h}=p^2/2M+V(\mathbf{x})$. (It is easy to see that the inequality is strict.) Since $2V=V$, $E(h)=E(\tilde{h})$ and thus (6) is reversed.

Helpful conversations with E. Witten and R. Askey are also gratefully acknowledged, as is the partial support of the U. S. National Science Foundation (PHY-8116101-A03).

[1]R. Bertlmann and A. Martin, Nucl. Phys. **B168**, 111 (1980).

[2]J. M. Richard and P. Taxil, Ann. Phys. (N.Y.) **150**, 267 (1983).

[3]S. Nussinov, Phys. Rev. Lett. **52**, 966 (1984).

[4]D. Weingarten, Phys. Rev Lett. **51**, 1830 (1983).

[5]E. Witten, Phys. Rev. Lett. **51**, 2351 (1983).

[6]J. P. Ader, J. M. Richard, and P. Taxil, Phys. Rev. D **25**, 2370 (1982).

[7]S. Nussinov, Phys. Rev. Lett. **51**, 2081 (1983).

[8]J. M. Richard, Phys. Lett. **139B**, 408 (1984).

[9]M. Gell-Mann, Phys. Rev. **125**, 1067 (1962).

[10]S. Okubo, Prog. Theor. Phys. **27**, 949 (1962).

[11]J. M. Richard and P. Taxil, Phys. Rev. Lett. **54**, 847 (1985).

[12]M. Reed and B. Simon, *Methods of Modern Mathematical Physics, Vol. 4* (Academic, New York, 1978).

[13]R. Askey, in *Harmonic Analysis on Homogeneous Spaces: Proceedings of the Symposia in Pure Mathematics, Vol. 26* (American Mathematics Society, Providence, 1973), pp. 335–338. In this paper Askey proves the sufficiency of (9) if a conjecture about Bessel functions holds. This he proves for d odd in R. Askey, Trans. Am. Math. Soc. **179**, 71 (1973). The even-d case was proved in J. Fields and M. Ismail, J. Math. Anal. **6**, 551 (1975).

[14]C. Quigg and J. Rosner, Phys. Rep. **56**, 167 (1979).

KINETIC ENERGY BOUNDS AND THEIR APPLICATION TO THE STABILITY OF MATTER

Elliott H. Lieb
Departments of Mathematics and Physics, Princeton University
P.O. Box 708, Princeton, NJ 08544

The Sobolev inequality on $\mathbf{R}^n, n \geq 3$ is very important because it gives a lower bound for the kinetic energy $\int |\nabla f|^2$ in terms of an L^p norm of f. It is the following.

$$\int_{\mathbf{R}^n} |\nabla f|^2 \geq S_n \left\{ \int_{\mathbf{R}^n} |f|^{2n/(n-2)} \right\}^{(n-2)/n} = S_n \|f\|^2_{2n/(n-2)}. \tag{1}$$

Applying Hölder's inequality to the right side we obtain the following modification of (1).

$$\int_{\mathbf{R}^n} |\nabla f|^2 \geq K_n^1 \left\{ \int_{\mathbf{R}^n} \rho^{(n+2)/n} \right\} / \left\{ \int_{\mathbf{R}^n} \rho \right\}^{2/n} = K_n^1 \|\rho\|^{(n+2)/n}_{(n+2)/n} \|\rho\|_1^{-2/n} \tag{2}$$

with $\rho(x) \equiv |f(x)|^2$. The superscript 1 on K_n^1 indicates that in (2) we are considering only one function, f. Hölder's inequality implies that $K_n^1 \geq S_n$ but, in fact, the ***sharp*** value of K_n^1 (which can be obtained by solving a nonlinear PDE) is larger than S_n. In particular, $K_n^1 > 0$ for *all* $n \geq 1$, even though $S_n = 0$ for $n < 3$.

Inequality (2), unlike (1) has the following important property: The non-linear term $\int \rho^{(n+2)/n}$ enters with the power 1 (and not $(n-2)/n$) and is therefore "extensive." The price we have to pay for this is the factor $\|f\|_2^{4/n} = \|\rho\|_1^{2/n}$ in the denominator, but since we shall apply (2) to cases in which $\|f\|_2 = 1$ (L^2 normalization condition) this is not serious.

Inequality (2) is equivalent to the following: Consider the Schrödinger operator on $\mathbf{R}^n$

$$H = -\Delta - V(x) \tag{3}$$

and let $e_1 = \inf \operatorname{spec}(H)$. (We assume H is self-adjoint.) Let $V_+(x) \equiv \max\{V(x), 0\}$. Then

$$e_1 \geq -L^1_{1,n} \int V_+(x)^{(n+2)/n} dx = -L^1_{1,n} \|V_+\|^{(n+2)/2}_{(n+2)/2} \tag{4}$$

with

$$L^1_{1,n} = \left(\frac{n}{2K_n^1} \right)^{n/2} \left(1 + \frac{n}{2} \right)^{-(n+2)/n}. \tag{5}$$

The reason for the subscript 1 in $L^1_{1,n}$ will be clarified in eq. (8).

Here is the proof of the equivalence. We have

$$e_1 \geq \inf_f \left\{ \int |\nabla f|^2 - \int \rho V_+ \mid \|f\|_2 = 1 \text{ and } \rho = |f|^2 \right\}.$$

Use (2) and Hölder to obtain (with $X = \|\rho\|_{(n+2)/n}$)

$$e_1 \geq \inf_X \left\{ K_n^1 X^{(n+2)/n} - \|V_+\|_{(n+2)/2} X \right\} \tag{6}$$

Minimizing (6) with respect to X yields (4). To go from (4) to (2), take $V = V_+ = \alpha |f|^{4/n} = \alpha \rho^{2/n}$ in (3). Then

$$-L_{1,n}^1 \alpha^{(n+2)/2} \int \rho^{(n+2)/n} \leq e_1 \leq (f, Hf) = \int |\nabla f|^2 - \alpha \int \rho^{(n+2)/n}.$$

Optimizing this with respect to α yields (2).

So far this is trivial, but now we turn to a more interesting question. Let $e_1 \leq e_2 \leq \ldots \leq 0$ be the negative spectrum of H (which may be empty). Is there a bound of the form

$$\sum e_i \geq -L_{1,n} \int V_+(x)^{(n+2)/2} dx \tag{7}$$

for some universal, V independent, constant $L_{1,n} > 0$ (which, of course, is $\geq L_{1,n}^1$)? The point is that the right side of (7) has the same form as the right side of (4). More generally, given $\gamma \geq 0$, does

$$\sum |e_i|^\gamma \leq L_{\gamma,n} \int V_+(x)^{\gamma + \frac{n}{2}} \tag{8}$$

hold for suitable $L_{\gamma,n}$? When $\gamma = 0, \sum |e_i|^0$ is interpreted as the number of $e_i \leq 0$.

The answer to these questions is yes in the following cases:

$\underline{n=1}$: All $\gamma > \frac{1}{2}$. The case $\gamma = 1/2$ is unsettled. For $\gamma < \frac{1}{2}$, examples show there can be no bound of the form (8).

$\underline{n=2}$: All $\gamma > 0$. There can be no bound when $\gamma = 0$.

$\underline{n \geq 3}$: All $\gamma \geq 0$.

The cases $\gamma > 0$ were first done in [10], [11]. The $\gamma = 0$ case for $n \geq 3$ was done in [3], [6], [14], with [6] giving the best estimate for $L_{0,n}$. For a review of what is currently known about these constants and conjectures about the sharp values of $L_{\gamma,n}$, see [8]. The proof of (8) is involved (especially when $\gamma = 0$) and will not be given here. It uses the Birman-Schwinger kernel, $V_+^{1/2}(-\Delta + \lambda)^{-1} V_+^{1/2}$.

There is a natural "guess" for $L_{\gamma,n}$ in terms of a semiclassical approximation (and which is not unrelated to the theory of pseudodifferential operators):

$$\sum |e_i|^\gamma \approx (2\pi)^{-n} \int_{\mathbf{R}^n \times \mathbf{R}^n, p^2 \leq V(x)} [V(x) - p^2]^\gamma dp dx \tag{9}$$

$$= L^c_{\gamma,n} \int V_+(x)^{\gamma+n/2} dx. \tag{10}$$

From (9),

$$L^c_{\gamma,n} = (4\pi)^{-n/2}\Gamma(\gamma+1)/\Gamma(1+\gamma+n/2). \tag{11}$$

It is easy to prove that

$$L_{\gamma,n} \geq L^c_{\gamma,n}. \tag{12}$$

The evaluation of the sharp $L_{\gamma,n}$ is an interesting open problem – especially $L_{1,n}$. In particular, for which γ, n is $L_{\gamma,n} = L^c_{\gamma,n}$? It is known [1] that for each fixed n, $L_{\gamma,n}/L^c_{\gamma,n}$ is nonincreasing in γ. Thus, if $L_{\gamma_0,n} = L^c_{\gamma_0,n}$ for some γ_0, then $L_{\gamma,n} = L^c_{\gamma,n}$ for all $\gamma > \gamma_0$. In particular, $L_{3/2,1} = L^c_{3/2,1}$ [11], so $L_{\gamma,1} = L^c_{\gamma,1}$ for $\gamma \geq 3/2$. No other sharp values of $L_{\gamma,n}$ are known. It is also known [11] that $L_{\gamma,1} > L^c_{\gamma,1}$ for $\gamma < 3/2$ and $L_{\gamma,n} > L^c_{\gamma,n}$ for $n = 2, 3$ and small γ.

Just as (4) is related to (2), inequality (7) is related to a generalization of (2). (The proof is basically the same.) Let $\phi_1, \ldots, \phi_N$ be any set of L^2 orthonormal functions on $\mathbf{R}^n (n \geq 1)$ and define

$$\rho(x) \equiv \sum_{i=1}^{N} |\phi_i(x)|^2. \tag{13}$$

$$T \equiv \sum_{i=1}^{N} \int |\nabla \phi_i|^2 \tag{14}$$

Then we have **The Main Inequality**

$$T \geq K_n \int \rho(x)^{1+2/n} dx \tag{15}$$

with K_n related to $L_{1,n}$ as in (5), i.e.

$$L_{1,n} = \left(\frac{n}{2K_n}\right)^{n/2} \left(1 + \frac{n}{2}\right)^{-(n+2)/2} \tag{16}$$

The best current value of K_n, for $n = 1, 2, 3$ is in [8]; in particular $K_3 \geq 2.7709$. We might call (15) *a Sobolev type inequality for orthonormal functions.* The point is that if the ϕ_i are merely normalized, but not orthogonal, then the best one could say is

$$T \geq N^{-2/n} K^1_n \int \rho(x)^{1+2/n} dx. \tag{17}$$

The orthogonality eliminates the factor $N^{-2/n}$, but replaces K_n^1 by the slightly smaller value K_n.

One should notice, especially, the N dependence in (15). The right side, loosely speaking, is proportional to $N^{(n+2)/n}$, whereas the right side of (17) appears, falsely, to be proportional to N^1, which is the best one could hope for without orthogonality. The difference is crucial for applications. In fact, if one is willing to settle for N^1 one can proceed directly from (1) (for $n \geq 3$). One then has (with $p = n/(n-2)$)

$$T \geq S_n \left\{ \int \rho(x)^p dx \right\}^{1/p}, \quad (n \geq 3). \tag{17a}$$

This follows from $\sum \|\phi_i^2\|_p \geq \left\| \sum |\phi_i|^2 \right\|_p$.

Eq. (11) gives a "classical guess" for $L_{1,n}$. Using that, together with (16), we have a "classical guess" for K_n, namely

$$\begin{aligned} K_n^c &= 4\pi n \Gamma \left(\frac{n+2}{2} \right)^{2/n} /(2+n) \\ &= \frac{3}{5} (6\pi^2)^{2/3} = 9.1156 \text{ for } n = 3. \end{aligned} \tag{18}$$

Since $L_{1,n} \geq L_{1,n}^c$, we have $K_n \leq K_n^c$. A conjecture in [11] is that $K_3 = K_3^c$, and it would be important to settle this.

Inequality (15) can be easily extended to the following: Let $\psi(x_1, \ldots, x_N) \in L^2((\mathbf{R}^n)^N)$, $x_i \in \mathbf{R}^n$. Suppose $\|\psi\|_2 = 1$ and ψ is *antisymmetric* in the N variables, i.e., $\psi(x_1, \ldots, x_i, \ldots, x_j, \ldots, x_N) = -\psi(x_1, \ldots x_j, \ldots, x_i, \ldots, x_N)$. Define

$$\rho_i(x) = \int \left| \psi(x_1, \ldots, x_{i-1}, x, x_{i+1}, \ldots, x_N) \right|^2 dx_1 \ldots \widehat{dx_i} \ldots dx_N \tag{19}$$

$$T_i(x) = \int |\nabla_i \psi|^2 dx_1 \ldots dx_N \tag{20}$$

$$\rho(x) = \sum_{i=1}^N \rho_i(x) \qquad T = \sum_{i=1}^N T_i. \tag{21}$$

(Note that $\rho(x) = N\rho_1(x)$ and $T = NT_1$ since ψ is antisymmetric, but the general form (19)-(21) will be used in the next paragraph.) **Then (15) holds with ρ and T given by (19)-(21)** (with the same K_n as in (15)). This is a generalization of (13)-(15) since we can take

$$\psi(x_1, \ldots, x_N) = (N!)^{-1/2} \det \{\phi_i(x_j)\}_{i,j=1}^N,$$

which leads to (13) and (14).

A variant of (15) is given in (52) below. It is a consequence of the fact that (17) and (17a) *also hold* with the definitions (19)-(21). *Antisymmetry of ψ is not required.* The proof of (17a) just uses (1) as before plus Minkowski's inequality, namely for $p \geq 1$

$$\int \left\{ \int |F(x,y)|^p dy \right\}^{1/p} dx \geq \left\{ \int \left\{ \int |F(x,y)| dx \right\}^p dy \right\}^{1/p} .$$

We turn now to some applications of these inequalities.

Application 1. Inequality (15) can be used to bound L^p **norms of Riesz and Bessel potentials of orthonormal functions** [7]. Again, $\phi_1, \ldots, \phi_N$ are L^2 orthonormal and let

$$u_i = (-\Delta + m^2)^{-1/2} \phi_i \tag{22}$$

$$\rho(x) \equiv \sum_{i=1}^{N} |u_i(x)|^2 . \tag{23}$$

Then there are constants L, B_p, A_n (independent of m) such that

$$n = 1: \quad \|\rho\|_\infty \leq L/m, \qquad m > 0 \tag{24}$$

$$n = 2: \quad \|\rho\|_p \leq B_p m^{-2/p} N^{1/p}, \qquad 1 \leq p < \infty, m > 0 \tag{25}$$

$$n \geq 3: \quad \|\rho\|_p \leq A_n N^{1/p}, \qquad p = n/(n-2), m \geq 0. \tag{26}$$

If the orthogonality condition is dropped then the right sides of (24)-(26) have to be multiplied by $N, N^{1-1/p}, N^{1-1/p}$ respectively. Possibly the absence of N in (24) is the most striking. Similar results can be derived [7] for $(-\Delta + m^2)^{-\alpha/2}$ in place of $(-\Delta + m^2)^{-1/2}$, with $\alpha < n$ when $m = 0$.

Inequality (15) also has applications in mathematical physics.

Application 2. (Navier-Stokes equation.) Suppose $\Omega \subset \mathbf{R}^n$ is an open set with finite volume $|\Omega|$ and consider

$$H = -\Delta - V(x)$$

on Ω with Dirichlet boundary conditions. Let $\lambda_1 \leq \lambda_2 \leq \ldots$ be the eigenvalues of H. Let $\bar{N}$ be the smallest integer, N, such that

$$E_N \equiv \sum_{i=1}^{N} \lambda_i \geq 0. \tag{27}$$

We want to find an upper bound for $\bar{N}$.

If $\phi_1, \phi_2, \ldots$ are the normalized eigenfunctions then, from (13)-(15) with $\phi_1, \ldots, \phi_N$,

$$E_N = T - \int \rho V \geq K_n \int \rho^{1+n/2} - \int V_+ \rho \geq G(\rho), \tag{28}$$

where (with $p = 1 + n/2$ and $q = 1 + 2/n$)

$$G(\rho) \equiv K_n \|\rho\|_p^p - \|V_+\|_q \|\rho\|_p. \tag{29}$$

Thus, for all N,

$$E_N \geq \inf\{G(\rho) |\ \|\rho\|_1 = N,\ \rho(x) \geq 0\}. \tag{30}$$

But $\|\rho\|_p |\Omega|^{1/q} \geq \|\rho\|_1 = N$ so, with $X \equiv \|\rho\|_p$,

$$E_N \geq \inf\{J(X) \mid X \geq N|\Omega|^{-1/q}\} \tag{31}$$

where

$$J(X) \equiv K_n X^p - \|V_+\|_q X. \tag{32}$$

Now $J(X) \geq 0$ for $X \geq X_0 = \{\|V_+\|_q / K_n\}^{1/(p-1)}$, whence we have the following implication:

$$N \geq |\Omega|^{1/q} \{\|V_+\|_q / K_n\}^{1/(p-1)} \Longrightarrow E_N \geq 0. \tag{33}$$

Therefore

$$\bar{N} \leq |\Omega|^{1/q} \{\|V_+\|_q / K_n\}^{1/(p-1)}. \tag{34}$$

The bound (34) can be applied [8] (following an idea of Ruelle) to the Navier-stokes equation. There, $\bar{N}$ is interpreted as the Hausdorff dimension of an attracting set for the N-S equation, while $V(x) \equiv \nu^{-3/2} \varepsilon(x)$, where $\varepsilon(x) = \nu |\nabla v(x)|^2$ is the average energy dissipation per unit mass in a flow v. ν is the viscosity.

Application 3. (Stability of matter.) This is the original application [10,11]. In the quantum mechanics of Coulomb systems (electrons and nuclei) one wants a lower bound for the Hamiltonian operator:

$$H = -\sum_{i=1}^{N} \Delta_i - \sum_{i=1}^{N} \sum_{j=1}^{K} z_j |x_i - R_j|^{-1} + \sum_{1 \leq i < j \leq N} |x_i - x_j|^{-1} + \sum_{1 \leq i < j \leq K} z_i z_j |R_i - R_j|^{-1} \tag{35}$$

on the L^2 space of *antisymmetric* functions $\psi(x_1, \ldots, x_N), x_i \in \mathbf{R}^3$. Here, N is the number of electrons (with coordinates x_i) and $R_1, \ldots, R_K \in \mathbf{R}^3$ are fixed vectors representing the locations of fixed nuclei of charges $z_1, \ldots, z_K > 0$. The desired bound is linear:

$$H \geq -A(N+K) \tag{36}$$

for some A independent of $N, K, R_1, \ldots, R_K$ (assuming all $z_i <$ some $\bar{z}$).

The main point is that antisymmetry of ψ is crucial for (36) and this is reflected in the fact that (15) holds with antisymmetry, but only (17) holds without it. Without the antisymmetry condition, H would grow as $-(N+K)^{5/3}$. This is discussed in Application 6 below. By using (15) one can eliminate the differential operators Δ_i. The functional $\psi \to (\psi, H\psi)$, with $(\psi,\psi) = 1$ can be bounded below using (15) by a functional (called the Thomas-Fermi functional) involving only $\rho(x)$ defined in (21). The minimization of this latter functional with respect to ρ is tractable and leads to (36).

Application 4. (Stellar structure.) Going from atoms to stars, we now consider N neutrons which attract each other gravitationally with a coupling constant $\kappa = Gm^2$, where G is the gravitational constant and m is the neutron mass. There are no Coulomb forces. Moreover, a "relativistic" form is assumed for the kinetic energy, which means that $-\Delta$ is replaced by $(-\Delta)^{1/2}$. Thus (35) is replaced by

$$H_N = \sum_{i=1}^{N} (-\Delta_i)^{1/2} - \kappa \sum_{1 \leq i < j \leq N} |x_i - x_j|^{-1} \tag{37}$$

(again on antisymmetric functions). One finds asymptotically for large N, that

$$\begin{aligned} \inf \operatorname{spec}(H_N) &= 0 \quad \text{if} \quad \kappa \leq CN^{-2/3} \\ &= -\infty \quad \text{if} \quad \kappa > CN^{-2/3} \end{aligned} \tag{38}$$

for some constant, C. Without antisymmetry, $N^{-2/3}$ must be replaced by N^{-1}. Equation (38) is proved in [12]. An important role is played by Daubechies's generalization [4] of (15) to the operator $(-\Delta)^{1/2}$ on $L^2(\mathbf{R}^N)$, namely (for antisymmetric ψ with $\|\psi\|_2 = 1$)

$$\left(\psi, \sum_{i=1}^{N} (-\Delta_i)^{1/2} \psi\right) \geq B_n \int \rho(x)^{1+1/n} \tag{39}$$

with ρ given by (19), (21). In general, one has

$$\left(\psi, \sum_{i=1}^{N} (-\Delta)^p \psi\right) \geq C_{p,n} \int \rho(x)^{1+2p/n} dx. \tag{40}$$

Recently [13] there has been considerable progress in this problem beyond that in [12]. Among other results there is an evaluation of the sharp asymptotic C in (38), i.e. if we first define $\kappa^c(N)$ to be the precise value of κ at which $\inf \operatorname{spec}(H_N) = -\infty$, we then define

$$C = \lim_{N\to\infty} N^{2/3}\kappa^c(N). \tag{41}$$

Let B_n^c be the "classical guess' in (39). This can be calculated from the analogue of (9) (using $|p|$ instead of p^2, and which leads to $\sum |e_i| \approx \widetilde{L}\int V_+^{1+n}$), and then from the analogue of (16), namely $\widetilde{L} = C_n B_n^{-n}$. One finds $B_3^c = (3/4)(6\pi^2)^{1/3}$ (cf. (18)). Using B_3^c, we introduce the functional

$$\mathcal{E}(\rho) = B_3^c \int \rho^{4/3} - \frac{1}{2}\kappa \int\int \rho(x)\rho(y)|x-y|^{-1}dxdy \tag{42}$$

for $\rho \in L^1(\mathbf{R}^3)\cap L^{4/3}(\mathbf{R}^3)$ and define the energy

$$E^c(N) = \inf\{\mathcal{E}(\rho) | \int \rho = N\}. \tag{43}$$

One finds there is a finite $\alpha^c > 0$ such that $E^c(N) = 0$ if $\kappa N^{2/3} \le \alpha^c$ and $E^c(N) > -\infty$ if $\kappa N^{2/3} > \alpha^c$. (This α^c is found by solving a Lane-Emden equation.)

Now (42) and (43) constitute the semiclassical approximation to H_N in the following sense. We expect that if we set $\kappa = \alpha N^{-2/3}$ in (37), with α fixed, then if $\alpha < \alpha^c$

$$\lim_{N\to\infty} \inf \operatorname{spec}(H_N) = 0 \tag{44}$$

while if $\alpha > \alpha^c$ there is an N_0 such that

$$\inf \operatorname{spec}(H_N) = -\infty \quad \text{if} \quad N > N_0. \tag{45}$$

Indeed, (44) and (45) are true [13], and thus α^c is the sharp asymptotic value of C in (38).

An interesting point to note is that Daubechies's B_3 in (39) is about half of B_3^c. The sharp value of B_3 is unknown. Nevertheless, with some additional tricks one can get from (37) to (42) with B_3^c and not B_3. Inequality (39) plays a role in [13], but it is not sufficient.

Application 5. (Stability of atoms in magnetic fields.) This is given in [9]. Here $\psi(x_1,\ldots,x_N)$ becomes a spinor-valued function, i.e. ψ is an antisymmetric function in $\bigwedge_1^N L^2(\mathbf{R}^3;\mathbf{C}^2)$. The operator H of interest is as in (35) but with the replacement

$$-\Delta \to \{\sigma\cdot(i\nabla - A(x))\}^2 \tag{46}$$

where $\sigma_1, \sigma_2, \sigma_3$ are the 2×2 Pauli matrices (i.e. generators of $SU(2)$) and $A(x)$ is a given vector field (called the magnetic vector potential). Let

$$E_0(A) = \inf \operatorname{spec}(H) \tag{47}$$

after the replacement of (46) in (35). As $A \to \infty$ (in a suitable sense), $E_0(A)$ can go to $-\infty$. The problem is this: Is

$$\widetilde{E}(A) \equiv E_0(A) + \frac{1}{8\pi}\int (\text{curl } A)^2 \tag{48}$$

bounded below for all A? In [9] the problem is resolved for $K = 1$, all N and $N = 1$, all K. It turns out that $\widetilde{E}(A)$ is bounded below in these cases if and only if all the z_i satisfy $z_i < z^c$ where z^c is some fixed constant independent of N and K. The problem is still open for all N and all K.

One of the main problems in bounding $\widetilde{E}(A)$ is to find a lower bound for the kinetic energy (the first term in (35) after the replacement given in (46)) for an antisymmetric ψ. First, there is the identity

$$\left(\psi, \sum_{i=1}^{N} \{\sigma \cdot (i\nabla - A(x_i)\}^2 \psi\right) = T(\psi, A) - \left(\psi, \sum_{i=1}^{N} \sigma \cdot B(x_i)\psi\right) \tag{49}$$

with $B = \text{curl } A$ being the magnetic field and

$$T(\psi, A) = \left(\psi, \sum_{i=1}^{N} |i\nabla - A(x)|^2 \psi\right). \tag{50}$$

The last term on the right side of (49) can be controlled, so it will be ignored here. The important term is $T(\psi, A)$. Since Pauli matrices do not appear in (50) we can now let ψ be an ordinary complex valued (instead of spinor valued) function.

It turns out that (8), and hence (15), hold with some $\widetilde{L}_{\gamma,n}$ which is *independent* of A. The T in (15) is replaced, of course, by the $T(\psi, A)$ of (50). To be more precise, the *sharp* constants $L_{\gamma,n}$ and $\widetilde{L}_{\gamma,n}$ are unknown (except for $\gamma \geq 3/2, n = 1$ in the case of $L_{\gamma,n}$) and conceivably $\widetilde{L}_{\gamma,n} > L_{\gamma,n}$. However, all the current bounds for $L_{\gamma,n}$ (see [8]) also hold for $\widetilde{L}_{\gamma,n}$. Thus, for $n = 3$ we have

$$T(\psi, A) \geq K_3 \int \rho^{5/3} \tag{51}$$

with K_3 being the value given in [8], namely 2.7709.

However, in [9] another inequality is needed

$$T(\psi, A) \geq C\left\{\int \rho^2\right\}^{2/3}. \tag{52}$$

It seems surprising that we can go from an $L^{5/3}$ estimate to an L^2 estimate, but the surprise is diminished if (17a) with its L^3 estimate is recalled. First note that (1) holds (with the same S_n) if $|\nabla f|^2$ is replaced by $|[i\nabla - A(x)]f|^2$. (By writing $f = |f|e^{i\theta}$ one finds that $|\nabla|f||^2 \leq |[i\nabla - A(x)]f|^2$.) Then (17a) holds since only convexity was used. Thus, using the mean of (15) and (17a),

$$T(\psi, A) \geq (S_n K_n)^{1/2} \|\rho\|_3^{1/2} \|\rho\|_{5/3}^{5/6}. \tag{53}$$

An application of Hölders inequality yields (52) with $C^2 = S_n K_n$.

Application 6. (Instability of bosonic matter.) As remarked in Application 3, dropping the antisymmetry requirement on ψ (the particles are now bosons) makes $\inf\operatorname{spec}(H)$ diverge as $-(N+K)^{5/3}$. The extra power 2/3, relative to (36) can be traced directly to the factor $N^{-2/3}$ in (17).

An interesting problem is to allow the positive particles also to be movable and to have charge $z_i = 1$. This should raise $\inf\operatorname{spec} H$, but by how much? For $2N$ particles the new H is

$$\widetilde{H} = -\sum_{i-1}^{2N} \Delta_i + \sum_{1 \leq i < j \leq 2N} e_i e_j |x_i - x_j|^{-1} \tag{54}$$

with $e_i = +1$ for $1 \leq i \leq N$ and $e_i = -1$ for $N+1 \leq i \leq 2N$. $\widetilde{H}$ acts on $L^2(\mathbf{R}^{3N})$ without any symmetry requirements.

Twenty years ago Dyson [5] proved, by a variational calculation, that $\inf\operatorname{spec}(\widetilde{H}) \leq -AN^{7/5}$ for some $A > 0$. Thus, stability (i.e. a linear law (36)) is not restored, but the question of whether the correct exponent is 7/5 or 5/3, or something in between, remained open. It has now been proved [2] that $N^{7/5}$ is correct, $\inf\operatorname{spec}(\widetilde{H}) \geq -BN^{7/5}$. The proof is much harder than for (36) because no simple semiclassical theory (like Thomas-Fermi theory) is a good approximation to $\widetilde{H}$. Correlations are crucial.

Application 7. (Stability of relativistic matter.) Let us return to Coulomb systems (electrons and nuclei) as in application 3, but with (35) replaced by

$$H = \sum_{i=1}^{N} \{(-\Delta_i + m^2)^{1/2} - m\} + \alpha V_c(x_1, \ldots, x_N; R_1, \ldots, R_K) \tag{55}$$

with $\alpha = e^2$ = electron charge squared (and $\hbar = c = 1$) and where

$$V_c = -\sum_{i=1}^{N}\sum_{j=1}^{K} z_j |x_i - R_j|^{-1} + \sum_{1 \leq i < j \leq N} |x_i - x_j|^{-1} + \sum_{1 \leq i < j \leq K} z_i z_j |R_i - R_j|^{-1} \tag{56}$$

is the Coulomb potential. The electron charge, $\alpha^{1/2}$, is explicitly displayed in (55) for a reason to be discussed presently. Also (55) differs from (35) in that the kinetic energy operator $-\Delta$ is replaced by the relativistic form $(-\Delta+m^2)^{1/2}-m$, where m is the electron mass. Since $-\Delta - m \le (-\Delta+m^2)^{1/2}-m \le -\Delta$, the difference of these two operators is a bounded operator and therefore, as far as the stability question is concerned, we may as well use the simplest operator $(-\Delta)^{1/2}$ in (55), which will be done henceforth. This, in fact, was already done in (37). We define

$$E_{N,K}(R_1,\dots,R_K) = \inf \operatorname{spec} H \tag{57}$$

$$E_{N,K} = \inf_{R_1,\dots,R_K} E_{NK}(R_1,\dots,R_K) \tag{58}$$

$$E = \inf_{N,K} E_{NK} \tag{59}$$

Under scaling (dilation of coordinates in $\mathbf{R}^{3N+3K}$) the operators $(-\Delta)^{1/2}$ and $|x|^{-1}$ behave the same (proportional to length)$^{-1}$ and hence we conclude that

$$E_{N,K} = 0 \qquad \text{or} \qquad -\infty. \tag{60}$$

The system is said to be **stable** if $E=0$. For simplicity of exposition let us take all z_j to be some common value, z.

For the hydrogenic atom $N=K=1$ the only constant that appears is the combination $z\alpha$. It is known that $E_{1,1}=0$ if and only if $z\alpha \le 2/\pi$. In the many-body case there are two constants (which can be taken to be $z\alpha$ and α) and the question is whether the system is stable *all the way up to* $z\alpha = 2/\pi$ for α less than some small, but fixed $\alpha_c > 0$. The answer will depend on q, the number of spin states allowed for the fermionic electrons. (Note: in application 3 we implicitly took $q=1$. In fact $q=2$ in nature. To say that there are q spin states means that under permutations $\psi(x_1,\dots,x_N)$ belongs to a Young's tableaux of q or fewer columns.)

This problem is resolved in [15] where it is shown that stability occurs if

$$q\alpha \le 1/47. \tag{61}$$

The kinetic energy bound (39) plays a crucial role in the proof (but, of course, many other inequalities are also needed). It is also shown in [15] that stability definitely fails to occur if

$$\alpha > 36 q^{-1/3} z^{2/3} \tag{62}$$

or if

$$\alpha > 128/15\pi. \tag{63}$$

If (63) holds then instability occurs for *every* $z>0$, no matter how small.

REFERENCES

[1] M. Aizenman and E.H. Lieb, On semiclassical bounds for eigenvalues of Schrödinger operators, Phys. Lett. *66*A, 427-429 (1978).

[2] J. Conlon, E.H. Lieb and H.-T. Yau, The $N^{7/5}$ law for bosons, Commun. Math. Phys. (submitted).

[3] M. Cwikel, Weak type estimates for singular values and the number of bound states of Schrödinger operators, Ann. Math. *106*, 93-100 (1977).

[4] I. Daubechies, Commun. Math. Phys. *90*, 511-520 (1983).

[5] F.J. Dyson, Ground state energy of a finite systems of charged particles, J. Math. Phys. *8*, 1538-1545 (1967).

[6] E.H. Lieb, The number of bound states of one-body Schrödinger operators and the Weyl problem, A.M.S. Proc. Symp. in Pure Math. *36*, 241-251 (1980). The results were announced in Bull. Ann. Math. Soc. *82*, 751-753 (1976).

[7] E.H. Lieb, An L^p bound for the Riesz and Bessel potentials of orthonormal functions, J. Funct. Anal. *51*, 159-165 (1983).

[8] E.H. Lieb, On characteristic exponents in turbulence, Commun. Math. Phys. *92*, 473-480 (1984).

[9] E.H. Lieb and M. Loss, Stability of Coulomb systems with magnetic fields: II. The many-electron atom and the one-electron molecule, Commun. Math. Phys. *104*, 271-282 (1986).

[10] E.H. Lieb and W.E. Thirring, Bounds for the kinetic energy of fermions which proves the stability of matter, Phys. Rev. Lett. *35*, 687-689 (1975). Errata *35*, 1116 (1975).

[11] E.H. Lieb and W.E. Thirring, "Inequalities for the moments of the eigenvalues of the Schrödinger Hamiltonian and their relation to Sobolev inequalities" in *Studies in Mathematical Physics* (E. Lieb, B. Simon, A. Wightman eds.) Princeton University Press, 1976, pp. 269-304.

[12] E.H. Lieb and W.E. Thirring, Gravitational collapse in quantum mechanics with relativistic kinetic energy, Ann. of Phys. (NY) *155*, 494-512 (1984).

[13] E.H. Lieb and H.-T. Yau, The Chandrasekhar theory of stellar collapse as the limit of quantum mechanics, Commun. Math. Phys. *112*, 147-174 (1987).

[14] G.V. Rosenbljum, Distribution of the discrete spectrum of singular differential operators. Dokl. Akad. Nauk SSSR *202*, 1012-1015 (1972). (MR 45 #4216). The details are given in Izv. Vyss. Ucebn. Zaved. Matem. *164*, 75-86 (1976). (English trans. Sov. Math. (Iz VUZ) *20*, 63-71 (1976).)

[15] E.H. Lieb and H.T. Yau, The stability and instability of relativistic matter, Commun. Math. Phys. *118*, 177-213 (1988). For a short summary see: Many body stability implies a bound on the fine structure constant, Phys. Rev. Lett. *61*, 1695-1697 (1988).

Adv. Theor. Math. Phys. **2** (1998) 719 – 731

A sharp bound for an eigenvalue moment of the one-dimensional Schrödinger operator

[1] **Dirk Hundertmark[a], Elliott H. Lieb[a], Lawrence E. Thomas[b]**

[a]Department of Physics and Mathematics
Jadwin Hall
Princeton University
P.O. Box 708
Princeton
New Jersey 08544

[b] Department of Mathematics
University of Virginia
Charlottesville
Virginia 22903

Abstract

We give a proof of the Lieb-Thirring inequality in the critical case $d = 1$, $\gamma = 1/2$, which yields the best possible constant.

e-print archive: http://xxx.lanl.gov/abs/hep-th/9806012
[1]On leave of NWF I - Mathematik, Universität Regensburg, D-93040 Regensburg

1 Introduction

There is a family of inequalities [9], [10] that has proved to be useful in various areas of mathematical physics, especially in the proofs of stability of matter. They state that given a Schrödinger operator

$$-\Delta + V \quad \text{on } \mathrm{L}^2(\mathbb{R}^d),$$

the sum of the moments of the negative eigenvalues $-E_1 < -E_2 \leq -E_3 \leq \ldots \leq 0$ (if any) of this operator is bounded by

$$\sum E_i^\gamma \leq L_{\gamma,d} \int (V_-(x))^{\gamma+d/2}\, dx, \tag{1}$$

with $V_-(x) := \max(-V(x), 0)$. These inequalities have been generalized in several directions, e.g. manifolds instead of $\mathbb{R}^d$. Here we are concerned with the case $d = 1$.

The cases originally shown to hold [10] are

$$d = 1,\ \gamma > \frac{1}{2}, \quad d = 2,\ \gamma > 0, \quad \text{and } d \geq 3,\ \gamma \geq 0.$$

When $d = 2$ there cannot be any bound for $\gamma = 0$ (meaning the number of negative eigenvalues) since at least one negative eigenvalue always exists for arbitrarily small negative perturbations of the free Laplacian in two dimensions [5, page 156-157], [15].

The critical case $d \geq 3$ and $\gamma = 0$ was open for a while and proved independently by Cwikel [4], Lieb [7], and Rozenbljum [11]. Still later, different proofs where given by Conlon [3] and Li and Yau [6]. The sharp constants are still not known, but the best one so far is in [7].

If $d = 1$ it is not hard to see that the inequality cannot hold for $\gamma < 1/2$. To prove this choose a sequence of aproximate δ-functions. They converge to zero in $\mathrm{L}^{\gamma+1/2}(\mathbb{R})$ but the limit may have a negative eigenvalue; see the discussion of a Dirac potential below. In the critical case $d = 1, \gamma = 1/2$, which concerns us here, it was not known until recently whether $L_{1/2,1}$ is finite. This case was settled by Timo Weidl [17] who showed that $L_{1/2,1} < 1.005$. Unfortunately his method of proof *cannot* be improved to yield the sharp constant as can be seen from the following argument: His method is also applicable for a half-line problem corresponding to a Schrödinger operator on $\mathbb{R}_+$ with Neumann boundary conditions at the origin; in fact he reduces the full problem (but not the determination of the sharp constant) to this case. Since in this half-line problem the trivial lower bound for the sharp constant is given by 1 his method cannot yield a better bound than 1 in the problem concerning us here.

Hence, the sharp constant $L_{1/2,1}$ remained undetermined, a tantalizing situation, since there is an obvious conjecture about the value of this constant [10]. In one dimension the potential can be a measure (thanks to the fact that $H^1(\mathbb{R}^1)$ functions are continuous) and when $\gamma = 1/2$ the right hand side of (1) is simply the total mass of this measure. In order to maximize the sum of the square roots of the eigenvalues it is reasonable to suppose that one should concentrate the potential at one point and the extreme case should hence correspond to a δ-function.

It is well-known that $-\partial_x^2 - c\delta$ is a well-defined closed quadratic form on the Sobolev space $H^1(\mathbb{R}^1)$ and the Hamiltonian corresponding to this form is used in textbooks as a simple solvable model in quantum mechanics. An exercise shows that the only bound state of this operator for positive c is given by $\psi(x) = \exp(-c|x|/2)$ with eigenvalue $-c^2/4$.

If it is true that this Dirac potential is the optimal case we conclude that the sharp constant in the Lieb-Thirring inequality for $d = 1, \gamma = 1/2$ is given by $L_{1/2,1} = 1/2$. The proof of this statement is the main result of this paper. A corollary of our result is that for the half-line problem with Neumann boundary conditions considered by Weidl, the sharp constant is 1.

Before turning to the proof let us note the corresponding – still unproved – conjecture when $1/2 < \gamma < 3/2$. The optimal potential should be given by

$$V(x) = -\frac{1}{\gamma^2 - 1/2}\left(\cosh\left(\frac{x}{\gamma^2 - 1/2}\right)\right)^{-2}$$

and the sharp constant is supposed to be [10]

$$L_{\gamma,1} = \pi^{-1/2}\frac{1}{\gamma - 1/2}\frac{\Gamma(\gamma+1)}{\Gamma(\gamma+1/2)}\left(\frac{\gamma-1/2}{\gamma+1/2}\right)^{\gamma+1/2} = 2L^c_{\gamma,1}\left(\frac{\gamma-1/2}{\gamma+1/2}\right)^{\gamma-1/2}$$

Here $L^c_{\gamma,1} := (2\pi)^{-1/2}\Gamma(\gamma+1)/\Gamma(\gamma+3/2)$ is its classical value. Unlike the case $\gamma < 3/2$ the optimal constant in one dimension and $\gamma \geq 3/2$ is known [1], [10] to be $L_{\gamma,1} = L^c_{\gamma,1}$. Using the fact proved in [1] that $L_{\gamma,1}/L^c_{\gamma,1}$ is monotone decreasing in γ and the sharp value for $L_{1/2,1}$ obtained here we conclude that $L_{\gamma,1} \leq 2L^c_{\gamma,1}$ for all $\gamma \geq 1/2$. As a last remark, let us note that our proof uses no special 1-D technique, except for the explicit form of the Birman-Schwinger kernel (3) in one dimension.

2 Proof of the main result for potentials

The principal result of this paper is

Theorem 1. *For a Schrödinger operator* $-\partial_x^2 + V$ *in one dimension the optimal constant* $L_{1/2,1}$ *is* $1/2$, *i.e.*

$$\sum_{E_i \le 0} \sqrt{E_i} \le \frac{1}{2} \int V_-(x)\, dx. \tag{2}$$

The inequality is strict if the negative part V_- *is a non-zero* L^1 *function.*

In this section we prove this theorem in the case the potential is an L^1 function. In the last section we extend the bound (2) to potentials that are (finite) measures and prove that the δ-function is the unique maximizer up to translations. By the minmax principle it suffices to investigate the operator $-\partial_x^2 - V_-$. We will henceforth assume $V = -U$ with U non-negative and integrable.

To study the bound states energies of a Schrödinger operator it is often useful to investigate another problem. To do so we need some more notation. For $E > 0$ let

$$\mathcal{K}_E(x,y) := \sqrt{U(x)}\, \frac{\exp(-\sqrt{E}\,|x-y|)}{2\sqrt{E}} \sqrt{U(y)}, \quad \text{for all } x,y \in \mathbb{R} \tag{3}$$

be the Birman-Schwinger kernel for the Schrödinger operator $-\partial_x^2 - U$ in $\mathrm{L}^2(\mathbb{R})$. $\mathcal{K}_E$ stands for the integral operator given by this kernel. The Birman-Schwinger principle [2, 13] states that $-E_n < 0$ is the nth eigenvalue of $-\partial_x^2 - U$ if and only if the nth eigenvalue of $\mathcal{K}_{E_n}$ equals one. The explicit expression (3) suggests that multiplying (3) by $\sqrt{E_n}$ will yield a still implicit but perhaps more flexible expression for $\sqrt{E_n}$. This is exactly what we are going to do. Let us define, for $\mu \ge 0$,

$$\mathcal{L}_\mu(x,y) := \sqrt{U(x)}\mathrm{e}^{-\mu|x-y|}\sqrt{U(y)}, \quad \text{for all } x,y \in \mathbb{R}. \tag{4}$$

Moreover, given some arbitrary non-negative locally finite Borel-measure κ on $\mathbb{R}$, we can generalize the kernel (4) to

$$\mathcal{L}^\kappa(x,y) := \sqrt{U(x)}\mathrm{e}^{-|\mathcal{J}(x)-\mathcal{J}(y)|}\sqrt{U(y)}, \quad \text{for all } x,y \in \mathbb{R}, \tag{5}$$

where the function $\mathcal{J}$ is given by

$$\mathcal{J}(x) := \int_0^x \kappa(dz). \tag{6}$$

Again $\mathcal{L}_\mu$ and $\mathcal{L}^\kappa$ are the corresponding integral operators. Of course, $\mathcal{L}_\mu$ in (4) corresponds to $\kappa(dz) = \mu dz$. Both $\mathcal{K}_E$ and $\mathcal{L}^\kappa$ are compact integral operators; their Hilbert-Schmidt norms are bounded by $(\int U(x)\,dx)^2/(2\sqrt{E})$ and $(\int U(x)\,dx)^2$, respectively. For a positive compact operator A we denote its ordered eigenvalues by $\lambda_1(A) \geq \lambda_2(A) \geq \ldots \geq 0$. With the help of the Fourier transform $(\exp(-\varepsilon|x|)/(2\varepsilon) = \int e^{ipx}/(p^2+\varepsilon^2)\,dp/(2\pi))$ one sees the following facts:

(i) $\mathcal{L}^\kappa$ and K_E are positive definite operators,

and hence the (ordered) eigenvalues $\lambda_j(\mathcal{L}^\kappa)$ obey

(ii) $\lambda_1(\mathcal{L}^\kappa) > \lambda_2(\mathcal{L}^\kappa) \geq \lambda_3(\mathcal{L}^\kappa) \geq \ldots \geq 0$

with a similar statement for $\lambda_j(\mathcal{K}_E)$. The strict inequality follows from the positivity of the integral kernel and the Perron-Frobenius theorem. The trace of $\mathcal{L}^\kappa$ is given by

(iii) $\operatorname{tr}\mathcal{L}^\kappa = \int U(x)\,dx$,

independent of κ, and

(iv) $\mathcal{L}^0 = \mathcal{L}_0$ is a rank one operator with eigenvalue $\int U(x)\,dx$.

The discussion above suggests that the sum of the square roots of the eigenvalues of the one dimensional Schrödinger operator is related to the sum of the eigenvalues of $\mathcal{L}_\mu$. Indeed we have the following bound:

Theorem 2 (Domination by $\mathcal{L}_\mu$). *Suppose $U \geq 0$ with $U \in L^1(\mathbb{R})$ and let $-E_1 < -E_2 \leq -E_3 \leq \ldots \leq 0$ be the negative eigenvalues counting multiplicity of the Schrödinger operator $-\partial_x^2 - U$ given by the minmax principle. Furthermore, we denote by $\lambda_j(\mathcal{L}_\mu)$ the eigenvalues of $\mathcal{L}_\mu$ in (4). Then, for all $n \in \mathbb{N}$ and $0 \leq E \leq E_n$*

$$2\sum_{i\leq n}\sqrt{E_i} \leq \sum_{i\leq n}\lambda_i(\mathcal{L}_{\sqrt{E}}) + \lambda_1(\mathcal{L}_{\sqrt{E_1}}) - \lambda_1(\mathcal{L}_{\sqrt{E_2}}). \tag{7}$$

In (7) we set $E_{j+1} = 0$ in case the Schrödinger operator happens to have only j negative eigenvalues.

Proof. As already mentioned, the Birman-Schwinger principle gives a one-to-one correspondence between negative eigenvalues of a Schrödinger operator and the eigenvalues of $\mathcal{K}_E$: $\lambda_i(\mathcal{K}_{E_i}) = 1$. Multiplying this equality by $2\sqrt{E_i}$ yields $2\sqrt{E_i} = 2\sqrt{E_i}\,\lambda_i(\mathcal{K}_{E_i}) = \lambda_i(\mathcal{L}_{\sqrt{E_i}})$ for all i such that $E_i > 0$. Note that $\lambda_i(\mathcal{L}_0) = 0$ if $i \geq 2$ since $\mathcal{L}_0$ is a rank one operator. Therefore we have

$$2\sum_{i\leq n}\sqrt{E_i} = \sum_{i\leq n}\lambda_i(\mathcal{L}_{\sqrt{E_i}}) \tag{8}$$

for arbitrary $n \in \mathbb{N}$. If the eigenvalues of $\mathcal{L}_\mu$ were monotonically decreasing as $\mu \geq 0$ increases this would immediately imply

$$2\sum_{i\leq n}\sqrt{E_i} \leq \sum_{i\leq n}\lambda_i(\mathcal{L}_{\sqrt{E_n}}) \leq \sum_{i\leq n}\lambda_i(\mathcal{L}_{\sqrt{E}}) \quad \text{for } 0 \leq E \leq E_n.$$

However, such a monotonicity *cannot* hold since the trace of $\mathcal{L}_\mu$ is *independent* of $\mu \geq 0$. Nevertheless, *the partial sums* $\sum_{i\leq n}\lambda_i(\mathcal{L}_{\sqrt{E}})$ *of its eigenvalues are monotone in* E even for the slightly more general operator $\mathcal{L}^\kappa$ given by (5). Lemma 4 below is the key lemma in our analysis. Assuming the monotonicity given in Lemma 4, the proof of the theorem follows immediately from (8): For $n = 1$ we have

$$\begin{aligned}2\sqrt{E_1} &= \lambda_1(\mathcal{L}_{\sqrt{E_1}}) = \lambda_1(\mathcal{L}_{\sqrt{E_2}}) + \lambda_1(\mathcal{L}_{\sqrt{E_1}}) - \lambda_1(\mathcal{L}_{\sqrt{E_2}}) \\ &\leq \lambda_1(\mathcal{L}_{\sqrt{E}}) + \lambda_1(\mathcal{L}_{\sqrt{E_1}}) - \lambda_1(\mathcal{L}_{\sqrt{E_2}}) \quad \text{for all } 0 \leq E \leq E_1\end{aligned}$$

where we take $E_2 = 0$ if the potential has only one negative eigenvalue. If there are two or more negative eigenvalues it follows by induction that

$$\begin{aligned}&2\sum_{i\leq n}\sqrt{E_i} + 2\sqrt{E_{n+1}} \\ &\quad\leq \sum_{i\leq n}\lambda_i(\mathcal{L}_{\sqrt{E_{n+1}}}) + \lambda_{n+1}(\mathcal{L}_{\sqrt{E_{n+1}}}) + \lambda_1(\mathcal{L}_{\sqrt{E_1}}) - \lambda_1(\mathcal{L}_{\sqrt{E_2}}) \\ &\quad\quad\leq \sum_{i\leq n+1}\lambda_i(\mathcal{L}_{\sqrt{E}}) + \lambda_1(\mathcal{L}_{\sqrt{E_1}}) - \lambda_1(\mathcal{L}_{\sqrt{E_2}})\end{aligned}$$

for all $0 \leq E \leq E_{n+1}$ and $n \in \mathbb{N}$. ∎

Before proving the Lemma, we note a simple consequence of this theorem which proves our main bound (2).

Corollary 3 (Sharp constant). *Under the hypotheses of Theorem 2 and for* $U \neq 0$

$$2\sum_{i\in\mathbb{N}}\sqrt{E_i} < \int U(x)\,dx.$$

Proof. From the theorem we get

$$\begin{aligned}2\sum_{i\in\mathbb{N}}\sqrt{E_i} &\leq \lambda_1(\mathcal{L}_0) + \lambda_1(\mathcal{L}_{\sqrt{E_1}}) - \lambda_1(\mathcal{L}_{\sqrt{E_2}}) \\ &= \int U(x)\,dx + \lambda_1(\mathcal{L}_{\sqrt{E_1}}) - \lambda_1(\mathcal{L}_{\sqrt{E_2}}),\end{aligned} \tag{9}$$

since $\mathcal{L}_0$ is a rank one operator with eigenvalue $\int U(x)\,dx$. To conclude the strict inequality also note that $\lambda_1(\mathcal{L}_{\sqrt{E}})$ is *strictly* monotone decreasing in $E \geq 0$ by Lemma 4. The Perron-Frobenius theorem [12, Theorem XIII.44] implies E_1 is simple and hence $\lambda_1(\mathcal{L}_{\sqrt{E_1}}) - \lambda_1(\mathcal{L}_{\sqrt{E_2}}) < 0$. ∎

Lemma 4 (Monotonicity). *For all $n \in \mathbb{N}$ the nth partial sum of the eigenvalues of the operator $\mathcal{L}^\kappa$ defined in (5) is monotonically decreasing in the sense that*

$$\sum_{i \leq n} \lambda_i(\mathcal{L}^{\kappa'}) \leq \sum_{i \leq n} \lambda_i(\mathcal{L}^{\kappa}) \tag{10}$$

if $\kappa'([s,t]) \geq \kappa([s,t])$ for all $s \leq t \in \mathbb{R}$. Moreover the largest eigenvalue $\lambda_1(\mathcal{L}^\kappa)$ is strictly monotone decreasing in κ.

Proof. To clarify the line of reasoning we consider first a toy-model given by an $(m+1)\times(m+1)$ matrix where the two variables x and y in (5) take on $m+1$ values $x_0 \leq \ldots \leq x_m$. With $a_i = \exp(-|J(x_i) - J(x_0|)) \leq 1$ (where J is defined in (6) and with $U = 1$ on $\{x_0, \ldots, x_m\}$ for simplicity) the operator given in (5) has the matrix

$$L(\{a_i\}) := \begin{pmatrix} 1 & a_1 & a_1a_2 & a_1a_2a_3 & \ldots & a_1\ldots a_m \\ a_1 & 1 & a_2 & a_2a_3 & \ldots & a_2\ldots a_m \\ \vdots & & & & & \vdots \\ a_1\ldots a_{m-1} & a_2\ldots a_{m-1} & \ldots & & 1 & a_m \\ a_1\ldots a_m & a_2\ldots a_m & \ldots & & a_m & 1 \end{pmatrix}.$$

Let $\lambda_1(\{a_i\}) \geq \lambda_2(\{a_i\}) \geq \ldots \geq \lambda_{m+1}(\{a_i\})$ be the ordered eigenvalues of $L(\{a_i\})$. We investigate the sum of the largest n eigenvalues in the cube given by $0 \leq a_k \leq 1$ for all $k \in \{1, \ldots, m+1\}$ and want to show that it is a (separately) monotone increasing function of each a_k in the interval $0 \leq a_k \leq 1$. Fix $k \in \{1, \ldots, m+1\}$ and $\{a_i\}_{i \neq k}$. For simplicity we write $L(a_k)$ for $L(\{a_i\}_{i\neq k}, a_k)$. The matrix L has the form

$$L(a_k) := L(\{a_i\}_{i\neq k}, a_k) = \begin{pmatrix} A & a_kW \\ a_kW^\dagger & B \end{pmatrix} =: L(0) + a_kT$$

with

$$L(0) := L(\{a_i\}_{i\neq k}, 0) = \begin{pmatrix} A & \mathbf{0} \\ \mathbf{0} & B \end{pmatrix} \quad \text{on } \mathbb{C}^k \oplus \mathbb{C}^{m+1-k} = \mathbb{C}^{m+1}$$

and the perturbation

$$T = \begin{pmatrix} \mathbf{0} & W \\ W^\dagger & \mathbf{0} \end{pmatrix}, \quad W : \mathbb{C}^{n+1-k} \to \mathbb{C}^k,$$

where A, B, and W are $k\times k$, $(m+1-k)\times(m+1-k)$, and $k\times(m+1-k)$ matrices respectively, depending only on $\{a_i\}_{i\neq k}$. This shows that the dependence of L on a_k (for fixed $\{a_i\}_{i\neq k}$) is affine-linear. Now the claimed monotonicity of the sum of the largest n eigenvalues in $0 \le a_k \le 1$ is easily seen by the usual quantum mechanics textbook arguments of perturbation theory, cf. [16, chapter 3.5]: The sum is given by

$$\begin{aligned}\sum_{i\le n} \lambda_i(L(a_k)) &= \sup_{0\le d\le 1,\ \mathrm{tr}\, d=n} \mathrm{tr}(dL(a_k)) \\ &= \sup_{0\le d\le 1,\ \mathrm{tr}\, d=n} \{\mathrm{tr}(dL(0) + a_k\, \mathrm{tr}(dT)\}\end{aligned}$$

where $d\colon \mathbb{C}^{m+1} \to \mathbb{C}^{m+1}$ is a density matrix. Consequently, being a supremum of affine-linear functions, it is convex. To conclude monotonicity in a_k it is enough to show that the derivative of the sum with respect to a_k at $a_k = 0$ is non-negative. If the eigenvalues of $L(0)$ are non-degenerate this follows immediately from the Feynmann-Hellman theorem of perturbation theory: Since $L(0)$ leaves the decomposition $\mathbb{C}^{n+1} = \mathbb{C}^k \oplus \mathbb{C}^{m+1-k}$ invariant its eigenvectors Φ_i live either in the subspace $\mathbb{C}^k$ or $\mathbb{C}^{m+1-k}$, so $\langle\Phi_i, T\Phi_i\rangle = 0$. Thus by the Feynman-Hellman formula each eigenvalue has derivative 0 at $a_k = 0$, and for this reason each partial sum has zero derivative at $a_k = 0$.

In the degenerate case a single eigenvalue might have a negative derivative at $a_k = 0$ but the partial sum of the *largest* n eigenvalues always has a non-negative derivative. Indeed, if the eigenvalues are degenerate we first have to diagonalize the perturbation T in the corresponding eigenspace h of $L(0)$. This eigenspace, however, can be decomposed into $h = h_1 \oplus h_2$, with $h_1 \subset \mathbb{C}^k$, $h_2 \subset \mathbb{C}^{m+1-k}$, h_1 or h_2 possibly empty. With P_i being the orthogonal projection onto h_i, $i = 1, 2$, the perturbation T restricted to the subspace h is again of the form $T|_h = P_h T P_h = \widetilde{W} + \widetilde{W}^\dagger$, i.e., $T|_h = \begin{pmatrix} 0 & \widetilde{W} \\ \widetilde{W}^\dagger & 0 \end{pmatrix}$ with $\widetilde{W} := P_{h_1} W P_{h_2} : h_2 \to h_1$. This gives $\mathrm{tr}_h T = \mathrm{tr}\, T|_h = 0$. The Feynman-Hellman formula tells us that the eigenvalues of the restricted perturbation $T|_h$ are the derivatives of the eigenvalue branches emerging from this degeneracy subspace at $a_k = 0$. Since even the perturbation restricted to the eigenspace h has trace zero, we conclude that the derivative of the sum at $a_k = 0$ is at most greater or equal to zero.

For the strict monotonicity of the largest eigenvalue $\lambda_1(L(\{a_i\}))$ in the cube $0 < a_i \le 1$, $i \in \{1, \ldots, m+1\}$ note that by the Frobenius–Perron theorem the corresponding eigenvector $\Phi(\{a_i\})$ has only positive entries. Consequently for $0 < a_i < a_i' \le 1$, all $i \in \{1, \ldots, m+1\}$, the minmax principle implies

$$\begin{aligned}\lambda_1(L(\{a_i\})) &= \langle \Phi(\{a_i\}), L(\{a_i\})\Phi(\{a_i\})\rangle \\ &< \langle \Phi(\{a_i\}), L(\{a_i'\})\Phi(\{a_i\})\rangle \\ &\leq \langle \Phi(\{a_i'\}), L(\{a_i'\})\Phi(\{a_i'\})\rangle = \lambda_1(\{a_i'\})\end{aligned}$$

Remark: The above reasoning for the toy model remains valid if L is replaced by MLM where M is a multiplication operator, i.e. a diagonal matrix, so that the partial sums of the eigenvalues for MLM are also monotone.

To apply this reasoning to our operator $\mathcal{L}_\mu$ it is enough to show the monotonicity (10) for finite discrete measures $\kappa = \sum c_j \delta_{x_j}$ and $\kappa' = \sum c_j' \delta_{x_j}$ with $c_j' \geq c_j$. Indeed, approximate κ and $\kappa' - \kappa$ by finite sums κ_m and Δ_m of δ-functions. This is possible since they are weakly dense in the set of locally finite Borel-measures. It is easy to see that the corresponding operators $\mathcal{L}^{\kappa_m}$ and $\mathcal{L}^{\kappa_m+\Delta_m}$ converge in Hilbert-Schmidt norm to $\mathcal{L}^\kappa$ and $\mathcal{L}^{\kappa'}$. Monotonicity of the partial sums of eigenvalues of $\mathcal{L}^\kappa$ for arbitrary κ then follows by approximation and, without loss of generality, we may assume

$$\kappa = \sum_{j=1}^m c_j \delta_{x_j}, \quad \kappa' = \sum_{j=1}^m c_j' \delta_{x_j} \qquad \text{for some } m \in \mathbb{N}$$

with $c_j' \geq c_j \geq 0$, $j \in \{1, \ldots, m\}$, and $-\infty < x_1 < \ldots < x_m < \infty$. For $x \leq y$ we infer

$$|J(x) - J(y)| = \int_x^y \kappa(dz) = \sum_{x \leq x_j \leq y} c_j$$

and

$$\begin{aligned}\mathcal{L}^{\kappa_m}(x,y) &= \sqrt{U(x)} \exp\Big(-\sum_{x \leq x_j \leq y} c_j\Big) \sqrt{U(y)} \\ &= \prod_{x \leq x_j \leq y} \mathrm{e}^{-c_j} \sqrt{U(x)}\sqrt{U(y)} \\ &= \prod_{x \leq x_j \leq y} a_j \sqrt{U(x)}\sqrt{U(y)}, \qquad a_j := \mathrm{e}^{-c_j},\ j = 1, \ldots, m \\ &=: \mathcal{L}(\{a_j\})(x,y). \end{aligned} \tag{11}$$

As in the matrix case the dependence of $\mathcal{L}(\{a_i\})$ on a single a_k (for fixed $\{a_j\}_{j\neq k}$) is affine-linear and decomposition of the Hilbert space is now given by $\mathrm{L}^2(\mathbb{R}) = \mathrm{L}^2(-\infty, x_k) \oplus \mathrm{L}^2(x_k, \infty)$. Hence we are in precisely the same situation as for our MLM toy–model, and we infer that the partial sums of the largest eigenvalues are monotone in κ for $\mathcal{L}^{\kappa m}$. By the above limiting argument therefor for $\mathcal{L}^{\kappa}$ and in particular for $\mathcal{L}_{\mu}$.

Strict monotonicity of the largest eigenvalue $\lambda_1(\mathcal{L}^{\kappa})$ in κ, i.e. $\lambda_1(\mathcal{L}^{\kappa'}) < \lambda_1(\mathcal{L}^{\kappa})$ if $\kappa' > \kappa$, follows from the Perron-Frobenius theorem, the minmax principle, and the strict monotonicity of the kernel (5) in κ. One can, however, avoid the minmax principle in this conclusion. The Perron-Frobenius theorem states that the eigenvectors Φ_1^{κ} and $\Phi_1^{\kappa'}$ corresponding to $\lambda_1(\mathcal{L}^{\kappa})$ and $\lambda_1(\mathcal{L}^{\kappa'})$ are non-negative and strictly positive on the support of the potential U. By definition

$$\lambda_1(\mathcal{L}^{\kappa})\Phi_1^{\kappa} = \mathcal{L}^{\kappa}\,\Phi_1^{\kappa}$$

and the same for κ'. From this we get

$$\lambda_1(\mathcal{L}^{\kappa'})\langle\Phi_1^{\kappa},\Phi_1^{\kappa'}\rangle - \lambda_1(\mathcal{L}^{\kappa})\langle\Phi_1^{\kappa'},\Phi_1^{\kappa}\rangle = \langle\Phi_1^{\kappa},\mathcal{L}^{\kappa'}\Phi_1^{\kappa'}\rangle - \langle\Phi_1^{\kappa'},\mathcal{L}^{\kappa}\Phi_1^{\kappa}\rangle. \quad (12)$$

since $\langle\Phi_1^{\kappa},\Phi_1^{\kappa'}\rangle > 0$ and the scalar products in (12) are real, hence symmetric, we get by interchanging the integration variables

$$\begin{aligned}\lambda_1(\mathcal{L}^{\kappa'}) - \lambda_1(\mathcal{L}^{\kappa}) &= \frac{1}{\langle\Phi_1^{\kappa},\Phi_1^{\kappa'}\rangle}\iint \Phi_1^{\kappa}(x)\Phi_1^{\kappa'}(y)(\mathcal{L}^{\kappa'}(x,y) - \mathcal{L}^{\kappa}(x,y))\,dxdy \\ &< 0\end{aligned}$$

by the strict monotonicity of the kernel $\mathcal{L}^{\kappa}(x,y)$ in κ and the strict positivity of Φ_1^{κ}, $\Phi_1^{\kappa'}$ on the support of U. This concludes the proof of the monotonicity lemma. ∎

3 Extension to 'potentials' that are measures

In this section we extend theorem 1 to measure perturbations of $-\partial_x^2$. As mentioned in the introduction the Sobolev inequality in one dimension, cf. [8][Theorem 8.5], ensures that a finite measure τ on $\mathbb{R}$ yields a quadratic form $\tau[\phi] := \int |\phi(x)|^2\,\tau(dx)$ that is infinitesimally form bounded with respect to the Laplacian in one dimension. The quadratic form

$$\begin{aligned}\langle\psi, H\phi\rangle &= \langle\psi, -\partial_x^2\phi\rangle + \langle\psi,\tau\phi\rangle \\ &:= \langle\partial_x\psi,\partial_x\phi\rangle + \int_{\mathbb{R}} \overline{\psi(x)}\phi(x)\,\tau(dx)\end{aligned} \quad (13)$$

is thus closed on the Sobolev space $H^1(\mathbb{R})$ and defines a unique self-adjoint operator $H = -\partial_x^2 + \tau$ on $\mathrm{L}^2(\mathbb{R})$. By the minmax principle for forms it is again enough to consider the case $\tau = -\nu$ for some positive bounded measure ν on $\mathbb{R}$. We will hence consider $H = -\partial_x^2 - \nu$. Our result is

Theorem 5. *Suppose ν is a non-negative measure with $\nu(\mathbb{R}) < \infty$ and let $-E_1 < -E_2 \le -E_3 \le \ldots \le 0$ be the negative eigenvalues counting multiplicity of the Schrödinger operator $-\partial_x^2 - \nu$ (if any) given by the corresponding quadratic form. Then*

$$\sum_{i=1}^{\infty} \sqrt{E_i} \le \frac{1}{2}\nu(\mathbb{R}) \tag{14}$$

with equality if and only if the measure ν is a single Dirac measure.

Proof. One obstacle in the proof of this theorem is to construct an analog of the Birman-Schwinger kernel (3) for measures. It is given by

$$\widetilde{K}_E[\nu](x,y) := \int \frac{1}{\sqrt{p^2+E}}(x,\zeta)\frac{1}{\sqrt{p^2+E}}(\zeta,y)\,\nu(d\zeta) \tag{15}$$

where we set $p^2 := -\partial_x^2$ for convenience. A given measure ν can be approximated by smooth functions by convoluting it with an approximate δ-function $\nu \to \nu_\varepsilon = \delta_\varepsilon * \nu$. Of course $\nu_\varepsilon \to \nu$ weakly and the operators $\widetilde{K}_E[\nu_\varepsilon]$ converge to $\widetilde{K}_E[\nu]$ for large E in Hilbert–Schmidt norm, hence in the usual operator norm, too. By Tiktopoulos' formula [14] this shows the norm convergence of the resolvents $(p^2-\nu_\varepsilon+E)^{-1}$ to $(p^2-\nu+E)^{-1}$ and thus any finite collection of eigenvalues of $p^2-\nu_\varepsilon$ converges to those of $p^2-\nu$. So, applying the results of the last section, we have for any partial sum, i.e. any $n \in \mathbb{N}$

$$\begin{aligned} 2\sum_{i\le n}\sqrt{E_i} &\le \lim_{\varepsilon\to 0}\int \nu_\varepsilon(x)\,dx + \lim_{\varepsilon\to 0}\left(\lambda_1(\mathcal{L}_{\sqrt{E_1}}[\nu_\varepsilon]) - \lambda_1(\mathcal{L}_{\sqrt{E_2}}[\nu_\varepsilon])\right) \\ &= \int \nu(dx) + \lim_{\varepsilon\to 0}\left(\lambda_1(\mathcal{L}_{\sqrt{E_1}}[\nu_\varepsilon]) - \lambda_1(\mathcal{L}_{\sqrt{E_2}}[\nu_\varepsilon])\right) \end{aligned}$$

where for $\mu \ge 0$ the operator $\mathcal{L}_\mu[\nu_\varepsilon]$ is defined by the right hand side of (4) with $U(x)$ replaced by $\nu_\varepsilon(x)$. For any positive bounded measure ν let $\widetilde{\mathcal{L}}_\mu[\nu] = 2\mu(p^2+\mu^2)^{-1/2}\nu(p^2+\mu^2)^{-1/2}$ be defined by its kernel

$$\widetilde{\mathcal{L}}_\mu[\nu](x,y) := 2\mu\int \frac{1}{\sqrt{p^2+\mu^2}}(x,\zeta)\frac{1}{\sqrt{p^2+\mu^2}}(\zeta,y)\,\nu(d\zeta).$$

Since the spectrum of an operator of the form $AA^\dagger$ is the same as that of $A^\dagger A$ except at zero we conclude for $\mu \ge 0$

$$\lambda_1(\mathcal{L}_\mu[\nu_\varepsilon]) = \lambda_1(\widetilde{\mathcal{L}}_\mu[\nu_\varepsilon]) \underset{\varepsilon\to 0}{\longrightarrow} \lambda_1(\widetilde{\mathcal{L}}_\mu[\nu])$$

since $\lambda_1(\mathcal{L}_\mu[\nu_\varepsilon]) > 0$ and the operators $\widetilde{\mathcal{L}}_\mu[\nu_\varepsilon]$ converge to $\widetilde{\mathcal{L}}_\mu[\nu]$ in Hilbert–Schmidt norm as $\varepsilon \to 0$. Thus the equivalent of (9) in the measure case is given by

$$2\sum_{i\in\mathbb{N}}\sqrt{E_i} \le \nu(\mathbb{R}) + \lambda_1(\widetilde{\mathcal{L}}_{\sqrt{E_1}}[\nu]) - \lambda_1(\widetilde{\mathcal{L}}_{\sqrt{E_2}}[\nu]) \tag{16}$$

By the Perron–Frobenius theorem for quadratic forms we know that the lowest negative eigenvalue $-E_1$ of $p^2-\nu$ is simple, ie. $E_1 > E_2$. So (14) will follow from (16) once we prove that $0 \le \mu \mapsto \lambda_1(\widetilde{\mathcal{L}}_\mu[\nu])$ is (strictly) monotone decreasing. The operator $\widetilde{\mathcal{L}}_\mu[\nu]$ is given by a strictly positive integral kernel and hence the eigenvector ϕ_μ corresponding to the largest eigenvalue is strictly positive. Rewriting $\widetilde{\mathcal{L}}_\mu[\nu]\phi_\mu = \lambda_1(\widetilde{\mathcal{L}}_\mu[\nu])\phi_\mu$ with $\psi_\mu = (p^2+\mu^2)^{1/2}\phi_\mu > 0$ we get $2\mu(p^2+\mu^2)^{-1}\nu\psi_\mu = \lambda_1(\widetilde{\mathcal{L}}_\mu[\nu])\psi_\mu$. Consequently for $0 \le \mu_1, \mu_2$

$$\lambda_1(\widetilde{\mathcal{L}}_{\mu_1}[\nu])\langle\psi_{\mu_2}, \nu\psi_{\mu_1}\rangle = 2\mu_1\langle\psi_{\mu_2}, \nu\frac{1}{p^2+\mu_1^2}\nu\psi_{\mu_1}\rangle$$

and similarly for $\lambda_1(\widetilde{\mathcal{L}}_{\mu_2}[\nu])$ with μ_1 and μ_2 interchanged. As in the end of the proof of Lemma 4 we can substract these equations and interchange the integration variables to arrive at

$$\begin{aligned}&\lambda_1(\widetilde{\mathcal{L}}_{\mu_1}[\nu]) - \lambda_1(\widetilde{\mathcal{L}}_{\mu_2}[\nu])\\ &= \frac{1}{\langle\psi_{\mu_1}, \nu\psi_{\mu_2}\rangle}\iint \nu(dx)\nu(dy)\psi_{\mu_1}(x)\psi_{\mu_2}(y)\left[\mathrm{e}^{-\mu_1|x-y|} - \mathrm{e}^{-\mu_2|x-y|}\right]\\ &< 0 \quad \text{for } 0 \le \mu_2 < \mu_1\end{aligned}$$

if ν is not concentrated at one point. ∎

Acknowledgment: D.H. and L.T. would like to thank the physics department of Princeton university for its warm hospitality and we thank Wolfgang Spitzer for discussions. — The authors also thank the following organizations for their support: Deutsche Forschungsgemeinschaft, grant Hu 773/1-1 (DH), and the U.S. National Science Foundation, grant PHY95-13072 A02 (EHL), and grant DMS 9801329 (LET).

References

[1] M. Aizenmann and E. H. Lieb: *On semi-classical bounds for eigenvalues of Schrödinger operators.* Phys. Lett. **66A** (1978), 427-429.

[2] M. S. Birman: *The spectrum of singular boundary problems.* Mat. Sb. **55** No.2 (1961), 125-174, translated in Amer. Math. Soc. Trans. (2), **53** (1966), 23-80.

[3] J. G. Conlon: *A new proof of the Cwikel-Lieb-Rosenbljum bound.* Rocky Mountain J. Math., **15**, no.1 (1985), 117–122.

[4] M. Cwikel: *Weak type estimates for singular values and the number of bound states of Schrödinger operators.* Trans. AMS, **224** (1977), 93-100.

[5] L. D. Landau and E. M. Lifshitz: *Quantum Mechanics. Non-relativistic theory.* Volume **3** of Course of Theoretical Physics, Pergamon Press (1958)

[6] P. Li and S.-T. Yau: *On the Schrödinger equation and the eigenvalue problem.* Comm. Math. Phys., **88** (1983), 309-318.

[7] E. H. Lieb: *The number of bound states of one body Schrödinger operators and the Weyl problem.* Bull. Amer. Math. Soc., **82** (1976), 751-753. See also Proc. A.M.S. Symp. Pure Math. **36** (1980), 241-252.

[8] E. H. Lieb and M. Loss: *Analysis.* Graduate Studies in Mathematics **14**, American Mathematical Society 1997.

[9] E. H. Lieb and W. Thirring: *Bound for the kinetic energy of fermions which proves the stability of matter.* Phys. Rev. Lett., **35** (1975), 687-689. Errata **35** (1975), 1116.

[10] E. H. Lieb and W. Thirring: *Inequalities for the moments of the eigenvalues of the Schrödinger Hamiltonian and their relation to Sobolev inequalities.* Studies in Math. Phys., Essays in Honor of Valentine Bargmann, Princeton (1976),

[11] G. V. Rozenbljum: *Distribution of the discrete spectrum of singular differential operators.* Dokl. AN SSSR, **202**, **N 5** 1012-1015 (1972), Izv. VUZov, Matematika, **N.1**(1976), 75-86.

[12] M. Reed and B. Simon: *Methods of modern mathematical physics IV: Analysis of operators.* Academic Press, New York 1978.

[13] J. Schwinger: *On the bound states of a given potential.* Proc. Nat. Acad. Sci. U.S.A. **47**, (1961), 122–129.

[14] B. Simon: *Quantum mechanics for Hamiltonians defined as quadratic forms.* Princeton Series in Physics, Princeton University press, New Jersey, 1971.

[15] B. Simon: *The bound state of weakly coupled Schrdinger operators in one and two dimensions.* Ann. Physics **97**, no. 2, (1976), 279–288.

[16] W. Thirring: *A course in mathematical physics. Vol. 3. Quantum mechanics of atoms and molecules.* Translated from the German by Evans M. Harrell. Lecture Notes in Physics, **141**. Springer-Verlag, New York-Vienna, 1981.

[17] T. Weidl: *On the Lieb-Thirring constants $L_{\gamma,1}$ for $\gamma \geq 1/2$.* Comm. Math. Phys., **178**, no. 1, (1996), 135–146.

Part IV

Coherent States

Commun. math. Phys. 31, 327—340 (1973)

The Classical Limit of Quantum Spin Systems

Elliott H. Lieb*

Institut des Hautes Etudes Scientifiques, Bures-sur-Yvette, France

Received February 28, 1973

Abstract. We derive a classical integral representation for the partition function, Z^Q, of a quantum spin system. With it we can obtain upper and lower bounds to the quantum free energy (or ground state energy) in terms of two classical free energies (or ground state energies). These bounds permit us to prove that when the spin angular momentum $J \to \infty$ (but after the thermodynamic limit) the quantum free energy (or ground state energy) is equal to the classical value. In normal cases, our inequality is $Z^C(J) \leqq Z^Q(J) \leqq Z^C(J+1)$.

I. Introduction

It is generally believed in statistical mechanics that if one takes a quantum spin system of N spins, each having angular momentum J, normalizes the spin operators by dividing by J, and takes the limit $J \to \infty$, then one obtains the corresponding classical spin system wherein the spin variables are replaced by classical vectors and the trace is replaced by an integration over the unit sphere. Indeed, Millard and Leff [1] have shown this to be true for the Heisenberg model *when N is held fixed*. Their proof is quite complicated and it is therefore not surprising that this goal was not achieved before 1971. Despite that success, however, the problem is not finished. One wants to show that one can interchange the limit $N \to \infty$ with the limit $J \to \infty$, i.e. is the classical system obtained if we first let $N \to \infty$ and then let $J \to \infty$? In the Millard-Leff proof the control over the N dependence of the error is not good enough to achieve this desideratum.

A more useful result, and one which would include the above, would be to obtain, for each J, upper and lower bounds to the quantum free energy in terms of the free energies of two classical systems such that those two bounds have a common classical limit as $J \to \infty$. In this paper we do just that, and the result is surprisingly simple: In most cases of interest (including the Heisenberg model), the classical upper bound is

* On leave from the Department of Mathematics, M.I.T., Cambridge, Mass. 02139, USA. Work partially supported by National Science Foundation Grant GP-31674X and by a Guggenheim Memorial Foundation Fellowship.

obtained by replacing the quantum spin by $(J+1)$ times the classical unit vector, while the lower bound is obtained by using J instead of $(J+1)$. Symbolically,

$$Z^C(J) \leqq Z^Q(J) \leqq Z^C(J+1). \tag{1.1}$$

In other cases the result is a little more complicated to state, but it is of the same nature. With an upper and lower bound in hand, it is then possible to derive rigorous bounds on expectation values, as we shall describe in Sections V and VI.

The main tool in our derivation will be what has been termed by Arrechi *et al.* [2] the Bloch coherent state representation. These states and some of their properties were obtained earlier [3, 4], but the most complete account is in Ref. [2]. Our lower bound is obtained by a variational calculation, while the upper bound is obtained from a representation of the quantum partition function that bears some similarity to the Wiener (or path) integral. Apart from its use in deriving the upper bound, the representation may be of theoretical value in proving other properties of quantum spin systems. In particular, it provides a sensible definition of the quantum partition function for all complex J, not just when J is half an integer, and one may discuss the existence or non-existence of a phase transition as a function of the continuous parameter J.

In a forthcoming paper [7] it will be shown how to apply the methods and bounds developed herein (using not only the Bloch states but the Glauber coherent photon states as well) to certain models of the interaction of atoms with a quantized radiation field, for example the Dicke Maser model.

II. Bloch Coherent States

In this section we recapitulate results derived in Refs. [2] and [3]. We consider a single quantum spin of fixed total angular-momentum and shall denote by $\boldsymbol{S} \equiv (S_x, S_y, S_z)$ the usual angular momentum operators:

$$[S_x, S_y] = iS_z, \quad \text{and cyclically,} \quad S_\pm = S_x \pm iS_y. \tag{2.1}$$

We denote by J the total angular momentum, i.e.

$$S^2 = S_x{}^2 + S_y{}^2 + S_z{}^2 = J(J+1). \tag{2.2}$$

The Hilbert space on which these operators act has dimension $2J+1$, i.e. it is $\mathbb{C}^{2J+1}$.

On the classical side, we denote by $\mathscr{S}$ the unit sphere in three dimensions:

$$\mathscr{S} = \{(x, y, z) \mid x^2 + y^2 + z^2 = 1\}, \tag{2.3}$$

and by $L^2(\mathscr{S})$ the space of square integrable functions on $\mathscr{S}$ with the usual measure

$$\Omega = (\theta, \varphi), \ 0 \leqq \theta \leqq \pi, \ 0 \leqq \varphi < 2\pi, \tag{2.4}$$

$$d\Omega = \sin\theta \, d\theta \, d\varphi, \tag{2.5}$$

$$x = \sin\theta \cos\varphi, \ y = \sin\theta \sin\varphi, \ z = \cos\theta. \tag{2.6}$$

(Note: In Ref. [2], but not Ref. [3] the "south pole", instead of the customary "north pole" corresponds to $\theta = 0$. Hence our formulas will differ from Ref. [2] by the replacement $\theta \to \pi - \theta$).

With $|J\rangle \in \mathbb{C}^{2J+1}$ being a normalized "spin up" state, $S_z|J\rangle = J|J\rangle$, one defines the Bloch state $|\Omega\rangle \in \mathbb{C}^{2J+1}$ by

$$\begin{aligned} |\Omega\rangle &= \exp\{\tfrac{1}{2}\theta[S_- e^{i\varphi} - S_+ e^{-i\varphi}]\} |J\rangle \\ &= [\cos\tfrac{1}{2}\theta]^{2J} \exp\{(\tan\tfrac{1}{2}\theta) e^{i\varphi} S_-\} |J\rangle \\ &= \sum_{M=-J}^{J} \binom{2J}{M+J}^{1/2} (\cos\tfrac{1}{2}\theta)^{J+M} (\sin\tfrac{1}{2}\theta)^{J-M} \exp[i(J-M)\varphi] |M\rangle \end{aligned} \tag{2.7}$$

where $|M\rangle$ is the normalized state

$$|M\rangle = \binom{2J}{M+J}^{-1/2} [(J-M)!]^{-1} (S_-)^{J-M} |J\rangle \tag{2.8}$$

such that

$$S_z|M\rangle = M|M\rangle. \tag{2.9}$$

It is clear from (2.7) that the set of states $|\Omega\rangle$ are complete in $\mathbb{C}^{2J+1}$. Their overlap is given by

$$\begin{aligned} K_J(\Omega', \Omega) &\equiv \langle \Omega' | \Omega \rangle \\ &= \{\cos\tfrac{1}{2}\theta \cos\tfrac{1}{2}\theta' + e^{i(\varphi-\varphi')} \sin\tfrac{1}{2}\theta \sin\tfrac{1}{2}\theta'\}^{2J} \end{aligned} \tag{2.10}$$

so that if we think of $K_J(\Omega', \Omega)$ as the kernel of a linear transformation on $L^2(\mathscr{S})$ it is selfadjoint and compact. In fact, it is positive semidefinite. We also have

$$|K_J(\Omega', \Omega)|^2 = [\cos\tfrac{1}{2}\Theta]^{4J}, \tag{2.11}$$

where

$$\cos\Theta = \cos\theta \cos\theta' + \sin\theta \sin\theta' \cos(\varphi - \varphi') \tag{2.12}$$

Commun. Math. Phys. 31, 327–340 (1973)

is the cosine of the angle between Ω and Ω'. In particular $|\Omega\rangle$ is normalized since $K_J(\Omega,\Omega)=1$.

Now let $\mathscr{M}^{2J+1}$ be the set of linear transformations on $\mathbb{C}^{2J+1}$ (i.e. operators on the spin space) and, for a given $G\in L^1(\mathscr{S})$, define $A_G\in\mathscr{M}^{2J+1}$ by

$$A_G=\frac{2J+1}{4\pi}\int d\Omega\, G(\Omega)\,|\Omega\rangle\langle\Omega|\,. \tag{2.13}$$

$\left(\text{Note: } \int d\Omega \text{ always means } \int_{\mathscr{S}} d\Omega\right)$. Since the Hilbert space is finite dimensional there is no problem in giving a meaning to (2.13). It is a remarkable fact that every operator in $\mathscr{M}^{2J+1}$ can be written in the form (2.13). In particular,

$$\mathbf{1}=\frac{2J+1}{4\pi}\int d\Omega\,|\Omega\rangle\langle\Omega|\,. \tag{2.14}$$

Thus, to every operator $A\in\mathscr{M}^{2J+1}$ there correspond two functions:

$$g(\Omega)=\langle\Omega|\,A\,|\Omega\rangle\,, \tag{2.15}$$

and the $G(\Omega)$ of (2.13). The former is, of course, unique, but the latter is not. However, it is always possible to choose $G(\Omega)$ to be infinitely differentiable. In Table 1 we list some function pairs for operators of common interest and useful formulas for calculation are given in Appendix A.

Table 1. Expectation values, $g(\Omega)$, and operator kernels, $G(\Omega)$, [cf. (2.13), (2.15)] for various operators commonly appearing in quantum spin Hamiltonians

Operator	$g(\Omega)$, (2.15)	$G(\Omega)$, (2.13)
S_z	$J\cos\theta$	$(J+1)\cos\theta$
S_x	$J\sin\theta\cos\varphi$	$(J+1)\sin\theta\cos\varphi$
S_y	$J\sin\theta\sin\varphi$	$(J+1)\sin\theta\sin\varphi$
$S_z{}^2$	$J(J-\frac{1}{2})(\cos\theta)^2+J/2$	$(J+1)(J+3/2)(\cos\theta)^2-\frac{1}{2}(J+1)$
$S_x{}^2$	$J(J-\frac{1}{2})(\sin\theta\cos\varphi)^2+J/2$	$(J+1)(J+3/2)(\sin\theta\cos\varphi)^2-\frac{1}{2}(J+1)$
$S_y{}^2$	$J(J-\frac{1}{2})(\sin\theta\cos\varphi)^2+J/2$	$(J+1)(J+3/2)(\sin\theta\cos\varphi)^2-\frac{1}{2}(J+1)$

We need three final remarks. The first is that if we consider $|\Omega\rangle\langle\Omega'|\in\mathscr{M}^{2J+1}$ then

$$\operatorname{Tr}|\Omega\rangle\langle\Omega'|=K_J(\Omega',\Omega) \tag{2.16}$$

(where Tr means Trace) as may be seen from (2.7). Hence, from (2.13)

$$\operatorname{Tr} A_G = \frac{2J+1}{4\pi} \int d\Omega\, G(\Omega)\,. \tag{2.17}$$

The second is that

$$\frac{2J+1}{4\pi} \int d\Omega\, K_J(\Omega', \Omega)\, K_J(\Omega, \Omega'') = K_J(\Omega', \Omega'')\,, \tag{2.18}$$

as may be seen from (2.14). Thus, K_J reproduces itself under convolution.

The third remark is that for any $A \in \mathcal{M}^{2J+1}$ we can use (2.14) to obtain

$$\begin{aligned} \operatorname{Tr} A &= \frac{2J+1}{4\pi} \int d\Omega\, \operatorname{Tr} |\Omega\rangle \langle \Omega|\, A \\ &= \frac{2J+1}{4\pi} \int d\Omega \sum_{M=-J}^{J} \langle M|\Omega\rangle \langle \Omega|\, A\, |M\rangle \\ &= \frac{2J+1}{4\pi} \int d\Omega\, \langle \Omega|\, A\, |\Omega\rangle\,. \end{aligned} \tag{2.19}$$

III. Lower Bound to the Quantum Partition Function

We consider a system of N quantum spins and shall label the operators and the angular momenta (which need not all be the same) by a superscript i, $i = 1, \ldots, N$. The Hamiltonian, H, can be completely general but, in any event, it can always be written as a polynomial in the $3N$ spin operators. The partition function is

$$Z^Q = \alpha_N \operatorname{Tr} \exp(-\beta H)\,, \tag{3.1}$$

where

$$\alpha_N = \prod_{i=1}^{N} (2J^i + 1)^{-1}\,. \tag{3.2}$$

[The normalization factor α_N is inessential; it is chosen to agree with the classical partition function when $\beta = 0$]. The Hilbert space is

$$\mathcal{H}_N = \bigotimes_{i=1}^{N} \mathcal{H}^i = \bigotimes_{i=1}^{N} \mathbb{C}^{2J^i+1}\,. \tag{3.3}$$

We denote by $|\Omega_N\rangle$ the complete, normalized set of states on $\mathcal{H}_N$ defined by

$$|\Omega_N\rangle = \bigotimes_{i=1}^{N} |\Omega^i\rangle\,, \tag{3.4}$$

Commun. Math. Phys. 31, 327–340 (1973)

by $\mathscr{S}_N$ the Cartesian product of N copies of the unit sphere, and by $d\Omega_N$ the product measure (2.4), (2.5) and (2.6) on $\mathscr{S}_N$. Using (2.19),

$$Z^Q = (4\pi)^{-N} \int d\Omega_N \langle \Omega_N | e^{-\beta H} | \Omega_N \rangle . \tag{3.5}$$

By the Peierls-Bogoliubov inequality, $\langle \psi | e^X | \psi \rangle \geqq \exp \langle \psi | X | \psi \rangle$ for any normalized $\psi \in \mathscr{H}_N$ and X selfadjoint. Thus,

$$Z^Q \geqq (4\pi)^{-N} \int d\Omega_N \exp\{-\beta \langle \Omega_N | H | \Omega_N \rangle\} . \tag{3.6}$$

Suppose, at first, that the polynomial, H, is linear in the operators $\boldsymbol{S}^i$ of each spin. That is we allow multiple site interactions of arbitrary complexity such as $S_x^1 S_y^2 S_y^3 S_z^4$, but do not allow monomials such as $(S_x^1)^2$ or $S_x^1 S_y^1$. In this case, which we shall refer to as *the normal case*, we see from (2.15) and Table 1 that the right side of (3.6) is precisely the classical partition function in which each $\boldsymbol{S}^i$ is replaced by J^i times a vector in $\mathscr{S}$. I.e.

$$\boldsymbol{S}^i \to J^i (\sin\theta^i \cos\varphi^i, \sin\theta^i \sin\varphi^i, \cos\theta^i) . \tag{3.7}$$

Thus, in the normal case,

$$Z^Q \geqq Z^C (J^1, \ldots, J^N) , \tag{3.8}$$

where Z^C means the classical partition function (with the normalization $(4\pi)^{-N}$).

In more complicated cases, (3.7) is not correct and S_z^1, for example, has to be replaced by $J^1 \cos\theta^1$ if it appears linearly in H, $(S_z^1)^2$ has to be replaced by $[J^1 \cos\theta^1]^2 + J^1 (\sin\theta^1)^2/2$ and so forth (see Table 1). However, to leading order in J^i, (3.7) is correct.

We note in passing that it is not necessary to use the Peierls-Bogoliubov inequality for all operators appearing in H. Thus, suppose the whole Hilbert space is $\mathscr{H}' = \mathscr{H} \otimes \mathscr{H}_N$ where $\mathscr{H}$ is the Hilbert space of some additional degrees of freedom (which may or may not themselves be spins) and H is selfadjoint on $\mathscr{H}'$. Then (by a generalized Peierls-Bogoliubov inequality)

$$\begin{aligned} Z^Q &= \alpha_N \operatorname{Tr}_{\mathscr{H}'} \operatorname{Tr}_{\mathscr{H}} \exp(-\beta H) \\ &\geqq \operatorname{Tr}_{\mathscr{H}} (4\pi)^{-N} \int d\Omega_N \exp\{-\beta \langle \Omega_N | H | \Omega_N \rangle\} \end{aligned} \tag{3.9}$$

where $\langle \Omega_N | H | \Omega_N \rangle$ is a partial expectation value and defines a selfadjoint operator on $\mathscr{H}$. We shall give an example of (3.9) in Appendix B. It is clear that if $\mathscr{H}$ is itself a spin space, then (3.9) gives a better bound than (3.6) applied to the full space $\mathscr{H}'$.

IV. Upper Bound to the Quantum Partition Function

Returning to the definitions (3.1) and (3.3) we note that

$$Z^Q = \lim_{n\to\infty} Z(n), \tag{4.1}$$

where

$$Z(n) = \alpha_N \operatorname{Tr}(\mathbf{1} - \beta n^{-1} H)^n . \tag{4.2}$$

Now, let H be represented by some $G(\Omega_N)$ as in (2.13), whence $\mathbf{1} - \beta n^{-1} H$ is represented by

$$F_n(\Omega_N) = 1 - \beta n^{-1} G(\Omega_N) . \tag{4.3}$$

Using (2.10), (2.13) and (2.16), we can represent Z_n as an nN fold integral:

$$Z(n) = \alpha_N \int d\Omega_N{}^1 \cdots \int d\Omega_N{}^n \prod_{j=1}^{n} F_n(\Omega_N{}^j)\, L_J(\Omega_N{}^j, \Omega_N{}^{j+1}) \tag{4.4}$$

with $n+1 \equiv 1$ in the last factor, and where

$$L_J(\Omega_N{}', \Omega_N) \equiv (4\pi)^{-N} \alpha_N{}^{-1} \prod_{i=1}^{N} K_{J^i}(\Omega'^i, \Omega^i) . \tag{4.5}$$

Thus

$$L_J(\Omega_N, \Omega_N) = (4\pi)^{-N} a_N{}^{-1} . \tag{4.6}$$

$$\int d\Omega_N\, L_J(\Omega_N{}', \Omega_N)\, L_J(\Omega_N, \Omega_N{}'') = L_J(\Omega_N{}', \Omega_N{}'') . \tag{4.7}$$

Equations (4.1) and (4.4) are our desired integral representation for Z^Q. To use them to obtain a bound, we think of F_n as a multiplication operator and of L_J as the kernel of a compact, selfadjoint operator on $L^2(\mathscr{S}_N)$. If $B(\Omega_N{}', \Omega_N)$ is such a kernel, then

$$\operatorname{Tr} B = \int d\Omega_N\, B(\Omega_N, \Omega_N) \tag{4.8}$$

is the trace on $L^2(\mathscr{S}_N)$. Thus,

$$Z(n) = \alpha_N \operatorname{Tr}(F_n L_J)^n . \tag{4.9}$$

In general, if $m = 2^j$, $j = 0, 1, 2, 3, \ldots$,

$$|\operatorname{Tr}(AB)^{2m}| \leqq \operatorname{Tr}(A^2 B^2)^m \leqq \operatorname{Tr} A^{2m} B^{2m} , \tag{4.10}$$

whenever A and B are selfadjoint. This follows from the Schwarz inequality (see Ref. [5] for details). Hence, if we take a sequence $n = 2^j$, $j = 1, 2, \ldots$ in (4.2) and use (4.7) n times and (4.6), we obtain, in the limit

Commun. Math. Phys. 31, 327–340 (1973)

$n\to\infty$,

$$Z^Q \leqq (4\pi)^{-N} \int d\Omega_N \exp[-\beta G(\Omega_N)]\,. \tag{4.11}$$

(4.11) is our desired classical upper bound. It is just like (3.6). In the normal case we see from Table 1 that $\boldsymbol{S}^i$ is replaced by (J^i+1) times a classical unit vector. In other cases, $G(\Omega_N)$ is a bit more complicated, but the same remarks as in Section III apply. Thus, in the normal case

$$Z^C(J^1, \ldots, J^N) \leqq Z^Q \leqq Z^C(J^1+1, \ldots, J^N+1)\,. \tag{4.12}$$

This inequality says that as J increases the quantum and classical free energies form two decreasing, *interlacing* sequences.

As in Section III, if $\mathscr{H}' = \mathscr{H} \otimes \mathscr{H}_N$ an inequality similar to (4.11) can be shown to hold, i.e.

$$Z^Q \leqq \mathrm{Tr}_{\mathscr{H}} (4\pi)^{-N} \int d\Omega_N \exp[-\beta H(\cdot, \Omega_N)]\,, \tag{4.13}$$

where $H(\cdot, \Omega_N)$ is a selfadjoint operator on $\mathscr{H}$ obtained by replacing each monomial in the spin operators in H by the appropriate $G(\Omega_N)$ function found in Table 1. We shall illustrate (4.13) in Appendix B. If $\mathscr{H}$ is a spin space then (4.13) gives a better bound than (4.11) applied to the full $\mathscr{H}'$.

V. Bounds on Expectation Values and the Ground State Energy

The expectation value of a quantum operator (observable), A, is

$$\langle A \rangle^Q = \mathrm{Tr}\, A \exp(-\beta H)/\mathrm{Tr} \exp(-\beta H)\,. \tag{5.1}$$

We can always assume A is selfadjoint (otherwise consider $A + A^\dagger$ and $iA - iA^\dagger$), in which case the Peierls-Bogoliubov inequality reads, for λ real,

$$\lambda \langle A \rangle^Q \geqq f(\lambda) - f(0)\,, \tag{5.2}$$

where

$$f(\lambda) = -\beta^{-1} \ln \mathrm{Tr} \exp[-\beta(H + \lambda A)]\,, \tag{5.3}$$

is a free energy. Hence, with $\lambda > 0$,

$$[f(0) - f(-\lambda)]/\lambda \geqq \langle A \rangle^Q \geqq [f(\lambda) - f(0)]/\lambda\,. \tag{5.4}$$

The upper and lower bounds to $f(\lambda)$ derived in the preceding two sections can be used to advantage in (5.4). In particular, we use (5.4) in the next section to derive $J\to\infty$ limits of quantum expectation values.

If we take the limit $\beta \to \infty$ in (3.1) we obtain bounds on the quantum ground state energy:

$$E_{-}^{C} \leqq E^{Q} \leqq E_{+}^{C}, \tag{5.5}$$

where E^C is the classical ground state energy (i.e. the minimum of the classical Hamiltonian over $\mathscr{S}_N$) and the + (resp. −) refers to the substitution of the appropriate $G(\Omega_N)$ (resp. $g(\Omega_N)$) functions from Table 1. In the normal case

$$E^{C}(J^{1}, \ldots, J^{N}) \geqq E^{Q} \geqq E^{C}(J^{1}+1, \ldots, J^{N}+1). \tag{5.6}$$

As ground state expectation values obey an inequality similar to (5.2), with f replaced by E, a bound similar to (5.4) holds for E. This is merely the variational principle.

The upper bound in (5.6) is easy to obtain directly by a variational calculation, but the lower bound is not. It is not easy to find a direct proof of it in a system consisting of three spins antiferromagnetically coupled to each other.

VI. The Thermodynamic Limit

A. The Free Energy

We shall, for simplicity, consider only the normal case here. The general case can be handled in a similar manner.

Let H_N be a Hamiltonian (polynomial) of N spins in which each spin has angular momentum one. Replace each spin operator $\boldsymbol{S}^i$ by $(J)^{-1}\boldsymbol{S}^i$ and let $\boldsymbol{S}^i$ now have angular momentum J. We shall denote this symbolically by $H_N^Q(J)$ and the partition function, (3.1), by $Z_N^Q(J)$. [It would equally be possible to allow different J values for different spins, but that is a needless complication. Also, the factor J^{-1} is not crucial. One could as well use $J^{-1/2}(J+1)^{-1/2}$]. Denoting the free energy per spin by $f_N(J) = -(N\beta)^{-1}\ln Z_N(J)$, the theorem to be proved is that

$$\lim_{J\to\infty}\lim_{N\to\infty} f_N^Q(J) = f^C \equiv \lim_{N\to\infty} f_N^C, \tag{6.1}$$

where f_N^C is the free energy per spin of the classical partition function in which each $\boldsymbol{S}^i$ is replaced by a classical unit vector. It is assumed that H_N is known to have a thermodynamic limit for the free energy per spin. We also want to prove an analogous formula for the ground state energy per spin. Our bounds are

$$f_N^C \geqq f_N^C(J) \geqq f_N^C(\delta_J), \tag{6.2}$$

Commun. Math. Phys. *31*, 327–340 (1973)

where the right side is the classical free energy per spin in which each vector is multiplied by $\delta_J \equiv (J+1)/J$.

If we think of δ_J as a variable, δ, then $H_N^C(\delta)$, the classical Hamiltonian as a function of δ, is continuous in δ. Moreover, $N^{-1} H_N^C(\delta)$ is equicontinuous in N, i.e. given any $\varepsilon > 0$ it is possible to find a $\gamma > 0$ such that $\|N^{-1}[H_N^C(\delta + x) - H_N^C(\delta)]\| \leqq \varepsilon$ for $|x| < \gamma$, independent of N, where $\| \ \|$ means the uniform on $\mathscr{S}_N$. Hence, the limit function

$$f^C(\delta) \equiv \lim_{N\to\infty} f_N^C(\delta) \tag{6.3}$$

is continuous in δ. This, together with (6.2), proves (6.1).

The same equicontinuity holds for the classical ground state energy. Thus, the analogue of (6.1) is also true for the ground state energy per spin:

$$\lim_{J\to\infty} \lim_{N\to\infty} N^{-1} E_N^Q(J) = \lim_{N\to\infty} E_N^C. \tag{6.4}$$

B. Expectation Values

We consider expectation values of intensive observables $N^{-1} A_N$. For example, A_N might be the Hamiltonian itself, in which case $\langle N^{-1} A_N \rangle$ is the energy per spin. Alternatively, A_N could be $\sum_{i=1}^{N} S_z^i$ so that $\langle N^{-1} A_N \rangle$ is the magnetization per spin. As before, we replace each $\boldsymbol{S}^i$ by $(J)^{-1}$ times a quantum spin of angular momentum J, *both* in the Hamiltonian and in A_N. Then, using inequality (5.4) and the bounds (6.2) we have, for each positive λ, fixed N and fixed J,

$$\begin{aligned} \lambda^{-1}[f_N^C(0;1) - f_N^C(-\lambda;\delta_J)] &\geqq N^{-1}\langle A_N \rangle^Q \\ &\geqq \lambda^{-1}[f_N^C(\lambda;\delta_J) - f_N^C(0;1)], \end{aligned} \tag{6.5}$$

where $f_N^C(\lambda;\delta)$ is the classical free energy per spin when the Hamiltonian is $H_N^C + \lambda A_N^C$ and where each classical spin unit vector in H_N^C and A_N^C is multiplied by δ. We are interested in $\delta_J = (J+1)/J$.

Now take the limit $N \to \infty$ and then the limit $J \to \infty$ in (6.5). By the same equicontinuity remark as in Section VI.A, for each $\lambda > 0$,

$$\limsup_{J\to\infty} \limsup_{N\to\infty} N^{-1}\langle A_N \rangle^Q \leqq \lambda^{-1}[f^C(0) - f^C(-\lambda)],$$

$$\liminf_{J\to\infty} \liminf_{N\to\infty} N^{-1}\langle A_N \rangle^Q \geqq \lambda^{-1}[f^C(\lambda) - f^C(0)]. \tag{6.6}$$

In (6.5), $f^C(\lambda)$ is the limiting classical free energy per spin for the Hamiltonian $H_N^C + \lambda A_N^C$ (with $\delta = 1$). It is easy to see that $f^C(\lambda)$ is concave in λ

and hence $\lim_{\lambda\downarrow 0}\lambda^{-1}[f^C(\lambda)-f^C(0)]\equiv G^+$ and $\lim_{\lambda\downarrow 0}\lambda^{-1}[f^C(0)-f^C(-\lambda)]$ $\equiv G^-$ exist everywhere. If $G^+=G^-$ (i.e. the right derivative equals the left derivative) then by a theorem of Griffiths (6)

$$\lim_{N\to\infty}\frac{d}{d\lambda}f_N^C(\lambda)=\frac{d}{d\lambda}f^C(\lambda). \tag{6.7}$$

This is the case in which the classical expectation value $N^{-1}\langle A_N\rangle^C$ has a well defined limit. Call it α. Then

$$\lim_{J\to\infty}\lim_{N\to\infty}N^{-1}\langle A_N\rangle^Q=\alpha, \tag{6.8}$$

as one sees by taking the limit $\lambda\to 0$ in (6.6). In other words, we have proved that *for intensive observables, as defined above, the quantum expectation value equals the classical expectation value after first taking the thermodynamic limit and then taking the classical limit* $J\to\infty$. If one takes the limits in the opposite order the theorem is trivially true and uninteresting. Note that we have not proved that the quantum thermodynamic limit, $\lim_{N\to\infty}N^{-1}\langle A_N\rangle^Q$ exists. It may not.

The same proof obviously goes through for ground state expectation values, as in Section VI.A, because the ground state energy is also concave in λ.

Acknowledgements. The author thanks the Institut des Hautes Etudes Scientifiques for its hospitality, as well as the Chemistry Laboratory III, University of Copenhagen where part of this work was done. The financial assistance of the Guggenheim Memorial Foundation is gratefully acknowleged. The author also acknowledges his gratitude to Dr. N. W. Dalton who suggested the problem to him in 1967.

Appendix A: Some Useful Formulas

The algebra $\mathcal{M}^{2J+1}$ has S_+, S_- and S_z as generators. Hence, the following generating function permits, by differentiation, easy calculation of $g(\Omega)$ in (2.15) or Table 1 for any operator. It is to be found, with appropriate modifications, in Ref. [2].

$$\langle\Omega|\exp(\gamma S_+)\exp(\beta S_z)\exp(\alpha S_-)|\Omega\rangle \tag{A.1}$$
$$=\{[e^{-i\varphi}\sin\tfrac{1}{2}\theta+\gamma\cos\tfrac{1}{2}\theta]\,[e^{i\varphi}\sin\tfrac{1}{2}\theta+\alpha\cos\tfrac{1}{2}\theta]\,e^{-\beta/2}+e^{\beta/2}[\cos\tfrac{1}{2}\theta]^2\}^{2J}.$$

Turning to (2.13), we calculate A_G for a sufficiently large class of functions $G(\Omega)$. Let

$$G(\Omega)=e^{im\varphi}(\cos\tfrac{1}{2}\theta)^p(\sin\tfrac{1}{2}\theta)^q \tag{A.2}$$

Commun. Math. Phys. *31*, 327–340 (1973)

where m is an integer and p and q are complex numbers. Defining $A(m, p, q) \equiv A_G$, the matrix elements of this operator can be calculated using (2.7) to be

$$\begin{aligned} A(m, p, q; M, M') = \delta(M - M' - m)\,\Gamma(J + \alpha + 1 + p/2)\,\Gamma(J - \alpha + 1 + q/2) \\ \cdot [(J + \alpha + m/2)!\,(J + \alpha - m/2)!\,(J - \alpha - m/2)!\,(J - \alpha + m/2)!]^{-1/2} \\ \cdot (2J + 1)!/\Gamma(2J + 2 + p/2 + q/2)\,, \end{aligned} \tag{A.3}$$

where δ is the Kroenecker delta function, Γ is the gamma function and $\alpha = (M + M')/2$. This formula has been used to calculate Table 1.

Appendix B: Application to the One Dimensional Heisenberg Chain

To illustrate the methods of this paper, we derive bounds for the free energy of a Heisenberg chain whose Hamiltonian is

$$H = -\sum_{i=1}^{N-1} \boldsymbol{S}^i \cdot \boldsymbol{S}^{i+1}\,. \tag{B.1}$$

Each spin is assumed to have angular momentum J. We have chosen the isotropic case for simplicity, but one could equally well handle the anisotropic Hamiltonian with a magnetic field. Note that $\beta > 0$ is the ferromagnetic case while $\beta < 0$ is the antiferromagnetic case.

The classical partition function is

$$Z_N^C(\beta, x) = (4\pi)^{-N} \int d\Omega_N \exp\left\{\beta x^2 \sum_{i=1}^{N-1} \boldsymbol{\Omega}^i \cdot \boldsymbol{\Omega}^{i+1}\right\} \tag{B.2}$$

with free energy per spin

$$f^C(\beta, x) = -\lim_{N\to\infty} (N|\beta|)^{-1} \ln Z_N^C(\beta, x)\,. \tag{B.3}$$

Our bounds are that

$$f^C(\beta, J) \geqq f^Q(\beta, J) \geqq f^C(\beta, J + 1)\,. \tag{B.4}$$

It is easy to evaluate (B.2) by the transfer matrix method. The normalized eigenfunction (of Ω) giving the largest eigenvalue is obviously the constant function $(4\pi)^{-1/2}$. Thus,

$$f^C(\beta, x) = -|\beta|^{-1} \ln A(\beta, x)\,, \tag{B.5}$$

where

$$A(\beta, x) = (4\pi)^{-1} \int d\Omega \exp\{\beta x^2 \boldsymbol{\Omega} \cdot \boldsymbol{\Omega}'\} \\ = (\beta x^2)^{-1} \sinh(\beta x^2), \tag{B.6}$$

and $A(\beta, x)$ is independent of Ω' as it should be. In this approximation, (B.4), one cannot distinguish between the ferro- and antiferromagnetic cases as far as the free energy is concerned.

To illustrate the idea mentioned at the ends of Sections III and IV, we suppose that the chain has $2N+1$ spins and we let $\mathscr{H}_N$ (resp. $\mathscr{H}$) be the Hilbert space for the odd (resp. even) numbered spins. $\mathscr{H}' = \mathscr{H} \otimes \mathscr{H}_N$ is the whole space. Our bounds are

$$g(\beta, J) \geqq f^Q(\beta, J) \geqq g(\beta, J+1), \tag{B.7}$$

where

$$g(\beta, x) = -\lim_{N\to\infty} (2N|\beta|)^{-1} \ln\{(2J+1)^{-N} \tilde{Z}_N(\beta, x)\}, \tag{B.8}$$

$$\tilde{Z}_N(\beta, x) = (4\pi)^{-N} \int d\Omega_N \operatorname{Tr} \exp\left\{\beta x \sum_{i=1}^{N} \boldsymbol{S}^{2i} \cdot (\boldsymbol{\Omega}^{2i-1} + \boldsymbol{\Omega}^{2i+1})\right\} \tag{B.9}$$

and where $d\Omega_N = d\Omega^1\, d\Omega^3 \dots d\Omega^{2N+1}$ and the trace is over the Hilbert space of $\boldsymbol{S}^2, \boldsymbol{S}^4, \dots, \boldsymbol{S}^{2N}$.

Since the remaining spin operators no longer interact, it is easy to calculate the trace. For a single spin:

$$\operatorname{Tr} \exp[b\boldsymbol{S} \cdot \boldsymbol{v}] = \sum_{M=-J}^{J} \exp[bMv] \tag{B.10}$$

where b is a constant and $\boldsymbol{v}$ is a vector of length v. Now we can do the integration over $\mathscr{S}_N$ by the transfer matrix method (with the same eigenvector $(4\pi)^{-1/2}$) and obtain

$$g(\beta, x) = -\tfrac{1}{2}|\beta|^{-1} \ln[A(\beta, x)/(2J+1)], \tag{B.11}$$

where

$$A(\beta, x) = (4\pi)^{-1} \int d\Omega \sum_{M=-J}^{J} \exp\{\beta x M |\boldsymbol{\Omega} + \boldsymbol{\Omega}'|\} \\ = 2\int_0^1 y\, dy \sinh[(2J+1)\beta x y]/\sinh[\beta x y]. \tag{B.12}$$

Again, no distinction between the ferro- and antiferromagnetic cases appears.

Commun. Math. Phys. *31*, 327–340 (1973)

References

1. Millard, K., Leff, H.: J. Math. Phys. **12**, 1000—1005 (1971).
2. Arecchi, F. T., Courtens, E., Gilmore, R., Thomas, H.: Phys. Rev. A**6**, 2211—2237 (1972).
3. Radcliffe, J. M.: J. Phys. A**4**, 313—323 (1971).
4. Kutzner, J.: Phys. Lett. A**41**, 475—476 (1972).
   Atkins, P. W., Dobson, J. C.: Proc. Roy. Soc. (London) A, A**321**, 321—340 (1971).
5. Golden, S.: Phys. Rev. B**137**, 1127—1128 (1965).
6. Griffiths, R. B.: J. Math. Phys. **5**, 1215—1222 (1964).
7. Hepp, K., Lieb, E. H.: The equilibrium statistical mechanics of matter interacting with the quantized radiation field. Preprint.

E. H. Lieb
I.H.E.S.
F-91440 Bures-sur-Yvette, France

Commun. math. Phys. 62, 35—41 (1978)

Communications in Mathematical Physics

Proof of an Entropy Conjecture of Wehrl

Elliott H. Lieb*

Departments of Mathematics and Physics, Princeton University, Princeton, New Jersey 08540, USA

Abstract. Wehrl has proposed a new definition of classical entropy, S, in terms of coherent states and conjectured that $S \geqq 1$. A proof of this is given. We discuss the analogous problem for Bloch coherent spin states, but in this case the conjecture is still open. An inequality for the entropy of convolutions is also given.

I. Introduction

In a recent paper [1], A. Wehrl introduced a new definition of the "classical" entropy corresponding to a quantum system, proved that it had several interesting properties that deserve to be studied further, and posed a conjecture about the minimum value of this "classical" entropy. The main purpose of this paper is to prove Wehrl's conjecture. It is somewhat surprising that while the conjecture appears to be almost obvious, the proof we give requires some difficult theorems in Fourier analysis. The conjecture may or may not be important physically, but it reveals an interesting feature of coherent states.

To briefly recapitulate Wehrl's analysis, consider a single particle in one dimension, so that the Hilbert space is $L^2(\mathbb{R})$. (The generalization to $\mathbb{R}^N$ is trivial.) For each $z=(p,q)\in\mathbb{R}^2$, define the normalized vector $|z\rangle$ in $L^2(\mathbb{R})$ by

$$|z\rangle \equiv (\pi\hbar)^{-1/4}\exp\{[-(x-q)^2/2+ipx]/\hbar\} \equiv R(x|p,q). \tag{1.1}$$

These vectors are the coherent states used by Schrödinger [2], Bargmann [3], Klauder [4], and Glauber [5]. If

$$P_z = |z\rangle\langle z| \tag{1.2}$$

is the orthogonal projection onto $|z\rangle$ then

$$\int \frac{dz}{\pi} P_z = I, \tag{1.3}$$

* Work partially supported by US National Science Foundation grant MCS 75-21684 A02

Commun. Math. Phys. *62*, 35–41 (1978)

where $dz/\pi \equiv dpdq/2\pi\hbar$ and I = identity. The integral in (1.3) can be defined as a weak integral and (1.2) is simply the Plancherel equality.

For a "density matrix" ϱ^Q (a positive semidefinite operator of trace 1) on $L^2(\mathbb{R})$, its quantum entropy is

$$S^Q(\varrho^Q) \equiv -\operatorname{Tr}\varrho^Q \ln \varrho^Q \geqq 0. \tag{1.4}$$

The right side of (1.4) is well defined, although it may be $+\infty$.

For a nonnegative function f on $\mathbb{R}^2$, with $\int f(z)dz/\pi = 1$, its classical entropy is

$$\tilde{S}(f) = -\int \frac{dz}{\pi} f(z) \ln f(z). \tag{1.5}$$

In general this integral may not be well defined, but even if it is it can be negative.

Given a quantum density matrix ϱ^Q, Wehrl defines the function

$$\varrho^{\mathrm{cl}}(z) \equiv \langle z|\varrho^Q|z\rangle, \tag{1.6}$$

whence $0 \leqq \varrho^{\mathrm{cl}}(z) \leqq 1$. Then

$$S^{\mathrm{cl}}(\varrho^Q) \equiv \tilde{S}(\varrho^{\mathrm{cl}}). \tag{1.7}$$

This is the *classical entropy* of ϱ^Q. [Note that by (1.3), $\int \varrho^{\mathrm{cl}}(z)dz/\pi = 1$.] Since $0 \leqq \varrho^{\mathrm{cl}}(z) \leqq 1$, the integral in (1.5) is now well defined, and $S^{\mathrm{cl}} \geqq 0$.

The positivity of S^{cl} is one advantage of Wehrl's definition. On the contrary, if, as is usual, $\varrho^Q = Z_Q^{-1} \exp[-\beta(-\hbar^2\Delta/2m + V(q))]$, the customary classical approximation is $f(z) = Z_{\mathrm{cl}}^{-1} \exp[-\beta(p^2/2m + V(q))]$. The difficulty with f is that $\tilde{S}(f)$ can be negative and, in general, $\tilde{S}(f) \to -\infty$ as $\beta \to \infty$.

A second advantage of Wehrl's definition is that S^{cl} is *monotonic*. If ϱ_{12}^Q is a density matrix on $L^2(\mathbb{R}) \otimes L^2(\mathbb{R})$, and $|z_1, z_2\rangle \equiv |z_1\rangle \otimes |z_2\rangle$, one defines

$$\varrho_{12}^{\mathrm{cl}}(z_1, z_2) = \langle z_1, z_2|\varrho_{12}^Q|z_1, z_2\rangle. \tag{1.8}$$

One can then define $\varrho_1^{\mathrm{cl}}(z_1)$ by partial trace on 2 (either first on ϱ_{12}^Q or else on the right side of (1.8); by (1.3) they are identical). Wehrl shows that the entropies satisfy

$$S_{12}^{\mathrm{cl}} \equiv \tilde{S}(\varrho_{12}^{\mathrm{cl}}) \geqq \tilde{S}(\varrho_1^{\mathrm{cl}}) \equiv S_1^{\mathrm{cl}}, \tag{1.9}$$

in an obvious notation. This property, which is obviously desirable physically, does not hold in general for either the quantum entropies or for ordinary classical continuous entropies (see [6] for further details). It does hold for these particular classical entropies.

Not only is $S^{\mathrm{cl}} \geqq 0$, but Wehrl proves [1]

$$S^{\mathrm{cl}}(\varrho^Q) > S^Q(\varrho^Q). \tag{1.10}$$

[To prove $\geqq$ note that $s(x) = -x\ln x$ is concave, so $s(\varrho^{\mathrm{cl}}(z)) \geqq \langle z|s(\varrho^Q)|z\rangle$. But $S^Q(\varrho^Q) = \int \langle z|s(\varrho^Q)|z\rangle dz/\pi$.] While the minimum of S^Q is zero (for any pure state, i.e. one dimensional projection) the minimum of S^{cl} is not zero. *Wehrl's conjecture* is the following:

Theorem 1. *The minimum of S^{cl} is 1 (independent of $\hbar$). This minimum occurs if $\varrho^Q = P_z$ for any z.*

Remarks. 1) There is no upper bound or lower bound (other than zero) for $S^{cl}(\varrho^Q) - S^Q(\varrho^Q)$.

2) It is easy to see from Theorem 1 that in $L^2(\mathbb{R}^N)$, the minimum of S^{cl} is N.

The proof of Theorem 1 will be given in Section II. An analogous conjecture can be posed for Bloch coherent spin states and this is discussed, but not proved, in Section III. In Section II an inequality (Theorem 3) on L^p norms is also presented. Section IV contains an inequality which may be of use for related problems.

II. Proof of Wehrl's Conjecture

From now on we set $\hbar = 1$. As a preliminary remark we note:

Lemma 2. *If ϱ^Q minimizes S^{cl}, ϱ^Q must be a pure state.*

Proof. If $\varrho^Q = \sum_i \lambda_i \pi_i$, the π_i being one dimensional orthogonal projections, $\lambda_i > 0$ and $\sum \lambda_i = 1$, then $\varrho^{cl}(z) = \sum \lambda_i \varrho_i(z)$ with $\varrho_i(z) = \langle z|\pi_i|z\rangle$. By concavity of $\tilde{S}$, $\tilde{S}(\varrho^{cl}(z)) \geqq \sum \lambda_i \tilde{S}(\varrho_i)$, with equality if and only if $\varrho_i(z) = \varrho_j(z)$ almost everywhere for all i,j. Suppose ϱ_i is a projection onto $\psi_i \in L^2(\mathbb{R})$. Let $w = q + ip \in \mathbb{C}$ and let $f_i(w) = \int \psi_i(x) \exp[-x^2/2 + wx]dx$, which is an entire analytic function of w [3]. Then equality almost everywhere implies that $|f_i(w)| = |f_j(w)|$, all w, and hence $f_i(w) = f_j(w)\exp(i\theta(w))$ and θ is real and analytic on the complement of the zeros of f_i. Hence, $\theta(w) = \text{const}$. By the uniqueness of the Fourier transform, $\psi_i = \alpha\psi_j$, with $|\alpha| = 1$, almost everywhere, and, hence $\pi_i = \pi_j$, which is a contradiction. □

Thus, to prove Theorem 1 we have to consider

$$f(p,q) = \int \psi(x) R(x|p,q)dx \tag{2.1}$$

with $\|\psi\|_2 = 1$, and show that

$$\tilde{S}(|f|^2) \geqq 1 \tag{2.2}$$

with equality if $\psi(x) = R(x|p,q)$ for some (p,q).

We will first prove Theorem 3 which concerns L^p norms of $f(p,q)$. Theorem 1 is a corollary of Theorem 3.

Theorem 3. *Let $\psi \in L^2(\mathbb{R})$ with $\|\psi\|_2 = 1$, and f given by* (2.1) *and* (1.1). *Then, for* $s \geqq 2$

$$I_s \equiv \int\int |(f(p,q)|^s dp dq/2\pi \leqq 2/s \tag{2.3}$$

with equality for $s > 2$ if $\psi(x) = \alpha R(x|p,q)$ for some p, q and $|\alpha| = 1$. For $s = 2$, (2.3) *is an equality for all ψ.*

To prove Theorem 3 we will require the following two lemmas (for $N = 1$). The first (best constant in the Hausdorff-Young inequality) was proved by Beckner [7] and the second (best constant in Young's inequality) simultaneously by Beckner [7] and Brascamp and Lieb [8].

Lemma 4. *Let $f \in L^p(\mathbb{R}^N)$, $1 \leqq p \leqq 2$, and $\hat{f}$ its Fourier transform ($\hat{f}(k) = \int f(x)e^{ikx}dx$). Then, with $1/p + 1/p' = 1$,*

$$\|\hat{f}\|_{p'} \leqq \{C_p(2\pi)^{1/p'}\}^N \|f\|_p, \tag{2.4}$$

Commun. Math. Phys. *62*, 35–41 (1978)

where

$$C_p^2 = p^{1/p}(p')^{-1/p'} \quad \textit{and} \quad C_1 = C_\infty = 1.$$

Remark. Equality holds in (2.4) if f is any Gaussian, i.e. $f(x) = a\exp\{-(x, Mx) + (x, b)\}$, $a \in \mathbb{C}$, $b \in \mathbb{C}^N$, and M positive definite.

Lemma 5. *Let* $f \in L^p(\mathbb{R}^N)$, $g \in L^q(\mathbb{R}^N)$ $1 \leqq p, q \leqq \infty$. *Then, with* $1 + 1/r = 1/p + 1/q$, $r \geqq 1$, *and* $* =$ *convolution,*

$$\|f * g\|_r \leqq \{C_p C_q / C_r\}^N \|f\|_p \|g\|_q. \tag{2.5}$$

Equality holds [8] *for* $r > 1$ *and* $N = 1$ *if and only if* $f(x) = a\exp[-p'(x-b)^2 + i\delta x]$ *and* $g(x) = \alpha\exp[-q'(x-\beta)^2 + i\delta x]$ *for some* $a, \alpha \in \mathbb{C}$ *and* $\delta, b, \beta \in \mathbb{R}$. *For* $r = 1$ *(all* N*),* $p = q = 1$ *and* (2.5) *is an equality for all positive* f, g.

Remark. In the classical inequalities, C_p is replaced everywhere by 1 in Lemmas 4 and 5.

Proof of Theorem 3. As a first step apply Lemma 4 (with $p' = s$) to the function $g_q(x) = \psi(x)\pi^{-1/4}\exp[-(x-q)^2/2]$, with q regarded as a parameter. ($g_q \in L^{s'}(\mathbb{R})$ by Hölder's inequality.) Thus,

$$\int |f(p,q)|^s dp/2\pi \leqq C_{s'}^s \pi^{-s/4} \phi_s(q)^{s/s'}, \tag{2.6}$$

where ϕ_s is the convolution

$$\phi_s = |\psi(x)|^{s'} * \exp[-s'x^2/2]. \tag{2.7}$$

The second step is to integrate (2.6) over q and use Lemma 5 with $p = q = 2/s'$ and $r = s/s'$. Since $\|\exp(-x^2/2)\|_2 = \pi^{1/4}$, $I_s \leqq \{C_{s'}(C_{2/s'})^{2/s'}(C_{s/s'})^{-1/s'}\}^s = 2/s$.

Equality holds in the first step if ψ is any Gaussian. In the second step, since $p = q = 2/s'$, equality holds for $s > 2$ if ψ is a Gaussian with the same variance as $\exp(-x^2/2)$, which is the condition stated in the theorem. When $s = s' = 2$, equality for all ψ is a simple consequence of the Plancherel formula. □

Proof of Theorem 1. We continue to use the notation of Theorem 3. Let $\varepsilon > 0$. Since $I_2 = 1$, $K_\varepsilon = \varepsilon^{-1}\{I_2 - I_{2(1+\varepsilon)}\} \geqq (1+\varepsilon)^{-1}$ by Lemma 5. Assuming $\tilde{S}(|f|^2) < \infty$ (otherwise, there is nothing to prove), we claim that $\lim_{\varepsilon \downarrow 0} K_\varepsilon = \tilde{S}(|f|^2)$, which proves that $\tilde{S}(|f|^2) \geqq 1$. To see this note that by Theorem 3 or by the Schwarz inequality, $|f(p,q)| \leqq \|f\|_\infty \leqq 1$, and hence $0 \leqq \varepsilon^{-1}|f|^2\{1 - |f|^{2\varepsilon}\} \leqq -|f|^2 \ln|f|^2$. Thus, $K_\varepsilon \to \tilde{S}(|f|^2)$ by dominated convergence. □

III. Bloch Coherent Spin States

Instead of $L^2(\mathbb{R})$, one can consider the finite dimensional vector space $\mathscr{H}_J = \mathbb{C}^{2J+1}$, $J = 1/2, 1, 3/2, \ldots$. The analogue of the vectors $|z\rangle$ are the Bloch coherent states [9–13] in $\mathscr{H}_J$. These have been used to prove the classical limit of quantum spin systems [13]. For each unit vector $\Omega \in \mathbb{R}^3$, the vector $|\Omega\rangle \in \mathscr{H}_J$ is defined as the normalized vector (unique up to the phase) satisfying

$$\Omega \cdot \mathbf{S}|\Omega\rangle = J|\Omega\rangle, \tag{3.1}$$

where $S=(S_x, S_y, S_z)$ are the usual angular momentum operators satisfying $[S_x, S_y] = iS_z$ and cyclically. An explicit representation is

$$|\Omega\rangle = \sum_{M=-J}^{J} A_M(\theta)\exp(-iM\phi)|M\rangle, \tag{3.2}$$

$$A_M(\theta) = \binom{2J}{M+J}^{1/2} [\cos(\theta/2)]^{J+M}[\sin(\theta/2)]^{J-M}, \tag{3.3}$$

where (θ, ϕ) are the polar coordinates of Ω. $|M\rangle$ is the normalized vector satisfying $S_z|M\rangle = M|M\rangle$ and whose phase is given by $|M\rangle =$(pos. const.) $(S_x - iS_y)^{J-M}|J\rangle$. With the measure

$$d\mu_J(\Omega) = (2J+1)\sin\theta d\theta d\phi/4\pi \tag{3.4}$$

on the unit sphere S_2, and

$$P_\Omega = |\Omega\rangle\langle\Omega| \tag{3.5}$$

the projection onto $|\Omega\rangle$, one has the analogue of (1.3):

$$\int d\mu_J(\Omega) P_\Omega = I. \tag{3.6}$$

Now given a density matrix ϱ^Q on $\mathscr{H}_J$ one can imitate the Wehrl construction:

$$\varrho^{\text{cl}}(\Omega) = \langle\Omega|\varrho^Q|\Omega\rangle \tag{3.7}$$

and $S^{\text{cl}}(\varrho^Q) = \tilde{S}(\varrho^{\text{cl}})$ with

$$\tilde{S}(f) = -\int f(\Omega)\ln f(\Omega) d\mu_J(\Omega). \tag{3.8}$$

The monotonicity of S^{cl} and the inequality $S^{\text{cl}} \geqq S^Q$ carry over to this case.

It is easy to compute that since [13] $\langle\Omega'|P_\Omega|\Omega'\rangle = [\cos\frac{1}{2}\Theta]^{4J}$, where Θ is the angle between Ω and Ω',

$$S^{\text{cl}}(P_\Omega) = 2J/(2J+1). \tag{3.9}$$

The analogue of Theorem 1 is then

Conjecture. $S^{\text{cl}}(\varrho^Q) \geqq 2J/(2J+1)$.

We will have to content ourselves with the following remarks.

Remark A. Suppose ϱ^Q is of the form

$$\varrho^Q = \int d\mu_J(\Omega) h(\Omega) P_\Omega \tag{3.10}$$

with $h(\Omega) \geqq 0$ and $\int h d\mu_J = 1$. Every ϱ^Q can be written in the form (3.10) with h real but, for $J \geqq 1$, not necessarily with $h \geqq 0$, even though ϱ^Q is positive. However P_Ω is of this form with h being a delta function. By (3.10)

$$\varrho^{\text{cl}}(\Omega) = \int d\mu_J(\Omega')[\cos\tfrac{1}{2}\Theta]^{4J} h(\Omega').$$

Since $\varrho^{\text{cl}}(\Omega)$ is then a convex combination of $|\langle\Omega|\Omega'\rangle|$'s, the concavity of $\tilde{S}$ leads to

$$S^{\text{cl}}(\varrho^Q) \geqq S^{\text{cl}}(P_{\Omega'}) = 2J(2J+1)$$

if $h(\Omega) \geqq 0$. The analogue of this remark would, of course, also hold for the original Wehrl problem.

Commun. Math. Phys. *62*, 35–41 (1978)

Remark B. Lemma 2 holds for the Bloch case as well. Thus we can assume ϱ^Q is a projection onto $\psi\in\mathscr{H}_J$. Then

$$\varrho^{cl}(\Omega)=|f(\Omega)|^2 \tag{3.12}$$

$$f(\Omega)=\sum_{M=-J}^{J} C_M A_M(\theta)e^{-iM\phi} \tag{3.13}$$

and $\sum|C_M|^2=1$.

If $J=1/2$, *every* $\psi=\alpha|\Omega\rangle$ for some $|\Omega\rangle$ and α. Thus the conjecture is manifestly true for $J=1/2$.

IV. An Inequality for Entropy of Convolutions

Lemmas 4 and 5 yielded a lower bound for S^{cl}. Lemma 5 alone yields the following entropy inequality which, while not strictly related to coherent states, may be useful for related problems.

We first remark that if f is a nonnegative function on $\mathbb{R}^N$ with $\int f(x)dx=1$, and if $f\in L^s(\mathbb{R}^N)$ for some $s>1$, then $\tilde{S}(f)$ is well defined in the sense that $\int_{f(x)>1} f(x)\ln f(x)dx<\infty$. $\tilde{S}(f)$ may be $+\infty$, however.

Theorem 6. *Suppose f and g are nonnegative functions on $\mathbb{R}^N$ with $\int f=\int g=1$ and $f,g\in L^s(\mathbb{R}^N)$ for some $s>1$. Then $f*g$ has the same properties and*

$$\exp[2\tilde{S}(f*g)/N]\geqq\exp[2\tilde{S}(f)/N]+\exp[2\tilde{S}(g)/N] \tag{4.1}$$

(4.1) *is equivalent to the following:*

$$2\tilde{S}(f*g)\geqq 2\lambda\tilde{S}(f)+2(1-\lambda)\tilde{S}(g) \\ -N\lambda\ln\lambda-N(1-\lambda)\ln(1-\lambda) \tag{4.2}$$

for all $\lambda\in[0,1]$.

Corollary. $\tilde{S}(f*g)\geqq\frac{1}{2}[\tilde{S}(f)+\tilde{S}(g)+N\ln 2]$.

Remark. (4.1) is an equality if f and g are any two Gaussians of the form $f(x)\sim\exp[-(x,Mx)+(b,x)]$, $g(x)\sim\exp[-\alpha(x,Mx)+(c,x)]$ with $\alpha>0$, $b,c\in\mathbb{R}^N$ and M positive definite.

Proof. By Lemma 5, $(f*g)\in L^p(\mathbb{R}^N)$ for $p=1$ and for $p=s(2-s)^{-1}$. Hence $\tilde{S}(f*g)$ is well defined.

(4.2)⇒(4.1): Choose

$$\lambda=\{\exp[2\tilde{S}(f)/N]+\exp[2\tilde{S}(g)/N]\}^{-1} \\ \cdot\exp[2\tilde{S}(f)/N].$$

(4.1)⇒(4.2): Geometric-arithmetic mean inequality. We now prove (4.2). In Lemma 5, choose $p'=r'/\lambda$, $q'=r'/(1-\lambda)$, so that $1+r^{-1}=p^{-1}+q^{-1}$. By convexity, $f\in L^s\cap L^1$ implies $f\in L^t$ for $1\leqq t\leqq s$ and $t\mapsto\|f\|_t$ is continuous for $t\in[0,s]$. For r close enough 1, $p,q<s$, so $f*g\in L^r$ and (2.5) holds. Furthermore, (2.5) is an equality for $r=p=q=1$ so one can take the right derivative at $r=1$. Without loss we can

assume $\tilde{S}(f)$ and $\tilde{S}(g)<\infty$, for otherwise $\tilde{S}(f*g)=\infty$ by concavity and there is nothing to prove. For the same reason, one can assume $\tilde{S}(f*g)<\infty$. Next, we claim that if $F\in L^1\cap L^s$, $s>1$, and $\tilde{S}(F)<\infty$ then $\lim_{\varepsilon\downarrow 0}\varepsilon^{-1}\int F(1-F^\varepsilon)=\tilde{S}(F)$. To see this, let $A=\{x|F(x)<1\}$. Then for $x\in A$, $0\leqq 1-F(x)^\varepsilon\leqq -\varepsilon\ln F(x)$. For $x\in A^c$ and

$$1\leqq\varepsilon\leqq s-1\,,\quad 0\leqq F(x)^\varepsilon-1\leqq\varepsilon(s-1)^{-1}\{F(x)^{s-1}-1\}\,.$$

The claim follows by dominated convergence. Thus, the right side of (2.5) is differentiable at $r=1$ and Theorem 6 follows by explicit calculation. This calculation can be avoided noting that as r varies, $p'/q'=\text{const}=(1-\lambda)/\lambda$. As noted in Lemma 5, (if $N=1$, and hence for all N) (2.5) is saturated for the Gaussians $f(x)=\exp(-x^2/\lambda)$, $g(x)=\exp(-x^2/(1-\lambda))$, *independent* of r. But these Gaussians also give equality in (4.1). □

References

1. Wehrl, A.: On the relation between classical and quantum-mechanical entropy. Rept. Math. Phys.
2. Schrödinger, E.: Naturwissenschaften **14**, 664—666 (1926)
3. Bargmann, V.: Commun. Pure Appl. Math. **14**, 187—214 (1961); **20**, 1—101 (1967)
4. Klauder, J. R.: Ann. Phys. (N.Y.) **11**, 123 (1960)
5. Glauber, R.J.: Phys. Rev. **131**, 2766 (1963)
6. Lieb, E.H.: Bull. Am. Math. Soc. **81**, 1—13 (1975)
7. Beckner, W.: Ann. Math. **102**, 159—182 (1975)
8. Brascamp, H.J., Lieb, E.H.: Advan. Math. **20**, 151—173 (1976)
9. Radcliffe, J.M.: J. Phys. A**4**, 313—323 (1971)
10. Kutzner, J.: Phys. Lett. A**41**, 475—476 (1972)
11. Atkins, P.W., Dobson, J.C.: Proc. Roy. Soc. (London) A**321**, 321—340 (1971)
12. Arrechi, F.T., Courtens, E., Gilmore, R., Thomas, H.: Phys. Rev. A**6**, 2211—2237 (1972)
13. Lieb, E.H.: Commun. math. Phys. **31**, 327—340 (1973)

Communicated by J. Glimm

Received May 12, 1978

Letters in Mathematical Physics **22**: 145–154, 1991. 145

Quantum Coherent Operators: A Generalization of Coherent States

ELLIOTT H. LIEB*
Departments of Mathematics and Physics, Princeton University, PO Box 708, Princeton, NJ 08544-0708, U.S.A.

and

JAN PHILIP SOLOVEJ**
School of Mathematics, Institute for Advanced Study, Princeton, NJ 08540, U.S.A.

(Received: 15 March 1991)

Abstract. We introduce a technique to compare different, but related, quantum systems, thereby generalizing the way that coherent states are used to compare quantum systems to classical systems in semiclassical analysis. We then use this technique to estimate the dependence of the free energy of the quantum Heisenberg model on the spin value, and to estimate the relation between the ferromagnetic and antiferromagnetic free energies.

AMS subject classifications (1991). 81R30, 81S30.

1. Introduction

Coherent states have been used since the origin of quantum mechanics as one possible approach to semiclassical analysis, i.e., to compare quantum systems to corresponding classical systems. A complete list of references would be enormous. To mention just a few, see [2, 3, 6, 9, 11, 12, 14] for applications to continuous systems and [1, 8, 13] for applications to spin systems. For reviews, see [5, 7, 10].

In this Letter, we introduce a technique that can be used to compare two different quantum systems in very much the same way as regular coherent states compare a quantum system to a classical system.

Coherent states can be introduced from several points of view. The first is simply to think of them as being an interesting parameterization, by points z in the classical phase space $\mathscr{C}$, of a complete set of vectors ψ_z in the Hillbert space, $\mathscr{H}$, describing the quantum system. In one dimension, $\mathscr{H} = L^2(\mathbb{R})$ and the classical phase space is $\mathbb{R}^2$. The latter can be identified with $\mathbb{C}$ through $z = q + ip$. The usual coherent states in $L^2(\mathbb{R})$ are then

$$\psi_z(x) = \pi^{-1/2} \exp(zx - \tfrac{1}{2}(x^2 + |z|^2)). \tag{1}$$

* Work supported in part by the U.S. National Science Foundation grant PHY-9019433.
** Work supported in part by the U.S. National Science Foundation grant DMS-9002416. Current address: Department of Mathematics, Princeton University, Fine Hall, Washington Road, Princeton, NJ 08544-1000, U.S.A.

Another way, mainly emphasized by Segal [12] and Bargmann [2, 3], is to consider the coherent states as defining an isometry between two Hilbert spaces, namely between $\mathscr{H}$ and a suitable subspace of the space of square integrable functions on $\mathscr{C}$. In the case of (1), we map $L^2(\mathbb{R})$ into the subspace of analytic functions of $L^2(\mathbb{C}\,;\pi^{-1}\,\mathrm{e}^{-|z|^2}\,\mathrm{d}^2z)$ by

$$f\mapsto\hat{f}(z)=\int\exp(\bar{z}x-\tfrac{1}{2}x^2)f(x)\,\mathrm{d}x.$$

Bargmann proved that this map is, in fact, a bijection.

A third point of view (see [4] and [8]), which is the one that will be generalized in this paper, emphasizes the one-dimensional projections $\Pi(z)=|\psi_z\rangle\langle\psi_z|$ onto the coherent states rather than the vectors ψ_z themselves. One point is that this suppresses the unimportant information of the phase of ψ_z. The completeness can now be written as

$$\int_{\mathscr{C}}\Pi(z)\,\mathrm{d}z=\mathbf{1}_{\mathscr{H}}\tag{2}$$

for a certain measure – proportional to the Liouville measure in the case of a Hamiltonian system – $\mathrm{d}z$ on $\mathscr{C}$. If $\mathscr{H}$ is finite-dimensional then $\int_{\mathscr{C}}\mathrm{d}z=\dim\mathscr{H}$. Here, $\mathbf{1}_{\mathscr{H}}$ denotes the identity on $\mathscr{H}$. The general philosophy is that all interesting operators A can be written as (or approximated by) operators that can be written in 'diagonal form' as $A=\int G(z)\Pi_z\,\mathrm{d}z$. The function G is called the **upper symbol** for the operator A. The function $g(z)=\langle\psi_z|A|\psi_z\rangle$ is called the **lower symbol**. It is the third point of view that is useful for proving the classical limit of quantum systems [8] and, generally, the Berezin–Lieb inequalities [4, 8]. We shall illustrate this technique in the case of spin systems.

Quantum spin systems are given by representation spaces $\mathscr{H}_J=\mathbb{C}^{2J+1}$ for SU(2), where the spin J is a half-integer. The corresponding classical phase space is $J\mathbb{S}^2$, namely vectors in $\mathbb{R}^3$ with length J. To a point $\Omega\in\mathbb{S}^2$ we associate the Bloch coherent state vector $|\Omega\rangle_J\in\mathscr{H}_J$ defined up to an arbitrary phase by $\Omega\cdot\mathbf{S}_J|\Omega\rangle_J=J|\Omega\rangle_J$, where $\mathbf{S}_J=(S_J^x,S_J^y,S_J^z)$ is the vector of spin operators on $\mathscr{H}_J$. The projector, $|\Omega\rangle_{JJ}\langle\Omega|$ is denoted by $\Pi_J(\Omega)$. The completeness relation (2) is here

$$\frac{2J+1}{4\pi}\int\Pi_J(\Omega)\,\mathrm{d}\Omega=\mathbf{1}_{\mathscr{H}_J},\tag{3}$$

where $\mathrm{d}\Omega$ is the normalized Euclidean measure on $\mathbb{S}^2$. In this case it is, indeed, true that all operators on $\mathscr{H}_J$ can be expressed in diagonal form.

EXAMPLE. Consider the Heisenberg model of interacting spins at inverse temperature β. We denote the free energy by $f(J,\beta)$ in the quantum case (for the precise definition see (19) below) or $f^{\mathrm{class}}(J,\beta)$ in the classical case. The Bloch coherent states can be used (see [8]) to prove that

$$f^{\mathrm{class}}(J,\beta)\geqslant f(J,\beta)\geqslant f^{\mathrm{class}}(J+1,\beta)\tag{4}$$

By using these inequalities twice together with the fact that the classical free energy depends on J through the simple scaling relation $f^{\text{class}}(J, \beta) = f^{\text{class}}(1, J^2\beta,)$, we can relate quantum spin systems with spins K and J. We get

$$f\left(K, \left(\frac{J}{K+1}\right)^2 \beta\right) \geqslant f(J, \beta) \geqslant f\left(K, \left(\frac{J+1}{K}\right)^2 \beta\right). \tag{5}$$

The point is now that the route through the classical system is not an optimal procedure to obtain inequalities like (5). An obvious drawback of Equation (5) is that it does not reduce to equalities when $K = J$.

Our new result here will be Equation (20) in Theorem 7. As an illustration, suppose we wish to compare spin 1 and spin 1/2. Then (5), with $K = 1, J = 1/2$, says

$$f(1, \tfrac{1}{16}\beta) \geqslant f(\tfrac{1}{2}, \beta) \geqslant f(1, \tfrac{9}{4}\beta),$$

whereas (20) gives the better bound

$$f(1, \tfrac{1}{4}\beta) \geqslant f(\tfrac{1}{2}, \beta) \geqslant f(1, \tfrac{9}{16}\beta).$$

The technique presented here is intrinsically quantum mechanical. In Theorem 8, we also compare the antiferromagnetic and ferromagnetic free energies on a bipartite lattice for the *same* spin values. Classically, there is no real distinction between antiferromagnets and ferromagnets. The free energies are the same by a simple change of variable. On the other hand, in quantum mechanics the two systems are not unitarily equivalent, and the free energies are, indeed, different. Our bounds delimit that difference.

To describe the framework of our generalization, consider two Hilbert spaces $\mathscr{H}_1$ and $\mathscr{H}_2$ either both finite dimensional or both infinite dimensional. A positive semi-definite operator Γ on $\mathscr{H}_1 \otimes \mathscr{H}_2$ is called a **quantum coherent operator** for the pair $(\mathscr{H}_1, \mathscr{H}_2)$ if it satisfies

$$\overline{\text{Tr}}_{\mathscr{H}_1}\Gamma = \mathbf{1}_{\mathscr{H}_2}, \tag{6}$$

$$\overline{\text{Tr}}_{\mathscr{H}_2}\Gamma = \mathbf{1}_{\mathscr{H}_1}, \tag{7}$$

where $\overline{\text{Tr}}_{\mathscr{H}_1}$ denotes the *normalized* partial trace, i.e.,

$$\overline{\text{Tr}}_{\mathscr{H}_1} = \begin{cases} \dfrac{1}{\dim_{\mathscr{H}_1}} \text{Tr}_{\mathscr{H}_1}, & \text{if } \mathscr{H}_1 \text{ is finite-dimensional,} \\ \text{Tr}_{\mathscr{H}_1}, & \text{if } \mathscr{H}_1 \text{ is infinite-dimensional.} \end{cases}$$

The definition of partial trace over $\mathscr{H}_1$, which gives an operator on $\mathscr{H}_2$, is well known. To make an analogy with (2) we can pretend that $\mathscr{H}_2 = \mathscr{H}$ and that '$\mathscr{H}_1$ is the classical phase space $\mathscr{C}$'. Then (6) is the same as (2), whereas (7) imitates the trace condition

$$\text{Tr}\,\Pi(z) = 1. \tag{8}$$

However, (6) and (7) bring out the symmetry between the two spaces $\mathscr{H}_1$ and $\mathscr{H}_2$.

If A is an operator on $\mathscr{H}_1$ we define an operator $\check{A}$ on $\mathscr{H}_2$ by

$$\check{A} = \overline{\mathrm{Tr}}_{\mathscr{H}_1}(\Gamma A). \tag{9}$$

(Really, A means $A \otimes \mathbf{1}_{\mathscr{H}_2}$ in (9).) Since Γ can be written as a linear combination of tensor products of an operator on $\mathscr{H}_1$ with an operator on $\mathscr{H}_2$, we see from the cyclicity of the $\mathscr{H}_1$-trace that $\overline{\mathrm{Tr}}_{\mathscr{H}_1}(\Gamma A) = \overline{\mathrm{Tr}}_{\mathscr{H}_1}(A\Gamma)$ and, hence, if A is Hermitean, then so is $\check{A}$. One virtue of this generalization is that it establishes a complete symmetry between the upper and lower symbols, i.e., $\check{A}$ is the (unique) lower symbol for A, and A is an upper symbol for $\check{A}$. The main comparison inequality that generalizes the Berezin–Lieb inequalities such as (4) is

THEOREM 1. *If A is a Hermitean operator on $\mathscr{H}_1$ and f is any convex function on the reals, then*

$$\overline{\mathrm{Tr}}_{\mathscr{H}_1} f(A) \geqslant \mathrm{Tr}_{\mathscr{H}_2} f(\check{A}).$$

If f is concave the inequality is reversed. An equivalent restatement of this is that if $\hat{A}$ is any upper symbol for A, then $\overline{\mathrm{Tr}}_{\mathscr{H}_2} f(\hat{A}) \geqslant \overline{\mathrm{Tr}}_{\mathscr{H}_1} f(A)$.

Proof.

$$\begin{aligned}\overline{\mathrm{Tr}}_{\mathscr{H}_1} f(A) &= \overline{\mathrm{Tr}}_{\mathscr{H}_2}\overline{\mathrm{Tr}}_{\mathscr{H}_1}(\Gamma f(A)) \\ &= \frac{1}{\dim \mathscr{H}_2} \sum_n \overline{\mathrm{Tr}}_{\mathscr{H}_1}(\langle v_n|\Gamma|v_n\rangle f(A)),\end{aligned}$$

(for the finite-dimensional case, and without $1/\dim \mathscr{H}_2$ in the infinite-dimensional case), where $\{v_n\}$ is the orthonormal basis in $\mathscr{H}_2$ consisting of eigenfunctions for $\check{A}$. $\langle v_n|\Gamma|v_n\rangle$ is a positive operator on $\mathscr{H}_1$ with

$$\overline{\mathrm{Tr}}_{\mathscr{H}_1}\langle v_n|\Gamma|v_n\rangle = \langle v_n|\mathbf{1}_{\mathscr{H}_1}|v_n\rangle = 1.$$

It then follows immediately from the spectral theorem and Jensen's inequality that

$$\begin{aligned}\overline{\mathrm{Tr}}_{\mathscr{H}_1} f(A) &\geqslant \frac{1}{\dim \mathscr{H}_2} \sum_n f(\langle v_n|\check{A}|v_n\rangle) \\ &= \overline{\mathrm{Tr}}_{\mathscr{H}_2} f(\check{A}),\end{aligned}$$

since all the v_n are eigenfunctions for $\check{A}$. □

Remarks. (1) An important open problem is to decide what condition on Γ, $\mathscr{H}_1$ and $\mathscr{H}_2$ will guarantee that *every* operator $\check{A}$ on $\mathscr{H}_2$ has an upper symbol A, satisfying (9). We can call this **operator completeness**. Obviously, $\dim \mathscr{H}_1 \geqslant \dim \mathscr{H}_2$ is needed for operator completeness, but conditions on Γ are also needed. If Γ is the identity operator on $\mathscr{H}_1 \otimes \mathscr{H}_2$ then $\Gamma = \mathbf{1}_{\mathscr{H}_1} \otimes \mathbf{1}_{\mathscr{H}_2}$ and, for any A, the $\check{A}$ of (9) is proportional to $\mathbf{1}_{\mathscr{H}_2}$. Operator completeness clearly fails in this case. A less trivial case of operator incompleteness, due to G. M. Graf, is mentioned in the acknowledgement at the end. As already mentioned, operator completeness holds in the case (3) above. For a proof of this see [7], pp. 29–34 or the remark after Theorem 5 below.

(2) If ρ_2 is a density matrix on $\mathscr{H}_2$, i.e., a positive semidefinite matrix with the somewhat unusual normalization $\overline{\mathrm{Tr}}_{\mathscr{H}_2}\rho_2 = 1$, we find that $\check{\rho}_2$ is a density matrix on $\mathscr{H}_1$. Furthermore, if H_1 is any operator on $\mathscr{H}_1$ we get

$$\overline{\mathrm{Tr}}_{\mathscr{H}_1}(\check{\rho}_2 H_1) = \overline{\mathrm{Tr}}_{\mathscr{H}_2}(\rho_2 \check{H}_1). \tag{10}$$

In the following section, we shall give interesting examples of quantum coherent operators, Γ, for spin systems. Moreover, they will be operator complete from the big space to the small one; see Theorem 5. We shall also study how coherent these operators are. More precisely, we shall define entropies related to these operators and estimate them in (13). In Section 3, we use these operators to give the stated estimates on the Heisenberg free energies.

2. Coherent Operators for Spin Systems

Let $\mathscr{H}_J = \mathbb{C}^{2J+1}$ and $\mathscr{H}_K = \mathbb{C}^{2K+1}$ be representation spaces for SU(2) corresponding to spin values $K \geqslant J$, where K and J are half-integers. Let P_L, $L = K-J, \ldots, K+J$ be the projection from $\mathscr{H}_K \otimes \mathscr{H}_J$ onto its subspace in which $(\mathbf{S}_K + \mathbf{S}_J)^2 = L(L+1)$. Here $\mathbf{S}_K = (S_K^x, S_K^y, S_K^z)$ is the vector of spin operators on $\mathscr{H}_K$ which we also identify as an operator on $\mathscr{H}_K \otimes \mathscr{H}_J$ (really $S_K^x \otimes \mathbf{1}_J, \ldots$). Define

$$\Gamma_L = \frac{(2K+1)(2J+1)}{2L+1} P_L.$$

LEMMA 2. *For $L = K-J, \ldots, K+J$, Γ_L is a quantum coherent operator for $(\mathscr{H}_K, \mathscr{H}_J)$.*

Proof. Let $U_J = U_J(R)$ and $U_K = U_K(R)$ denote the action of some $R \in \mathrm{SU}(2)$ in the representation spaces $\mathscr{H}_J$ and $\mathscr{H}_K$. Then

$$\begin{aligned} U_J[\overline{\mathrm{Tr}}_{\mathscr{H}_K} P_L] U_J^{-1} &= \overline{\mathrm{Tr}}_{\mathscr{H}_K}[U_K^{-1} U_K U_J P_L U_J^{-1} U_K^{-1} U_K] \\ &= \overline{\mathrm{Tr}}_{\mathscr{H}_K}[U_K^{-1} P_L U_K] = \overline{\mathrm{Tr}}_{\mathscr{H}_K} P_L. \end{aligned}$$

Here we have used the cyclicity of the trace on $\mathscr{H}_K$ and the fact that P_L commutes with $U_K U_J = U_K(R) U_J(R)$. Since this holds for all $R \in \mathrm{SU}(2)$ and since both representations U_J and U_K are irreducible, it follows from Schur's Lemma that $\overline{\mathrm{Tr}}_{\mathscr{H}_K} P_L$ is a multiple of the identity. The same is true, of course, with K and J interchanged. The lemma then follows from $\mathrm{Tr}_{\mathscr{H}_K \otimes \mathscr{H}_J} P_L = 2L+1$, from which the relevant constants can be computed. □

The next question is how Γ_L transforms the regular spin operators according to (9).

LEMMA 3. *We have*

$$\check{\mathbf{S}}_K \equiv \overline{\mathrm{Tr}}_{\mathscr{H}_K}(\Gamma_L \mathbf{S}_K) = \frac{L(L+1) - K(K+1) - J(J+1)}{2J(J+1)} \mathbf{S}_J,$$

and the same identity with J and K interchanged.

Proof. Using the same argument as in Lemma 2 we find that

$$U_J \check{\mathbf{S}}_K U_J^{-1} = \overline{\mathrm{Tr}}_{\mathcal{H}_K}[P_L U_K \mathbf{S}_K U_K^{-1}]. \tag{11}$$

If we now set $U_J = \exp[itS^z]$ in (11) and take the derivative with respect to t at $t=0$, we infer the commutation relation $[S_J^z, \check{S}_K^+] = \check{S}_K^+$. As usual $S^\pm = S^x \pm iS^y$. Likewise, we infer $[S_J^+, \check{S}_K^+] = 0$. It follows easily that $\check{S}_K^+ = \kappa S_J^+$ for some scalar κ. Since we can interchange x, y and z, we have shown that $\check{\mathbf{S}}_K = \kappa \mathbf{S}_J$. We compute κ from

$$\begin{aligned}\kappa J(J+1) &= \mathbf{S}_J \cdot \check{\mathbf{S}}_K = \overline{\mathrm{Tr}}_{\mathcal{H}_K}(\mathbf{S}_J \cdot \mathbf{S}_K \Gamma_L)\\ &= \tfrac{1}{2}(L(L+1) - J(J+1) - K(K+1)).\end{aligned}$$

□

The most interesting cases are $L_{\max} = K + J$ and $L_{\min} = K - J$. We denote the respective operators by $\Gamma_{\max}$ and $\Gamma_{\min}$. Then

$$\begin{aligned}&\overline{\mathrm{Tr}}_{\mathcal{H}_K}(\mathbf{S}_K \Gamma_{\min}) = -\frac{K+1}{J+1}\mathbf{S}_J \quad \text{and} \quad \overline{\mathrm{Tr}}_{\mathcal{H}_J}(\mathbf{S}_J \Gamma_{\min}) = -\frac{J}{K}\mathbf{S}_K,\\ &\overline{\mathrm{Tr}}_{\mathcal{H}_K}(\mathbf{S}_K \Gamma_{\max}) = \frac{K}{J+1}\mathbf{S}_J \quad \text{and} \quad \overline{\mathrm{Tr}}_{\mathcal{H}_J}(\mathbf{S}_J \Gamma_{\max}) = \frac{J}{K+1}\mathbf{S}_K.\end{aligned} \tag{12}$$

Equation (12) has a surprising lack of symmetry between J and K; but here the condition $K \geqslant J$ has to be remembered.

We see that $\Gamma_{\min}$, except for a minus sign, gives a scaling of the spin. A natural question is now whether we can find a coherent operator that acts like $\Gamma_{\min}$ but without the minus sign. In the remark after Theorem 8 below, we shall see that such an operator does not exist.

We now prove in a very precise sense that $\Gamma_{\min}$ is a better coherent operator than $\Gamma_{\max}$. In fact, if we are given a density matrix ρ_J on $\mathcal{H}_J$ we can compute entropies relative to $\mathcal{H}_K$ as $\sigma_K(\check{\rho}_J^{\min})$ and $\sigma_K(\check{\rho}_J^{\max})$, where

$$\sigma_K(\rho) = \frac{1}{\dim_{\mathcal{H}_K}} \overline{\mathrm{Tr}}_{\mathcal{H}_K} f(\rho) \quad \text{and} \quad f(t) = -t \ln t.$$

This definition is similar to the definition of Wehrl's [15] classical entropy given in [9], i.e.,

$$\sigma_{cl}(\rho) = \frac{1}{4\pi}\int f(\langle \Omega | \rho | \Omega \rangle_J)\, \mathrm{d}\Omega.$$

Notice that we are using our unconventional normalization of always working with normalized traces.

THEOREM 4.

$$\sigma_J(\rho_J) \leqslant \sigma_K(\check{\rho}_J^{\min}) \leqslant \sigma_{cl}(\rho_J) \leqslant \sigma_K(\check{\rho}_J^{\max}). \tag{13}$$

Proof. The first inequality follows from Theorem 1 and the fact that $f(t)$ is a concave function. The second follows from the inequality $\sigma_K(\rho) \leqslant \sigma_{cl}(\rho)$ in [9] together with the fact that

$$\langle\Omega|\check{\rho}_J^{\min}|\Omega\rangle_K = \langle -\Omega|\rho_J|-\Omega\rangle_J. \tag{14}$$

This identity follows from

$$\overline{\mathrm{Tr}}_{\mathscr{H}_K}(\Gamma_{\min}\Pi_K(\Omega)) = \frac{2J+1}{2K+1}\Pi_J(-\Omega), \tag{15}$$

which, since it is rotation invariant, can be checked by choosing $|\Omega\rangle_K$ to be the maximal weight vector $|K\rangle_K$ (i.e., $S_K^z|K\rangle_K = K|K\rangle_K$). Before proving the last inequality in (13), we first notice that it follows from the theory of coherent states that

$$\Gamma_{\max} = \frac{(2K+1)(2J+1)}{4\pi}\int \Pi_K(\Omega)\otimes\Pi_J(\Omega)\,\mathrm{d}\Omega, \tag{16}$$

because $|\Omega\rangle_K \otimes |\Omega\rangle_J$ is a coherent state in the subspace on which $P_{\max}$ projects. From (16) we easily conclude that $\check{\rho}_J^{\max}$ has the following simple representation in terms of Bloch coherent states.

$$\check{\rho}_J^{\max} = (2K+1)\int \langle\Omega|\rho_J|\Omega\rangle_J \Pi_K(\Omega)\,\mathrm{d}\Omega. \tag{17}$$

The last inequality in (13) now follows from a proof almost identical to the proof of Theorem 1. □

If ρ_J is a pure state then $\sigma_J(\rho_J) = -\ln(2J+1)$ which is the smallest possible value with our normalization. It is now clear that $\check{\rho}_J^{\min}$ is not a pure state if $K > J$ because $\sigma_K(\check{\rho}_J^{\min}) \geqslant -\ln(2J+1)$ by (13), whereas $\sigma_K(\rho) = -\ln(2K+1)$ for any pure state ρ.

We shall now show operator completeness for $\Gamma_{\max}$ and $\Gamma_{\min}$.

THEOREM 5. *If $K \geqslant J$, then the coherent operators $\Gamma_{\max}$ and $\Gamma_{\min}$ are operator complete from $\mathscr{H}_K$ to $\mathscr{H}_J$.*

Proof. We have to show that we can get all operators (matrices) in End($\mathscr{H}_J$) as $\check{A}$ in (9) with A an operator in End($\mathscr{H}_K$). In the case of $\Gamma_{\min}$ this follows from (15), since we know from the operator completeness of (3) that the projections $\Pi_J(\Omega)$ span all operators. For $\Gamma_{\max}$ we note that the group SU(2) acts on operators in End($\mathscr{H}_J$) or End($\mathscr{H}_K$) through the adjoint representation $\mathrm{ad}_{U_K}A = U_K A U_K^{-1}$. As in Lemma 3, we see that for all the coherent operators Γ_L

$$(\mathrm{ad}_{U_K}A)^\vee = \mathrm{ad}_{U_J}(\check{A}).$$

Thus, End($\mathscr{H}_K$) is a vector space on which SU(2) acts, and it is clear that End($\mathscr{H}_K$) can be written as a direct sum of irreducible representations for SU(2) corresponding to spins $M = 0, \ldots, 2K$. The map $A \mapsto \check{A}$ will map the subspaces corresponding to representations with $M = 0, \ldots, 2J$, into the corresponding subspaces of $\mathrm{End}(\mathcal{H}_J)$, while the subspaces with $M = 2J+1, \ldots, 2K$ are in the kernel. To see that the map is onto, we have to show that none of the representation subspaces with $M = 0, \ldots, 2J$ in $\mathrm{End}(\mathcal{H}_K)$ are mapped to zero, and thus, from irreducibility, conclude that they are disjoint from the kernel. In the adjoint representation, the generators $\mathbf{S}_K$ act via commutators, and from the identities

$$[S_K^x, [S_K^x, (S_K^+)^M]] + [S_K^y, [S_K^y, (S_K^+)^M]] + [S_K^z, [S_K^z, (S_K^+)^M]] \\ = M(M+1)(S_K^+)^M,$$

and

$$[S_K^z, (S_K^+)^M] = M(S_K^+)^M,$$

we see that $(S_K^+)^M$ is a heighest weight vector in the irreducible subspace of $\mathrm{End}(\mathcal{H}_K)$ with spin M. It is therefore enough to show that $(S_K^+)^M$ is not mapped to zero. From [8] formula (A.1) we can calculate

$$\langle \Omega | (S_K^\pm)^M | \Omega \rangle = C_K^{(M)} (\Omega_x \pm i\Omega_y)^M,$$

where $C_K^{(M)} > 0$ for $M = 0, \ldots, 2K$. Using these lower symbols we get from (16) that if $M = 0, \ldots, 2J$

$$\overline{\mathrm{Tr}}_{\mathcal{H}_J} \overline{\mathrm{Tr}}_{\mathcal{H}_K} [\Gamma_{\max} (S_K^+)^M (S_J^-)^M] = C_K^{(M)} C_J^{(M)} (4\pi)^{-1} \int |\Omega_x + i\Omega_y|^{2M} \, \mathrm{d}\Omega > 0,$$

from which the theorem follows in the case of $\Gamma_{\max}$. □

Remark. In view of (16) operator completeness in the case of $\Gamma_{\max}$ is clearly a stronger statement than operator completeness in the classical case (3). Therefore, *the above proof for $\Gamma_{\max}$ automatically gives an alternative proof of completeness in the classical case*, and shows, moreover, that *one can always choose an upper symbol from the subspace of all lower symbols* (see [7], pp. 29–34).

From the above proof we also immediately get the following corollary.

COROLLARY 6. *If $K \geqslant J$ then the subspace of* $\mathrm{End}(\mathcal{H}_K)$ *consisting of matrices $\check{A}$ with $A \in$* $\mathrm{End}(\mathcal{H}_J)$ *(for both $\Gamma_{\min}$ and $\Gamma_{\max}$) is the direct sum of irreducible subspaces under the adjoint representation with spin values $M = 0, \ldots, 2J$.*

3. Free Energy of the Heisenberg Model

In this section, we shall use the method described in the previous sections to estimate the free energy of the Heisenberg model of interacting spins. For simplicity, we take the same spin J on each site, but this is not necessary.

Let Λ denote a finite collection of $|\Lambda|$ points and define the Heisenberg Hamiltonian $H(J)$ on $\mathcal{H}_J(\Lambda) = \otimes_{i \in \Lambda} \mathcal{H}_J$ by

$$H(J) = -\sum_{i,j \in \Lambda} E_{ij} \mathbf{S}_J(i) \cdot \mathbf{S}_J(j), \tag{18}$$

where E_{ij} are real numbers. No assumption is made about the sign of the E_{ij}. The partition function is defined to be

$$Z_\Lambda(J, \beta) = (2J+1)^{-|\Lambda|} \mathrm{Tr}_{\mathscr{H}_J(\Lambda)} \, \mathrm{e}^{-\beta H(J)}.$$

The normalized free energies are

$$f(J, \beta) = -(J^2\beta)^{-1} \ln Z_\Lambda(J, \beta). \tag{19}$$

The operator

$$\Gamma(\Lambda) = \otimes_{i \in \Lambda} \Gamma_{\min}(i)$$

on $\mathscr{H}_K(\Lambda) \otimes \mathscr{H}_J(\Lambda)$ is a coherent operator for $(\mathscr{H}_K(\Lambda), \mathscr{H}_J(\Lambda))$. We get from (12) for $K \geqslant J$

$$\check{H}(J) = \left(\frac{J}{K}\right)^2 H(K) \quad \text{and} \quad \check{H}(K) = \left(\frac{K+1}{J+1}\right)^2 H(K).$$

Thus, Theorem 1 implies that

$$Z_\Lambda(J, \beta) \geqslant Z_\Lambda\left(K, \left(\frac{J}{K}\right)^2 \beta\right), \qquad Z_\Lambda(K, \beta) \geqslant Z_\Lambda\left(J, \left(\frac{K+1}{J+1}\right)^2 \beta\right).$$

Using (19) we arrive at Theorem 7

THEOREM 7. *If* $K \geqslant J$

$$f\left(K, \left(\frac{J}{K}\right)^2 \beta\right) \geqslant f(J, \beta) \geqslant f\left(K, \left(\frac{J+1}{K+1}\right)^2 \beta\right), \tag{20}$$

$$f\left(J, \left(\frac{K+1}{J+1}\right)^2 \beta\right) \geqslant f(K, \beta) \geqslant f\left(J, \left(\frac{K}{J}\right)^2 \beta\right). \tag{21}$$

Inequalities (20) and (21) are the same, but both are given here for the sake of clarity.

Finally, let us compare the free energy for the Hamiltonian H given in (18) with that of $-H$, i.e., we reverse the sign of all the E_{ij}. (Recall that the sign of each E_{ij} is arbitrary but fixed.) With an application to ferro- and antiferromagnets in mind, we shall call the former case (with H) the ferromagnet and shall call the latter case (with $-H$) the antiferromagnet. Subscripts a and f will denote the two cases. One important new assumption must now be made, however. We assume that Λ is **bipartite**. This means that $\Lambda = A \cup B$ with $A \cap B$ empty and with $E_{ij} = 0$ whenever $i \in A$ and $j \in A$ or else $i \in B$ and $j \in B$. The coherent operator to be used is

$$\Gamma' = \otimes_{i \in A} \Gamma_{\min}(i) \otimes_{j \in B} \Gamma_{\max}(j),$$

on $\mathscr{H}_J(\Lambda) \otimes \mathscr{H}_J(\Lambda)$. Note the combination of min and max used here. We get

$$\check{H}_f(J) = \frac{J}{J+1} H_a(J) \quad \text{and} \quad \check{H}_a(J) = \frac{J}{J+1} H_f(J).$$

Thus

THEOREM 8. (*ferro- and antiferromagnetic comparison*).

$$f_a\left(J, \frac{J}{J+1}\beta\right) \leqslant f_f(J, \beta) \leqslant f_a\left(J, \frac{J+1}{J}\beta\right),$$

$$f_f\left(J, \frac{J}{J+1}\beta\right) \leqslant f_a(J, \beta) \leqslant f_f\left(J, \frac{J+1}{J}\beta\right).$$

Remark. If we could find a coherent operator inducing the same transformation on the basic spin operators as $\Gamma_{\min}$, but without the minus sign, then the above proof would give $f_a = f_f$, which is, of course, wrong.

Acknowledgements

We are grateful to G. M. Graf for pointing out to us that Γ_L need not be operator complete for every J, K,and L. By Lemma 3 we see that $\mathbf{S}_K$ is mapped to zero when $K = J = 2$ and $L = 3$; from this we conclude that $\mathbf{S}_J$ cannot be in the range of the Γ_L transform in this case.

References

1. Arecchi, F. T., Courtens, E., Gilmore, R., and Thomas, H., Atomic coherent states in quantum optics, *Phys. Rev. A* **6**, 2211–2237 (1972).
2. Bargmann, V., On a Hilbert space of analytic functions and an associated integral transform, part 1, *Comm. Pure Appl. Math.* **14**, 187–214 (1961).
3. Bargmann, V., On a Hilbert space of analytic functions and an associated integral transform, part 2, *Comm. Pure Appl. Math.* **20**, 1–101 (1967).
4. Berezin, F. A., *Izv. Akad. Nauk SSSR Ser. Mat.* **36**(5), 1134–1167 (1972). English translation: Covariant and contravariant symbols of operators. *Math. USSR-Izv.* **6**(5), 1117–1151 (1972) and F. A. Berezin. General concept of quantization, *Comm. Math. Phys.* **40**, 153–174 (1975).
5. Feng, D. H., Gilmore, R., and Zhang, W-M., Coherent states: Theory and some applications, *Rev. Mod. Phys.* **62**, 867–927 (1990).
6. Klauder, J. R., The action option and a Feynman quantization of spinor fields in terms of ordinary *c*-numbers, *Ann. Phys.* **11**, 123 (1960).
7. Klauder, J. R., and Skagerstam, B-S., *Coherent States*, World Scientific, Singapore, 1985.
8. Lieb, E. H., The classical limit of quantum spin systems, *Comm. Math. Phys.* **31**, 327–340 (1973).
9. Lieb, E. H., Proof of an entropy conjecture of Wehrl, *Comm. Math. Phys.* **62**, 35–41 (1978).
10. Perelomov, A., *Generalized Coherent States and their Applications*, Springer-Verlag, New York, Berlin, Heidelberg, 1986.
11. Schrödinger, E., Der stetige übergang von der Mikro-zur Makromechanik, *Naturwiss.* **14**, 664–666 (1926).
12. Segal, I. E., Mathematical characterizations of the physical vacuum for the linear Bose–Einstein fields, *Illinois J. Math.* **6**, 500–523 (1962).
13. Simon, B., The classical limit of quantum partition functions, *Comm. Math. Phys.* **71**, 247–276 (1980).
14. Thirring, W. E., A lower bound with the best possible constant for Coulomb hamiltonians, *Comm. Math. Phys.* **79**, 1–7 (1981).
15. Wehrl, A., On the relation between classical and quantum-mechanical entropy. *Rep. Math. Phys.* **12**, 385 (1977).

COHERENT STATES AS A TOOL FOR OBTAINING RIGOROUS BOUNDS

Elliott H. Lieb*
Departments of Mathematics and Physics, Princeton University
Princeton, NJ 08544 USA

ABSTRACT

This talk reviews some of the ways in which coherent states can be used to give rigorous bounds for quantities of physical interest and, in certain cases, can yield exact asymptotic formulas. Three main topics will be discussed.

1. The Berezin-Lieb inequalities which yield upper and lower bounds to quantum mechanical free energies in terms of classical free energies.
2. Coherent states (combined with a variational principle and correlation inequality) can generate upper and lower bounds for the ground state energies of atoms and other Coulomb systems.
3. Wehrl's conjecture about the entropy of coherent states.

0. Introduction

This talk reviews some of the ways in which coherent states can be used to give rigorous bounds to quantities of physical interest and, in certain cases, can yield exact asymptotic formulas. Three main topics will be discussed.

1. The Berezin-Lieb inequalities, which yield upper and lower bounds to quantum mechanical free energies in terms of classical free energies: Some applications are (a) upper and lower bounds to the free energy of quantum spin systems in terms of the corresponding classical spin systems and (b) the exact evaluation of the ground state energy and free energy of the Dicke laser model. (The generalization to bounds of one quantum system in terms of another quantum system are given in J.P. Solovej's talk.)

* Work partially supported by U.S. National Science Foundation grant PHY 9019433 A02.

2. Coherent states (combined with a variational principle and a correlation inequality) can generate upper and lower bounds for the ground state energies of atoms and other Coulomb systems. In the limit $Z \to \infty$ these bounds coincide and thereby establish the asymptotic exactness of Thomas-Fermi theory.
3. Wehrl's conjecture about the entropy of coherent states, its resolution for Glauber states (which also leads to L^p bounds for Wigner distribution functions), and the open conjecture about $SU(2)$ (or Bloch) coherent states.

1. Free Energies

It is helpful to have an example in mind, and for this it is convenient to take the Heisenberg model in which each spin has a value S (which can be $1/2, 1, 3/2, \ldots$) and the interaction is $\mathbf{S}_i \cdot \mathbf{S}_j$ for nearest neighbor pairs $\langle i,j\rangle$. Thus, the Hamiltonian is

$$H(S) = \frac{1}{S^2} \sum_{\langle i,j\rangle} \mathbf{S}_i \cdot \mathbf{S}_j. \tag{1.1}$$

The normalization $1/S^2$ is taken for convenience so that $H(S)$ has, in some sense to be determined, a nice limit as $S \to \infty$.

We are interested in the partition function

$$Z^Q(S) = (2S+1)^{-N} \mathrm{Tr}\, e^{-\beta H(S)} = e^{-\beta F(S)} \tag{1.2}$$

and will try to bound it in terms of a classical partition function Z^{cl} given by

$$Z^{cl} = (4\pi)^{-N} \int \exp[-\beta H^{cl}(\boldsymbol{\Omega}_1, \ldots, \boldsymbol{\Omega}_N)] d\Omega_1 \ldots d\Omega_N \tag{1.3}$$

$$H^{cl} = \sum_{\langle i,j\rangle} \boldsymbol{\Omega}_i \cdot \boldsymbol{\Omega}_j \tag{1.4}$$

with $|\boldsymbol{\Omega}_i| = 1$. The N integrations in (1.3) are each over the unit sphere S^2.

The general situation is the following. We are given a Hilbert space $\mathcal{H}$ and a family of coherent states $|z\rangle$, parametrized symbolically by z (and which might, in fact, be $\Omega \in S^2$ as a particular case), satisfying

$$\langle z|z\rangle = 1 \qquad \text{(normalization)} \tag{1.5}$$

and

$$\int |z\rangle\langle z| dz = \mathbf{1} \qquad \text{(resolution of identity)}, \tag{1.6}$$

for a suitable measure dz on the parameter space. For each operator H on $\mathcal{H}$ we can define the **lower symbol,** $\check{H}(z)$, which is a function on the parameter space, by

$$\check{H}(z) = \langle z|H|z\rangle. \tag{1.7}$$

Usually there is at least one **upper symbol,** $\widehat{H}(z)$, which is a measurable function satisfying

$$H = \int \widehat{H}(z)|z\rangle\langle z|dz. \tag{1.8}$$

Such a function may not exist (it does not exist for the Coulomb potential, $|x|^{-1}$, for example, using Glauber coherent states, or even generalized states of the type given in (2.8)) and if it exists it is not always unique. In the finite dimensional case (i.e., spins) it always exists, but is never unique.

An important — and frustrating — point is that while the lower symbol $\check{H}$ is always a positive function when H is positive, the upper symbol $\widehat{H}$ need not be positive. For example, with Glauber coherent states, the lower symbol for the oscillator Hamiltonian $H = a^\dagger a$ is $p^2 + q^2$ while the upper symbol is $p^2 + q^2 - 1$.

The Berezin-Lieb inequalities[1,2] are as follows.

THEOREM:

$$Z^Q := \frac{\operatorname{Tr} e^{-\beta H}}{\operatorname{Tr} \mathbf{1}} \begin{cases} \leq \int \exp[-\beta \widehat{H}(z)]dz \\ \geq \int \exp[-\beta \check{H}(z)]dz \end{cases} \tag{1.9}$$

This relates a quantum partition function to a classical partition function. Recent developments, with J.P. Solovej[3], relate one quantum Z to *another quantum* Z. Note that this inequality also holds for any convex function of H, not just for the exponential function.

Returning now to our example with spins, we take $|\boldsymbol{\Omega}\rangle$ to be the Bloch coherent state, i.e., the vector in $\mathbf{C}^{2S+1}$ defined (up to a phase) by

$$(\mathbf{S}\cdot\boldsymbol{\Omega})|\boldsymbol{\Omega}\rangle = S|\boldsymbol{\Omega}\rangle. \tag{1.10}$$

The measure on S^2 for (1.6) and (1.8) is $(4\pi)^{-1}d\Omega$. We then have the following upper and lower symbols for the three spin operators $\mathbf{S} = (S^x, S^y, S^z)$.

$$\widehat{\mathbf{S}}(\boldsymbol{\Omega}) = (S+1)\boldsymbol{\Omega}, \qquad \check{\mathbf{S}}(\boldsymbol{\Omega}) = S\boldsymbol{\Omega}. \tag{1.11}$$

Recalling the $1/S^2$ normalization convention in (1.1), the Berezin-Lieb inequalities yield

$$Z^{cl}(\beta) \leq Z^{Q}(S,\beta) \leq Z^{cl}\left(\left(\frac{(S+1)}{S}\right)^2 \beta\right). \tag{1.12}$$

With these upper and lower bounds, which are *uniform in the size of the system*, it is a trivial matter to prove the *classical limit of the quantum spin systems*, i.e., if we define $f(S) = N^{-1}F(S)$ to be the free energy per spin, then

$$\lim_{S\to\infty}\lim_{N\to\infty} f^{Q}(S) = f^{cl} = \lim_{N\to\infty} -\beta^{-1}\ln Z^{cl}. \tag{1.13}$$

Another example of the use of these inequalities to evaluate a quantum free energy is the **Dicke laser model**, defined by the Hamiltonian

$$H = a^{\dagger}a + \varepsilon S^{z} + N^{-1/2}(a+a^{\dagger})S^{x} \tag{1.14}$$

with

$$\mathbf{S} = \sum_{j=1}^{N} \mathbf{S}_j. \tag{1.15}$$

Each $\mathbf{S}_j$ is a spin 1/2 particle (in reality, a 2-level atom with "spin-up" being an excited state and "spin down" being the ground state). The operators a and $a^{\dagger}$ are annihilation and creation operators for a single photon made in a cavity. The first term in H is the photon energy. The last term in H is the atom-photon interaction. There are N atoms and we want to take the thermodynamic limit $N \to \infty$ and find the free energy per atom.

It turns out that this system has a phase transition (in the thermodynamic limit) as a function of β from a low β state in which the average photon number $\langle a^{\dagger}a\rangle$ is $O(1)$ to a high β state in which $\langle a^{\dagger}a\rangle$ is $O(N)$. K. Hepp and I were initially able to prove this only with great difficulty[4]. Later we realized[5] it could be done with ease using a *variant* of Theorem 1.

This variant noted in Ref. 2 and also in Ref. 5 is that when $\mathcal{H}$ is a tensor product the inequalities (1.9) hold if we replace only some of the operators by their symbols and then take the proper trace over the remaining operators. Thus, when the total spin $\mathbf{S}$, which is a conserved quantity, has the value S, we can replace the spin operators by their lower symbols, for example, and we have (with π_S being the projection onto spin S)

$$\mathrm{Tr}\pi_S e^{-\beta H} \geq \nu_S \frac{2S+1}{4\pi}\int \mathrm{Tr}_a \exp\{-\beta[a^{\dagger}a+\varepsilon S\Omega^{z}+(a+a^{\dagger})N^{-1/2}S\Omega^{x}]\}d\Omega. \tag{1.16}$$

Here ν_S is the number of ways of getting spin S with N spin 1/2 particles and Tr_a is the trace over the photon field. A similar upper bound is obtained using upper symbols. The photon field trace in the right side of (1.16) is easy to compute because it is just the partition function of a displaced oscillator, namely $(1-e^{-\beta})^{-1}\exp\{-\beta[\varepsilon S\Omega^z - S^2N^{-1}(\Omega^x)^2]\}$. The Ω integration can then be done by steepest descent as N tends to ∞. Finally, the expression in (1.16) has to be maximized with respect to S.

Alternatively, we can get upper/lower bounds to the partition function by replacing $a^\dagger a$ and $a + a^\dagger$ by their upper/lower symbols with respect to Glauber coherent states and then taking the trace over the spin operators.

Either way, the upper bounds and the lower bounds converge as $N \to \infty$ and the earlier results in Ref. 4 appear in a simple way.

2. Coulomb Systems

We are interested in computing the ground state energy, $E^Q(N,Z)$ of an atom consisting of N electrons and a nucleus of charge Z; units in which $e = \hbar = 2m = 1$ will be used. The nucleus is assumed to be infinitely massive and fixed at the origin in $\mathbf{R}^3$. The well known non-relativistic Hamiltonian is

$$H = \sum\nolimits_{j=1}^{N} h_j + \sum\nolimits_{i\le i<j\le N} |\mathbf{x}_i - \mathbf{x}_j|^{-1} \tag{2.1}$$

with the one-body operator h given by

$$h = -\Delta - Z/|\mathbf{x}|. \tag{2.2}$$

(What follows can also be generalized to a "relativistic" Hamiltonian in which $-\Delta$ is replaced by $\sqrt{-c^2\Delta + m^2c^4}$.)

Our goal is to show that when $\lambda = N/Z$ is fixed (usually $\lambda = 1$, which is the neutral case) and $Z \to \infty$ then

$$\lim_{Z\to\infty} E^Q(N,Z)/E^{\mathrm{TF}}(N,Z) = 1 \tag{2.3}$$

where $E^{\mathrm{TF}}(N,Z)$ is the Thomas-Fermi (TF) energy of an atom.

The TF energy is defined by minimizing the following functional of a nonnegative density $\rho(\mathbf{x})$, $\mathbf{x} \in \mathbf{R}^3$, under the condition $\int \rho = N$.

$$\mathcal{E}(\rho) = \tfrac{3}{5}(3\pi^2)^{2/3}\int \rho(\mathbf{x})^{5/3}d^3x - Z\int \rho(\mathbf{x})|\mathbf{x}|^{-1}d^3x + D(\rho,\rho), \tag{2.4}$$

where $D(f,g) := \frac{1}{2}\int\int f(\mathbf{x})g(\mathbf{y})|\mathbf{x}-\mathbf{y}|^{-1}d^3x d^3y$ is the Coulomb repulsion functional. We denote the unique minimizer by $\rho_{N,Z}^{\mathrm{TF}}(\mathbf{x})$.

There is no space here to go into the details of TF theory, but it is a fact that (2.3) is true, as shown by Lieb and Simon[6] using a rather involved proof. The point here is that coherent states offer a much easier route. The following ideas can be found in the review article[7]. A lower bound to E^Q using coherent states was given at the same time by Thirring[8]. We note that E^{TF} has the form

$$E^{\mathrm{TF}} = C(\lambda)Z^{7/3}. \tag{2.5}$$

An upper bound to E^Q of the form $C(\lambda)Z^{7/3} + O(Z^2)$ and a lower bound to E^Q of the form $C(\lambda)Z^{7/3} - O(Z^{7/3-1/30})$ can be derived using coherent states.

The following is a very sketchy derivation of the upper and lower bounds. For details see Ref. 7. For simplicity of exposition here I shall deal with spinless electrons, which means that $3\pi^2$ must be replaced by $6\pi^2$ in (2.4).

2.1 Upper bound to $E^Q(N,Z)$

One necessary input is the following variational principle[9] (whose proof was later simplified by Bach[10]). The one-particle reduced density matrix, $\gamma(x,y)$, $x \in$ $\mathbf{R}^3$,
$y \in \mathbf{R}^3$ of any N-fermion density matrix satisfies

$$0 \le \gamma \le 1 \qquad \text{(as an operator)} \tag{2.6a}$$

$$\mathrm{Tr}\,\gamma = N \tag{2.6b}$$

Given any γ satisfying (2.6) (which we call *admissible*), the following inequality (or variational principle) holds:

$$E^Q(N,Z) \le \mathrm{Tr}\,\gamma h + \int\int |\mathbf{x}-\mathbf{y}|^{-1}[\gamma(\mathbf{x},\mathbf{x})\gamma(\mathbf{y},\mathbf{y})) - |\gamma(\mathbf{x},\mathbf{y})|^2]d^3x d^3y. \tag{2.7}$$

The right side of (2.7) is well known for a Hartree-Fock γ (i.e., a γ that is a projection); the interesting point is that it holds for *all admissible* γ's.

Next, we introduce the family of coherent states parametrized by $p \in \mathbf{R}^3$, $q \in \mathbf{R}^3$,

$$f_{\mathbf{p},\mathbf{q}}(\mathbf{x}) = g(\mathbf{x}-\mathbf{q})e^{i\mathbf{p}\cdot\mathbf{x}}, \qquad \int |g(\mathbf{x})|^2 d^3x = 1. \tag{2.8}$$

(g will not be a Gaussian as in the Glauber states; in fact it is convenient to let g have compact support.) We define our variational γ by

$$\gamma(\mathbf{x},\mathbf{y}) = \int\int M(\mathbf{p},\mathbf{q}) f_{\mathbf{p},\mathbf{q}}(\mathbf{x}) f_{\mathbf{p},\mathbf{q}}(\mathbf{y}) d^3p d^3q \tag{2.9}$$

with

$$M(\mathbf{p},\mathbf{q}) = \Theta([6\pi^2 \rho_{N,Z}^{\mathrm{TF}}(\mathbf{q})]^{2/3} - p^2). \tag{2.10}$$

Here, Θ is the step-function ($\Theta(t) = 1$ if $t > 0$ and $\Theta(t) = 0$ if $t \leq 0$).

Since the function M satisfies $0 \leq M(\mathbf{p},\mathbf{q}) \leq 1$, it follows that γ satisfies (2.6a). Since $\int \rho_{N,Z}^{\mathrm{TF}}(\mathbf{x}) d^3x = N$, it follows that γ satisfies (2.6b).

This construction, (2.9) is simple and effective. It generates a useful *admissible* density matrix without having to construct a Hartree-Fock determinantal wave function. The construction eliminates what used to be called "The orthogonality problem (or catastrophe)".

If we substitute (2.9) into (2.7) and do some computations we find that

$$E^Q \leq E^{\mathrm{TF}} + O(Z^2), \tag{2.11}$$

as required.

2.2 Lower Bound to $E^Q(N,Z)$

Let $\psi(\mathbf{x}_1,\ldots,\mathbf{x}_N)$ be any normalized function satisfying the Pauli principle (i.e., ψ is antisymmetric) and define the one-body density matrix

$$\gamma(\mathbf{x},\mathbf{y}) := N \int \psi(\mathbf{x},\mathbf{x}_2,\ldots,\mathbf{x}_N)\psi(\mathbf{y},\mathbf{x}_1,\ldots,\mathbf{x}_N)^* d^3x_2 \ldots d^3x_N, \tag{2.12}$$

and the one-body density $\rho(\mathbf{x}) := \gamma(\mathbf{x},\mathbf{x})$.

A second input we shall need is the following inequality, which controls the difference between the true Coulomb repulsion and the classical analogue, $D(\rho,\rho)$.

$$\langle\psi| \sum_{1\leq i<j\leq N} |\mathbf{x}_i - \mathbf{x}_j|^{-1}|\psi\rangle \geq D(\rho,\rho) - (1.68)\int \rho^{4/3}(\mathbf{x}) d^2x \tag{2.13}$$

By Schwarz's inequality for the Coulomb potential we have

$$D(\rho,\rho) \geq 2D(\rho_{N,Z}^{\mathrm{TF}},\rho) - D(\rho_{N,Z}^{\mathrm{TF}},\rho_{N,Z}^{\mathrm{TF}}). \tag{2.14}$$

Thus

$$\langle\psi|H|\psi\rangle \geq \operatorname{Tr}\gamma\widetilde{h} - D(\rho_{N,Z}^{\mathrm{TF}}, \rho_{N,Z}^{\mathrm{TF}}) - 1.68\int \rho^{4/3}(\mathbf{x})d^3x. \tag{2.15}$$

where $\widetilde{h}$ is the one-body Schroedinger operator

$$\widetilde{h} = -\Delta - \phi^{\mathrm{TF}}(\mathbf{x})$$

with

$$\phi^{\mathrm{TF}}(\mathbf{x}) = -\frac{Ze^2}{|\mathbf{x}|} - \frac{1}{|\mathbf{x}|} * \rho^{\mathrm{TF}}.$$

We now use coherent states (2.8) to find a lower bound to $\operatorname{Tr}\gamma\widetilde{h}$. Define $0 \leq \widetilde{M}(p,q) \leq 1$ by

$$\widetilde{M}(\mathbf{p},\mathbf{q}) := \langle f_{\mathbf{p},\mathbf{q}}|\,\gamma\,|f_{\mathbf{p},\mathbf{q}}\rangle \leq 1. \tag{2.16}$$

We then have

$$\operatorname{Tr}[-\nabla^2\gamma] = \int\int \widetilde{M}(\mathbf{p},\mathbf{q})\; p^2 d^3p d^3q - N\int |\nabla g(\mathbf{x})|^2 d^3x \tag{2.17}$$

$$\operatorname{Tr}[\phi^{\mathrm{TF}}\gamma] = \int\int \widetilde{M}(\mathbf{p},\mathbf{q})\; \phi_{N,Z}^{\mathrm{TF}}(\mathbf{q})\; d^3p d^3q + \text{“controllable error”}. \tag{2.18}$$

The last term, “controllable error”, is a bit mysterious and a bit hard to compute, but it can be evaluated and shown to be $O(Z^{7/3-1/30})$ when combined with the other “errors” $-N\int|\nabla g|^2$ and $(-1.68)\int\rho^{4/3}$ in (2.13).

The quantity $\operatorname{Tr}\gamma\widetilde{h}$ is bounded below by the sum of the negative eigenvalues of $\widetilde{h}$, i.e. $-\operatorname{Tr}g(\widetilde{h})$. Here $g(t) = -t\theta(-t)$ is a convex function. If $\widetilde{h}$ had an upper symbol we could get a lower bound to (2.15) by using the Berezin-Lieb inequality. The “controllable error” in (2.18) is due to the fact that the TF potential, like the Coulomb potential, does not have an upper symbol and, therefore, must be approximated by a potential that does have one.

The sum of the two main terms in (2.17) and (2.18) can now be minimized with respect to *all* functions $\widetilde{M}(\mathbf{p},\mathbf{q})$ satisfying $O \leq \widetilde{M}(\mathbf{p},\mathbf{q}) \leq 1$. The minimizer is found, with the help of the TF equation, to be the same as in (2.10). If this is substituted into (2.17), (2.18) and if $D(\rho_{N,Z}^{\mathrm{TF}}, \rho_{N,Z}^{\mathrm{TF}})$ is subtracted from this (as in (2.15)), the result is precisely the anticipated quantity $E^{\mathrm{TF}}(N,Z)$.

3. Wehrl's Entropy

In 1977 Wehrl[12] proposed a classical interpretation of quantum-mechanical entropy. While it differs from both the usual “classical entropy” and from the true quantum-mechanical entropy, it remedies some deficiencies of both.

We begin with the density matrix of a system with Hamiltonian H,

$$\Gamma := e^{-\beta H}/\mathrm{Tr}e^{-\beta H}. \tag{3.1}$$

Next, we define the function on phase space

$$\rho(z) := \langle z|\Gamma|z\rangle \tag{3.2}$$

with $|z\rangle$ being a family of coherent states, as in (1.5)-(1.7). Thus, $\rho(z)$ is the lower symbol of Γ, and we note that

$$0 \le \rho(z) \le 1, \tag{3.3}$$

$$\int \rho(z)dz = 1 \tag{3.4}$$

since $\mathrm{Tr}\,\Gamma = 1$. ***Wehrl's entropy*** is given by

$$S^{W}(\Gamma) := -\int \rho(z)\ln\rho(z)dz, \tag{3.5}$$

i.e., S^{W} is the ordinary classical or Shannon entropy of the density ρ.

The entropy S^{W} is *not* the classical entropy of Γ. That quantity is usually defined by attributing a classical function $\widetilde{H}(z)$ to H (say, $\check{H}(z)$ as in (1.7)) and then defining

$$\rho^{cl}(z) := \exp[-\beta\widetilde{H}(z)]/\int \exp[-\beta\widetilde{H}(z)]dz, \tag{3.6}$$

and then setting

$$S^{cl}(\Gamma) := -\int \rho^{cl}(z)\ln\rho^{cl}(z)dz. \tag{3.7}$$

The major drawback to (3.7) is that $\rho^{cl}(z)$ can easily exceed 1 for some large β. Indeed as $\beta\to\infty$, $S^{cl}(\Gamma)\to-\infty$. This is in stark contradiction with the fact that the quantum entropy

$$S^{Q}(\Gamma) = -\mathrm{Tr}\,\Gamma\ln\Gamma \tag{3.8}$$

is always nonnegative. As Wehrl points out[12], this simple observation shows the impossiblity (for all β) of S^{cl} tending to S^{Q} as Planck's constant tends to zero. On the other hand, since $\rho(z)$ in (3.2) never exceeds 1, $S^{W}(\Gamma) > 0$ for all β, and thus S^{W} behaves better than S^{cl} from this perspective. Indeed, Wehrl proved that

$$S^{W}(\Gamma) > S^{Q}(\Gamma) \ge 0. \tag{3.9}$$

The left side of (3.9) is the Berezin-Lieb inequality applied to the convex function $x \ln x$.

The quantum entropy S^Q also has a serious defect — this time a physical one. If our Hilbert space $\mathcal{H}$ is the tensor product of two spaces, $\mathcal{H} = \mathcal{H}_1 \otimes \mathcal{H}_2$, and Γ is an operator on $\mathcal{H}$, we have the entropy $S^Q(\Gamma)$ which we shall call S_{12}. Using the partial trace we can also define Γ_1 by $\Gamma_1 := \mathrm{Tr}_{\mathcal{H}_2}\Gamma$ as an operator on $\mathcal{H}_1$, with corresponding entropy $S^Q(\Gamma_1) := S_1$. Likewise, S_2 is defined. If we did this with classical discrete densities and replaced partial traces by sums, we would have the inequalities

$$S_1 \leq S_{12} \leq S_1 + S_2. \tag{3.10}$$

It turns out that the subadditivity inequality $S_{12} \leq S_1 + S_2$ does hold for the quantum entropies and classical continuous entropy (in which sums become integrals) but *the monotonicity, $S_1 \leq S_{12}$ fails in general for quantum systems*! While it does hold for classical discrete systems, *monotonicity also fails for classical continuous systems* (cf. Ref. 13). Thus, the universe could be in a pure state, and hence have zero entropy, while S_1, the entropy of Earth, is quite large.

An advantage of the Wehrl entropy is that both parts of (3.10) hold! This means that we define

$$\rho_{12}(z_1, z_2) = \langle z_1, z_2 | \Gamma | z_1, z_2 \rangle, \tag{3.11}$$

where $|z_1, z_2\rangle$ is the ordinary tensor product of two coherent states on $\mathcal{H}_1$ and $\mathcal{H}_2$. We can then define

$$\rho_1(z_1) = \int \rho_{12}(z_1, z_2) dz_2 = \langle z_1 | \mathrm{Tr}_{\mathcal{H}_2}\Gamma | z_1 \rangle, \tag{3.12}$$

noting that the two possible definitions of ρ_1 are, in fact, the same.

In addition to (3.10) the Wehrl entropy also satisfies all the other nice properties of entropy such as concavity in Γ and strong subadditivity[13].

Returning now to (3.9) we can ask for the minimum (with respect to all Γ's) of the value of $S^W(\Gamma)$. By concavity, it is easy to prove that a minimizing Γ must be a pure state, i.e., $\Gamma = |\phi\rangle\langle\phi|$ for some normalized vector $|\phi\rangle$ in the Hilbert space. In case that $\mathcal{H} = L^2(\mathbf{R}^n)$ and the coherent states are the Glauber coherent states (i.e. (2.8) and with $g(x) = \exp[-x^2]$), Wehrl conjectured that the minimizing $|\phi\rangle$ must itself be a coherent state, i.e., $|\phi\rangle$ is an $f_{\mathrm{p,q}}$ in (2.8). Anyone will do. An easy computation would then show

$$\min_{\Gamma} S^W(\Gamma) = 1. \tag{3.13}$$

This conjecture was proved[14], but the strange fact was that two deep theorems in harmonic analysis had to be used — the sharp constant in the Hausdorff-Young inequality and the sharp constant in Young's inequality. In view of the Heisenberg group lying behind Glauber coherent states (which are minimal weight vectors), it is tempting to suppose that a much simpler proof, perhaps group theoretical, of Wehrl's conjecture is possible. This is an *interesting open mathematical problem.*

Another interesting mathematical problem concerns the obvious analog of Wehrl's conjecture, made in Ref.14, for the spin S Bloch coherent spin states $|\boldsymbol{\Omega}\rangle$ used in Sect. 1. If, as the conjecture states, the minimum Wehrl entropy occurs when $|\phi\rangle$ is an $|\boldsymbol{\Omega}\rangle$ the entropy, which is independent of $|\boldsymbol{\Omega}\rangle$ and easy to calculate, is

$$\min_{\Gamma} S^{W}(\Gamma) = \frac{2S}{2S+1}. \tag{3.14}$$

For $S = 1/2$ the proof is trivial since *all* vectors in $\mathbf{C}^2$ are coherent states. But no proof exists for any other S value, even though many attempts have been made to find one. *It would be very nice if someone could solve this 15 year old problem!* Clearly, we do not know everything there is to be known about $SU(2)$.

A final remark about the Wehrl conjecture for Glauber states is its generalization[14] in $\mathbf{R}^n$:

$$(2\pi)^{-n}\int |\langle\phi|f_{\mathbf{p},\mathbf{q}}\rangle|^{2r} d^n p d^n q \geq r^{-n} \tag{3.15}$$

for all $|\phi\rangle$ satisfying $\langle\phi|\phi\rangle = 1$ and all $r \geq 1$. Since (3.15) is always an equality when $r = 1$, we can deduce (3.13) from (3.15) by differentiating (3.15) at $r = 1$.

Further generalizations, related to radar signal analysis, wavelets and Wigner distribution functions, were also obtained[15]. Among them there is the following. Let the $f_{\mathbf{p},\mathbf{q}}$ in (3.15) be given by (2.18) but with an *arbitrary* normalized $g(\mathbf{x}-\mathbf{q})$, so that the left side of (3.15) now involves two arbitrary functions ϕ and g. The inequality (3.15) remains true!

References

1. F.A. Berezin, *Covariant and contravariant symbols of operators*, Izv. Akad. SSSR Ser. Mat. **6** (1972) 1134-1167.
2. E.H. Lieb, *The classical limit of quantum spin systems*, Commun. Math. Phys. **31** (1973) 327-340.
3. E.H. Lieb and J.P. Solovej, *Quantum coherent operators: A generalization of coherent states*, Lett. Math. Phys. **22** (1991) 145-154.
4. K. Hepp and E.H. Lieb, *On the superradiant phase transition for molecules in a quantized radiation field*, Ann. of Phys. (NY) **76** (1973) 360-404.

5. K. Hepp and E.H. Lieb, *The equilibrium statistical mechanics of matter interacting with the quantized radiation field*, Phys. Rev. **A8** (1973) 2517-2525.
6. E.H. Lieb and B. Simon, *Thomas-Fermi theory of atoms, molecules and solids*, Adv. in Math. **23** (1977) 22-116.
7. E.H. Lieb, *Thomas-Fermi and related theories of atoms and molecules*, Rev. Mod. Phys. **53** (1981) 603-641; errata **54** (1981) 311. See Sect. V.
8. W. Thirring, *A lower bound with the best possible constants for Coulomb Hamiltonians*, Commun. Math. Phys. **79** (1981) 1-7.
9. E.H. Lieb, *A variational principle for many-fermion systems*, Phys. Rev. Lett. **46** (1981) 457-459; errata **47** (1981) 69.
10. V. Bach, *Error bounds for the Hartree-Fock energy of atoms and molecules*, Commun. Math. Phys. **147** (1992) 527-548.
11. E.H. Lieb and S. Oxford, *An improved lower bound on the indirect Coulomb energy*, Int. J. Quant. Chem. **19** (1981) 427-439.
12. A. Wehrl, *On the relation between classical and quantum-mechanical entropy*, Rept. Math. Phys. **16** (1979) 353-358.
13. E.H. Lieb, *Some convexity and subadditivity properties of entropy*, Bull. Amer. Math. Soc. **81** (1975) 1-13.
14. E.H. Lieb, *Proof of an entropy conjecture of Wehrl*, Commun. Math. Phys. **62** (1978) 35-41.
15. E.H. Lieb, *Integral bounds for radar ambiguity functions and Wigner distributions*, J. Math. Phys. **31** (1990) 594-599.

Part V

Brunn-Minkowski Inequality and Rearrangements

Reprinted from JOURNAL OF FUNCTIONAL ANALYSIS
Vol. 17, No. 2, October 1974

Printed in Belgium

A General Rearrangement Inequality for Multiple Integrals

H. J. BRASCAMP*§

The Institute for Advanced Study, Princeton, New Jersey 08540

ELLIOTT H. LIEB§

Departments of Mathematics and Physics, Massachusetts Institute of Technology, Cambridge, Massachusetts 02139

AND

J. M. LUTTINGER†

Department of Physics, Columbia University, New York, New York 10027

Communicated by Irving Segal

Received March 21, 1974

In this paper we prove a rearrangement inequality that generalizes inequalities given in the book by Hardy, Littlewood and Pólya[1] and by Luttinger and Friedberg.[2] The inequality for an integral of a product of functions of one variable is further extended to the case of functions of several variables.

I. INTRODUCTION

Rearrangement inequalities were studied by Hardy, Littlewood and Pólya in the last chapter of their book "Inequalities." Let us start by recapitulating the definition of the symmetric decreasing rearrangement of a function, and the integral inequalities following from that definition. Our new results are contained in Theorems 1.2 and 3.4.

In the following, measure always means Lebesgue measure and is denoted by μ.

DEFINITION 1.1. Let f be a nonnegative measurable function on $\mathbf{R}$,

* Work partially supported by National Science Foundation Grant GP-16147 A#1.
§ Work partially supported by National Science Foundation Grant GP-31674 X.
† Work partially supported by a grant from the National Science Foundation.

let $K_y^f = \{x \mid f(x) \geqslant y\}$ and let $M_y^f = \mu(K_y^f)$. Assume that $M_a^f < \infty$ for some $a < \infty$. If f^* is another function on $\mathbf{R}$ with the same properties as f and, additionally,

$$\text{(a)}\quad f^*(x) = f^*(-x),\ \forall x,$$

$$\text{(b)}\quad 0 < x_1 < x_2 \Rightarrow f^*(x_2) \leqslant f^*(x_1),$$

$$\text{(c)}\quad M_y^{f^*} = M_y^f, \quad \forall y > 0,$$

then f^* is called a *symmetric decreasing rearrangement* of f.

Remarks. (1) If g and h are two symmetric decreasing rearrangements of f, then

$$g(x) = h(x) \quad \text{a.e.}$$

(2) If χ is the characteristic function of a measurable set, we can define $\chi^*(x) = 1$ if $2\,|\,x\,| \leqslant \int \chi$ and $\chi^*(x) = 0$, otherwise. For a general function f, define $\chi_y(x) = 1$ if $f(x) \geqslant y$ and $\chi_y(x) = 0$, otherwise. Then

$$f(x) = \int_0^\infty dy \chi_y(x),$$

and

$$f^*(x) = \int_0^\infty dy \chi_y^*(x)$$

is a symmetric decreasing rearrangement of f. The fact that $M_a^f < \infty$ implies that $f^*(x) < \infty$, $\forall x \neq 0$.

(3) In the following theorems we shall always be dealing with integrals. Consequently, by remark (1), f^* is unique for our purposes. Trivially, $f \in L^1(\mathbf{R})$ iff $f^* \in L^1(\mathbf{R})$ and $\int f = \int f^*$.

The inequalities to be found in [1] are

$$\int_{\mathbf{R}} dx f(x)\, g(x) \leqslant \int_{\mathbf{R}} dx f^*(x)\, g^*(x);$$

$$\int_{\mathbf{R}^2} dx_1 dx_2 f(x_1)\, g(x_2)\, h(x_1 - x_2) \leqslant \int_{\mathbf{R}^2} dx_1 dx_2 f^*(x_1)\, g^*(x_2)\, h^*(x_1 - x_2),$$

the latter being due to Riesz [3].

A generalization due to Luttinger and Friedberg [2] reads

$$\int_{\mathbf{R}^n} d^n x \prod_{j=1}^{n} f_j(x_j)\, h_j(x_j - x_{j+1}) \leqslant \int_{\mathbf{R}^n} d^n x \prod_{j=1}^{n} f_j^*(x_j)\, h_j^*(x_j - x_{j+1}),$$

where $x_{n+1} \equiv x_1$. This formula was derived for the purpose of physical applications (inequalities for Green's functions, Luttinger [4]).

In the present paper we give a further generalization, one which was already conjectured in [2].

THEOREM 1.2. *Let f_j, $1 \leqslant j \leqslant k$, be nonnegative measurable functions on* $\mathbf{R}$, *and let a_{jm}, $1 \leqslant j \leqslant k$, $1 \leqslant m \leqslant n$, be real numbers. Then*

$$\int_{\mathbf{R}^n} d^n x \prod_{j=1}^{k} f_j \left(\sum_{m=1}^{n} a_{jm} x_m \right) \leqslant \int_{\mathbf{R}^n} d^n x \prod_{j=1}^{k} f_j^* \left(\sum_{m=1}^{n} a_{jm} x_m \right).$$

Remark. Theorem 1.2 is nontrivial only for $k > n$. If $k < n$, both integrals diverge. If $k = n$ and $\det | a_{jm} | = 0$, both integrals diverge. If $k = n$ and $\det | a_{jm} | \neq 0$, equality holds (change variables to $y_j = \sum_{m=1}^{n} a_{jm} x_m$ and then use the fact that $\int f_j = \int f_j^*$).

A proof of Theorem 1.2 is given in Section 2. An important tool is Brunn's part of the Brunn–Minkowski theorem, which we recall here (see e.g., [5] Section 11.48). Note that every convex set in $\mathbf{R}^n$ is measurable.

LEMMA 1.3. *Let C be a convex set in $\mathbf{R}^{n+1}$, let $\varphi \in \mathbf{R}^{n+1}$, and let $V(t)$ be the family of planes $\langle \varphi, x \rangle = t$, $-\infty < t < \infty$. Let $S(t)$ be the n-dimensional volume of the convex set $V(t) \cap C$. Then $S(t)^{1/n}$ is a concave function of t in the interval where $S(t) > 0$.*

COROLLARY 1.4. *Let C, φ and $S(t)$ be as in Lemma 1.3 and, in addition, let C be balanced (i.e., $x \in C \Rightarrow -x \in C$). Then $S(t) = S(-t)$ and $S(t_2) \leqslant S(t_1)$ for $t_2 \geqslant t_1 \geqslant 0$.*

In Section 3 we generalize Theorem 1.2 to the Schwarz symmetrization (Definition 3.3) of functions of several variables. An auxiliary lemma that we need for this purpose is given in the Appendix.

II. PROOF OF THEOREM 1.2

Although in general $f \to f^*$ is not linear, by Remark (2) following Definition 1.1 it is sufficient to assume that each f_j is the characteristic function of some measurable set. By standard approximation arguments we may assume this set to be a finite union of disjoint compact intervals (cf. [1], Section 10.14).

We start by assuming that each f_j is the characteristic function of one interval.

LEMMA 2.1. *Let f_j, $1 \leqslant j \leqslant k$, be the characteristic functions of the intervals*

$$b_j - c_j \leqslant x \leqslant b_j + c_j,$$

and define

$$f_j(x \mid t) = f_j(x + b_j t).$$

Then

$$I(t) = \int_{\mathbf{R}^n} d^n x \prod_{j=1}^{k} f_j \left(\sum_{m=1}^{n} a_{jm} x_m \mid t \right)$$

is a nondecreasing function of $t \in [0, 1]$.

Remark. Note, that $f_j(x \mid 0) = f_j(x)$ and $f_j(x \mid 1) = f_j^*(x)$, so Lemma 2.1 includes a special case of Theorem 1.2.

Proof of Lemma 2.1. $I(t)$ is the volume of the intersection of the k strips

$$S_j = \left\{ x \in \mathbf{R}^n \mid b_j(1-t) - c_j \leqslant \sum_{m=1}^{n} a_{jm} x_m \leqslant b_j(1-t) + c_j \right\}.$$

In $\mathbf{R}^{n+1}$, consider the set

$$C = \bigcap_{1 \leqslant j \leqslant k} \left\{ x \in \mathbf{R}^{n+1} \mid -c_j \leqslant \sum_{m=1}^{n} a_{jm} x_m - b_j x_{n+1} \leqslant c_j \right\}.$$

$I(t)$ is the volume of the intersection of C with the plane $x_{n+1} = 1 - t$. Since C is convex and balanced, $I(t)$ is nondecreasing for $t \in [0, 1]$ by Corollary 1.4. Q.E.D.

We now conclude the proof of Theorem 1.2 with the following lemma.

LEMMA 2.2. *Theorem* 1.2 *holds under the restriction, that each f_j is the characteristic function of a finite union of disjoint compact intervals.*

Proof. Let f_j be the characteristic function of n_j intervals. We prove the lemma by induction on $N = \{n_1, n_2, \ldots, n_k\}$, with fixed k.

We say that $M \prec N$ if $m_j \leqslant n_j$, $1 \leqslant j \leqslant k$, and $m_i < n_i$ for some i.

Lemma 2.2 is true for $N = \{1, 1,..., 1\}$ by Lemma 2.1. Now assume that Lemma 2.2 is true for all $M \prec N$.

Let $f_j(x)$ be the characteristic function of

$$\bigcup_{1 \leqslant p \leqslant n_j} \{x \in \mathbf{R} \mid b_{jp} - c_{jp} \leqslant x \leqslant b_{jp} + c_{jp}\},$$

with

$$b_{jp} + c_{jp} < b_{j,p+1} - c_{j,p+1}, \; 1 \leqslant p \leqslant n_j - 1, \; 1 \leqslant j \leqslant k.$$

Further define $f_j(x \mid t)$ to be the characteristic function of

$$\bigcup_{1 \leqslant p \leqslant n_j} \{x \in \mathbf{R} \mid b_{jp}(1-t) - c_{jp} \leqslant x \leqslant b_{jp}(1-t) + c_{jp}\}$$

for $0 \leqslant t \leqslant \tau$, where

$$\tau = \min_{j,p}[1 - (b_{j,p+1} - b_{jp})^{-1}(c_{j,p+1} + c_{jp})] > 0.$$

For $0 \leqslant t < \tau$, the intervals belonging to each function f_j remain disjoint; at $t = \tau$ at least two intervals coalesce for some j.

Since each f_j is a positive sum of characteristic functions of single intervals of the type stated in the hypothesis of Lemma 2.1, we can apply that lemma interval by interval and find

$$\int_{\mathbf{R}^n} d^n x \prod_{j=1}^{k} f_j \left(\sum_{m=1}^{n} a_{jm} x_m \right) \leqslant \int_{\mathbf{R}^n} d^n x \prod_{j=1}^{k} f_j \left(\sum_{m=1}^{n} a_{jm} x_m \mid \tau \right).$$

At $t = \tau$, the family of functions $\{f_j(x \mid \tau\}$ satisfies the hypothesis of Lemma 2.2, except that N has been reduced to some $M \prec N$. Therefore, by assumption

$$\int_{\mathbf{R}^n} d^n x \prod_{j=1}^{k} f_j \left(\sum_{m=1}^{n} a_{jm} x_m \mid \tau \right) \leqslant \int_{\mathbf{R}^n} d^n x \prod_{j=1}^{k} f_j^* \left(\sum_{m=1}^{n} a_{jm} x_m \right),$$

because $f_j(\cdot \mid \tau)$ and $f_j(\cdot)$ have the same symmetric decreasing rearrangement. This proves Lemma 2.2 and at the same time Theorem 1.2

III. Generalization to Functions of Several Variables

In this section we indicate how to generalize Theorem 1.2 to functions of several variables (Lemma 3.2 and Theorem 3.4). The intuitive idea was given in [4], *p.* 1450.

Let f be a nonnegative, measurable function on $\mathbf{R}^p$, and let V be a $p-1$ dimensional plane through the origin of $\mathbf{R}^p$. Choose an orthogonal coordinate system in $\mathbf{R}^p$ such that the x^1-axis is perpendicular to V.

DEFINITION 3.1. A nonnegative, measurable function $f^*(x \mid V)$ on $\mathbf{R}^p$ is called a Steiner-symmetrization with respect to V of the function $f(x)$, if $f^*(x^1, x^2,..., x^p)$ is a symmetric decreasing rearrangement with respect to x^1 of $f(x^1, x^2,..., x^p)$ for each fixed $x^2,..., x^p$.

Remark. The notion of Steiner symmetrization is usually reserved for sets; for any $y > 0$, the set $\{x \in \mathbf{R}^p \mid f^*(x \mid V) \geqslant y\}$ is a Steiner symmetrization with respect to V of the set $\{x \in \mathbf{R}^p \mid f(x) \geqslant y\}$ (see e.g., Pólya and Szegö [6], Note *A*).

LEMMA 3.2. *Let $f_j(x)$, $1 \leqslant j \leqslant k$, be nonnegative measurable functions on $\mathbf{R}^p$, let a_{jm}, $1 \leqslant j \leqslant k$, $1 \leqslant m \leqslant n$, be real numbers, and let V be any plane through the origin of $\mathbf{R}^p$. Then*

$$\int_{\mathbf{R}^{np}} d^{np}x \prod_{j=1}^{k} f_j\left(\sum_{m=1}^{n} a_{jm}x_m\right) \leqslant \int_{\mathbf{R}^{np}} d^{np}x \prod_{j=1}^{k} f_j^*\left(\sum_{m=1}^{n} a_{jm}x_m \mid V\right),$$

where $\mathbf{R}^{np} \ni x = (x_1,..., x_n)$, $x_m \in \mathbf{R}^p$.

Proof. Choose appropriate orthogonal coordinates in $\mathbf{R}^p \ni x = (x^1,..., x^p)$ as above, so that the x^1-axis is orthogonal to V. Then, by Theorem 2.1, the inequality already holds for the integration over $x_1{}^1, x_2{}^1,..., x_n{}^1$ for any fixed $x_m{}^q$, $1 \leqslant m \leqslant n$, $2 \leqslant q \leqslant p$. Q.E.D.

DEFINITION 3.3. Let f be a nonnegative measurable function on $\mathbf{R}^p$, let $K_y{}^f = \{x \mid f(x) \geqslant y\}$ and let $M_y{}^f = \mu(K_y{}^f)$. Assume that $M_a{}^f < \infty$ for some $a < \infty$. If f^{**} is another function on $\mathbf{R}^p$ with the same properties as f and, additionally,

(a) $f^{**}(x_1) = f^{**}(x_2)$ when $|x_1| = |x_2|$,

(b) $0 < |x_1| < |x_2| \Rightarrow f^{**}(x_2) \leqslant f^{**}(x_1)$,

(c) $M_y^{f^{**}} = M_y{}^f$, $\forall y > 0$,

then f^{**} is called a Schwarz symmetrization of f.

Remarks. (1) The remarks after Definition 1.1 apply, mutatis mutandis, to Schwarz symmetrization.

(2) The notion of Schwarz symmetrization is usually reserved for sets; the set in $\mathbf{R}^{p+1}$ under the graph of $y = f^{**}(x)$ is the Schwarz symmetrization with respect to the y-axis of the set under the graph of $y = f(x)$ (see [5], Note A).

It is intuitively clear, that the Schwarz symmetrization can be obtained as the $L^1(\mathbf{R}^p)$ limit of a sequence of Steiner symmetrizations with respect to different planes. That fact will be proved in the Appendix for the characteristic function of a bounded measurable set (Lemma A1). For the moment we use it, together with Lemma 3.2 and the remarks at the beginning of Section 2, to conclude our main theorem, which is the following.

THEOREM 3.4. *Under the assumptions of Lemma* 3.2,

$$\int_{\mathbf{R}^{np}} d^{np}x \prod_{j=1}^{k} f_j \left(\sum_{m=1}^{n} a_{jm}x_m\right) \leqslant \int_{\mathbf{R}^{np}} d^{np}x \prod_{j=1}^{k} f_j^{**} \left(\sum_{m=1}^{n} a_{jm}x_m\right).$$

APPENDIX

We give the lemma that suffices to establish Theorem 3.4. For two sets A and B, $A \Delta B \equiv (A \cup B)\backslash(A \cap B)$. μ_p denotes Lebesgue measure in $\mathbf{R}^p$.

LEMMA A.1. *Let K be a bounded measurable set in $\mathbf{R}^p$, and let S be the ball centered at the origin with $\mu_p(S) = \mu_p(K)$. Then there exists a sequence of sets K_n, where $K_0 = K$ and where K_{n+1} is obtained from K_n by Steiner symmetrization with respect to some $(p - 1)$-dimensional subspace of $\mathbf{R}^p$, such that*

$$\lim_{n\to\infty} \mu_p(K_n \Delta S) = 0.$$

Remark. There exist various theorems stating the convergence of K_n to S in the Hausdorff metric ([7], Section 21 for compact convex sets; [8], Section 4.5.3 and [9], Section 2.10.31 for general compact sets).

Let us first give a precise definition of the Steiner symmetrization for arbitrary measurable sets (cf. the Remark following Definition 3.1).

DEFINITION A.2. Let K be a bounded measurable set in $\mathbf{R}^p$, and let V be a $(p - 1)$-dimensional subspace of $\mathbf{R}^p$. Then the set $K_V{}^*$ is called a Steiner symmetrization of K with respect to V, if, for every straight line L perpendicular to V with $K \cap L$ measurable in $\mathbf{R}$, $K_V{}^* \cap L$ is a segment (open or closed) with center in V and

$$\mu_1(K_V{}^* \cap L) = \mu_1(K \cap L).$$

Remarks. (1) Let K be open (resp. closed) and take for $K_V{}^* \cap L$ in Definition A2 the open (resp. closed) segments. Then $K_V{}^*$ is open resp. closed).

To prove this, choose coordinates $x = (x^1,..., x^p) \in \mathbf{R}^p$ with x^1 in the direction orthogonal to V. Let χ_K be the characteristic function of K. Then the statement is true if the function $\mathbf{R}^{p-1} \ni y \to \int dx^1\, \chi_K(x^1, y)$ is lower (resp. upper) semicontinuous. But this follows from the fact that $\chi_K(x^1, y)$ is lower (resp. upper) semicontinuous in $\mathbf{R}^p$.

(2) For arbitrary measurable K, all Steiner symmetrizations are measurable and satisfy (Fubini's theorem)

$$\mu_p(K_V{}^*) = \mu_p(K).$$

Two Steiner symmetrizations can only differ by a set of measure zero. All this is readily seen by sandwiching K between closed sets from within and open sets from without.

(3) If K and M are measurable sets, Lemma 3.2 gives that

$$\mu_p(K_V{}^* \cap M_V{}^*) \geqslant \mu_p(K \cap M),$$

and therefore

$$\mu_p(K_V{}^* \mathbin{\Delta} M_V{}^*) \leqslant \mu_p(K \mathbin{\Delta} M).$$

In particular, if K and M differ only by a set of measure zero, so do $K_V{}^*$ and $M_V{}^*$.

(4) In view of Remarks 2 and 3, we shall further speak of *the* Steiner symmetrization of a measurable set, which in fact associates with each equivalence class of measurable sets a unique equivalence class of measurable sets.

PROPOSITION A.3. *Let K and S be as in Lemma* A.1. *Then*

$$\mu_p(K_V{}^* \mathbin{\Delta} S) \leqslant \mu_p(K \mathbin{\Delta} S)$$

and the equality holds for all subspaces V iff $K = S$.

Proof. The $\leqslant$ inequality holds by Remark 3 above, since $S_V{}^* = S$.

Denote by $L(v)$ the straight line perpendicular to V through $v \in V$. Let $K(v) = K \cap L(v)$, and let $\pi_V(K)$ be the projection of K on V,

$$\pi_V(K) = \{v \in V \mid \mu_1(K(v)) > 0\}.$$

Now let $K \neq S$ so that $\mu_p(K \backslash S) = \mu_p(S \backslash K) > 0$. It can be shown by a tedious but trivial argument that there exists a subspace V such that $P = \pi_V(K \backslash S) \cap \pi_V(S \backslash K)$ has positive μ_{p-1} measure. If $v \in P$, neither $K(v) \subset S(v)$ nor $S(v) \subset K(v)$; therefore

$$\mu_1(K_V{}^*(v) \mathbin{\Delta} S(v)) = \mid \mu_1(K(v)) - \mu_1(S(v)) \mid < \mu_1(K(v) \mathbin{\Delta} S(v)).$$

for all $v \in P$. Because, generally, for all $v \in V$

$$\mu_1(K_V{}^*(v) \mathbin{\Delta} S(v)) \leqslant \mu_1(K(v) \mathbin{\Delta} S(v)),$$

we have for the particular subspace V under consideration

$$\mu_p(K_V{}^* \mathbin{\Delta} S) < \mu_p(K \mathbin{\Delta} S).$$

This proves Proposition A.3.

Let us now specify the sequence of sets in Lemma A.1. Given K_n, choose a subspace V_1, such that

$$\mu_p(K_{nV_1}^* \mathbin{\Delta} S) < \inf_V \mu_p(K_{nV}^* \mathbin{\Delta} S) + n^{-1}$$

Then construct K_{n+1} from K_n by p consecutive Steiner symmetrizations with respect to a set of $p-1$ dimensional subspaces $V_1, V_2, \ldots, V_p$ (beginning with V_1 specified above) whose orthogonal complements are pairwise orthogonal. In that way,

$$\mu_p(K_{n+1} \mathbin{\Delta} S) < \mu_p(K_{nW}^* \mathbin{\Delta} S) + n^{-1}$$

for all n and for all subspaces W.

Proposition A.4. *There exist a subsequence K_{n_j} and a measurable set M such that*

$$\lim_{j \to \infty} \mu_p(K_{n_j} \mathbin{\Delta} M) = 0$$

Proof. Express a point $x \in \mathbf{R}^p$ in coordinates $(x^1, x^2, \ldots, x^p)$ corresponding to the planes used to construct K_n. Then, it is not difficult to show that for $n > 0$ (i.e., after the first set of p orthogonal symmetrizations), $x \in K_n$ implies $y \in K_n$ if $\mid y^m \mid \leqslant \mid x^m \mid$, $m = 1, \ldots, p$. Therefore, if χ_n is the characteristic function of K_n

$$\int_{\mathbf{R}} dx^m \mid \chi_n(x^1, \ldots, x^m + y^m, \ldots, x^p) - \chi_n(x^1, \ldots, x^m, \ldots, x^p) \mid \leqslant 2 \mid y^m \mid.$$

Note that by assumption K is contained in some ball B of radius R centered at the origin; then also $K_n \subset B$. This implies that

$$\int_{\mathbf{R}^p} d^p x \mid \chi_n(x+y) - \chi_n(x) \mid \leqslant 2(2R)^{p-1} \sum_{m=1}^{p} \mid y^m \mid .$$

In other words

$$\lim_{y \to 0} \int_{\mathbf{R}^p} d^p x \mid \chi_n(x+y) - \chi_n(x) \mid = 0$$

uniformly in n. Hence the family of functions $\{\chi_n\}$ is conditionally compact in $L^1(\mathbf{R}^p)$ (Dunford and Schwartz [10], Theorem IV, 8.21). Q.E.D.

Propositions A.3. and A.4. immediately give the following.

COROLLARY A.5. *$\mu_p(K_n \Delta S)$ decreases monotonously to $\mu_p(M \Delta S)$.*

Let us now conclude the proof of Lemma *A*.1. Assume that $M \neq S$; we shall show that this leads to a contradiction.

Let $\mu_p(M \Delta S) = \delta > 0$. Then there exist a $p-1$ dimensional subspace W and an $\epsilon > 0$ such that

$$\mu_p(M_W{}^* \Delta S) = \delta - \epsilon.$$

by Proposition A.3. Also

$$\lim_{j \to \infty} \mu_p(K^*_{n_j W} \Delta M_W{}^*) = 0,$$

so

$$\lim_{j \to \infty} \mu_p(K^*_{n_j W} \Delta S) = \delta - \epsilon.$$

Then there exists an n_k such that $n_k > 2\epsilon^{-1}$ and

$$\mu_p(K^*_{n_k W} \Delta S) < \delta - \epsilon/2.$$

But by the construction of the sequence K_n,

$$\mu_p(K_{n_k+1} \Delta S) < \mu_p(K^*_{n_k W} \Delta S) + n_k^{-1} < \delta,$$

which contradicts Corollary A.5.

Thus we find that $M = S$; then by Corollary A.5., $\mu_p(K_n \Delta S)$ decreases monotonously to zero. This proves Lemma A.1.

References

1. G. E. Hardy, J. E. Littlewood, and G. Pólya, Inequalities, Cambridge University Press, London and New York (1952).
2. J. M. Luttinger and R. Friedberg, Preprint, A New Rearrangement Inequality for Multiple Integrals (1973).
3. F. Riesz, Sur une Inégalité Intégrale, *J. L.M.S.* **5** (1930), 162–168.
4. J. M. Luttinger, Generalized isoperimetric inequalities, *J. Math. Phys.* **14** (1973), 586–593, 1444–1447, 1448–1450.
5. T. Bonnesen and W. Fenchel, Theorie der Konvexen Körper, Chelsea, New York (1948).
6. G. Pólya and G. Szegö, Isoperimetric Inequalities in Mathematical Physics, Princeton Univ. Press, Princeton (1951).
7. W. Blaschke, Kreis und Kugel, Veit and Comp., Leipzig (1916).
8. H. Hadwiger, Vorlesungen über Inhalt, Oberfläche und Isoperimetrie, Springer, Berlin-Göttingen-Heidelberg (1957).
9. H. Federer, Geometric Measure Theory, Springer, New York (1969).
10. N. Dunford and J. T. Schwartz, Linear Operators, Part I, Interscience, New York and London (1958).

1. Some inequalities for Gaussian measures and the long-range order of the one-dimensional plasma

H. J. BRASCAMP AND E. H. LIEB

1.1. Introduction

THE following is a preliminary report on some recent work, the full details of which will be published elsewhere. We have come across some inequalities about integrals and moments of log concave functions which hold in the multidimensional case and which are useful in obtaining estimates for multidimensional modified Gaussian measures. By making a small jump (we shall not go into the technical details) from the finite to the infinite dimensional case, upper and lower bounds to certain types of functional integrals can be obtained. As a non-trivial application of the latter we shall, for the first time, prove that the one-dimensional one-component quantum-mechanical plasma has long-range order when the interaction is strong enough. In other words, the Wigner lattice can exist, in one dimension at least. As another application we shall prove a log concavity theorem about the fundamental solution (Green's function) of the diffusion equation.

1.2. Basic concavity theorem

We begin with a theorem (Theorem 1.1) which, to the best of our knowledge, is new and which constitutes the basis of all our other inequalities.

DEFINITION 1.1. A function F from $\mathbf{R}^n$ to $\mathbf{R}$ is a *log concave function* if $F(x) \geq 0$, $\forall x \in \mathbf{R}^n$, and $F(x)^\lambda F(y)^{1-\lambda} \leq F[\lambda x + (1-\lambda)y]$, $\forall x, y \in \mathbf{R}^n$ and $\lambda \in (0, 1)$. If the inequality is reversed, we say that F is *log convex*. We shall sometimes write $F(x) = \mathrm{e}^{f(x)}$ and f is concave, but it then is understood that f can take on the value $-\infty$. We say that F is *even* if $F(x) = F(-x)$, $\forall x$.

Two important examples of log concave functions are:

(a) $F(x) = \exp[-(x, Ax)]$, where A is any symmetric real positive-semidefinite quadratic form on $\mathbf{R}^n$.

(b) Let C be any convex set in $\mathbf{R}^n$ and let $\chi_C(x) = 1$ for $x \in C$, $\chi_C(x) = 0$ for $x \notin C$ be the characteristic function of C. Then χ_C is a log concave function. χ_C is even if and only if C is *balanced*, i.e. $x \in C \Rightarrow -x \in C$.

THEOREM 1.1. *Let F be a log concave function on $\mathbf{R}^{m+n}$ and F: $(x, y) \mapsto F(x, y)$ for $x \in \mathbf{R}^m$, $y \in \mathbf{R}^n$. Then $G(x) \equiv \int_{\mathbf{R}^n} F(x, y)\, \mathrm{d}y$ is a log concave function on $\mathbf{R}^m$.*

We have four different proofs of this theorem, one of which is the following.

Proof. It is sufficient to prove the theorem when $m=n=1$; the general case follows by Fubini's theorem and induction. Choose two points x and x' such that $G(x)\neq 0$ and $G(x')\neq 0$. We may assume that

$$\sup_y\{F(x,y)\}=\sup_y\{F(x',y)\},$$

for otherwise we can replace $F(x,y)$ by $e^{bx}F(x,y)$ with b suitably chosen. For each $z\geq 0$, define

$$C(z)=\{(x,y)|F(x,y)\geq z\}\subset \mathbf{R}^2,$$

$C(x,y)=\{y|F(x,y)\geq z\}\subset \mathbf{R}$ and $g(x,z)=\text{meas}\{C(x,z)\}$.
Then

(*i*) $C(z)$ is convex and thus $C(x,z)$ is an interval;
(*ii*) $G(x)=\int_0^\infty g(x,z)\,dz$;
(*iii*) for all $0\leq\lambda\leq 1$, $g(\lambda x+(1-\lambda)x',z)\geq\lambda g(x,z)+(1-\lambda)g(x',z)$.

This last fact follows easily from the convexity of $C(z)$; it is also the Brunn–Minkowski theorem which, in one dimension, is trivial. Thus

$$G(\lambda x+(1-\lambda)x')\geq\lambda G(x)+(1-\lambda)G(x')\geq G(x)^\lambda G(x')^{1-\lambda}.$$

Q.E.D.

Theorem 1.1 should not be confused with the following theorem, which is much simpler and which follows directly from Hoelder's inequality.

THEOREM 1.2. *Let* $F:\mathbf{R}^{m+n}\to\mathbf{R}$ and, for $x\in\mathbf{R}^m$, $y\in\mathbf{R}^n$, *let* $F(x,y)$ *be log convex in* x *for each fixed* y. *Then* $G(x)\equiv\int_{\mathbf{R}^n}F(x,y)\,dy$ *is log convex on* $\mathbf{R}^m$.

An immediate consequence of Theorem 1.1 is the following.

THEOREM 1.3. *The convolution of two log concave functions on* $\mathbf{R}^n$ *is log concave.*

Proof. $H(x)=\int_{\mathbf{R}^n}F(x-y)G(y)\,dy$ is log concave since $F(x-y)G(y)$ is jointly log concave in $(x,y)\in\mathbf{R}^{n+m}$.

Q.E.D.

REMARK. In the case of $\mathbf{R}$, Theorem 1.3 is known [1].

1.3. Application of Theorem 1.1 to Gaussian measures

A Gaussian measure on $\mathbf{R}^n$ is given by an (unnormalized) density function $W(x)=\exp[-(x,Ax)/2]$, $A>0$. The expectation value of a real-valued function H, on $\mathbf{R}^n$, is given by

$$\langle H\rangle_0=\frac{\int H(x)W(x)\,dx}{\int W(x)\,dx}.$$

Now suppose that $W(x)$ is replaced by $W_F(x) = W(x)F(x)$, where F is a log concave function. With respect to the new weight we define $\langle H\rangle_F$ as above. How does $\langle H\rangle_F$ compare with $\langle H\rangle_0$?

THEOREM 1.4. *The covariance matrix* M^F, *whose elements are* $M^F_{ij} = \langle x_i x_j\rangle_F - \langle x_i\rangle_F\langle x_j\rangle_F$, *satisfies*

$$M^F \leqslant M^0 = A^{-1}$$

in the sense of forms, i.e. $M^0 - M^F$ *is positive-semidefinite.*

Proof. Consider the function T on $\mathbf{R}^{n+m}$ defined by

$$\begin{aligned} T(x, y) &= W(x)F(x)\exp[-(y, A^{-1}y)/2 + (y, x)] \\ &= W(x - A^{-1}y)F(x). \end{aligned}$$

Then T is log concave and, by Theorem 1.1, $U(y) = \int dx T(x, y)$ is log concave on $\mathbf{R}^n$. Thus, the matrix $\partial^2 \ln U(y)/\partial y_i \partial y_j|_{y=0} = M^F - M^0$ is negative-semidefinite.

Q.E.D.

THEOREM 1.5. *If, in the above, we replace F by a log convex function then* $M^F \geqslant M^0 = A^{-1}$.

Proof. Write $U(y) = \int_{\mathbf{R}^n} W(x)F(x + A^{-1}y)\,dx$ and use Theorem 1.2.

Q.E.D.

As an application of Theorem 1.4, consider an Ising model with Boltzmann factor $B(\sigma) = \exp[\frac{1}{2}(\sigma, A\sigma)]$, $\sigma_i = \pm 1$, $i = 1, \ldots, n$. By adding an unimportant multiple of the identity to A, we can always assume $A > 0$. Since

$$B(\sigma) = (2\pi)^{-n/2}[\det A]^{-\frac{1}{2}} \int_{\mathbf{R}^n} \exp[-(x, A^{-1}x)/2 + (x, \sigma)]\,dx$$

we find that the covariance matrix of the σs, $N_{ij} \equiv \langle\sigma_i\sigma_j\rangle_B$, is simply related to the covariance matrix M^F, introduced above (with A replaced by A^{-1}), by

$$N = A^{-1}M^F A^{-1} - A^{-1},$$

where

$$F(x) = \sum_\sigma e^{(x,\sigma)} = \prod_{i=1}^n 2\cosh x_i.$$

Now $F(x)$ is log convex, so Theorem 1.5 states that $M^F \geqslant A$, which implies that $N \geqslant 0$—hardly an interesting result. Note, however, that

$$G(x) \equiv F(x)\exp\left(-\frac{1}{2}\sum_{i=1}^n x_i^2\right)$$

is log concave. Therefore, provided $A^{-1}>I$ (equivalently, $A<I$) we can write

$$\exp[-(x, A^{-1}x)/2]F(x)=\exp[-(x,(A^{-1}-I)x)/2]G(x)$$

and Theorem 1.4 states that $M^F\leqslant(A^{-1}-I)^{-1}$ and $N\leqslant(I-A)^{-1}$. In the physical situation, A is a matrix whose eigenvalues are of O(1) independent of n and $A<I$ occurs for sufficiently high temperature, independently of n. Hence, for high temperature, the eigenvalues of N are O(1); this means there is no long-range order. Although previously there existed elementwise bounds on N for special choices of A([2] and [3]: inequalities), our result is the first case of a quadratic form inequality on N.

We now quote an assortment of theorems, to indicate some of the directions in which Theorem 1.4 can be generalized.

THEOREM 1.6. *Consider the weight $W_F(x)=W(x)F(x)$ with F log concave, as in Theorem 1.4, and let F be even. Let L be any symmetric, real, n-square matrix. Then*

$$\langle(x, Lx)^2\rangle_F-\langle(x, Lx)\rangle_F^2\leqslant 2\langle(x, LA^{-1}Lx)\rangle_F. \tag{1.1}$$

Proof. We consider the case in which $A=I$; the general case can be handled by the change of variables $x\to A^{-\frac{1}{2}}x$. Let $Z=\int_{\mathbf{R}^n}e^{-(x,x)/2}F(x)\,dx$. Then

$$2\Delta\equiv 2\langle(x, Lx)^2\rangle_F-2\langle(x, Lx)\rangle_F^2$$

$$=Z^{-2}\int_{\mathbf{R}^n}\int_{\mathbf{R}^n}e^{-[(x,x)+(y,y)]/2}F(x)F(y)[(x, Lx)-(y, Ly)]^2\,dx\,dy$$

$$=Z^{-2}\int_{\mathbf{R}^n}\int_{\mathbf{R}^n}e^{-[(u,u)+(v,v)]/2}F[2^{-\frac{1}{2}}(u+v)]\,F(2^{-\frac{1}{2}}(u-v))4(u, Lv)^2\,du\,dv$$

after the change of variables $x=2^{-\frac{1}{2}}(u+v)$, $y=2^{-\frac{1}{2}}(u-v)$. Now do the v integration and recall that $\{\langle v_i, v_j\rangle\}_{i,j=1}^n\leqslant I$ for each u, by Theorem 1.4. Thus, $2\Delta\leqslant 4\langle(u, L^2u)\rangle$. Returning to the original x, y variables, one notes that $2(u, L^2u)=(x, L^2x)+(y, L^2y)+2(x, L^2y)$.

Finally $\langle(x, L^2x)\rangle=\langle(y, L^2y)\rangle=\langle(x, L^2x)\rangle_F$ and $\langle x_iy_j\rangle=0$.

Q.E.D.

REMARKS. (*i*) If F is log convex, the inequality in Theorem 1.6 is reversed.

(*ii*) The significance of Theorem 1.6 is that if L and A are of the order of I, the left side of (1.1) is the difference of two terms of $O(n^2)$, while the right side is $O(n)$. Choosing $L=A$, the left side of (1.1) is like n times a specific heat, while the right side is like n times an internal energy—to use the language of statistical mechanics. Usually, it is difficult to obtain an upper bound on a specific heat.

COROLLARY 1.7. *Let A and L be symmetric, n-square matrices with A non-singular, let F be even and log concave and let λ be real. Then*

$$Z(\lambda) \equiv \int_{\mathbf{R}^n} \exp[-(x, A\, e^{\lambda L} A x)] F(x)\, dx$$

is log concave in λ.

Proof. Compute $d^2 \ln Z/d\lambda^2$ and compare with Theorem 1.6.

Q.E.D.

THEOREM 1.8. *Let $W_F(x) = e^{-x^2/2} F(x)$ be a weight in **R** with F log concave. Define $\langle\,.\,\rangle_F$ and $\langle\,.\,\rangle_0$ as before. Then*

$$\langle |x - \langle x\rangle_F|^\alpha\rangle_F \leq \langle |x - \langle x\rangle_0|^\alpha\rangle_0$$

for $\alpha \geq 1$.

The proof of Theorem 1.8 is lengthy and will not be given here. The theorem says that multiplying a Gaussian weight on **R** by a log concave function may, if the function is not even, shift the mean, but all moments, higher than the first, with respect to the new mean are decreased.

We present next a theorem which will play an important role in the next section.

THEOREM 1.9. *Let A be a real positive-definite $(n+m)-$square matrix partitioned as $A = \begin{bmatrix} \alpha & \beta \\ \beta^T & \gamma \end{bmatrix}$, where α is n-square, γ is m-square, β is n × m, and T means transpose. Let F be a log concave function on $\mathbf{R}^{n+m}$ and form the unnormalized weight on $\mathbf{R}^{n+m}$: $W_F(x) = W(x)F(x)$, $W(x) = \exp[-(x, Ax)/2]$. Denoting, as before, a point $x \in \mathbf{R}^{m+n}$ as $x = (y, z)$, $y \in \mathbf{R}^n$, $z \in \mathbf{R}^m$, define the unnormalized weight V on $\mathbf{R}^n$ by*

$$V(y) = \int_{\mathbf{R}^m} W_F(y, z)\, dz.$$

If we define $G\colon \mathbf{R}^n \to \mathbf{R}$ by

$$V(y) = \exp[-(y, By)/2] G(y),$$

with $B = \alpha - \beta\gamma^{-1}\beta^T > 0$, then G is log concave.

Proof. Note that the $(n+m)$-square matrix $C \equiv A - \begin{bmatrix} B & 0 \\ 0 & 0 \end{bmatrix}$ is positive-semidefinite, since A is positive-definite. Hence $U_F(x) = \exp[-(x, Cx)/2] F(x)$ is log concave on $\mathbf{R}^{n+m}$. Since

$$V(y) = \exp[-(y, By)/2] \int_{\mathbf{R}^m} U_F(y, z)\, dz,$$

Theorem 1.9 follows from Theorem 1.1.

Q.E.D.

REMARKS. (*i*) *Mutatis mutandis*, if F is replaced by a log convex function, then G is log convex on $\mathbf{R}^n$.

(*ii*) If $F(x)$ is a constant, then $G(y)$ is also a constant. Thus, Theorem 1.9 states that if one does a partial integration over a Gaussian weight times a log concave function, the result is the Gaussian weight one would have obtained without the log concave multiplier times a new log concave function.

To pursue the ideas of Theorem 1.9 a bit further, let us formulate the Brunn–Minkowski theorem for Gaussian measures. We recall the classical Brunn–Minkowski theorem [4].

THEOREM 1.10. *Let C_0, C_1 be non-empty convex sets in $\mathbf{R}^n$, and let*

$$C_\lambda = \lambda C_1 + (1-\lambda)C_0, \qquad 0 \leqslant \lambda \leqslant 1.$$

Denote by $|C|$ the n-dimensional Lebesgue measure of C. Then

$$|C_\lambda|^{1/n} \geqslant \lambda |C_1|^{1/n} + (1-\lambda)|C_0|^{1/n}.$$

REMARK. If $C_0 = \{0\}$, then $C_\lambda = \lambda C_1$.

In the case of Gaussian measures we have the following.

THEOREM 1.11. *Let C_0, C_1 and C_λ be as in Theorem 1.10, and let A be a real, positive-definite, n-square matrix. Let*

$$\mu_G(C) = \int_C \exp[-(x, Ax)/2]\, dx.$$

Then

$$\mu_G(C_\lambda) \geqslant \mu_G(C_1)^\lambda \mu_G(C_0)^{1-\lambda}.$$

Proof. Define the convex set

$$D = \{(\lambda, x) | 0 \leqslant \lambda \leqslant 1, x \in C_\lambda\} \subset \mathbf{R} \times \mathbf{R}^n.$$

Then

$$\mu_G(C_\lambda) = \int_{\mathbf{R}} \chi_D(\lambda, x) \exp[-(x, Ax)/2]\, dx.$$

Since the integrand is log concave in (λ, x), $\mu_G(C_\lambda)$ is log concave by Theorem 1.1.

Q.E.D.

As a corollary to Theorem 1.11 we quote a theorem of L. Gross [5]. The Gaussian measure μ_n on $\mathbf{R}^n$ defines the measure of a Borel set $B \subset \mathbf{R}^n$ to be

$$\mu_n(B) = (2\pi)^{-n/2} \int_B \exp[-(x, x)/2]\, \mathrm{d}x.$$

THEOREM 1.12. (L. Gross) *Let C be a convex, balanced set in $\mathbf{R}^n \times \mathbf{R}^m$, let D be the intersection of C with $\mathbf{R}^m$, and let E be the projection of C on $\mathbf{R}^n$. Then*

$$\mu_{n+m}(C) \leqslant \mu_n(E)\mu_m(D).$$

Proof. Let C_x be the intersection of C with the plane parallel to $\mathbf{R}^m$ through $x \in \mathbf{R}^n$; in particular, $C_0 = D$. By the symmetry of C and Theorem 1.11, $x \to \mu_m(C_x)$ is log concave and even on $\mathbf{R}^n$, and hence $\mu_m(C_x)$ is maximal for $x = 0$. Thus

$$\mu_{n+m}(C) = \int_E \mu_m(C_x)\, \mathrm{d}\mu_n(x) \leqslant \mu_m(D) \int_E \mathrm{d}\mu_n(x) = \mu_n(E)\mu_m(D).$$

Q.E.D.

Let us return to the Brunn-Minkowski theorem (1.11) for Gaussians. By passing to the limit $n \to \infty$, the same theorem obviously remains true for infinite-dimensional Gaussian measures, for example, the Wiener measure. In that case we should deal with measurable, convex sets of Wiener paths. We shall consider here particular convex sets of paths, namely those passing through a convex set $C_\lambda \subset \mathbf{R}^n$ for all t.

With C_λ, $0 \leqslant \lambda \leqslant 1$, defined as in Theorem 1.10, consider the fundamental solution $G_\lambda(x, y; t)$, $t \geqslant 0$, of the diffusion equation with potential V, in $\mathbf{R}^n$, defined by

$$\left(\frac{\partial}{\partial t} - \frac{1}{2}\Delta_x + V(x)\right) G_\lambda(x, y; t) = 0;$$

$$G_\lambda(x, y; 0) = \delta(x - y), \qquad x, y \in C_\lambda;$$

$$G_\lambda(x, y; t) = 0, \qquad x \in \partial C_\lambda, \qquad \forall y, t;$$

$$\equiv 0, \qquad x \notin C_\lambda \quad \text{or} \quad y \notin C_\lambda.$$

THEOREM 1.13. *Let $V(x)$ be a convex function. Then $G_\lambda(x, y; t)$ is log concave in $(x, y, \lambda) \in \mathbf{R}^n \times \mathbf{R}^n \times [0, 1]$.*

Proof. Use the Trotter product formula with $x_0 = x$, $x_N = y$:

$$G_\lambda(x, y; t) = \lim_{N \to \infty} (2\pi t/N)^{-nN/2} \int_{\mathbf{R}^{n(N-1)}} \mathrm{d}x_1 \ldots \mathrm{d}x_{N-1} \times$$
$$\times \prod_{j=1}^{N} \left\{ \exp\left[-\frac{N}{2t}(x_j - x_{j-1})^2 - \frac{t}{N} V(x_j) \right] \chi_{C_\lambda}(x_j) \right\}.$$

The integrand is log concave in $(x, x_1, \ldots, x_{N-1}, y, \lambda)$. Finally the pointwise limit of a sequence of log concave functions is log concave.

Q.E.D.

COROLLARY 1.14. *In addition to the hypotheses of Theorem 1.13, either let C_0 and C_1 be compact or let* $\exp(-tV)$ *be in* $L^1(\mathbf{R}^n)$, $\forall t>0$. *Define*

$$Z_\lambda(t)=\int_{C_\lambda} G_\lambda(x, x; t)\,\mathrm{d}x=\operatorname{tr} \mathrm{e}^{-tH},$$

with $H=-\frac{1}{2}\Delta+V$. *Then* $Z_\lambda(t)<\infty$ *and* $Z_\lambda(t)$ *is log concave in* λ.

Proof. That $Z_\lambda(t)$ is finite is a standard result and can be proved from the Trotter product formula above using Hoelder's inequality. The log concavity of $Z_\lambda(t)$ follows from Theorems 1.1 and 1.13.

Q.E.D.

COROLLARY 1.15. *Let $V(x)$ and C_λ be as in Corollary* 1.14. *Let $\varepsilon_0(\lambda)$ be the lowest eigenvalue of the equation*

$$[-\tfrac{1}{2}\Delta+V(x)]f(x)=\varepsilon_0(\lambda)f(x),$$

with $f(x)=0$ for $x\in\partial C_\lambda$. Then $\varepsilon_0(\lambda)$ is a convex function of $\lambda\in[0,1]$.

Proof. Since e^{-tH} is trace class, $Z_\lambda=\sum_{j=0}^{\infty}\exp[-t\varepsilon_j(\lambda)]$, $\varepsilon_{j+1}(\lambda)\geqslant\varepsilon_j(\lambda)$ and each $\varepsilon_j(\lambda)$ has finite multiplicity. Then

$$\varepsilon_0(\lambda)=-\lim_{t\to\infty} t^{-1}\ln Z_\lambda(t),$$

and, since the pointwise limit of a sequence of convex functions is convex, Corollary 1.15 is proved.

Q.E.D.

1.4. The one-dimensional plasma

In this section we apply the previous theorems to an old problem in physics, namely, to the one-dimensional, one-component plasma in a neutralizing background. We shall consider both the classical and quantum-mechanical cases. The latter requires the introduction of the Wiener integral, and thus provides another example of the application of our theorems to functional integrals. The object of our investigations is to show that long-range order exists for sufficiently large coupling constant, i.e. that the one-particle distribution function is a non-constant periodic function. The occurrence of this phenomenon was first predicted by Wigner [6].

Let $\boldsymbol{x}=(x_{-n},\ldots,x_n)$ be the coordinates of $(2n+1)$ one-dimensional particles, each having a negative charge of one unit. The one-dimensional Coulomb potential between two unit charges separated by a distance $|x|$ is $-|x|$. Then the total potential energy of $(2n+1)$ particles in a 'box' $[-L, L]$ with a fixed uniform positive charge background of density ρ is

$$\phi(\boldsymbol{x})=-\sum_{-n\leq i<j\leq n}|x_i-x_j|+\rho\sum_{i=-n}^{n}\int_{-L}^{L}|x_i-x|\,\mathrm{d}x-\frac{\rho^2}{2}\int_{-L}^{L}\int_{-L}^{L}|x-y|\,\mathrm{d}x\,\mathrm{d}y. \tag{1.2}$$

We shall further assume that the total charge is zero, i.e.

$$2L\rho=2n+1.$$

Since ϕ is symmetric in the x_i, it is sufficient to consider the convex domain

$$C=\{\mathbf{x}|-L\leq x_{-n}\leq x_{-n+1}\ldots\leq x_n\leq L\}. \tag{1.3}$$

In C,

$$\phi(\mathbf{x})=\rho\sum_{j=-n}^{n}\left(x_j-\frac{j}{\rho}\right)^2, \tag{1.4}$$

where a constant term in the potential has been neglected. Our methods are capable of handling the domain C as it stands, but then the function we wish to calculate, $\rho(x)$, will not be strictly periodic, except in the thermodynamic limit $n\to\infty$, $\rho=$ constant. To circumvent this difficulty we extend C to the larger convex domain

$$D=\{x|x_{-n}\leq x_{-n+1}\ldots\leq x_n\leq x_{-n}+2L\}. \tag{1.5}$$

The domain D no longer confines the particles to the box, and we shall cheat a little by supposing that the expression (1.4) for ϕ is valid in all of D. The original walls of the box are still visible in ϕ.

Remark then that the domain D and the potential ϕ are invariant under the linear transformations R(reflection) and T(translation), defined by

$$(\mathrm{R}\mathbf{x})_j=-x_{-j} \tag{1.6}$$

$$(\mathrm{T}\mathbf{x})_j=x_{j+1}-\frac{1}{\rho},\qquad -n\leq j\leq n-1;$$

$$(\mathrm{T}\mathbf{x})_n=x_{-n}+\frac{2n}{\rho}. \tag{1.7}$$

1.4.1. The classical case

The Gibbs distribution function of the jth particle is defined by (β is the reciprocal temperature)

$$\rho_j^{(n)}(x)=\frac{\int_D\delta(x-x_j)\exp[-\beta\phi(\mathbf{x})]\,\mathrm{d}\mathbf{x}}{\int_D\exp[-\beta\phi(\mathbf{x})]\,\mathrm{d}\mathbf{x}}.$$

The symmetry properties (1.6) and (1.7) imply that

$$\rho_0^{(n)}(x) = \rho_0^{(n)}(-x); \tag{1.8}$$

$$\rho_j^{(n)}(x) = \rho_0^{(n)}\left(x - \frac{j}{\rho}\right), \quad -n \leq j \leq n. \tag{1.9}$$

Since D is a convex domain in $\mathbf{R}^{2n+1}$, direct application of Theorem 1.9 gives that

$$\rho_0^{(n)}(x) = \exp(-\beta\rho x^2) F^{(n)}(x), \tag{1.10}$$

where $F^{(n)}$ is log concave; by eqn (1.8), $F^{(n)}(x)$ is also even.

We shall not go into the details of the existence of the limiting distribution functions

$$\rho_j(x) \equiv \lim_{n\to\infty} \rho_j^{(n)}(x)$$

in the thermodynamic limit ($n \to \infty$, $L \to \infty$, $2n+1 = 2L\rho$). Obviously, properties (1.8)–(1.10) remain true in the limit. It is also fairly clear, that the use of domain C instead of domain D would give the same distribution functions in the limit.

Thus far we have established part (i) of the following theorem:

THEOREM 1.16. (i) *The one-particle distribution functions of the classical one-dimensional plasma computed in D satisfy*

$$\rho_0(x) = \rho_0(-x),$$

$$\rho_j(x) = \rho_0\left(x - \frac{j}{\rho}\right),$$

$$\rho_0(x) = \exp[-\beta\rho x^2] F(x),$$

where $F(x)$ is a log concave, even function.

(ii)
$$\int_{\mathbf{R}} |x|^\alpha \rho_0(x)\,\mathrm{d}x \leq \left(\frac{\pi}{\beta\rho}\right)^{-\frac{1}{2}} \int_{\mathbf{R}} |x|^\alpha \exp[-\beta\rho x^2]\,\mathrm{d}x$$

for $\alpha \geq 1$.

(iii) *For large values of β/ρ, the total density*

$$\rho(x) = \sum_{j=-\infty}^{\infty} \rho_j(x)$$

is non-trivially periodic.

Proof. (ii) Follows from Theorem 1.8, (iii) will be proved in § 1.4.3, Theorem 1.18.

REMARKS. (*i*) Theorem 1.16 (*ii*) states that the moments of the one-particle distribution functions are smaller than they would be without the restriction $x_j \leqslant x_{j+1}$.

(*ii*) The interpretation of Theorem 1.16 (*iii*) is that the plasma is in a crystalline state. The specific position of the crystal is a consequence of the hard walls that were imposed at $\pm L = \pm(n+\frac{1}{2})\rho$. This fact is reflected not only in the domain D (eqn (1.5)) but also in the expressions (1.4) for $\phi(\mathbf{x})$. Hard walls at positions $\pm L+\delta$ would translate the crystal through a distance δ.

(*iii*) The fact that $\rho(x)$ is not a constant has recently been proved by Kunz [7] who, by other methods, showed that to be true for all β,ρ except possibly for a countable number of values of β/ρ.

1.4.2. The quantum-mechanical case

The quantum-mechanical Hamiltonian of the system defined by equation (1.2), with $\hbar^2/m = 1$, is

$$H = -\tfrac{1}{2}\Delta + \phi(\mathbf{x}),$$

where

$$\Delta = \sum_{j=-n}^{n} \frac{\partial^2}{\partial x_j^2}.$$

We consider the case that the particles are spinless fermions. This means that H acts on square integrable, antisymmetric functions. As is well known, an equivalent statement in one dimension is that H acts on square integrable functions defined on $E = \{\mathbf{x} | x_{-n} \leqslant x_{-n+1}, \ldots, \leqslant x_n\}$ which vanish on the boundary of E. The 'box' condition requires that the functions vanish on the boundary of $C \subset E$. As in the classical case, we shall use the larger domain D instead of C.

The distribution function of the jth particle is then

$$\rho_j^{(n)}(x) = \frac{\operatorname{tr}_D \mathrm{e}^{-\beta H}\delta_j(x)}{\operatorname{tr}_D \mathrm{e}^{-\beta H}},$$

where $\delta_j(x)$ is the operator of multiplication by $\delta(x-x_j)$. Since H and D are invariant under the transformations R and T (eqns (1.6), (1.7)), the distribution functions again have the symmetry properties ((1.8), (1.9)).

To find the analogue of eqn (1.10), use the Trotter product formula for $\exp(-\beta H)$, which gives, with $\mathbf{x}^0 = \mathbf{x}^N$,

$$\begin{aligned}\operatorname{Tr}_D \mathrm{e}^{-\beta H}\delta_0(x) = \lim_{N\to\infty}\left(\frac{2\pi\beta}{N}\right)^{-(2n+1)N/2} \int_{D^N} \mathrm{d}\mathbf{x}^1 \ldots \mathrm{d}\mathbf{x}^N \times \\ \times \prod_{k=1}^{N} \exp\left[-\frac{N}{2\beta}\sum_{j=-n}^{n}(x_j^k - x_j^{k-1})^2 - \frac{\beta\rho}{N}\sum_{j=-n}^{n}\left(x_j^k - \frac{j}{\rho}\right)^2\right]\delta(x_0^1 - x).\end{aligned} \tag{1.11}$$

Since D^N is convex in $\mathbf{R}^{(2n+1)N}$, we can apply Theorem 1.9 to conclude that

$$\rho_0^{(n)}(x)=\exp(-\gamma x^2)H^{(n)}(x), \tag{1.12}$$

where $H^{(n)}$ is log concave, and where $\exp(-\gamma x^2)$ is, up to multiplication by an x-independent constant, what eqn (1.11) would give if D were replaced by $\mathbf{R}^{2n+1}$. But in that case the integrations separate into $(2n+1)$ independent integrations over $\mathbf{R}^N$. Therefore, $\exp(-\gamma x^2)$ is proportional to $G(x, x; \beta)$, where $G(x, y; t)$ is the fundamental solution (Green's function) of the differential equation, for $t>0$,

$$\left(\frac{\partial}{\partial t}-\frac{1}{2}\frac{\partial^2}{\partial x^2}+\rho x^2\right)G(x, y; t)=0;$$

$$G(x, y; 0)=\delta(x-y).$$

Using the well-known expression [8] for G, we obtain

$$\gamma=(2\rho)^{\frac{1}{2}}\tanh\beta(\rho/2)^{\frac{1}{2}}. \tag{1.13}$$

The analogue of Theorem 1.16 is now immediate.

THEOREM 1.17: *Theorem 1.16 is correct for the quantum-mechanical plasma of spinless fermions except that in part* (*ii*) $\beta\rho$ *is replaced by* γ (*eqn* (1.13)) *and in part* (*iii*) β/ρ *is replaced by* γ/ρ^2.

Remarks (*i*) and (*ii*) after Theorem 1.16 also apply here.

We turn next to the demonstration that parts (*iii*) of Theorems 1.16 and 1.17 follow from parts (*i*) of those theorems.

1.4.3. *Can modified theta functions be constant?*

Let $f(x)=\exp(-\lambda x^2)F(x)$ with $F(x)$ even and log concave and $\lambda>0$. Consider

$$\rho(x)=\sum_{j=-\infty}^{\infty} f(x-j). \tag{1.14}$$

The question to which we address ourselves here is whether or not F can be chosen so that $\rho(x)$ is constant. The answer, surprisingly, depends on λ. As Theorem 1.18 shows, $\rho(x)$ cannot be constant when λ is large, and thus parts (*iii*) of Theorems 1.16 and 1.17 are proved.

Define the Fourier transform by

$$\hat{f}(k)=\int_{\mathbf{R}} e^{2\pi ikx}f(x)\,dx. \tag{1.15}$$

Then, by the Poisson summation formula,

$$\rho(x)=\sum_{j=-\infty}^{\infty} \hat{f}(j)\,e^{-2\pi ijx}. \tag{1.16}$$

Therefore $\rho(x)$ is constant iff $\hat{f}(j)=0$ for $j=\pm1, \pm2, \ldots$.

THEOREM 1.18. *Let $\rho(x)$ be defined as in eqn* (1.14). *Then there exists a* λ_0, $0<\lambda_0<\infty$, such that

(a) *For all $\lambda>\lambda_0$ and for all F even and log concave, $\rho(x)$ is not constant*;

(b) *For all $\lambda<\lambda_0$, $\lambda>0$ there exists an even, log concave F such that $\rho(x)=$ constant.*

Proof. (*i*) Existence of λ_0: If, for some λ, there is an $F(x)$ that leads to a constant $\rho(x)$, then, for $\mu<\lambda$, the log concave function $F(x)\exp[(\mu-\lambda)x^2]$ gives the same constant $\rho(x)$.

(*ii*) $\lambda_0<\infty$: Normalize to $F(0)=1$. Then $\rho(0)\geq 1$, and

$$\rho(\tfrac{1}{2})\leq\sum_{j=-\infty}^{\infty}\exp[-\lambda(j-\tfrac{1}{2})^2]\leq 2\,e^{-\lambda/4}\sum_{j=0}^{\infty}e^{-2\lambda j}=2\,e^{-\lambda/4}(1-e^{-2\lambda})^{-1}.$$

This gives the simple estimation $\lambda_0<3$.

(*iii*) $\lambda_0>0$: We indicate how to construct an example of constant ρ for λ sufficiently small. Choose a non-constant, even, log concave function G, and normalize it so that $g(x)\equiv\exp(-\lambda x^2)G(x)$ satisfies

$$\int_{\mathbf{R}} g(x)\,dx=\hat{g}(0)=1.$$

Define

$$\hat{f}(k)=\prod_{j=1}^{\infty}\hat{g}(k/j), \tag{1.17}$$

which is the Fourier transform of the convolution

$$f(x)=\prod_{j=1}^{\infty}{}^{*j}\exp(-\lambda j^2x^2)G(jx). \tag{1.18}$$

The infinite product (1.17) is defined and $\hat{g}(k)>0$ in a neighbourhood of $k=0$, since

$$1>\hat{g}(k/j)\approx 1+\tfrac{1}{2}(k/j)^2\hat{g}''(0)>0\quad\text{for}\quad |k/j|\ll 1.$$

Equation (1.18) then follows from the Lebesgue dominated convergence theorem, and $f\not\equiv 0$. Now Theorem 1.9 applied to eqn (1.18) gives

$$f(x)=\exp(-\alpha\lambda x^2)F(x),$$

where $F(x)$ is log concave and even and $\alpha=(\sum_{j=1}^{\infty}j^{-2})^{-1}=6\pi^{-2}$.

It is now sufficient to determine λ and G such that $\hat{g}(\pm 1)=0$; then, by eqn (1.17),

$$\hat{f}(j)=0\quad\text{for all integers}\quad j\neq 0,$$

and we are done. Take $G(x)=[1+\sqrt{(\lambda/\pi)}]\chi(x)$, where χ is the characteristic function of $[-\frac{3}{4},\frac{3}{4}]$. Then

$$\lim_{\lambda\to 0}\hat{g}(k)=(\pi k)^{-1}\sin\tfrac{3}{2}\pi k;\ \lim_{\lambda\to\infty}\hat{g}(k)=1.$$

Therefore λ can be chosen such that $\hat{g}(\pm 1)=0$.

Q.E.D.

Acknowledgements

This work has been partially supported by U.S. National Science Foundation Grants GP-31674X and GP-16147A #1.

References

1. SCHOENBERG, I. J. On Pólya frequency functions I: The totally positive functions and their Laplace transforms. *J. Anal. math.* **1**, 331–74 (1951).
2. GRIFFITHS, R. B. Correlations in Ising ferromagnets, I, II, III. *J. Math. Phys.* **8**, 478–83, 484–9 (1967); *Commun. Math. Phys.* **6**, 121–7 (1967); KELLY, D. and SHERMAN, S. General Griffiths inequalities on correlations in Ising ferromagnets. *J. Math. Phys.* **9**, 466–84 (1968).
3. FORTUIN, C. M., KASTELEYN, P. W., and GINIBRE, J. Correlation functions on some partially ordered sets. *Commun. Math. Phys.* **22**, 89–103 (1971).
4. BONNESEN, T. and FENCHEL, W. *Theorie der Konvexen Koerper.* Chelsea, New York (1948).
5. GROSS, L. Measurable functions on Hilbert space. *Trans. Am. math. Soc.* **105**, 372–90 (1962); see also DUDLEY, R. M., FELDMAN, J., and LE CAM, L. On seminorms and probabilities, and abstract Wiener spaces. *Ann. Math.* **93**, Ser. 2, 390–408 (1971).
6. WIGNER, E. P. Effects of the electron interaction on the energy levels of electrons in metals. *Trans. Faraday Soc.* **34**, 678–85 (1938).
7. KUNZ, H. Equilibrium properties of the one-dimensional classical electron gas. Preprint E.P.F.L., Lausanne (1974).
8. MERZBACHER, E. *Quantum mechanics*, Chapter 8. Wiley, New York (1961).

Note

Since this work was completed we have found that Theorem 1.1 has been proved independently by A. Prekopa, *Acta Math. Szeged* **32**, 301–15 (1971), and Y. Rinott, Thesis, Weizmann Institute, Rehovoth, Israel (Nov. 1973).

ADVANCES IN MATHEMATICS **20,** 151–173 (1976)

Best Constants in Young's Inequality, Its Converse, and Its Generalization to More than Three Functions

HERM JAN BRASCAMP*

Department of Physics, Princeton University, Princeton, New Jersey 08540

AND

ELLIOTT H. LIEB*

Departments of Mathematics and Physics, Princeton University, Princeton, New Jersey 08540

The best possible constant D_{pqt} in the inequality $|\iint dx\,dy\, f(x)g(x-y)\,h(y)| \leqslant D_{pqt} \| f \|_p \| g \|_q \| h \|_t$, $p, q, t \geqslant 1$, $1/p + 1/q + 1/t = 2$, is determined; the equality is reached if f, g, and h are appropriate Gaussians. The same is shown to be true for the converse inequality ($0 < p, q < 1$, $t < 0$), in which case the inequality is reversed. Furthermore, an analogous property is proved for an integral of k functions over n variables, each function depending on a linear combination of the n variables; some of the functions may be taken to be fixed Gaussians. Two applications are given, one of which is a proof of Nelson's hypercontractive inequality.

1. INTRODUCTION

The classical inequality of Young is that

$$\| f * g \|_r \leqslant \| f \|_p \| g \|_q , \tag{1.1}$$

where $*$ means convolution, $1/p + 1/q = 1 + 1/r$, $p, q, r \geqslant 1$ and f and g are functions on $\mathbb{R}$. Alternatively, (1.1) is equivalent to

$$I = \left| \iint f(x)\, g(x-y)\, h(y)\, dx\, dy \right| \leqslant \| f \|_p \| g \|_q \| h \|_t , \tag{1.2}$$

when $1/p + 1/q + 1/t = 2$. Unlike Hölder's inequality,

$$\| fg \|_r \leqslant \| f \|_p \| g \|_q$$

* Work partially supported by National Science Foundation Grant Number GP-31674X.

($1/p + 1/q = 1/r$), the best possible constant in (1.1) is not unity. About a year ago, Beckner [1] showed that Gaussians give the best constant when $1 \leqslant p, q, t \leqslant 2$, by finding the best constant in the Hausdorff–Young inequality. The latter result is very deep, but will not play a role in this paper. Thus the conjecture was raised that for all p, q, r, Gaussians give the best constant in (1.1) and (1.2). This fact was proved simultaneously by Beckner [1] and us by different methods. We report our method here because it also leads to a generalization of (1.2), namely to integrals involving k functions instead of 3 and to integrations over n variables instead of 2. This is contained in Theorem 1 and the explicit constant for (1.2) is in Eqs. (2.19) and (2.20).

In Section 3, we also find the best constant in the converse of Young's inequality (Eq. (1.1) with the reversed inequality, for $0 < p, q, r < 1$), first shown by Leindler [2]. In particular, we rederive the Prékopa–Leinder inequality [3–5]. In Section 4 we show that, as far as Young's inequality and its converse are concerned, the equality holds uniquely for Gaussians. We are not able to show this for the general inequality of Theorem 1; this remains an open question. Section 5 contains two applications of Theorem 1: Nelson's hypercontractivity theorem and an inequality in statistical mechanics.

An amusing example of Theorem 1 which shows how Gaussians arise, and which can be done by elementary methods is the following: let

$$J = \iint f(x)\, g(y)\, h(x - y)\, k(x + y)\, dx\, dy.$$

Thus, using the Schwarz inequality,

$$\begin{aligned} |J| &\leqslant \left[\iint |f(x)\, g(y)|^2 dx\, dy\right]^{1/2} \left[\iint |h(x - y)\, k(x + y)|^2\, dx\, dy\right]^{1/2} \\ &= 2^{-1/2} \| f \|_2 \| g \|_2 \| h \|_2 \| k \|_2 . \end{aligned}$$

Equality holds if $f(x)\, g(y)$ is proportional to $h(x - y)\, k(x + y)$. But this is true for the Gaussians $f(x) = g(x) = \exp(-2x^2)$, $h(x) = k(x) = \exp(-x^2)$. Thus, $2^{-1/2}$ is the best constant.

The general case is not as simple as this example. The idea behind our proof is that I^M can trivially be written as an integral over $\mathbb{R}^M \times \mathbb{R}^M$. However, by the rearrangement inequality [6–9], $| I^M |$ can be increased by replacing $f(x_1, ..., x_M) = |f(x_1)| \cdots |f(x_M)|$ by its spherically symmetric, decreasing rearrangement, F, and similarly for g and h. This rearrangement does not affect the L^p norms. The main fact is that for large M, all spherically symmetric, decreasing functions look like Gaussians in some sense. The proof is concluded by letting $M \to \infty$.

2. The Main Theorem

In this section we prove the following theorem.

Theorem 1. *Let n and k be integers with $1 \leqslant n \leqslant k$. Let p_j, $1 \leqslant j \leqslant k$ be real numbers such that $1 \leqslant p_j \leqslant \infty$, $\sum_{j=1}^{k} 1/p_j = n$. Let f_j, $1 \leqslant j \leqslant k$, be complex-valued functions on $\mathbb{R}$, and let $f_j \in L^{p_j}(\mathbb{R})$. Let a^j, $1 \leqslant j \leqslant k$, be vectors in $\mathbb{R}^n$, and let*

$$I(\{f_j\}) = \int_{\mathbb{R}^n} d^n x \prod_{j=1}^{k} f_j(\langle a^j, x\rangle). \tag{2.1}$$

Then

$$| I(\{f_j\})| \leqslant D \prod_{j=1}^{k} \| f_j \|_{p_j}, \tag{2.2}$$

where

$$D = \sup\{I(\{\phi_j\}) | \phi_j \in G, \| \phi_j \|_{p_j} = 1,\ j = 1,..., k\}, \tag{2.3}$$

the supremum being taken over the class G of all Gaussian functions with maximum at the origin.

The value of D will be exhibited in Section 2.3.

2.1 *Auxiliary Remarks*

Let us first pave the way for the proof of Theorem 1 with some remarks and propositions.

Obviously it is sufficient to consider only non-negative functions f_j, since taking the absolute values of all f_j increases $| I(\{f_j\})|$ and does not change the L^p norms.

In the same way, one can restrict oneself to symmetric decreasing functions. To see this, let us introduce the symmetric decreasing rearrangement f^* of a non-negative function f, [6]: f^* is the symmetric decreasing function that is equimeasurable to f, i.e., the sets $\{x \in \mathbb{R} \mid f(x) \geqslant z\}$ and $\{x \in \mathbb{R} \mid f^*(x) \geqslant z\}$ have equal Lebesgue measures for all $z > 0$. Obviously, f and f^* have equal p-norms. Also, according to a theorem proved by Luttinger and ourselves [9],

$$\int_{\mathbb{R}^n} d^n x \prod_{j=1}^{k} f_j(\langle a^j, x\rangle) \leqslant \int_{\mathbb{R}^n} d^n x \prod_{j=1}^{k} f_j^*(\langle a^j, x\rangle). \tag{2.4}$$

We shall further need a generalization of the inequality (2.4) to functions of several variables, also given in [9]. Given a non-negative function $f(x)$, $x \in \mathbb{R}^M$, its Schwarz symmetrization f^* (spherically symmetric decreasing rearrangement) is defined as the spherically symmetric function which is decreasing in radial directions and which is equimeasurable to f. Then the inequality reads

$$\begin{aligned} &\int_{\mathbb{R}^{nM}} d^n x_1 \cdots d^n x_M \prod_{j=1}^{k} f_j(\langle a^j, x_1\rangle, \ldots, \langle a^j, x_M\rangle) \\ &\quad \leqslant \int_{\mathbb{R}^{nM}} d^n x_1 \cdots d^n x_M \prod_{j=1}^{k} f_j^*(\langle a^j, x_1\rangle, \ldots, \langle a^j, x_M\rangle). \end{aligned} \tag{2.5}$$

The derivation in [9] of Eq. (2.5) from Eq. (2.4) follows Sobolev's method [8].

Now, restricting ourselves to non-negative, symmetric decreasing functions f_j, each of those functions can be approximated pointwise from below by functions of the type

$$f_j^K(x) = \sum_{m=1}^{K} g_j^m \chi^m(x). \tag{2.6}$$

Here, the χ^m are characteristic functions of symmetric intervals $[-l^m, l^m]$, with $l^m > l^{m+1}$. Note that the function $f_j^K(x)$ takes only K different positive values, namely, $h_j^1 = g_j^1$, $h_j^2 = g_j^1 + g_j^2, \ldots, h_j^K = g_j^1 + \cdots + g_j^K$.

As $K \to \infty$, $f_j^K(x) \uparrow f_j(x)$ for all $x \in \mathbb{R}$, and hence by monotone convergence

$$\| f_j^K \|_{p_j} \uparrow \| f_j \|_{p_j};\ I(\{f_j^K\}) \uparrow I(\{f_j\}). \tag{2.7}$$

The latter remains true if $I(\{f_j\}) = \infty$.

As a consequence of Eq. (2.7), it suffices now to prove Theorem 1 for step functions of the form given in Eq. (2.6). We conclude this subsection with two useful propositions.

PROPOSITION 2. *Let* ψ_j, $1 \leqslant j \leqslant K$, *be non-negative functions in* $L^p(\mathbb{R}^M)$, $p \geqslant 1$. *Then*

$$\left\| \sum_{j=1}^{K} \psi_j \right\|_p \geqslant K^{-1/q} \sum_{j=1}^{K} \| \psi_j \|_p ,$$

where $1/p + 1/q = 1$.

Proof. Hölder's inequality applied to a finite sum yields

$$\sum_{j=1}^{K} \| \psi_j \|_p \leqslant K^{1/q} \left(\sum_{j=1}^{K} \| \psi_j \|_p^p \right)^{1/p} = K^{1/q} \left(\int dx \sum_{j=1}^{K} \psi_j(x)^p \right)^{1/p}.$$

However,

$$\sum_{j=1}^{K} \psi_j(x)^p \leqslant \left(\sum_{j=1}^{K} \psi_j(x) \right)^p. \qquad \text{Q.E.D.}$$

PROPOSITION 3. *Let* η_a *be the characteristic function of the ball* $\{x \in \mathbb{R}^M : |x| \leqslant a\}$, *and let*

$$\phi_a(x) = \exp[(1 - x^2/a^2)\, M/2p],$$

so that $\eta_a(x) \leqslant \phi_a(x)$. *Then*

$$\| \phi_a \|_p \leqslant \| \eta_a \|_p (3\sqrt{M})^{1/p}.$$

Proof. If Ω_M is the surface area of a unit sphere in M dimensions,

$$\| \eta_a \|_p^p = \Omega_M a^M / M;$$

$$\| \phi_a \|_p^p = \Omega_M e^{M/2} \int_0^\infty dx\, x^{M-1} e^{-Mx^2/2a^2} = \tfrac{1}{2}\, \Omega_M e^{M/2} a^M (2/M)^{M/2}\, \Gamma(M/2).$$

Hence, by Stirling's formula

$$\begin{aligned} (\| \phi_a \|_p / \| \eta_a \|_p)^p &= \Gamma(M/2 + 1)/[M/(2e)]^{M/2} \\ &< (\pi M)^{1/2}\, e^{1/(6M)} < 3\sqrt{M}. \end{aligned} \qquad \text{Q.E.D.}$$

These foundations being laid, we can turn to the proof of Theorem 1.

2.2 *Proof of Theorem* 1

As explained above, each of the functions f_j can be taken to be a step function of the form (2.6), all with the same number of steps, say K. Now write

$$I(\{f_j\})^M = \int_{\mathbb{R}^{nM}} d^n x_1 \cdots d^n x_M \prod_{j=1}^{k} \prod_{m=1}^{M} f_j(\langle a^j, x_m \rangle).$$

Let $F_j(x_1, ..., x_M)$ be the $\mathbb{R}^M$ Schwarz symmetrization of $\prod_{m=1}^M f_j(x_m)$. Then by Eq. (2.5) the last integral is not larger than

$$\int_{\mathbb{R}^{nM}} d^n x_1 \cdots d^n x_M \prod_{j=1}^{k} F_j(\langle a^j, x_1 \rangle, ..., \langle a^j, x_M \rangle).$$

Now notice that $f_j(x)$ only takes K positive values, say $h_j^1, ..., h_j^K$. Then $\prod_{m=1}^M f_j(x_m)$ and F_j take the values $(h_j^1)^{\alpha_1}(h_j^2)^{\alpha_2} \cdots (h_j^K)^{\alpha_K}$, with $\alpha_m \in \{0, ..., M\}$, and $\sum_{m=1}^K \alpha_m = M$. The number of values taken by F_j is thus certainly smaller than $(M+1)^K$. We can write

$$F_j = \sum_{m=1}^{(M+1)^K} H_j^m \eta^m, \tag{2.8}$$

where the η^m are characteristic functions of M-dimensional balls centered at the origin. Then, by Proposition 2

$$\| f_j \|_{p_j}^M = \| F_j \|_{p_j} \geqslant (M+1)^{-K/q_j} \sum_m H_j^m \| \eta^m \|_{p_j},$$

where $1/p_j + 1/q_j = 1$. Altogether, this gives

$$\Big[I(\{f_j\}) \Big/ \prod_{j=1}^{k} \| f_j \|_{p_j}\Big]^M \leqslant (M+1)^{(k-n)K}$$

$$\times \frac{\sum_{m_1} \cdots \sum_{m_k} H_1^{m_1} \cdots H_k^{m_k} \int_{\mathbb{R}^{nM}} d^n x_1 \cdots d^n x_M \prod_{j=1}^k \eta^{m_j}(\langle a^j, x_1 \rangle, ..., \langle a^j, x_M \rangle)}{\sum_{m_1} \cdots \sum_{m_k} H_1^{m_1} \cdots H_k^{m_k} \prod_{j=1}^k \| \eta^{m_j} \|_{p_j}}. \tag{2.9}$$

Now pick any k-tuple of characteristic functions $\eta_1, ..., \eta_k$ of balls with radii $b_1, ..., b_k$. Define

$$\phi_j = \exp[(1 - x^2/b_j^2)\, M/2p_j].$$

Then, by Proposition 3

$$\frac{\int_{\mathbb{R}^{nM}} d^n x_1 \cdots d^n x_M \prod_{j=1}^k \eta_j(\langle a^j, x_1\rangle, \ldots, \langle a^j, x_M\rangle)}{\prod_{j=1}^k \| \eta_j \|_{p_j}}$$

$$\leqslant (3\sqrt{M})^n \frac{\int_{\mathbb{R}^{nM}} d^n x_1 \cdots d^n x_M \prod_{j=1}^k \phi_j(\langle a^j, x_1\rangle, \ldots, \langle a^j, x_M\rangle)}{\prod_{j=1}^k \| \phi_j \|_{p_j}}. \quad (2.10)$$

Since each ϕ_j is a product of M one-dimensional Gaussians, the quotient on the right side of Eq. (2.10) is at most D^M, by the very definition of D, Eq. (2.3). Using this together with Eqs. (2.9) and (2.10), we get

$$I(\{f_j\}) \leqslant [(M+1)^{(k-n)K}(3\sqrt{M})^n]^{1/M} D \prod_{j=1}^k \| f_j \|_{p_j} .$$

The desired result is obtained by letting M go to ∞. Q.E.D.

2.3 *Computation of the Best Constant, D*

We now proceed to compute the supremum D in Eq. (2.2). Let $\phi_j(x) = \exp(-z_j x^2)$. Then

$$\int_{\mathbb{R}^n} d^n x \prod_{j=1}^k \phi_j(\langle a^j, x\rangle) = \pi^{n/2} (\det A)^{-1/2},$$

where the $n \times n$ matrix A is given by

$$A_{uv} = \sum_{j=1}^k z_j a_u{}^j a_v{}^j, \qquad 1 \leqslant u, \quad v \leqslant n.$$

PROPOSITION 4.

$$\det A = \sum_{|S|=n} J_S z_S ,$$

where

$$z_S = \prod_{j\in S} z_j , \quad (2.11)$$

and J_S, $S = \{j_1, \ldots, j_n\}$, *is defined as*

$$J_S = [\det(a^{j_1} \cdots a^{j_n})]^2. \quad (2.12)$$

Proof. Note that $\det A$ is homogeneous of degree n in the z_j. Further

$$\frac{\partial}{\partial z_j} \det A = \sum_{u,v=1}^{n} a_u{}^j \det \tilde{A}_{u;v} a_v{}^j (-1)^{u+v},$$

where the $(n-1) \times (n-1)$ matrix $\tilde{A}_{u;v}$ is obtained from A by removing the uth row and vth column. Repeating this procedure, one gets

$$\frac{\partial^2}{\partial z_j\, \partial z_i} \det A = \sum_{u \neq w} \sum_{v \neq t} a_u{}^j a_w{}^i \eta(u, w) \det \tilde{A}_{uw;vt} a_v{}^j a_t{}^i \eta(v, t)(-1)^{u+v+w+t},$$

where

$$\begin{aligned} \eta(u, w) = 1, &\quad \text{if} \quad u < w, \\ -1, &\quad \text{if} \quad u > w. \end{aligned}$$

In particular, $(\partial^2/\partial z_j{}^2) \det A = 0$. Differentiating n times, one ends up with

$$\left(\partial^n \Big/ \prod_{j \in S} \partial z_j\right) \det A = J_S. \qquad \text{Q.E.D.}$$

Since $\| \phi_j \|_{p_j}^2 = (\pi / z_j p_j)^{1/p_j}$, we get

$$D^2 = \sup_{z_1,\ldots,z_k > 0} \prod_{j=1}^{k} (z_j\, p_j)^{1/p_j} \Big/ \sum_S J_S z_S.$$

Now consider the function

$$\psi(z_1, \ldots, z_k) = \prod_{j=1}^{k} z_j^{1/p_j} \Big/ \sum_S J_S z_S,$$

defined on $W = (\mathbb{R}^+)^k$. By Schwarz's inequality,

$$\psi((z_1 t_1)^{1/2}, \ldots, (z_k t_k)^{1/2})^2 \geqslant \psi(z_1, \ldots, z_k)\, \psi(t_1, \ldots, t_k). \tag{2.13}$$

In other words, $\log \psi$ is a concave function of the $\log z_j$. Therefore, if the variational equations

$$1/p_j = \sum_{S \ni j} J_S z_S \Big/ \sum_S J_S z_S$$

have a solution in W, ψ reaches its absolute maximum there.

We show now that the variational equations have a unique solution (modulo the trivial rescaling $z_j \to cz_j$) if $J_S > 0$, $1 < p_j < \infty$. Firstly, the equality sign in Eq. (2.13) holds iff $J_S z_S = (\text{const})\ J_S t_S$. If $J_S > 0$, this implies that $z_j = ct_j$. Thus, modulo rescaling, $\log \psi$ is strictly concave.

Secondly, let z_j approach the boundary of W; say $z_j \sim N^{\alpha_j}$ with $N \to \infty$ and any real α_j. Let

$$\gamma = \sum_{j=1}^{k} \alpha_j/p_j - \max_{|S|=n} \sum_{j \in S} \alpha_j .$$

If $J_S > 0$, $\psi \sim N^{\gamma}$; moreover, if $1 < p_j < \infty$, $\gamma < 0$ and $\psi \to 0$ (unless $\alpha_j = \text{const.}$ for all j, which again corresponds to the rescaling).

The results are summarized in the following theorem.

THEOREM 5. *Under the assumptions of Theorem* 1, *and with the notation of Eqs.* (2.11, 12), *let the equations*

$$1/p_j = \sum_{S \ni j} J_S z_S \Big/ \sum_{S} J_S z_S , \qquad j = 1,..., k \tag{2.14}$$

have a solution for $0 < z_j < \infty$. *Then the constant* D *in Theorem* 1 *is given by*

$$D^2 = \prod_{j=1}^{k} (z_j p_j)^{1/p_j} \Big/ \sum_{S} J_S z_S , \tag{2.15}$$

and the equality sign in Eq. (2.2) *holds for*

$$f_j(x) = \exp(-z_j x^2). \tag{2.16}$$

If $1 < p_j < \infty$ *and* $J_S > 0$, *the Eqs.* (2.14) *have a unique solution satisfying* $0 < z_j < \infty$ (*modulo the rescaling* $z_j \to cz_j$).

Remark. If $J_S = 0$ for some S, Eqs. (2.14) may or may not have a solution and D may be finite or infinite. If $J_S > 0$ and some $p_j = 1$, Eq. (2.14) formally leads to $z_j = \infty$. If $J_S > 0$ and some $p_j = \infty$, Eq. (2.14) formally leads to $z_j = 0$. In both cases this gives the right value for D.

An important consequence of Theorem 5 is this: Normally one would apply Theorem 1 with fixed values of $p_1 ,..., p_k$, but then the determination of $z_1 ,..., z_k$ from Eq. (2.14) may not be easy to do when k is large. It may be much easier to fix the values of the z_j, whence the p_j are trivially given by Eq. (2.14). Eq. (2.15) then correctly gives the value of D for those p_j. Examples of such usage are given in Section 5.

2.4 *A Generalization of Theorem* 1

THEOREM 6. *Let m, n, k be integers with $0 \leqslant m \leqslant n$, $1 \leqslant n \leqslant k+m$. Let p_j, $1 \leqslant j \leqslant k$, satisfy*

$$1 \leqslant p_j \leqslant \infty, \qquad n - m \leqslant \sum_{j=1}^{k} 1/p_j \leqslant n.$$

Let f_j, a^j, $1 \leqslant j \leqslant k$, be as in Theorem 1. *Finally, let B be a nonnegative real, $n \times n$ matrix of rank m:*

$$B_{uv} = \sum_{j=k+1}^{k+m} z^j a_u{}^j a_v{}^j.$$

Then

$$\int_{\mathbb{R}^n} d^n x \prod_{j=1}^{k} f_j(\langle a^j, x\rangle) \exp(-\langle x, Bx\rangle) \leqslant E \prod_{j=1}^{k} \| f_j \|_{p_j}, \tag{2.17}$$

where the optimal constant E can be determined by restricting the f_j to be Gaussians.

If the equations

$$1/p_j = \sum_{S \ni j} J_S z_S \Big/ \sum_S J_S z_S, \qquad 1 \leqslant j \leqslant k \tag{2.18}$$

(with S running over the n-point subsets of $\{1,..., k+m\}$) have a solution satisfying $0 < z_j < \infty$ for $1 \leqslant j \leqslant k$, E is given by

$$E^2 = \prod_{j=1}^{k} (\pi z_j \, p_j)^{1/p_j} \Big/ \pi^n \sum_S J_S z_S,$$

and the equality sign in Eq. (2.17) *holds if*

$$f_j(x) = \exp(-z_j x^2), \qquad 1 \leqslant j \leqslant k.$$

Proof. If Eqs. (2.18) have a solution, one can define p_j, $k+1 \leqslant j \leqslant k+m$, by Eq. (2.18) extended to $k+1 \leqslant j \leqslant k+m$. Then Theorem 6 reduces to Theorems 1 and 5.

In the general case, Theorem 6 can be proved in the same way as Theorem 1, following the lines of Sections 2.1 and 2.2. During that operation, $\exp(-\langle x, Bx\rangle)$ is kept fixed. Q.E.D.

2.5 *Young's Inequality*

Theorems 1 and 5 contain the following special case, which gives the best possible improvement to Young's inequality.

$$\left|\int_{\mathbb{R}^2} dx\, dy\, f(x)\, g(x-y)\, h(y)\right| \leqslant C_p C_q C_t \| f \|_p \| g \|_q \| h \|_t\,, \tag{2.19}$$

$$C_p{}^2 = p^{1/p}/p'^{1/p'}, \tag{2.20}$$

where $1 \leqslant p, q, t \leqslant \infty$, $1/p + 1/q + 1/t = 2$, $1/p + 1/p' = 1$. [Throughout the remainder of this paper we use the convention $1/p' = 1 - 1/p$] The equality sign holds if

$$f(x) = \exp(-p'x^2), \qquad g(x) = \exp(-q'x^2), \qquad h(x) = \exp(-t'x^2). \tag{2.21}$$

Eqs. (2.20, 21) can be immediately read off from Eq. (2.14-16). In Section 4 we shall show that (2.21) is essentially the only choice to obtain equality in (2.19).

An equivalent form of Eq. (2.19) is

$$\| f * g \|_r \leqslant C_p C_q C_{r'} \| f \|_p \| g \|_q\,. \tag{2.22}$$

Repeated application of the last equation gives

$$\| f_1 * \cdots * f_n \|_r \leqslant C_{r'} \prod_{j=1}^{n} C_{p_j} \| f \|_{p_j}\,, \tag{2.23}$$

where $1 \leqslant p_j \leqslant \infty$, $\sum_{j=1}^{n} 1/p_j = n - 1 + 1/r$. The constant in Eq. (2.23) is the best possible, the equality sign holding for

$$f_j(x) = \exp(-p_j{}'x^2).$$

In Section 3 we shall show that the inequality (2.23) is reversed, if the exponents $p_1, \ldots, p_n$ lie between 0 and 1.

2.6 *A Multi-Dimensional Version of Theorem* 1

Theorem 1 has been stated and proved for functions f_j from $\mathbb{R}^1 \to \mathbb{C}$. We now state a generalization of that theorem for functions from $\mathbb{R}^M \to \mathbb{C}$.

THEOREM 7. *With the same assumptions as in Theorem* 1, *let* $f_j \in L^{p_j}(\mathbb{R}^M)$, $1 \leqslant j \leqslant k$. *Let* $\{a_i^j\}$, $1 \leqslant i \leqslant M$, $1 \leqslant j \leqslant k$ *be vectors in* $\mathbb{R}^n$. *Then*

$$\int_{\mathbb{R}^{nM}} d^n x_1 \cdots d^n x_M \prod_{j=1}^{k} f_j(\langle a_1{}^j, x_1\rangle, \ldots, \langle a_M{}^j, x_M\rangle) \leqslant D \prod_{j=1}^{k} \| f_j \|_{p_j},$$

and D is determined by taking the supremum over Gaussian functions.

Proof. We note that the rearrangement inequality, cf. Eq. (2.5), is not true for such integrals. However, the theorem easily follows from Theorem 1 by integrating first over x_1, then over x_2, etc. In this way one finds that the optimal Gaussians are of the form

$$\phi_j(x_1, \ldots, x_M) = \exp\left[-\sum_{i=1}^{M} C_i x_i{}^2\right].$$

Q.E.D.

An open question is the following: Let $B^1, \ldots, B^k$ be k linear maps from $\mathbb{R}^N$ to $\mathbb{R}^M$ and let $f_1, \ldots, f_k$ be functions in $L^{p_j}(\mathbb{R}^M)$. Let

$$I = \int_{\mathbb{R}^N} d^N x \prod_{j=1}^{k} f_j(B^j(x)).$$

When can I be bounded by a constant times $\prod_{j=1}^{k} \| f_j \|_{p_j}$ and when is the optimal set of f's Gaussian?

3. THE CONVERSE INEQUALITY

This section is devoted to the following theorem.

THEOREM 8. *Let* p_j, $1 \leqslant j \leqslant n$, *and* r *satisfy* $0 < p_j, r \leqslant 1$. *Let* $\sum_{j=1}^{n} 1/p_j = n - 1 + 1/r$. *With* $1/p + 1/p' = 1$, *let*

$$C_p{}^2 = | p |^{1/p} / | p' |^{1/p'}. \tag{3.1}$$

Finally, let f_j, $1 \leqslant j \leqslant n$, *be non-negative functions in* $L^{p_j}(\mathbb{R})$. *Then*

$$\| f_1 * \cdots * f_n \|_r \geqslant C_{r'} \prod_{j=1}^{n} C_{p_j} \| f \|_{p_j}.$$

The equality sign holds (for $p_j \neq 1$) if

$$f_j(x) = \exp(p_j' x^2).$$

3.1 *Preliminary Remarks*

It is sufficient to prove Theorem 8 for $n = 2$ ($0 < p, q, r \leqslant 1$, $1/p + 1/q = 1 + 1/r$):

$$\| f * g \|_r \geqslant C_p C_q C_{r'} \| f \|_p \| g \|_q ; \tag{3.2}$$

the general case then follows by repeated application.

A weaker form of Eq. (3.2) was found by Leindler [4]:

$$\| f * g \|_r \geqslant \| f \|_p \| g \|_q . \tag{3.3}$$

If $p = 1$, $q = r$ and Eq. (3.2) is the same as Eq. (3.3). Thus we shall further restrict ourselves to $0 < p, q < 1$.

As in Section 2, we shall need a rearrangement inequality.

PROPOSITION 9. *Let $f, g\colon \mathbb{R}^M \to \mathbb{R}^+$ and let $0 < r \leqslant 1$. Then*

$$\| f * g \|_r \geqslant \| f^* * g^* \|_r . \tag{3.4}$$

Proof. If $r = 1$, Eq. (3.4) is a trivial equality. For $0 < r < 1$ and $f, h \geqslant 0$, Hölder's inequality becomes

$$\int f(x)\, h(x)\, dx \geqslant \| f \|_r \| h \|_{r'} .$$

Hence

$$\| f * g \|_r = \inf \left\{ \int f(x-y)\, g(y)\, h(x)\, d^M x\, d^M y \,\middle|\, h(x) \geqslant 0, \| h \|_{r'} = 1 \right\}. \tag{3.5}$$

Note that $r' < 0$. Define the symmetric increasing rearrangement ${}^*h$ of h by

$${}^*h = [(h^{-1})^*]^{-1}.$$

Then $\| {}^*h \|_{r'} = \| h \|_{r'}$. For $A > 0$, let

$$h^A(x) = \min[A, h(x)]; \qquad k^A(x) = A - h^A(x).$$

Then, as $A \to \infty$

$$h^A(x) \uparrow h(x); \quad A - k^{A*}(x) \uparrow {}^*h(x). \tag{3.6}$$

We can assume, that $\| f * g \|_r < \infty$. But in that case, Leindler's inequality (3.3) implies that $f, g \in L^1 \cap L^r$. Then by the rearrangement inequality (2.5) and Eq. (3.6) together with monotone convergence, we have

$$\begin{aligned} &\int f(x-y)\, g(y)\, h(x)\, d^Mx\, d^My \\ &\quad = \lim_{A\to\infty} \left[A \| f \|_1 \| g \|_1 - \int f(x-y)\, g(y)\, k^A(x)\, d^Mx\, d^My \right] \\ &\quad \geqslant \lim_{A\to\infty} \left[A \| f^* \|_1 \| g^* \|_1 - \int f^*(x-y)\, g^*(y)\, k^{A*}(x)\, d^Mx\, d^My \right] \\ &\quad = \int f^*(x-y)\, g^*(y)^*\, h(x)\, d^Mx\, d^My. \end{aligned}$$

Eq. (3.4) now follows from Eq. (3.5). Q.E.D.

A consequence of Proposition 9 is, that we can restrict ourselves to symmetric decreasing functions in proving Eq. (3.2). Then we can find sequences of simple step functions as in Eq. (2.6) such that

$$\| f^K \|_p \to \| f \|_p, \qquad \| g^K \|_q \to \| g \|_q; \qquad f^K(x) \leqslant f(x), \qquad g^K(x) \leqslant g(x).$$

This means that it suffices to prove Eq. (3.2) for step functions of the form given in Eq. (2.6).

We need the analogue of Proposition 3, which reads

PROPOSITION 10. *Let ψ_j, $1 \leqslant j \leqslant K$, be non-negative functions in $L^p(\mathbb{R}^M)$, $0 < p < 1$. Then*

$$\sum_{j=1}^K \| \psi_j \|_p \leqslant \Big\| \sum_{j=1}^K \psi_j \Big\|_p \leqslant K^{-1/p'} \sum_{j=1}^K \| \psi_j \|_p,$$

where $1/p + 1/p' = 1$.

Proof. The first inequality follows from the fact that $\| \psi \|_p$ can be written as an infimum (cf. Eq. (3.5)); the second one is proved as in Proposition 3, where it should be noted that both inequalities encountered in that proof are reversed, Q.E.D.

We also need some comparison between characteristic functions of balls and Gaussians, as in Proposition 4. It is true that Proposition 4 remains valid for $0 < p < 1$; however the direction of the inequality signs makes it quite useless here. In fact, no such simple trick seems to be available now, and we are obliged to make a brutal computation of the volume of the intersection of two balls.

PROPOSITION 11. *Let* η_a *be the characteristic function of the ball* $\{x \in \mathbb{R}^M : |x| \leqslant a\}$, *and let* (*with* $0 < p, q, r < 1$, $1/p + 1/q = 1 + 1/r$)

$$\psi_M(a/b) = \| \eta_a * \eta_b \|_r / \| \eta_a \|_p \| \eta_b \|_q .$$

Then

$$\psi_M(a/b) \geqslant (C_p C_q C_{r'})^M.$$

Proof. Note that by the rearrangement inequality in Proposition 9

$$\psi_{MN}(a/b) \leqslant [\psi_M(a/b)]^N;$$

hence it suffices to show that

$$\lim_{M\to\infty} [\psi_M(a/b)]^{1/M} \geqslant C_p C_q C_{r'} .$$

The intersection in $\mathbb{R}^M$ of a ball with radius a centered at the origin and a ball with radius b centered at the point x can be thought of as the union of $M - 1$-dimensional balls, each centered on the line connecting the origin with x. The greatest radius $h(x)$ occurring among these balls is

$$h(x) = \min(a, b), \qquad 0 \leqslant x \leqslant |a^2 - b^2|^{1/2};$$

$$h(x) = [-x^4 + 2(a^2 + b^2)\, x^2 - (a^2 - b^2)^2]^{1/2}/2x, \ |a^2 - b^2|^{1/2} \leqslant x \leqslant a + b;$$

$$h(x) = 0, \qquad x \geqslant a + b.$$

Then

$$(\eta_a * \eta_b)(x) \sim \Omega_M h(x)^M$$

(i.e., the Mth root of the ratio of both members goes to 1 as $M \to \infty$). In the same way

$$\| \eta_a * \eta_b \|_r \sim \Omega_M^{1+1/r} [\max_{x>0} \{x^{1/r} h(x)\}]^M.$$

The maximum on the right side is reached for $|a^2 - b^2|^{1/2} < x < a + b$; hence

$$\lim_{M\to\infty} [\psi_M(a/b)]^{1/M}$$
$$= \tfrac{1}{2} \max_x \{(x/a)^{1/p} (x/b)^{1/q} [-1 + 2(a^2/x^2 + b^2/x^2) - (a^2/x^2 - b^2/x^2)^2]^{1/2}\}.$$

607/20/2-5

Let

$$\phi(A, B) = A^{-1/p}B^{-1/q}[-1 + 2(A + B) - (A - B)^2].$$

Straightforward calculation gives that the unique solution to

$$\frac{\partial \phi}{\partial A} = \frac{\partial \phi}{\partial B} = 0$$

is given by

$$A = rr'/pp'; \qquad B = rr'/qq'.$$

Since $[\psi_M(a/b)]^{1/M} \to \infty$ for $a/b \to 0$ or $a/b \to \infty$, substitution of these values for A and B into $\phi(A, B)$ must lead to the minimum over a/b of $\lim_{M\to\infty} [\psi_M(a/b)]^{1/M}$. The result is

$$\min_{a/b} \lim_{M\to\infty} [\psi_M(a/b)]^{1/M} = C_p C_q C_{r'} .$$

Q.E.D.

3.2 *Proof of Theorem* 8

Theorem 8 is now proved along the same lines as Theorem 1. Given step functions f, g as in Eq. (2.6), define $F(x_1 ,..., x_M)$ [resp. $G(x_1 ,..., x_M)$] as the Schwarz symmetrization of $\prod_{m=1}^M f(x_m)$ [resp. $\prod_{m=1}^M g(x_m)$]; then F, G are as in Eq. (2.8). We have, by Proposition 9,

$$(\| f * g \|_r)^M \geqslant \| F * G \|_r .$$

Hence, by Propositions 10 and 11,

$$\begin{aligned} &[\| f * g \|_r / \| f \|_p \| g \|_q]^M \\ &\quad \geqslant (M+1)^{(1/p'+1/q')K} \frac{\sum_m \sum_n H_1{}^m H_2{}^n \| \eta^m * \eta^n \|_r}{\sum_m \sum_n H_1{}^m H_2{}^n \| \eta^m \|_p \| \eta^n \|_q} \\ &\quad \geqslant (M+1)^{(1/p'+1/q')K} (C_p C_q C_{r'})^M. \end{aligned}$$

The proof is again concluded by taking the Mth root. Q.E.D.

The proof given above does not allow for a generalization of Theorem 8 to a full analogue of Theorem 1, concerning k functions and n variables. In fact, a converse rearrangement inequality as in Proposition 9 only seems true if $k = n + 1$. (cf. the proof of Proposition 9).

3.3 *A Limiting Case of Theorem* 8

Theorem 8 allows us to rederive a theorem due to Prékopa [3, 5] and Leindler [4].

THEOREM 12 (Prékopa–Leindler). *Let f, $g \geqslant 0$, f, $g \in L^1(\mathbb{R})$, and let $\lambda \in (0, 1)$. Let*

$$h(x) \equiv \operatorname*{ess\,sup}_y f\Big(\frac{x-y}{\lambda}\Big)^{\lambda} g\Big(\frac{y}{1-\lambda}\Big)^{1-\lambda}.$$

Then h is measurable and

$$\| h \|_1 \geqslant \| f \|_1^{\lambda} \, \| g \|_1^{1-\lambda}.$$

Proof. The measurability of h is proved in [10]. Let $f^{(n)}$ (resp. $g^{(n)}$) be a sequence of bounded functions of compact support which approach f (resp. g) in L^1 norm and such that $f^{(n)}(x) \leqslant f(x)$, $g^{(n)}(x) \leqslant g(x)$, $\forall x$. Defining $h^{(n)}$ using $f^{(n)}$ and $g^{(n)}$, one has that $\| h^{(n)} \|_1 \leqslant \| h \|_1$, and hence it is sufficient to prove the theorem for bounded functions of compact support. For such functions

$$h(x) = \lim_{R\to\infty} h_R(x), \qquad h_R(x) = \Big[\int dy\, f\Big(\frac{x-y}{\lambda}\Big)^{\lambda R} g\Big(\frac{y}{1-\lambda}\Big)^{(1-\lambda)R}\Big]^{1/(R-1)},$$

$$\| h \|_1 = \lim_{R\to\infty} \| h_R \|_1 .$$

The interchange of the R limit and the integral is allowed by dominated convergence since the h_R are uniformly bounded and their supports lie in some common compact set.

Now for $R > \max(\lambda^{-1}, (1-\lambda)^{-1})$, let $1/p = \lambda R$, $1/q = (1-\lambda)R$, $1/r = R - 1$, $1/r' = 2 - R$. Using (3.2) one has, with $t = R(R-1)^{-1}$,

$$\| h_R \|_1 \geqslant (C_p C_q C_{r'})^r [\lambda \| f \|_1]^{\lambda t} [(1-\lambda) \| g \|_1]^{(1-\lambda)t}.$$

When $R \to \infty$, $t \to 1$ and $(C_p C_q C_{r'})^r \to \lambda^{-\lambda} (1-\lambda)^{-(1-\lambda)}$. Q.E.D.

Note, that Prékopa and Leindler proved a slightly weaker form of Theorem 12, concerning sup instead of ess sup. Variants of their theorem were later found by Rinott [11] and ourselves [12]. Much simpler proofs are possible without using Theorem 8 and these will be published in the *Journal of Functional Analysis*.

4. Uniqueness

In this section we show that Eqs. (2.22) and (3.2) hold as equalities only if f and g are Gaussians.

Theorem 13. *Let* $f(x) \in L^p(\mathbb{R})$, $g(x) \in L^q(\mathbb{R})$, *with either* $1 < p$, $q < \infty$ *or* $0 < p$, $q < 1$. *In the latter case, let* $f(x) \geqslant 0$, $g(x) \geqslant 0$. *Let* $1 + 1/r = 1/p + 1/q$, *and let*

$$\| f * g \|_r = C_p C_q C_{r'} \| f \|_p \| g \|_q . \tag{4.1}$$

Then

$$\begin{aligned} f(x) &= A \exp[-\gamma \mid p' \mid (x - \alpha)^2 + i\,\delta x], \\ g(x) &= B \exp[-\gamma \mid q' \mid (x - \beta)^2 + i\,\delta x], \end{aligned} \tag{4.2}$$

with constants $\gamma \in \mathbb{R}^+$, $\alpha, \beta, \delta \in \mathbb{R}$, *and* A, $B \in \mathbb{R}^+$, $\delta = 0$ *if* $0 < p$, $q < 1$; A, $B \in \mathbb{C}$ *if* p, $q > 1$.

Proof. If p, $q > 1$, the equality (4.1) implies that $f \geqslant 0$, $g \geqslant 0$ (apart from arbitrary multiplicative constants (see note added in proof)). Eq. (4.1) holds, iff there exists a function $h \in L^{r'}(\mathbb{R})$ such that

$$\int_{\mathbb{R}^2} dx\, dy\, f(x - y)\, g(y)\, h(x) = C_p C_q C_{r'} \| f \|_p \| g \|_q \| h \|_{r'} . \tag{4.3}$$

In fact, by Hölder's inequality the only possible choice for h is

$$h = (\text{const})(f * g)^{r/r'}. \tag{4.4}$$

Now let Eq. (4.3) be satisfied for the triples f, g, h and f_1, g_1, h_1. Then

$$\int_{\mathbb{R}^4} dx\, dy\, du\, dv\, f(x - y) f_1(u - v - x + y)\, g(y)\, g_1(v - y)\, h(x)\, h_1(u - x)$$
$$= (C_p C_q C_{r'})^2 \| f \|_p \| f_1 \|_p \| g \|_q \| g_1 \|_q \| h \|_{r'} \| h_1 \|_{r'} .$$

Now first integrate over (x, y) and then over (u, v). Using Eq. (2.22), resp. Eq. (3.2), twice, this implies that, for almost all (u, v), Eq. (4.3) is satisfied for the triple $f(x) f_1(u - v - x)$, $g(x)\, g_1(v - x)$, $h(x)\, h(u - x)$. Therefore, this triple must satisfy an equation of the form (4.4), with the constant depending on (u, v).

As a special choice, take

$$f_1(x) = \exp(- \mid p' \mid x^2/2),\; g_1(x) = \exp(- \mid q' \mid x^2/2),\; h_1(x) = \exp[-r(\operatorname{sgn} r')x^2/2].$$

Define

$$F(x) = f(x) \exp(-|p'| x^2/2),$$
$$G(x) = g(x) \exp(-|q'| x^2/2),$$
$$H(x) = h(x)^{r'/r} \exp(-|r'| x^2/2).$$

Then, for almost all (u, v), we have for almost all x

$$H(x) \exp(r'ux) = K(u, v) \int_{\mathbb{R}} dy\, F(x-y) \exp[p'(u-v)(x-y)]\, G(y) \exp(q'vy). \tag{4.5}$$

Define the two-sided Laplace transform by

$$\hat{A}(s) = \int_{\mathbb{R}} dx\, A(x)\, e^{-sx}.$$

Since F, G, and H contain a Gaussian factor, their Laplace transforms are defined and analytic in the whole complex s-plane. Eq. (4.5) becomes

$$\hat{H}(s - r'u) = K(u, v) \hat{F}(s - p'(u - v))\, \hat{G}(s - q'v).$$

By a shift $s \to s + r'u$, this becomes

$$\hat{H}(s) = K(u, v) \hat{F}(s + p't)\, \hat{G}(s - q't), \tag{4.6}$$

with

$$t = v - ur'/q'.$$

Since $\hat{H}$ does not depend on u and v, $K(u, v)$ can only depend on t. Since $\hat{F}$, $\hat{G}$ and $\hat{H}$ are entire functions and are strictly positive for real arguments, one can take the second logarithmic derivative with respect to s and t of (4.6). One then finds that

$$\hat{F}(s) = D \exp(\mu s^2/p' + \delta s),$$
$$\hat{G}(s) = E \exp(\mu s^2/q' + \epsilon s),$$

with constants D, E, μ, δ, ϵ. With the inverse Laplace transform, this leads to Eq. (4.2). Q.E.D.

Remarks. Obviously, the uniqueness of the Gaussians can be proved in the same way for multiple convolutions, as in Eq. (2.23) and Theorem 8. However, the above proof fails for the general case of Theorem 1, if

$k \geqslant n+2$. Then introduction of the Laplace transform in an equality like (4.5) does not lead to a simple product, as in Eq. (4.6).

Theorem 14 does not extend to the case in which p or q is one.

5. Applications

5.1 *A Theorem of Nelson*

We are now in the position to give a simple proof of Nelson's hypercontractivity theorem [13]. On $\mathbb{R}$, consider the Gaussian measure

$$d\mu(x) = (2\pi)^{-1/2}\, e^{-x^2/2}\, dx,$$

with the corresponding spaces $L^p(\mathbb{R}, \mu)$. If $f \in L^q(\mathbb{R}, \mu)$, the map $\Gamma(c)$, $0 \leqslant c \leqslant 1$, is defined by

$$(\Gamma(c)f)(x) = [2\pi(1-c^2)]^{-1/2} \int_{\mathbb{R}} \exp\left[-\frac{(y-cx)^2}{2(1-c^2)}\right] f(y)\, dy.$$

THEOREM 14. *Let* $1 \leqslant q \leqslant p \leqslant \infty$. *Then* $\Gamma(c)$ *is a contraction from* $L^q(\mathbb{R}, \mu)$ *to* $L^p(\mathbb{R}, \mu)$ *if*

$$c \leqslant [(q-1)/(p-1)]^{1/2}. \tag{5.1}$$

The contraction constant is 1.

Proof. It has to be shown that

$$(2\pi)^{-1}(1-c^2)^{-1/2} \int_{\mathbb{R}^2} g(x) \exp\left[-\tfrac{1}{2}x^2 - \frac{(y-cx)^2}{2(1-c^2)}\right] f(y)\, dx\, dy$$
$$\leqslant \|f\|_{q,\mu} \|g\|_{p',\mu}, \tag{5.2}$$

with $1/p + 1/p' = 1$. If we write

$$F(x) = f(x)\exp(-x^2/2q), \qquad G(x) = g(x)\exp(-x^2/2p'),$$

we are in the situation of Theorem 6; however, for that theorem to apply, the quadratic form

$$-\frac{y^2}{2q} + \frac{1}{2}x^2 + \frac{(y-cx)^2}{2(1-c^2)} - \frac{x^2}{2p'}$$

must be non-negative definite. This is equivalent to the condition (5.1).

If it fails, one can choose Gaussians for F and G such, that the left side of Eq. (5.2) diverges.

Now let us assume (5.1) to hold. If we put $f(x) = \exp(-\alpha x^2/2)$, $g(x) = \exp(-\beta x^2/2)$, the ratio of the left and right sides of Eq. (5.2) is

$$\{(\alpha p' + 1)^{1/p'}(\beta q + 1)^{1/q}[\alpha\beta(1 - c^2) + \alpha + \beta + 1]^{-1}\}^{1/2}. \tag{5.3}$$

This expression reaches an extremum if

$$(\alpha p' + 1)^{-1} = [\beta(1 - c^2) + 1][\alpha\beta(1 - c^2) + \alpha + \beta + 1]^{-1};$$
$$(\alpha q + 1)^{-1} = [\alpha(1 - c^2) + 1][\alpha\beta(1 - c^2) + \alpha + \beta + 1]^{-1}.$$

It is ensured by the general concavity argument in Section 3.3 that any solution to these equations gives the absolute maximum; hence we can take $\alpha = \beta = 0$ and the maximal ratio of the left and right sides of Eq. (5.2) is 1. Q.E.D.

5.2 *The Anharmonic Crystal in Statistical Mechanics*

We consider a d-dimensional crystal of size L. This means that we have $N = L^d$ particles. The equilibrium position of the nth particle is the vector $n = \{n_1, \ldots, n_d\} \in \mathbb{Z}^d$, with $0 \leqslant n_j \leqslant L - 1$, $j = 1, \ldots, d$. The vector n labels the particles and the n's are distinct. We assume that each particle has a one-dimensional motion with coordinate x_n. Neighboring particles interact through a potential $\phi(x_n - x_m)$, $\phi(x) = \phi(-x)$. Let us take periodic boundary conditions, that is, particles numbered $(n_1, \ldots, L - 1, \ldots, n_d)$ and $(n_1, \ldots, 0, \ldots, n_d)$ interact. Fixing the center of mass, we define the partition function

$$Z_N(\phi) = \int_{\mathbb{R}^N} d^N x \, \delta\left(N^{-1/2} \sum_n x_n\right) \exp\left[- \sum_{\langle n,m \rangle} \phi(x_n - x_m)\right],$$

where the summation in the exponential extends over all pairs of nearest neighbours. We now apply Theorem 6, with the δ-function playing the role of a fixed Gaussian. We get

$$Z_N(\phi) \leqslant \sup_\gamma \{\| e^{-\phi} \|_p^{Nd} / \| e^{-\gamma x^2} \|_p^{Nd}\} \, Z_N(\gamma x^2). \tag{5.4}$$

Note, that we have chosen all exponents p_j in Theorem 6 to be equal; then, by symmetry, the Gaussians giving the best constant E can all be taken the same. The condition on the p_j in Theorem 6 now becomes

$$N - 1 \leqslant Nd/p \leqslant N.$$

However, the expression on the right side of Eq. (5.4) contains $\gamma^{-(N-1)/2+Nd/2p}$ as a factor; hence, the supremum is finite only if

$$Nd/p = N - 1,$$

so that

$$Z_N(\phi) \leqslant \{\| e^{-\phi} \|_{Nd/(N-1)} / \| e^{-x^2} \|_{Nd/(N-1)}\}^{Nd} Z_N(x^2).$$

The partition function for the harmonic crystal, $Z_N(x^2)$, is easily computed by going over to normal modes; one has

$$\sum_{\langle n,m \rangle} (x_n - x_m)^2 = \sum_k \omega_k \mid q_k \mid^2,$$

where

$$q_k = N^{-1/2} \sum_n x_n \exp[2\pi i \langle n, k \rangle / L],$$

$$\omega_k = 2 \left[d - \sum_{j=1}^{d} \cos(2\pi k_j / L) \right],$$

$$k = \{k_1, \ldots, k_d\} \in \mathbb{Z}^d, \qquad 0 \leqslant k_j \leqslant L - 1.$$

Hence

$$Z_N(x^2) = \prod_{k \neq 0} (\pi / \omega_k)^{1/2}.$$

For the free energy,

$$f(\phi) = -\lim_{N \to \infty} N^{-1} \log Z_N(\phi),$$

this gives the lower bound

$$f(\phi) \geqslant -d \log \| e^{-\phi} \|_d - \tfrac{1}{2} \log(d/2)$$
$$+ \tfrac{1}{2} \int_0^1 dk_1 \cdots \int_0^1 dk_d \log \left(d - \sum_{j=1}^{d} \cos 2\pi k_j \right).$$

Note added in proof. In the original version of Theorem 13 we did not include the term $i\delta x$ in (4.2). We are indebted to J. Fournier for pointing out this oversight to us.

References

1. W. Beckner, Inequalities in Fourier analysis, *Ann. of Math.* **102** (1975), 159–182.
2. L. Leindler, On a certain converse of Hölder's inequality, *In* "Linear Operators and Approximation," Proceedings of the 1971 Oberwolfach Conference, Birkhäuser Verlag, Basel-Stuttgart, 1972.

3. A. Prékopa, Logarithmic concave measures with application to stochastic programming, *Acta Sci. Math. Szeged* **32** (1971), 301–315.
4. L. Leindler, On a certain converse of Hölder's inequality. II, *Acta Sci. Math. Szeged* **33** (1972), 217–223.
5. A. Prékopa, On logarithmic measures and functions, *Acta Sci. Math. Szeged* **34** (1973), 335–343.
6. G.E. Hardy, J.E. Littlewood, and G.Pólya, "Inequalities," Cambridge University Press, London and New York, 1952.
7. F. Riesz, Sur une Inéqualité Intégrale, *J. L.M.S.* **5** (1930), 162–168.
8. S. Sobolev, On a theorem of functional analysis, *Mat. Sb. (N.S.)* **4** (1938), 471–497; *Amer. Math. Soc. Transl.* **34**, 2 (1963), 39–68.
9. H. J. Brascamp, E. H., Lieb, and J. M. Luttinger, A general rearrangement inequality for multiple integrals, *J. Funct. Anal.* **17** (1974), 227–237.
10. P. R. Chernoff, Advanced problems and solutions, *Amer. Math. Monthly* **81** (1974), 1038–1039.
11. Y. Rinott, Thesis, Tel Aviv, 1973.
12. H. J. Brascamp and E. H. Lieb, Some inequalities for Gaussian measures and the long range order of the one-dimensional plasma, *in* "Functional Integration and Its Applications" (A. M. Arthurs, Ed.), Clarendon, Oxford, 1975.
13. E. Nelson, The free Markoff field, *J. Funct. Anal.* **12** (1973), 211–227.

JOURNAL OF FUNCTIONAL ANALYSIS **22**, 366–389 (1976)

On Extensions of the Brunn–Minkowski and Prékopa–Leindler Theorems, Including Inequalities for Log Concave Functions, and with an Application to the Diffusion Equation

HERM JAN BRASCAMP*

Department of Physics, Princeton University, Princeton, N.J. 08540

AND

ELLIOTT H. LIEB*

Departments of Mathematics and Physics, Princeton University, Princeton, N.J. 08540

Communicated by the Editors

We extend the Prékopa–Leindler theorem to other types of convex combinations of two positive functions and we strengthen the Prékopa–Leindler and Brunn–Minkowski theorems by introducing the notion of essential addition. Our proof of the Prékopa–Leindler theorem is simpler than the original one. We sharpen the inequality that the marginal of a log concave function is log concave, and we prove various moment inequalities for such functions. Finally, we use these results to derive inequalities for the fundamental solution of the diffusion equation with a convex potential.

1. INTRODUCTION

In this paper we give various extensions of the Brunn–Minkowski and Prékopa–Leindler theorems. The Brunn–Minkowski theorem for the convex addition $D = \lambda A + (1 - \lambda)B = \{x \in \mathbf{R}^n \mid x = \lambda y + (1 - \lambda)z, y \in A, z \in B\}$ of two nonempty, measurable sets $A, B \subset \mathbf{R}^n$ reads [1, 2]

$$\mu_n(D)^{1/n} \geqslant \lambda\mu_n(A)^{1/n} + (1 - \lambda)\, \mu_n(B)^{1/n}, \tag{1.1}$$

* Work partially supported by National Science Foundation Grant MPS71-03375 A03 at M.I.T.

where μ_n means Lebesgue measure in $\mathbf{R}^n$. The requirement that A and B are nonempty is crucial.

The Prékopa–Leindler theorem [3, 4, 5] reads

$$\| k \|_1 \geqslant \| f \|_1^{\lambda} \| g \|_1^{1-\lambda}, \tag{1.2}$$

where

$$k(x \mid f, g) = \sup_{y \in \mathbf{R}^n} f\left(\frac{x-y}{\lambda}\right)^{\lambda} g\left(\frac{y}{1-\lambda}\right)^{1-\lambda}, \tag{1.3}$$

and f, g are nonnegative, measurable functions on $\mathbf{R}^n$. If f and g are the characteristic functions of A and B, respectively, k is the characteristic function of D. Thus, Eq. (1.2) states that $\mu_n(D) \geqslant 1$ if $\mu_n(A) = \mu_n(B) = 1$. By the scaling property, $\mu_n(\lambda A) = \lambda^n \mu_n(A)$. Thus Eq. (1.2) implies Eq. (1.1). In that sense, the Prékopa–Leindler theorem can be viewed as an extension of the Brunn–Minkowski theorem.

These theorems are extended here in the following ways.

EXTENSION 1. The sup in Eq. (1.3) is replaced by ess sup:

$$h(x \mid f, g) = \operatorname*{ess\,sup}_{y \in \mathbf{R}^n} f\left(\frac{x-y}{\lambda}\right)^{\lambda} g\left(\frac{y}{1-\lambda}\right)^{1-\lambda}. \tag{1.4}$$

The Prékopa–Leindler theorem strengthened in this way is contained in Theorems 3.2 and 3.3.

Our new version really is stronger than the old; generally, $\| h \|_1 \leqslant \| k \|_1$, and there are functions f and g such that h differs greatly from k. It is a fact, however, established in the Appendix, that f and g can always be replaced by functions f^* and g^* which differ only by null functions from f and g such that

$$h(x \mid f, g) = h(x \mid f^*, g^*) = k(x \mid f^*, g^*).$$

Thus, once one knows how to construct f^* and g^*, the strengthened Prékopa–Leindler theorem follows from the known one.

However, we prefer to work with the essential supremum h, because (1) $h(x)$ is unaltered if null functions are added to f and g, and (2) $h(x)$ is lower semicontinuous for any measurable f and g. The supremum k has neither property.

By taking characteristic functions for f and g, a stronger form of the Brunn–Minkowski theorem results; as above, it can be derived from the known theorem (see the Appendix). The proof given here of the Prékopa–Leindler theorem is based on the Brunn–Minkowski theorem; it is simpler than the original proof by Prékopa and Leindler. The idea of our proof is already contained in [6]. Another (rather involved) proof of the strengthened Prékopa–Leindler theorem is given by us in [7].

EXTENSION 2. Other types of convex combinations, h_α, of two functions, f and g are defined for $\alpha \in [-\infty, \infty]$; see Eqs. (2.1–2.3). The convex combination in Eq. (1.4) is the case $\alpha = 0$.

In Section 3 theorems of the Prékopa–Leindler type are given for general α (Theorems 3.1–3.3). A Brunn–Minkowski-like version of these theorems is contained in Corollary 3.4. For the case $\alpha = 0$ and with sup instead of ess sup, it was first given by Prékopa [2, 4]. A much simpler proof for that case was found by Rinott [8]; his proof is completely different from ours. Rinott also found the case $\alpha = -1/n$ in Corollary 3.4. Moreover, he found a converse of Corollary 3.4, saying that Eq. (3.8) for all A, B implies the existence of a log concave density function.

In Section 4 we consider log concave functions. A corollary of the Prékopa–Leindler theorem is that $\int F(x, y)\, dy$ is log concave in x if $F(x, y)$ is log concave in (x, y). This result is sharpened in Theorem 4.2. In Theorem 4.1 a Sobolev-type inequality for log concave measures is given.

Some theorems on log concave functions have counterparts for log convex functions (Theorems 4.3, 5.1, and 6.1). However, these counterparts are comparatively trivial; they essentially follow from the usual convexity arguments (Hölder's inequality). We stress that the log concave theorems and other Brunn–Minkowski and Prékopa–Leindler-like theorems do not follow trivially from Hölder's inequality.

In Section 5 we give inequalities for the moments of a Gaussian distribution, compared with the moments of the same distribution perturbed by a log concave (or log convex) function (Theorem 5.1).

In Section 6 we give an application to the diffusion equation in $\mathbf{R}^n$ with convex potential. More applications (the Ising model, the one dimensional Coulomb plasma) are given in [6].

2. NOTATION

Given nonnegative measurable functions $f(x)$, $g(x)$ on $\mathbf{R}^n$, we shall introduce various convex combinations of them, parametrized by the real number $\alpha \in [-\infty, \infty]$. With $0 < \lambda < 1$, we define

$$h_\alpha(x \mid f, g) = \operatorname*{ess\,sup}_{y \in \mathbf{R}^n} \left\{ \lambda f\left(\frac{x-y}{\lambda}\right)^\alpha \oplus (1-\lambda) g\left(\frac{y}{1-\lambda}\right)^\alpha \right\}^{1/\alpha}. \quad (2.1)$$

The symbol $\oplus$ differs from the ordinary addition $+$ in that for

$$f = 0 \quad \text{or} \quad g = 0, \quad \{\lambda f^\alpha \oplus (1-\lambda)\, g^\alpha\}^{1/\alpha} = 0. \tag{2.2}$$

Otherwise, $\oplus$ and $+$ are the same: For $f > 0$ and $g > 0$,

$$\begin{aligned} &\{\lambda f^\alpha \oplus (1-\lambda)\, g^\alpha\}^{1/\alpha} \\ &\quad = \{\lambda f^\alpha + (1-\lambda)\, g^\alpha\}^{1/\alpha}, && \text{if} \quad -\infty < \alpha < 0, \quad 0 < \alpha < \infty; \\ &\quad = \min(f, g), && \text{if} \quad \alpha = -\infty; \\ &\quad = \max(f, g), && \text{if} \quad \alpha = \infty; \\ &\quad = f^\lambda g^{1-\lambda}, && \text{if} \quad \alpha = 0. \end{aligned} \tag{2.3}$$

Note, that $\oplus$ and $+$ are completely identical for $\alpha \leqslant 0$; however, for $\alpha > 0$ Eq. (2.2) makes them essentially different. Note further that

$$h_\alpha(x) \leqslant h_\beta(x), \qquad \text{if} \quad \alpha < \beta.$$

We shall often write $h_\alpha(f, g)$, $h_\alpha(x)$ or h_α if the dependence of $h_\alpha(x \mid f, g)$ on x, f and g, or both is obvious. The dependence on λ is not displayed, λ being held fixed.

As a particular case, take for f and g characteristic functions of measurable sets $A, B \subset \mathbf{R}^n$: $f = \chi_A$, $g = \chi_B$. Then by Eqs. (2.2, 2.3),

$$\{\lambda f^\alpha \oplus (1-\lambda)\, g^\alpha\}^{1/\alpha} = 0 \qquad \text{or} \quad 1,$$

independent of α. Hence, there is a set C such that

$$h_\alpha(\chi_A, \chi_B) = \chi_C, \qquad \forall \alpha.$$

We shall use the notation

$$C = \operatorname{ess}\{\lambda A + (1-\lambda) B\}.$$

To stress the difference with the ordinary Brunn–Minkowski addition we give appropriate definitions:

$$\begin{aligned} \lambda A + (1-\lambda) B &= \{x \in \mathbf{R}^n \mid (x - \lambda A) \cap (1-\lambda) B \neq \varnothing\}; \\ \operatorname{ess}\{\lambda A + (1-\lambda) B\} &= \{x \in \mathbf{R}^n \mid \mu_n[(x - \lambda A) \cap (1-\lambda) B] > 0\}. \end{aligned} \tag{2.4}$$

The ordinary addition results, if ess sup in Eq. (2.1) is replaced by sup. The ordinary and the essential additions may differ considerably, as can be seen by taking for A a single point. However, there always exist sets A^* and B^* which differ from A and B by null sets and such that

$$A^* + B^* = \text{ess}(A^* + B^*) = \text{ess}(A + B) \tag{2.5}$$

(see the Appendix). Equation (2.5) and the Brunn–Minkowski theorem, Eq. (1.1), immediately imply the strengthened Brunn–Minkowski theorem

$$\mu_n(C)^{1/n} \geqslant \lambda\mu_n(A)^{1/n} + (1-\lambda)\,\mu_n(B)^{1/n}, \tag{2.6}$$

if $\mu_n(A) > 0$, $\mu_n(B) > 0$.

In the next section we show how Eq. (2.6) extends to inequalities for $\| h_\alpha \|_1$ in terms of $\| f \|_1$ and $\| g \|_1$.

3. Inequalities for $\| h_\alpha \|_1$

The following theorem is basic.

THEOREM 3.1. *Let f, g be nonnegative, measurable functions on* $\mathbf{R}$ *and define $h_{-\infty}$ as in Eqs.* (2.1–2.3):

$$h_{-\infty}(x) = \operatorname*{ess\,sup}_{y\in\mathbf{R}} \min \left\{ f\left(\frac{x-y}{\lambda}\right), g\left(\frac{y}{1-\lambda}\right)\right\}.$$

Let $\| f \|_\infty = \| g \|_\infty \equiv m$. Then

$$\| h_{-\infty} \|_1 \geqslant \lambda \| f \|_1 + (1-\lambda) \| g \|_1 .$$

Proof. For $z > 0$, define the sets

$$A(z) = \{x \in \mathbf{R} \mid f(x) > z\},$$
$$B(z) = \{x \in \mathbf{R} \mid g(x) > z\},$$
$$D(z) = \{x \in \mathbf{R} \mid h_{-\infty}(x) > z\}.$$

Then

$$D(z) \supset \text{ess}\{\lambda A(z) + (1-\lambda)\, B(z)\},$$

by the definitions of $h_{-\infty}$ and of the essential addition.

If $z < m$, $\mu_1(A(z)) > 0$ and $\mu_1(B(z)) > 0$. Thus, by Eq. (2.6)

$$\mu_1(D(z)) \geqslant \lambda\mu_1(A(z)) + (1-\lambda)\,\mu_1(B(z)).$$

Note, further, that $\mu_1(D(z)) = \mu_1(A(z)) = \mu_1(B(z)) = 0$ for $z \geqslant m$, and that

$$\| f \|_1 = \int_0^\infty \mu_1(A(z))\, dz, \quad \text{etc.}$$

This gives the desired result. Q.E.D.

By a simple rescaling, Theorem 3.1 immediately leads to

THEOREM 3.2. *Let f, g be nonnegative measurable functions on* **R** *and define h_α as in Eqs.* (2.1–2.3). *Let $\| f \|_1 > 0$, $\| g \|_1 > 0$. Then, for $\alpha \geqslant -1$,*

$$\| h_\alpha \|_1 \geqslant \{\lambda \| f \|_1^\beta + (1 - \lambda) \| g \|_1^\beta\}^{1/\beta}, \tag{3.1}$$

with $\beta = \alpha/(1 + \alpha)$. In particular,

$$\| h_0 \|_1 \geqslant \| f \|_1^\lambda \, \| g \|_1^{1-\lambda}. \tag{3.2}$$

Proof. It is sufficient to consider bounded functions f and g, since any f, g can be approximated from below in L^1 by bounded functions. Now define

$$F(x) = f(x)/\| f \|_\infty\,; \qquad G(x) = g(x)/\| g \|_\infty\,.$$

Let us first consider the case $\alpha \neq 0$. Then

$$\begin{aligned} h_\alpha(x \mid f, g) &= \operatorname*{ess\,sup}_{y \in \mathbf{R}} \left\{ \lambda \| f \|_\infty^\alpha F\left(\frac{x - y}{\lambda}\right)^\alpha \oplus (1 - \lambda) \| g \|_\infty^\alpha G\left(\frac{y}{1 - \lambda}\right)^\alpha \right\}^{1/\alpha} \\ &= [\lambda \| f \|_\infty^\alpha + (1 - \lambda) \| g \|_\infty^\alpha]^{1/\alpha} \\ &\quad + \operatorname*{ess\,sup}_{y \in \mathbf{R}} \left\{ \theta F\left(\frac{x - y}{\lambda}\right)^\alpha \oplus (1 - \theta)\, G\left(\frac{y}{1 - \lambda}\right)^\alpha \right\}^{1/\alpha}, \end{aligned}$$

with the obvious meaning of θ, $0 < \theta < 1$. Thus

$$h_\alpha(x \mid f, g) \geqslant [\lambda \| f \|_\infty^\alpha + (1 - \lambda) \| g \|_\infty^\alpha]^{1/\alpha}\, h_{-\infty}(x \mid F, G),$$

and by Theorem 1

$$\| h_\alpha \|_1 \geqslant [\lambda \| f \|_\infty^\alpha + (1 - \lambda) \| g \|_\infty^\alpha]^{1/\alpha} \left[\lambda \frac{\| f \|_1}{\| f \|_\infty} + (1 - \lambda) \frac{\| g \|_1}{\| g \|_\infty} \right] \tag{3.3}$$

Now Eq. (3.1) for $-1 \leqslant \alpha < 0$ or $0 < \alpha \leqslant \infty$ follows by Hölder's inequality.

For $\alpha = 0$,

$$h_0(f, g) = \|f\|_\infty^\lambda \|g\|_\infty^{1-\lambda} h_0(F, G) \geqslant \|f\|_\infty^\lambda \|g\|_\infty^{1-\lambda} h_{-\infty}(F, G).$$

Then Theorem 1 gives

$$\|h_0\|_1 \geqslant \|f\|_\infty^\lambda \|g\|_\infty^{1-\lambda} \left[\lambda \frac{\|f\|_1}{\|f\|_\infty} + (1-\lambda) \frac{\|g\|_1}{\|g\|_\infty}\right], \tag{3.4}$$

and Eq. (3.2) follows by the arithmetic-geometric mean inequality. Q.E.D.

Remarks. 1. Equation (3.3) (supplemented with Eq. (3.4) for $\alpha = 0$) holds for all $\alpha \in [-\infty, \infty]$. The restriction $\alpha \geqslant -1$ arises from the final application of Hölder's inequality.

2. Theorem 3.2 does not hold if $\alpha > 0$, $\|f\|_1 = 0$, $\|g\|_1 > 0$; in that case $h_\alpha = 0$. Analogously, the extended Brunn–Minkowski theorem [Eq. (2.6)] is not true if A or B has measure zero.

The n-dimensional version of Theorem 3.2 reads thus.

THEOREM 3.3. *Let f, g be nonnegative measurable functions on $\mathbf{R}^n$ and define h_α as in Eqs.* (2.1–2.3). *Let* $\|f\|_1 > 0$, $\|g\|_1 > 0$. *Then for* $\alpha \geqslant -1/n$,

$$\|h_\alpha\|_1 \geqslant \{\lambda \|f\|_1^\gamma + (1-\lambda)\|g\|_1^\gamma\}^{1/\gamma}, \tag{3.5}$$

with $\gamma = \alpha/(1 + n\alpha)$. *In particular,*

$$\|h_0\|_1 \geqslant \|f\|_1^\lambda \|g\|_1^{1-\lambda}.$$

Proof. Write $\mathbf{R}^n \ni x = (y, z)$, with $y \in \mathbf{R}$, $z \in \mathbf{R}^{n-1}$. Define

$$F(z) = \int dy\, f(y, z); \qquad G(z) = \int dy\, g(y, z). \tag{3.6}$$

Since

$$h_\alpha(y, z \mid f, g) = \operatorname*{ess\,sup}_{w \in \mathbf{R}^{n-1}} \operatorname*{ess\,sup}_{v \in \mathbf{R}} \left\{\lambda f\left(\frac{y-v}{\lambda}, \frac{z-w}{\lambda}\right)^\alpha \oplus (1-\lambda) g\left(\frac{v}{1-\lambda}, \frac{w}{1-\lambda}\right)^\alpha\right\}^{1/\alpha},$$

it follows from Theorem 3.2 that

$$\int dy\, h_\alpha(y, z \mid f, g) \geqslant h_\beta(z \mid F, G), \tag{3.7}$$

with $\beta = \alpha/(\alpha + 1)$. Note, that we used that

$$\int dy \operatorname*{ess\,sup}_w \geqslant \operatorname*{ess\,sup}_w \int dy.$$

Note further, that Theorem 3.2 does not apply, if z and w are such that $F((z - w)/\lambda) = 0$ or $G(z/(1 - \lambda)) = 0$. However, Eq. (3.7) is saved by the $\oplus$ sign in the definition of h_β [cf. Eq. (2.2)].

If we assume Theorem 3.3 to be true for $n - 1$, we have that

$$\| h_\beta(F, G)\|_1 \geqslant \{\lambda \| F \|_1^\gamma + (1 - \lambda) \| G \|_1^\gamma\}^{1/\gamma},$$

with $\gamma = \beta/[1 + (n - 1)\beta] = \alpha/(1 + n\alpha)$.

With Eqs. (3.6, 3.7) and Fubini's theorem, this leads to Eq. (3.5). Thus Theorem 3.3 is proved by induction. Q.E.D.

As an introduction to two corollaries of Theorem 3.3, let us define the classes of functions $K_\alpha(\mathbf{R}^n)$.

DEFINITION. $K_\alpha(\mathbf{R}^n)$ consists of the nonnegative, measurable functions F on $\mathbf{R}^n$ such that for all $\lambda \in (0, 1)$

$$F = h_\alpha(F, F) \quad \text{a.e.}$$

In more pedestrian terms, this means that F has the following convexity properties (apart from null functions).

$\alpha = -\infty$: F is unimodal, i.e., the sets $\{z \mid F(x) > z\}$ are convex.

$-\infty < \alpha < 0$: F^α is convex.

$\alpha = 0$: F is logarithmically concave, i.e.,

$$F(\lambda x + (1 - \lambda) y) \geqslant F(x)^\lambda F(y)^{1-\lambda}.$$

$0 < \alpha < \infty$: F^α is concave on a convex set, and $F(x) = 0$ outside this set.

$\alpha = \infty$: $F(x) =$ const. on a convex set, and $F(x) = 0$ outside this set.

Note, that $K_\alpha \subset K_\beta$ if $\alpha > \beta$. This follows from Jensen's inequality.

COROLLARY 3.4. *Let A, B be measurable sets in $\mathbf{R}^n$ of positive measure, and let*

$$C = \text{ess}\{\lambda A + (1 - \lambda)B\}.$$

Let $F \in K_\alpha(\mathbf{R}^n)$, $\alpha \geqslant -1/n$, *and let*

$$\mu_F(A) = \int_A F(x)\, dx.$$

Then, with $\gamma = \alpha/(1 + n\alpha)$,

$$\mu_F(C) \geqslant \{\lambda \mu_F(A)^\gamma + (1 - \lambda)\, \mu_F(B)^\gamma\}^{1/\gamma}.$$

In particular, if F is log concave,

$$\mu_F(C) \geqslant \mu_F(A)^\lambda\, \mu_F(B)^{1-\lambda}. \tag{3.8}$$

Proof. Let $f = F\chi_A$ and $g = F\chi_B$. Then $h_\alpha(f, g) \leqslant \chi_C h_\alpha(F, F) = \chi_C F$. Apply Theorem 3.3 to complete the proof. Q.E.D.

EXAMPLES. (1) Let $F(x) \equiv 1 \in K_\infty$. Then $\gamma = 1/n$ and we recover the Brunn–Minkowski theorem, Eq. (2.6).

(2) Let $G(x) = \exp(-x^2) \in K_0$. Then in any $\mathbf{R}^n$

$$\mu_G(C) \geqslant \mu_G(A)^\lambda\, \mu_G(B)^{1-\lambda}.$$

(3) Let $L(x) = (1 + x^2)^{-1} \in K_{-1/2}$. Then

$$\mu_L(C) \geqslant \{\lambda \mu_L(A)^{-1} + (1 - \lambda)\, \mu_L(B)^{-1}\}^{-1}, \qquad \text{in } \mathbf{R},$$

$$\mu_L(C) \geqslant \min\{\mu_L(A), \mu_L(B)\}, \qquad \text{in } \mathbf{R}^2.$$

COROLLARY 3.5. *Let* $F(x, y) \in K_\alpha(\mathbf{R}^{m+n})$, $x \in \mathbf{R}^m$, $y \in \mathbf{R}^n$. *Let*

$$G(x) = \int_{\mathbf{R}^n} F(x, y)\, dy.$$

Then $G \in K_\gamma(\mathbf{R}^m)$, $\gamma = \alpha/(1 + n\alpha)$. *In particular, if F is log concave, so is G.*

Proof. Since $F(x, y) > 0$ on a convex set in $\mathbf{R}^{m+n}$, $G(x) > 0$ on a convex set in $\mathbf{R}^m$. Now fix points x_0, x_1 in this set, and define $f(y) = F(x_1, y)$, $g(y) = F(x_0, y)$. Then

$$F(\lambda x_1 + (1 - \lambda)\, x_0, y) \geqslant h_\alpha(y \mid f, g).$$

Now apply Theorem 3.3 to $h_\alpha(y \mid f, g)$. Q.E.D.

4. Log Concave Functions and Measures

In this section we prove a Sobolev-type inequality (Theorem 4.1) for log concave measures (i.e., measures given by a log concave density function). We shall write $F(x) \equiv \exp[-f(x)]$, $x \in \mathbf{R}^n$; $F(x)$ is log concave iff $f(x)$ is convex. If $f(x)$ is twice continuously differentiable, this means that the second derivatives matrix, f_{xx}, is nonnegative.

It is often convenient to write $\mathbf{R}^{n+m} \ni x = (y, z)$, $y \in \mathbf{R}^m$, $z \in \mathbf{R}^n$. The matrix f_{xx} is then partitioned in an obvious way as

$$f_{xx} = \begin{pmatrix} f_{yy} & f_{yz} \\ f_{zy} & f_{zz} \end{pmatrix}. \tag{4.1}$$

We shall often encounter

$$G(y) = \exp[-g(y)] \equiv \int_{\mathbf{R}^n} F(y, z)\, dz. \tag{4.2}$$

Then $G(y)$ is log concave by Corollary 3.5. A sharper form of this result will be given in Theorem 4.2.

With F as a density function, define

$$\begin{aligned} \langle A \rangle &= \int_{\mathbf{R}^n} A(x)\, F(x)\, dx \Big/ \int_{\mathbf{R}^n} F(x)\, dx, \\ \operatorname{var} A &= \langle | A - \langle A \rangle |^2 \rangle, \\ \operatorname{cov}(A, B) &= \langle (\bar{A} - \langle \bar{A} \rangle)(B - \langle B \rangle) \rangle. \end{aligned} \tag{4.3}$$

If $x = (y, z)$, $y \in \mathbf{R}^m$, $z \in \mathbf{R}^n$, we write

$$\begin{aligned} \langle A \rangle_z (y) &= \int_{\mathbf{R}^n} A(y, z)\, F(y, z)\, dz \Big/ \int_{\mathbf{R}^n} F(y, z)\, dz, \\ \langle B \rangle_y &= \int_{\mathbf{R}^m} B(y)\, G(y)\, dy \Big/ \int_{\mathbf{R}^m} G(y)\, dy, \end{aligned} \tag{4.4}$$

so that $\langle A \rangle = \langle \langle A \rangle_z \rangle_y$. In analogy with Eq. (4.3), var_y, cov_y, var_z, and cov_z are defined.

Theorem 4.1. *Let $F(x) = \exp[-f(x)]$, $x \in \mathbf{R}^n$, let f be twice continuously differentiable and let f be strictly convex. Let f have a minimum, so that F decreases exponentially in all directions; then*

$$\int_{\mathbf{R}^n} F(x)\, dx < \infty.$$

Let $h \in C^1(\mathbf{R}^n)$, *and let* $\operatorname{var} h < \infty$. *Then*

$$\operatorname{var} h \leqslant \langle (h_x, (f_{xx})^{-1} h_x) \rangle, \tag{4.5}$$

where the inner product is with respect to $\mathbf{C}^n$, *and* h_x *denotes the gradient of* h.

It is convenient to postpone the proof of Theorem 4.1 a moment. We prefer to give an immediate corollary first.

THEOREM 4.2. *Let* $F(x) = F(y, z) = \exp[-f(y, z)]$, $y \in \mathbf{R}^m$, $z \in \mathbf{R}^n$, *satisfy the assumptions of Theorem* 4.1. *Moreover, let the integrals*

$$\int_{\mathbf{R}^n} (\phi, f_{yy}\phi) F \, dz, \qquad \int_{\mathbf{R}^n} (\phi, f_y)^2 F \, dz \tag{4.6}$$

converge uniformly in y *in a neighborhood of a given point* $y_0 \in \mathbf{R}^m$, *for all vectors* $\phi \in \mathbf{R}^m$. *Then, with the notation of Eqs.* (4.1, 4.2, 4.4), $g(y)$ *is twice continuously differentiable near* y_0, *and*

$$g_{yy} \geqslant \langle f_{yy} - f_{yz}(f_{zz})^{-1} f_{zy} \rangle_z \tag{4.7}$$

as a matrix inequality.

Proof. We denote differentiation in a direction t at y_0 by a subscript t. Then Eq. (4.7) is equivalent to saying that for all directions t

$$g_{tt} \geqslant \langle f_{tt} - f_{tz}(f_{zz})^{-1} f_{zt} \rangle_z .$$

By differentiating $g(y) = \log \int F(y, z) \, dz$, one gets

$$g_{tt} = \langle f_{tt} \rangle_z - \operatorname{var}_z f_t . \tag{4.8}$$

The differentiation can be done under the integral sign by the uniform convergence of the integrals (4.6), which also ensures the continuity of g_{tt}.

The result (4.7) follows by applying Theorem 4.1 with $h(z) = f_t(y_0, z)$. Q.E.D.

Remark. Even though F is assumed to be a log concave function, decreasing exponentially in all directions, the convergence of the integrals (4.6) does not follow automatically. For example, define the convex function $\phi(x)$, $x \in \mathbf{R}$, by $\phi(0) = \phi'(0) = 0$, and

$$\phi''(x) = \sum_{n \neq 0} a_n \delta(x - n), \qquad a_n > 0, \quad a_n = a_{-n} .$$

Then

$$\int \phi''(x) \exp[-\phi(x)]\, dx = 2 \sum_{n=1}^{\infty} a_n \exp\left[-\sum_{k=1}^{n-1} (n-k)\, a_k\right],$$

which can be made divergent by an appropriate recursive definition of a_n. If we take

$$f(y, z) = y^2 + \phi(y + z), \qquad y, z \in \mathbf{R},$$

the integrals (4.6) obviously diverge for all y.

The function ϕ can be approximated by a C^2 function without changing the conclusion.

Proof of Theorem 4.1. We can obviously restrict h to be real valued. Let us first give the proof for $\mathbf{R}^1$. If $f(x)$ has its unique minimum at $x = a$, write

$$h(x) - h(a) = f'(x)\, k(x).$$

Then $k(x)$ is continuously differentiable, except possibly at $x = a$. However, if we set $k(a) = h'(a)/f''(a)$, k is continuous at $x = a$. Now

$$\begin{aligned}\int (h')^2/f''F\, dx &= \int [(k'f')^2/f'' + 2kk'f' + k^2f'']F\, dx\\ &= \int [(k'f')^2/f'' + (kf')^2]F\, dx + [k^2f'F]_{-\infty}^{a} + [k^2f'F]_a^{\infty}\\ &\geqslant \int [h(x) - h(a)]^2\, F(x)\, dx.\end{aligned}$$

Equation (4.5) follows by noting that

$$\operatorname{var} h \leqslant \langle [h - h(a)]^2 \rangle.$$

Now assume that Theorem 4.1 has been proved for $x \in \mathbf{R}^{n-1}$. Hence we also have Theorem 4.2 for $z \in \mathbf{R}^{n-1}$ at our disposition.

Write $\mathbf{R}^n \ni x = (y, z)$, $y \in \mathbf{R}$, $z \in \mathbf{R}^{n-1}$. Then

$$\operatorname{var} h = \langle \operatorname{var}_z h \rangle_y + \operatorname{var}_y \langle h \rangle_z,$$

with the notation of Eqs. (4.3, 4.4).

Let us first restrict ourselves to functions h with compact support. This has the advantage that F can be modified outside the support of h in such a way, that it satisfies all the assumptions of Theorem 4.2

for all y. Then $G(y) = \int F(y, z)\, dz$ satisfies the assumptions of Theorem 4.1, so that

$$\operatorname{var}_y \langle h \rangle_z \leqslant \langle ((d/dy)\langle h \rangle_z)^2 / g'' \rangle_y .$$

Now all differentiations can be carried out under the integral signs, since h has compact support and F has been appropriately modified. Thus we find (cf. Eq. (4.8))

$$\operatorname{var} h \leqslant \langle B \rangle_y ,$$

$$B = \operatorname{var}_z h + \frac{[\langle h_y \rangle_z - \operatorname{cov}_z(h, f_y)]^2}{\langle f_{yy} \rangle_z - \operatorname{var}_z f_y} . \tag{4.9}$$

Applying Theorem 4.1 for $z \in \mathbf{R}^{n-1}$, with fixed $y \in \mathbf{R}$, we have

$$\operatorname{var}_z H \leqslant \langle (H_z , f_{zz}^{-1} H_z) \rangle_z .$$

Since this is true for

$$H = \lambda h + \mu f_y$$

with arbitrary λ and μ, we get

$$B \leqslant \langle (h_z , f_{zz}^{-1} h_z) \rangle_z + \frac{\langle h_y - (h_z , f_{zz}^{-1} f_{zy}) \rangle_z^2}{\langle f_{yy} - (f_{yz} , f_{zz}^{-1} f_{zy}) \rangle_z} .$$

Since f is convex, the denominator above is positive and we can use Schwarz's inequality to obtain

$$B \leqslant \left\langle (h_z , f_{zz}^{-1} h_z) + \frac{[h_y - (h_z , f_{zz}^{-1} f_{zy})]^2}{f_{yy} - (f_{yz} , f_{zz}^{-1} f_{zy})} \right\rangle_z$$

$$= \langle (h_x , f_{xx}^{-1} h_x) \rangle_z . \tag{4.10}$$

Eq. (4.5) follows by combining Eqs. (4.9) and (4.10).

Now only the restriction that h has compact support remains to be removed. As an intermediate step, let us show that for all h and F satisfying the assumptions of Theorem 4.1

$$\operatorname{var}_S h \leqslant \langle (h_x , f_{xx}^{-1} h_x) \rangle_S , \tag{4.11}$$

where the averages are taken over a ball with radius S centered at the origin, instead of over all $\mathbf{R}^n$.

Modify h outside the ball smoothly to a function k with compact support, and let

$$f^{(N)}(x) = f(x), \qquad \text{if } |x| \leqslant S;$$

$$f^{(N)}(x) = f(x) + N(|x| - S)^4, \qquad \text{if } |x| \geqslant S.$$

By our results until now, we have that

$$\operatorname{var}_N k \leqslant \langle (k_x, (f_{xx}^{(N)})^{-1} k_x) \rangle_N,$$

with averages with respect to the weight $\exp[-f^{(N)}(x)]$. Equation (4.11) is proved by taking the limit $N \to \infty$ and using the monotone convergence theorem.

Now let $S \to \infty$ in Eq. (4.11). Then $\operatorname{var}_S h \to \operatorname{var} h$, and

$$\int_S (h_x, f_{xx}^{-1} h_x) F\, dx$$

increases (it may actually increase to ∞). This concludes the proof. Q.E.D.

EXAMPLES. 1. Let $M_{ij} = \operatorname{cov}(x_i, x_j)$. Then we have the matrix inequality

$$M \leqslant \langle (f_{xx})^{-1} \rangle, \tag{4.12}$$

as can be seen by taking $h(x) = (\phi, x)$ for any $\phi \in \mathbf{R}^n$ in Theorem 4.1. As a curiosity, compare (4.12) with the one dimensional inequality

$$\operatorname{var} x \geqslant \langle f'' \rangle^{-1}, \tag{4.13}$$

which holds for general weights F. The proof is

$$1 = [\operatorname{cov}(x, f')]^2 \leqslant \operatorname{var} f' \operatorname{var} x = \langle f'' \rangle \operatorname{var} x,$$

with Schwarz's inequality and two integrations by parts.

2. For the Gaussian weight $F(x) = \exp[-(x, Ax)]$,

$$\operatorname{var} h \leqslant \langle (h_x, (2A)^{-1} h_x) \rangle. \tag{4.14}$$

In particular, if $F(x) = \exp[-(x, x)/2]$,

$$\operatorname{var} h \leqslant \langle |h_x|^2 \rangle. \tag{4.15}$$

3. If $F(x) = \exp[-(x, Ax)]$, $M = (2A)^{-1}$, and thus the inequality in (4.12) holds as an equality.

4. The analog of Example 3 in the setting of Theorem 4.2 concerns the Gaussian

$$\Phi(x, y) = \exp\left[-(x, y)\begin{pmatrix} A & B \\ B^* & C \end{pmatrix}\begin{pmatrix} x \\ y \end{pmatrix}\right], \qquad (x, y) \in \mathbf{R}^m \times \mathbf{R}^n, \quad (4.16)$$

with a real, positive matrix $\left(\begin{smallmatrix} A & B \\ B^* & C \end{smallmatrix}\right)$. Then

$$\int \Phi(x, y)\, dy = \text{const.} \exp[-(x, Dx)], \qquad (4.17)$$

with

$$D = B - BC^{-1}B^*. \qquad (4.18)$$

Thus for Gaussians the equality sign in Eq. (4.7) holds.

THEOREM 4.3. *With the notation of Eqs.* (4.16–4.18), *let* $G(x)$ *be defined by*

$$\int \Phi(x, y)\, F(x, y)\, dy = G(x) \exp[-x, Dx)].$$

Then, if $F(x, y)$ *is log concave,* $G(x)$ *is log concave; if* $F(x, y)$ *is log convex,* $G(x)$ *is log convex.*

Proof. Write

$$\Phi(x, y) = \exp[-(x, Dx) - (y', Cy')],$$
$$y' = y + C^{-1}B^*x.$$

Then

$$G(x) = \int \exp[-(y, Cy)]\, F(x, y - C^{-1}B^*x)\, dy. \qquad (4.19)$$

If $F(x, y)$ is log concave, the integrand in Eq. (4.19) is log concave. Then $G(x)$ is log concave by Corollary 3.5. If $F(x, y)$ is log convex, the integrand is log convex in x for all fixed y. Then $G(x)$ is log convex by Hölder's inequality. Q.E.D.

Note, that the log concave part of Theorem 4.3 also follows from Theorem 4.2.

5. Moment Inequalities

Theorem 5.1. *Let $F(x)$ be a nonnegative function on* $\mathbf{R}^n$, *and let A be a real, positive definite, $n \times n$ matrix. Assume* $\exp[-(x, Ax)]\, F(x) \in L^1$ *and define*

$$\langle k \rangle_F = \int k(x) \exp[-(x, Ax)]\, F(x)\, dx \Big/ \int \exp[-(x, Ax)]\, F(x)\, dx.$$

If $F(x) \equiv 1$ we write $\langle \cdot \rangle_1$. Let $\phi \in \mathbf{R}^n$, $a \in \mathbf{R}$. Then

$$\langle |(\phi, x) - \langle(\phi, x)\rangle_F|^\alpha \rangle_F \leqslant \langle |(\phi, x)|^\alpha \rangle_1,$$

when F is log concave and $\alpha \geqslant 1$;

$$\langle |(\phi, x) - a|^\alpha \rangle_F \leqslant \langle |(\phi, x)|^\alpha \rangle_1, \qquad \text{if} \quad \alpha > 0,$$

$$\langle |(\phi, x) - a|^\alpha \rangle_F \geqslant \langle |(\phi, x)|^\alpha \rangle_1, \qquad \text{if} \quad -1 < \alpha < 0,$$

when F is log convex.

Proof. By a linear transformation such that $(\phi, x) \to x_1$ and by Theorem 4.3 it suffices to prove Theorem 5.1 for the one-dimensional case. This will be done in Lemmas 5.2 and 5.3. Q.E.D.

Lemma 5.2. *Let $F(x)$ be a log convex function on $\mathbf{R}$, and let the averages $\langle \cdot \rangle_F$ and $\langle \cdot \rangle_1$ be computed with the weights* $\exp(-x^2)\, F(x)$ *and* $\exp(-x^2)$, *respectively. Let $a \in \mathbf{R}$. Then*

$$\langle |x - a|^\alpha \rangle_F \geqslant \langle |x|^\alpha \rangle_1, \qquad \text{if} \quad \alpha > 0; \tag{5.1}$$

$$\langle |x - a|^\alpha \rangle_F \leqslant \langle |x|^\alpha \rangle_1, \qquad \text{if} \quad -1 < \alpha < 0. \tag{5.2}$$

Proof. Note that

$$\langle |x - a|^\alpha \rangle_F = \langle |x|^\alpha \rangle_G = \langle |x|^\alpha \rangle_H,$$

where

$$G(x) = F(x + a) \exp(-2ax),$$

$$H(x) = G(x) + G(-x).$$

Since F is log convex, G and H are log convex; moreover, H is even. Thus, for $\alpha > 0$, it has to be shown that

$$\langle x^\alpha H(x) \rangle \geqslant \langle x^\alpha \rangle \langle H(x) \rangle, \tag{5.3}$$

with the averages computed over $x > 0$ with the weight $\exp(-x^2)$. But this is equivalent to the inequality

$$\int_0^\infty \int_0^\infty dx\, dy \exp(-x^2 - y^2)[H(x) - H(y)](x^\alpha - y^\alpha) \geqslant 0, \tag{5.4}$$

which is obvious, since $H(x)$ and x^α are increasing functions for $x > 0$.

If $-1 < \alpha < 0$, x^α is decreasing for $x > 0$, and hence

$$\langle x^\alpha H(x)\rangle \leqslant \langle x^\alpha\rangle\langle H(x)\rangle.$$

This proves Eq. (5.2). Q.E.D.

LEMMA 5.3. *Let $F(x)$ be a log concave function on* **R**. *Then, with the notation of Lemma* 5.2,

$$\langle | x - \langle x\rangle_F |^\alpha\rangle_F \leqslant \langle | x |^\alpha\rangle_1, \qquad \text{if} \quad \alpha \geqslant 1. \tag{5.5}$$

Proof. Write

$$\langle | x - \langle x\rangle_F |^\alpha\rangle_F = \langle | x |^\alpha\rangle_G,$$

with

$$G(x) = F(x + \langle x\rangle_F) \exp(-2x\langle x\rangle_F).$$

Then $G(x)$ is log concave, and $\langle x\rangle_G = 0$. By approximation, it is sufficient to assume $G \in C^1$. Hence

$$\int dx \exp(-x^2)\, G'(x) = 2 \int dx\, x \exp(-x^2)\, G(x) = 0. \tag{5.6}$$

Moreover, there must exist a number K such that $G(x)$ is increasing for $x < K$; decreasing for $x > K$. By Eq. (5.6) K must be finite and we can assume that $K \geqslant 0$, say. Then $G'(x) \geqslant 0$ for $x < 0$, and Eq. (5.6) implies that

$$\int_0^\infty dx \exp(-x^2)\, G'(x) \leqslant 0. \tag{5.7}$$

It has to be shown that

$$\langle x^\alpha[G(x) + G(-x)]\rangle \leqslant \langle x^\alpha\rangle\langle G(x) + G(-x)\rangle, \tag{5.8}$$

where the averages are with respect to $\exp(-x^2)$, $x > 0$.

We assumed, that $G'(x) \geqslant 0$ for $x < 0$, and thus (cf. Eqs. (5.3, 5.4)]

$$\langle x^\alpha G(-x)\rangle \leqslant \langle x^\alpha\rangle\langle G(-x)\rangle.$$

We wish to show the same inequality for the $G(x)$ part in Eq. (5.8), which is equivalent to

$$\int_0^\infty dx \int_0^x dy \exp(-x^2 - y^2)[G(x) - G(y)](x^\alpha - y^\alpha) \leqslant 0. \tag{5.9}$$

If we write

$$G(x) - G(y) = \int_y^x G'(z)\, dz,$$

Eq. (5.9) becomes

$$\int_0^\infty dz\, \psi(z) \exp(-z^2)\, G'(z) \leqslant 0, \tag{5.10}$$

$$\psi(z) = \exp(z^2) \int_z^\infty dx \int_0^z dy \exp(-x^2 - y^2)(x^\alpha - y^\alpha). \tag{5.11}$$

If we manage to show that $\psi(z)$ is an increasing function for $z > 0$, Eq. (5.10) follows from Eq. (5.7) and the fact that $G'(x) \geqslant 0$ for $0 < x < K$; $G'(x) \leqslant 0$ for $x > K$, and Lemma 5.3 is proved.

After some manipulation, we find that

$$\begin{aligned}\psi'(z) = {} & \int_z^\infty dx \exp(-x^2)(x^\alpha - z^\alpha) \\ & + z \exp(z^2) \int_z^\infty dx \int_0^\infty dy \exp(-x^2 - y^2)[(\alpha - 1)\, x^{\alpha-2} + y^\alpha x^{-2}].\end{aligned}$$

Thus, if $\alpha \geqslant 1$, $\psi'(z) > 0$. Q.E.D.

Remark. Here, as well as in Theorem 4.3, the log convex case is much simpler than the log concave case. We leave as an open question, the correct generalization of Eq. (5.5) when $-1 < \alpha < 1$. If $F(x)$ is symmetric decreasing, which implies that $\langle x \rangle_F = 0$ but does not imply that F is log concave, then Eq. (5.5) trivially generalizes to

$$\begin{aligned} \langle | x |^\alpha \rangle_F &\leqslant \langle | x |^\alpha \rangle_1, \qquad \text{if} \quad \alpha > 0; \\ \langle | x |^\alpha \rangle_F &\geqslant \langle | x |^\alpha \rangle_1, \qquad \text{if} \quad -1 < \alpha < 0; \end{aligned}$$

cf. the proof of Lemma 4.2 and especially Eqs. (5.3, 5.4).

THEOREM 5.4. *Under the assumptions of Theorem* 5.1, *let M be the covariance matrix*

$$M_{ij} = \langle x_i x_j \rangle_F - \langle x_i \rangle_F \langle x_j \rangle_F .$$

Then

$$\begin{aligned} M &\leqslant \langle (2A + f_{xx})^{-1} \rangle_F \leqslant (2A)^{-1}, \qquad \textit{if} \quad F \equiv \exp(-f) \textit{ is log concave}; \\ M &\geqslant (2A)^{-1}, \qquad \textit{if} \quad F \textit{ is log convex}. \end{aligned} \tag{5.12}$$

Proof. Setting $\alpha = 2$ in Theorem 5.1 leads to $M \leqslant (2A)^{-1}$ resp. $M \geqslant (2A)^{-1}$. The stronger inequality (5.12) is obtained from Theorem 4.1 by taking $h(x) = (\phi, x)$ and replacing the weight $F(x)$ by $\exp[-(x, Ax)]\, F(x)$. Q.E.D.

6. The Diffusion Equation

Consider the diffusion equation in $\mathbf{R}^n$

$$\partial\psi/\partial t = -H_A\psi \tag{6.1}$$

with the Hamiltonian

$$(H_A\psi)(x) = -\tfrac{1}{2}(\Delta\psi)(x) + V(x)\,\psi(x), \tag{6.2}$$

defined on an open, connected region $A \subset \mathbf{R}^n$, with zero boundary conditions. The potential $V(x)$ is assumed to be convex; in particular, $V(x)$ may be ∞ outside a convex set D. Further we assume the region A to be such that

$$\int_A \exp[-tV(x)]\, dx < \infty, \qquad \forall t > 0. \tag{6.3}$$

(This means that A is bounded in the directions, for which $V(x)$ does not go to ∞ as $|x| \to \infty$.)

The fundamental solution $G_A(x, y; t)$ of Eq. (6.1) is defined by

$$\begin{aligned}
((\partial/\partial t) - H_{A,x})\, G_A(x, y; t) &= 0, && x, y \in A \cap D,\ t > 0;\\
G_A(x, y; 0) &= \delta(x - y), && x, y \in A \cap D;\\
G_A(x, y; t) &= 0, && x \in \partial(A \cap D);\\
G_A(x, y; t) &= 0, && x \notin A \cap D \text{ or } y \notin A \cap D.
\end{aligned}$$

We could, of course, replace A by $A \cap D$ without changing G_A, but the point is that in Theorem 6.2 we want to vary A while keeping D fixed.

Using the Trotter product formula, we can write

$$G_A(x, y; t) = \lim_{M\to\infty} \left(\frac{2\pi t}{M}\right)^{-nM/2} \int_A dx_1 \cdots \int_A dx_{M-1} \times \prod_{j=1}^{M} \exp\left[-\frac{M}{2t}(x_j - x_{j-1})^2 - \frac{t}{M} V(x_j)\right], \tag{6.4}$$

where $x_0 = x$, $x_M = y$.

Define the partition function by

$$Z_A(t) = \operatorname{Tr} \exp(-tH_A) = \int_A G_A(x, x; t)\, dx. \tag{6.5}$$

Then Eq. (6.3) guarantees, that $Z_A(t) < \infty$ for all $t > 0$, so that H_A has a pure point spectrum. In fact, Hölder's inequality applied to Eqs. (6.4, 6.5) gives that

$$\begin{aligned} Z_A(t) &\leqslant \int_A G^0(x, x; t) \exp[-tV(x)]\, dx \\ &= (2\pi t)^{-n/2} \int_A \exp[-tV(x)]\, dx, \end{aligned}$$

where G^0 is the fundamental solution of Eq. (6.1) with $V(x) = 0$. Moreover the ground state is nondegenerate and the corresponding eigenfunction is nonnegative [9].

THEOREM 6.1. *Let $A = \mathbf{R}^n$, and let the potential be of the form*

$$V(x) = \tfrac{1}{2}\omega^2 x^2 + W(x), \qquad \omega \geqslant 0, \tag{6.6}$$

with a convex function $W(x)$. Then the ground state wave function $\psi_0(x)$ is of the form

$$\psi_0(x) = \exp(-\tfrac{1}{2}\omega x^2)\, \phi(x),$$

where $\phi(x)$ is log concave.

Proof. Let $G_\omega(x, y; t)$ be the fundamental solution of Eq. (6.1) for $V(x) = \frac{1}{2}\omega^2 x^2$. Then the fundamental solution for the potential (6.6) is of the form

$$G(x, y; t) = G_\omega(x, y; t)\, H(x, y; t),$$

where $H(x, y, t)$ is log concave in (x, y) for all t. This follows directly from Theorem 4.3 applied to Eq. (6.4).

If ϵ is the ground state energy,

$$\psi_0(x)\, \psi_0(y) = \lim_{t\to\infty} G(x, y; t) \exp(\epsilon t).$$

Since the pointwise limit of log concave functions is log concave, the theorem follows. Q.E.D.

Remark. If $W(x)$ is concave instead of convex, (but such that Eq. (6.3) still holds), the log convex part of Theorem 4.3 implies in the same way as above that $\phi(x)$ is log convex.

THEOREM 6.2. *Let A and B be open, connected regions, let $C = \lambda A + (1-\lambda)B$, and let $V(x)$ be convex. Then*

$$Z_C(t) \geqslant Z_A(t)^\lambda \, Z_B(t)^{1-\lambda}; \tag{6.7}$$

$$\epsilon_C \leqslant \lambda\epsilon_A + (1-\lambda)\,\epsilon_B, \tag{6.8}$$

where $\epsilon_A(\epsilon_B, \epsilon_C)$ is the ground state energy of $H_A(H_B, H_C)$.

Proof. Equations (6.4, 6.5) together give an expression for the partition function. We note, that we can apply Corollary 3.4 to the sets A^M, B^M, and C^M. This proves Eq. (6.7). Further

$$\epsilon_A = -\lim_{t\to\infty} t^{-1} \log Z_A(t),$$

which gives Eq. (6.8). Q.E.D.

APPENDIX

THEOREM A.1. *For measurable sets A and $B \subset \mathbf{R}^n$, define the essential sum $C = \text{ess}\{A+B\}$ as in Eq. (2.4). Then C is open, and*

$$\mu_n(C)^{1/n} \geqslant \mu_n(A)^{1/n} + \mu_n(B)^{1/n}. \tag{A.1}$$

THEOREM A.2. *For nonnegative, measurable functions $f(x)$ and $g(x)$ on $\mathbf{R}^n$, define*

$$H_\alpha(x \mid f, g) = \operatorname*{ess\,sup}_{y\in\mathbf{R}^n}\{f(x-y)^\alpha \oplus g(y)^\alpha\}^{1/\alpha}, \tag{A.2}$$

cf. Eqs. (2.1–2.3). *Then $H_\alpha(x)$ is lower semicontinuous in x for all α.*

Proof of Theorem A.1. All the above facts are based on the following observation: For an arbitrary measurable set $A \subset \mathbf{R}^n$, define

$$A^* = \{x \in \mathbf{R}^n \mid \mu_n[A \cap V(\epsilon, x)]/W_n(\epsilon) \to 1 \quad \text{for} \quad \epsilon \downarrow 0\}, \tag{A.3}$$

where $V(\epsilon, x)$ is the open ball of radius ϵ centered at x, and $W_n(\epsilon)$ is its volume. Then A^* is measurable and $\mu_n(A^* \varDelta A) = 0$, where $\varDelta$ means symmetric difference [2, Theorem 2.9.11]. Hence

$$\text{ess}(A+B) = \text{ess}(A^* + B^*), \tag{A.4}$$

and it is sufficient to prove the theorem when A and B are replaced by A^* and B^*.

Let $x \in A^* + B^*$, i.e., there is a point $y \in A^* \cap (x - B^*)$. Notice, that $A^{**} = A^*$; thus for some $\epsilon > 0$,

$$\mu_n[A^* \cap V(\epsilon, y)] \geqslant \tfrac{3}{4} W_n(\epsilon);$$

$$\mu_n[(x - B^*) \cap V(\epsilon, y)] \geqslant \tfrac{3}{4} W_n(\epsilon).$$

Hence, $\mu_n[A^* \cap (v - B^*)] > 0$ for all v in some neighborhood $V(\delta, x)$, which implies that $A^* + B^*$ is open, and that

$$A^* + B^* = \text{ess}(A^* + B^*). \tag{A.5}$$

Equation (A.1) now follows from Eqs. (A.4, A.5) and the Brunn–Minkowski theorem, Eq. (1.1). Q.E.D.

Proof of Theorem A.2. For a nonnegative, measurable function f, let

$$A_f = \{(x, z) \in \mathbf{R}^{n+1} \mid 0 < z < f(x)\}. \tag{A.6}$$

Define $A_f{}^*$ as in (A.3). If $(x, z) \in A_f{}^*$, $(x, t) \in A_f{}^*$ for all t, $0 < t < z$. Thus it makes sense to define

$$f^*(x) = \sup\{z \mid (x, y) \in A_f{}^*\}. \tag{A.7}$$

The supremum over the empty set is taken to be zero. Given f^*, define A_{f^*} according to definition (A.6). Clearly A_f, $A_f{}^*$ and f^* are all measurable. By (A.6) and (A.7), $A_f{}^* \supset A_{f^*}$. Since

$$A_f{}^* \backslash A_{f^*} \subset G \equiv \{(x, f^*(x)) \mid x \in \mathbf{R}^n\},$$

and since $\mu_{n+1}(G) = 0$, it follows that $\mu_{n+1}(A_f{}^* \backslash A_{f^*}) = 0$. In general, $\int p = \mu_{n+1}(A_p)$. Therefore

$$\begin{aligned} \int |f^* - f| \, dx &= \mu_{n+1}(A_{f^*} \Delta A_f) \\ &= \mu_{n+1}(A_f{}^* \Delta A_f) = 0. \end{aligned} \tag{A.8}$$

As a consequence of (A.8),

$$H_\alpha(f, g) = H_\alpha(f^*, g^*). \tag{A.9}$$

Now consider the function

$$K_\alpha(x \mid f, g) = \sup_{y \in \mathbf{R}^n} \{f(x - y)^\alpha \oplus g(y)^\alpha\}^{1/\alpha}. \tag{A.10}$$

Note that generally $K_\alpha(x) \geqslant H_\alpha(x)$. Let

$$D(z) = \{x \in \mathbf{R}^n \mid K_\alpha(x \mid f^*, g^*) > z\}, \qquad z \geqslant 0. \tag{A.11}$$

Choose $z \geqslant 0$, $x \in D(z)$. By definitions (A.10) and (A.11), there is a $y \in \mathbf{R}^n$, and numbers b, $c > 0$ such that

$$z \leqslant (b^\alpha + c^\alpha)^{1/\alpha},$$
$$f^*(x-y) > b, g^*(y) > c.$$

In other words

$$\beta \equiv (x-y, b) \in A_{f^*}, \qquad \gamma \equiv (y, c) \in A_{g^*}.$$

Then for all $\delta > 0$ there exist balls $V(\epsilon, \beta)$ and $V(\epsilon, \gamma)$ in $\mathbf{R}^{n+1}$ such that, in the notation of (A.3),

$$\mu_{n+1}(A_{f^*} \cap V(\epsilon, \beta)) \geqslant (1-\delta)\, W_{n+1}(\epsilon),$$
$$\mu_{n+1}(A_{g^*} \cap V(\epsilon, \gamma)) \geqslant (1-\delta)\, W_{n+1}(\epsilon).$$

If δ is small enough, it follows that the sets

$$\{v \in V(\epsilon, x-y) \mid f^*(v) > b\},$$
$$\{w \in V(\epsilon, y) \mid g^*(w) > c\}$$

have measure at least equal to $\frac{3}{4} W_n(\epsilon)$. This implies (1) that $H_\alpha(x \mid f^*, g^*) > z$, so that in fact

$$H_\alpha(f^*, g^*) = K_\alpha(f^*, g^*), \tag{A.12}$$

and (2) that $D(z)$ contains a neighborhood of x, such that $D(z)$ is open. Hence $K_\alpha(f^*, g^*)$ is lower semicontinuous. By Eqs. (A.9, A.12), so is $H_\alpha(f, g)$. Q.E.D.

References

1. L. Lusternik, Die Brunn-Minkowskische Ungleichung für beliebige messbare Mengen, *C. R. Dokl. Acad. Sci. URSS* No. **3**, 8 (1935), 55–58.
2. M. Federer, "Geometric Measure Theory," Springer, New York, 1969.
3. A. Prékopa, Logarithmic concave measures with application to stochastic programming, *Acta Sci. Math. (Szeged)*, **32** (1971), 301–315.
4. L. Leindler, On a certain converse of Hölder's inequality II, *Acta Sci. Math. (Szeged)* **33** (1972), 217–223.

5. A. Prékopa, On logarithmic concave measures and functions, *Acta Sci. Math. (Szeged)* **34** (1973), 335–343.
6. H. J. Brascamp and E. H. Lieb, Some inequalities for Gaussian measures, *in* "Functional Integral and its Applications" (A. M. Arthurs, Ed.), Clarendon Press, Oxford, 1975.
7. H. J. Brascamp and E. H. Lieb, Best constants in Young's inequality, its converse and its generalization to more than three functions, *Advances in Math.* **20** (1976).
8. Y. Rinott, On convexity of measures, Thesis, Weizmann Institute, Rehovot, Israel, November 1973, to appear.
9. B. Simon and R. Høegh-Krohn, Hypercontractive semigroups and two-dimensional self-coupled Bose fields, *J. Functional Analysis* **9** (1972), 121–180.

Note added in proof. After this paper was submitted for publication we discovered that Corollary 3.4 and its converse were proved by C. Borell:

C. Borell, Convex measures on locally convex spaces, *Ark. Mat.* **12** (1974), 239–252.

C. Borell, Convex set functions, *Period. Math. Hungar.* **6** (1975), 111–136.

Studies in Appl. Math. *57*, 93–105 (1977)

Existence and Uniqueness of the Minimizing Solution of Choquard's Nonlinear Equation

*By Elliott H. Lieb**

The equation dealt with in this paper is

$$\left\{-\Delta-2\int|\phi(y)|^2|x-y|^{-1}dy\right\}\phi=e\phi \qquad \text{in three dimensions.}$$

It comes from minimizing the functional

$$\mathcal{E}(\phi)=\int|\nabla\phi|^2dx-\int\int|\phi(x)|^2|x-y|^{-1}|\phi(y)|^2dx\,dy,$$

which, in turn, comes from an approximation to the Hartree-Fock theory of a plasma. It describes an electron trapped in its own hole. The interesting mathematical aspect of the problem is that $\mathcal{E}$ is not convex, and usual methods to show existence and uniqueness of the minimum do not apply. By using symmetric decreasing rearrangement inequalities we are able to prove existence and uniqueness (modulo translations) of a minimizing ϕ. To prove uniqueness a strict form of the inequality, which we believe is new, is employed.

I. Introduction

We consider the functional

$$\mathcal{E}(\phi)=\int|\nabla\phi(x)|^2dx-\int\int|\phi(x)|^2|x-y|^{-1}|\phi(y)|^2dx\,dy \tag{1.1}$$

on $W^1(\mathbf{R}^3)$, the space of functions on $\mathbf{R}^3$ such that $\|\nabla\phi\|_2$ and $\|\phi\|_2$ are finite.

*Work supported by U.S. National Science Foundation grant MCS 75-21684.

Published by Elsevier North-Holland, Inc.

Studies in Appl. Math. *57*, 93–105 (1977)

This functional arises in a certain approximation to Hartree-Fock theory for a one component plasma. Ph. Choquard proposed it for investigation at the Symposium on Coulomb Systems, Lausanne, July, 1976.

If one defines

$$E(\lambda)=\inf\{\mathcal{E}(\phi)|\phi\in W^1(\mathbf{R}^3),\ \|\phi\|_2\leqslant\lambda\}, \tag{1.2}$$

intuition suggests that:

(i) $E(\lambda)$ is finite.

(ii) There is a minimizing ϕ for $E(\lambda)$ which satisfies the nonlinear Schrödinger equation

$$\{-\Delta+V_\phi(x)\}\phi(x)=e\phi(x) \tag{1.3}$$

with

$$V_\phi(x)\equiv-2\int|\phi(y)|^2|x-y|^{-1}dy. \tag{1.4}$$

(iii) The minimizing ϕ is unique except for translations (i.e., $\phi(x)\to\phi(x+a)$, $a\in\mathbf{R}^3$) and $\|\phi\|_2=\lambda$. Furthermore, ϕ is infinitely differentiable. Thus,

$$E(\lambda)=\inf\{\mathcal{E}(\phi)|\phi\in W^1(\mathbf{R}^3),\ \|\phi\|_2=\lambda\}. \tag{1.5}$$

These facts will be proved in this paper. The mathematical difficulty of the problem stems from the minus sign in $\mathcal{E}$, which precludes the conventional arguments about convex functionals. To overcome the lack of convexity, the theory of symmetric decreasing functions will be employed. This is reviewed in Sec. III. To prove uniqueness of the minimum a strict form of the inequality is used. This we believe to be new, and it is given in the appendix.

The uniqueness proof is technically the hardest, if not the most novel, part of this paper. The proof in Sec. VI would not have been possible without important insights generously contributed by S. Parter and M. Steuerwalt.

For the uniqueness proof we rely heavily on the fact that the kernel in (1.1) is $|x-y|^{-1}$; in particular, the kernel yields a useful scaling relation. On the other hand, for the existence proof we only use the fact that $|x|^{-1}$ is a symmetric decreasing function. Thus, our method should be applicable in a wider context. For example, the existence proof is applicable for the functional $\mathcal{E}(\phi)-\int R(x)|\phi(x)|^2dx$, where R is a symmetric decreasing function in $L^{3/2}(\mathbf{R}^3)+L^\infty(\mathbf{R}^3)$. This latter functional arises in the Hartree-Fock theory of the helium atom.

II. Boundedness of $\mathcal{E}(\phi)$

We use the notation

$$T(\phi)=\int|\nabla\phi(x)|^2dx=\|\nabla\phi\|_2^2,$$

$$W(\phi)=\int\int|\phi(x)|^2|x-y|^{-1}|\phi(y)|^2dx\,dy. \tag{2.1}$$

Sobolev's inequality in $\mathbf{R}^3$ states that for $\phi\in W^1$,

$$T(\phi)\geqslant K\|\phi\|_6^2, \tag{2.2}$$

with the best possible K given by [1, 5, 7]

$$K=3(\pi/2)^{4/3}\approx5.478. \tag{2.3}$$

If we define

$$\rho_\phi(x)\equiv|\phi(x)|^2, \tag{2.4}$$

then $\phi\in W^1$ implies $\rho_\phi\in L^3$ and

$$T(\phi)\geqslant K\|\rho_\phi\|_3. \tag{2.5}$$

To discuss $E(\lambda)$ we also assume $\rho_\phi\in L^1$ with $\|\rho_\phi\|_1\leqslant\lambda^2$.
The function $|x|^{-1}$ can be written

$$|x|^{-1}=h_1(x)+h_2(x) \tag{2.6}$$

with $h_1\in L^{3/2}$ and $h_2\in L^\infty$, where $h_1(x)=|x|^{-1}$ for $|x|\leqslant R$, $h_1(x)=0$ otherwise. For any $\lambda>0$ we can choose R such that

$$\|h_1\|_{3/2}=K\lambda^{-2}/2, \tag{2.7}$$

and we then define

$$b(\lambda)=\|h_2\|_\infty=\text{const }\lambda^2. \tag{2.8}$$

By Young's inequality,

$$\int\int\rho(x)h_1(x-y)\rho(y)\,dx\,dy\leqslant\|h_1\|_{3/2}\|\rho\|_3\|\rho\|_1 \tag{2.9}$$

$$\int\int\rho(x)h_2(x-y)\rho(y)\,dx\,dy\leqslant\|h_2\|_\infty\|\rho\|_1^2. \tag{2.10}$$

Studies in Appl. Math. 57, 93–105 (1977)

From the above facts we can conclude
LEMMA 1. *If* $\phi \in W^1$ *and* $\|\phi\|_2 \leqslant \lambda$, *then*

$$\mathcal{E}(\phi) \geqslant -b(\lambda)\lambda^4. \tag{2.11}$$

Furthermore,

$$\begin{aligned}&\text{(i) } E(\lambda) < 0;\\ &\text{(ii) } \textit{if } \mathcal{E}(\phi) \leqslant E(\lambda)+1, \textit{ then}\end{aligned} \tag{2.12}$$

$$T(\phi) < 2\left[1 + b(\lambda)\lambda^4\right]. \tag{2.13}$$

Proof: (2.11) follows from (2.5), (2.7), (2.9) and (2.10). To prove (2.12) it is sufficient to find some $\phi \in W^1$ such that $\mathcal{E}(\phi) < 0$ and $\|\phi\|_2 \leqslant \lambda$. A Gaussian $\phi(x) = a\exp(-bx^2)$ will do this. For (2.13) we note that for $\|\phi\|_2 \leqslant \lambda$,

$$\begin{aligned} W(\phi) &\leqslant \tfrac{1}{2}K\|\rho_\phi\|_3 + b(\lambda)\lambda^4 \\ &\leqslant \tfrac{1}{2}T(\phi) + b(\lambda)\lambda^4. \end{aligned} \tag{2.14}$$

Hence, $1 > E(\lambda) + 1 \geqslant \mathcal{E}(\phi) \geqslant T(\phi)/2 - b(\lambda)\lambda^4$. ■

COROLLARY 2. *If* $\phi \in W^1$, $\|\phi\|_2 \leqslant \lambda$, *and* $\mathcal{E}(\phi) = E(\lambda)$, *then* $\|\phi\|_2 = \lambda$.

Proof: If $\|\phi\|_2 = \gamma < \lambda$, let $\psi = a\phi$ with $a = \lambda/\gamma > 1$. Then $\mathcal{E}(\psi) = (\lambda/\gamma)^2[T(\phi) - (\lambda/\gamma)^2 W(\phi)] < (\lambda/\gamma)^2 E(\lambda) < E(\lambda)$. ■

III. Symmetric decreasing rearrangements

It is necessary to use some inequalities about the symmetric decreasing rearrangement of a function, and we therefore briefly review some of the main facts. (See [2] for details and generalizations.)

Let

$$S = \left\{ f : \mathbf{R}^3 \to [0,\infty] \,|\, f(x) \leqslant f(y) \text{ if } |x| \geqslant |y| \right\} \tag{3.1}$$

be the symmetric decreasing functions, and let

$$S' = \left\{ f : \mathbf{R}^3 \to [0,\infty] \,|\, f(x-v) = g(x) \text{ a.e. for some } v \in \mathbf{R}^3 \text{ and } g \in S \right\} \tag{3.2}$$

be the translates (a.e.) of functions in S. The functions in S' are Lebesgue measurable. Two functions f and h in S' are said to be *equicentered* if the same v can be chosen for f and h in (3.2).

If χ is the characteristic function of a measurable set in $\mathbf{R}^3$, we define χ^* by

$$\begin{aligned} \chi^*(x) &= 1 \quad \text{if} \quad 4\pi|x|^3/3 \leqslant \|\chi\|_1, \\ &= 0 \quad \text{otherwise.} \end{aligned} \tag{3.3}$$

Clearly $\chi^* \in S$ and $\|\chi^*\|_1 = \|\chi\|_1$. Given $f: \mathbf{R}^3 \to [0, \infty]$, let $\chi_a^f(x) = 1$ if $f(x) \geqslant a$, $\chi_a^f(x) = 0$ otherwise. Then

$$f(x) = \int_0^\infty \chi_a^f(x)\, da, \tag{3.4}$$

and we define

$$f^*(x) \equiv \int_0^\infty \chi_a^{f^*}(x)\, da. \tag{3.5}$$

If $f: \mathbf{R}^3 \to \mathbf{C}$ we define

$$f^* = |f|^*. \tag{3.6}$$

Clearly $f^* \in S$, and for all a, $\mu\{x | f^*(x) \geqslant a\} = \mu\{x || f(x)| \geqslant a\}$, where μ is Lebesgue measure. This implies that for all p,

$$\|f^*\|_p = \|f\|_p. \tag{3.7}$$

It is easy to check that for $\alpha > 0$,

$$(f^\alpha)^* = (f^*)^\alpha \qquad \text{pointwise.} \tag{3.8}$$

The inequality of Riesz [4] states that for any three measurable functions on $\mathbf{R}^3$,

$$\left| \int\int f(x) g(x-y) h(y)\, dx\, dy \right| \leqslant \int\int f^*(x) g^*(x-y) h(y)^*\, dx\, dy. \tag{3.9}$$

To prove uniqueness we will need the following strict version of (3.9) (see the appendix).

Lemma 3. *If $g \in S$ and g is positive and strictly decreasing (i.e., $|x| < |y| \Rightarrow g(x) > g(y) > 0$), then* (3.9) *is a strict inequality when the right side is finite unless f and h are equicentered functions in S'.*

Since $g(x) = |x|^{-1}$ satisfies the hypothesis of Lemma 3, and since $|\phi| \in S' \Leftrightarrow |\phi|^2 \in S'$, we have

Corollary 4. *If $|\phi| \notin S'$, then $W(\phi) < W(\phi^*)$.*

Next we turn to $T(\phi)$.

Lemma 5. *If $\phi \in W^1$, then $\phi^* \in W^1$ and $T(\phi) \geqslant T(\phi^*)$.*

This lemma is well known, but what we believe to be an original and simple proof is given in the appendix. Probably a strict version of Lemma 5 is true, but we do not need it, since we have the strict inequality for $W(\phi)$.

The results of this section can be summarized as follows:

Studies in Appl. Math. *57*, 93–105 (1977)

LEMMA 6.

(a) *There exists a sequence of symmetric decreasing functions* $\phi^{(j)} \in W^1$ *such that* $\|\phi^{(j)}\|_2 = \lambda$ *and* $\mathcal{E}(\phi^{(j)}) \to E(\lambda)$.

(b) *If* $\phi \in W^1$, $\|\phi\|_2 = \lambda$ *and* $\mathcal{E}(\phi) = E(\lambda)$, *then* $\phi \in S'$.

Proof: If $\{\phi^{(j)}\}_{j=1}^{\infty}$ is a minimizing sequence for $\mathcal{E}(\phi)$ and if $\phi^{(j)}$ is replaced by $\psi^{(j)} \equiv \phi^{(j)*}$, then (a) follows from Corollaries 2 and 4 and Lemma 5. (b) follows from Corollary 4. ■

Remark: Part (b) is crucial for the uniqueness question, because it is then sufficient to prove uniqueness among the functions in S'.

IV. Existence of a minimum and its properties

THEOREM 7. *There exists a* $\phi \in S$ *with* $\|\phi\|_2 = \lambda$ *such that* $\mathcal{E}(\phi) = E(\lambda)$.

Proof: Let $\phi^{(j)} \in S$ be a minimizing sequence for $E(\lambda)$. W^1 is a Hilbert space with norm $\|\phi\| = \|\phi\|_2 + \|\nabla\phi\|_2$, and $\|\nabla\phi\|_2$ is bounded by Lemma 1. By the Banach-Alaoglu theorem there exists a W^1-weakly convergent subsequence which we shall denote by $\phi^{(j)}$. If ϕ is the weak limit then $\liminf_{j\to\infty} T(\phi^{(j)}) \geqslant T(\phi)$ and $\|\phi\|_2 \leqslant \lambda$.

Now consider $\rho^{(j)}(x) = \phi^{(j)}(x)^2$. We abuse notation by writing $\rho(r)$ with $r = |x|$ for spherically symmetric functions. Since $\rho^{(j)} \in S$ and $\|\rho^{(j)}\|_1 = \lambda^2$, we have, for any $R \geqslant 0$, $\rho^{(j)}(R) 4\pi R^3/3 \leqslant 4\pi \int_0^R \rho^{(j)}(s) s^2\, ds \leqslant \|\rho^{(j)}\|_1 = \lambda^2$. Likewise, $\|\rho^{(j)}\|_3 \leqslant C$ by (2.2) and Lemma 1, and hence $[\rho^{(j)}(R)]^3 4\pi R^3/3 \leqslant C^3$. Thus $\rho^{(j)}(R) < f(R)$, where

$$
\begin{aligned}
f(r) &= Ar^{-1} \quad \text{for} \quad r \leqslant 1,\\
&= Ar^{-3} \quad \text{for} \quad r \geqslant 1.
\end{aligned}
\tag{4.1}
$$

By a trivial generalization of Helly' theorem [3], we can find a further subsequence such that $\rho^{(j)}(r) \to \rho(r) \leqslant f$ *pointwise* for $r > 0$. Hence $\phi^{(j)} \to \tilde{\phi} \equiv \rho^{1/2}$ pointwise on $\mathbf{R}^3 \setminus \{0\}$. We also know that $\phi^{(j)} \to \phi$ in weak L^2. Since $\phi^{(j)} \leqslant f^{1/2} \in L^2_{\text{loc}}$, it is easy to see that $\tilde{\phi} = \phi$. [Proof: If $g \in C_0^\infty$, then $\int g(\phi^{(j)} - \phi) \to 0$ by the weak convergence, while $\int g(\phi^{(j)} - \tilde{\phi}) \to 0$ by dominated convergence. Hence $\int g(\phi - \tilde{\phi}) = 0$ for all $g \in C_0^\infty$, which implies that $\phi = \tilde{\phi}$.]

Since $\rho^{(j)} \to \rho \equiv \phi^2$ pointwise, and $\rho \leqslant f$, we have, by dominated convergence, that $W(\phi^{(j)}) \to W(\phi)$ provided $W(f^{1/2}) < \infty$. This latter fact is easy to verify.

In summary, $E(\lambda) = \liminf_{j\to\infty} \mathcal{E}(\phi^{(j)}) \geqslant \mathcal{E}(\phi)$, so ϕ is a minimum for $E(\lambda)$. ■

We turn next to some properties of any minimizing function ϕ. In the next theorem we do *not* use the fact that ϕ is symmetric decreasing.

THEOREM 8. *If* $\phi \in W^1$, $\|\phi\|_2 = \lambda$ *and* $\mathcal{E}(\phi) = E(\lambda)$, *then* ϕ *satisfies the (distributional) equation*

$$
\{-\Delta + V_\phi(x)\}\phi(x) = e\phi(x) \tag{1.3}
$$

for some $e<0$. V_ϕ *is given in* (1.4). *If* $\phi\in W^1$ *is any complex valued function (not necessarily minimizing) satisfying* (1.3) *in the distributional sense for any e then*:

(ia) $V_\phi\in L^p$, $4\leqslant p\leqslant\infty$.
(ib) $\phi V_\phi\in L^p$, $\frac{4}{3}\leqslant p\leqslant 6$.
(ii) V_ϕ *is a continuous function which goes to zero at infinity.*
(iii) *If* $e<0$, *then* $\phi\in C^\infty$ *and goes to zero at infinity, and hence* ϕ *is a strong solution of* (1.3).

Proof: The proof of (1.3) is standard. Simply replace ϕ by $\phi+\lambda g$, $g\in\mathcal{S}$ (Schwarz space), and compute the derivative at $\lambda=0$ of $\mathcal{E}(\phi+\lambda g)$. To see that $e<0$ for a minimizing ϕ, multiply (1.3) by $\overline{\phi(x)}$ and integrate. Then

$$\mathcal{E}(\phi)=E(\lambda)=e\lambda^2+W(\phi) \tag{4.2}$$

and $E(\lambda)<0$, while $W(\phi)>0$.

For the second part we write $|x|^{-1}=h_1(x)+h_2(x)$ as in (2.6) and note that $h_1\in L^{3/2}$ and $h_2\in L^4$. By Young's inequality, if $f\in L^p$, $g\in L^q$, $p^{-1}+q^{-1}=1+r^{-1}$, then $f*g\in L^r$. (ia) follows from the fact that $\rho_\phi=|\phi|^2\in L^q$ for $1\leqslant q\leqslant 3$, so that $h_1*\rho_\phi\in L^p$, $\frac{3}{2}\leqslant p\leqslant\infty$ and $h_2*\rho_\phi\in L^p$, $4\leqslant p\leqslant\infty$. For (ib) we note that $fg\in L^r$ with $r^{-1}=p^{-1}+q^{-1}$. (ii) follows from the fact [6] that if $f\in L^p$, $g\in L^q$, $p^{-1}+q^{-1}=1$ and $1<p<\infty$, then $f*g$ is a continuous function which goes to zero at infinity.

For (iii) we write $\phi=Y_e*(\phi V_\phi)$, where $Y_e(x)=(4\pi|x|)^{-1}\exp(-|e|^{1/2}|x|)$ is a kernel for $(-\Delta-e)^{-1}$. As $Y_e\in L^p$, $p=2$, (ib) implies that ϕ is a continuous function which goes to zero at infinity. Now fix $x_0\in\mathbf{R}^3$, and let $\psi\in C_0^\infty$ be a function which is 1 near x_0. Let $\phi_1(x)=\psi(x)\phi(x)$, $\phi_2=\phi-\phi_1$. Let $\phi=\phi_a+\phi_b$, where $\phi_a=-Y_e*(V_\phi\phi_2)$. Since ϕ_2 vanishes near x_0, ϕ_a is C^∞ near x_0. Assuming that ϕ is C^k $(k\geqslant 0)$ in a neighborhood of x_0, we shall prove that ϕ is C^{k+1} near x_0. Write $\rho_\phi=\rho^1+\rho^2$, where $\rho^1=|\phi_1|^2$ and $\rho^2=|\phi_2|^2+\bar{\phi}_1\phi_2+\phi_1\bar{\phi}_2$. Since ρ_2 is zero near x_0, $(|x|^{-1})*\rho^2$ is harmonic and hence C^∞ near x_0. Since ρ^1 has compact support, it is in all L^p, and ρ^1 is C^k near x_0. Then $(|x|^{-1})*\rho^1$ is C^k near x_0. Therefore $\phi_1 V_\phi$ is C^k near x_0 and has compact support. Hence $\phi_b=-Y_e*(\phi_1 V_\phi)$ is C^{k+1} near x_0. ■

V. Scaling properties

In this section we shall exploit the fact that the kernel in (1.1) is $|x|^{-1}$. For $z>0$ consider the functional

$$\mathcal{E}_z(\phi)=T(\phi)-zW(\phi) \tag{5.1}$$

on W^1 and

$$E(\lambda,z)=\inf\{\mathcal{E}_z(\phi)|\phi\in W^1,\ \|\phi\|_2=\lambda\}. \tag{5.2}$$

The results of the previous sections carry through *mutatis mutandis*.

THEOREM 9. *Let $\phi(x;\lambda,z)$ be a minimizing function for $E(\lambda,z)$, and let $e(\lambda,z)$ be the eigenvalue in the analogue of* (1.3), *i.e.,*

$$\{-\Delta+zV_\phi(x)\}\phi(x;\lambda,z)=e(\lambda,z)\phi(x;\lambda,z). \tag{5.3}$$

Let $\phi_1(x)$, E_1 and e_1 denote such a triplet when $\lambda=z=1$. Then for every solution to the (λ,z) problem there is a solution to the $(1,1)$ problem and conversely. These are related by

$$\phi(x;\lambda,z)=z^{3/2}\lambda^4\phi_1(z\lambda^2x),$$

$$zV_{\phi(\cdot;\lambda,z)}(x)=z^2\lambda^4V_{\phi_1}(z\lambda^2x),$$

$$E(\lambda,z)=z^2\lambda^6E_1,$$

$$e(\lambda,z)=z^2\lambda^4e_1,$$

$$T(\phi(\cdot;\lambda,z))=z^2\lambda^6T(\phi_1),$$

$$zW(\phi(\cdot;\lambda,z))=z^2\lambda^6W(\phi_1).$$

Proof: Trivial. ■

VI. Uniqueness of the minimum

If ϕ minimizes $\mathcal{E}(\phi)$ subject to $\|\phi\|_2=\lambda$, Lemma 6 asserts that $\phi\in S'$.

THEOREM 10. *If ϕ is minimizing for $E(\lambda)$ and $\phi\in S$, then ϕ is unique.*

Proof: By Theorem 9, we know that if we prove uniqueness for any $\lambda>0$, then we have uniqueness for all $\lambda>0$. If ϕ is minimizing for some λ_0, then, by Theorem 9, for every $\lambda>0$ there is a scaled copy of ϕ that minimizes for λ. Consider $I(\lambda)\equiv\int|\phi(x;\lambda)|^2|x|^{-1}dx$. Since $\phi\in L^2\cap L^6$, $I(\lambda)$ is finite. By scaling, $I(\lambda)=A\lambda^4$ and $e(\lambda)=$[eigenvalue in (5.3)]$=B\lambda^4$. By Newton's 1687 theorem, and using the fact that ϕ is spherical, we can conveniently express V_ϕ in polar coordinates as

$$\begin{aligned}V_\phi(x)&=-8\pi r^{-1}\int_0^r|\phi(s;\lambda)|^2s^2ds-8\pi\int_r^\infty|\phi(s;\lambda)|^2s\,ds\\&=\int_0^rK(r,s)|\phi(s;\lambda)|^2ds-2I(\lambda),\end{aligned} \tag{6.1}$$

where $r=|x|$ and, for $r\geqslant s$,

$$K(r,s)=8\pi s^2(s^{-1}-r^{-1})\geqslant0. \tag{6.2}$$

Thus, (5.3) reads

$$\{-\Delta+U_\phi(x)\}\phi(x)=(e+2I)\phi(x) \tag{6.3}$$

and

$$U_\phi(r)=\int_0^r K(r,s)|\phi(s)|^2\,ds. \tag{6.4}$$

(6.3) is a Schrödinger equation with potential U_ϕ. As $U_\phi \geqslant 0$, we see that $e+2I$ must be positive. Since $e+2I=(B+2A)\lambda^4$, $B+2A>0$.

Now choose λ such that $(B+2A)\lambda^4=1$. Then we have the following *canonical form* of (1.3) for spherical functions:

$$\left\{-\frac{d^2}{dr^2}-\frac{2}{r}\frac{d}{dr}+U_\phi(r)\right\}\phi(r)=\phi(r). \tag{6.5}$$

In other words, every W^1 spherically symmetric solution of (1.3), whether minimizing or not and whether $e<0$ or not, obeys (6.5) after a suitable scale transformation. Our goal will be to show that (6.5) has only one (non-null) nonnegative solution in W^1, for this will imply that the minimum is unique (modulo translations).

An advantage of (6.5) is that the parameter λ appears nowhere. Another advantage of (6.5), especially for numerical work, is that $U_\phi(r)$ depends upon $\phi(s)$ only for $s<r$. Hence one can integrate from $r=0$ outwards.

Given any solution ϕ of (6.5), we can reconstruct the original problem by

$$\tilde\lambda^2=4\pi\int_0^\infty|\phi(r)|^2r^2\,dr, \tag{6.6}$$

$$I(\tilde\lambda)=4\pi\int_0^\infty|\phi(r)|^2r\,dr, \tag{6.7}$$

$$T(\phi)=4\pi\int_0^\infty\left|\frac{d[r\phi(r)]}{dr}\right|^2dr, \tag{6.8}$$

$$W(\phi)=-2\pi\int_0^\infty|\phi(r)|^2U_\phi(r)r^2\,dr+\tilde\lambda^2I(\tilde\lambda), \tag{6.9}$$

$$\phi(x;\lambda)=(\lambda/\tilde\lambda)^4\phi\big((\lambda/\tilde\lambda)^2x\big). \tag{6.10}$$

If ϕ is minimizing,

$$E(\tilde\lambda)=T(\phi)-W(\phi) \tag{6.11}$$

Studies in Appl. Math. *57*, 93–105 (1977)

and, for any $\lambda>0$,

$$E(\lambda)=(\lambda/\tilde{\lambda})^{6}E(\tilde{\lambda}). \tag{6.12}$$

We turn next to the uniqueness proof for (6.5). Suppose $\phi\in W^1$ is a nonnegative solution of (6.5). ϕ and U_ϕ can also be thought of as functions on $\mathbf{R}^3$. Consider the following functional on $W^1(\mathbf{R}^3)$:

$$A_\phi(\psi)\equiv\int|\nabla\psi(x)|^2dx+\int|\psi(x)|^2U_\phi(x)\,dx. \tag{6.13}$$

Let $\Gamma_\phi=\inf\{A_\phi(\psi)|\psi\in W^1,\ \|\psi\|_2=1\}$. It follows by standard arguments [or, since $U_\phi(x)$ is symmetric increasing and $U_\phi(x)$ is bounded, by the methods of Theorem 7] that there is a minimizing function for Γ_ϕ. This function satisfies (6.5) and is positive and (also by standard arguments) unique. Therefore it must be proportional to ϕ itself. Hence, $A_\phi(\phi)=\|\phi\|_2^2$ and

$$A_\phi(\psi)\geqslant\|\psi\|_2^2 \tag{6.14}$$

for any $\psi\in W^1$, and equality holds in (6.14) when $\psi=\phi$.

Suppose there are two different, non-null, nonnegative solutions ϕ_1 and ϕ_2 of (6.5). Denote the potentials simply by U_1 and U_2 and the functionals in (6.13) by A_1 and A_2. Consider first the case that $\phi_1(r)\geqslant\phi_2(r)$, all $r\geqslant 0$. Then $U_1(r)\geqslant U_2(r)$, all $r\geqslant0$. It is easy to check that $B\equiv\int[U_1(x)-U_2(x)]\phi_1(x)^2dx>0$. Then $\|\phi_1\|_2^2\leqslant A_2(\phi_1)=A_1(\phi_1)-B<\|\phi_1\|_2^2$, and this is a contradiction.

Next, suppose that $\psi=\phi_1-\phi_2$ is not of one sign. It is easy to see by the methods of Theorem 8 that ϕ_1 and ϕ_2, and hence ψ, are continuous. There are two cases: (i) $\psi(0)\neq0$, in which case we can assume $\psi(0)>0$; (ii) there exists an $R>0$ such that $\psi(r)=0$ for $0\leqslant r\leqslant R$ and ψ is not identically zero in any open interval of the form $I_\varepsilon=(R,R+\varepsilon)$. In case (ii) we write $\phi_i(r)=u_i(r)/r$, and (6.5) can be solved for $r\in I_\varepsilon$ as

$$u_i(r)=u_i(R)+a_i(r-R)+T(r,u_i), \tag{6.15}$$

where

$$T(r,f)=\int_R^r(r-s)\big[\tilde{U}_f(s)-1\big]f(s)\,ds, \tag{6.16}$$

and $\tilde{U}_f$ is given by (6.4) with $\phi(r)=\phi_1(r)=\phi_2(r)$ for $r\leqslant R$ and $\phi(r)=f(r)/r$ for $r\in I_\varepsilon$. Let D_ε be the set of continuous functions on I_ε and which are bounded by $\sup\{u_i(r)|r\in I_\varepsilon,\ i=1,2\}$ and such that $f(R)=u_1(R)=u_2(R)$. Equip D_ε with the usual sup norm. For sufficiently small $\varepsilon>0$, $T(r,\cdot)$ is a strict contraction on D_ε, and hence $u_1=u_2$ in I_ε if $a_1=a_2$. If $a_1>a_2$, then $u_1(r)>u_2(r)$ for some (possibly smaller) open interval I_ε.

Thus we can say, in either case, that there exists an $R>0$ such that $\psi(R)=0$ $\psi(r)\geqslant0$ for $r\in I\equiv[0,R]$ and $\psi(r)$ does not vanish identically in I. This implies that the function $F(r)\equiv\frac{1}{2}[U_1(r)-U_2(r)]\geqslant 0$ in I. Define $\tilde{\psi}\in W^1$ to be the function $\tilde{\psi}(x)=\psi(x)$, $|x|\leqslant R$, $\tilde{\psi}(x)=0$, $|x|\geqslant R$. From (6.5), $\tilde{\psi}$ satisfies the following equation for $|x|\leqslant R$:

$$\left[-\Delta+\tfrac{1}{2}(U_1+U_2)\right]\tilde{\psi}=\tilde{\psi}-F(\phi_1+\phi_2).$$

Multiplying this by $\tilde{\psi}$ and integrating by parts yields

$$K\equiv\tfrac{1}{2}\left[A_1(\tilde{\psi})+A_2(\tilde{\psi})\right]=\|\tilde{\psi}\|_2^2-L,$$

where $L=\int\tilde{\psi}(x)F(x)[\phi_1(x)+\phi_2(x)]\,dx$. It is easy to see that $L>0$. On the other hand, $K\geqslant\|\tilde{\psi}\|_2^2$ by (6.14). This is a contradiction. ■

Appendix: Proof of two theorems on symmetric decreasing rearrangements

Given a nonnegative (respectively complex valued) function, f, on $\mathbf{R}^n$, then f^* denotes the symmetric decreasing rearrangement of f (respectively $|f|$).

We turn first to the strong form of the Riesz theorem [4].

LEMMA 3. *Suppose g is a positive spherically symmetric decreasing function on* $\mathbf{R}^n$ *and g is strictly decreasing (i.e.,* $|x|<|y|\Rightarrow g(x)>g(y)>0$*). For any two nonnegative functions* $f\in L^p(\mathbf{R}^n)$, $h\in L^q(\mathbf{R}^n)$ *define*

$$I(f,g,h)=\int\int f(x)g(x-y)h(y)\,dx\,dy. \tag{A.1}$$

Then

$$I(f,g,h)\leqslant I(f^*,g,h^*) \tag{A.2}$$

with strict inequality whenever $I(f^*,g,h^*)<\infty$, *unless the following holds: For some* $v\in\mathbf{R}^n$, $f(x-v)=f^*(x)$ *and* $h(x-v)=h^*(x)$ a.e.

Proof: The Riesz theorem which gives $\leqslant$ in (A.2) will be assumed; our problem will be to prove "less than". By subtracting positive constants, if necessary, from f, g, and h we can suppose without loss of generality that f^*, h^*, and $g=g^*$ go to zero at infinity. It can also be assumed that neither f nor h are null functions. We first prove the lemma for R^1. g can be written as

$$g(x)=\int\chi_r(x)\,d\mu(r)$$

where μ is a positive measure on $A=[0,\infty]$ and χ_r is the characteristic function of the interval $[-r,r]$. The hypothesis about g implies that $\mu((a,b))>0$ for every open interval (a,b) in A. Now suppose that f and h are characteristic functions of two sets F and H of finite measure. Then f^* (resp. g^*) is the characteristic function of the closed interval $[-c,c]$(resp.$[-d,d]$), where $2c=\text{meas}(F)$ ($2d=$

Studies in Appl. Math. *57*, 93–105 (1977)

meas(H)). Let $B=[-c-d,c+d]$. If $m\equiv f*h$, then m is continuous (by the remark in the proof of Theorem 8, [5]), and supp(m)$\subset B$ if and only if F and H are equicentered intervals. Let $[-R,R]$ be the smallest symmetric interval that contains supp(m). For any f, g, and h, $I(f,g,h)=\int g(x)m(x)\,dx$. Suppose that F and H are not equicentered intervals. Then for $r\in(c+d,R)$ we have that $J(r)\equiv I(f,\chi_r,h)<\int m(x)\,dx=\int f\int h$ while $K(r)\equiv I(f^*,\chi_r,h^*)=\int f^*\int h^*=\int f\int h$. For all $r\geqslant 0$, $J(r)\leqslant K(r)$ by the Riesz theorem. Therefore

$$I(f^*,g,h^*)-I(f,g,h)\geqslant\int_{c+e}^{R}[J(r)-K(r)]\,d\mu(r)>0,$$

and this proves the lemma for characteristic functions. For arbitrary f and h we can write

$$f(x)=\int_0^\infty \chi_a^f(x)\,da, \tag{A.3}$$

where χ_a^f is the characteristic function of the set $B_a^f=\{x\,|\,f(x)\geqslant a\}$, and similarly for h. By Fubini's theorem, to have equality in (A.2) we must have for almost all (a,b) (in the sense of $\mathbf{R}^2$ Lebesgue measure) that there exists a $v\in R^1$ such that $\chi_a^f(\cdot-v)$ and $\chi_b^h(\cdot-v)$ are (a.e.) the characteristic functions of symmetric intervals. This v, if it exists, cannot depend on a or b. [To see this, choose an a such that χ_a^f is not null. Then the v such that $\chi_a^f(\cdot-v)$ is symmetric is unique, and hence cannot depend on b.] Hence, for equality, there exists a fixed v such that $\chi_a^f(\cdot-v)$ $[\chi_b^h(\cdot-v)]$ is symmetric (a.e.) for almost all a (in the $\mathbf{R}^1$ sense) [almost all b]. By (A.3), f and h then satisfy the last line of the lemma.

Next we turn to $\mathbf{R}^{n+1}$ and suppose the lemma to be true for $\mathbf{R}^n$. f and h can be assumed to be Borel measurable. If $x=(x_1,\ldots,x_n)\in\mathbf{R}^n$ and $y\in\mathbf{R}^1$, consider $F_y(x)\equiv f(x_1,\ldots,x_n,y)$ to be a function on $\mathbf{R}^n$. G_y and H_y are defined similarly. G_y satisfies the hypothesis of the lemma for each y. In (A.1) first do the integral over $x_1,\ldots,x_n$ and $y_1,\ldots,y_n$ holding x_{n+1} and y_{n+1} fixed. By induction, equality holds in (A.2) only if F_y and H_z are equicentered functions in S' for almost all (y,z) (in the $\mathbf{R}^2$ sense). By the same argument as given above for the $\mathbf{R}^1$ case, the displacement $v\in\mathbf{R}^n$ must be independent of y and z. If the argument is repeated holding some other coordinate [not necessarily orthogonal to the $(n+1)$th] fixed, we conclude that for equality there exists $w\in\mathbf{R}^{n+1}$ such that the two translated functions $f'\equiv f(\cdot-w)$ and $h'\equiv(\cdot-w)$ have the following property: Let P_t be any family of parallel n-dimensional hyperplanes in $\mathbf{R}^{n+1}$ parametrized by the distance t from the origin, and let f_t be f' restricted to P_t. Then for almost all t, f_t can be modified on a set of measure zero such that f_t is symmetric decreasing. By standard but tedious arguments (see the appendix of [2] for details), this implies that the last line of the lemma holds. ■

The next theorem concern the behavior of the W^1 norm under rearrangement of a function.

LEMMA 5. *If* $\phi\in W^1(\mathbf{R}^n)$, *then* $\phi^*\in W^1(\mathbf{R}^n)$ *and* $\|\nabla\phi\|_2\geqslant\|\nabla\phi^*\|_2$.

Proof: Let $t>0$ and consider the following function on $\mathbf{R}^n$: $G_t(x)=(4\pi t)^{-n/2}\exp(-x^2/4t)$. G_t is a kernel for $e^{t\Delta}$, the fundamental solution of the heat equation. G_t is in all the L^p spaces, so

$$I_t(\phi)=t^{-1}\left\{\int|\phi(x)|^2dx-\int\bar{\phi}(x)G_t(x-y)\phi(y)\,dx\,dy\right\} \tag{A.4}$$

is well defined. By Riesz's rearrangement theorem (3.9), $I_t(\phi)\geqslant I_t(\phi^*)$, since $G_t(x)$ is symmetric decreasing. $\phi^*\in L^2(\mathbf{R}^n)$, since $\phi\in L^2(\mathbf{R}^n)$. To complete the proof we have to show that for any $f\in L^2(\mathbf{R}^n)$,

$$\lim_{t\downarrow 0} I_t(f)=\|\nabla f\|_2^2 \quad \text{if } f\in W^1, \tag{A.5}$$

$$\lim_{t\downarrow 0} I_t(f)=\infty \quad \text{if } f\notin W^1. \tag{A.6}$$

Recall that for $f\in L^2(\mathbf{R}^n)$, $\|\nabla f\|_2^2=\int k^2|\hat{f}(k)|^2dk$ by definition, where $\hat{f}$ is the Fourier transform of f. We can rewrite (A.4) as

$$I_t(f)=\int|\hat{f}(k)|^2\left\{t^{-1}\left[1-\exp(-k^2t)\right]\right\}dk. \tag{A.7}$$

Suppose $f\in W^1$. Since $1-e^{-x}\leqslant x$, $t^{-1}[1-\exp(-k^2t)]\leqslant k^2$ and (A.5) is true by dominated convergence. Suppose $f\notin W^1$. Since $1-e^{-x}\geqslant 1-(1+x)^{-1}=x(1+x)^{-1}$, $t^{-1}[1-\exp(-k^2t)]\geqslant k^2(1+k^2t)^{-1}$. (A.6) follows from this. ∎

References

1. T. Aubin, Problèmes isoperimetrique et espaces de Sobolev, *C. R. Acad. Sci. Paris* **280**, 279–281 (1975).
2. H. J. Brascamp, E. H. Lieb, and J. M. Luttinger, A general rearrangement inequality for multiple integrals, *J. Funct. Anal.* **17**, 227–237 (1974).
3. W. Feller, *An Introduction to Probability Theory and its Applications*, Vol. 2, Wiley, New York, 1966, p. 261.
4. F. Riesz, Sur une inégalité intégrale, *J. LMS* **5**, 162–168 (1930).
5. G. Rosen, Minimum value for c in the Sobolev inequality $\|\phi\|_3\leqslant c\|\nabla\phi\|_2$, *SIAM J. Appl. Math.* **21**, 30–32 (1971).
6. W. Rudin, *Fourier Analysis on Groups*, Interscience, New York, 1962.
7. G. Talenti, Best constant in the Sobolev inequality, to be published.

Princeton University

(Received November 15, 1976)

BULLETIN (New Series) OF THE
AMERICAN MATHEMATICAL SOCIETY
Volume 20, Number 2, April 1989

SYMMETRIC DECREASING REARRANGEMENT CAN BE DISCONTINUOUS

FREDERICK J. ALMGREN, JR. AND ELLIOTT H. LIEB

Suppose $f(x^1, x^2) \geq 0$ is a continuously differentiable function supported in the unit disk in the plane. Its symmetric decreasing rearrangement is the rotationally invariant function $f^*(x^1, x^2)$ whose level sets are circles enclosing the same area as the level sets of f. Such rearrangement preserves $\mathbf{L}^p$ norms but decreases convex gradient integrals, e.g. $\|\nabla f^*\|_p \leq \|\nabla f\|_p$ $(1 \leq p < \infty)$. Now suppose that $f_j(x^1, x^2) \geq 0$ $(j = 1, 2, 3, \ldots)$ is a sequence of infinitely differentiable functions also supported in the unit disk which converge uniformly together with first derivatives to f. The symmetrized functions also converge uniformly. The real question is about convergence of the derivatives of the symmetrized functions. We announce that *the derivatives of the symmetrized functions need not converge strongly*, e.g. it can happen that $\|\nabla f_j^* - \nabla f^*\|_p \nrightarrow 0$ for every p. We further characterize exactly those f's for which convergence is assured and for which it can fail.

The rearrangement map $\mathscr{R}: f \to f^*$ in general dimensions also decreases gradient norms. For this reason alone, rearrangement has long been a basic tool in the calculus of variations and in the theory of those PDE's that arise as Euler-Lagrange equations of variational problems; it permits one to concentrate attention on radial, monotone functions and thereby reduces many problems to simple one dimensional ones. Some examples are (i) the lowest eigenfunction of the Laplacian in a ball is symmetric decreasing; (ii) the body with smallest capacity for a given volume is a ball **[PS]**; (iii) the optimal functions for the Sobolev and Hardy-Littlewood-Sobolev inequalities are symmetric decreasing and can be explicitly calculated **[LE]**. Other examples are given in **[KB]**.

Obviously $\mathscr{R}$ is highly nonlocal, nonlinear, and nonintuitive, but the property of decreasing gradient norms would lead one to surmise that $\mathscr{R}$ is a smoothing operator in some sense. Thus when W. Ni and L. Nirenberg asked, some years ago, whether $\mathscr{R}$ is continuous in the $\mathbf{W}^{1,p}$ topology the answer appeared to be that it should be so (it is easy to prove that $\mathscr{R}$ is always a contraction in $\mathbf{L}^p$). Indeed, by an elegant analysis Coron **[CJ]** proved this in $\mathbf{R}^1$. An affirmative answer to this question would have meant that the mountain-pass lemma could be used to establish spherically symmetric solutions of certain PDE's, and Coron's result led to just such an application **[RS]**. Our result is that $\mathscr{R}$ is not continuous in $\mathbf{W}^{1,p}(\mathbf{R}^n)$ for $n \geq 2$ and it is surprising, to us at least. Since almost all applications

Received by the editors October 17, 1988 and, in revised form, November 29, 1988.
1980 *Mathematics Subject Classification* (1985 *Revision*). Primary 46E35; Secondary 26B99, 47B38.

of $\mathscr{R}$—apart from the mountain-pass application—do not rely on continuity, our result does not have much immediate impact on applications. It reveals, however, an unexpected subtlety about the geometry of level sets of functions and shows that intuition can be very wrong. More precisely, our analysis has led us to isolate a property of $\mathbf{W}^{1,p}$ functions on their critical sets which we call **co-area regularity**, in terms of which we prove **[AL]**.

MAIN THEOREM. *The rearrangement map $\mathscr{R}$ is $\mathbf{W}^{1,p}(\mathbf{R}^n)$ continuous at a function f if and only if f is co-area regular.*

Each $\mathbf{W}^{1,p}$ function on the line is automatically co-area regular. In higher dimensions both the regular and irregular functions are dense in $\mathbf{W}^{1,p}$.

The symmetric decreasing rearrangment of a vector valued function f is defined by setting $f^* = |f|^*$. One can also replace $\mathbf{W}^{1,p}$ norms by gradient integrals of other convex integrands $\psi\colon \mathbf{R}^+ \to \mathbf{R}^+$, i.e. $\|\nabla f\|_p^p = \int |\nabla f|^p \, d\mathscr{L}^n$ ($\mathscr{L}^n$ is Lebesgue measure) is replaced by $\int \psi(|\nabla f|)\, d\mathscr{L}^n$. Our *conclusions about continuity remain the same.* However, for each $0 < \alpha < 1$, *each $p \geq 1$, and each $n \geq 1$ we show that the rearrangement map $\mathscr{R}$ is continuous everywhere on the fractional Sobolev space $\mathbf{W}^{\alpha,p}(\mathbf{R}^n)$.* We thus have the curious fact that co-area regularity plays a role for $\mathbf{W}^{\alpha,p}$ only when $\alpha = 1$.

DEFINITION. Suppose $f\colon \mathbf{R}^n \to \mathbf{R}^+$ and set

$$\mathscr{G}_f(y) = \int \chi_{\{f>y\}} \chi_{\{\nabla f = 0\}} \, d\mathscr{L}^n;$$

for each positive number y; here χ_A denotes the characteristic function of the set A. Since $\mathscr{G}_f\colon \mathbf{R}^+ \to \mathbf{R}^+$ is nonincreasing, its distribution first derivative $\mathscr{G}_f'$ is a (negative) measure. Our function f is called **co-area regular** if and only if the measure $\mathscr{G}_f'$ is purely singular with respect to $\mathscr{L}^1$. Otherwise f is called **co-area irregular**.

The term co-area regular was suggested by H. Federer's "co-area formula" for the absolutely continuous function

$$y \to \int \chi_{\{f>y\}} \chi_{\{\nabla f \neq 0\}} \, d\mathscr{L}^n$$

which is complementary to our $\mathscr{G}_f$.

We also announce

THEOREM. *For each $n \geq 2$ and each $0 < \lambda < 1$, there is (by construction) a positive constant C and a function $f\colon \mathbf{R}^n \to [0,1]$ in $\mathbf{C}^{n-1,\lambda}$ whose support is the unit cube $\mathbf{Q}$ such that $\mathscr{G}_f(y) = C(1-y)$ for each $0 < y \leq 1$. In particular, the measure $\mathscr{G}_f'$ is absolutely continuous with respect to $\mathscr{L}^1$; thus f is a co-area irregular function.*

Each f in $\mathbf{C}^{n-1,1}$ turns out to be co-area regular and both the regular and the irregular functions are dense in $\mathbf{W}^{1,p}$ for $n \geq 2$.

The idea behind the construction above is to decompose $\mathbf{Q}$ into 2^n cubes $\mathbf{Q}_j$ of half the size, then decompose each of those into 2^n $\mathbf{Q}_{jk}$'s and so on. We first set $f(x) = \sum_{i=1}^{\infty} a_i(x)2^{-ni}$ where $a_i(x)$ equals $(l-1)$ when x belongs to the cube $\mathbf{Q}_{...l...}$ and $l \in \{1, \ldots, 2^n\}$ is the index in the ith position. This f is not continuous but its range is uniformly spread over $(0, 1)$. The second step is to "smooth" this f in such a way that it belongs to $\mathbf{C}^{n-1,\lambda}$ and $\mathscr{L}^n\{x: \nabla f = 0\} > 0$.

A fuller statement of failure of continuity is the following.

THEOREM (DISCONTINUITY AT CO-AREA IRREGULAR FUNCTIONS). *Suppose $n \geq 2$ and f is a co-area irregular function belonging to $\mathbf{W}^{1,p}(\mathbf{R}^n)$. Then there is a sequence $f_1, f_2, f_2, \ldots$ of infinitely differentiable functions in $\mathbf{W}^{1,p}(\mathbf{R}^n)$ such that $f_j \to f$ in $\mathbf{W}^{1,p}(\mathbf{R}^n)$ as $j \to \infty$ but $f_j^* \nrightarrow f^*$ in $\mathbf{W}^{1,p}(\mathbf{R}^n)$.*

The basic idea behind the proof is the following. Let U_j be a suitable smooth approximation of $\chi_{\{\nabla f=0\}}$ and set

$$f_j(x) = f(x) + \frac{1}{2j} U_j(x) \sin(jf(x))$$

for each x. We confirm that $f_j \to f$ in $\mathbf{W}^{1,p}$ as $j \to \infty$. Defining sets $K_{(j)}(y) = \{x: f_{(j)}(x) > y\}$ for each y, we check for integers m that $K(y) = K_j(y)$ when $y = (2m)(\pi/j)$ while $K(y)$ is generally a proper subset of $K_j(y)$ when $0 < \sigma < 1$ and $y = (2m+\sigma)(\pi/j)$. Since K_j and K define f_j^* and f^* one can estimate (by using the Schwarz inequality several times and a simple Sobolev inequality) that $\|\nabla f_j^* - \nabla f^*\|_p \geq (\text{constant}) \int h^{1/2} \, d\mathscr{L}^1$, where $\mathscr{L}^1 \wedge h$ denotes the absolutely continuous part of $-\mathscr{G}_f'$.

Now, suppose that $f_j \to f$ in $\mathbf{W}^{1,p}$ and that f is co-area regular. As a further part of our Main Theorem we will indicate why $f_j^* \to f^*$ in $\mathbf{W}^{1,p}$. We infer, using Federer's co-area formula and dominated convergence, that

$$\omega_f(y) = \int_{f^{-1}\{y\}} \frac{1}{|\nabla f|} \, d\mathscr{H}^{n-1}$$

is well defined and finite for $\mathscr{L}^1$ almost every y ($\mathscr{H}^{n-1}$ denotes Hausdorff measure) and that

$$\int \omega_f \, d\mathscr{L}^1 = \int \chi_{\{\nabla f \neq 0\}} \, d\mathscr{L}^n.$$

The co-area formula fails to give information about the set $\{x: \nabla f = 0\}$. This missing information is contained in $\mathscr{G}_f$.

To compute f^* we must compute $\sigma_f(y) = \int \chi_{\{f>y\}} \, d\mathscr{L}^n$ and we have $\sigma_f'(y) = \mathscr{L}^1 \wedge \omega_f(y) + \mathscr{G}_f'(y)$. We show that $\liminf_{j\to\infty} \omega_{f_j}(y) \geq \omega_f(y)$ for almost every y. Let $\mathscr{L}^1 \wedge \delta_{(j)}(y)$ be the absolutely continuous part of $\sigma_{f_{(j)}}'(y)$. Since $\mathscr{G}_f'$ is purely singular (**this is where co-area regularity is used**) we infer that

$$\liminf_{j\to\infty} \delta_j(y) \geq \delta(y). \tag{1}$$

To prove the convergence of ∇f_j^* to ∇f^* we prove the convergence of arc length of the one dimensional graphs representing these functions in polar coordinates (this is the geometrically invariant notion). It turns out (using several involved convexity arguments) that if we use the $\mathbf{L}^p$ convergence of f_j^* to f^* (so that the graphs converge pointwise) only the absolutely continuous pieces $\delta_{(j)}$ are needed and that (1) suffices for our purposes.

References

[AL] F. Almgren and E. Lieb, *Symmetric decreasing rearrangement is sometimes continuous* (submitted).

[CJ] J-M. Coroń, *The continuity of the rearrangement in* $\mathbf{W}^{1,p}(\mathbf{R})$, Ann. Scuola Norm. Sup. Pisa Sér 4 **11** (1984), 57–85.

[KB] B. Kawohl, *Rearrangements and convexity of level sets in PDE*, Lecture Notes in Math., vol. 1150, Springer-Verlag, Berlin and New York, 1985, 134 pp.

[LE] E. Lieb, *Sharp constants in the Hardy-Littlewood-Sobolev and related inequalities*, Ann. of Math. (2) **118** (1983), 349–374.

[PS] G. Pólya and G. Szegö, *Isoperimetric inequalities in mathematical physics*, Ann. of Math. Studies no. 27, Princeton Univ. Press, Princeton, N. J., 1952.

[RS] B. Ruf and S. Solimini, *On a class of superlinear Sturm-Liouville problems with arbitrarily many solutions*, SIAM J. Math. Anal. **17** (1986), 761–771.

Department of Mathematics, Princeton University, Princeton, New Jersey 08544

THE (NON) CONTINUITY OF SYMMETRIC DECREASING REARRANGEMENT

FREDERICK J. ALMGREN JR. - ELLIOTT H. LIEB

Abstract. *The operation* $\mathcal{R}$ *of symmetric decreasing rearrangement maps* $W^{1,p}(\mathbb{R}^n)$ *to* $W^{1,p}(\mathbb{R}^n)$. *Even though it is norm decreasing we show that* $\mathcal{R}$ *is not continuous for* $n \geq 2$. *The functions at which* $\mathcal{R}$ *is continuous are precisely characterized by a new property called* **co-area regularity**. *Every sufficiently differentiable function is co-area regular, and both the regular and the irregular functions are dense in* $W^{1,p}(\mathbb{R}^n)$.

1. INTRODUCTION

Suppose $f(x^1, x^2) \geq 0$ is a continuously differentiable function supported in the unit disk in the plane. Its rearrangement is the rotationally invariant function $f^*(x^1, x^2)$ whose level sets are circles enclosing the same area as the level sets of f, i.e.

$$\mathcal{L}^2\{(x^1, x^2) : f(x^1, x^2) > y\} = \mathcal{L}^2\{(x^1, x^2) : f^*(x^1, x^2) > y\}$$

for each positive height y ($\mathcal{L}^n$ denotes Lebsgue over $\mathbb{R}^n$). Such rearrangement preserves L^p norms, i.e.

$$\int |f^*|^p d\,\mathcal{L}^2 = \int |f|^p d\,\mathcal{L}^2 \quad (1 \leq p < \infty)$$

but decreases convex gradient integrals, e.g.

$$\int |\nabla f^*|^p d\,\mathcal{L}^2 \leq \int |\nabla f|^p d\,\mathcal{L}^2 .$$

Now suppose that $f_j(x^1, x^2) \geq 0$ $(j = 1, 2, 3, \ldots)$ is a sequence of continuosly differentiable functions also supported in the unit disk which converge uniformly together with first derivatives to f, i.e.

$$f_j(x^1, x^2) \to f(x^1, x^2) \quad \text{and} \quad \nabla f_j(x^1, x^2) \to \nabla f(x^1, x^2)$$

uniformly in (x^1, x^2) as $j \to \infty$. It is not difficult to check that the symmetrized functions also converge uniformly. The real question is about convergence of the derivatives of the symmetrized functions. It is certainly plausible that they should converge strongly (we believed it for some time). Our principal new result is that *the derivatives of the symmetrized functions need not converge strongly,* e.g. for special f's and f_j's satisfying our conditions above it can happen that for every p

$$\lim_{j\to\infty} \inf \int |\nabla f_j^* - \nabla f^*|^p \mathrm{d}\mathcal{L}^2 > 0 .$$

Furthermore, we are able to characterize exactly those f's for which convergence is assured and for which it can fail.

The general notion of the symmetric decreasing rearrangement f^* of a function $f : \mathbf{R}^n \to \mathbf{R}^+$ is important in various parts of analysis. For example, various rotationally invariant variational integrals (like the gradient norms mentioned above) are not increased by symmetrization of competing functions. One is then free to search for a minimum among rotationally invariant decreasing functions (which are much easier to analyze since they are essentially functions of a single independent variable). A particular application of this technique has been in the computation of optimal constants for Sobolev inequalities.

Some years ago W. Ni and L. Nirenberg raised the question whether the rearrangement map $\mathcal{R} : f \to f^*$ is strongly continuous in the $W^{1,p}(\mathbf{R}^n)$ topology for all $1 \leq p < \infty$ (this would facilitate application of the «mountain pass lemma», for example). J-M. Coron [CJ] showed such strong continuity (and more) to be true in case $n = 1$, and we, at least, were led to the «obvious» conjecture that continuity holds for all n. We have settled this question [AL] – **rearrangement is not continuous in dimensions larger than one.** As indicated above, we can also identify precisely those f's at which the map $\mathcal{R}$ is continuous and those at which it is not. Our analysis has led us to isolate a property of functions which we call **co-area regularity** which deals with the behavior of functions on their critical sets. For $W^{1,p}$ functions our main result is

THEOREM 1. [AL] *For each $1 \leq p < \infty$ the rearrangement map $\mathcal{R}$ is $W^{1,p}(\mathbf{R}^n)$ continuous at a function f if and only if f is co-area regular.*

Each $W^{1,p}$ function on the line turns out to be necessarily co-area regular so that our theorem is consistent with Coron's result. For higher dimensional domains, however, there are always functions which are not co-area regular. In particular, in $\mathbf{R}^n (n \geq 2)$ there are irregular functions in $C^{n-1,\lambda}$ for each $0 < \lambda < 1$ (i. e. f's which are $n-1$ times continuosly differentiable with $(n-1)^{th}$ derivatives which are Hölder continuous with exponent λ). In fact these irregular functions are dense in $W^{1,p}(\mathbf{R}^n)$. However, each f with Lipschitz $(n-1)^{th}$ derivatives (i.e. $\lambda = 1$) is co-area regular.

In this note we shall briefly review symmetric rearrangement, introduce co-area regularity, sketch the construction of a co-area irregular function, give the reason that co-area irregularity implies lack of continuity of $\mathcal{R}$ in $W^{1,p}$, and finally sketch the reason that co-area regularity implies continuity of $\mathcal{R}$. Our proof of continuity discussed here uses the theory of rectifiable currents in an essential way. The version in [AL] uses more traditional functional analysis instead.

REMARK. One sometimes defines the symmetric decresing rearrangement of vector valued function $f : \mathbf{R}^n \to \mathbf{R}^m$ (as well as functions $\mathbf{R}^n \to \mathbf{R}^+$) by setting $f^* = |f|^*$. Sometimes it is also of interest to replace $W^{1,p}$ norms by gradient energies associated with integrals of other convex integrands $\psi : \mathbf{R}^+ \to \mathbf{R}^+$, i.e. $||\nabla f||_p^p = \int |\nabla f|^p d\,\mathcal{L}^n$ is replaced by $\int \psi(|\nabla f|)\,d\,\mathcal{L}^n$. These two generalizations are carried out in [AL] but are omitted here for simplicity. *The conclusions about continuity remain the same.*

It is worth pointing out that although the map $\mathcal{R}$ is not continuous for $W^{1,p}$ norms we show [AL].

THEOREM 2. *For each* $0 < \alpha < 1$, *each* $1 \leq p < \infty$, *and each* $n \geq 1$, *the rearrangement map* $\mathcal{R}$ *is continuous on the fractional Sobolev space* $W^{\alpha,p}(\mathbf{R}^n)$.

For $0 < \alpha < 1$ the norm $||f||_{W^{\alpha,p}}$ is given by

$$||f||^p_{W^{\alpha,p}} = ||f||^p_p + \int\int |f(x) - f(y)|^p\, |x-y|^{-n-p\alpha} d\,\mathcal{L}^n x\, d\,\mathcal{L}^n y.$$

We have the curious conclusion that co-area regularity plays a role for $W^{\alpha,p}$ only when $\alpha = 1$. Fractional derivatives, of course, are not a local construct.

2. REARRANGEMENTS AND CO-AREA REGULARITY

2.1. Rearrangements

We review the definition and basic properties of the symmetric decreasing rearrangement $f^* = \mathcal{R}f$ of a function $f : \mathbf{R}^n \to \mathbf{R}^+$. It is convenient to use the notation $\chi_{\{A\}} : \mathbf{R}^n \to \{0,1\}$ symbolically to denote a function which takes value 1 when the test A is passed and takes value 0 otherwise; e.g. $\chi_{\{f>y\}}(x)$ equals 1 in case $f(x) > y$ and equals 0 otherwise. Also we associate to a fixed function f a radius function $R : \mathbf{R}^+ \to \mathbf{R}^+$ defined by requiring

$$\alpha(n)R(y)^n = \int \chi_{\{f>y\}} d\mathcal{L}^n \tag{2.1}$$

for each y; here $\alpha(n)$ is the volume of the unit ball in $\mathbf{R}^n$. We further denote by $\chi_R : \mathbf{R}^n \to \{0,1\}$ the characteristic function of the open ball centered at the origin and of radius R. Finally, our **rearranged function**

$$f^* : \mathbf{R}^n \to \mathbf{R}^+$$

is defined by setting

$$f^*(x) = \int_{y>0} \chi_{R(y)}(x) d\mathcal{L}^1 y \tag{2.2}$$

for each x. It is immediate to check that f^* *is symmetric and decreasing,* i.e. $f(x) = f(z)$ if $|z| = |x|$ and $0 \leq f(x) \leq f(z)$ if $|x| \geq |z|$. It is also clear that f^* *is equimeasurable with* f, i.e.

$$\mathcal{L}^n(\{x : f(x) > y\}) = \mathcal{L}^n(\{x : f^*(x) > y\}) \tag{2.3}$$

for each $y > 0$.

Equation (2.3) implies immediately that *rearrangement preserves L^p norms,* i.e.

$$||f||_p = ||f^*||_p. \tag{2.4}$$

Moreover [CG], *rearrangement is a contraction on L^p,* i.e.

$$||f - g||_p \geq ||f^* - g^*||_p, \quad \text{whenever } f, g \in L^p. \tag{2.5}$$

In particular, R *is a continuous map from L^p into L^p.*

The function space $W^{1,p}(\mathbf{R}^n)$ consists of those functions f which belong to $L^p(\mathbf{R}^n)$ and whose distribution gradients ∇f are functions belonging to $L^p(\mathbf{R}^n, \mathbf{R}^n)$. It has long been known [B] [BZ] [H] [K] [L] [PS] [S1] [S2] [T] that $\mathcal{R}$ is $W^{1,p}$ *norm non-increasing,* i.e.

$$||\nabla f||_p \geq ||\nabla f^*||_p. \tag{2.6}$$

This implies that $\mathcal{R}f$ *also belongs to $W^{1,p}$*. (Actually, when $p = 1$ it is not obvious that f^* is in $W^{1,1}$ and not merely in BV; this was proved by Hilden

[H].) However, $\mathcal{R}$ *is not a contraction mapping.* Indeed, $||\nabla f - \nabla g||_p$ can be arbitrarily large compared to $||\nabla f^* - \nabla g^*||_p$. To see why this can happen, suppose that $f, g : \mathbf{R} \to \mathbf{R}^+$ are smooth functions with $f(x) = g(x)$ for $x \leq 0$ and $f(x) > g(x)$ for $x > 0$. Suppose also, for $x \leq 0$, that both ∇f (and hence ∇g) are very large in L^p norm while, for $x > 0$, both ∇f and ∇g are of order 1 in L^p norm. Then $||\nabla f - \nabla g||_p$ is of order one because of the cancellation for $x \leq 0$. On the other hand it is easy to arrange things so that the rearrangement destroys this cancellation so that $||\nabla f^* - \nabla g^*||_p$ will be large.

These facts suggest some of the subtlely of questions about the continuity of $\mathcal{R}$ on $W^{1,p}$. We can phrase our question in the following way.

Given $f, f_1, f_2, \ldots$ in $W^{1,p}$ with $f_j \to f$ in $W^{1,p}$, is it true that $A_j = ||\nabla f_j^* - \nabla f^*||_p$ ultimately converges to 0 as $j \to \infty$ even though A_j may be large for very many j's?

2.2. Co-area Regularity

Instead of the integral in (2.1) representing the full crossectional area at height y of the subgraph of our function f, consider the integral

$$\mathcal{G}_f(y) = \int \chi_{\{f>y\}} \chi_{\{\nabla f = 0\}} d\mathcal{L}^n \tag{2.7}$$

which, for each y, represents that part of the crossection of the subgraph associated with critical points of f. Since our function $\mathcal{G}_f : \mathbf{R}^+ \to \mathbf{R}^+$ is nonincreasing its distribution first derivative $\mathcal{G}'_f$ is a (negative) measure. Since a smooth function must be constant on any connected open set on which its gradient vanishes, there are many functions f for which the contributions to the integral in (2.7) come only from flat parts of the graph corresponding to those positive numbers y for which the set $\{x : f(x) = y\}$ has positive measure. Since there can be at most countably many such y's, the measure $\mathcal{G}'_f$ would then be singular with respect to $\mathcal{L}^1$ on $\mathbf{R}^+$. This situation is not the most general one, however, and there are «irregular» smooth functions f for which the measure $\mathcal{G}'_f$ has an absolutely continuous piece as well. Indeed, we have the following theorem.

THEOREM 3. [AL] *For each $n \geq 2$ and each $0 < \lambda < 1$, there is (by construction) a positive constant C and a function $f : \mathbf{R}^n \to [0,1]$ withe the following properties.*

(1) The function f belongs to $C^{n-1,\lambda}(\mathbf{R}^n)$ and has support equal to the cube

$$Q = \{x : |x^i| \leq 1 \text{ for each } i = 1, \ldots, n\}$$

of side length 2.

(2) For each $0 < y \leq 1$,

$$\mathcal{G}_f(y) = C(1-y).$$

In particular, the measure $\mathcal{G}'_f$ *is absolutely coninuous with respect to* $\mathcal{L}^1$. *Thus* f *is co-area irregular. (See Definition below).*

It can be difficult to picture such a function. Somehow its gradient vanishes on a set of positive $\mathcal{L}^n$ measure containing no open subsets or flat spots, i.e. $\mathcal{L}^n(\{x : f(x) = y\}) = 0$ for every y. Furthermore, the image of the critical set is distributed uniformly over all y values in the range [0, 1].

Theorem 3 also tells us that the following definition is not an empty one.

DEFINITION. *A function* f *in* $W^{1,p}$ *is called* **co-area regular** *if and only if the measure* $\mathcal{G}'_f$ *(see (2.7)) is purely singular with respect to* $\mathcal{L}^1$. *Otherwise* f *is called* **co-area irregular**.

The term co-area in these definitions was suggested by H. Federer's «co-area formula» which gives an integral representation of the absolutely continuous function

$$y \mapsto \int \chi_{\{f>y\}}\chi_{\{\nabla f \neq 0\}} \mathrm{d}\,\mathcal{L}^n.$$

A mild generalization [AL] of the Morse-Sard-Federer theorem shows that each f belonging to $C^{n-1,1}$ is automatically co-area regular. An easy argument then shows

THEOREM 4. [AL] *For each* $n \geq 2$ *and each* $p \geq 1$, *the co-area regular and the co-area irregular functions are each dense in* $W^{1,p}(\mathbf{R}^n)$.

Questions of the behavior of functions on their critical sets have a substantial mathematical heritage both in theory and in examples. We here sketch the construction of a function f as in Theorem 3 when $n = 2$. First set $f(x) = 0$ for $x \notin Q$. For $x \in Q$ we will use 4-adic notation to express the values of our f, i.e. we will write

$$f(x) = \sum_{\ell=1}^{\infty} 4^{-\ell} a_\ell(x) \qquad \text{with } a_\ell(x) \in \{0,1,2,3\}.$$

First divide Q in the obvious way into four squares each of side length 1 and label these squares $S_0^{(1)}, S_1^{(1)}, S_2^{(1)}, S_3^{(1)}$ in clockwise order. Set $a_1(x) = j$ if $x \in S_j^{(1)}$ (don't worry about the boundaries of the $S_j^{(1)}$'s). Next, divide each $S_j^{(1)}$ into four squares each of side length $\frac{1}{2}$ and label these $S_{jk}^{(2)}$ (with $k = 0,1,2,3,$) in the same clockwise order. Set $a_2(x) = k$ if $x \in S^{(2)}_{jk}$. The construction continues in the obvious way ulimately to define an f. For each $0 < a < b < 1$ we have $\mathcal{L}^2(f^{-1}(a,b)) = 4(b-a)$. At present our f is not even continuous much less smooth. We fix this up by modifying this construction. We replace each a_ℓ by a carefully constructed smooth function b_ℓ in our sum above. The support of each b_ℓ is contained within the $4^{\ell-1}$ squares on which $b_{\ell-1}$ assumes constant values, and b_ℓ assumes constant values on 4^ℓ squares nested within the $b_{\ell-1}$ constant value squares. The subgraph then resembles a union of step pyramids (like Zhoser not Cheops) with those at the ℓ-th level having bases on the tops of those at the $\ell-1$-th level. With some effort one can construct the b_ℓ's so that $f \in C^{1,\lambda}$ and $\{x : \nabla f = 0\}$ has positive measure. As expected the measure of the set $\{x : \nabla f = 0\}$ goes to zero as λ approaches 1.

3. REARRANGEMENT IS DISCONTINUOUS AT CO-AREA IRREGULAR FUNCTIONS

THEOREM 5. [AL] *Suppose $n \geq 2$ and f is a co-area irregular function belonging to $W^{1,p}(\mathbf{R}^n)$. Then there is a sequence $f_1, f_2, f_3, \ldots$ of functions in $W^{1,p}(\mathbf{R}^n)$ such that $f_j \to f$ in $W^{1,p}(\mathbf{R}^n)$ as $j \to \infty$ but $f_j^* \not\to f^*$. Moreover, for each $\epsilon > 0$, the f_j's can be chosen with the following properties.*

(1) The sequence of differences $f_j - f$ converges to zero in $L^\infty(\mathbf{R}^n)$.

(2) There is a positive number Y such that

$$f_j(x) = f(x) \qquad \textit{whenever} \quad f(x) < Y \text{ or } f(x) > Y + \epsilon$$

and

$$Y \leq f_j(x) \leq Y + \epsilon \qquad \textit{whenever} \quad Y \leq f(x) \leq Y + \epsilon.$$

(3) For $\mathcal{L}^n$ almost every x,

$$|\nabla f_j(x)| \leq \frac{3}{2}|\nabla f(x)| + \epsilon$$

(4) The measure of the set

$$\{x : \nabla f(x) \neq 0 \textit{ and } \nabla f_j(x) \neq \nabla f(x)\}$$

converges to zero as $j \to \infty$.

If we do not require properties (2), (3), (4) then the difference $f_j - f$ can be chosen to belong to C^∞. If we drop all four properties then each f_j can be chosen to belong to C^∞. The basic idea behind the proof of Theorem 5 (omitting refinements

(1), (2), (3), (4)) is the following. Let W be the characteristic function of the critical set of f, i.e. the set for which $\nabla f = 0$, and set

$$f_j(x) = f(x) + \frac{1}{2j} W(x) \sin(jf(x)) \tag{3.1}$$

for each x. Then clearly $f_j \to f$ in L^p as $j \to \infty$. For the gradients we compute formally

$$\begin{aligned}\nabla f_j(x) - \nabla f(x) = {} & \frac{1}{2} W(x) \nabla f(x) \cos(jf(x)) \\ & + \frac{1}{2j} \nabla W(x) \sin(jf(x)).\end{aligned} \tag{3.2}$$

The first term on the right side is zero since W vanishes when ∇f does not vanish. The second term on the right side in (3.2) is a bit problematic since ∇W is not p-th power summable. This defect, however, can be remedied (with some effort) by mollifying W in a j-dependent way so that $\|\nabla W\|_\infty < j^{1/2}$. This establishes the L^p convergence of ∇f_j to ∇f.

Now define sets

$$K_j(y) = \{x : f_j(x) > y\} \quad (j = 1, 2, 3, \ldots) \quad \text{and} \quad K(y) = \{x : f(x) > y\}$$

for each y. Since the function $t \mapsto t + \frac{1}{2j} \sin(tj)$ is increasing (check the derivative) we infer that $K_j(y) = K(y)$ whenever m is an integer and $y = 2m\pi/j$. For these special y values we infer that the radius functions are equal, i.e. $R_j(y) = R(y)$ (recall (2.1)). On the other hand, if $0 < \sigma < 1$ and $y = (2m + \sigma)(\pi/j)$, then $R_j(y) \geq R(y)$ and, in general, $R_j(y) \geq R(y)$.

Think of the graphs of f_j^* and f^* parametrized by the height y instead of the radius $|x|$. When $y = 2m\pi/j$ the graphs intersect. When $y = (2m+\sigma)(\pi/j)$ and $0 < \sigma < 1$, the graph of f_j^* lies to the right of the graph of f^*. For our purposes it sufficies to show that the numbers $B_j \equiv \|\nabla f_j^* - \nabla f^*\|_1$ are bounded away from zero. We then try to estimate the B_j's in terms of the distribution $\mathcal{G}_f$ from (2.7). Using the Schwarz inequality several times and a simple Sobolev inequality we are able to estimate

$$B_j \geq (\text{constant}) \int |h|^{1/2} \mathrm{d}\mathcal{L}^1; \tag{3.3}$$

here $\mathcal{L}^1 \wedge h$ denotes the absolutely continuous part of our $\mathcal{G}'_f$.

It is reassuring that the bound (3.3) above involves $|h|^{1/2}$ instead of $|h|$. This is so because «the square root of a *singular* measure is zero»; by this we mean that if the singular part of $\mathcal{G}'_f$ (which cannot contribute to the lack of convergence, as we assert in the next section) is approximated by absolutely continuous measures $\mathcal{L}^1 \wedge \tilde{h}^{(k)}$ $(k = 1, 2, 3, \ldots)$, then $\int |\tilde{h}^{(k)}|^{1/2} \mathrm{d}\mathcal{L}^1$ converges to zero as $k \to \infty$.

4. REARRANGEMENT IS CONTINUOUS AT CO-AREA REGULAR FUNCTIONS

The proof [AL] that the co-area regularity of f implies $W^{1,p}$ continuity of $\mathcal{R}$ at f is quite technical. We will attempt to outline some of the main ideas. In our proof in [AL] sections 4.2 and 4.3 below are replaced by more traditional methods in functional analysis.

4.1. Reduction to $W^{1,1}$

Our first step is to establish the fact that continuity of $\mathcal{R}$ in $W^{1,p}$ is implied by continuity of $\mathcal{R}$ in $W^{1,1}$. This may seem surprising since ordinarily nothing can be inferred about $||\nabla f_j^* - \nabla f^*||_p$ from information about $||\nabla f_j^* - \nabla f^*||_1$. In the present case, however, our rearrangement operator $\mathcal{R}$ acts independently on slabs $\{x : Y_1 < f(x) < Y_2\}$. We can then surgically remove small, well chosen slabs from the f_j and f on which $|\nabla f_j^*|$ or $|\nabla f^*|$ is large. On these slabs we can control $||\nabla f_j^* - \nabla f^*||_p$ in terms of $||\nabla f_j^*||_p$ and $||\nabla f^*||_p$ and these quantities can, in turn, be controlled by $||\nabla f_j||_p$ and $||\nabla f||_p$ with use of the basic inequality (2.6). After these small slabs are removed, the f_j and f effectively have bounded gradients and then $W^{1,1}$ convergence implies $W^{1,p}$ convergence.

4.2. The co-area formula and co-area regularity

The basic tool in our second step is H. Federer's co-area formula as extended by J. Brothers and W. Ziemer [BZ]. Suppose $f \in W^{1,1}(\mathbf{R}^n)$ and g is a nonnegative Borel function. Then the slice integral

$$A(y) \equiv \int_{f^{-1}\{y\}} g \, \mathrm{d}\mathcal{H}^{n-1} \tag{4.1}$$

exists for $\mathcal{L}^1$ almost every positive number y and we have the *co-area formula*

$$\int_{y>0} A \, \mathrm{d}\mathcal{L}^1 = \int g|\nabla f| \, \mathrm{d}\mathcal{L}^n; \tag{4.2}$$

here $\mathcal{H}^{n-1}$ denotes Hausdorff's $(n-1)$-dimensional measure over $\mathbf{R}^n$. In one application of (4.2) we replace $f(x)$ by $F_t(x) = \max\{f(x), t\}$ (with $t > 0$), then replace $g(x)$ by $(|\nabla f(x)| + \delta)^{-1}$, then let $\delta \to 0+$, and finally use Lebesgue's monotone convergence theorem applied to each side of (4.2) to infer

$$\int_{y>0} \omega_f(y) \, \mathrm{d}\mathcal{L}^1 y = \int \chi_{\{f>t\}} \chi_{\{\nabla f \neq 0\}} \mathrm{d}\mathcal{L}^n \equiv \gamma_f^{(t)} \tag{4.3}$$

where we have written

$$\omega_f(y) = \int_{f^{-1}\{y\}} |\nabla f|^{-1} d\mathcal{H}^{n-1} \tag{4.4}$$

for each y. In other words, the basic distribution integral on the right side of (2.1) (call it $\sigma_f(y)$) breaks up naturally into two pieces

$$\sigma_f(y) = \gamma_f(y) + \mathcal{G}_f(y) \tag{4.5}$$

and (4.3) states that γ_f is absolutesly continuous with derivative $-\omega_f$. The KEY POINT is: *the only absolutely continuous part of the measure $-\sigma'_f$ is ω_f if and only if f is co-area regular.*

4.3. Currents and the lower semicontinuity of slice integrals

Suppose that we have a sequence f_j converging to f in $W^{1,1}$ and that f is co-area regular. Henceforth we will omit the subscript f (e.g. σ_f will be denoted σ) when referring to f, and will use a subscript j when referring to f_j (e.g. σ_{f_j} will be denoted σ_j). We assert that

$$\liminf_{j\to\infty} \omega_j(y) \geq \omega(y) \qquad \mathcal{L}^1 \text{ almost every } y. \tag{4.6}$$

To show this it sufficies to prove that

$$\lim_{j\to\infty} \int_{f_j^{-1}\{y\}} g\, d\mathcal{H}^{n-1} = \int_{f^{-1}\{y\}} g\, d\mathcal{H}^{n-1} \qquad \text{for } \mathcal{L}^1 \text{ almost every } y \tag{4.7}$$

whenever $g \in L^\infty$. An approximation argument shows it is sufficient to prove (4.7) for $g \in C_0^\infty$. It is here that we need to utilize the inherent current structure of the graph and subgraph of f and the f_j's and the inherent convergence as currents. To do this we form the $n+1$ dimensional current

$$Q = E^{n+1} \llcorner \{(x,y) : x \in \mathbf{R}^n, y < f(x)\}$$

whose boundary $T = \partial Q$ is the current associated with the graphs of f. The current T can then be sliced by the coordinate function y to obtain an $n-1$ dimensional slice current $T(y)$ corresponding to the level set f^{-1} for $\mathcal{L}^1$ almost every y. Likewise, we define $Q_j, T_j, T_j(y)$ for the various j's and further set $S_j = Q - Q_j$ with associated slice currents $S_j(y)$. Since «slicing commutes with boundaries» in the current setting we infer $\partial S_j(y) = T(y) - T_j(y)$ for almost every y.

Since the mass M of a current corresponds to its volume, we readily check that

$$M(S_j) = M(Q - Q_j) = \|f - f_j\|_1 \to 0 \qquad \text{as } j \to \infty. \tag{4.8}$$

Since $M(S_j) = \int M(S_j(y))\mathrm{d}\mathcal{L}^1 y$ for each j, there will be a subsequence (still denoted by j's) such that

$$\lim_{j\to\infty} M(S_j(y)) = 0 \qquad \text{for almost every } y. \tag{4.9}$$

Since $\partial S_j(y) = T(y) - T_j(y)$ we conclude the convergence of the $T_j(y)$'s to $T(y)$ for almost every y. The lower semicontinuity of mass under such convergence then implies

$$\liminf_{j\to\infty} M(T_j(y)) \geq M(T(y)) \quad \text{for } \mathcal{L}^1 \text{ almost every } y. \tag{4.10}$$

Using, for example, J. Michael's [M] Lipschitz approximation theorem we readily infer

$$M(T_{(j)}(y)) = \mathcal{H}^{n-1}(f_{(j)}^{-1}(y)) \qquad \text{for } \mathcal{L}^1 \text{ almost every } y; \tag{4.11}$$

here (j) denotes either j or no j. We use the co-area formula again to infer

$$\int M(T_{(j)}(y))\mathrm{d}\mathcal{L}^1 y = \int |\nabla f_{(j)}|\mathrm{d}\mathcal{L}^n. \tag{4.12}$$

However, $\int |\nabla f_j|\mathrm{d}\mathcal{L}^n \to \int |\nabla f|\mathrm{d}\mathcal{L}^n$ by the assumed L^1 convergence of ∇f_j to ∇f.

The following is a general lemma. Suppose μ is a measure and $h, h_1, h_2, h_3, \ldots$ are nonnegative, summable functions such that $\liminf_j h_j(x) \geq h(x)$ for μ almost every x. In case $\int h_j \mathrm{d}\mu \to \int h \mathrm{d}\mu$ as $j \to \infty$ then there is a subsequence $j(k)$ of the j's such that $h_{j(k)}(x) \to h(x)$ as $k \to \infty$ for μ almost every x.

We apply this lemma to the case at hand to infer that, for a further subsequence,

$$\liminf_{j\to\infty} M(T_j(y)) = M(T(y)) \qquad \text{for } \mathcal{L}^1 \text{ almost every } y. \tag{4.13}$$

Equation (4.13), with a little more work, then leads to (4.7).

As an application of (4.7) we return to (4.4) and prove that

$$\liminf_{j\to\infty} \omega_j(y) \geq \omega(y) \qquad \text{for } \mathcal{L}^1 \text{ almost every } y. \tag{4.14}$$

This result is crucial for us. To prove it, we use (4.7) with $g_{(j)}(y) = (|\nabla f_{(j)}|+\delta)^{-1}$ (as in the proof of (4.4)) and then let $\delta \to 0$.

4.4. Graph arc length as an invariant measure

The last main step in our proof is to combine (4.14), the co-area regularity of f, and the $W^{1,1}$ convergence of the f_j's to f to show that the ∇f_j^*'s convergence to ∇f^* in L^1. Since $f_{(j)}^*(x)$ is really only a function of $r = |x|$, our considerations are essentially one-dimensional. (It is true that the real measure is $r^{n-1}\,dr$ and not dr, but this is merely a nuisance which one can handle). Let us suppose then $n = 1$ and we will denote d/dr by a prime.

Think of the graph of f^* (or f_j^*) which is a curve in $\mathbb{R}^2$. The geometrically invariant notion is not $f^{*\prime}$ (which is the quantitativity in which we are really interested) but rather the arc length derivative $(1+(f^{*\prime})^2)^{1/2}$. The arc length can be computed in two different ways. The first way is to use the height y as parameter. Then the arc length of the graph of $f_{(j)}^*$ equals

$$\int (1+(\rho_{(j)}(y))^2)^{\frac{1}{2}}\mathrm{d}\mathcal{L}^1 y + \int \mathrm{d}\nu_{(j)};$$

here $\nu_{(j)}$ is the singular part of the measure $-(\mathrm{d}\sigma_{(j)}/\mathrm{d}y)$ while $\mathcal{L}^1 \wedge \rho_{(j)}$ is the absolutely continuous part of $-(\mathrm{d}\sigma_{(j)}/\mathrm{d}y)$. *The crucial point is the following: The co-area regularity of f implies that $\rho(y) = \omega(y)$*. For f_j^*, all we can say is that $\rho_j(y) \geq \omega_j(y)$; but this is of no concern since, from (4.14), we have

$$\text{(4.15)} \qquad \liminf_{j\to\infty} \rho_j(y) \geq \rho(y) \qquad \text{for } \mathcal{L}^1 \text{ almost every } y.$$

Concerning the singular components ν_j and ν one knows nothing. However, by the L^1 convergence of f_j^* to f^* (see (2.5)) we can infer that the arcs convergence pointwise, i.e. for any $0 < a < b$

$$\text{(4.16)} \qquad \int_a^b \rho_j \mathrm{d}\mathcal{L}^1 + \int_a^b \mathrm{d}\nu_j \to \int_a^b \rho\mathrm{d}\mathcal{L}^1 + \int_a^b \mathrm{d}\nu.$$

It is then a simple exercise to show that (4.15), (4.16) alone imply *arc length convergence*, i.e.

$$\text{(4.17)} \qquad \int (1+\rho_j^2)^{1/2}\mathrm{d}\mathcal{L}^1 + \int \mathrm{d}\nu_j \to \int (1+\rho^2)^{1/2}\mathrm{d}\mathcal{L}^1 + \int \mathrm{d}\nu.$$

Now think about this arc length convergence (4.17) in terms of the radius parameterization, i.e.

$$\text{(4.18)} \qquad \int (1+(f_j^{*\prime}(r))^2)^{1/2}\mathrm{d}\mathcal{L}^1 r \to \int (1+(f^{*\prime}(r))^2)^{1/2}\mathrm{d}\mathcal{L}^1 r.$$

There is no singular part of the measure (since $f_j^{*\prime}$ is a function). Intuitively, it is clear (by drawing a few graphical examples) that arc length convergence implies L^1 convergence of $f_j^{*\prime}$ to $f^{*\prime}$ because the function $t \mapsto (1+t^2)^{1/2}$ is strictly convex. This is indeed correct as the following general theorem [AL] states.

THEOREM 6. *Suppose $\psi : \mathbf{R}^n \to \mathbf{R}^+$ is a convex function. Suppose also that $f, f_1, f_2, f_3, \ldots$ are functions in $L^1_{\mathrm{loc}}(\mathbf{R}^n, \mathbf{R})$ having distributional gradients which are functions in $L^1_{\mathrm{loc}}(\mathbf{R}^n, \mathbf{R})$. Suppose that $\psi(\nabla f), \psi(\nabla f_1), \psi(\nabla f_2), \psi(\nabla f_3), \ldots$ also are functions in $L^1(\mathbf{R}^n)$ and that $f_j - f \to 0$ in $L^1(\mathbf{R}^n)$ as $j \to \infty$. Then (as has been known for some time [SJ])*

$$\text{(1)} \qquad \liminf_{j\to\infty} \int \psi(\nabla f_j)\, d\mathcal{L}^n \geq \int \psi(\nabla f)\, d\mathcal{L}^n.$$

(2) Suppose further that equality holds in (1) and that ψ is strictly convex (i.e. $\psi(x) + \psi(y) > 2\psi\left(\frac{x+y}{2}\right)$ whenever $x \neq y$). Uniform convexity is not assumed. Then $\psi(\nabla f_j) \to \psi(\nabla f)$ in $L^1(\mathbf{R}^n)$ as $j \to \infty$. Furthermore, there is a subsequence $j(1), j(2), j(3), \ldots$ of $1, 2, 3, \ldots$ such $\nabla f_{j(k)}(x) \to \nabla f(x)$ for $\mathcal{L}^n$ almost every x as $k \to \infty$.

(3) Finally, suppose $\psi(\xi) \to \infty$ as $|\xi| \to \infty$ (e.g. our function $\xi \mapsto (1 + |\xi|^2)^{1/2}$). Then, for every measurable subset Ω of $\mathbf{R}^n$ of finite measure, $\nabla f_j \to \nabla f$ in $L^1(\Omega, \mathbf{R}^n)$.

REFERENCES

[AL] F. ALMGREN and E. LIEB: Symmetric decresing rearrangement is sometimes continuous, J. Amer. Math. Soc. **2**, 683-773 (1989).

[B] C. BANDLE: Isoperimetric inequalities and applications. Pitman (Boston, London, Melboune), 1980.

[BZ] J. BROTHERS and W. ZIEMER: Minimal rearrangements of Sobolev functions. *Journ. Reine Angew. Math.* **384**, 153-179 (1988).

[CG] G. CHITI: Rearrangements of functions and convergence in Orlicz spaces. *Appl. Anal.* **9**, 23-27 (1979).

[CJ] J-M. CORON: The continuity of the rearrangement in $W^{1,p}(\mathbb{R})$. *Ann. Scuol. Norm. Sup. Pisa, Ser 4*, **11**, 57-85 (1984).

[H] K. HILDEN: Symmetrization of functions in Sobolev spaces and the isoperimetric inequality. *Manuscr. Math.* **18**, 215-235 (1976).

[K] B. KAWOHL: Rearrangements and convexity of level sets in partial differential equations. *Lect. Notes in Math.* **1150**, Springer (Berlin, Heidelberg, New York), 1985.

[L] E. LIEB: Existence and uniqueness of the minimizing solution of Choquard's nonlinear equation. *Stud. Appl. Math.* **57**, 93-105 (1977). See appendix.

[M] J. MICHAEL: Lipschitz approximations to summable functions. *Acta Math.* **111**, 73-94 (1964).

[PS] G. POLYA and G. SZEGÖ: Isoperimetric inequalities in mathematical physics. *Ann. Math. Stud.* **27**, Princeton University Press (Princeton) (1951).

[SJ] J. SERRIN: On the definition and properties of certain variational integrals. *Trans. Amer. Math. Soc.* **101**, 139-167, (1961).

[S1] E. SPERNER: Zur symmetrisierung von Funktionen auf Sphären. *Math. Z.* **134**, 317-327 (1973).

[S2] E. SPERNER: Symmetrisierung für Funktionen mehrerer reeller Variablen. *Manuscr. Math.* **11**, 159-170 (1974).

[T] G. TALENTI: Best constant in Sobolev inequality. *Ann. Pura Appl.* **110**, 353-372 (1976).

ADVANCES IN MATHEMATICS **117**, 193–207 (1996)
ARTICLE NO. 0008

On the Case of Equality in the Brunn–Minkowski Inequality for Capacity

LUIS A. CAFFARELLI*

School of Mathematics, Institute for Advanced Study, Princeton, New Jersy, 08540; and Courant Institute for the Mathematical Sciences, New York, New York 10012-1110

DAVID JERISON*

Department of Mathematics, Massachusetts Institute of Technology, Cambridge, Massachusetts 02139-4307

AND

ELLIOTT H. LIEB*

Departments of Mathematics and Physics, Jadwin Hall, Princeton University, P.O. Box 708, Princeton, New Jersey 08544-0708

Received July 28, 1995

Suppose that Ω_0 and Ω_1 are convex, open subsets of $\mathbf{R}^N$. Denote their convex combination by

$$\Omega_t = (1-t)\,\Omega_0 + t\Omega_1 = \{(1-t)\,x + ty : x \in \Omega_0 \text{ and } y \in \Omega_1\}.$$

The Brunn–Minkowski inequality says that

$$(\mathrm{vol}\,\Omega_t)^{1/N} \geqslant (1-t)\,\mathrm{vol}\,\Omega_0^{1/N} + t\,\mathrm{vol}\,\Omega_1^{1/N}$$

for $0 \leqslant t \leqslant 1$. Moreover, if there is equality for some t other than an endpoint, then the domains Ω_1 and Ω_0 are translates and dilates of each other.

Borell proved an analogue of the Brunn–Minkowski inequality with capacity (defined below) in place of volume. Borell's theorem [B] says

* The work of the first author was partially supported by NSF Grant DMS-9101324. The work of the second author was partially supported by NSF Grant DMS-9401355. The work of the third author was partially supported by NSF Grant PHY90-19433 A04.

THEOREM A. *Let $\Omega_t = t\Omega_1 + (1-t)\Omega_0$ be a convex combination of two convex subsets of $\mathbf{R}^N$, $N \geq 3$. Then*

$$(\operatorname{cap} \Omega_t)^{1/(N-2)} \geq (1-t) \operatorname{cap} \Omega_0^{1/(N-2)} + t \operatorname{cap} \Omega_1^{1/(N-2)}$$

for $0 \leq t \leq 1$.

The main purpose of this note is to prove.

THEOREM B. *There is equality in the inequality of Theorem A if and only if Ω_1 is a translate and dilate of Ω_0.*

The case of equality in the classical Brunn–Minkowski inequality can be used to prove uniqueness in the Minkowski problem described below. In particular, it implies that any two convex bodies with the same Gauss curvature (as a function of the unit normal) are translates of each other. Theorem B will be used to prove uniqueness for an analogous problem associated to the first variation of capacity [J1, J2]. There is a similar theory in the case $N = 2$ in which the capacity is replaced by the transfinite diameter (the exponential of the logarithmic capacity).

1. The Minkowski Problem and Its Variational Formulation

Let g denote the Gauss map, that is, the map from $\partial\Omega$ to S^n, $n = N-1$, that sends a point X to $\partial\Omega$ to the outer unit normal to Ω at X. The mapping g is defined almost everywhere with respect to surface measure $d\sigma$ on $\partial\Omega$. We define a measure μ_Ω on S^n by $d\mu_\Omega = g_*(d\sigma)$, i.e.,

$$\mu_\Omega(E) = \sigma(g^{-1}(E))$$

for every Borel subset E of S^n, is a measure on S^n. The Minkowski problem asks under what conditions on μ one can find a convex, bounded open set Ω such that $\mu_\Omega = \mu$. In the case of measures that consist of a finite number of point masses, each mass corresponds to the area of a face of a convex polyhedron and the location on S^n of the point mass is the unit normal to the face. Thus the problem is to find a convex polyhedron given the areas of its faces and the normals to the faces. In the case the measure μ has a smooth positive density with respect to the uniform measure $d\xi$ on the sphere, $d\mu = (1/K)\, d\xi$, the function K is the Gauss curvature of $\partial\Omega$ and the problem can be restated as the problem of finding a convex body given its Gauss curvature as a function of the unit normal.

Here are the basic existence and uniqueness theorems:

THEOREM 1.1. *Let μ be a positive Borel measure on S^n, $n=N-1$. There exists a bounded, convex open set $\Omega \subset \mathbf{R}^N$ such that $\mu_\Omega = \mu$ if and only if*

(a) $\mu(\{\xi\colon \xi\cdot e>0\})>0$ *for every* $e\in S^n$ *and*

(b) $\int_{S^n} e\cdot\xi \, d\mu(\xi)=0$ *for every* $e\in S^n$.

THEOREM 1.2. $\mu_{\Omega_0}=\mu_{\Omega_1}$ *if and only if* Ω_0 *and* Ω_1 *are translates of each other.*

The Minkowski problem can be solved variationally. Let Ω be a convex domain. The support function μ_Ω of Ω is the function defined for $\xi\in S^n$ by

$$u_\Omega(\xi)=\sup\{X\cdot\xi : X\in\Omega\}.$$

The support function determines Ω because

$$\Omega=\{X\in\mathbf{R}^N : X\cdot\xi<u_\Omega(\xi) \text{ for all } \xi\in S^n\}.$$

Consider the functional

$$\lambda=\inf\left\{\int_{S^n} u_\Omega \, d\mu : \text{convex } \Omega \text{ such that vol } \Omega\geqslant 1\right\}. \tag{$*$}$$

THEOREM 1.3. *If μ is a finite positive measure satisfying* (a) *and* (b) *of Theorem* 1.1, *then $\lambda>0$ and a minimizer Ω of* $(*)$ *exists. Moreover, it is unique up to translation, and it solves $\mu_\Omega=N\lambda^{-1}d\mu$. One then recovers the solution of Theorem* 1.1 *by dilation.*

The Lagrange multiplier factor $N\lambda^{-1}$ arises from the volume constraint and the relation

$$\text{vol } \Omega=\frac{1}{N}\int_{S^n} u_\Omega \, d\mu_\Omega. \tag{1.4}$$

The proofs of Theorems 1.1–1.3 are contained in [BF, CY].

In parallel with the Minkowski problem there is a problem of prescribing the first variation of capacity [J2]. To define capacity, let $N\geqslant 3$ and let Ω be a bounded, convex, open subset of $\mathbf{R}^N$. The equilibrium potential of Ω is the continuous function U defined in $\bar{\Omega}'=\mathbf{R}^N\backslash\Omega$ satisfying

$$\Delta U=0 \quad \text{in } \Omega' \qquad \text{and} \qquad U=1 \quad \text{on } \partial\Omega'$$

and such that U tends to zero at infinity. The electrostatic capacity of Ω is defined as the constant $\gamma=\text{cap }\Omega$ such that

$$U(x)=\gamma a_N\,|x|^{2-N}+O(|x|^{1-N}) \qquad \text{as} \quad x\to\infty$$

where the dimensional constant a_N is chosen according to the fundamental solution of Laplace's equation

$$\Delta(-a_N\,|x|^{2-N})=\delta_0.$$

By a theorem of Dahlberg, $|\nabla U|^2$ is defined almost everywhere on $\partial\Omega$ and integrable with respect to surface measure. Define ν_Ω by

$$d\nu_\Omega=g_*(|\nabla U|^2\,d\sigma).$$

The analogous problem is to find a convex domain Ω such that $\nu_\Omega=\nu$. The associated functional is

$$m=\inf\left\{\int_{S^n}u_\Omega\,d\nu : \text{convex } \Omega \text{ such that } \operatorname{cap}\Omega\geqslant 1\right\}. \qquad (\dagger)$$

The results analogous to Theorems 1.1 through 1.3 are

THEOREM 1.5. *Let $N\geqslant 4$, $n=N-1$. Suppose that ν is a positive measure on S^n. There exists a bounded, convex, open set $\Omega\subset\mathbf{R}^N$ such that $\nu_\Omega=\nu$ if and only if*

(a) $\nu(\{\xi:\xi\cdot e>0\})>0$ *for every* $e\in S^n$ *and*

(b) $\int_{S^n}e\cdot\xi\,d\nu(\xi)=0$ *for every* $e\in S^n$.

When $N=3$, conditions (a) *and* (b) *hold if and only if there exists a number $c>0$ and a bounded, convex, open set Ω that $\nu_\Omega=c\nu$.*

THEOREM 1.6. *Let $N\geqslant 4$. Then $\nu_{\Omega_0}=\nu_{\Omega_1}$ if and only if Ω_0 and Ω_1 are translates of each other. When $N=3$, $\nu_{\Omega_0}=\nu_{\Omega_1}$ if and only if Ω_0 and Ω_1 are translates and dilates of each other.*

THEOREM 1.7. *If $N\geqslant 4$, and ν is a finite, positive measure satisfying* (a) *and* (b) *of Theorem* 1.5, *then $m>0$ and a minimizer Ω of* ($\dagger$) *exists. Moreover, it is unique up to translation, and it solves $g_*(|\nabla U|^2\,d\sigma)=(N-2)\,m^{-1}\,d\nu$. When $N=3$, the result is the same except that Ω is unique up to translation and dilation.*

When $N\geqslant 4$, a dilation of the minimizer given in Theorem 1.7 solves the equation in Theorem 1.5. But when $N=3$, ν_Ω is dilation invariant. Therefore the statements of the theorems must be modified theorems must be modified as indicated. When $N=3$, there is exactly one constant c, $c=m^{-1}$, for which the equation $\nu_\Omega=c\nu$ has a solution.

The uniqueness statements in Theorems 1.2 and 1.3 are not logically equivalent, although this is a problematic distinction to make between two true statements. The distinction is that the uniqueness in Theorem 1.2 applies to any stationary point of the functional, whereas the one in Theorem 1.3 refers only to minimizers. (This distinction is a trivial one because it follows from convexity of the functional that all stationary points are minimizers; see Proposition 5.2.) More important to the present article, the fact that the minimizer of $(*)$ is unique up to translation follows from Theorem 1.2 only after one proves the variational equation $\mu_\Omega = N\lambda^{-1}\mu$ for the minimizing body Ω. The situation in the case of the capacity theorems is less complete than it appears. Although it is not hard to show directly that the minimizer of $(\dagger)$ exists, we cannot confirm directly that it satisfies the equation $\nu_\Omega = (N-2)\, m^{-1}\nu$. Instead, we will prove Theorem 1.7 using Theorem B and Theorem 1.5. Theorem 1.5 is proved in [J], using a mixture of variational and limiting techniques. It would be nice to have a direct proof of Theorem 1.7. This problem will be discussed again at the end of the paper.

We will frequently identify the boundary of Ω with the unit sphere by the Gauss map. In particular, we will abuse notation by considering the support function as a function on $\partial\Omega$: $u(x) = u(g(x))$ is defined almost everywhere on $\partial\Omega$.

2. First and Second Variations of Capacity

The analogue of formula 1.4 for capacity is [J2]

$$\operatorname{cap} \Omega = \frac{1}{N-2} \int_{S^n} u_\Omega \, d\nu_\Omega. \tag{2.1}$$

The following first variation formula, proved by Poincaré in the smooth case, says that $|\nabla U|^2 d\sigma$ is the first variation of capacity in the same sense in which $d\sigma$ is the first variation of volume.

Proposition 2.2 [J]. *Let u_0 and u_1 be support functions for convex domains Ω_0 and Ω_1 respectively. Let $\nu_0 = \nu_{\Omega_0}$, then*

$$\text{(a)} \quad \frac{d}{dt} \operatorname{cap}(\Omega_0 + t\Omega_1)|_{t=0^+} = \int_{S^n} u_1 \, d\nu_0$$

and

$$\text{(b)} \quad \frac{d}{dt} \operatorname{cap}((1-t)\,\Omega_0 + t\Omega_1)|_{t=0^+} = \int_{S^n} (u_1 - u_0) \, d\nu_0.$$

Next, we describe the second variation, that is the Fréchet derivative of the mapping $\Omega \to v_\Omega$. Following [J2, J1, CY], we write this only in the smooth strictly convex case and express it in terms of the variation of the support function u_Ω. Let $e_1, ..., e_n$ be an orthonormal frame for S^n, and let covariant derivatives with respect to this frame be denoted ∇_i and ∇_{ij}. Denote $\mathscr{U} = \{u \in C^\infty(S^n): \nabla_{ij}u + u\delta_{ij} \gg 0\}$. It can be shown that the correspondence $\Omega \to u_\Omega$ is a one-to-one correspondence between C^∞ convex domains with strictly positive Gauss curvature and functions of $\mathscr{U}$. Let $b \in \mathbf{R}^N$. Translation of the domain Ω to $\Omega + b$ corresponds to the change in u to $u + b \cdot \xi$. Denote the N-dimensional space $\mathscr{P}_1 = \text{span}\{\xi_1, ..., \xi_N\}$. The Gauss mapping g is a diffeomorphism and we denote the inverse mapping by $F: S^n \to \partial\Omega$. It is given by the formula $F = \nabla \bar{u}$, where $\bar{u}$ is the extension of u from S^n to $\mathbf{R}^N$ as homogeneous function of degree 1: $\bar{u}(r\xi) = ru(\xi)$ for all $\xi \in S^n$. The Gauss curvature K can be defined as a function of the unit normal by $g_*(d\sigma) = (1/K(\xi))\, d\xi$, where $d\xi$ is the uniform measure on the Gauss sphere. The density $1/K$ can be computed in terms of u and written

$$1/K(\xi) = \det(\nabla_{ij}u(\xi) + u(\xi)\,\delta_{ij}) \tag{2.3}$$

K is unchanged by translation of Ω. In fact, each individual entry of the matrix whose determinant is $1/K$ is unchanged by translation: if $v \in \mathscr{P}_1$, then

$$(\nabla_{ij}v(\xi) + v(\xi)\,\delta_{ij}) = 0 \qquad \text{for all } i, j. \tag{2.4}$$

Define the coefficients c_{ij} of the cofactor matrix of $\nabla_{ij}u + u\delta_{ij}$ by

$$c_{ij}(\nabla_{jl}u + u\,\delta_{jl}) = \delta_{il}\det(\nabla_{pq}u + u\,\delta_{pq}) = \delta_{il}/K. \tag{2.5}$$

Here and in subsequent formulas we follow the convention that repeated indices are summed. Define the density $S \in C^\infty(S^n)$ by $g_*(|\nabla U|^2\, d\sigma) = S\, d\xi$, define the mapping $\mathscr{F}: \mathscr{U} \to C^\infty(S^n)$ by $\mathscr{F}(u) = S$. We have the formula

$$S(\xi) = h(F(\xi))^2/K(\xi), \tag{2.6}$$

where $h(x) = |\nabla U(x)|$ for $x \in \partial\Omega$.

Let $f \in C^\infty(S^n)$ and let w be the harmonic function in Ω' that vanishes at infinity and has boundary values $f(\xi)$ at $x = F(\xi)$ on $\partial\Omega'$. Define the operator Λ acting on $C^\infty(S^n)$ by $\Lambda(f) = \xi \cdot \nabla w(F(\xi))$, the normal derivative of the harmonic extension. Let $v \in C^\infty(S^n)$. For t sufficiently small, $u + tv \in \mathscr{U}$. Furthermore, if v is the support function of a domain Ω_1, then $u + tv$ is the support function of $\Omega + t\Omega_1$.

PROPOSITION 2.7 [J]. *The directional derivative of $\mathscr{F}$ is given by*

$$\frac{d}{dt}\mathscr{F}(u+tv)|_{t=0} = Lv,$$

where $L = L_u$ *is defined as*

$$Lv = \nabla_i(h^2 c_{ij}\nabla_j v) - (2/K)\, h\Lambda(hv) - h^2\, Tr(c_{ij})\, v.$$

Green's formula implies that Λ is selfadjoint on $L^2(\partial\Omega, d\sigma)$. It follows that

Remark 2.8. L is selfadjoint on $L^2(S^n, d\xi)$.

3. UNIQUENESS FOR SMALL PERTURBATIONS OF THE SPHERE

We analyze the second variation L to deduce uniqueness for small perturbations of the sphere.

LEMMA 3.1. *Let Ω_0 be the domain with support function u. If $u-1$ has sufficiently small $C^{2N}(S^n)$ norm, and $N \geqslant 4$, then the null space of L is $\mathscr{P}_1$, and there is an orthonormal basis for the orthogonal complement of the null space of the form $\{\phi_k\}$, $k=0, 1, \ldots$ with*

$$L\phi_0 = \alpha_0\phi_0 \qquad \text{and} \qquad L\phi_k = -\alpha_k\phi_k, \qquad k = 1, 2, \ldots$$

and $\alpha_k \geqslant 1$ for all $k = 0, 1, \ldots$ and $\alpha_k = \mathcal{O}(k^{2/n})$. In the case $N=3$, the null space is the span of $\mathscr{P}_1$ and the additional vector u. Furthermore, the complement of the null space has a basis $\{\phi_k\}$, with $k = 1, 2, \ldots$, that is, all the rest of the eigenvalues are strictly negative.

Proof. Denote $\mathscr{F}(u) = S$. Dilation gives $\mathscr{F}((1+t)u) = (1+t)^{N-3}S$, so that

$$Lu = (N-3)\, S. \tag{3.2}$$

Translation gives $\mathscr{F}(u+v) = S$ for all $v \in \mathscr{P}_1$, so that

$$Lv = 0 \qquad \text{for all} \quad v \in \mathscr{P}_1. \tag{3.3}$$

Thus the null space contains $\mathscr{P}_1$ (and u in the case $N=3$). The asymptotic size of the eigenvalues follows from standard elliptic estimates. The fact that there are no other zero eigenvalues and the uniform lower bound on the eigenvalues follows from perturbation and the explicit calculation of the case of the unit sphere that follows.

In the case of the sphere, $u \equiv 1$, $U(x) = |x|^{2-N}$, $h = N-2$, $K = 1$, and $c_{ij} = \delta_{ij}$. The operator Λ can be computed from the observation that if $P_k(x)$ is homogeneous harmonic polynomial of degree k, then its extension to the exterior of the ball is given by $w(x) = |x|^{2-N} P_k(x/|x|^2)$ which is homogeneous of degree $2-N-k$. Thus,

$$\Lambda(P_k) = (2-N-k)\, P_k.$$

The Laplace–Beltrami operator on the sphere satisfies

$$c_{ij} \nabla_{ij} P_k = -k(k+N-2)\, P_k.$$

Therefore, if L_1 denotes the operator for $u \equiv 1$,

$$L_1 P_k = -(N-2)^2\, (k(k+n-2) - 2(N+k-2) + (N-1))\, P_k.$$

In particular,

$$L_1 P_0 = (N-2)^2\, (N-3)\, P_0 \qquad \text{and} \qquad L_1 P_1 = 0,$$

and the remaining eigenvalues are negative integers strictly less than -1.

Let $L = L_u$. Standard perturbation theory implies that for u sufficiently close to 1, all the small eigenvalues of L are within, say, unit distance of corresponding eigenvalues of L_1. The asymptotic estimate from above and below $\alpha_k = \mathcal{O}(k^{2/n})$ follows from standard theory of elliptic theory. This proves all the assertions of Lemma 3.1 provided we can show that the null space of L is the space V defined by $V = \mathscr{P}_1$ if $N \geqslant 4$ and $V = \operatorname{span}\{\xi_1, \xi_2, \xi_3, u\}$ when $N = 3$. The null space of L_1 is $\mathscr{P}_1$ when $N \geqslant 4$ and $\operatorname{span}\{\xi_1, \xi_2, \xi_3, 1\}$ when $N = 3$. Let T_1 denote the projection onto the null space of L_1. Let A be the partial inverse of L_1 with the same null space as L_1 and satisfying $AL_1 = 1 - T_1$. Let $\|\cdot\|$ denote the norm of $L^2(S^n, d\xi)$. For u sufficiently close to 1,

$$\|A(L-L_1)\, w\| < \|w\|/4.$$

If w is orthogonal to V, then for u sufficiently close to 1,

$$\|T_1 w\| < \|w\|/4$$

(3.2) and (3.3) imply that V is contained in the null space of L. In order to show that V is the null space of L, consider a function w that is orthogonal to V and satisfies $Lw = 0$. Then

$$0 = ALw = A(L-L_1)\, w + AL_1 w = A(L-L_1)\, w + w - T_1 w.$$

Therefore,

$$\|w\| \leqslant \|A(L-L_1)\, w\| + \|T_1 w\| < \|w\|/2$$

and we conclude that $w=0$. This proves Lemma 3.1.

PROPOSITION 3.4. *Let Ω_0 be the domain with support function u and let Ω_1 be the domain with support function v. If $u-1$ and $v-1$ have small $C^{2N}(S^n)$ norm, and*

$$\operatorname{cap}((1-t)\,\Omega_0 + t\Omega_1)^{1/(N-2)} = (1-t)\operatorname{cap}(\Omega_0)^{1/N-2} + t\operatorname{cap}(\Omega_1)^{1/(N-2)}$$

for some $t \in (0,1)$, then v is a linear combination of u and a first-order spherical harmonic: $v(\xi) = au(\xi) + b\cdot\xi$ for some $a>0$ and some $b \in \mathbf{R}^N$. In other words, Ω_1 is a translate and dilate of Ω_0.

Proof. Let

$$f(t) = \operatorname{cap}(\Omega_0 + t\Omega_1),$$
$$m(t) = \operatorname{cap}((1-t)\,\Omega_0 + t\Omega_1)^{1/(N-2)} = (1-t)\, f(t/(1-t))^{1/(N-2)}.$$

Formula 2.1 implies

$$f(t) = \frac{1}{N-2}\int_{S^n} (u+tv)\,\mathscr{F}(u+tv)\,d\xi.$$

Proposition 2.2 implies

$$f'(t) = \int_{S^n} v\mathscr{F}(u+tv)\,d\xi.$$

Consequently, Proposition 2.7 implies

$$f''(0) = \int_{S^n} vLv\,d\xi.$$

Since m is concave and agrees with a linear function at 0, t and 1, it must be linear. Thus $m''=0$. We can calculate

$$m''(0) = (N-2)^{-2} f(0)^{-2+1/(N-2)}\,[(N-2)\,f(0)\,f''(0) - (N-3)\,f'(0)^2].$$

Thus $m''(0)=0$ implies

$$(N-2)\,f(0)\,f''(0) = (N-3)\,f'(0)^2. \tag{3.5}$$

Denote by (,) the inner product on $L^2(S^n, d\xi)$, then

$$f(0) = (u, S)/(N-2), \qquad f'(0) = (v, S), \qquad f''(0) = (v, Lv).$$

When $N=3$, (3.5) implies $(v, Lv)=f''(0)=0$. By Lemma 3.1, this implies $Lv=0$ and v belongs to the span of u and ξ_1, ξ_2, and ξ_3. The case $N\geqslant 4$ is more complicated because L has a positive eigenvalue α_0. Rewrite equation (3.5) as

$$(u, Lu)(v, Lv)=(v, Lu)^2. \tag{3.6}$$

If we let $a_k=(u, \phi_k)$ and $b_k=(v, \phi_k)$, then $u-\sum a_k\phi_k \in \mathscr{P}_1$ and $v-\sum b_k\phi_k \in \mathscr{P}_1$ and (3.6) can be restated as

$$\left(-\alpha_0 a_0^2+\sum_{k=1}^{\infty} \alpha_k a_k^2\right)\left(-\alpha_0 b_0^2+\sum_{k=1}^{\infty} \alpha_k b_k^2\right)=\left(-\alpha_0 a_0 b_0+\sum_{k=1}^{\infty} \alpha_k a_k b_k\right)^2. \tag{3.7}$$

Given $\varepsilon>0$, we can choose u and v sufficiently close to 1, that $|a_0-1|<\varepsilon$ and $|b_0-1|<\varepsilon$ and

$$\sum_{k=1}^{\infty} k^2(|a_k|^2+|b_k|^2)<\varepsilon.$$

If we recall that $\alpha_0\geqslant 1$ and $\beta_0\geqslant 1$, then, in particular we can choose ε small enough that

$$\sum_{k=1}^{\infty} \alpha_k a_k^2<\alpha_0 a_0^2 \qquad \text{and} \qquad \sum_{k=1}^{\infty} \alpha_k b_k^2<\alpha_0 b_0^2. \tag{3.8}$$

Our goal is to deduce from (3.7) and (3.8) that $a_k/a_0=b_k/b_0$ for all $k=1, 2, \ldots$. It then follows that $v/b_0-u/a_0 \in \mathscr{P}_1$, which is what we want to prove. Let

$$x_k=a_k\sqrt{\alpha_k}/a_0\sqrt{\alpha_0}, \qquad y_k=b_k\sqrt{\alpha_k}/b_0\sqrt{\alpha_0}.$$

Then (3.7) and (3.8) are rephrased as

$$(-1+|x|^2)(-1+|y|^2)=(-1+(x, y))^2 \tag{3.7$'$}$$

$$|x|<1 \qquad \text{and} \qquad |y|<1 \tag{3.8$'$}$$

where $|\cdot|$ denotes the norm on ℓ^2. The conclusion that we wish to draw is that $x=y$.

To prove this, let $A=|x|$ and $B=|y|$. Then x/AB and y/AB have length greater than 1. If $x\neq y$, then

$$|x/A-y/B|^2<|x/AB-y/AB|^2=|x-y|^2/A^2B^2.$$

Furthermore,

$$AB-(x,y)=\frac{AB}{2}\,|x/A-x/B|^2.$$

Therefore,

$$\begin{aligned}A^2B^2-(x,y)^2&=(AB+(x,y))(AB-(x,y))\\&=(AB+(x,y))\frac{AB}{2}\,|x/A-y/B|^2\\&\leqslant A^2B^2\,|x/A-y/B|^2<|x-y|^2.\end{aligned}$$

The equation in the hypothesis can be written as

$$|x|^2+|y|^2-2(x,y)=|x|^2\,|y|^2-(x,y)^2.$$

Thus,

$$|x-y|^2=A^2B^2-(x,y)^2<|x-y|^2.$$

This is a contradiction, so it must be that $x=y$.

4. Analytic Continuation

We can now prove the main result, Theorem B. Note that

$$\operatorname{cap}(s\Omega)=s^{N-2}\operatorname{cap}\Omega.$$

Consider the regions Ω_0 and Ω_1 of Theorem B. After dilation, one can assume without loss of generality that $\operatorname{cap}\Omega_0=\operatorname{cap}\Omega_1$. Furthermore, if equality holds for one value of t, then it holds for all values because a concave function that agrees with a linear function at three points is linear. Let U_t be the equilibrium potential of Ω_t, $0\leqslant t\leqslant 1$. Let $\Omega_t(\lambda)=\{x\in\Omega'\colon U_t(x)>\lambda\}\cup\bar{\Omega}$. The equilibrium potential of $\Omega_t(\lambda)$ is U_t/λ, so that

$$\operatorname{cap}\Omega_t(\lambda)=\lambda\operatorname{cap}\Omega_t. \tag{4.1}$$

Therefore, the hypothesis of Theorem B implies

$$(\operatorname{cap}\Omega_t(\lambda))^{1/(N-2)}=(1-t)\operatorname{cap}\Omega_0(\lambda)^{1/(N-2)}+t\operatorname{cap}\Omega_1(\lambda)^{1/(N-2)}. \tag{4.2}$$

Borell's inequality and 4.2 imply

$$\begin{aligned}(\operatorname{cap}(1-t)\,\Omega_0(\lambda)+t\Omega_1(\lambda))^{1/(N-2)}\\ &\geqslant(1-t)\operatorname{cap}\Omega_0(\lambda)^{1/(N-2)}+t\operatorname{cap}\Omega(\lambda)^{1/(N-2)}\\ &=(\operatorname{cap}\Omega_t(\lambda))^{1/(N-2)}.\end{aligned}$$

Borell [B], in the process of proving Theorem A, shows that if $0\leqslant t\leqslant 1$ and $x_t=(1-t)\,x_0+tx_1$, then

$$U_t(x_t)\geqslant\min(U_0(x_0),\,U_1(x_1))\qquad\text{for all}\quad x_0\in\Omega_0,\,x_1\in\Omega_1.$$

This can be rephrased as

$$\Omega_t(\lambda)\supset(1-t)\,\Omega_0(\lambda)+t\Omega_1(\lambda).$$

On the other hand, the capacity of the smaller set is at least as large as the larger, so

$$\Omega_t(\lambda)=(1-t)\,\Omega_0(\lambda)+t\Omega_1(\lambda)$$

holds for all $\lambda<1$ and all t, $0\leqslant t\leqslant 1$. Furthermore,

$$\begin{aligned}\operatorname{cap}((1-t)\,\Omega_0(\lambda)+t\Omega_1(\lambda))^{1/(N-2)}\\ &=(1-t)\operatorname{cap}\Omega_0(\lambda)^{1/(N-2)}+t\operatorname{cap}\Omega_1(\lambda)^{1/(N-2)}.\end{aligned}\tag{4.3}$$

We will show that as λ tends to 0, the domains $\Omega_0(\lambda)$ and $\Omega_1(\lambda)$ approach spheres. We will then be able to apply Proposition 3.4. Let $\lambda=cs^{N-2}$, where $c=a_N\operatorname{cap}\Omega_0=a_N\operatorname{cap}\Omega_1$. For z a unit vector in $\mathbf{R}^N$, define $\rho(z,s)$ implicitly by

$$U_0(s^{-1}\rho(z,s)\,z)=cs^{N-2}.$$

(There is a unique value of ρ because the radial derivative of U_0 is negative.) There is a harmonic function ϕ defined in the image of Ω' by the mapping $x\to x/|x|^2$ satisfying $\phi(0)=c$ and

$$U_0(x)=|x|^{2-N}\,\phi(x/|x|^2).$$

The equation for ρ can be written

$$\rho^{2-N}\phi(s\rho^{-1}z)=c.$$

The implicit function theorem shows that ρ is a real analytic function of (z,s) near $s=0$ and that $\rho(z,0)=1$ for all z and $\rho(z,s)$ tends to the function 1 on S^n in the C^∞ topology as s tends to 0. Thus a suitable dilate of $\Omega_0(\lambda)$ is very close to the unit ball:

$$s\Omega_0(\lambda) = \{rz : z \in \mathbf{R}^N, |z| = 1, 0 < r < \rho(z, s)\}.$$

Similarly, $s\Omega_1(\lambda)$ tends to the unit ball as s and λ tend to 0. Finally (4.3) implies,

$$\begin{aligned} &\operatorname{cap}((1-t)\, s\Omega_0(\lambda) + ts\Omega_1(\lambda))^{1/(N-2)} \\ &\quad = (1-t) \operatorname{cap} s\Omega_0(\lambda)^{1/(N-2)} + t \operatorname{cap} s\Omega_1(\lambda)^{1/(N-2)}. \end{aligned} \tag{4.4}$$

Fix s and λ sufficiently small that both $s\Omega_0(\lambda)$ and $s\Omega_1(\lambda)$ are close to the unit ball. Then Proposition 3.4 says that they are translates and dilates of each other. In fact, since we have normalized the capacities to be equal, they are translates of each other. Therefore, the same is true of $\Omega_0(\lambda)$ and $\Omega_1(\lambda)$. It follows that there is a vector $b \in \mathbf{R}^N$ such that

$$U_1(x-b) = U_0(x)$$

provided $U_0(x) \leqslant \lambda$. By analytic continuation, this equality holds for all x and we are done.

5. Applications to Existence and Uniqueness in the Variational Problem

We can now deduce Theorem 1.6. Consider first the case $N \geqslant 4$. Suppose that Ω_0 and Ω_1 are two domains satisfying $v = v_{\Omega_0} = v_{\Omega_1}$. Let u_0 and u_1 denote the support functions of Ω_0 and Ω_1, and denote $\Omega_t = (1-t)\,\Omega_0 + t\Omega_1$. Then Proposition 2.2 and formula 2.1 imply

$$\frac{d}{dt} \operatorname{cap} \Omega_t|_{t=0^+} = \int_{S^n} (u_1 - u_0)\, dv = (N-2)(\operatorname{cap} \Omega_1 - \operatorname{cap} \Omega_0). \tag{5.1}$$

Denote $m(t) = \operatorname{cap} \Omega_t^{1/(N-2)}$. Then

$$\begin{aligned} m'(0^+) &= \operatorname{cap} \Omega_0^{-1+1/(N-2)}(\operatorname{cap} \Omega_1 - \operatorname{cap} \Omega_0) \\ &= m(0)^{3-N}\,(m(1)^{N-2} - m(0)^{N-2}). \end{aligned}$$

Because m is concave, $m'(0^+) \geqslant m(1) - m(0)$. This can be rewritten as

$$m(1)^{N-3} \geqslant m(0)^{N-3}.$$

By symmetry we have the opposite inequality. Therefore $m(0) = m(1)$ and $\operatorname{cap} \Omega_1 = \operatorname{cap} \Omega_0$. Consequently, $m'(0^+) = 0$; and since m is concave, it must be constant. Therefore, by Theorem B, Ω_1 is a translate and dilate of Ω_0. Since the capacities are the same, Ω_1 must be a translate of Ω_0.

Next, suppose that $N=3$, and suppose that Ω_0 and Ω_1 are two domains with corresponding measures $c_0 \nu$ and $c_1 \nu$. Formula 5.1 yields

$$m'(0^+) = \int_{S^2} (u_1 - u_0)\, c_0\, d\nu = (c_0/c_1) \operatorname{cap} \Omega_1 - \operatorname{cap} \Omega_0.$$

Because $m(t)$ is concave,

$$(c_0/c_1) \operatorname{cap} \Omega_1 - \operatorname{cap} \Omega_0 = m'(0^+) \geqslant m(1) - m(0) = \operatorname{cap} \Omega_1 - \operatorname{cap} \Omega_0$$

so that $c_0 \geqslant c_1$. By symmetry, $c_0 = c_1$. It follows that $m'(0^+) = m(1) - m(0)$, and since m is concave, it must be linear. Finally, Theorem B implies that Ω_1 is a translate and dilate of Ω_0.

Next let us deduce Theorem 1.7 from Theorem 1.5 and Theorem B. Fix a positive measure ν on S^n satisfying the two necessary conditions (a) and (b). Theorem 1.5 implies that there is a bounded, convex open set Ω_0 and a positive number c such that Ω_0 has capacity 1 and induced measure $d\nu_{\Omega_0} = c\, d\nu$. Formula 2.1 implies

$$1 = \operatorname{cap} \Omega_0 = \frac{1}{N-2} \int_{S^n} u_0 c\, d\nu$$

where u_0 is the support function of Ω_0. Let

$$\mathscr{C} = \{\Omega : \Omega \text{ is convex and open, } \operatorname{cap} \Omega \geqslant 1\}.$$

Denote

$$F(\Omega) = \int_{S^n} u_\Omega\, d\nu.$$

PROPOSITION 5.1. *Ω_0 is the unique minimizer of F in the class $\mathscr{C}$, up to translation.*

Proof. Let $\Omega_1 \in \mathscr{C}$, and let u_1 be the support function of Ω_1. Suppose that

$$F(\Omega_1) \leqslant F(\Omega_0).$$

From Proposition 2.2 we obtain

$$\frac{d}{dt} \operatorname{cap} \Omega_t |_{t=0^+} = \int_{S^n} (u_1 - u_0)\, c\, d\nu = c(F(\Omega_1) - F(\Omega_0)) \leqslant 0.$$

Since $(\operatorname{cap} \Omega_t)^{1/(N-2)}$ is concave and equals 1 at $t=0$ and $t=1$, it must be constant. Therefore by Theorem B, Ω_1 is a translate and dilate of Ω_0. (No dilation is needed because the two convex bodies have the same capacity.)

It is probably possible to carry out a direct variational proof of Theorem 1.5, but we did not do so because we have not proved that minimizers satisfy the natural Euler–Lagrange equation. To make this remark more precise, consider an arbitrary positive, continuous function u on S^n, which need not be the support function of a convex domain. Denote

$$\Omega[u] = \{X : X \cdot \xi < u(\xi) \text{ for all } \xi \in S^n\}.$$

If u^* is the support function of $\Omega[u]$, then $0 < u^* \leqslant u$, but there may not be equality. It would be possible to make a direct argument for existence if the variational formula

$$\frac{d}{dt} \operatorname{cap} \Omega[u + tv]|_{t=0} = \int_{S^n} v \, dv_{\Omega[u]} \tag{5.3}$$

were proved for all $v \in C(S^n)$ and all support functions u. In (5.1), this is proved for $t = 0^+$ provided the function v is also a support function of a convex domain. The corresponding identity for volume is true for all continuous v. The proof is not immediate, but follows from the fact that the Gauss mapping g is continuous almost everywhere with respect to $d\sigma$.

References

[BF] T. Bonnesen and W. Fenchel, "Theory of Convex Bodies," BCS Associates, Moscow, ID, 1987, Translation from German.

[B] C. Borell, Capacitary inequalities of the Brunn–Minkowski type, *Math. Ann.* **263** (1983), 179–184.

[CY] S.-Y. Cheng and S.-T. Yau, On the regularity of the solution of the n-dimensional Minkowski problem, *Comm. Pure Appl. Math.* **29** (1976), 495–516.

[D] B. E. J. Dahlberg, Estimates for harmonic measure, *Arch. Rational Mech. Anal.* **65** (1977), 275–283.

[J1] D. Jerison, Prescribing harmonic measure on convex domains, *Invent. Math.* **105** (1991), 375–400.

[J2] D. Jerison, A Minkowski problem for electrostatic capacity, *Acta Math.*, to appear.

Part VI

General Analysis

J. Funct. Anal. *51*, 159–165 (1983)

Reprinted from JOURNAL OF FUNCTIONAL ANALYSIS
Vol. 51, No. 2, April 1983

An L^p Bound for the Riesz and Bessel Potentials of Orthonormal Functions

ELLIOTT H. LIEB*

Institute for Advanced Study, Princeton, New Jersey 08540

Communicated by Irving Segal

Received September 14, 1982

Let $\psi_1,..., \psi_N$ be orthonormal functions in $\mathbb{R}^d$ and let $u_i = (-\Delta)^{-1/2}\psi_i$ or $u_i = (-\Delta + 1)^{-1/2}\psi_i$, and let $\rho(x) = \sum |u_i(x)|^2$. L^p bounds are proved for ρ, an example being $\|\rho\|_p \leqslant A_d N^{1/p}$ for $d \geqslant 3$, with $p = d(d-2)^{-1}$. The unusual feature of these bounds is that the orthogonality of the ψ_i yields a factor $N^{1/p}$ instead of N, as would be the case without orthogonality. These bounds prove some conjectures of Battle and Federbush (a Phase Cell Cluster Expansion for Euclidean Field Theories, I, 1982, preprint) and of Conlon (*Comm. Math. Phys.*, in press).

The genesis of this paper is a problem posed by Battle and Federbush [2] in connection with their new approach to Euclidean φ^4 quantum field theory. The problem is related to the proof of stability in [2, Sect. 7].

Let $\psi_1,..., \psi_N$ be N orthonormal complex valued functions in $L^2(\mathbb{R}^d)$ and define

$$u_i = (-\Delta + m^2)^{-1/2}\psi_i \tag{1}$$

$$\rho(x) = \sum_{i=1}^{N} |u_i(x)|^2, \tag{2}$$

where Δ is the Laplacian and $m \geqslant 0$. Battle and Federbush prove that for $d = 1$, 2, and 3 and $m > 0$ there is a universal constant W such that

$$\|\rho\|_2 \leqslant W m^{-2+d/2} N^{1/2}. \tag{3}$$

They ask whether (3) also holds for $d = 4$, and also point out that no such bound can hold for $d \geqslant 5$.

Equation (3) looks like a Sobolev inequality—and it is—except for one important innovation. The standard Sobolev inequality would have a factor N, not $N^{1/2}$, in (3). What makes (3) interesting is that the orthogonality of

* Permanent address: Jadwin Hall, Princeton University, P. O. Box 708, Princeton, N.J. 08544. Work partially supported by National Science Foundation Grant PHY-8116101.

J. Funct. Anal. *51*, 159–165 (1983)

the ψ_i yields $N^{1/2}$. If the ψ_i were normalized, but not orthogonal, the best estimate in (3) would have N. Indeed, (3) is easily seen to hold with a factor N by the standard Sobolev inequality, even for $d=4$.

Here we shall not only prove (3) for $d=4$ but will generalize the inequality (for all $d \geqslant 1$) to what is essentially the best possible—in the sense that any proposed strengthening will fail even for $N=1$.

The main results are the following, but generalizations are given in Theorems 3 and 4:

THEOREM 1. *Let $\psi_1,\ldots,\psi_N$ be orthonormal in $L^2(\mathbb{R}^d)$ with u_i and ρ given by* (1) *and* (2). *Then*

(i) $d=1$. *For $m>0$, $\rho \in C^{0,1/2}$ (the Hölder continuous functions with exponent $\frac{1}{2}$) and $\rho \in L^\infty$. There is a universal constant L such that*

$$\|\rho\|_\infty \leqslant L/m. \tag{4}$$

(ii) $d=2$. *For $m>0$ and all $1 \leqslant p < \infty$, $\rho \in L^p$ and*

$$\|\rho\|_p \leqslant B_p m^{-2/p} N^{1/p}, \tag{5}$$

where B_p is a universal constant. (ρ is not necessarily in L^∞.)

(iii) $d \geqslant 3$. *For all $m \geqslant 0$ (including $m=0$), and with $p=d(d-2)^{-1}$, $\rho \in L^p$ and*

$$\|\rho\|_p \leqslant A_d N^{1/p}, \tag{6}$$

where A_d is a universal constant (independent of m).

Remark 1. If the orthogonality (but not the normality) of the ψ_i is omitted, then the right side of (4) has to be multiplied by N, and $N^{1/p}$ has to be replaced by N on the right sides of (5) and (6). In some sense the effect of orthogonality is most striking in (4).

Remark 2. The theorem yields (3) for $d=4$. For $d=1$, 2, or 3, the theorem also yields (3) via the Hölder inequality and the obvious fact (for all d), which follows by taking Fourier transforms, that

$$\|\rho\|_1 \leqslant Nm^{-2}. \tag{7}$$

Proof of Theorem 1. Let us first study the situation for $N=1$. The operator $(-\Delta)^{-1/2} \equiv I$ is the Riesz operator while $(-\Delta+m^2)^{-1/2} \equiv J_m$ is the Bessel operator. We refer to [7, Chap. V] for a discussion and definitions. What will concern us here is

(a) For $d \geqslant 3$, I is a bounded map from L^2 to L^r and from L^s to L^2 with $r=2d(d-2)^{-1}$ and $s=r'=2d(d+2)^{-1}$.

(b) For all d, J_m is given by an integral kernel of the form

$$G_m(x-y) = m^{d-1}\, G(m(x-y)). \tag{8}$$

The m dependence given by (8) accounts trivially (by scaling) for the m dependence in (4)–(7). Henceforth we shall take $m=1$ and drop the m subscript. J has the same properties as I in (a) for $d \geqslant 3$.

(c) For all d and $1 < p < \infty$, the spaces $J(L^p(\mathbb{R}^d))$ and $L_1^p(\mathbb{R}^d)$ are identical ($L_1^p = \{f \mid f \text{ and } \nabla f \in L^p\}$). For $d=2$, $L_1^2 \subset L^p$, all $2 \leqslant p < \infty$. For $d=1$, if $f \in L^2$, then $Jf \in L^\infty$ and $Jf \in C^{0,1/2}$ (see [1, p. 98]).

Conditions (a)–(c) account for all the statements in Theorem 1 except for the N dependence in Eqs. (4)–(6), which we prove next.

(i) $d=1$. Let V be a nonnegative function in $L^1 \cap L^\infty$ which we think of as a multiplication operator. Let

$$H = V^{1/2} J, \qquad H^* = J V^{1/2}. \tag{9}$$

By (c), H is a bounded operator from L^2 to L^2 and H^* is its adjoint. Let $K = H^*H$. K is compact, for

$$\operatorname{Tr} K = \iint v(x)\, R(x-y)\, dx dy \leqslant C_1 \| V \|_1. \tag{10}$$

The last inequality comes from the fact that $R(x) = G * G(x) = C_2 \exp(-|x|)$. Let $\lambda_1 \geqslant \lambda_2 \geqslant \cdots$ be the eigenvalues (including multiplicity) of K. Then for any N orthonormal functions ψ_i,

$$\sum_{i=1}^{N} \| H\psi_i \|^2 \leqslant \sum_{i=1}^{N} \lambda_i \leqslant \operatorname{Tr} K. \tag{11}$$

However, the left side of (11) is just $\int \rho V$. Since $\int \rho V \leqslant C_1 \| V \|_1$ for any $V \in L^1 \cap L^\infty$, (4) is proved.

(ii) $d=2$. The $p=1$ case is given in (7), so assume $1 < p < \infty$. Again, consider the operator K with $V = \rho^{p-1} \in L^r$ with $r = p'$. By (c), H is bounded from L^2 to L^2 and H^* is its adjoint. Let $T = \operatorname{Tr} K^r$. Then $T^{1/r} \leqslant C_3 \| V \|_r$. (To prove this we can appeal to a general result of Cwikel [3] (see also Theorem 2 and [6, Theorem 4.1]) that $|\!|\!| f(x)\, g(-i\nabla) |\!|\!|_{2r} \leqslant C_4 \| f \|_{2r} \| g \|_{2r}$, where $|\!|\!| \cdot |\!|\!|_q$ is the trace norm.) Using the same variational principle as in (i) we have that

$$\int \rho V \leqslant \sum_{i=1}^{N} \lambda_i \leqslant N^{1/p} \left\{ \sum_{i=1}^{\infty} \lambda_i^r \right\}^{1/r} \leqslant C_3 N^{1/p} \| V \|_r. \tag{12}$$

Since $V = \rho^{p-1}$, $\| V \|_r = \| \rho \|_p^{p-1}$.

(iii) $d \geqslant 3$. First consider $m > 0$. For reasons of clarity we reintroduce the parameter m, namely, $H = V^{1/2} J_m$. With $V^{1/2} \in L^d$, H and H^* are bounded from L^2 to L^2 by (a). If we try to imitate the $d = 2$ proof (with $V = \rho^{p-1}$) we would have, as in (12), $\int \rho V \leqslant C_4 N^{1/p} |||K|||_t$ with $t = p' = d/2$. However, $|||K|||_t$ need not be finite; it is certainly not bounded by $\|V\|_t$.

A new idea is needed and this is provided by the Cwikel–Lieb–Rosenbljum bound [3–5]. (This bound was proved by these authors by completely independent methods. The Cwikel and Rosenbljum methods extend to a wider class of operators, but for the operator of interest K, Lieb's bound gives the best constant of the three.)

First, K is compact. The nonzero eigenvalues of K are, of course, the same as those of $B = HH^*$. B is called the Birman–Schwinger kernel [6]. Second, let $n(V)$ denote the number of eigenvalues of K which are $\geqslant 1$. Then

$$n(V) \leqslant C_5^{d/2} \int V^{d/2}. \tag{13}$$

Here C_5 is *independent* of m (as it must be). Since K is linear in V, (13) can be inverted to read

$$\lambda_j \leqslant C_5 \|V\|_{d/2} j^{-2/d}. \tag{14}$$

(Simply consider V/λ_j and $n(V/\lambda_j) = j$ in (13).)

Now we can imitate (12). Take $V = \rho^{p-1}$ whence

$$\int V\rho \leqslant \sum_{j=1}^{N} \lambda_j \leqslant C_5 \|V\|_{d/2} \sum_{j=1}^{N} j^{-2/d} \leqslant C_6 \|V\|_{d/2} N^{1-2/d}. \tag{15}$$

This completes the proof for $m > 0$.

For $m = 0$ we take $H = V^{1/2} I$, $H^* = I V^{1/2}$. By (a), these are bounded from L^2 to L^2 with a bound $C_1 \|V\|_{d/2}$. Bound (13) continues to hold, and (15) is again true. Alternatively, we can note that for fixed ψ, $u_m = J_m \psi$ converges pointwise a.e. to $u = I\psi$ as $m \to 0$ by dominated convergence using the explicit integral kernels for J_m and I (see [7, Chap. V, Theorem 1a]). Then (6) follows by Fatou's lemma. ∎

Variations on the Theme

An obvious generalization is to replace (1) by

$$u_i = (-\Delta + m^2)^{-\alpha/2} \psi_i \tag{16}$$

with $\alpha > 0$, and with $\alpha < d$ if $m = 0$ (see [7]). Here ρ is still defined by (2). Equation (7) becomes

$$\|\rho\|_1 \leqslant N m^{-2\alpha}. \tag{17}$$

Cwikel's theorem [3], the first half of which was mentioned just before (12), will be needed. See [6, Theorems 4.1 and 4.2].

THEOREM 2 (*Cwikel*). (i) *If $f, g \in L^q(\mathbb{R}^d)$ with $2 \leqslant q < \infty$ then, with* $|||X|||_q \equiv \{\mathrm{Tr}|X|^q\}^{1/q}$,

$$|||f(x)\, g(-i\nabla)|||_q \leqslant (2\pi)^{-d/q} \|f\|_q \|g\|_q. \tag{18}$$

(ii) *If $f \in L^q(\mathbb{R}^d)$* and $g \in L^q_w(\mathbb{R}^d)$, *then for $2 < q < \infty$ there is a finite constant $C_{q,d}$ such that*

$$|||f(x)\, g(-i\nabla)|||_{q,w} \leqslant C_{q,d} \|f\|_q \|g\|_{q,w}. \tag{19}$$

By definition, $\|g\|_{q,w} = \sup_{t>0} t \,\mathrm{meas}\{x \mid t \leqslant |g(x)|\}^{1/q}$, and $|||O|||_{q,w} = \sup_n \lambda_n n^{1/q}$, where $\lambda_1 \geqslant \lambda_2 \geqslant \cdots$ are the eigenvalues of the (compact) operator $(O^*O)^{1/2}$. Note that the nonzero eigenvalues of O^*O and OO^* are the same.

In our application

$$g(k) = (k^2 + m^2)^{-\alpha/2}$$

and

$$\|g\|_q = R_{q,d,\alpha}\, m^{-\alpha + d/q}, \qquad \text{if} \quad \alpha q > d \quad \text{and} \quad m > 0. \tag{20}$$

$$\|g\|_{q,w} = T_{d,\alpha}, \qquad \text{if} \quad q\alpha = d \quad \text{and} \quad m \geqslant 0. \tag{21}$$

With this information, and by imitating the proof of Theorem 1, case $d \geqslant 3$, we have the following generalization of Theorem 1:

THEOREM 3. (i) *For all d, $m > 0$, and $\alpha > 0$ a finite universal constant $B_{p,d,\alpha}$ exists such that*

$$\|\rho\|_p \leqslant B_{p,d,\alpha} M^{d-2\alpha-d/p} N^{1/p} \tag{22}$$

provided that

$$\begin{aligned} &1 \leqslant p \leqslant \infty && \textit{when} \quad 2\alpha > d,\\ &1 \leqslant p < \infty && \textit{when} \quad 2\alpha = d,\\ &1 \leqslant p \leqslant d(d-2\alpha)^{-1} && \textit{when} \quad 2\alpha < d. \end{aligned}$$

J. Funct. Anal. *51*, 159–165 (1983)

(ii) *For all* d, $m=0$, $0<\alpha<d/2$, *and* $p=d(d-2\alpha)^{-1}$, *a finite universal constant* $A_{d,\alpha}$ *exists such that*

$$\|\rho\|_p \leqslant A_{d,\alpha} N^{1/p}. \tag{23}$$

Note. The q in (20), (21) is chosen to be $2p(p-1)^{-1}$ when $p>1$. Also, $V=\rho^{p-1}$.

In the foregoing, the operators H and H^*, given by (9), were used with $V=\rho^r$ for some suitable r. Now let us consider the following problem as suggested by Conlon [8]:

Consider the operator L given by the kernel

$$L(x,y)=\sum_{i=1}^{N} \bar{\psi}_i(x)\, G_{m,\alpha}(x-y)\, \psi_i(y), \tag{24}$$

where $G_{m,\alpha}$ is the kernel for $(-\Delta+m^2)^{-\alpha/2}$ with $\alpha>0$, and with $\alpha<d$ if $m=0$. Again, the $\{\psi_i\}$ are an orthonormal set.

For $d=3$, $m=0$, and $\alpha=2$, Conlon [8] proved that when $1/r+1/s=\frac{2}{3}$, $2<r$, $s<6$, $|(f,Lg)|\leqslant(\text{const})\, N^{1/2}\,\|f\|_r\,\|g\|_s$ (with $(v,u)=\int \bar{v}u$). In this case, the operator L is the *exchange Coulomb energy operator* of Hartree–Fock theory. Conlon [8] suggested that the exponent $\frac{1}{2}$ could be improved to $\frac{1}{3}$ by using the results of Cwikel [3]. Subsequently, Conlon (private communication) was able to prove the $N^{1/3}$ bound for $r=s=3$ by a completely different method from that given below. The general case is contained in

THEOREM 4. *With L given by* (24) *and* $f\in L^r(\mathbb{R}^d)$, $g\in L^s(\mathbb{R}^d)$, *there are universal constants C, independent of the* $\{\psi_i\}$, *such that*

(i) *For all d and all* $m\geqslant 0$,

$$|(f,Lg)|\leqslant C_{r,s,d,\alpha}\, N^{1-\alpha/d}\, \|f\|_r\, \|g\|_s, \tag{25}$$

when $1/r+1/s=\alpha/d$ *and* $2<r,s<\infty$.

(ii) *For all d and* $m>0$

$$|(f,Lg)|\leqslant C_{r,s,d,\alpha}\, m^{-\alpha+d/r+d/s}\, N^{1-1/r-1/s}\, \|f\|_r\, \|g\|_s$$

when $1/r+1/s<\alpha/d$ *and* $2\leqslant r,s<\infty$.

Proof. Let $H_f=(-\Delta+m^2)^{-\beta/2} f$ and $H_g=(-\Delta+m^2)^{-\gamma/2} g$ with $\beta+\gamma=\alpha$. Then $|(\psi_i, H_f^* H_g \psi_i)|\leqslant \|H_g\psi_i\|_2 \|H_f\psi_i\|_2$ and $|(f,Lg)|=|\sum_{i=1}^N(\psi_i, H_f^* H_g\psi_i)|\leqslant\{\sum_{i=1}^N\|H_g\psi_i\|_2^2\sum_{i=1}^N\|H_f\psi_i\|_2^2\}^{1/2}$. For (i), use Theorem 2(ii) with $r\beta=d$, $s\gamma=d$. For (ii), use Theorem 2(i) with $r\beta>d$, $s\gamma>d$. For part (i), one mimics the proof of Theorem 1, $d\geqslant 3$. For both parts, it is necessary to note that the orthonormality of the $\{\psi_i\}$ implies that $\sum_{i=1}^{N} \|H\psi_i\|_2^2 \leqslant \sum_{i=1}^{N} \lambda_i$, where $\lambda_1 \leqslant \lambda_2 \leqslant \cdots$ are the eigenvalues of H^*H. ∎

Acknowledgments

I thank Professor Paul Federbush for drawing my attention to the $d = 4$ problem contained in [2] and I thank Professor Joseph Conlon for valuable discussions about the problem raised in [8]. The Institute for Advanced Study is thanked for its hospitality and support.

References

1. R. A. Adams, "Sobolev Spaces," Academic Press, New York, 1975.
2. G. A. Battle III and P. Federbush, A Phase Cell Cluster Expansion for Euclidean Field Theories, Part I, 1982, preprint.
3. M. Cwikel, Weak type estimates for singular values and the number of bound states of Schroedinger operators, *Ann. of Math.* **106** (1977), 93–102.
4. E. H. Lieb, The number of bound states of one-body Schroedinger operators and the Weyl problem, Proc. A.M.S. Symposia in Pure Math., Vol. 36, pp. 241–252, 1980; these results were announced in Bounds on the eigenvalues of the Laplace and Schroedinger operators, *Bull. Amer. Math. Soc.* **82** (1976), 751–753.
5. G. V. Rosenbljum, Distribution of the discrete spectrum of singular differential operators, *Dokl. Akad. Nauk SSSR* **202** (1972), 1012–1015 (MR **45**, No. 4216); the details are given in Distribution of the discrete spectrum of singular differential operators, *Izv. Vyss. Učebn. Zaved. Matematika* **164** (1976), 75–86; English trans., *Soviet Math.* (*Iz. VUZ*) **20** (1976), 367–380.
6. B. Simon, "Trace Ideals and their Applications," Cambridge Univ. Press, Cambridge, 1979.
7. E. M. Stein, "Singular Integrals and Differentiability Properties of Functions," Princeton Univ. Press, Princeton, N.J., 1970.
8. J. G. Conlon, Semi-classical limit theorems for Hartree–Fock theory, *Comm. Math. Phys.*, in press.

PROCEEDINGS OF THE
AMERICAN MATHEMATICAL SOCIETY
Volume 88, Number 3, July 1983

A RELATION BETWEEN POINTWISE CONVERGENCE OF FUNCTIONS AND CONVERGENCE OF FUNCTIONALS

HAÏM BREZIS AND ELLIOTT LIEB[1]

ABSTRACT. We show that if $\{f_n\}$ is a sequence of uniformly L^p-bounded functions on a measure space, and if $f_n \to f$ pointwise a.e., then $\lim_{n\to\infty}\{\|f_n\|_p^p - \|f_n - f\|_p^p\} = \|f\|_p^p$ for all $0 < p < \infty$. This result is also generalized in Theorem 2 to some functionals other than the L^p norm, namely $\int |j(f_n) - j(f_n - f) - j(f)| \to 0$ for suitable $j\colon \mathbf{C} \to \mathbf{C}$ and a suitable sequence $\{f_n\}$. A brief discussion is given of the usefulness of this result in variational problems.

1. Introduction. Let (Ω, Σ, μ) be a measure space and let $\{f_n\}_{n=1}^{\infty}$ be a sequence of complex valued measurable functions which are uniformly bounded in $L^p = L^p(\Omega, \Sigma, \mu)$ for some $0 < p < \infty$. Suppose that $f_n \to f$ pointwise almost everywhere (a.e.). What can be said about $\|f\|_p$?

The simplest tool for estimating $\|f\|_p$ is Fatou's lemma, which yields

$$\|f\|_p \leq \liminf_{n\to\infty} \|f_n\|_p.$$

The purpose of this note is to point out that much more can be said, namely

$$\lim_{n\to\infty}\left\{\|f_n\|_p^p - \|f_n - f\|_p^p\right\} = \|f\|_p^p. \tag{1}$$

More generally, if $j\colon \mathbf{C} \to \mathbf{C}$ is a continuous function such that $j(0) = 0$, then, when $f_n \to f$ a.e. and $\int |j(f_n(x))|\, d\mu(x) \leq C < \infty$, it follows that

$$\lim_{n\to\infty} \int [j(f_n) - j(f_n - f)] = \int j(f) \tag{2}$$

under suitable conditions on j and/or $\{f_n\}$.

Heuristically, (2) says the following. If we write $f_n = f + g_n$ with $g_n \to 0$ a.e., then, for large n, $\int j(f + g_n)$ decouples into two parts, namely $\int j(f)$ and $\int j(g_n)$.

Equation (1) is not merely an idle exercise, but it is actually useful in the calculus of variations to prove the existence of maximizing (resp. minimizing) functions in some cases in which compactness is not available. In fact (1) was first used by one of us (E. Lieb), but with a different notion of convergence than pointwise convergence of $f_n \to f$, to solve a variational problem [**1**]. Later, it was also used in another variational problem [**2**]. At the end of this note we shall give a brief account of how (1) can be used.

Received by the editors August 9, 1982 and, in revised form, November 17, 1982.
1980 *Mathematics Subject Classification*. Primary 28A20, 35J60, 46E30.
Key words and phrases. Convergence of functionals, pointwise convergence, L^p spaces.
[1]Work partially supported by U.S. National Science Foundation Grant PHY-8116101.

Two theorems will be stated: (i) the L^p case ($0 < p < \infty$), (ii) the general case (2). Although (i) is a corollary of (ii) we state it separately because it is an important special case and because the assumptions are especially transparent.

E. Lieb is most grateful to the Institute for Advanced Study for its support and hospitality. Both authors thank the Summer Research Institute for bringing them together in Melbourne, Australia, where this note had its origin.

2. The L^p case ($0 < p < \infty$).

THEOREM 1. *Suppose $f_n \to f$ a.e. and $\| f_n \|_p \leq C < \infty$ for all n and for some $0 < p < \infty$. Then the limit in* (1) *exists and the equality in* (1) *holds.*

REMARKS. (i) By Fatou's lemma, $f \in L^p$.

(ii) In case $0 < p \leq 1$, and if we assume that $f \in L^p$, then we do not need the hypothesis that $\| f_n \|_p$ is uniformly bounded. [This follows from the inequality $|\,|f_n|^p - |f_n - f|^p\,| \leq |f|^p$ and the dominated convergence theorem.] However, when $1 < p < \infty$, the hypothesis that $\| f_n \|_p$ is uniformly bounded is really necessary (even if we assume that $f \in L^p$) as a simple counterexample shows.

(iii) When $1 < p < \infty$, the hypotheses of Theorem 1 imply that $f_n \rightharpoonup f$ weakly in L^p. [By the Banach-Alaoglu theorem, for some subsequence, f_{n_k} converges weakly to some g; but $g = f$ since $f_n \to f$ a.e.] However, weak convergence in L^p is insufficient to conclude that (1) holds, except in the case $p = 2$. When $p \neq 2$ it is easy to construct counterexamples to (1) under the assumption only of weak convergence. When $p = 2$ the proof of (1) is trivial under the assumption of weak convergence.

3. The general case. In order to prove (2), some conditions are needed on the function j and the sequence $\{f_n\}$. To make this point clear we shall later give an example for which (2) fails. On the other hand, we shall not attempt to find the most general conditions for which (2) holds but shall, instead, content ourselves here with conditions which are reasonably simple, yet general enough to cover many examples.

Let $j\colon \mathbf{C} \to \mathbf{C}$ be a continuous function with $j(0) = 0$. In addition let j satisfy the following hypothesis:

For every sufficiently small $\varepsilon > 0$ there exist two continuous, nonnegative functions φ_ε and ψ_ε such that

$$|j(a+b) - j(a)| \leq \varepsilon \varphi_\varepsilon(a) + \psi_\varepsilon(b) \tag{3}$$

for all $a, b \in \mathbf{C}$.

THEOREM 2. *Let j satisfy the above hypothesis and let $f_n = f + g_n$ be a sequence of measurable functions from Ω to $\mathbf{C}$ such that:*

(i) $g_n \to 0$ *a.e.*

(ii) $j(f) \in L^1$.

(iii) $\int \varphi_\varepsilon(g_n(x))\, d\mu(x) \leq C < \infty$, *for some constant C, independent of ε and n.*

(iv) $\int \psi_\varepsilon(f(x))\, d\mu(x) < \infty$ *for all* $\varepsilon > 0$.

Then, as $n \to \infty$,

$$\int |j(f + g_n) - j(g_n) - j(f)|\, d\mu \to 0. \tag{4}$$

REMARKS. (i) It is not assumed that $j(f_n)$ or $j(g_n)$ are separately in L^1.

(ii) Note that the convergence in (4) is in the strong L^1 topology. This is a stronger statement than (2).

PROOF OF THEOREM 2. Fix $\varepsilon > 0$ and let

$$W_{\varepsilon,n}(x) = \left[|j(f_n(x)) - j(g_n(x)) - j(f(x))| - \varepsilon\varphi_\varepsilon(g_n(x))\right]_+,$$

where $[a]_+ = \max(a, 0)$. As $n \to \infty$, $W_{\varepsilon,n}(x) \to 0$ a.e. On the other hand,

$$\begin{aligned} |j(f_n) - f(g_n) - j(f)| &\leq |j(f_n) - j(g_n)| + |j(f)| \\ &\leq \varepsilon\varphi_\varepsilon(g_n) + \psi_\varepsilon(f) + |j(f)|. \end{aligned}$$

Therefore, $W_{\varepsilon,n} \leq \psi_\varepsilon(f) + |j(f)| \in L^1$. By dominated convergence, $\int W_{\varepsilon,n}\, d\mu \to 0$ as $n \to \infty$. However,

$$|j(f_n) - j(g_n) - j(f)| \leq W_{\varepsilon,n} + \varepsilon\varphi_\varepsilon(g_n)$$

and, thus,

$$I_n \equiv \int |j(f_n) - j(g_n) - j(f)|\, d\mu \leq \int [W_{\varepsilon,n} + \varepsilon\varphi_\varepsilon(g_n)]\, d\mu.$$

Consequently, $\limsup_{n\to\infty} I_n \leq \varepsilon C$. Now let $\varepsilon \to 0$. □

EXAMPLES. (a) $j(t) = |t|^p$, $0 < p < \infty$. Here (3) is satisfied with $\varphi_\varepsilon(t) = |t|^p$ and $\psi_\varepsilon(t) = C_\varepsilon |t|^p$ for some C_ε sufficiently large. Therefore hypotheses (ii)—(iv) are simply that $f \in L^p$ and the $\{g_n\}$ are uniformly bounded in L^p. This proves Theorem 1.

(b) Suppose that j is a continuous, convex function from $\mathbf{C}$ to $\mathbf{R}$ with $j(0) = 0$. Choose some number $k > 1$. Then (3) holds for $\varepsilon k < 1$ with

$$\varphi_\varepsilon(t) = j(kt) - kj(t) \quad \text{and} \quad \psi_\varepsilon(t) = |j(C_\varepsilon t)| + |j(-C_\varepsilon t)|,$$

with $1/C_\varepsilon = \varepsilon(k-1)$. This is proved in Lemma 3 below. Therefore, the hypotheses of Theorem 2 are satisfied if there is some fixed $k > 1$ such that $[j(kg_n) - kj(g_n)]$ is uniformly bounded in L^1, and if $j(Mf)$ is in L^1 for every real M.

(c) The condition in example (b) that $j(kg_n) - kj(g_n)$ is uniformly bounded in L^1 for some constant $k > 1$ can be essential, not only for the hypotheses of Theorem 2 but for the conclusion as well. Let $\Omega = [0, 1]$, $j(t) = e^{|t|} - 1$, $d\mu = dx$, $f(x) = 1$, $g_n(x) = \ln(1 + n)$ if $0 < x < 1/n$, and $g_n(x) = 0$ otherwise. Then $\int j(f_n) = 2e - 1$, $\int j(g_n) = 1$ and $\int j(f) = e - 1$. In this example we see that (2) does *not* hold even though $j(g_n)$ is uniformly bounded in L^1 and $j(Mf) \in L^1$ for all real M. Note that for this sequence $\{g_n\}$, $j(kg_n)$ is *not* uniformly bounded when $k > 1$. However since $j(t)$ is convex, (b) above tells us that the conclusion of Theorem 2 would be valid for any other sequence $\tilde{g}_n$ such that $j(k\tilde{g}_n)$ is uniformly bounded in L^1 for some $k > 1$.

LEMMA 3. *Let j: $\mathbf{C} \to \mathbf{R}$ be convex and let $k > 1$. Then*

$$|j(a + b) - j(a)| \leq \varepsilon[j(ka) - kj(a)] + |j(C_\varepsilon b)| + |j(-C_\varepsilon b)|$$

for all $a, b \in \mathbf{C}$, $0 < \varepsilon < 1/k$ and $1/C_\varepsilon = \varepsilon(k-1)$.

PROOF. Let $\alpha = 1 - k\varepsilon$, $\beta = \varepsilon$, $\gamma = (k-1)\varepsilon$. Then $\alpha + \beta + \gamma = 1$ and $(a+b) = \alpha a + \beta(ka) + \gamma(C_\varepsilon b)$. By convexity,

$$j(a+b) \leqslant \alpha j(a) + \beta j(ka) + \gamma j(C_\varepsilon b).$$

This implies that

$$j(a+b) - j(a) \leqslant \varepsilon[j(ka) - kj(a)] + |j(C_\varepsilon b)|.$$

For the reverse inequality let

$$\alpha = 1/(1+k\varepsilon), \quad \beta = \varepsilon/(1+k\varepsilon), \quad \gamma = \varepsilon(k-1)/(1+k\varepsilon),$$

whence $a = \alpha(a+b) + \beta(ka) + \gamma(-C_\varepsilon b)$. Then

$$j(a) - j(a+b) \leqslant \varepsilon[j(ka) - kj(a)] + \varepsilon(k-1)j(-C_\varepsilon b). \quad \square$$

4. Applications. In the calculus of variations an oft-met problem is to show that an infimum or supremum is achieved. We shall outline by two examples how Theorem 1 can be used for this purpose.

(A) If K is the sharp constant in the inequality $\|Af\|_q \leqslant K\|f\|_p$, where A is a bounded linear operator from L^p to L^q, can one find f such that equality holds? We shall assume that $\infty > q \geqslant p \geqslant 1$. In fact, the problem in **[1]** that motivated Theorem 1 was the Hardy-Littlewood-Sobolev inequality on $L^p(\mathbf{R}^n, dx)$. Namely, A is the integral kernel $A(x,y) = |x-y|^{-\lambda}$, $0 < \lambda < n$ and $p^{-1} + \lambda/n = 1 + q^{-1}$. Let $K = \sup\{R(f) \mid f \in L^p,\ f \neq 0\}$, where $R(f) = \|Af\|_q / \|f\|_p$. The problem we address here is to prove the existence of a maximizing f, i.e. $R(f) = K$. Suppose that an L^p-bounded sequence $\{f_n\}$ can be found such that (i) $R(f_n) \to K$, (ii) $f_n \to f$ a.e., (iii) $f \neq 0$. (For the H.L.S. inequality, this can be done by using a rearrangement inequality.) The difficulty that one faces is to show $R(f) = K$. This difficulty can be overcome by Theorem 1 if we make the additional assumption that $Af_n \to Af$ a.e. (This can also be verified for the H.L.S. problem.) With these assumptions we have that

$$K^p = \lim_{n\to\infty} \frac{\|Af_n\|_q^p}{\|f_n\|_p^p} = \lim_{n\to\infty} \frac{\{\|Af\|_q^q + \|Ag_n\|_q^q\}^{p/q}}{\{\|f\|_p^p + \|g_n\|_p^p\}}$$

with $f_n = f + g_n$ as before. Since $p/q \leqslant 1$ and $(a+b)^t \leqslant a^t + b^t$ for $a, b \geqslant 0$ and $t \leqslant 1$, and since $\|Ag_n\|_q \leqslant K\|g_n\|_p$ (by definition), it follows that $K^p \leqslant \|Af\|_p^p / \|f\|_p^p$. Thus f is maximizing, as desired. For further details see **[1]**.

(B) This is taken from **[2]**. Let $\Omega \subset \mathbf{R}^n$, $n \geqslant 3$, be a bounded domain. Let $\lambda \geqslant 0$ and let

$$R_\lambda(f) = \frac{\int|\nabla f|^2 - \lambda\int|f|^2}{\|f\|_p^2} \quad \text{with } p = \frac{2n}{n-2}.$$

The problem is to show that $K_\lambda = \inf\{R_\lambda(f) \mid f \in H_0^1(\Omega), f \neq 0\}$ is achieved.

Suppose that we know that $K_\lambda < K_0$ (this is indeed the case for every $\lambda > 0$ when $n \geqslant 4$, and for λ sufficiently large when $n = 3$; see **[2]**); then K_λ is achieved.

To prove this, let $\{f_n\}$ be a minimizing sequence with $\|f_n\|_p = 1$. Since f_n is bounded in $H^1(\Omega)$ we may assume that $f_n \rightharpoonup f$ weakly in H^1, $f_n \to f$ strongly in L^2 and

$f_n \to f$ a.e. We have

$$\int |\nabla f_n|^2 - \lambda \int |f_n|^2 = K_\lambda + o(1),$$

and since $\int |\nabla f_n|^2 \geq K_0 \| f_n \|_p^2 = K_0$ (by definition of K_0), it follows that $\lambda \int |f|^2 \geq K_0 - K_\lambda > 0$. Therefore $f \neq 0$. On the other hand, let $g_n = f_n - f$. We have

$$\int |\nabla f_n|^2 - \lambda \int |f_n|^2 = K_\lambda \|f_n\|_p^2 + o(1),$$

and since $g_n \rightharpoonup 0$ weakly in H^1, we obtain

$$\int |\nabla f|^2 + \int |\nabla g_n|^2 - \lambda \int |f|^2 = K_\lambda \|f_n\|_p^2 + o(1).$$

Consequently,

$$\int |\nabla f|^2 + K_0 \|g_n\|_p^2 - \lambda \int |f|^2 \leq K_\lambda \|f_n\|_p^2 + o(1).$$

On the other hand, it follows from Theorem 1 that

$$\|f_n\|_p^p = \|f\|_p^p + \|g_n\|_p^p + o(1).$$

Since $p \geq 2$ we deduce that

$$\|f_n\|_p^2 \leq \|f\|_p^2 + \|g_n\|_p^2 + o(1).$$

If $K_\lambda \geq 0$, we conclude that

$$K_\lambda \|f_n\|_p^2 \leq K_\lambda \|f\|_p^2 + K_0 \|g_n\|_p^2 + o(1)$$

and, therefore,

$$\int |\nabla f|^2 - \lambda \int |f|^2 \leq K_\lambda \|f\|_p^2 + o(1),$$

i.e. f is minimizing, as desired.

If $K_\lambda < 0$, we have

$$\int |\nabla f|^2 - \lambda \int |f|^2 \leq K_\lambda + o(1) \leq K_\lambda \|f\|_p^2 + o(1)$$

since $\| f \|_p \leq 1$. Here again, f is minimizing, as desired. For further details see [**2**].

References

1. E. H. Lieb, *Sharp constants in the Hardy-Littlewood-Sobolev and related inequalities*, Ann. of Math. (to appear).

2. H. Brezis and L. Nirenberg, *Positive solutions of nonlinear elliptic equations involving critical Sobolev exponents*, Comm. Pure Appl. Math. (to appear).

Département de Mathématiques, Université Paris VI, 4, pl. Jussieu 75230 Paris Cedex 05, France

The Institute for Advanced Study, Princeton, New Jersey 08540

Permanent address: Departments of Mathematics and Physics, Princeton University, Jadwin Hall, P.O.B. 708, Princeton, New Jersey 08544

Annals of Mathematics, **118** (1983), 349–374

Sharp constants in the Hardy-Littlewood-Sobolev and related inequalities

By Elliott H. Lieb*

Abstract

A maximizing function, f, is shown to exist for the HLS inequality on $\mathbf{R}^n$: $\| |x|^{-\lambda} * f\|_q \le N_{p,\lambda,n}\|f\|_p$, with N being the sharp constant and $1/p + \lambda/n = 1 + 1/q$, $1 < p, q, n/\lambda < \infty$. When $p = q'$ or $p = 2$ or $q = 2$, f and N are explicitly evaluated. A maximizing f is also shown to exist for other inequalities:

(i) The Okikiolu, Glaser, Martin, Grosse, Thirring inequality: $K_{n,p}\|\nabla f\|_2 \ge \| |x|^{-b} f\|_p$, $n \ge 3$, $0 \le b < 1$, $p = 2n/(2b + n - 2)$. (This was known before, but the proof here has certain simplifications.)

(ii) The doubly weighted HLS inequality of Stein and Weiss:

$$\left\| \int V(x,y) f(y)\, dy \right\|_q \le P_{\alpha,\beta,p,\lambda,n}\|f\|_p$$

with $V(x,y) = |x|^{-\beta}|x-y|^{-\lambda}|y|^{-\alpha}$, $0 \le \alpha < n/p'$, $0 \le \beta < n/q$, $1/p + (\lambda + \alpha + \beta)/n = 1 + 1/q$.

(iii) The weighted Young inequality: $\| |x|^{\gamma} f\|_p^m \ge Q_{p,m,n}\|f^{(m)}\|_\infty$, where $f^{(m)}(x)$ is the m-fold convolution of f with itself, $m \ge 3$, $m/(m-1) \le p < m$, $\gamma/n + 1/p = (m-1)/m$. When $p = m/(m-1)$ or $p = 2$, f and Q are explicitly evaluated.

I. Introduction

A classical inequality, due to Hardy and Littlewood [15], [16] and Sobolev [26] (see also [11]) states that

$$\left| \int\int f(x)|x-y|^{-\lambda} g(y)\, dx\, dy \right| \le N_{p,\lambda,n}\|f\|_p\|g\|_t \tag{1.1}$$

for all $f \in L^p(\mathbf{R}^n)$, $g \in L^t(\mathbf{R}^n)$, $1 < p$, $t < \infty$, $1/p + 1/t + \lambda/n = 2$ and $0 < \lambda < n$. (By notational definition, $N_{p,\lambda,n}$ is the best, or sharp constant in (1.1).)

*Work partially supported by the U.S. National Science Foundation under grant no. PHY-8116101.

Annals of Math. *118*, 349–374 (1983)

The main purpose of this paper is two-fold. In Section II, it is shown that a maximizing pair f, g exists for (1.1), i.e., a pair that gives equality in (1.1). This will require the use of two rearrangement inequalities and a new compactness technique (Lemma 2.7) for maximizing sequences. From the point of view of general methodology, this is perhaps the most interesting part of this work. In Section III, N and f, g are explicitly computed for the case $t = p$ and, as a corollary, for the cases $t = 2$ or $p = 2$. This part is amusing for the following reason: one can guess what the pair f, g ought to be and verify that they satisfy the integral equations (Euler-Lagrange equations) for (1.1). But these equations are nonlinear and it is far from clear that this choice is actually maximal. The proof that it is so requires use of stereographic projection from $\mathbf{R}^n$ to the sphere $\mathbf{S}^n$ and exploitation of the symmetry of f, g given by the rearrangement inequalities used in Section II.

Additional examples are given of the techniques of Section II. Section IV contains a comparatively simple proof of the existence of a maximizing f for the Sobolev inequality

$$K_n \| \nabla f \|_2 \geq \| f \|_{2n/(n-2)}, \qquad n \geq 3 \tag{1.2}$$

and its generalization due to Okikiolu [21], Glaser, Martin, Grosse and Thirring [14]:

$$K_{n,p} \| \nabla f \|_2 \geq \| \, |x|^{-b} f \|_p, \qquad n \geq 3 \tag{1.3}$$

for $0 \leq b < 1$ and $p = 2n/(2b + n - 2)$. Of course (1.2) has been treated before by Aubin [2] and Talenti [31] and (1.3) in [14], but the directness of the proofs given here may be of some value.

Section V uses the techniques of Section II to prove the existence of a maximizing f, g for the doubly weighted HLS inequality of Stein and Weiss [29]:

$$\left| \int\int g(x) V(x,y) f(y)\, dx\, dy \right| \leq P_{\alpha,\beta,p,\lambda,n} \| f \|_p \| g \|_t \tag{1.4}$$

with $V(x,y) = |x|^{-\beta} |x - y|^{-\lambda} |y|^{-\alpha}$, $0 \leq \alpha < n/p'$, $0 \leq \beta < n/t'$, $1/p + 1/t + (\lambda + \alpha + \beta)/n = 2$. Finally, the weighted Young inequality is shown to have a maximizing f:

$$Q_{p,m,n} \| f^{(m)} \|_\infty \leq \| \, |x|^\gamma f \|_p^m, \qquad m \geq 3 \tag{1.5}$$

where $f^{(m)}(x)$ is the m-fold convolution of f with itself and $m/(m-1) \leq p < m$, $\gamma/n + 1/p = (m-1)/m$. Moreover, Q can be evaluated in two cases: $p = m/(m-1)$ and $p = 2$. The latter case turns out, by Fourier transformation, to be (1.1) in disguise with $p = t$. Thus, the evaluation of the sharp constant in (1.5) for $p = 2$ brings the work full circle.

My indebtedness to Alan Sokal is profound. He stimulated this investigation by suggesting that (1.5) was true for $m = 4$, $p = 2$, a case which arose in his study of quantum field theory [27]. Later he proposed the general case of (1.5). He also suggested that the techniques of Section II would work for (1.4). Throughout the course of this work he was a constant source of encouragement and stimulation. I am also indebted to Henri Berestycki for his encouragement. I thank Haïm Brezis for pointing out the last part of Lemma 2.7 and I thank the referee for many helpful remarks, in particular for drawing my attention to [3]. I am most grateful to the Institute for Advanced Study for its support and hospitality.

A technical remark can be made about (1.1) in the context of *weak* L^p *spaces*, $|x|^{-\lambda} \in L_w^{n/\lambda}(\mathbf{R}^n)$. There are two definitions of what is meant by $\|h\|_{q,w}$ for $1 < q < \infty$. One is

$$\|h\|_{q,w}^{*} = (n/\sigma_n)^{1/q} \sup_{\alpha>0} \alpha\mu\{x|\,|h(x)| > \alpha\}^{1/q} \tag{1.6}$$

where σ_n is the area of the unit sphere, (2.13), and μ is Lebesgue measure. This is not a true norm (the triangle inequality is not satisfied), but it is convenient and it is equivalent to the following, due to Calderón, which is a true norm:

$$\|h\|_{q,w}^{**} = (1/q')(n/\sigma_n)^{1/q} \sup_A \mu(A)^{-1/q'} \int_A |h(x)|\, dx \tag{1.7}$$

for $0 < \mu(A) < \infty$. Clearly $|x|^{-\lambda}$ has unit norm in both definitions ($q = n/\lambda$). The generalization of (1.1) is

$$\left| \int\int f(x)h(x-y)g(y)\, dx\, dy \right| \le N_{p,\lambda,n} \|f\|_p \|g\|_t \|h\|_{q,w} \tag{1.8}$$

with $q = n/\lambda$, $1/p + 1/t + 1/q = 2$, $1 < p, t, q < \infty$, and *the same* $N_{p,\lambda,n}$ is sharp in (1.8) as in (1.1). *Either one* of the two definitions, (1.6) or (1.7), may be used in (1.8), and the same N is sharp for both.

The justification for (1.8) is the following: if we replace f, g, h by their symmetric decreasing rearrangement, f^*, g^*, h^* (see Section II), the left side of (1.8) does not decrease. All the norms on the right side of (1.8) are invariant. The maximizing f, g for (1.1) satisfies $f = f^*$, $g = g^*$ (Section II). We note that $|x|^{-\lambda} = \sup\{h(x)|\, \|h\|_{p,w}^{*} \le 1, h = h^*\}$. Thus, (1.8) holds with (1.6). The proof for (1.7) is trickier. Let $b = f * g$. Clearly, $b = b^*$ since $f = f^*$, $g = g^*$. Thus, $b(x) = \int_0^\infty da\, \chi_a(x)$ where χ_a is the characteristic function of $T_a = \{x|b(x) > a\}$, which is a ball of some radius R_a. Assume that $\|h\|_{q,w}^{**} = 1$. The left side of

(1.8) is

$$\int b(x)h(x)\,dx = \int_0^\infty da \int \chi_a(x)h(x)\,dx \le \int_0^\infty da\, q'(\sigma_n/n)^{1/q}|T_a|^{1/q'}$$

$$= \int_0^\infty da\, q'(\sigma_n/n)R_a^{n-\lambda} = \int_0^\infty da \int \chi_a(x)|x|^{-\lambda}\,dx$$

$$= \int b(x)|x|^{-\lambda}\,dx.$$

II. Existence of a maximizing function

Here we shall establish the existence of a maximizing pair of functions f, g giving equality in (1.1). This means finding $f \in L^p$, $p^{-1} + \lambda/n = 1 + q^{-1}$, such that if

$$R(F) \equiv \| |x|^{-\lambda} * F\|_q / \|F\|_p, \qquad F \neq 0, \text{ then} \tag{2.1}$$

$$R(f) = N_{p,\lambda,n} \equiv \sup\{R(F) | F \in L^p, F \neq 0\}. \tag{2.2}$$

Some remarks might be helpful to explain the difficulties to be faced in finding this f. First, the usual way to find f is by a compactness argument. But $R(F)$ is not upper-semicontinuous in the L^p weak topology. Second, $R(F)$ is invariant under the conformal group of dilations, rotations and translations, namely,

$$F(x) \to F(\gamma\mathcal{R}x + y), \qquad \gamma > 0, \quad \mathcal{R} \in O(n), \quad y \in \mathbf{R}^n. \tag{2.3}$$

Furthermore, if $q = p'$, a case for which we shall explicitly find f, an inversion symmetry also exists, i.e.,

$$F(x) \to |x|^{\lambda - 2n} F(x|x|^{-2}). \tag{2.4}$$

The existence of this large invariance group means that a maximizing f cannot be unique and also that it is easy for a weakly convergent maximizing sequence $\{f_n\}$ to converge to zero.

Third, if the kernel $|x-y|^{-\lambda}$ is changed slightly, a maximizing f need not exist. Explicitly, let $K(r)$, for $r > 0$, be any positive function such that $r^\lambda K(r)$ is strictly monotone increasing. Consider $K(|x-y|)$ in place of $|x-y|^{-\lambda}$ as a kernel in (2.1). The Fourier transform of the Bessel potential, $(1 + |x-y|^2)^{-\lambda/2}$, is a good example; it is even positive definite. For any $f \in L^p$, let $F(x) = |f(x/2)|$. It is easy to check that $R(F) > R(f)$, and hence, that no maximum can exist.

One of the key tools we shall use (several times, in fact) is Riesz's rearrangement inequality [24] (for a generalization see [7]) in the strong form given by Lieb [20]. It is recalled in Lemma 2.1.

Definition 1. Let f: $\mathbf{R}^n \to \mathbf{C}$ satisfy $\mu_f(a) = \mu\{x \mid |f(x)| > a\} < \infty$ for all $a > 0$. (Here, μ is Lebesgue measure.) f^*: $\mathbf{R}^n \to [0, \infty)$ is a *symmetric decreasing rearrangement* of f if $f^*(x)$ depends only on $|x|$, and $f^*(x_1) \geq f^*(x_2) \geq 0$ if $|x_1| \leq |x_2|$, and $\mu_{f^*}(a) = \mu_f(a)$, for all $a > 0$.

It is easy to check that f^* always exists and it is defined uniquely almost everywhere (see [20]). Henceforth, notation will be abused in the sense that any function $f(x)$ that depends only on $|x|$ will sometimes be written as $f(|x|)$.

It is convenient to introduce the following sets of functions from $\mathbf{R}^n \to [0, \infty)$ (where T denotes "translate"):

SD $= \{f \mid f$ is symmetric decreasing, i.e., $f = f^*\}$;
SSD $= \{f \mid f \in$ SD and f is strictly monotone decreasing$\}$;
TSD $= \{f \mid f_y(x) = f(x+y)$ and $f_y \in$ SD for some $y \in \mathbf{R}^n\}$;
TSSD $= \{f \mid f \in$ TSD and $f_y \in$ SSD$\}$.

LEMMA 2.1. *Let f, g, h be functions on $\mathbf{R}^n$ satisfying the conditions of Definition* 1 *and let*

$$I(f, g, h) = \int\int f(x) g(x-y) h(y)\, dx\, dy.$$

Then (i) $I(f^*, g^*, h^*) \geq |I(f, g, h)|$.

If, in addition, $g^* \in$ SSD *then*

(ii) $I(f^*, g^*, h^*) > |I(f, g, h)|$ *unless* $f(x) = f^*(x+y)$ *and* $h(x) = h^*(x+y)$ *for some (common)* $y \in \mathbf{R}^n$.

The first part of this lemma has been generalized to more than three functions and more than two variables in [7].

Another closely related fact that will be needed later is Lemma 2.2. We omit the easy proof (which mimics the proof of Lemma 2.1); it can also be obtained from Lemma 2.1 by suitable choice of h.

LEMMA 2.2. *Let* $g = g^*$, $f = f^*$. *Suppose the convolution* $k \equiv g * f$ *satisfies* $k(x) < \infty$ *for all* $x \neq 0$. *Then*

(i) $k = k^*$ *and*
(ii) $k \in$ SSD *if* $g \in$ SSD.

The existence theorem to be proved is

THEOREM 2.3. *Let* $1/p + \lambda/n = 1 + 1/q$ *with* $0 < \lambda < n$, $1 < p, q < \infty$. *Then*

(i) $N_{p,\lambda,n}$ *in* (2.2) *is finite and there exists an* $f \in L^p$ *that maximizes* R, *i.e.,* $R(f) = N_{p,\lambda,n}$.

Annals of Math. *118*, 349–374 (1983)

(ii) *After multiplication by a suitable complex constant, every maximizing* f *is in* TSSD *and satisfies the pair of equations*

$$|x|^{-\lambda} * f = g^{t-1}, \qquad |x|^{-\lambda} * g = f^{p-1} \tag{2.5}$$

for some $g \in L^t$ *and* $g \in$ TSSD. (*Here* $t = q' = q/(q-1)$.) *After a common translation*, $f, g \in$ SSD.

(iii) *When* $q' = p = t$, *then* $g = f$.

(iv) *Let* $q' = p = t$ *and let* f *be translated so that* $f = f^*$. *Then, possibly after a dilation* $f(r) \to \gamma^{n/p} f(\gamma r)$, f *has the inversion symmetry of* (2.4):

$$f(1/r) = r^{2n/p} f(r). \tag{2.6}$$

In the following, irrelevant positive constants will all be denoted by the common symbol C.

Proof of (ii) *and* (iii): These two parts are easy in view of Lemma 2.1. N in (2.2), which is here assumed to be finite, can be written as

$$N_{p,\lambda,n} = \sup_{f,g} \int\int f(x) g(y) K(x-y)\, dx\, dy / \|f\|_p \|g\|_t, \tag{2.7}$$

where $f \in L^p$, $g \in L^t$ and $K(x) = |x|^{-\lambda}$ and $1/t + 1/p + \lambda/n = 2$. Since the rearrangements $f \to f^*$, $g \to g^*$ do not change the norms $\|f\|_p$, $\|g\|_t$, and since $K = K^* \in$ SSD, Lemma 2.1 (ii) shows that $f, g \in$ TSD (possibly after the multiplication by constants). The equations (2.3) are then easy to derive by letting $f \to f + \varepsilon\varphi$, $g \to g + \delta\psi$ and setting the derivatives at $\varepsilon = \delta = 0$ equal to zero. (Again, it may be necessary to multiply f and g by constants to get unity on the right side of (2.5).) By Lemma 2.2, equations (2.5) imply that (after a translation), $f, g \in$ SSD. (iii) follows from (2.7) and the fact that $K(x-y)$ is positive definite. In fact, $|x|^{-\lambda} = C|x|^{-(n+\lambda)/2} * |x|^{-(n+\lambda)/2}$. (See (3.6).) □

Note that (2.7) implies

$$N_{p,\lambda,n} = N_{t,\lambda,n}, \qquad 1/p + 1/t + \lambda/n = 2. \tag{2.8}$$

Beginning of Proof of (i). Let $\{f_j\}$ be a maximizing sequence, i.e., $R(f_j) \to N$. Assume $\|f_j\|_p = 1$. Since $f_j \in L^p$, f_j^* exists and $\|f_j^*\|_p = 1$. By Lemma 2.1 (see the proof of (ii)), $R(f_j^*) \geq R(f_j)$, so we can henceforth assume that $f_j = f_j^*$. Now

$$\begin{aligned} 1 = C\int_0^\infty r^{n-1} f_j(r)^p\, dr &\geq C\int_0^R r^{n-1} f_j(r)^p\, dr \geq CR^n f_j(R)^p \\ &\Rightarrow 0 \leq f_j(r) \leq Cr^{-n/p}. \end{aligned} \tag{2.9}$$

By passing to a subsequence, we can then assume

$$f_j(r) \to f(r) \quad \text{and} \quad 0 \leq f(r) \leq Cr^{-n/p} \tag{2.10}$$

for all rational r. Since $f_j(r)$ is non-increasing in r, it is easy to see that $f_j(r)$ converges for almost all $r > 0$ and therefore that $f = f^*$. (This is essentially Helly's theorem.) By Fatou's lemma, $f \in L^p$.

The problem we face is that f could easily be zero because of the dilation subgroup mentioned in (2.3). Even if $f \neq 0$ it is not obvious that $R(f) \geq N_{p,\lambda,n}$, but this fact will be proved with the help of the following lemma.

LEMMA 2.4. *Let $1/p + \lambda/n = 1 + 1/q$, with $0 < \lambda < n$. Suppose $f \in L^p(\mathbf{R}^n)$ is spherically symmetric and $|f(r)| \leq \varepsilon r^{-n/p}$ for all $r > 0$. There is a constant, C_n, independent of f and ε such that $\||x|^{-\lambda} * f\|_q \leq C_n \|f\|_p^{p/q} \varepsilon^{1-p/q}$. (Note that $p < q$.)*

Remarks. (i) Lemma 2.4 and (2.9) obviously imply that $N < \infty$.

(ii) Lemma 2.4 follows from known results about the Lorentz spaces $L(p, q)$ (see [22], [9], [28]). We give our own proof, which is based on a transformation to *logarithmic radial variables*, for two reasons: (a) In conjunction with Lemma 2.1 it provides an alternative strategy for proving many known facts about $L(p, q)$ spaces; (b) The formulation given in our proof will be needed later in order to establish (and hence to exploit) the inversion symmetry for $q = p'$ given in Theorem 2.3 (iv).

Proof. Define F: $\mathbf{R} \to \mathbf{R}$ by

$$F(u) = e^{un/p} f(e^u), \quad \text{whence} \tag{2.11}$$

$$\sigma_n^{1/p} \|F\|_{L^p(\mathbf{R})} = \|f\|_{L^p(\mathbf{R}^n)}. \tag{2.12}$$

Here, σ_n is the area of the unit sphere in $\mathbf{R}^n$,

$$\sigma_n = 2\pi^{n/2}/\Gamma(n/2). \tag{2.13}$$

Without loss of generality, we can assume $f(r) \geq 0$. Define $h = |x|^{-\lambda} * f$, which is spherically symmetric, and H: $\mathbf{R} \to \mathbf{R}$ by $H(u) = e^{un/q} h(e^u)$. As in (2.11), $\sigma_n^{1/q} \|H\|_q = \|h\|_q$. An explicit form for H, which is easily obtained by integrating $d^n x$ over angles in $\mathbf{R}^n$, is the following:

$$H(u) = \int_{-\infty}^{\infty} L_n(u - v) F(v)\, dv, \quad \text{where} \tag{2.14}$$

$$L_n(u) = 2^{-\lambda/2} \exp\{u(n/q - \lambda/2)\} Z_n(u), \tag{2.15}$$

$$\begin{aligned} Z_n(u) &= \sigma_{n-1} \int_0^{\pi} [\cosh u - \cos\theta]^{-\lambda/2} (\sin\theta)^{n-2}\, d\theta, \qquad n \geq 2, \\ &= [\cosh u + 1]^{-\lambda/2} + [\cosh u - 1]^{-\lambda/2}, \qquad n = 1. \end{aligned} \tag{2.16}$$

Now, $L_n \in L^1(\mathbf{R})$ and $F \in L^p(\mathbf{R})$ and $\|F\|_\infty < \varepsilon$. (Note that $|n/q - \lambda/2| < \lambda/2$, since $p > 1$, and that the singularity, if any, in $Z_n(u)$ for u near zero is no worse

Annals of Math. *118*, 349–374 (1983)

than $|u|^{-\lambda/n}$.) By Young's inequality, $\|H\|_p \leq C\|F\|_p$ and $\|H\|_\infty \leq C\|F\|_\infty \leq C\varepsilon$. Since $q > p$, the lemma follows from Hölder's inequality. □

Before returning to the proof of Theorem 2.3(i), let us draw two conclusions from the construction, (2.11)–(2.16).

First, the original problem (2.2) is equivalent to the one-dimensional problem

$$N_{p,\lambda,n} = \sigma^{1/q-1/p}\sup\left\{ \|L_n * F\|_q/\|F\|_p \,\middle|\, 0 \neq F \in L^p(\mathbf{R})\right\}. \tag{2.17}$$

In particular, $L_n *$ is a bounded operator from $L^p(\mathbf{R})$ to $L^q(\mathbf{R})$.

The second conclusion is

Proof of Theorem 2.3 (iv). Make the change of variables given in (2.10) and note that $n/q - \lambda/2 = 0$ in (2.14) when $q = p'$. Thus, $L_n = K_n \in \mathrm{SSD}(\mathbf{R})$. From (2.17) and by the same proof as for Theorem 2.3(ii), $F \in \mathrm{TSSD}(\mathbf{R})$. Translating F, namely $F(u) \to F(u+y)$, is the same as dilation of f. With $F \in \mathrm{SSD}(\mathbf{R})$, inverting (2.11) gives the desired result. □

It is worth noting that the strong rearrangement inequality had to be used *twice* to prove Theorem 2.3(iv).

Lemma 2.4 and (2.9) not only imply that $N < \infty$, they also imply that, after a suitable j-dependent dilation, we can assume that *the limit*, f, *in* (2.9) *is not zero*. To see this, let

$$a_j = \sup_r r^{n/p} f_j(r).$$

By Lemma 2.4, $a_j \nrightarrow 0$, for otherwise $\||x|^{-\lambda} * f_j\|_q \to 0$ while $\|f_j\|_p = 1$, which would mean that $\{f_j\}$ is not a maximizing sequence. Thus, $a_j > 2\beta > 0$. Replace $f_j(r)$ by $\gamma_j^{n/p} f_j(\gamma_j r)$, which does not change the norm of f_j. We can now choose γ_j so that $f_j(1) > a_j/2 > \beta > 0$. Therefore, $f(1) > \beta$ and, since $f \in \mathrm{SD}$, $f(r) > \beta$ for $r \leq 1$. Thus, f is not zero.

Let us briefly review the situation about Theorem 2.3(i). We have a maximizing sequence $\{f_j\}$ of non-negative symmetric decreasing functions which converge pointwise, almost everywhere to $f \neq 0$. By Fatou's lemma, $\|f\|_p \leq \lim\|f_j\|_p = 1$; therefore f will be maximizing if $I(f_j) \to I(f)$, where

$$I(g) \equiv \||x|^{-\lambda} * g\|_q. \tag{2.18}$$

The convergence of $I(f_j)$ to $I(f)$ will be proved, but only after we first prove that $R(f_j) \to R(f)$ and $\|f\|_p = 1$. Before doing so let us first consider a related problem which is interesting in its own right, for which it is easy to establish that $I(f_j) \to I(f)$. This other problem and its solution are stated as the following theorem.

THEOREM 2.5. *Let* $1/p + 1/t + \lambda/n = 2$ *with* $0 < \lambda < n$ *and* $1 < p$, $t < \infty$ *as before, and consider the ratio in* (2.6) *but with g restricted to be f; i.e.,*

$$\tilde{N} = \sup_f \int\int f(x)f(y)|x-y|^{-\lambda}\,dx\,dy/\|f\|_p\|f\|_t \tag{2.19}$$

with $f \in L^p \cap L^t$ *and* $f \neq 0$. *(Naturally,* $\tilde{N} \leq N$ *and* $\tilde{N} = N$ *when* $t = p = q'$ *as stated in Theorem* 2.3(iii).) *Let* $t \neq p$. *Then there exists a maximizing f for* $\tilde{N}$. *Furthermore, after multiplication by a constant, a dilation and a translation (i.e.,* $f(x) \to cf(\gamma x + y)$ *this f is in* SSD *and satisfies*

$$|x|^{-\lambda} * f = f^{p-1} + f^{t-1}. \tag{2.20}$$

Proof. All of the argument is as before, but with one additional fact at our disposal. We can (after dilation and multiplication by a constant) assume that $\|f_j\|_p = \|f_j\|_t = 1$. From (2.9) the limit f satisfies $f(r) \leq Cr^{-n/p}$ and $f(r) \leq Cr^{-n/t}$ (same C). Let $h(x) = C\min\{|x|^{-n/p}, |x|^{-n/t}\}$. Although h is neither in L^p nor in L^t, the function $h(x)h(y)|x-y|^{-\lambda} \in L^1(\mathbf{R}^n \times \mathbf{R}^n)$. (To see this, note that $h \in L^s$ when $\min(t,p) < s < \max(t,p)$. Choose s so that $1/s + 1/s + \lambda/n = 2$. But we already proved that $h \to |x|^{-\lambda} * h$ is bounded from L^s to $L^{s'}$.) Therefore, if $I(f)$ denotes the integral in (2.19), we have that $I(f_j) \to I(f)$ by dominated convergence. □

Returning to Theorem 2.3(i), we see that establishing the convergence of $I(f_j)$ to $I(f)$ is more delicate than in Theorem 2.5, even if $p \neq t$, because all we know is that $f \in L^p$, and not necessarily in L^t. Therefore, the dominated convergence argument cannot be used.

To control the convergence of $R(f_j)$ to $R(f)$, the following lemma due to Brezis and myself [8] is useful.

LEMMA 2.6. *Let* $0 < p < \infty$. *Let* (M, Σ, μ) *be a measure space and let* $\{f_j\}$ *be a uniformly norm-bounded sequence in* $L^p(M, \Sigma, \mu)$ *that converges pointwise, almost everywhere to f. (By Fatou's lemma* $f \in L^p$.) *Then the following limit exists and equality holds.*

$$\lim_{j\to\infty} \int \big||f_j(x)|^p - |f(x) - f_j(x)|^p - |f(x)|^p\big|\,d\mu(x) = 0.$$

Remarks. Lemma 2.6 says more than that $\|f_j\|_p^p - \|f - f_j\|_p^p \to \|f\|_p^p$. It improves Fatou's lemma which says that $\liminf \|f_j\|_p^p \geq \|f\|_p^p$. In [8] a similar theorem is proved for functionals of the form $f \to \int J(f)\,d\mu$. The conclusion of Lemma 2.6 does not hold (except when $p = 2$) if pointwise is replaced by weak convergence. Note that the lemma holds even for $0 < p < 1$. Lemma 2.6 can be

Annals of Math. *118*, 349–374 (1983)

proved simply without using the general results in [8]. Note that

$$\big||f_j|^p - |f - f_j|^p - |f|^p\big|$$
$$\leq \begin{cases} 2|f|^p, & 0 < p \leq 1 \\ p2^{p-1}\{|f - f_j|^{p-1}|f| + |f - f_j||f|^{p-1}\}, & 1 \leq p < \infty. \end{cases}$$

The lemma follows from the first inequality for $0 < p < 1$ by dominated convergence; for $1 \leq p < \infty$ it follows from the second by Egorov's theorem.

The utility of Lemma 2.6 for problems in the calculus of variations is given in the next lemma.

LEMMA 2.7. *Let (M, Σ, μ) and (M', Σ', μ') be measure spaces and let X (resp. Y) be $L^p(M, \Sigma, \mu)$ (resp. $L^q(M', \Sigma', \mu')$) with $1 \leq p \leq q < \infty$. Let A be a bounded linear operator from X to Y. For $f \in X$, $f \neq 0$ let*

$$R(f) = \|Af\|_Y/\|f\|_X \quad \text{and} \quad N = \sup\{R(f)|f \neq 0\}.$$

Let $\{f_j\}$ be a uniformly norm-bounded maximizing sequence for N and suppose that $f_j \to f \neq 0$ and that $Af_j \to Af$ pointwise almost everywhere. Then f maximizes, i.e., $R(f) = N$. Moreover, if $p < q$ and if $\lim\|f_j\|_X = C$ exists, then $\|f\|_X = C$, and hence $\|Af_j\|_Y \to \|Af\|_Y$.

Proof. By Lemma 2.6, $\|f_j\|_X^p = \|f - f_j\|_X^p + \|f\|_X^p + o(1)^p$ (where $o(1)$ denotes something that goes to zero as $j \to \infty$) and $\|Af_j\|_Y^q = \|Af - Af_j\|_Y^q + \|Af\|_Y^q + \tilde{o}(1)^q$. If $a, b, c \geq 0$, then $(a^q + b^q + c^q)^{p/q} \leq a^p + b^p + c^p$. Thus,

$$R(f_j)^p \leq \{\|Af\|_Y^p + \|A(f - f_j)\|_Y^p + o(1)^p\}/\{\|f\|_X^p + \|f - f_j\|_X^p + \tilde{o}(1)^p\}.$$

Now $\|A(f - f_j)\|_Y \leq N\|f - f_j\|_X$ for every j, and $o(1), \tilde{o}(1) \to 0$, and $R(f_j)^p \to N^p$. Since $f \neq 0$, we must have that $\|Af\|_Y \geq N\|f\|_X$, and hence $\|Af\|_Y = N\|f\|_X$. We also have that $\lim\|A(f - f_j)\|_Y - N\|f - f_j\|_X = 0$. For the last part, let $p < q$ and $\lim\|f_j\|_X = C$. Then $0 < \|f\|_X \leq C$. Suppose $0 < \|f\|_X = D < C$. Then $\lim\|f - f_j\|_X^p = C^p - D^p > 0$. However, $(a^q + b^q)^{p/q} < a^p + b^p$ unless $a = 0$ or $b = 0$. Thus, $\lim R(f_j)^p < N^p$, which is a contradiction. □

The last sentence of Lemma 2.7, and the proof, were pointed out to me by Haïm Brezis.

Conclusion of Proof of Theorem 2.3(i). Let us return to the one-dimensional equivalent formulation given in (2.11)–(2.17). Given the maximizing sequence $f_j \in \mathrm{SD}$ with $\|f_j\|_p = 1$, (2.11) defines a sequence F_j: $\mathbf{R} \to \mathbf{R}$ with $\|F_j\|_p = \sigma_n^{-1/p}$ and $\|F_j\|_\infty \leq C$ by (2.9). Also, $f_j \to f \neq 0$ pointwise, so $F_j \to F \neq 0$ pointwise.

Lemma 2.7 can now be applied to finish the proof provided the operator $A = L_n *: L^p(\mathbf{R}) \to L^q(\mathbf{R})$ satisfies $AF_j \to AF$ pointwise. But since $L_n \in L^1(\mathbf{R})$ and since $\|F_j - F\|_\infty < C$, $L_n * (F_j - F)(x) \to 0$ everywhere by dominated convergence. The fact that $I(f_j) \to I(f)$ is contained in the last sentence of Lemma 2.7. In our case, $p < q$ since $\lambda/n < 1$. □

III. The maximizing function when $p' = q$ or $p = 2$ or $q = 2$

In certain cases the equations (2.5) for the maximizing f can be solved and the constant N computed explicitly. In these cases a solution to (2.5) can be easily guessed and verified. The difficult part will be to prove that this f is actually maximizing. To prove this it will be necessary to use stereographic projection to recast (2.5) as an equation on the sphere $\mathbf{S}^n$.

Recall that $1/p + 1/t + \lambda/n = 2$ or $1/p + \lambda/n = 1 + 1/q$ with $t = q' = q/(q-1)$ and $0 < \lambda < n$, $1 < t, p, q < \infty$.

THEOREM 3.1. *When $t = p = 2n/(2n - \lambda)$, $q = 2n/\lambda$ (or $p' = q$), the maximizing f for (2.1) is (after multiplication by a constant and dilation) uniquely*

$$f(x) = \left(1 + |x|^2\right)^{-n/p} \quad \textit{and} \tag{3.1}$$

$$N_{p,\lambda,n} = \pi^{\lambda/2} \frac{\Gamma(n/2 - \lambda/2)}{\Gamma(n - \lambda/2)} \left\{ \frac{\Gamma(n/2)}{\Gamma(n)} \right\}^{-1+\lambda/n}. \tag{3.2}$$

COROLLARY 3.2. (i) *Let $q = t = 2$ and $p = 2n/(3n - 2\lambda)$, which requires $n < 2\lambda < 2n$. The maximizing f for (2.1) is (after multiplication by a constant and dilation) uniquely*

$$f(x) = \left(1 + |x|^2\right)^{-n/p} \quad \textit{and} \tag{3.3}$$

$$N_{p,\lambda,n} = \pi^{\lambda/2} \frac{\Gamma(n/2 - \lambda/2)}{\Gamma(\lambda/2)} \left\{ \frac{\Gamma(\lambda - n/2)}{\Gamma(3n/2 - \lambda)} \right\}^{1/2} \left\{ \frac{\Gamma(n/2)}{\Gamma(n)} \right\}^{-1+\lambda/n}. \tag{3.4}$$

(ii) *Let $p = 2$, $t = 2n/(3n - 2\lambda)$, $q = 2n/(2\lambda - n)$, which requires $n < 2\lambda < 2n$. The maximizing f for (2.1) is (after multiplication by a constant and dilation) uniquely*

$$f = |x|^{-\lambda} * \left(1 + |x|^2\right)^{-n/t} \tag{3.5}$$

and $N_{2,\lambda,n}$ is given by the right side of (3.4).

Proof of Corollary 3.2. Note that for $0 < \lambda, \gamma < n$ and $\lambda + \gamma > n$,

$$|x|^{-\lambda} * |x|^{-\gamma} = D(\lambda,\gamma)|x|^{n-\lambda-\gamma}, \tag{3.6}$$

$$D(\lambda,\gamma) = \pi^{n/2}\Gamma(n/2-\lambda/2)\Gamma(n/2-\gamma/2)\Gamma(\lambda/2+\gamma/2-n/2) \times\{\Gamma(\lambda/2)\Gamma(\gamma/2)\Gamma(n-\lambda/2-\gamma/2)\}^{-1}. \tag{3.7}$$

This follows from the fact [28, Theorem IV. 4.1] that the Fourier transform $w_\lambda(k) = \int|x|^{-\lambda}\exp(ik\cdot x)\,dx$ is

$$w_\lambda(k) = |k|^{\lambda-n}\pi^{n/2}2^{n-\lambda}\Gamma(n/2-\lambda/2)/\Gamma(\lambda/2). \tag{3.8}$$

When $q = 2$,

$$N^2 = \||x|^{-\lambda} * f\|_2^2/\|f\|_p^2 = D(\lambda,\lambda)(f, |x|^{n-2\lambda} * f)/\|f\|_p^2.$$

But this maximization problem is, by Theorem 2.3(iii) and (2.7), the same as in Theorem 3.1 (but with $\lambda \to 2\lambda - n$). When $p = 2$ the proof is similar, using (2.5). See (2.8). □

Beginning of Proof of Theorem 3.1. By Theorem 2.3, the f we seek *must have* (after dilation, etc.) two properties:

(a) The inversion symmetry $f(1/r) = r^{2n/p}f(r)$ and

(b) It satisfies (2.5) up to a constant, namely

$$|x|^{-\lambda} * f = Bf^{p-1}. \tag{3.9}$$

Clearly, (a) holds for (3.1). The fact that (3.1) satisfies (3.9) can be seen in several ways. One way is to note that

$$f_\mu(x) \equiv (1+|x|^2)^{-\mu} \tag{3.10}$$

has the Fourier transform

$$\begin{aligned}\hat{f}_\mu(k) &= (2\pi)^{n/2}\int_0^\infty (|k|r)^{1-n/2}r^{n-1}J_{-1+n/2}(|k|r)f_\mu(r)\,dr \\ &= \pi^{n/2}2^{1-\mu+2/n}\Gamma(\mu)^{-1}|k|^{\mu-n/2}K_{\mu-n/2}(|k|).\end{aligned} \tag{3.11}$$

Here, K is a Bessel function and satisfies $K_\nu(x) = K_{-\nu}(x)$. If we set $\mu = n/p = n - \lambda/2$ and use (3.8), we find that

$$|x|^{-\lambda} * f_{n/p} = B_\lambda f_{\lambda/2} = B_\lambda(f_{n/p})^{p-1}, \tag{3.12}$$

$$B_\lambda = \pi^{n/2}\Gamma(n/2-\lambda/2)/\Gamma(n-\lambda/2). \tag{3.13}$$

It follows that $R(f_{n/p})$ in (2.1) is

$$\begin{aligned}\||x|^{-\lambda} * f_{n/p}\|_q/\|f_{n/p}\|_p &= B_\lambda[\hat{f}_n(0)]^{1/q-1/p} \\ &= \text{right side of (3.2)}.\end{aligned} \tag{3.14}$$

The calculation just given is slightly formal, but it can easily be made rigorous. The real problem that faces us, however, is this: (3.12) shows that f in (3.1) (hereafter we shall denote $f_{n/p}$ by f) satisfies (3.9). It also has the correct inversion symmetry. Is this f maximizing? Is it the unique maximizer (up to a constant)? We do not know that (3.9) has an (essentially) unique solution—even if we restrict to the SSD category—and we shall offer no proof of this kind of uniqueness. *This is an open problem*! But it will be shown that f is (essentially) unique in the category of maximizers. In the course of this proof, (3.12)–(3.14) will be rederived in a simpler way. For the proof, a change of variables will be required, namely stereographic projection.

Stereographic Projection. Consider the sphere $\mathbf{S}^n$ in $\mathbf{R}^{n+1}$, $\mathbf{S}^n = \{\Omega \in \mathbf{R}^{n+1} \mid |\Omega| = 1\}$. Consider the invertible map $\Sigma\colon \mathbf{R}^n \to \mathbf{S}^n \setminus (0,\ldots,0,-1)$

$$\Sigma(x) = (\rho, \xi) = \left(2x/(1+|x|^2), (1-|x|^2)/(1+|x|^2)\right) \tag{3.15}$$

where $\rho \in \mathbf{R}^n$. Conversely, if $\rho \in \mathbf{R}^n$, $\xi \in (-1, 1]$ and $(\rho, \xi) \in \mathbf{S}^n$,

$$\Sigma^{-1}((\rho, \xi)) = \rho/(1+\xi). \tag{3.16}$$

Apart from trivial constants this is the usual stereographic projection with $0 \in \mathbf{R}^n \to$ "north pole".

Let $x_1, x_2 \in \mathbf{R}^n$ and $\Omega_i = \Sigma(x_i)$ with $\Omega_i = (\rho_i, \xi_i)$. Then

$$|x_1 - x_2| = |\Omega_1 - \Omega_2|\{(1+\xi_1)(1+\xi_2)\}^{-1/2}. \tag{3.17}$$

Here, $|\Omega_1 - \Omega_2|$ means Euclidean distance in $\mathbf{R}^{n+1}$, not geodesic distance on $\mathbf{S}^n$.

Let $d\Omega$ be the rotation invariant measure on $\mathbf{S}^n$ with the normalization

$$\int d\Omega = \sigma_{n+1} = 2\pi^{(n+1)/2}\Gamma(n+1)/2)^{-1} \tag{3.18}$$

which is the area of the unit sphere in $\mathbf{R}^{n+1}$ (see (2.13)). Then the Jacobian of Σ is given by

$$d\Omega = d\rho/|\xi| = 2^n(1+|x|^2)^{-n}\, dx = (1+\xi)^n\, dx. \tag{3.19}$$

With any $f\colon \mathbf{R}^n \to \mathbf{C}$ we associate $F\colon \mathbf{S}^n \to \mathbf{C}$ (denoted by $f \leftrightarrow F$) by

$$\begin{aligned} F(\Omega) &= (1+\xi)^{-\mu} f(\Sigma^{-1}(\Omega)), \\ f(x) &= 2^{-n}(1+|x|^2)^{\mu} F(\Sigma(x)), \end{aligned} \tag{3.20}$$

with $\mu = n/p = n - \lambda/2$. ($\lambda$ enters at this point.) Clearly,

$$\|F\|_p = \|f\|_p. \tag{3.21}$$

In particular, f given by (3.1) corresponds to $F(\Omega) = \text{constant} = 2^{-\mu}$.

Annals of Math. *118*, 349–374 (1983)

From (3.15)–(3.18) we have that when $f \leftrightarrow F$,

$$\{|x|^{-\lambda} * F\}(x) \leftrightarrow (1+\xi)^{\lambda-n}\{|\Omega|^{-\lambda} * F\}(\Omega), \quad \text{where} \tag{3.22}$$

$$(|\Omega|^{-\lambda} * F)(\Omega) = \int d\Omega' |\Omega - \Omega'|^{-\lambda} F(\Omega'). \tag{3.23}$$

(Again, note that Euclidean distance in $\mathbf{R}^{n+1}$ is used.) Equation (3.9) then takes the following simple form (with the same B):

$$|\Omega|^{-\lambda} * F = BF^{p-1}. \tag{3.24}$$

This, together with (3.20), (3.21), gives another equivalent form for N (cf. (2.17)):

$$N_{p,\lambda,n} = \sup\left\{ \|\Omega^{-\lambda} * F\|_q / \|F\|_p \mid F \in L^p(\mathbf{S}^n),\, F \neq 0\right\}. \tag{3.25}$$

As stated before, $F = 2^{-\mu} \leftrightarrow f$ in (3.1). To check (3.12)–(3.14) we must compute

$$\begin{aligned} I &= \int d\Omega' |\Omega - \Omega'|^{-\lambda} \\ &= \sigma_n \int_0^{\pi} d\theta (\sin\theta)^{n-1} (2 - 2\cos\theta)^{-\lambda/2} \\ &= \sigma_n 2^{n-\lambda} \int_0^{\pi/2} d\varphi (\cos\varphi)^{n-1} (\sin\varphi)^{n-1-\lambda} \\ &= \sigma_n 2^{n-\lambda-1} \Gamma(n/2)\Gamma(n/2 - \lambda/2)/\Gamma(n - \lambda/2). \end{aligned} \tag{3.26}$$

Thus, $B = I2^{\lambda - n} = B_\lambda$ in (3.13). Furthermore,

$$\begin{aligned} \||\Omega|^{-\lambda} * F\|_q / \|F\|_p &= BF^{p-2}\sigma_{n+1}^{1/q - 1/p} \\ &= \text{right side of (3.2)}, \end{aligned} \tag{3.27}$$

if we use the duplication formula for the gamma function.

Equation (3.24) has one very great advantage over (3.9). The $O(n+1)$ rotation invariance of (3.24) allows us to generate new solutions from old ones. This fact will eventually permit us to conclude that (3.1) is the (essentially) unique maximizer.

As an interesting aside before returning to the proof of Theorem 3.1, let us consider some other solutions to (3.9) and (3.24). (Irrelevant constants will be suppressed here.)

(a) We have $f(x) = (1 + |x|^2)^{-\mu} \leftrightarrow F(\Omega) = 1$. However, by translation and dilation of f in $\mathbf{R}^n$, there is an $n+1$ parameter family of solutions as follows:

$$f(x) = \left[b^2 + |x - z|^2\right]^{-\mu} \leftrightarrow F(\Omega) = \left[1 + (w, \Omega)\right]^{-\mu}, \tag{3.28}$$

where $b \in \mathbf{R}$, $z \in \mathbf{R}^n$, $w \in \mathbf{R}^{n+1}$. All b, z and w are allowed, except for the condition $|w| < 1$. The SSD category corresponds to $z = 0$ and $w = (0, \ldots, 0, c)$ with $|c| < 1$.

These other solutions are interesting for the following reason. Since $F =$ const. satisfies (3.24) and since this solution has the maximum possible $O(n+1)$ symmetry, it might have been supposed that $F =$ const. is the unique maximizer. But we see from (3.28) that there are other equivalent solutions with less symmetry. This is indeed surprising.

(b) $f(x) = |x - z|^{-\mu}$ with $z \in \mathbf{R}^n$ also satisfies (3.9). It is not allowed since it is not in L^p. This is an n parameter family and the correspondence is

$$f(x) = |x - z|^{-\mu} \leftrightarrow F(\Omega) = (1 + \xi)^{-\mu/2}|\Omega - \Omega'|^{-\mu} \tag{3.29}$$

with $\Omega' = \Sigma(z)$. Even more solutions can be obtained from (3.29) by applying an $O(n+1)$ rotation to Ω and Ω'. The function $2(1+\xi)$ becomes $|\Omega - \Omega''|^2$, with $\Omega \neq \Omega'$. Thus we have a $2n$ parameter family of solutions:

$$F(\Omega) = |\Omega - \Omega'|^{-\mu}|\Omega - \Omega''|^{-\mu} \leftrightarrow f(x) = |x - z'|^{-\mu}|x - z''|^{-\mu} \tag{3.30}$$

with $\Omega', \Omega'', z', z''$ arbitrary except that $\Omega' \neq \Omega''$ and $z' \neq z''$. This is amusing because the rotation invariance of (3.24) allowed us to generate a nontrivial $2n$ parameter family of solutions starting from $|x|^{-\mu}$.

Conclusion of Proof of Theorem 3.1. The f given in (3.1) satisfies (3.9) and we want to show that it maximizes and that it is (essentially) unique.

Let f be any maximizer. By Theorem 2.3 we can assume three things (after translation and dilation): (a) f satisfies (3.9). This means, in particular, that $f(x)$ is defined for *all* x by the left side of (3.9), not for just almost every x; (b) $f \in$ SSD; (c) $f(1/r) = r^{2\mu}f(r)$. Let $F \leftrightarrow f$. By (b), $F(\Omega)$ depends only on ξ: $F(\Omega) = \varphi(\xi)$. By (c), $\varphi(\xi) = \varphi(-\xi)$. Thus, $f(x) = (1 + |x|^2)^{-\mu}\varphi((1 - |x|^2)/(1 + |x|^2))$. Let $R \in O(n+1)$ be the following rotation:

$$R\colon (\rho_1,\ldots,\rho_n,\xi) \to (\rho_1\cos\theta - \xi\sin\theta, \rho_2,\ldots,\rho_n, \xi\cos\theta + \rho_1\sin\theta).$$

By the rotation invariance of (3.24),

$$\begin{aligned} f_R(x) &= (1 + |x|^2)^{-\mu}F(R\Sigma(x)) \\ &= (1 + |x|^2)^{-\mu}\varphi([(1 - |x|^2)\cos\theta + 2x_1\sin\theta]/[1 + |x|^2]) \end{aligned}$$

also maximizes. Therefore, by Theorem 2.3, there exists a unique $y \in \mathbf{R}^n$ such that $f_R(x + y) \in$ SSD. This y is the *unique* solution to $f_R(y) = \max_x f_R(x)$. By the $O(n-1)$ rotation invariance of f_R in $(x_2,\ldots,x_n)$ we see that $y_2 = \cdots = y_n = 0$. Since $f_R(x + y) \in \mathrm{SSD}(\mathbf{R}^n)$, $g(x_1) \equiv f_R(x_1 + y_1, x_2,\ldots,x_n) \in \mathrm{SSD}(\mathbf{R}^1)$ for any fixed $x_2,\ldots,x_n$. But $f_R((1,1,\ldots,1)) = f_R((-1,1,\ldots,1))$. Therefore $y_1 = 0$ also, and $y = 0$. This means that $\varphi([\ ]/[\ ])$ is spherically symmetric in x for *all* θ and I claim that φ must then be a constant. To see this, let $u_+, u_- \in [-1, 1]$ and

Annals of Math. *118*, 349–374 (1983)

let $x_{\pm} = (\pm b, 0, \ldots, 0)$. Since $f_R(x_+) = f_R(x_-)$, we shall have $\varphi(u_+) = \varphi(u_-)$ if we can find b and θ such that $u_{\pm} = [(1 - b^2)\cos\theta \pm 2b\sin\theta]/[1 + b^2]$. Let $b = -\tan(\psi/2)$. These equations then read $u_{\pm} = \cos(\psi \pm \theta)$, and we see that a solution is trivial. □

It should be noted that the proof above used the strong rearrangement inequality (Lemma 2.1) and Lemma 2.2 *twice*: Once to show that any maximizer, f, is in TSSD and once to show that $f(1/r) = r^{2\mu} f(r)$. For the latter, the formulation in (2.17) was essential.

The following is another way to conclude the proof of Theorem 3.1. It uses the rearrangement inequality on $\mathbf{S}^n$ of Baernstein and Taylor [3] which is a generalization of the inequality on $\mathbf{S}^1$ of Friedberg and Luttinger [13]. The inequality is:

(3.31)

$$\int\int F(\Omega)K(\Omega \cdot \Omega')G(\Omega')\, d\Omega\, d\Omega' \leq \int\int F^*(\Omega)K(\Omega \cdot \Omega')G^*(\Omega')\, d\Omega\, d\Omega',$$

where $K\colon [-1,1] \to \mathbf{R}$ is non-decreasing, and where F^* is equimeasurable with F, $F^*(\rho, \xi)$ depends only on ξ and is non-increasing in ξ (and likewise for G^*). Unfortunately, Baernstein and Taylor do not prove a strong inequality analogous to Lemma 2.1 (ii), which, it may be conjectured, exists. If it did hold, then the proof could be simplified.

In our case $K(\Omega \cdot \Omega') = |\Omega - \Omega'|^{-\lambda}$ which is strictly increasing. Let $F(\Omega) = \varphi(\xi)$ be the maximizer that satisfies $\varphi(\xi) = \varphi(-\xi)$. If the strong version of (3.31) held, we could immediately conclude that $\varphi(\xi)$ is either non-increasing or non-decreasing, in which case $\varphi(\xi) =$ constant and the proof would be finished. In the absence of this fact let $F^*(\Omega) = \psi(\xi)$ with $\psi(\xi)$ non-increasing. By (3.31), F^* also maximizes. Using (3.20), $F^* \leftrightarrow h$ with $h(x) = h(|x|)$. By the strong rearrangement inequality on $\mathbf{R}^n$, for some $\gamma > 0$, $h_\gamma(x) \equiv h(\gamma x)$ satisfies $h_\gamma(1/r) = r^{2\mu} h_\gamma(r)$. In general (without assuming any symmetry), if $F \leftrightarrow f$, then $F_\gamma \leftrightarrow f_\gamma$, where $F_\gamma(\rho, \xi) = F(2\lambda\rho/w, \nu/w)$ with $w = 1 + \xi + \gamma^2(1 - \xi)$ and $\nu = 1 + \xi - \gamma^2(1 - \xi)$. In our case $(F^*)_\gamma(\Omega) = \psi_\gamma(\xi) = \psi(\nu/w)$ and $\psi_\gamma(\xi) = \psi_\gamma(-\xi)$. Setting $\xi = 1$ we conclude that $\psi(1) = \psi(-1)$. Since ψ is non-increasing, $\psi =$ constant and the proof is completed.

IV. The Sobolev inequality

As another application of the method of Section II, we shall prove here the existence of a maximizing function f on $\mathbf{R}^n$ for the sharp constant in the Sobolev inequality

$$K_n \|\nabla f\|_2 \geq \|f\|_{2^*}, \qquad 2^* = 2n/(n-2), \quad n \geq 3. \tag{4.1}$$

I thank H. Brezis and P. L. Lions for suggesting to me that there should be a simple, direct "rearrangement inequality proof" of the existence of this f. Existence proofs already exist, of course (see [2], [31] and also [25]). (A generalization of (4.1) using Lorentz space norms was given by Alvino [1].) What is offered below, it is hoped, is a more direct and simpler argument.

A generalization of (4.1), useful in the theory of the Schroedinger equation, was given in [14] for $n = 3$:

$$K_{n,p}\|\nabla f\|_2 \geq \||x|^{-b}f\|_p, \qquad n \geq 3, \tag{4.2}$$

for $0 \leq b < 1$ and $p = 2n/(2b + n - 2)$.

In [14], an interesting extension of (4.2) is also given. If $1 - n/2 < b < 0$, no inequality of this type is possible for all f. But if f is restricted to be *spherically symmetric* (*not necessarily* symmetric decreasing), then a bound as in (4.2) holds and there is also a maximizing f. This is also given in Theorem 4.3.

Generalizations of (4.2) can be found in [12] and [21]. Flett [12] gives [21], Theorem 6.5.8, as the earliest reference to (4.2). Glaser, Martin, Grosse and Thirring [14] were unaware of this, but seem to have been the first to compute the sharp constant in (4.2). Generalizations of (4.1) can be found in [32], [33]. See also [5], [16].

It will be recalled that the rearrangement inequality, Lemma 2.1, was used *twice* in the proof of Theorem 2.3. First, it was used to show that a maximizing sequence could be sought in the SD($\mathbf{R}^n$) category. Second, when $p' = q$, it was used in the one-dimensional formulation (2.11)–(2.17) to deduce (2.6). The dual usage will also be needed here because we are faced with the same problem as that outlined in the beginning of Section II: The variational problem posed by (4.1) and (4.2) is invariant under the same conformal group (including inversion).

We begin with the fact that for $f \in W^{1,p}(\mathbf{R}^n) = \{f \mid f \text{ and } \nabla f \in L^p(\mathbf{R}^n)\}$, $\|\nabla f\|_p \geq \|\nabla f^*\|_p$ where f^* is the symmetric decreasing rearrangement of $|f|$. This fact has been known for a long time (see [10], [23], for example), but all proofs of it seem to be complicated. There is one case, suitable for our purposes here, in which the following simple proof can be given [20], and it would be desirable to be able to extend this argument to the $W^{1,p}$ case, $p \neq 2$. It would also be desirable to have a strict inequality as in Lemma 2.1(ii).

LEMMA 4.1. *Let $f \in W^{1,2}(\mathbf{R}^n) = H^1(\mathbf{R}^n)$. Then $f^* \in H^1(\mathbf{R}^n)$ and $\|\nabla f\|_2 \geq \|\nabla f^*\|_2$.*

Proof. Let $t > 0$ and $g_t(x - y) = e^{t\Delta}(x, y)$ be the kernel of the heat equation semigroup. Let $f \in L^2$. It is easy to see that if $A(f, t) \equiv (1/t)[\|f\|_2^2 - (f, g_t * f)]$, then $\lim_{t\to 0} A(f, t) = \|\nabla f\|_2^2$ or $+\infty$ according as $\nabla f \in L^2$ or not. (See [20] for a proof of this fact.) For each $t > 0$, $g_t(\cdot)$ is a

Annals of Math. *118*, 349–374 (1983)

Gaussian and hence in SSD($\mathbf{R}^n$). Therefore, $(f, g_t * f) \le (f^*, g_t * f^*)$ by Lemma 2.1. Since $\|f\|_2 = \|f^*\|_2$, the lemma is proved. □

With this preparation we can now prove the following about $\mathbf{R}^1$:

THEOREM 4.2. *Let $F \in H^1(\mathbf{R})$ and $2 < p < \infty$. For $F \neq 0$ let*

$$T(F) = \|F\|_p^2/\{\|F'\|_2^2 + \|F\|_2^2\} = \{\|F\|_p/\|F\|_{H^1}\}^2, \tag{4.3}$$

$$M_p = \sup\{T(F) | F \in H^1, F \neq 0\}. \tag{4.4}$$

Then M_p is finite and there exists a maximizing $F \in \mathrm{SD} \cap H^1$, i.e., $T(F) = M_p$. This F is unique (up to a constant and to translation). With $r = 2/(p-2)$,

$$F(x) = (\text{const.})\{\cosh(x/r)\}^{-r}, \tag{4.5}$$

$$M_p = \{(2r+1)\Gamma(2r)/r\Gamma(r)^2\}^{1-2/p}(r/4)^{2/p}(r+1)^{-1}. \tag{4.6}$$

Proof. By Lemma 4.1, $T(F^*) \ge T(F)$, so henceforth we can restrict attention to $F \in \mathrm{SD}$. Then $F \in L^\infty$ since $F(x) \to 0$ as $x \to -\infty$ and $F(x)^2 = 2\int_{-\infty}^x F'(y)F(y)\,dy \le 2\|F'\|_2\|F\|_2$. Let $\{F_n\}$ be a maximizing sequence for T and we can assume $\|F_n\|_2^2 + \|F_n'\|_2^2 = 1$. By (2.9), $F_n(x) \le C|x|^{-1/2}$. By the L^∞ bound just given, $F_n(x) \le C$. Therefore,

$$F_n(x) \le h(x) \equiv \min(C, C|x|^{-1/2}) \in L^p$$

since $p > 2$. As in Theorem 2.3, we can assume $F_n \to F \in \mathrm{SD}$ pointwise. We can also assume (Banach-Alaoglu theorem) that $F_n' \to G'$ and $F_n \to G$ weakly in L^2. Clearly, $G = F$. Then

$$\liminf \|F_n'\|_2^2 + \|F_n\|_2^2 \ge \|F'\|_2^2 + \|F\|_2^2.$$

It remains to show that $M_p = \lim\|F_n\|_p = \|F\|_p$, which will also prove the crucial fact that $F \neq 0$. This follows by dominated convergence since $F_n(x) \le h(x)$.

This maximizing F can easily be found as follows: By letting $F \to F + \varepsilon\varphi$, $\varphi \in C_0^\infty$, and equating the derivative at $\varepsilon = 0$ to zero,

$$F'' = F - F^{p-1}/M_p \tag{4.7}$$

in the distributional sense. By standard ODE methods, there is only one solution to (4.5) that vanishes as $|x| \to \infty$. (Recall that $F(x) = F(-x)$ and $\|F\|_p = 1$.) This solution is (4.5), (4.6). □

It should be noted that the last step—the calculation of F and M_p—was very easy compared to the proof of Theorem 3.1. Here, it is easy to verify that (4.5) is the (essentially) unique positive solution to (4.7). In Theorem 3.1, on the other hand, it was difficult to verify that (3.1) is the desired maximizing solution to (3.9); the apparatus of stereographic projection had to be used.

Next we turn to the problem posed by (4.1) and (4.2).

THEOREM 4.3. *Let $n \geq 3$ and let $f \in W^{1,2}(\mathbf{R}^n) = H^1(\mathbf{R}^n)$. Let $1 - n/2 < b < 1$ and $p = 2n/(2b + n - 2)$, so that $\infty > p > 2$. Let*

$$R(f) = \||x|^{-b} f\|_p / \|\nabla f\|_2, \qquad f \neq 0, \tag{4.8}$$

$$K_{n,p} = \sup\{R(f) | f \in H^1, f \neq 0\}. \tag{4.9}$$

(i) *If $1 > b \geq 0$, $K_{n,p}$ is finite and a maximizing $f \in$ SD exists, i.e., $R(f) = K_{n,p}$.*

$$f(x) = \{1 + |x|^{2t/r}\}^{-r}, \tag{4.10}$$

$$K_{n,p} = \sigma_n^{-1/2 + 1/p} t^{-1/2 - 1/p} M_p^{1/2}, \tag{4.11}$$

with $r = 2/(p - 2)$, $t = -1 + n/2$, M_p in (4.6), σ_n in (2.13).

$$K_{n,p} = [\pi n(n - 2)]^{-1/2} [\Gamma(n)/\Gamma(n/2)]^{1/n}, \qquad \textit{when } b = 0.$$

(ii) *If $1 - n/2 < b < 0$, $R(f)$ is unbounded on H^1, but $R(f)$ restricted to spherically symmetric functions (not necessarily decreasing) in H^1 (denoted by H^1_R) is bounded. If $K^R_{n,p} = \sup\{R(f) | f \in H^1_R, f \neq 0\}$ then there is a maximizing $f \in$ SD, $R(f) = K^R_{n,p}$, given by (4.10) and $K^R_{n,p}$ is given by the right side of (4.11).*

Note. When $n > 4$, the f in (4.10) is in $H^1(\mathbf{R}^n)$. When $n = 3$ or 4, this $f \notin H^1(\mathbf{R}^n)$ but $R(f)$ is well defined.

Proof. (i) Since $|x|^{-b} \in$ SD, we have that $\||x|^{-b} f\|_p \leq \||x|^{-b} f^*\|_p$. By Lemma 4.1, $\|\nabla f\|_2 \geq \|\nabla f^*\|_2$. Thus, we can henceforth restrict our attention to $f \in$ SD. As in (2.11), let $F: \mathbf{R} \to \mathbf{R}$ be given by

$$F(tu) = e^{tu} f(e^u) \tag{4.12}$$

with $t = -1 + n/2 > 0$. Then

$$(\sigma_n/t)^{1/p} \|F\|_p = \||x|^{-b} f\|_p, \tag{4.13}$$

$$(\sigma_n t)^{1/2} \|(F' - F)\|_2 = \|\nabla f\|_2 \tag{4.14}$$

where σ_n is given by (2.13). Since $f \in L^2$, as in (2.9) we have $F(u) \leq C e^{-u/t}$.

Now assume $f \in L^\infty$ in addition to $f \in H^1 \cap$ SD, whence $F(u) \leq C \exp(-|u|/2t)$. Then $\int F'F = 0$, and thus

$$R(f)^2 = \sigma_n^{-1 + 2/p} t^{-1 - 2/p} T(F), \tag{4.15}$$

with $T(F)$ given by (4.3). Since $\|\nabla f\|_2 < \infty$, $F \in H^1(\mathbf{R}^1)$. Thus, for $f \in L^\infty$,

Annals of Math. *118*, 349–374 (1983)

Theorem 4.2 completes the proof. (Note that (4.5) and (4.12) are consistent with $f \in L^\infty$.) For $f \notin L^\infty$ we use the fact that $L^\infty \cap H^1$ is dense in H^1. Thus, there exists a sequence $\{g_n\}$ in $L^\infty \cap H^1$ such that $\|\nabla g_n\|_2 \to \|\nabla f\|_2$ and $\|g_n\|_2 \to \|f\|_2$. By passing to a subsequence, $g_n \to f$ pointwise almost everywhere and hence, $\||x|^{-b} f\|_p \le \liminf \||x|^{-b} g_n\|_p$. Therefore,

$$R(f) \le \sup\{R(f) | f \in L^\infty \cap H^1, f \neq 0\}.$$

(ii) If $b < 0$ we cannot say that we can restrict our attention to $f \in \mathrm{SD}$. But for $f \in H^1_R(\mathbf{R}^n)$ we can make the same change of variables as in (4.12)–(4.14). The proof proceeds essentially as before. □

It is worth remarking about Theorem 4.3 as $b \to 1$ and $p \to 2$. From (4.11) we are led to believe that $K_{n,2}$ is finite, but that there is no maximizing f since (4.10) tends to unity as $p \to 2$. This is indeed correct (see [19, Lemma 2.7] where the authors attribute the result to Karlson [18] and to Herbst [17]), and $K_{n,2}$ is given by the limit of (4.11) as $p \to 2$, namely

$$(4.16) \qquad (-1 + n/2)\||x|^{-1} f\|_2 \le \|\nabla f\|_2, \qquad n \ge 3,$$

and this constant is the best possible.

Another remark concerns the relation of Theorem 4.3 with $b = 0$ and Corollary 3.2. With $p = 2n/(n-2)$, $\|f\|_p \le K_{n,p}\|\nabla f\|_2 = K_{n,p}\|(-\Delta)^{1/2} f\|_2$. Formally, this is equivalent to $\|(-\Delta)^{-1/2} g\|_p \le K_{n,p}\|g\|_2$. But

$$2\pi^{(n+1)/2}(-\Delta)^{-1/2} g = \Gamma(n/2 - 1/2)|x|^{-\lambda} * g$$

with $\lambda = n - 1$, [30, p. 117]. Thus, we should have

$$(4.17) \qquad 2\pi^{(n+1)/2} K_{n,2n/(n-2)} = \Gamma(n/2 - 1/2) N_{2,n-1,n},$$

which is confirmed by (3.4) and (4.11).

V. Doubly weighted HLS inequality and weighted Young inequality

Two more illustrations will be given of the use of the methods of Section II. The first is the *doubly weighted Hardy-Littlewood-Sobolev inequality* [15], [29] which generalizes the HLS inequality considered before.

THEOREM 5.1. *Let* $0 < \lambda < n$, $1 < p \le q < \infty$, $0 \le \alpha < n/p'$ *(with* $1/p + 1/p' = 1$), $0 \le \beta < n/q$ *and* $1/p + (\lambda + \alpha + \beta)/n = 1 + 1/q$. *Let*

$$(5.1) \qquad V(x, y) = |x|^{-\beta}|x - y|^{-\lambda}|y|^{-\alpha}$$

be an integral kernel on $\mathbf{R}^n$. *Then* $f \to Vf$, $(Vf)(x) = \int V(x, y) f(y)\, dy$, *is a bounded map from* $L^p(\mathbf{R}^n)$ *to* $L^q(\mathbf{R}^n)$. *Moreover, if* $p < q$,

$$R(f) = \|Vf\|_q/\|f\|_p, \qquad f \neq 0, \quad \text{and} \tag{5.2}$$

$$P_{\alpha,\beta,p,\lambda,n} = \sup\{R(f) | f \in L^p(\mathbf{R}^n), f \neq 0\} \tag{5.3}$$

then there is a maximizing $f \in \mathrm{SD} \cap L^p$, *i.e.*, $R(f) = P_{\alpha,\beta,p,\lambda,n}$.

Remarks. (i) In [29] the condition $0 \le \alpha, \beta$ is relaxed to $\alpha + \beta \ge 0$. However, the stronger condition is needed here in order to use rearrangement inequalities.

(ii) Obviously, (5.2), (5.3) are equivalent to

$$P_{\alpha,\beta,p,\lambda,n} = \sup \int\int g(x)|x-y|^{-\lambda} f(y)\,dx\,dy / \||x|^\alpha f\|_p \||x|^\beta g\|_{q'}. \tag{5.4}$$

(iii) When $p = q$ a maximizing f cannot be expected to exist. See the remark at the end of Section IV which corresponds to the case $p = q = 2$, $\lambda = n - 1$, $\beta = 1$, $\alpha = 0$, $n \ge 3$. See also [17].

An extension of Lemma 2.4 is needed.

LEMMA 5.2. *Let the hypothesis be the same as in Theorem* 5.1 *except that the condition* $0 \le \alpha, \beta$ *is eliminated. Let* $f \in L^p(\mathbf{R}^n)$ *be spherically symmetric and* $|f(r)| \le \varepsilon r^{-n/p}$ *for all* $r > 0$. *Then* $\|Vf\|_q \le C_n \|f\|_p^{p/q} \varepsilon^{1-p/q}$ *for some* $C_{n,\alpha,\beta}$ *independent of* f *and* ε.

Proof. This is the same as the proof of Lemma 2.4 except that (2.15) changes to

$$L_{n,\alpha,\beta}(u) = 2^{-\lambda/2} \exp\{u(n/q - \lambda/2 - \beta)\} Z_n(u). \tag{5.5}$$

The hypothesis guarantees that $|n/q - \lambda/2 - \beta| < \lambda/2$ so that $L_{n,\alpha,\beta} \in L^1(\mathbf{R})$. □

Proof of Theorem 5.1. Since $|x|^{-\alpha}$ and $|x|^{-\beta} \in \mathrm{SD}$ (here we use the fact that $\alpha, \beta \ge 0$) the generalization of the Riesz inequality given in [7] implies that a maximizing sequence $\{f_j\}$ can be taken in SD. Lemma 5.2 implies that $R(f)$ is bounded if we take $\|f_j\|_p = 1$ so that $f_j(r) \le Cr^{-n/p}$ as in (2.9). As in (2.10), we can assume $f_j(r) \to f(r) \le Cr^{-n/p}$ almost everywhere. If $q > p$ we can use Lemma 5.1 to dilate each f_j so that $f \neq 0$ (see the remarks after the proof of Theorem (2.3)(iv)). The final step is as in the conclusion of the proof of Theorem (2.3)(i), using Lemma 2.7. □

The second illustration is what A. Sokal has called *the weighted Young inequality*. Let f: $\mathbf{R}^n \to \mathbf{R}^+$ and let

$$f^{(m)} = f * f * \cdots * f \qquad (m \text{ factors}) \tag{5.6}$$

Annals of Math. *118*, 349–374 (1983)

be the convolution of f with itself m times. We consider $m \geq 3$. Now $f^m(0)$ makes sense, even if f is defined only almost everywhere, because

$$f^{(m)}(0) = \int f(-x_{m-1}) f(x_{m-1} - x_{m-2}) \cdots f(x_2 - x_1) f(x_1)\, dx_1 \cdots dx_{m-1}. \tag{5.7}$$

Let p and γ satisfy the conditions

$$m/(m-1) \leq p \leq m \quad \text{and} \quad \gamma/n + 1/p = (m-1)/m \geq 0. \tag{5.8}$$

Our interest will be in the ratio

$$R(f) = |f^{(m)}(0)| / \| |x|^\gamma f \|_p^m, \qquad f \neq 0, \tag{5.9}$$

$$Q_{p,m,n} = \sup\{ R(f) \mid |x|^\gamma f \in L^p(\mathbf{R}^n), f \neq 0 \}. \tag{5.10}$$

By the generalization of the Riesz rearrangement inequality in [7], $f^{(m)}(x) \leq (f^*)^{(m)}(0)$ and $\| |x|^\gamma f^* \|_p \leq \| |x|^\gamma f \|_p$. Thus,

$$Q_{p,m,n} = \sup\left\{ \|f^{(m)}\|_\infty / \| |x|^\gamma f \|_p^m \;\middle|\; |x|^\gamma f \in L^p(\mathbf{R}^n), f \neq 0 \right\}. \tag{5.11}$$

The idea that $R(f)$ should be bounded was suggested to me by A. Sokal. Initially, he was interested in the case $p = 2$, $m = 4$, $\gamma = n/4$ for use in a problem in quantum field theory [27]. As will shortly be seen, the $p = 2$ case reduces to the HLS inequality itself (with $p = 2$). But that is a case for which the sharp constant was derived in Section III, and thus we shall be able to compute Q when $p = 2$. Another case for which Q can be found is $p = m/(m-1)$ and this is given in (5.12).

THEOREM 5.3. *Assuming* (5.8), $Q_{p,m,n} < \infty$ *for all* p, n *and* $m \geq 3$. *Moreover, if* $m/(m-1) \leq p < m$, *there is a maximizing function*, $f \in$ SD, *i.e.*, $R(f) = Q_{p,m,n}$.

Remarks. (i) When $\gamma = 0$, $p = m/(m-1)$, (5.9) is one of the generalized Young inequalities treated in [6]. The ordinary Young inequality shows that $R(f)$ is bounded. In [6] it was shown that a maximizing f exists and that it is a Gaussian, $f(x) = \exp(-|x|^2)$. Then $Q_{p,m,n}$ can be easily computed in this case:

$$Q_{p,m,n} = \{ p^{m-1}/m \}^{n/2}, \qquad p = m/(m-1). \tag{5.12}$$

An alternative derivation [6] of (5.12) can be obtained from the sharp constants in the ordinary Young inequality, which was also derived in [4]. If that inequality is iterated $(m-2)$ times, one obtains $Q_{p,m,n} \leq$ (right side of (5.12)). However, the explicit choice of a Gaussian for f gives a lower bound which establishes (5.12).

(ii) A generalization of Theorem 5.3 (at least the first part) is obviously possible, namely

$$(f_1 * f_2 * \cdots * f_m)(0) \le C\prod_{j=1}^{m} \||x|^{\gamma_j} f_j\|_{p_j},$$

$\sum_{j=1}^{m}\gamma_j + n/p_j = n(m-1)$ and $\gamma_j \ge 0$, $p_j \le m$, for all j. This can easily be proved by imitating the following proof.

Proof. Let $\{f_j\}$ be a maximizing sequence and let $g_j(x) = |x|^\gamma f_j(x)$. The denominator in $R(f_j)$ is $\|g_j\|_p^m$. Then $f_j^{(m)}(0)$ is an integral over a product of $g_j(x_i - x_k)$ and $|x_i - x_k|^{-\gamma}$ factors. By the general rearrangement inequality in [7], the numerator does not decrease if we replace g_j by g_j^*. Henceforth, assume $g_j = g_j^*$ and $\|g_j\|_p = 1$. As in (2.10), we can assume $g_j(r) \to g(r) < Cr^{-n/p}$ almost everywhere and $g_j(r) < Cr^{-n/p}$. Let $g_j(r) = r^{-n/p}h_j(r)$, so that $\|h_j\|_\infty < C$ and $A_j \equiv \int h_j(x)^p|x|^{-n}\,dx < C$.

First we show that $R(f_j)$ is bounded. Substitute $f_j(x) = h_j(x)|x|^{-\gamma-n/p}$ in (5.7). By Hölder's inequality, $f_j^m(0) \le \prod_{k=1}^m I_k(h_j)^{1/m}$ where $I_k(h_j)$ is (5.7) with $h_j(x)^m|x|^{-\gamma-n/p}$ in the kth position and $|x|^{-\gamma-n/p}$ in the other $(m-1)$ positions. It is easy to do the trivial integrals and one finds that all I_k have the common value $C\int h_j(x)^m|x|^{-n}\,dx \le CA_j\|h_j\|_\infty^{m-p}$. This shows not only that $f_j^m(0)$ is bounded but it also shows that when $p < m$, $\|h_j\|_\infty$ cannot go to zero as $j \to \infty$. Therefore there are dilations of g_j so that $g \ne 0$ (see the remarks after the proof of Theorem (2.3)(iv)).

It remains to show that $f(r) = r^{-\gamma}g(r)$ maximizes. Write $f_j = f + b_j$ with $b_j \to 0$ almost everywhere. I claim that

$$f_j^{(m)}(0) = f^{(m)}(0) + b_j^{(m)}(0) + o(1) \tag{5.13}$$

when $p < m$. This will complete the proof (using Lemma 2.6) by the strategy of Lemma 2.7. One merely sets $p/q = p/m$ in the last part of the proof of Lemma 2.7.

To prove (5.13), we have to show that when $f + b_j$ is inserted in (5.7) and expanded out into 2^m terms, those terms that contain at least one f and one b_j factor vanish as $j \to \infty$. Write $f = r^{-(\gamma+n/p)}\varphi$ and $b_j = r^{-(\gamma+n/p)}\beta_j$. We shall use a Hölder inequality as in the proof that $f^{(m)}(0)$ is bounded, but with a slight change. Consider a term, I, in $f_j^m(0)$ that has φ m_1 times and β_j m_2 times with $m_1 + m_2 = m$ and $0 < m_1 < m$. All orderings of these functions give the same integral (by changing variables). Let $a = (m-p)/(m-1) > 0$. Then $I \le I_1^{m_1/m}I_2^{m_2/m}$ where I_1 has φ^p once, φ^a $(m_1 - 1)$ times and β_j^a m_2 times. I_2 has φ^a m_1 times, β_j^p once and β_j^a $(m_2 - 1)$ times. First consider I_1. We know that φ and

Annals of Math. *118*, 349–374 (1983)

β_j are bounded by a constant C (since $f_j r^{(n/p+\gamma)}$ is so bounded). Suppose the integration variable of φ^p is x_1. Then do all the other $(m-2)$ integrations and call the result $z_j(x_1)$. If we replace all the other $(m-1)$ functions by C, the $(m-2)$ integrals are finite for all $x_1 \neq 0$, namely $|x_1|^{-n}$. Therefore, by dominated convergence, $z_j(x_1) \to 0$ as $j \to \infty$ for every $x_1 \neq 0$. (Note: It is important that there is at least one factor of β_j^a and that $a > 0$.) Furthermore, $z_j(x_1)$ has the form $|x_1|^{-n} w_j(x_1)$ with $w_j(x_1)$ uniformly bounded (in x_1 and j). Thus, the final integral is $I_1 = \int \varphi(x_1)^p |x_1|^{-n} w_j(x_1)\,dx_1$. Since $|x|^{-n}\varphi(x)^p \in L^1$ and $w_j(x) < C$, $I_1 \to 0$ as $j \to \infty$ by dominated convergence. I_2 is uniformly bounded since

$$I_2 \leq C \int \beta_j(x)^p |x|^{-n}\,dx \leq C\|g_j\|_p^p = C.$$

Therefore $I \to 0$ as $j \to \infty$ and (5.13) is proved. □

The value of $Q_{p,m,n}$ has already been given in (5.12) when $p = m/(m-1)$. Let us conclude by evaluating Q when $p = 2$. The function $g(x) = |x|^\gamma f(x)$ is in L^2. Let G be the Fourier transform of g. Then $\|g\|_2 = (2\pi)^{-n/2}\|G\|_2$. The Fourier transform of $|x|^{-\gamma}$ is $w_\gamma(k)$ in (3.8) and has the form $E_{\gamma,n}|k|^{\gamma-n}$. Thus, the Fourier transform of f is $F = (2\pi)^{-n}E_{\gamma,n}|k|^{\gamma-n} * G$ and $f^{(m)}(0) = (2\pi)^{-n}\int F(k)^m\,dk$. Hence

$$R(f)^{1/m} = (2\pi)^{-n(1/2+1/m)} E_{\gamma,n} \left| \int [|k|^{\gamma-n} * G]^m \right|^{1/m} \Big/ \|G\|_2. \tag{5.14}$$

Comparing this with (2.1) (with $p = 2$, $q = m$, $\lambda = n - \gamma$ and $f \to G$), we see that apart from a constant, the two expressions are almost the same. The one difference is that $\|\,|k|^{-\lambda} * G\|_m$ is replaced by the integral in (5.14). However, the maximizing f for (2.1) is non-negative by Corollary 3.2(ii). It can be used as G in (5.14) and the two expressions are then the same. The maximizing G is thus

$$G = |k|^{\gamma-n} * (1 + |k|^2)^{-\gamma-n/2}. \tag{5.15}$$

This is *unique* (up to dilations, etc.).

This paper is thus brought full circle by the identification of the HLS problem and Theorem 5.3 for $p = 2$. *The maximizing f for* (5.9) *is unique* (*up to dilations, etc.*) *and is*

$$f(x) = |x|^{-\gamma} K_\gamma(|x|), \qquad p = 2, \tag{5.16}$$

where K_γ is a Bessel function (see (3.11)). Equation (5.16) should be compared with the $p = m/(m-1)$ case in which f is a Gaussian (see (5.12)). Q can be computed from (5.14), (3.4) and (3.8).

$$Q_{2,m,n} = \pi^{n(m-2)/4} \left\{ \frac{\Gamma(n/m)}{\Gamma(n - n/m)} \right\}^{m/2} \left\{ \frac{\Gamma(n)}{\Gamma(n/2)} \right\}^{m/2-1}. \tag{5.17}$$

THE INSTITUTE FOR ADVANCED STUDY, PRINCETON, NJ

PERMANENT ADDRESS:

PRINCETON UNIVERSITY, PRINCETON, NJ
(MATHEMATICS & PHYSICS DEPARTMENTS)

REFERENCES

[1] A. ALVINO, Sulla disguaglianza di Sobolev in spazi di Lorentz, Boll. Unione Mat. Ital. **14A** (1977), 148–156.

[2] T. AUBIN, Problèmes isopérimétriques et espaces de Sobolev, Compt. Rend. Acad. Sci. Paris **280A** (1975), 279–281. See also J. Diff. Geom. **11** (1976), 573–598.

[3] A. BAERNSTEIN II and B. A. TAYLOR, Spherical rearrangements, sub-harmonic functions and $*$-functions in n-space, Duke Math. J. **43** (1976), 245–268.

[4] W. BECKNER, Inequalities in Fourier analysis, Ann. of Math. **102** (1975), 159–182.

[5] G. A. BLISS, An integral inequality, J. London Math. Soc. **5** (1930), 40–46.

[6] H. J. BRASCAMP and E. H. LIEB, Best constants in Young's inequality, its converse, and its generalization to more than three functions, Adv. in Math. **20** (1976), 151–173.

[7] H. J. BRASCAMP, E. H. LIEB and J. M. LUTTINGER, A general rearrangement inequality for multiple integrals, J. Funct. Anal. **17** (1974), 227–237.

[8] H. BREZIS and E. H. LIEB, A relation between pointwise convergence of functions and convergence of functionals, Proc. A. M. S. **88** (1983), 486–490.

[9] H. BREZIS and S. WAINGER, A note on limiting cases of Sobolev embeddings and convolution inequalities, Commun. Part. Diff. Eq. **5** (1980), 773–789.

[10] G. F. D. DUFF, A general integral inequality for the derivative of an equimeasurable rearrangement, Can. J. Math. **28** (1976), 793–804.

[11] N. DU PLESSIS, Some theorems about the Riesz fractional integral, Trans. A. M. S. **80** (1955), 124–134.

[12] T. M. FLETT, On a theorem of Pitt, J. London Math. Soc. **7** (1973), 376–384.

[13] R. FRIEDBERG and J. M. LUTTINGER, Rearrangement inequality for periodic functions, Arch. Rat. Mech. Anal. **61** (1976), 35–44.

[14] V. GLASER, A. MARTIN, H. GROSSE and W. THIRRING, A family of optimal conditions for the absence of bound states in a potential, in *Studies in Mathematical Physics*, E. H. Lieb, B. Simon and A. S. Wightman, eds., Princeton University Press (1976), 169–194.

[15] G. H. HARDY and J. E. LITTLEWOOD, Some properties of fractional integrals (1), Math. Zeitschr. **27** (1928), 565–606.

[16] ______, On certain inequalities connected with the calculus of variations, J. London Math. Soc. **5** (1930), 34–39.

[17] I. W. HERBST, Spectral theory of the operator $(p^2 + m^2)^{1/2} - Ze^2/r$, Commun. Math. Phys. **53** (1977), 285–294.

[18] B. KARLSSON, Self adjointness of Schroedinger operators, Inst. Mittag-Leffler Report no. 6 (1976).

[19] V. F. KOVALENKO, M. A. PERLMUTTER and YA. A. SEMENOV, Schroedinger operators with $L_w^{l/2}(\mathbf{R}^l)$ potentials, J. Math. Phys. **22** (1981), 1033–1044.

[20] E. H. LIEB, Existence and uniqueness of the minimizing solution of Choquard's nonlinear equation, Studies in Appl. Math. **57** (1977), 93–105.

[21] G. O. OKIKIOLU, *Aspects of the Theory of Bounded Integral Operators in L^p-Spaces*. Academic Press, N.Y., 1971.

[22] R. O'NEIL, Convolution operators and $L(p, q)$ spaces, Duke Math. J. **30** (1963), 129–142.

[23] G. PÓLYA and G. SZEGÖ, *Isoperimetric Inequalities in Mathematical Physics*, Princeton University Press, 1951.

Annals of Math. *118*, 349–374 (1983)

[24] F. Riesz, Sur une Inégalité Intégral, J. London Math. Soc. **5** (1930), 162–168.
[25] G. Rosen, Minimum value for c in the Sobolev inequality $\|\phi\|_6 \le c\|\nabla\phi\|_2$, SIAM J. Appl. Math. **21** (1971), 30–32.
[26] S. L. Sobolev, On a theorem of functional analysis, Mat. Sb. (N.S.) **4** (1938), 471–479. A. M. S. Transl. Ser. 2, **34** (1963), 39–68.
[27] A. Sokal, Improved upper bound for the renormalized four-point coupling, in preparation.
[28] E. M. Stein and G. Weiss, *Introduction to Fourier Analysis on Euclidean Spaces*, Princeton University Press, 1971.
[29] ______, Fractional integrals in n-dimensional Euclidean space, J. Math. Mech. **7** (1958), 503–514.
[30] E. M. Stein, *Singular Integrals and Differentiability Properties of Functions*, Princeton University Press, 1970.
[31] G. Talenti, Best constant in Sobolev inequality, Ann. di Matem. Pura ed Appl. **110** (1976), 353–372.
[32] M. Weinstein, Nonlinear Schrödinger equations and sharp interpolation estimates, Comm. Math. Phys. **87** (1983), 567–576.
[33] F. Weissler, Logarithmic Sobolev inequalities for the heat-diffusion semigroup, Trans. A. M. S. **237** (1978), 255–269.

(Received January 31, 1983)

Invent. Math. *74*, 441–448 (1983)

Invent. math. 74, 441–448 (1983)

Inventiones mathematicae

On the lowest eigenvalue of the Laplacian for the intersection of two domains

Elliott H. Lieb*

Departments of Mathematics and Physics, Princeton University, P.O.B. 708, Princeton, NJ 08544, USA

Abstract. If A and B are two bounded domains in $\mathbb{R}^n$ and $\lambda(A)$, $\lambda(B)$ are the lowest eigenvalues of $-\Delta$ with Dirichlet boundary conditions, then there is some translate, B_x, of B such that $\lambda(A \cap B_x) < \lambda(A) + \lambda(B)$. A similar inequality holds for $\lambda_p(A) = \inf\{\|\nabla f\|_p^p / \|f\|_p^p \mid f \in W_0^{1,p}(A)\}$. There are two corollaries of this theorem: (i) A lower bound for $\sup_x$ {volume $(A \cap B_x)$} in terms of $\lambda(A)$, when B is a ball; (ii) A compactness lemma for certain sequences in $W^{1,p}(\mathbb{R}^n)$.

1. The main theorem

The chief purpose of this paper is to prove Theorem 1 which contains, as a corollary, the answer to a geometric question about domains in $\mathbb{R}^n$. Theorem 1 is generalized to Theorem 3 in Sect. 2, and this leads to the compactness lemma of Sect. 3, which was another motivation for Theorem 1.

Let us begin with a discussion of the geometric question. If A is an open set in $\mathbb{R}^n$ (bounded or unbounded), let $\lambda(A)$ denote the lowest eigenvalue of $-\Delta$ in A with Dirichlet boundary conditions. $\lambda(A) \equiv \infty$ if A is empty. To be precise, $\lambda(A)$ is defined by (1.7), (1.8). Intuitively, if $\lambda(A)$ is small then A must be large in some sense. One well known result in this direction is the inequality of Faber [5] and Krahn [7] which states that among all domains with a given volume $|A|$, the ball has the smallest λ. Thus,

$$\lambda(A) \geqq \beta_n |A|^{-2/n} \tag{1.1}$$

where β_n is the lowest eigenvalue of a ball of unit volume.

Equation (1.1) clearly does not tell the whole story. If $\lambda(A)$ is small then A must not only have a large volume, it must also be "fat" in some sense. One might suppose that there is a constant α_n such that $\lambda(A) \geqq \alpha_n R^{-2}$, where R is the radius of the largest ball contained in A. Unfortunately, this is not general-

* Work partially supported by U.S. National Science Foundation grant PHY-8116101 A01. AMS(MOS) Classification: 35P15

Invent. Math. 74, 441–448 (1983)

ly true. The situation is the following:

n=1: $\alpha_1 = \pi^2/4$

n=2: In general, there is no universal lower bound to α_2. Hayman [6] showed that $\alpha_2 \geqq 1/900$ *provided* A is simply connected. Osserman [8] improved this to $\alpha_2 \geqq 1/4$ (see also Osserman [9], [10]). Osserman [8] extended Hayman's result to $\alpha_2 \geqq k^{-2}$ for domains of connectivity $k \geqq 2$; Croke [4] improved this for $k \geqq 2$ to $\alpha_2 \geqq (2k)^{-1}$. (Earlier, Taylor [11] and Cheng [3] had found bounds of the Croke type but with worse constants.)

A related result is that of Cheeger [2], valid for $n \geqq 2$, namely $\lambda(A) \geqq \inf S^2/4V^2$, where the infimum is over all relatively compact subdomains of A of surface area S and volume V. However, Cheeger's result does not imply any universal lower bound to α_n. We shall return to this quantity, $\inf S/V$, later in (2.6).

n≧3: No such inequality is possible, even if topological properties are taken into account. Hayman [6] points out that if A is a ball with many narrow, inward pointing spikes removed from it, then $\lambda(A) \approx \lambda(\text{ball})$ but $R \approx 0$. In some special cases, however, a lower bound can be given for α_n. Hayman [6] shows that this can be done if every boundary point, x, of A has the property that every ball centered at x has a fixed fraction of its volume outside A. Another example is Osserman's result [10] that $\lambda(A) \geqq (2R)^{-2}$ for convex domains, based on Cheeger's result [2] and a result of Brascamp and Lieb [1] about the level sets of the lowest eigenfunction.

What these result shows, in a word, is that when $n>1$, A need not contain any ball of fixed radius R no matter how small $\lambda(A)$ may be. Small holes and spikes do not influence $\lambda(A)$ very much but they do have a great effect on the ability to insert a ball.

Nevertheless, the intuition persists that if $\lambda(A)$ is small then A contains "most of" a ball of radius $R \sim \lambda^{-1/2}$. The holes and spikes cannot prevent this. More precisely, for each fraction $\psi < 1$ there should be a constant $\sigma_n(\psi)$, with $\sigma_n(\psi) \to 0$ as $\psi \to 1$, such that

$$\lambda(A) \geqq \sigma_n(\psi) R^{-2} \tag{1.2}$$

where R is the largest radius such that $|A \cap B_R| \geqq \psi |B_R|$ for some ball B_R of radius R. This is the content of Corollary 2.

Equation (1.2) is the aforementioned geometric motivation. The following is the key to proving it.

Theorem 1. *Let A and B be non-empty open sets in $\mathbb{R}^n$, $n \geqq 1$, and let $\lambda(A)$, $\lambda(B)$ be the lowest eigenvalue of $-\Delta$ with Dirichlet boundary conditions. Let B_x denote B translated by $x \in \mathbb{R}^n$. Let $\varepsilon > 0$. Then there exists an x such that*

$$\lambda(A \cap B_x) < \lambda(A) + \lambda(B) + \varepsilon. \tag{1.3}$$

If A and B are both bounded then there is an x such that

$$\lambda(A \cap B_x) < \lambda(A) + \lambda(B). \tag{1.4}$$

Equivalently

$$\lambda(A) \geqq \inf_x \lambda(A \cap B_x) - \lambda(B) \tag{1.3'}$$

$$\lambda(A) > \inf_x \lambda(A \cap B_x) - \lambda(B), \qquad A, B \text{ bounded.} \tag{1.4'}$$

Moreover, $x \mapsto \lambda(A \cap B_x)$ *is upper-semicontinuous, so that the set of* x's *for which* (1.3) *or* (1.4) *holds is open.*

Remark 1. No assumption is made about the smoothness of the boundaries of A and B.

Remark 2. (1.3) and (1.4) are, in one sense, best possible as the following examples show.

Example 1. In $\mathbb{R}^2$, let A be the strip $A = \{(x,y) | 0 < x < \pi/s, -\infty < y < \infty\}$ and let B be the perpendicular strip $B = \{(x,y) | -\infty < x < \infty, 0 < y < \pi/t\}$. Then $\lambda(A) = s^2$, $\lambda(B) = t^2$, $\lambda(A \cap B_x) = s^2 + t^2$ for all $x \in \mathbb{R}^2$.

Example 2. If we now cut off the strips A and B above so that their lengths are finite (but larger than $\max(\pi/s, \pi/t)$) then $\lambda(A) > s^2$, $\lambda(B) > t^2$ and $\lambda(A \cap B_x) = s^2 + t^2$ for suitable x.

Example 3. In $\mathbb{R}^2$, let B be a ball of radius $2^{1/2}$. Let $\mathbb{Z}^2$ be the lattice with integer points, and for each $y \in \mathbb{Z}^2$ let b_y be the ball $b_y = \{x | |x-y| \leqq r_y\}$. Let $A = \mathbb{R}^2 \setminus \bigcup b_y$. If $r_y \to 0$ as $|y| \to \infty$ then $\lambda(A) = 0$. For $|x|$ sufficiently large $\lambda(A \cap B_x) < 0 + \lambda(B) + \varepsilon$. However, $\lambda(A \cap B_x) > 0 + \lambda(B)$ for every x

Similar examples hold in $\mathbb{R}^n$, $n > 2$.

Corollary 2. *Let A be a non-empty (bounded or unbounded) open set in $\mathbb{R}^n$, $n \geqq 1$, and let B_r be a ball of radius r of volume $|B_r| = r^n/\delta_n$, $\delta_n = [n\Gamma(n/2)/2]\pi^{-n/2}$. Let β_n be as given in* (1.1). *Let $0 < \psi < 1$ be fixed and let*

$$\sigma_n(\psi) = \beta_n \delta_n^{2/n} [\psi^{-2/n} - 1] > 0. \tag{1.5}$$

Suppose that for $0 < R \leqq \infty$

$$\lambda(A) \leqq \sigma_n(\psi) R^{-2}. \tag{1.6}$$

Then for every $0 < r < R$ there exists an $x \in \mathbb{R}^n$ and a ball $B_{r,x}$ of radius r centered at x such that

$$|A \cap B_{r,x}| > \psi |B_r| = \psi r^n/\delta_n$$

Remark. This corollary has an obvious analogue for domains B other than balls.

Proof of Corollary. Let $r < R$ and choose $\varepsilon = (r^{-2} - R^{-2}) \sigma_n(\psi)$. By Theorem 1 there is an x such that $\sigma_n(\psi) R^{-2} \geqq \lambda(A) > \lambda(A \cap B_{r,x}) - \lambda(B_r) - \varepsilon$. However, $\lambda(B_r) = \beta_n |B_r|^{-2/n}$ and, by (1.1), $\lambda(A \cap B_{r,x}) \geqq \beta_n |A \cap B_{r,x}|^{-2/n}$. □

We turn now to the proof of Theorem 1. The basic idea is really very simple and is most clearly displayed in the proof of the first part, (1.3). In the proof of the second part, (1.4), the basic idea is obscured by technicalities. I am

Invent. Math. *74*, 441–448 (1983)

indebted to Haim Brezis for helpful ideas about the second part. First, let us define

$$\lambda(A)=\inf\{J(f)|f\in H_0^1(A), f\neq 0\}$$
$$=\inf\{J(f)|f\in C_0^\infty(A), f\neq 0\}. \quad (1.7)$$

$$J(f)=\int|\nabla f|^2/\int|f|^2. \quad (1.8)$$

Proof of (1.3). There exists $f\in C_0^\infty(A)$, $g\in C_0^\infty(B)$, $f,g\neq 0$ such that $J(f)<\lambda(A)+\varepsilon/2$, $J(g)<\lambda(B)+\varepsilon/2$. Here, we think of f as defined on all of $\mathbb{R}^n$ with $f(x)=0$ for $x\notin A$ and similarly for g. Without loss f and g can be assumed to be real valued. The normalization $\int f^2=\int g^2=1$ is imposed. The functions h_x given by $h_x(y)=f(y)g(y-x)$ satisfy $h_x\in C_0^\infty(A\cap B_x)$. Let

$$T(x)=\int|\nabla h_x|^2, \quad D(x)=\int|h_x|^2. \quad (1.9)$$

Clearly,

$$\int D(x)\,dx=\iint f(y)^2 g(y-x)^2\,dy\,dx=1 \quad (1.10)$$

We now compute

$$|\nabla h_x|^2(y)=|\nabla f|^2(y)g^2(y-x)+f^2(y)|\nabla g|^2(y-x)$$
$$+(\nabla f^2)(y)\cdot(\nabla g^2)(y-x)/2. \quad (1.11)$$

The last term can be written as $-(\nabla f^2)(y)\cdot\nabla_x g^2(y-x)/2$. Thus the integral (over x) of this term vanishes and

$$\int T(x)\,dx=\int|\nabla f|^2+\int|\nabla g|^2<\lambda(A)+\lambda(B)+\varepsilon\equiv\Lambda. \quad (1.12)$$

Therefore, $\int dx[T(x)-\Lambda D(x)]<0$ and hence $\Lambda D(x)>T(x)\geqq 0$ on a set of positive measure. (1.3) then follows from the Definition (1.7), (1.8).

To prove that $x\mapsto\lambda(A\cap B_x)$ is upper-semicontinuous, we note the easily proved fact that $C_0^\infty(A\cap B_x)=\{f(\cdot)g(\cdot-x)|f\in C_0^\infty(A), g\in C_0^\infty(B)\}$. For any such product function, $T(x)$ and $D(x)$ in (1.9) are clearly continuous. The function j given by $j(x)=T(x)/D(x)$ if $D(x)>0$ and $J(x)=\infty$ otherwise is thus upper-semicontinuous. Equation (1.7) then gives $\lambda(A\cap B_x)$ as the infimum (over f,g) of upper-semicontinuous functions. □

Proof of (1.4). Since A and B are bounded there exist $F\in H_0^1(A)$ and $G_0^1\in H_0^1(B)$ such that $J(F)=\lambda(A)$ and $J(G)=\lambda(B)$, with $\int F^2=\int G^2=1$. This is a simple consequence of the Rellich-Kondrachov compactness theorem. (Again, we extend F to all of $\mathbb{R}^n$ with $F(x)=0$, $x\notin A$, and similarly for G; it is easy to see that $F,G\in H_0^1(\mathbb{R}^n)$.) Define $H_x(y)\equiv F(y)G(y-x)$. Since $F,G\in L^2$, $H_x\in L^1$. Likewise, since ∇F and ∇G are L^2 functions, $W_x(y)\equiv(\nabla F)(y)G(y-x)+F(y)\nabla G(y-x)\in L^1$. It is easy to see (by approximating F, G by C_0^∞ functions) that $\nabla H_x=W_x$ in the sense of distributions.

It is not a-priori clear that H_x or $W_x\in L^2$. (They are, in fact, in L^2 because F and $G\in L^\infty$. However, this latter fact is not elementary and we prefer to avoid using it. For our purpose it suffices to show that H_x and $W_x\in L^2$ for almost all x, and this can be done by the following elementary argument.) We note that (by Fubini) $D(x)=\int H_x^2$ satisfies $\int D(x)\,dx=1$ as in (1.10). Thus, $D(x)<\infty$ a.e.

and $H_x \in L^2$ a.e. (dx). Likewise, repeating the argument in (1.11), (1.12) for $T(x) = \int |W_x|^2$

$$\int T(x)\,dx = \lambda(A) + \lambda(B) \equiv \Lambda. \tag{1.13}$$

(Here, one has to note as above that $\nabla F(y)\,G(y-x) \in L^2(dy)$ a.e. (dx) and $F(y)\nabla G(y-x) \in L^2(dy)$ a.e. (dx). By the Schwartz inequality, $Z_x(y) \equiv F(y)\,G(y-x)\nabla F(y)\cdot\nabla G(y-x) \in L^1(dy)$ a.e. (dx) and $Z_x(y) \in L^1(dx\,dy)$. Finally, $2G\nabla G = \nabla G^2$ in the distributional sense, whence $\iint (Z_x(y)\,dx\,dy = 0.)$. Thus, for almost all x, $H_x \in H_0^1(dy)$ and $\nabla H_x = W_x$.

We have that $\int (T - \Lambda D) = 0$. The remainder of the proof is as before, except that in order to prove the strict inequality (1.4) we must show that $T(x) = \Lambda D(x)$ cannot hold a.e. To see this let K_A (resp. K_B) be the characteristic function of A (resp. B). Then $(K_A * K_B)(x) \equiv K(x) = |A \cap B_x|$ is a continuous function of compact support. For $\varepsilon > 0$ there is an open set C such that $0 < K(x) < \varepsilon$ for $x \in C$. By the strong maximum principle, both $F(y)$ and $G(y-x)$ are positive in the open set $A \cap B_x$, $x \in C$. (Here we assume that A and B are connected; if not, it suffices to replace A and B by connected components on which $F > 0$, $G > 0$.) Thus, $D(x) > 0$, $x \in C$. If $T(x) = \Lambda D(x)$, $x \in C$, then $\lambda(A \cap B_x) \leqq \Lambda$, but this is impossible for sufficiently small ε by the Faber-Krahn inequality (1.1). □

2. Some extensions of theorem 1

Instead of the ordinary eigenvalue given in (1.7), (1.8) we can consider the $W^{1,p}$ eigenvalue, $1 \leqq p < \infty$, given by

$$\begin{aligned} \lambda_p(A) &= \inf\{J(f) \mid f \in W_0^{1,p}(A), f \neq 0\} \\ &= \inf\{J(f) \mid f \in C_0^\infty(A), f \neq 0\} \end{aligned} \tag{2.1}$$

$$J(f) = \int |\nabla f|^p / \int |f|^p. \tag{2.2}$$

As in (1.1) there is an isoperimetric inequality for $\lambda_p(A)$, which is also proved by rearrangement inequalities. Given $|A|$, the minimum is achieved for a ball:

$$\lambda_p(A) \geqq \beta_{n,p} |A|^{-p/n}, \tag{2.3}$$

where $\beta_{n,p}$ is λ_p for a ball of unit volume.

The analogue of Theorem is

Theorem 3. *Let A and B be non-empty open sets in $\mathbb{R}^n$, $n \geqq 1$. Let $1 \leqq p < \infty$ and let $\lambda_p(A)$, $\lambda_p(B)$ be given by* (2.1), (2.2). *Let $\varepsilon > 0$. Then there exists an $x \in \mathbb{R}^n$ such that*

$$\lambda_p(A \cap B_x)^{1/p} < \lambda_p(A)^{1/p} + \lambda_p(B)^{1/p} + \varepsilon. \tag{2.4}$$

If A and B are bounded and $1 < p < \infty$, then for some x

$$\lambda_p(A \cap B_x)^{1/p} < \lambda_p(A)^{1/p} + \lambda_p(B)^{1/p}. \tag{2.5}$$

(2.4), (2.5) *are equivalent to*

$$\lambda_p(A)^{1/p} \geqq \inf_x \lambda_p(A \cap B_x)^{1/p} - \lambda_p(B)^{1/p} \tag{2.4'}$$

$$\lambda_p(A)^{1/p} \not> \inf_x \lambda_p(A \cap B_x)^{1/p} - \lambda_p(B)^{1/p}. \tag{2.5'}$$

For $1 \leqq p < \infty$, $x \mapsto \lambda_p(A \cap B_x)$ *is upper-semicontinuous so that the set of* x's *for which* (2.4) *or* (2.5) *holds is open.*

Proof. A few remarks about the necessary changes in the proof should suffice.

(i) For (2.4) we again use C_0^∞ approximants. The exponents $1/p$ in (2.4), (2.5) (which a clever reader might be able to eliminate) result from the fact that $|\nabla h_x|^p$ is more complicated than in (1.11). All we can say is that $|g \nabla f + f \nabla g|^p \leqq \gamma^{1-p} |g \nabla f|^p + (1-\gamma)^{1-p} |f \nabla g|^p$. The choice $\gamma^{-1} = 1 + [\lambda(B)/\lambda(A)]^{1/p}$ is optimal and gives (2.4), (2.5).

(ii) For (2.5) we note (by the Rellich-Kondrachov theorem) that when A is bounded and $1 < p < \infty$ there is a minimizing function $F \in W_0^{1,p}$ for (2.1). Likewise, there is a $G \in W_0^{1,p}$ for B. $\nabla F, \nabla G, F, G \in L^p$, but unfortunately this does not automatically imply that combinations such as $F(y)\,G(y-x)$ or $\nabla F(y)\,G(y-x)$, etc. are in $L^1(dy)$. (We could use the fact that $F, G \in L^\infty$ but, as before, we prefer to give an elementary proof.) The earlier proof (see (1.13)) does show, however, that they are in $L^p(dy)$ a.e. (dx). That they are also in $L^1(dy)$ a.e. (dx) follows from this and the fact that their supports are bounded for all x. Hence $W_x(y) = \nabla F(y)\; G(y-x) + F(y)\, \nabla G(y-x)$ makes sense as a distribution a.e. (dx) and it is then easy to see that $W_x(y) = \nabla_y\, F(y)\, G(y-x)$ a.e. (dx). □

Remark. When $p=1$ there is no minimizing function, F, for (2.1), (2.2), even if A is bounded. However [12],

$$\lambda_1(A) = \inf S(D)/|D|, \tag{2.6}$$

where the infimum is over all relatively compact subdomains, D, of A with surface area $S(D)$ and volume $|D|$.

Corollary 2 has the following analogue.

Corollary 4. *Let A be a non-empty open set in* $\mathbb{R}^n$, $n \geqq 1$, *let* $1 \leqq p < \infty$ *and let* B_r *be a ball of radius* r *with* $|B_r| = r^n/\delta_n$. *Let* $0 < \psi < 1$ *be fixed and let*

$$\sigma_{n,p}(\psi) = \beta_{n,p}\, \delta_n^{p/n} [\psi^{-1/n} - 1]^p. \tag{2.7}$$

If $\lambda(A) \leqq \sigma_{n,p} R^{-p}$, $0 < R \leqq \infty$, *then for every* $0 < r < R$ *there exists an* $x \in \mathbb{R}^n$ *and a ball* $B_{r,x}$ *of radius* r *such that*

$$|A \cap B_{r,x}| > \psi\, |B_r| = \psi\, r^n/\delta_n.$$

Proof. The proof imitates that of Corollary 2, using Theorem 4, with the choice $\varepsilon = \sigma_{n,p}^{1/p} [r^{-1} - R^{-1}]$. □

Another variation on Theorem 1 was suggested to me by S.T. Yau. One can consider manifolds other than $\mathbb{R}^n$ and symmetry groups other than the translation group. As an illustration, consider the sphere S^n and the rotation group $O(n+1)$. If $B\subset S^n$, $R\in O(n+1)$ then $B_R=\{x\in S^n|x=Ry,\ y\in B\}$. Let $d\mu(R)$ denote normalized Haar measure on $O(n+1)$. The analogue of Theorems 1 and 3 is the following.

Theorem 5. *Let A and B be non-empty open sets in S^n, $n\geqq 1$, and let $\lambda_p(A)$, $\lambda_p(B)$, $1\leqq p<\infty$, be given by (2.1), (2.2). Let $\varepsilon>0$. Then there exists an R in $O(n+1)$ such that*

$$\lambda_p(A\cap B_R)^{1/p}<\lambda_p(A)^{1/p}+\lambda_p(B)^{1/p}, \qquad 1<p<\infty \tag{2.8}$$

$$\lambda_p(A\cap B_R)^{1/p}<\lambda_p(A)^{1/p}+\lambda_p(B)^{1/p}+\varepsilon, \qquad 1\leqq p<\infty \tag{2.9}$$

If $p=2$ then the exponent $1/p$ can be replaced by 1 in (2.8), (2.9). The map $R\to\lambda_p(A\cap B_R)$ is upper-semicontinuous.

The proof is a before provided we note that for all $y\in S^n$ and $f\colon S^n\to\mathbb{C}$

$$\int d\mu(R)f(Ry)=|S_n|^{-1}\int_{S^n} f(y)\,dy.$$

3. A compactness lemma

One of the motivations, in addition to the geometric one mentioned in Sect. 1, for proving Theorems 1 and 3 was to prove the following compactness lemma. It is useful in the calculus of variations to show that, under some condition, a bounded sequence of functions in $W^{1,p}(\mathbb{R}^n)$ can, after suitable translations, be assumed to have a weak limit that is not zero. (See [13, 14] for example.)

Lemma 6. *Let $1<p<\infty$ and let $\{f_j\}_{j=1}^\infty$ be a uniformly bounded sequence of real valued functions in $W^{1,p}$ with the property that $E_j=\{x|f_j(x)>\varepsilon\}$ satisfies $|E_j|\geqq C$ for some fixed ε, $C>0$. Then there exists a sequence of translations $\{\tau_j\}$ of $\mathbb{R}^n$, $\tau_j\colon y\mapsto y+x_j$, $F_j(y)\equiv f_j(\tau_j y)=f_j(y+x_j)$, such that $F_{n_j}\rightharpoonup F$ weakly in $W^{1,p}$ and $F\neq 0$, for some subsequence $\{n_j\}$.*

Proof. By density, we can assume that $f_j\in C_0^\infty$ so that E_j is open and bounded (replace ε by $\varepsilon/2$ if necessary). Let $g_j(y)=\max(f_j(y)-\varepsilon/2,0)\in W^{1,p}$ and $\|g_j\|\leqq\|f_j\|\leqq D=\text{const}$. Let $A_j=\{y|g_j(y)>0\}\supset E_j$, which is also open and bounded. Then $g_j\in W_0^{1,p}(A_j)$ and $\int|\nabla g_j|^p<D^p$, $\int|g_j|^p\geqq(\varepsilon/2)^p|E_j|\geqq(\varepsilon/2)^pC$. Thus $\lambda_p(\underset{\sim}{A}_j)\leqq\sigma_{n,p}(1/2)R^{-p}$ for some fixed $0<R<\infty$. Cf. (2.7). By Corollary 4, there is an x_j and a ball B_{r,x_j} of radius $r=R/2$ such that $|A_j\cap B_{r,x_j}|\geqq r^n/2\delta_n$. Choose τ_j: $y\to y+x_j$. Let $\beta\in L^{p'}$ denote the characteristic function of $B_{r,0}$, so that $\int F_j\beta\geqq\varepsilon r^n/4\delta_n\equiv K$. By the Banach-Alaoglu theorem there is a subsequence such that $F_j\rightharpoonup F$. But $F\neq 0$ since $\int F\ =K$. $\square$

Motivated by the foregoing, H. Brezis (private communication) found another proof that does not use Corollary 4.

Brezis' proof of Lemma 6. We start with a simple remark. Let $u\in L^p_{\text{loc}}$ with $\nabla u\in L^p$ and $\|\nabla u\|_p\leqq 1$. Set $k=1+\|u\|_p^{-p}$ (for $\|u\|_p\leqq\infty$). Let B_x denote the unit ball in $\mathbb{R}^n$ centered at x and let β_x be its characteristic function. Clearly there is some x such that $\int|\nabla u|^p\beta_x<k\int|u|^p\beta_x$, and thus

$$\int(|\nabla u|^p+|u|^p)\beta_x<(k+1)\int|u|^p\beta_x. \tag{3.1}$$

On the other hand, by Sobolev's inequality we have

$$\int (|\nabla u|^p + |u|^p)\, \beta_x \geqq S\, \|u\, \beta_x\|_q^p \tag{3.2}$$

where $q^{-1}+n^{-1}=p^{-1}$ if $p<n$ and, if $p\geqq n$, q is arbitrary with $p<q<\infty$. $S>0$ depends only on p, q. Combining (3.1), (3.2) and Hölder's inequality we obtain

$$S<(k+1)|B_x \cap \operatorname{supp} u|^{1-p/q}. \tag{3.3}$$

Let us apply the previous remark to $u=\max(f_j-\varepsilon/2, 0)$. For simplicity, we assume that $\|\nabla f_j\|_p \leqq 1$ so that $\|\nabla u\|_p \leqq 1$. From the assumptions of Lemma 6 we have $\|u\|_p^p \geqq (\varepsilon/2)^p |E_j| \geqq (\varepsilon/2)^p C$, and thus $k \leqq 1+(2/\varepsilon)^p/C$. From (3.3) we deduce that there exists some x_j such that

$$|B_{x_j} \cap \{x \mid f_j(x) > \varepsilon/2\}| \geqq K$$

for some constant K depending only on p, q, ε, C. The conclusion follows as in the previous proof. □

References

1. Brascamp, H.J., Lieb, E.H.: Some inequalities for Gaussian measures and the long-range order of the one-dimensional plasma. In: Functional Integration and its Applications, Arthurs, A.M. (ed.) pp. 1-14. Oxford: Clarendon Press 1975
2. Cheeger, J.: A lower bound for the smallest eigenvalue of the Laplacian. In: Problems in Analysis, a Symposium in Honor of Salomon Bochner, Gunning, R.C. (ed.) pp. 145—199. Princeton, N.J.: Princeton University Press 1970
3. Cheng, S.Y.: On the Hayman-Osserman-Taylor inequality, (preprint).
4. Croke, C.B.: The first eigenvalue of the Laplacian for plane domains. Proc. Amer. Math. Soc. **81**, 304-305 (1981)
5. Faber, C.: Beweis das unter allen homogenen Membranen von gleicher Fläche und gleicher Spannung die kreisförmige den tiefsten Grundton gibt, Sitzungsber. Bayer. Akad. der Wiss. Math. Phys., Munich 1923, pp. 169-172
6. Hayman, W.K.: Some bounds for principle frequency. Applic. Anal. **7**, 247-254 (1977/1978)
7. Krahn, E.: Über eine von Rayleigh formulierte Minimaleigenschaft des Kreises. Math. Ann. **94**, 97-100 (1925)
8. Osserman, R.: A note on Hayman's theorem on the bass note of a drum. Comment. Math. Helv. **52**, 545-555 (1977)
9. Osserman, R.: The isoperimetric inequality, Bull. Amer. Math. Soc. **84**, 1182-1238 (1978)
10. Osserman, R.: Bonnesen-style isoperimetric inequalities. Amer. Math. Monthly **86**, 1-29 (1979)
11. Taylor, M.: Estimate on the fundamental frequency of a drum. Duke Math. J. **46**, 447-453 (1979)
12. Yau, S.T.: Isoperimetric constants and the first eigenvalue of a compact manifold. Ann. Sci. Ecole Norm. Sup. **8**, 487-507 (1975)
13. Lieb, E.H.: Some vector field equations. In: Proceedings of the March 1983 University of Alabama, Birmingham International Conference on Partial Differential Equations, Knowles, I. (ed.), North-Holland (in press)
14. Brezis, H., Lieb, E.H.: Minimum action solutions to some vector field equations (in preparation)

Oblatum 22-IV-1983

Commun. Math. Phys. 96, 97–113 (1984)

Communications in Mathematical Physics

Minimum Action Solutions of Some Vector Field Equations

Haim Brezis[1] and Elliott H. Lieb[2,*]

1 Département de Mathématiques, Université Paris VI, 4, Place Jussieu, F-75230 Paris, Cédex 05, France

2 Departments of Mathematics and Physics, Princeton University, P.O. Box 708, Princeton, NJ 08544, USA

Abstract. The system of equations studied in this paper is $-\Delta u_i = g^i(u)$ on $\mathbb{R}^d$, $d \geqq 2$, with $u: \mathbb{R}^d \to \mathbb{R}^n$ and $g^i(u) = \partial G/\partial u_i$. Associated with this system is the action, $S(u) = \int \{\frac{1}{2}|\nabla u|^2 - G(u)\}$. Under appropriate conditions on G (which differ for $d=2$ and $d \geqq 3$) it is proved that the system has a solution, $u \not\equiv 0$, of finite action and that this solution also minimizes the action within the class $\{v$ is a solution, v has finite action, $v \not\equiv 0\}$.

I. Introduction

The purpose of this paper is to demonstrate the existence of solutions to a class of systems of partial differential equations that arises in several branches of mathematical physics (e.g. calculating lifetimes of metastable states, estimates of large order behavior of perturbation theory, Ginzburg-Landau theory, density of states in disordered systems). The systems to be considered are of the form

$$-\Delta u_i(x) = g^i(u(x)), \qquad i = 1, \ldots, n. \tag{1.1}$$

Furthermore, it will be shown that among the nonzero solutions to (1.1) there is one that minimizes the action, $S(u)$, associated with (1.1).

The meaning of the quantities in (1.1) is the following: $u \equiv (u_1, \ldots, u_n) \in \mathbb{R}^n$ and each $u_i: \mathbb{R}^d \to \mathbb{R}$ with $d \geqq 2$. We require that $u_i(x) \to 0$ as $|x| \to \infty$ in a weak sense described below (namely $u \in \mathscr{C}$). (Note: In some applications it is required that $u(x) \to c =$ constant as $|x| \to \infty$ but, by redefining $u \to u - c$ and by redefining g^i, the problem can be reduced to the $u(x) \to 0$ case.) The n functions $g^i: \mathbb{R}^n \to \mathbb{R}$ are the gradients of some function $G \in C^1(\mathbb{R}^n \backslash \{0\})$, namely

$$\begin{aligned} g^i(u) &= \partial G(u)/\partial u_i, \qquad && u \neq 0, \\ g^i(u) &= 0, && u = 0, \end{aligned} \tag{1.2}$$

* Work partially supported by U.S. National Science Foundation Grant PHY-81-16101-A02

and G satisfies certain properties described in Sect. II ($d \geqq 3$) and Sect. III ($d=2$). In particular, we emphasize that $G(u)$ need not be differentiable at $u=0$ so that, for example, $G(u)$ could be $-|u|$ near $u=0$.

The *Action* associated with (1.1) is

$$S(u) \equiv K(u) - V(u)\,, \tag{1.3}$$

$$K(u) \equiv \tfrac{1}{2}\int |\nabla u(x)|^2\,dx \equiv \tfrac{1}{2}\sum_i \int |\nabla u_i(x)|^2\,dx\,, \tag{1.4}$$

$$V(u) \equiv \int G(u(x))\,dx\,. \tag{1.5}$$

In general, $S(u)$ is not bounded below, and one of our goals is to show that, under suitable conditions, $S(u) > -\infty$ if u satisfies (1.1) and that $S(u)$ actually has a minimum in the set of non-trivial solutions to (1.1). The word non-trivial (meaning $u \not\equiv 0$) is important; it will be shown later that when $d=2$ the function $u=0$ satisfies (1.1) and minimizes $S(u)$, but the non-trivial solutions to (1.1) all have $S(u)>0$. When $d \geqq 3$, the $u \equiv 0$ solution never has the minimum action.

The class of functions to which we shall restrict our investigation of (1.1) as an equation in $\mathscr{D}'$ is

$$\mathscr{C} = \{u \mid u \in L^1_{\mathrm{loc}}(\mathbb{R}^d),\ \nabla u \in L^2(\mathbb{R}^d),\ G(u) \in L^1(\mathbb{R}^d),\ \mu([|u| > \alpha]) < \infty \text{ for all } \alpha > 0\}. \tag{1.6}$$

Here, the symbol $[f > \alpha]$ denotes the set $\{x \mid f(x) > \alpha\}$. The same symbol, $[f > \alpha]$, will also be used to denote the characteristic function of this set. Lebesgue measure is denoted by μ. The set

$$\mathscr{E} = \{u \mid u \in \mathscr{C},\ g(u) \in L^1_{\mathrm{loc}}(\mathbb{R}^d),\ u \text{ satisfies (1.1) in } \mathscr{D}',\ u \not\equiv 0\} \tag{1.7}$$

is the subset of $\mathscr{C}$ which we shall prove is non-empty and in which there is a $\bar{u}$ such that

$$S(\bar{u}) \leqq S(u)\,, \quad \text{all} \quad u \in \mathscr{E}\,. \tag{1.8}$$

The solution of this problem was reported in 1983 and an outline of the proof was given [13]. The purpose of the present paper is to present all the details of the proof and certain additional refinements.

Probably the earliest general treatment of existence of finite action solutions to (1.1) was by Strauss [20] for $n=1$, $d \geqq 1$. (The case $n=1$ is called the scalar case.) While this work was very important because it introduced new techniques, it imposed severe restrictions on the function G. Moreover, Strauss did not explicitly consider the question of whether or not his solution to (1.1) minimized the action. Strauss and Vázquez extended this work to the vector case and to the "zero mass" case [22]. The next step was taken by Coleman et al. [10] who made an important contribution to the problem by their "constrained minimum method" which not only yields a solution for $d \geqq 3$ but also yields a minimum action solution. They discovered almost optimal assumptions on G so that the problem has a solution, but their method for finding a minimum action solution was restricted in an essential way to $d \geqq 3$ and $n=1$. A detailed treatment of the Coleman, Glaser, Martin method, together with some improvements and other theorems useful in the analysis of this and related problems was given by Berestycki and Lions [5, 6]. Then, in the same generality, Berestycki and Lions [6] went on to prove the existence of infinitely many finite action solutions to (1.1). Strauss [20] also had results in this direction. (Infinitely many solutions for the so-called zero mass case was done in [7].) As before, all of this was for $n=1$, $d \geqq 3$.

In view of the aforementioned work, two natural extensions suggest themselves. One is to $d=2$ and the other is to $n>1$. We thank Ian Affleck for suggesting both problems to one of us (E.L.). Affleck was interested in the $d=2$ case for physical applications [1]. Some results for both $d=2$ and for $n \geqq 1$ were obtained by one of us (E.L.) in 1983, and these were subsequently strengthened in collaboration with H.B. to the level of generality given here and in [13]. Independently, in 1982, Berestycki, Gallouet and Kavian had solved the $d=2$, $n=1$ case (with stronger hypotheses than in the present paper; in particular they do not treat the zero mass case) and this was published recently [3] (see also [4]). [However, they also showed there are infinitely many solutions of (1.1) for $d=2$, $n=1$.]

The proofs for $n=1$ all relied on the fact that one could look for minima in the class of radial functions (by rearrangement inequalities), and that these functions have certain compactness properties [20, 6]. For $n>1$, one can still restrict attention to radial solutions, although it is not known whether the minimum action solution lies in this class (because rearrangement inequalities are not applicable). Berestycki and Lions [5] showed how to prove the existence of radial solutions that minimize the action among all *radial* solutions of (1.1).

The extension to $n>1$ requires a new compactness device. In this paper, the heart of the matter is contained in Lemma 2.2. It should be noted that Lions has developed a general compactness principle [15, 16] which allows him to deal with the cases $d \geqq 2$, $n \geqq 1$.

II. The Case of Three or More Dimensions

A. The Minimization Problem

Let $G: \mathbb{R}^n \to \mathbb{R}$ be continuous with $G(0)=0$. In this subsection we shall consider a minimization problem that leads to (1.1) if G happens to be differentiable, but here we shall make no assumptions about the differentiability of G. Here, and henceforth, $C>0$ will denote an inessential, positive constant. G satisfies the following four conditions (2.2)–(2.5). [Note $G(u)$, not $|G(u)|$ in (2.2), (2.3).]

$$\limsup_{|u| \to \infty} |u|^{-p} G(u) \leqq 0, \tag{2.2}$$

where, for $d \geqq 3$, p always denotes

$$p = 2^* = 2d/(d-2),$$

$$\limsup_{|u| \to 0} |u|^{-p} G(u) \leqq 0, \tag{2.3}$$

$$G(u_0) > 0 \quad \text{for some} \quad u_0 \in \mathbb{R}^n, \tag{2.4}$$

$$\left.\begin{array}{l} \text{For all } \gamma>0 \text{ there exists } C_\gamma \text{ such that for all } u, w \in \mathbb{R}^n \\ |G(u+w)-G(u)| \leqq \gamma[|G(u)|+|u|^p] + C_\gamma[|G(w)|+|w|^p+1]. \end{array}\right\} \tag{2.5}$$

Remark 1. Condition (2.5) looks awkward, but it holds in several cases such as (2.6) or (2.7) or (2.8):

$$\lim_{|u|\to\infty} |u|^{-p}|G(u)|=0\,, \tag{2.6}$$

$$\left.\begin{array}{ll} G\in C^1(\mathbb{R}^n\backslash\{0\}) & \text{and } g=\nabla G \text{ satisfies} \\ |g(u)|\leqq C+C|u|^{p-1}\,, & \text{all } u\neq 0\,. \end{array}\right\} \tag{2.7}$$

$$\left.\begin{array}{ll} G\in C^1(\mathbb{R}^n\backslash\{0\}) \ \text{ and } & g=\nabla G \text{ satisfies} \\ |g(u)|\leqq C+C|u|^{q-1}\,, & \text{all } u\neq 0 \text{ and} \\ |u|^p+|G(u)|\geqq \alpha|u|^q-C\,, & \text{all } u\in\mathbb{R}^n \\ \quad\text{for some } q\geqq p, \ \text{ and } & C,\,\alpha>0\,. \end{array}\right\} \tag{2.8}$$

The main result of this section is the solution to the following minimization problem. We define

$$T=\inf\{\tfrac{1}{2}\textstyle\int|\nabla u|^2\,|\,u\in\mathscr{C},\int G(u)\geqq 1\}\,. \tag{2.9}$$

Theorem 2.1. *Assume* (2.2)–(2.5). *Then there exists* $v\in\mathscr{C}$ *such that*

$$\tfrac{1}{2}\textstyle\int|\nabla v|^2=T, \tag{2.10}$$

and

$$\textstyle\int G(v)=1\,. \tag{2.11}$$

Remark 2. Using (2.4) it is easy to see that there is some $u\in\mathscr{C}$ such that $\int G(u)=1$.

Remark 3. Let $u\in L^1_{\text{loc}}$ and $\nabla u\in L^2$, such that $u\to 0$ as $|x|\to\infty$ in the weak sense of (1.6), namely $\mu([|u|>\alpha])<\infty$, all $\alpha>0$. Then $u\in L^p$ and $\|u\|_p\leqq C\|\nabla u\|_2$. Thus, the class $\mathscr{C}$ in (1.6) can be characterized (for $d\geqq 3$) as

$$\mathscr{C}=\{u|u\in L^p(\mathbb{R}^d),\,\nabla u\in L^2(\mathbb{R}^d),\,G(u)\in L^1(\mathbb{R}^d)\}\,.$$

To prove this, let $\chi_n(x)=\chi(x/n)$, where $\chi\in C_0^\infty$ and $\chi\equiv 1$ near 0. Let $\varepsilon>0$ be fixed. Assume, provisionally, that $u\in\mathscr{C}$ and also $u\in L^\infty$. By Sobolev's inequality,

$$\begin{aligned}\|\chi_n(|u|-\varepsilon)_+\|_p&\leqq C\|\nabla\chi_n(|u|-\varepsilon)_+\|_2\\ &\leqq C\{\|\nabla u\|_2+[\textstyle\int_A|\nabla\chi_n|^2]^{1/2}\}\\ &\leqq C\|\nabla u\|_2+C_\varepsilon/n\,,\end{aligned}$$

where $A=[|u|>\varepsilon]$ and C_ε is some constant depending on ε. We conclude (in this L^∞ case) by letting $n\to\infty$ and then $\varepsilon\to 0$. If $u\in\mathscr{C}$ but $u\notin L^\infty$, we may truncate u, then use the foregoing, and then remove the truncation by Fatou's lemma.

In the following, $\{u^j\}$ denotes a minimizing sequence for (2.9).

Lemma 2.1. *There exist* $\varepsilon,\,\delta>0$ *such that for all* j, $\mu([|u^j|>\varepsilon])\geqq\delta$.

Proof. Since ∇u^j is bounded in L^2, Sobolev's inequality implies that

$$\|u^j\|_p^p\leqq C\,. \tag{2.12}$$

Let $\gamma=1/(2C)$. By (2.2), (2.3) there exists $1>\varepsilon>0$ such that

$$G(v)\leqq\gamma|v|^p \quad\text{for}\quad |v|\leqq\varepsilon \quad\text{or}\quad |v|\geqq 1/\varepsilon\,. \tag{2.13}$$

Thus, we have that

$$1 \leqq \int G(u^j) \leqq \gamma \int |u^j|^p \{[|u^j| \leqq \varepsilon] + [|u^j| \geqq 1/\varepsilon]\} + \int G(u^j)\,[\varepsilon < |u^j| < 1/\varepsilon]$$
$$\leqq \tfrac{1}{2} + C_\varepsilon \mu([|u^j| > \varepsilon]) .$$

This implies the lemma with $\delta = 1/(2C_\varepsilon)$. □

Next, we recall the following [14]:

Lemma 2.2. *Let v be a function such that $v \in L^1_{\mathrm{loc}}$, $\nabla v \in L^2$, $\|\nabla v\|_2 \leqq C$ and $\mu([|v| > \varepsilon]) \geqq \delta > 0$. Then, there exists a shift $T_y v(x) = v(x+y)$ such that, for some constant $\alpha = \alpha(C, \delta, \varepsilon) > 0$, $\mu(B \cap [|T_y v| > \varepsilon/2]) > \alpha$, where $B = \{x \in \mathbb{R}^d \,|\, |x| \leqq 1\}$.*

Using Lemmas 2.1 and 2.2 we can shift each u^j in such a way that $\mu(B \cap [|T_{y_j} u^j| > \varepsilon/2]) \geqq \alpha$, where $\alpha > 0$ is independent of j. Thus, we may assume without loss of generality that $\mu(B \cap [|u^j| > \varepsilon/2]) \geqq \alpha$. After extracting a subsequence we may also assume that (cf. (2.12))

$$u^j \rightharpoonup u \text{ weakly in } L^p,\ \nabla u^j \rightharpoonup \nabla u \text{ weakly in } L^2 ,$$
$$u^j \to u \text{ a.e. on } \mathbb{R}^d,\ \mu(B \cap [|u| \geqq \varepsilon/2]) \geqq \alpha .$$

Finally, we have $G(u) \in L^1$. To prove this, let us write

$$G = G_+ - G_- \quad \text{with} \quad G_+ = \operatorname{Max}\{G, 0\} \quad \text{and} \quad G_- = \operatorname{Max}\{-G, 0\} .$$

We have

$$\int G_+(u^j) \leqq \gamma \int |u^j|^p \{[|u^j| \leqq \varepsilon] + [|u^j| \geqq 1/\varepsilon]\}$$
$$+ \int G_+(u^j)\,[\varepsilon < |u^j| < 1/\varepsilon] \leqq \text{const} .$$

(The last integral is uniformly bounded because $\|u^j\|_p < C \Rightarrow \mu([|u^j| > \varepsilon]) < (C/\varepsilon)^p$; moreover, $G_+(\cdot)$ is bounded on $(\varepsilon, 1/\varepsilon)$ since G_+ is continuous.) We also have $\int G_-(u^j) \leqq \int G_+(u^j) - 1$. Hence, $\int |G(u^j)| \leqq \text{const}$, and we deduce from Fatou's lemma that $G(u) \in L^1$. Thus, $u \in \mathscr{C}$.

We conclude the proof of Theorem 2.1 with

Lemma 2.3. *The limit function satisfies $\int G(u) = 1$ and $\frac{1}{2}\int |\nabla u|^2 = T$, where T is defined in* (2.9).

Remark 4. It follows from Lemma 2.3 that in fact $\nabla u^j \to \nabla u$ strongly in L^2 and thus $u^j \to u$ strongly in L^p.

Proof of Lemma 2.3. It is easily seen by scaling [i.e. $v(x) \to v(\lambda x)$] that

$$\tfrac{1}{2}\int |\nabla v|^2 \geqq T\left[\int G(v)\right]^{(d-2)/d}, \quad \text{all } v \in \mathscr{C} \text{ with } \int G(v) > 0 . \tag{2.14}$$

Let $\phi \in L^p$ with $G(\phi) \in L^1$ and with ϕ having compact support. We claim that, as $j \to \infty$

$$\int G(u^j + \phi) \geqq 1 + \int G(u + \phi) - \int G(u) + o(1) . \tag{2.15}$$

[Note that the integrals in (2.15) make sense because of (2.5).]

Verification of (2.15). Let $K = \operatorname{Supp}\phi$; we have

$$\int G(u^j+\phi) = \int_K G(u^j+\phi) + \int_{\sim K} G(u^j) \geqq 1 + \int_K [G(u^j+\phi) - G(u^j)]$$
$$= 1 + \int_K [G(u+\phi) - G(u)] + o(1).$$

The last equality follows from Egorov's (or Vitali's) lemma. Indeed, given $\varepsilon > 0$ we fix $\gamma > 0$ small enough so that

$$\gamma \int \{|G(u^j)| + |u^j|^p\} < \varepsilon/2.$$

By (2.5) we have that

$$\int_A |G(u^j+\phi) - G(u^j)| \leqq \varepsilon/2 + C_\gamma \int_A [|G(\phi)| + |\phi|^p + 1] \leqq \varepsilon$$

for any set $A \subset K$ with $\mu(A) < \delta$ (δ small enough). Thus we have established (2.15).

Suppose now that $\phi \in H^1$, ϕ has compact support, $G(\phi) \in L^1$ and

$$\int [G(u+\phi) - G(u)] > -1. \tag{2.16}$$

For j large enough we may insert $v = u^j + \phi$ in (2.14) and, in the limit, we find that

$$T + \int \nabla u \cdot \nabla\phi + \tfrac{1}{2}\int |\nabla\phi|^2 \geqq T[1 + \int G(u+\phi) - \int G(u)]^{1-2/d}.$$

That is,

$$T + \tfrac{1}{2}\int |\nabla(u+\phi)|^2 - \tfrac{1}{2}\int |\nabla u|^2 \geqq T[1 + \int G(u+\phi) - \int G(u)]^{1-2/d}. \tag{2.17}$$

Let $\lambda > 0$ be fixed. We can find a mapping $S: \mathbb{R}^d \to \mathbb{R}^d$, bijective with S and S^{-1} smooth such that

$$S(x) = \begin{cases} \lambda x, & |x| < 1 \\ x, & |x| > R \end{cases}$$

(for some R depending on λ). Set $S_n(x) = nS(x/n)$ and $\phi_n(x) = u(S_n(x)) - u(x)$, so that $\phi_n \in H^1$ and ϕ_n has compact support and $G(\phi_n) \in L^1$. [The last assertion is obtained by choosing $u = \phi_n(x)$, $w = u(x)$ in (2.5).] We claim that as $n \to \infty$

$$\int G(u+\phi_n) = \int G(u(\lambda x))dx + o(1) = \lambda^{-d}\int G(u) + o(1), \tag{2.18}$$

and

$$\int |\nabla(u+\phi_n)|^2 = \int |\nabla[u(\lambda x)]|^2 dx + o(1) = \lambda^{2-d}\int G(u) + o(1). \tag{2.19}$$

Indeed we write

$$\int G(u+\phi_n) = \int G(u(S_n(x)))dx = \int G(u(y))J_n(y)dy,$$

where J_n denotes the Jacobian determinant of the mapping $y \to S_n^{-1}(y)$; it is easy to see that $|J_n| \leqq C$, C independent of n, and $J_n(y) \to \lambda^{-d}$ as $n \to \infty$ for all y. Thus (2.18) follows by dominated convergence. The same argument applies to (2.19). We fix $\lambda > 0$ with $|\lambda - 1|$ so small that $(\lambda^{-d} - 1)\int G(u) > -1$. Thus $\phi = \phi_n$ satisfies (2.16) for n large enough. Hence (2.17) holds for $\phi = \phi_n$ and in the limit (as $n \to \infty$) we obtain

$$T + \tfrac{1}{2}(\lambda^{2-d} - 1)\int |\nabla u|^2 \geqq T[1 + (\lambda^{-d} - 1)\int G(u)]^{1-2/d}. \tag{2.20}$$

Finally we choose $\lambda=1\pm\varepsilon$ in (2.20) and, as $\varepsilon\to 0$, we see that $\frac{1}{2}\int|\nabla u|^2=T\int G(u)$. Since $u\not\equiv 0$ we have $\int G(u)>0$, and we deduce from (2.14) (applied to $v=u$) that $\int G(u)\geqq 1$. On the other hand, since $\nabla u^j\rightharpoonup\nabla u$ weakly in L^2, we obtain, by lower semicontinuity, that $\frac{1}{2}\int|\nabla u|^2\leqq T$. Therefore $\int G(u)=1$ and $\frac{1}{2}\int|\nabla u|^2=T$. This concludes the proof of Theorem 2.1. □

B. Further Properties of u

Throughout this section we assume that G is differentiable on $\mathbb{R}^n\backslash\{0\}$. More precisely, let $G:\mathbb{R}^n\to\mathbb{R}$ be continuous (on all of $\mathbb{R}^n$) with $G(0)=0$. Assume that G satisfies (2.2)–(2.4) and $G\in C^1(\mathbb{R}^n\backslash\{0\})$. We set

$$g(v)=\begin{cases}\nabla G(v) & \text{if } v\neq 0\\ 0 & \text{if } v=0\,.\end{cases}$$

We assume (2.8). For every $v\in\mathscr{C}$ we define its action to be $S(v)=\frac{1}{2}\int|\nabla v|^2-\int G(v)$.

Theorem 2.2. *Let u be given by Theorem 2.1. Then after some appropriate scaling,*

$$\underline{u}(x)=u(\theta x),\ (\theta>0),\ \underline{u}\text{ satisfies}$$
$$-\Delta\underline{u}=g(\underline{u})\text{ in }\mathscr{D}'. \tag{2.21}$$

Moreover,

$$0<S(\underline{u})\leqq S(v),\ \text{all } v\in\mathscr{C}\cap L^\infty_{\mathrm{loc}},\ v\not\equiv 0,\ -\Delta v=g(v)\text{ in }\mathscr{D}'.$$

[In some cases, any solution v is automatically in L^∞_{loc} (see Theorem 2.3).]

Proof. Fix $\phi\in C_0^\infty$. We see easily by dominated convergence that as $t\to 0$

$$\int[G(u+t\phi)-G(u)][u\neq 0]=t\int[g(u)\cdot\phi][u\neq 0]+o(t)\,. \tag{2.22}$$

Here we use (2.8). Also, we have that

$$\int|G(u+t\phi)-G(u)|[u=0]\leqq Ct\int|\phi|[u=0]+o(t)\,. \tag{2.23}$$

From (2.14), and using (2.22) and (2.23) we deduce that, for $|t|$ small enough,

$$\tfrac{1}{2}\int|\nabla(u+t\phi)|^2\geqq T\{\textstyle\int G(u+t\phi)\}^{1-2/d}\geqq T\{1+t\int g(u)\cdot\phi-Ct\int|\phi|[u=0]+o(t)\}^{1-2/d}$$

$$\geqq T\left\{1+t\left(\frac{d-2}{d}\right)\int g(u)\cdot\phi-Ct\left(\frac{d-2}{d}\right)\int|\phi|[u=0]+o(t)\right\}.$$

Consequently,

$$\left|\int\nabla u\cdot\nabla\phi-T\left(\frac{d-2}{d}\right)\int g(u)\cdot\phi\right|\leqq C\int|\phi|[u=0]\,,\qquad\text{all }\phi\in C_0^\infty\,.$$

We deduce from the Riesz representation theorem that there exists some $h\in L^\infty$ such that

$$-\Delta u=T\left(\frac{d-2}{d}\right)g(u)+h[u=0]\text{ in }\mathscr{D}'.$$

Finally, we have $u\in L^q_{\mathrm{loc}}$ (by (2.8)) and $g(u)\in L^{q/(q-1)}_{\mathrm{loc}}$. We deduce from the elliptic regularity theory that $u\in W^{2,q/(q-1)}_{\mathrm{loc}}$ [since $q/(q-1)>1$]. Therefore $\Delta u=0$ a.e. on the set $[u=0]$ (see [19] or [11]). Hence we have proved that

$$-\Delta u = T\left(\frac{d-2}{d}\right) g(u) \text{ in } \mathscr{D}',$$

and therefore that $\underline{u}(x)=u(\theta x)$ satisfies (2.21) with $\theta^2 = d[(d-2)T]^{-1}$. In order to complete the proof of Theorem 2.1 we must establish Pohozaev's identity [18] in a setting slightly more general than usual.

Lemma 2.4. *Assume $G \in C^1(\mathbb{R}^n \setminus \{0\})$ and let $v \in \mathscr{C} \cap L^\infty_{\mathrm{loc}}$ be any solution of* (1.1) *in $\mathscr{D}'$. Then*

$$\int |\nabla v|^2 = \frac{2d}{d-2} \int G(v). \tag{2.24}$$

Proof. Since $v \in L^\infty_{\mathrm{loc}}$, it follows from (1.1) and the elliptic regularity theory that $v \in W^{2,t}_{\mathrm{loc}}$, all $t<\infty$. Note that $\partial G(v)/\partial x_i = g(v) \cdot \partial v/\partial x_i$ in $\mathscr{D}'$. Indeed, choose a smoothing sequence G_k for G so that $G_k \to G$ uniformly on compact sets of $\mathbb{R}^n$ and $g_k = \nabla G_k$ tends to g pointwise on $\mathbb{R}^n \setminus \{0\}$. We have $\partial G_k(v)/\partial x_i = g_k(v) \cdot \partial v/\partial x_i$, and thus, for $\phi \in C_0^\infty$,

$$\int \frac{\partial}{\partial x_i}(G_k(v))\phi = -\int G_k(v) \frac{\partial \phi}{\partial x_i} \to \int G(v) \frac{\partial \phi}{\partial x_i}$$

and

$$\int g_k(v) \cdot \frac{\partial v}{\partial x_i} \phi \to \int g(v) \cdot \frac{\partial v}{\partial x_i} \phi$$

by dominated convergence (recall that $\partial v/\partial x_i = 0$ a.e. on the set $[v=0]$). Next we multiply Eq. (1.1) by $\phi \sum_i x_i \partial v/\partial x_i$, where $\phi \in C_0^\infty$. Note that

$$\int \phi g(v) \cdot \sum_i x_i \frac{\partial v}{\partial x_i} = -d \int G(v)\phi - \int G(v) \sum_i x_i \frac{\partial \phi}{\partial x_i},$$

while

$$-\int \phi \Delta v \cdot \sum_i x_i \frac{\partial v}{\partial x_i} = \left(1 - \frac{d}{2}\right) \int \left\{ \phi |\nabla v|^2 + \sum_{i,j} \frac{\partial \phi}{\partial x_j} x_i \frac{\partial v}{\partial x_i} \cdot \frac{\partial v}{\partial x_j} \right\} - \frac{1}{2} \int \sum_i \frac{\partial \phi}{\partial x_i} x_i |\nabla v|^2.$$

Finally we choose $\phi(x) = \phi_n(x) = \phi_1(x/n)$, where ϕ_1 is any function in C_0^∞ such that $\phi_1(x)=1$ for $|x|<1$ and $\phi_1(x)=0$ for $|x|>2$. As $n \to \infty$ we obtain (2.24). □

Proof of Theorem 2.2 Concluded. We have

$$\int |\nabla \underline{u}|^2 = \frac{2d}{(d-2)} \int G(\underline{u}), \quad \int |\nabla v|^2 = \frac{2d}{(d-2)} \int G(v),$$

and on the other hand we also have

$$\tfrac{1}{2} \int |\nabla \underline{u}|^2 = T\left[\int G(\underline{u})\right]^{1-2/d}, \quad \tfrac{1}{2} \int |\nabla v|^2 \geqq T\left[\int G(v)\right]^{1-2/d}.$$

Combining these relations we see that $\int G(v) \geqq \int G(\underline{u})$. However,

$$S(v) = \tfrac{1}{2}\int |\nabla v|^2 - \int G(v) = \frac{2}{(d-2)} \int G(v),$$

$$S(\underline{u}) = \tfrac{1}{2}\int |\nabla \underline{u}|^2 - \int G(\underline{u}) = \frac{2}{(d-2)} \int G(\underline{u}),$$

and we obtain $0 < S(\underline{u}) \leqq S(v)$ for all $v \in \mathscr{C} \cap L^\infty_{\text{loc}}$, $v \not\equiv 0$, $-\Delta v = g(v)$ in $\mathscr{D}'$.

C. Regularity and Behavior at Infinity

In this section we shall only assume that $g: \mathbb{R}^n \to \mathbb{R}^n$ is any mapping bounded on bounded sets and such that $g(v) \cdot v \leqq C|v| + C|v|^p$, all $v \in \mathbb{R}^n$.

Theorem 2.3. *Let $v \in \mathscr{C}$ with $g(v) \in L^1_{\text{loc}}$ be any solution of $-\Delta v = g(v)$ in $\mathscr{D}'$. Then*

$$v \in W^{2,q}_{\text{loc}}, \text{ all } q < \infty \text{ (and consequently } v \in C^{1,\alpha}_{\text{loc}} \text{ for all } \alpha < 1) \tag{2.25}$$

and

$$v \in L^\infty \quad \text{with} \quad \lim_{|x| \to \infty} v(x) = 0 . \tag{2.26}$$

Assume, in addition, that

$$g(v) \cdot v \leqq -C|v|^r, \qquad \text{all } v \in \mathbb{R}^n \text{ with } |v| < \delta, \tag{2.27}$$

for some constants $C > 0$, $\delta > 0$, $1 \leqq r \leqq 2$. Then

$$\text{if } r = 2, \ v(x) \text{ decays exponentially as } |x| \to \infty, \tag{2.28}$$

$$\text{if } 1 \leqq r < 2, \ v(x) \text{ has compact support}. \tag{2.29}$$

Proof. By Kato's inequality (see Kato [12]) we have

$$\Delta |v| \geqq \Delta v \cdot \frac{v}{|v|} \text{ in } \mathscr{D}',$$

and thus

$$-\Delta |v| \leqq -\Delta v \cdot \frac{v}{|v|} = \frac{g(v) \cdot v}{|v|} \leqq C + C|v|^{p-1}.$$

Therefore

$$-\Delta |v| + |v| \leqq C + C|v|^{p-1} \leqq C + A|v|, \tag{2.30}$$

where

$$A = C|v|^{p-2}[|v| \geqq 1],$$

so that $A \in L^{d/2}$ and $\mu(\operatorname{Supp} A) < \infty$. We deduce from (2.30) that

$$|v| \leqq Y * (C + A|v|) \leqq C + [Y * (A|v|)],$$

where Y denotes the Yukawa potential (recall that $Y \in L^1$).

Applying Lemma A.1 (in the appendix) with $\alpha = d/(d-2)$ and $\beta = 2d/(d-2)$, we see that

$$\int_B |v|^q < \infty\,, \qquad \text{all } B \text{ with } \mu(B) < \infty, \text{ all } q < \infty\,. \tag{2.31}$$

In order to prove (2.26) we note that

$$-\Delta |v|^2 = -2v \cdot \Delta v - 2|\nabla v|^2 \leqq 2v \cdot g(v)\,. \tag{2.32}$$

Given $\varepsilon > 0$, we have, for some $\delta > 0$,

$$-\Delta |v|^2 + |v|^2 \leqq C|v| + C|v|^p \leqq \varepsilon + C|v|^p[|v| > \delta]\,,$$

and thus

$$|v|^2 \leqq C\varepsilon + (Y * \phi)\,, \tag{2.33}$$

where $\phi = C|v|^p[|v| > \delta]$.

From (2.31) we deduce that $\phi \in L^q$, all $q < \infty$. Since, on the other hand, $Y \in L^t$ for all $1 \leqq t < d/(d-2)$, it follows that $(Y * \phi)(x) \to 0$ as $|x| \to \infty$. Using (2.33) we obtain $v \in L^\infty$ and

$$\limsup_{|x| \to \infty} |v(x)|^2 \leqq C\varepsilon\,.$$

This implies (2.26) since ε is arbitrary.

Therefore we have $g(v) \in L^\infty$ and consequently $v \in W^{2,q}_{\text{loc}}$ for all $q < \infty$.

Finally we assume (2.27). Combining (2.26), (2.27), and (2.32) we see that

$$-\Delta |v|^2 + 2C|v|^r \leqq 0 \quad \text{for} \quad |x| > R\,, \tag{2.34}$$

(R large enough). We easily deduce (2.28) and (2.29) from (2.34) by comparison with radial supersolutions. (When $r = 2$ this is standard, when $1 \leqq r < 2$, see e.g. Benilan-Brezis-Crandall [2]. A systematic survey of available methods for proving compact support can be found in the book of Diaz [23].) □

III. The Two-Dimensional Case

Let $G : \mathbb{R}^n \to \mathbb{R}$ be continuous with $G(0) = 0$, and $G \in C^1(\mathbb{R}^n \setminus \{0\})$. We set

$$g(v) = \begin{cases} \nabla G(v) & \text{if} \quad v \neq 0 \\ 0 & \text{if} \quad v = 0\,. \end{cases}$$

We make the following assumptions

$$G(v) < 0 \quad \text{for} \quad 0 < |v| \leqq \varepsilon \quad \text{for some} \quad \varepsilon > 0\,, \tag{3.1}$$

$$G(v_0) > 0 \quad \text{for some } v_0\,, \tag{3.2}$$

$$|g(v)| \leqq C + C|v|^{p-1}\,, \quad \text{for all } v, \text{ for some } 1 < p < \infty\,. \tag{3.3}$$

The class $\mathscr{C}$ of functions is given in (1.6).

Theorem 3.1. *Assume* (3.1), (3.2), (3.3). *Then*

$$T = \operatorname{Inf}\{\tfrac{1}{2}\textstyle\int |\nabla v|^2 \,|\, v \in \mathscr{C},\ v \not\equiv 0,\ \int G(v) \geqq 0\} \tag{3.4}$$

is achieved by some $u\in\mathscr{C}$, $u\not\equiv 0$ *such that* $\int G(u)=0$. *Moreover u satisfies*

$$-\Delta u=g(u) \text{ in } \mathscr{D}', \tag{3.5}$$

and

$$0<S(u)\leqq S(v), \tag{3.6}$$

for all $v\in\mathscr{C}$ *such that* $v\not\equiv 0$ *and* $-\Delta v=g(v)$ *in* $\mathscr{D}'$.

It is important to note that Theorem 3.1 states that the unique minimizer of $S(u)$ in the set of functions that are in $\mathscr{C}$ and that satisfy (1.1) is, in fact, $u\equiv 0$. The existence of this trivial solution of (1.1) of lowest action is special to $d=2$. It is the chief difficulty in the two-dimensional case for the obvious reason that the minimum of $\int|\nabla u|^2$ with $\int G(u)=0$ would be $u\equiv 0$. Therefore we must impose the extra condition $u\not\equiv 0$. (Independently, Keller [21] introduced the $u\not\equiv 0$ constraint, but for $d\geqq 3$. Berestycki et al. [3, 4] used it for $d=2$.)

We do not have a general result (as in the $d\geqq 3$ case) for the existence of a minimum in (3.4) without assuming the differentiability of G on $\mathbb{R}^n\setminus\{0\}$. However, if we assume that for some $\varepsilon>0$,

$$\sup_{0\leqq t\leqq 1}|G(tv)|\leqq C|G(v)|$$

for all $|v|<\varepsilon$, then we can prove the existence of a minimizing u for (3.4) under the assumptions (3.1), (3.2) and $|v|^{-p}G(v)\to 0$, as $|v|\to\infty$ for some fixed $p<\infty$.

Proof. Let $\{u^j\}$ be a minimizing sequence for (3.4). Note that $\mu([|u^j|>\varepsilon])>0$, since $u^j\not\equiv 0$ and $\int G(u^j)\geqq 0$. On the other hand the expression $\int|\nabla u|^2$ is invariant under scaling. Thus we may always assume that

$$\mu([|u^j|>\varepsilon])=1\,. \tag{3.7}$$

Also, after a shift, we may assume that

$$\mu(B\cap[|u^j|>\varepsilon/2])\geqq\alpha>0\,, \tag{3.8}$$

where B is the unit ball (the argument is the same as for $d\geqq 3$).

The following lemma is needed in the proof of Theorem 3.1.

Lemma 3.1. *We have that*

$$\int|u^j|^q[|u^j|>\varepsilon]\leqq C_q,\qquad \text{all } q<\infty,\ \text{all } j\,. \tag{3.9}$$

Proof. First we claim that

$$\begin{gathered}\|\phi\|_q\leqq C_q\|\nabla\phi\|_2\,\mu(\operatorname{Supp}\phi)^{1/q},\quad \text{for all}\quad 1\leqq q<\infty\,,\\ \text{all } \phi\in L^1_{\text{loc}},\ \nabla\phi\in L^2,\ \mu(\operatorname{Supp}\phi)<\infty\,.\end{gathered} \tag{3.10}$$

The conclusion of Lemma 3.1 follows by choosing $\phi=(|u^j|-\varepsilon)_+$ in (3.10), and we obtain $\|(|u^j|-\varepsilon)_+\|_q\leqq C_q\|\nabla u^j\|_2$, which implies (3.9). For the proof of (3.10) we argue as follows:

Step (i). We start with the well-known inequality

$$\|\phi\|_2\leqq C\|\nabla\phi\|_1\,,\qquad \text{all } \phi\in C^1_0\,. \tag{3.11}$$

(See e.g. Nirenberg [17].)

Step (ii). Inserting $\phi^2, \phi^3, \dots, \phi^n, \dots$ in (3.11) and interpolating, we find

$$\|\phi\|_q \leqq C_q \|\nabla\phi\|_1^\alpha \|\nabla\phi\|_2^{1-\alpha}, \tag{3.12}$$

with $\alpha = 2/q$, $q < \infty$, all $\phi \in C_0^1$.

Step (iii). Smoothing by convolution, we see that (3.12) holds when $\phi \in L^\infty$, $\nabla\phi \in L^2$ and ϕ has compact support.

Step (iv). Inequality (3.12) still holds when $\phi \in L^\infty$, $\nabla\phi \in L^2$ and $\mu(\text{Supp}\phi) < \infty$. Indeed use step (iii) with $(\chi_n\phi)$, where $\chi_n(x) = \chi_1(x/n)$ and $\chi_1 \in C^1$ with $\chi_1(x) = 1$ for $|x| < 1$. Note that $\|\phi\nabla\chi_n\|_1 \to 0$ and $\|\phi\nabla\chi_n\|_2 \to 0$.

Step (v). Inequality (3.12) is valid for $\phi \in L^1_{\text{loc}}$, $\nabla\phi \in L^2$ and $\mu(\text{Supp}\phi) < \infty$. Indeed, we can use Step (iv) on truncated ϕ's.

Step (vi). We obtain (3.10) from (3.12) by the Cauchy-Schwarz inequality. □

Returning now to Theorem 3.1, we deduce from Lemma 3.1 that $\|u^j\|_{L^q(Q)} \leqq C(q, Q)$, all $q < \infty$, all Q bounded. We may now extract a subsequence (still denoted by $\{u^j\}$) such that

$$u^j \to u \text{ in } L^q(Q), \text{ all } q < \infty, \text{ all } Q \text{ bounded},$$
$$u^j \to u \text{ a.e. on } \mathbb{R}^2,$$
$$\nabla u^j \rightharpoonup \nabla u \text{ weakly in } L^2(\mathbb{R}^2),$$
$$\mu(B \cap [|u| \geqq \varepsilon/2]) \geqq \alpha > 0 \text{ (in particular } u \not\equiv 0),$$
$$\mu([|u| \geqq \varepsilon]) \leqq 1.$$

Moreover, we have $G(u) \in L^1$. Indeed, writing $G = G_+ - G_-$, we have that

$$\int G_+(u^j) = \int G_+(u^j)[|u^j| > \varepsilon] \leqq \int C|u^j|^p[|u^j| > \varepsilon] \leqq C$$

[here we use (3.3)]. On the other hand $\int G_-(u^j) \leqq \int G_+(u^j)$, and therefore $\int |G(u^j)| \leqq C$, which implies, by Fatou, that $G(u) \in L^1$.

We also deduce that $u \in \mathscr{C}$ since $\mu([|u| > \varepsilon]) \leqq 1$ and $G(u) \in L^1$ [here we use assumption (3.1)].

Note that for any set B of finite measure we have

$$\int_B |u^j|^q \leqq C(q, |B|), \qquad \text{all } q < \infty, \tag{3.13}$$

and, also in the limit

$$\int_B |u|^q \leqq C(q, |B|), \qquad \text{all } q < \infty. \tag{3.14}$$

Indeed we may write

$$\int_B |u^j|^q \leqq \int |u^j|^q [|u^j| \geqq \varepsilon] + \int_B \varepsilon^q \leqq C + \varepsilon^q \mu(B).$$

Let us introduce the class of functions from $\mathbb{R}^2 \to \mathbb{R}^n$,

$$\mathscr{K} = \{\phi | \phi \in L^1_{\text{loc}}, \nabla\phi \in L^2, \mu(\text{Supp}\phi) < \infty\}.$$

We recall that by (3.10) $\phi \in \mathscr{K}$ implies $\phi \in L^q(\mathbb{R}^2)$, all q, $1 \leqq q < \infty$.

Lemma 3.2. *Let $\phi \in \mathscr{K}$ be such that*

$$\int G(u+\phi)-G(u)>0. \tag{3.15}$$

Then

$$\int \nabla u \cdot \nabla \phi + \tfrac{1}{2}\int |\nabla \phi|^2 \geqq 0. \tag{3.16}$$

Note that $\int G(u+\phi)$ makes sense for $\phi \in \mathscr{K}$; indeed if $B = \operatorname{Supp}\phi$ then

$$\int |G(u+\phi)| \leqq \int_B |G(u+\phi)| + \int_{\sim B} |G(u)| \leqq \int_B (C + C|u+\phi|^p) + \int |G(u)| < \infty .$$

Proof. We have, with $B = \operatorname{Supp}\phi$,

$$\int G(u^j+\phi) = \int_B G(u^j+\phi) + \int_{\sim B} G(u^j) \geqq \int_B [G(u^j+\phi) - G(u^j)] \to \int_B G(u+\phi) - G(u).$$

For the last assertion we note that

$$G(u^j+\phi) - G(u^j) \to G(u+\phi) - G(u) \text{ a.e.}.$$

On the other hand, if $A \subset B$ we have

$$\int_A |G(u^j+\phi) - G(u^j)| \leqq \int_A (C + C|u^j|^p + C|\phi|^p) \leqq C\mu(A) + C\mu(A)^{1/2}\left(\int_B |u^j|^{2p} + \int_B |\phi|^{2p}\right)^{1/2}$$
$$\leqq C\mu(A) + C\mu(A)^{1/2},$$

and the last term can be made arbitrarily small by choosing $\mu(A)$ small enough. We conclude the proof by Egorov's or Vitali's lemma. Thus, if (3.15) holds, we have $\int G(u^j+\phi)>0$ for j large enough, and therefore $\frac{1}{2}\int |\nabla(u^j+\phi)|^2 \geqq T$. Since $\frac{1}{2}\int |\nabla u^j|^2 \to T$, we obtain (3.16) in the limit. □

Lemma 3.3. *There is a constant C_1 (depending only on G) such that if $\phi \in \mathscr{K}$, then*

$$\left(\int g(u)\cdot\phi - C_1 \int |\phi|\,[u=0] > 0\right) \Longrightarrow \left(\int \nabla u \cdot \nabla \phi \geqq 0\right). \tag{3.17}$$

Note that $\int g(u)\phi$ makes sense since $g(u) \in L^2(B)$ and $\phi \in L^2(B)$ (here $B = \operatorname{Supp}\phi$). On the other hand, $\int |\phi|\,[u=0]$ also makes sense since $\phi \in L^1(B)$.

Proof. By dominated convergence we have, as $t \to 0$,

$$\begin{aligned}\int [G(u+t\phi) - G(u)]\,[u \neq 0] &= t\int g(u)\cdot\phi[u\neq 0] + o(t)\\ &= t\int g(u)\cdot\phi + o(t).\end{aligned} \tag{3.18}$$

On the other hand we have

$$\left|\int G(t\phi)\,[u=0]\right| \leqq C_1 t \int |\phi|\,[u=0] + o(t) \tag{3.19}$$

[here we use assumption (3.3) to deduce that $|G(v)| \leqq C_1|v| + C|v|^p$, all v]). Let $\phi \in \mathscr{K}$ be such that

$$\int g(u)\cdot\phi - C_1\int |\phi|\,[u=0] > 0. \tag{3.20}$$

We deduce from (3.18) and (3.19) that $\int [G(u+t\phi) - G(u)] > 0$ for $t>0$ and small enough. Therefore, by Lemma 3.2

$$\int \nabla u \cdot \nabla \phi + \frac{t}{2}\int |\nabla \phi|^2 \geqq 0.$$

As $t\to 0$ we have $\int \nabla u \cdot \nabla \phi \geqq 0$, which is precisely (3.17). □

Lemma 3.4. *Consider the linear functional $L(\phi)=\int \nabla u \cdot \nabla \phi$. There is some $\phi \in \mathscr{K}$ such that $L(\phi)\neq 0$ and $\phi=0$ on $[u=0]$.*

Proof. Assuming the contrary, we should have $L(\phi)=0$, for all $\phi\in\mathscr{K}$ such that $\phi=0$ on $[u=0]$. In particular, taking $\phi=(\phi_1,0,\ldots,0)$, we should have $\int \nabla u_1\cdot\nabla\phi_1=0$, for all $\phi\in\mathscr{K}$ such that $\phi_1=0$ on $[u_1=0]$. We choose $\phi_1=(u_1-\delta)_+$, $\delta>0$. Then, from the above, $\int|\nabla(u_1-\delta)_+|^2=0$, which implies $(u_1-\delta)_+=C$, which in turn implies $(u_1-\delta)_+\equiv 0$ (since $u_1\to 0$ at infinity in the weak sense). Hence, $u_1\leqq\delta$, and thus $u_1\leqq 0$. The same argument applied to each component leads to $u\equiv 0$, which is a contradiction. □

Lemma 3.5. *There is a constant $k\geqq 0$ such that*

$$|\int g(u)\phi-kL(\phi)|\leqq C_1\int|\phi|\,[u=0]\,,\quad \textit{for all}\quad \phi\in\mathscr{K}\,. \tag{3.21}$$

Proof. We fix some $\phi_0\in\mathscr{K}$ such that $L(\phi_0)=-1$ and $\phi_0=0$ on $[u=0]$. (See Lemma 3.4.) Given $\phi\in\mathscr{K}$, note that

$$\psi=\phi+L(\phi)\phi_0+\alpha\phi_0\,,\qquad \alpha>0\,,$$

satisfies

$$L(\psi)=-\alpha<0$$

and, by Lemma 3.3, we have that

$$\int g(u)\cdot[\phi+L(\phi)\phi_0+\alpha\phi_0]-C_1\int|\phi|\,[u=0]\leqq 0$$

(since $\phi_0=0$ on $[u=0]$). As $\alpha\to 0$ we find that

$$\int g(u)\cdot\phi-kL(\phi)-C_1\int|\phi|\,[u=0]\leqq 0\,,\quad \text{for all}\quad \phi\in\mathscr{K}\,,$$

where $k=-\int g(u)\cdot\phi_0\geqq 0$ [by (3.17)]. By considering the two choices $\pm\phi$, we obtain (3.21). □

Lemma 3.6. *For $v\in H^1_{\text{loc}}$ we have that*

$$\partial G(v)/\partial x_i=g(v)\cdot\partial v/\partial x_i\quad \textit{in } \mathscr{D}'\,. \tag{3.22}$$

Note that $g(v)\cdot\partial v/\partial x_i\in L^1_{\text{loc}}$ and $G(v)\in L^1$, so that (3.22) makes sense in $\mathscr{D}'$.

Proof. Choose a smoothing sequence G_k for G so that $G_k\to G$ pointwise on $\mathbb{R}^n$, $g_k=\nabla G_k$ tends to g pointwise on $\mathbb{R}^n\backslash\{0\}$. Moreover,

$$|G_k(v)|\leqq C|v|+C|v|^p\quad \text{and}\quad |g_k(v)|\leqq C+C|v|^{p-1}\,.$$

We have that $\partial G_k(v)/\partial x_i=g_k(v)\cdot\partial v/\partial x_i$, and thus, for $\phi\in C_0^\infty$,

$$\int\phi\partial G_k(v)/\partial x_i=-\int G_k(v)\partial\phi/\partial x_i\to-\int G(v)\partial\phi/\partial x_i\,,$$

and

$$\int\phi g_k(v)\cdot\partial v/\partial x_i\to\int\phi g(v)\cdot\partial v/\partial x_i\,,$$

by dominated convergence (recall that $\partial v/\partial x_i=0$ a.e. on the set $[v=0]$). □

Proof of Theorem 1 Concluded. The linear functional $M(\phi)=\int g(u)\cdot\phi-kL(\phi)$, $\phi\in C_0^\infty$, satisfies, by (3.21), $|M(\phi)|\leqq C_1\|\phi\|_{L^1}$. Thus by the Riesz representation theorem there is some function $h\in L^\infty(\mathbb{R}^2)$, $h:\mathbb{R}^2\to\mathbb{R}^n$, such that $M(\phi)=\int h\cdot\phi$, for all $\phi\in C_0^\infty$. Moreover, by (3.21), we have $|\int h\cdot\phi|\leqq C\int|\phi|\,[u=0]$, for all $\phi\in C_0^\infty$, and hence for all $\phi\in L^1$. Thus, $h=0$ for $u\neq 0$ and, therefore, $g(u)+k\Delta u=[u=0]h$. It follows that $k\neq 0$ (and thus $k>0$), for otherwise $k=0\Rightarrow g(u)\equiv 0\Rightarrow \partial G(u)/\partial x_i=0$ by Lemma 3.6 $\Rightarrow G(u)=C\Rightarrow G(u)\equiv 0$ (since $G(u)\in L^1$) $\Rightarrow$ for a.e. x we have either $u(x)=0$ or $|u(x)|\geqq\varepsilon$. On the other hand, $|u|\in H^1_{\mathrm{loc}}$, and thus it has a mean value property; therefore we would have either $u\equiv 0$ a.e. on $\mathbb{R}^2$ or $|u|\geqq\varepsilon$ a.e. on $\mathbb{R}^2$. Both cases are excluded (since $u\not\equiv 0$). Hence we have proved that $k>0$ and u satisfies $-\Delta u=\frac{1}{k}g(u)+[u=0]h'$ for some $h'\in L^\infty$. It follows from the elliptic regularity theory that $u\in W^{2,q}_{\mathrm{loc}}$, all $q<\infty$, and therefore $\Delta u=0$ a.e. on $[u=0]$. Consequently $h'=0$ a.e. on $[u=0]$, i.e. we have

$$-\Delta u=g(u)/k \quad \text{for some} \quad k>0. \tag{3.23}$$

When $d=2$, Pohozaev's identity (the proof of which is similar to Lemma 2.4) states that $\int G(u)=0$. On the other hand, since $\nabla u^j\rightharpoonup\nabla u$ weakly in L^2, we have, by lower semicontinuity, $\frac{1}{2}\int|\nabla u|^2\leqq T$. Thus, in fact, $\frac{1}{2}\int|\nabla u|^2=T$ and u is a minimizer for (3.4). After scaling we can always assume that u also satisfies $-\Delta u=g(u)$. Finally, if $v\in\mathscr{C}$ satisfies $-\Delta v=g(v)$ in $\mathscr{D}'$, then $v\in L^q_{\mathrm{loc}}$, all $q<\infty$, $\Rightarrow g(v)\in L^q_{\mathrm{loc}}\Rightarrow v\in L^\infty_{\mathrm{loc}}$. By Pohozaev's identity we have $\int G(v)=0$, and thus if $v\not\equiv 0$ we obtain $\frac{1}{2}\int|\nabla v|^2\geqq T$. Therefore,

$$S(v)=\tfrac{1}{2}\int|\nabla v|^2\geqq T=\tfrac{1}{2}\int|\nabla u|^2=S(u). \quad \square$$

Behavior at Infinity

Here we assume only that $g:\mathbb{R}^n\to\mathbb{R}^n$ is any mapping such that for some $p<\infty$,

$$|g(v)|\leqq C+C|v|^{p-1}, \quad \text{for all} \quad v\in\mathbb{R}^n.$$

Theorem 3.2. *Let $v\in\mathscr{C}$ be any solution of (1.1) in $\mathscr{D}'$. Then*

$$\lim_{|x|\to\infty} v(x)=0.$$

Assume in addition that $g(v)\cdot v\leqq -C|v|^r$ for all $v\in\mathbb{R}^n$ with $|v|<\delta$ for some constants $C>0$, $\delta>0$, $1\leqq r\leqq 2$. Then

(i) *if $r=2$, $v(x)$ decays exponentially as $|x|\to\infty$,*

(ii) *if $1\leqq r<2$, $v(x)$ has compact support.*

Proof. For any $\delta>0$ the function $\phi=(|v|-\delta)_+$ satisfies $\phi\in L^1_{\mathrm{loc}}$, $\nabla\phi\in L^2$, $\mu(\mathrm{Supp}\,\phi)<\infty$, and thus, by (3.10), $\phi\in L^q(\mathbb{R}^2)$ for all $q<\infty$. Hence $\int|v|^q\,[|v|>\delta]<\infty$, for all $q<\infty$, all $\delta>0$. We note that

$$-\Delta|v|^2=-2v\cdot\Delta v-2|\nabla v|^2\leqq 2v\cdot g(v)\leqq C|v|+C|v|^p\cdot$$

Given any $\alpha>0$, we have, for some $\delta>0$,

$$-\Delta|v|^2+|v|^2\leqq C|v|+|v|^2+C|v|^p\leqq\alpha+C(|v|^2+|v|^p)\,[|v|>\delta],$$

and thus $|v|^2 \leqq C\alpha + (Y * \psi)$, where Y is the Yukawa (or Bessel) potential and $\psi = C(|v|^2 + |v|^p)[|v| > \delta]$. Thus $\psi \in L^q$, all $1 \leqq q < \infty$. On the other hand $Y \in L^t$ for all $1 \leqq t < \infty$. It follows that $(Y * \psi) \to 0$ as $|x| \to \infty$. Therefore,

$$\limsup_{|x| \to \infty} |v(x)|^2 \leqq C,$$

which implies that $v(x) \to 0$ as $|x| \to \infty$, since α is arbitrary.

The rest of the proof is the same as in Theorem 2.3. □

Appendix

Lemma A.1. *Let* $1 < \alpha < \beta < \infty$ *and* $1/\alpha + 1/\alpha' = 1$. *Let* $Y \in L^\alpha_w$ *(weak* L^α*),* $A \in L^{\alpha'}$ *with* $\mu(\operatorname{Supp} A) < \infty$, *and* $f \in L^\beta$ *with* $A, Y, f \geqq 0$. *We assume that* $f \leqq 1 + Y * (Af)$, *where* $*$ *denotes convolution. Then*

$$\int_B |f|^q < \infty,$$

for all $q < \infty$ *and all* B *of finite measure.*

Proof. Let χ be the characteristic function of $B \cup \operatorname{Supp} A$. We have $\chi f \leqq \chi + \chi[Y * A\chi f]$. Let $g = \chi f$, whence $g \leqq \chi + \chi[Y * (Ag)]$, and $g \in L^\beta$ with $\mu(\operatorname{Supp} g) < \infty$. Let $Q: \phi \mapsto Y * (A\phi)$. Note that Q is a well defined bounded operator from L^γ into L^γ for all $\alpha < \gamma < \infty$, with norm $\leqq C_\gamma \|Y\|_{\alpha,w} \|A\|_{\alpha'}$. We introduce β_1 defined by

$$1/\beta_1 = \begin{cases} 1/\beta - 1/\alpha', & \text{if } \beta < \alpha', \\ 1/(\beta+1), & \text{if } \beta \geqq \alpha'. \end{cases}$$

Note that $\beta_1 > \beta$. We shall prove that $g \in L^\beta \Rightarrow g \in L^{\beta_1}$. Iterating this fact with β replaced by β_1 we find that $g \in L^{\beta_k}$ for an increasing sequence $\beta_k \to \infty$. This will prove the lemma. Write $A = A_1 + A_2$ with $A_1 \in L^\infty$ and A_2 such that $K: \phi \mapsto Y * (A_2 \phi)$ is a bounded operator from L^β into L^β and L^{β_1} into L^{β_1} with norm < 1. We have that

$$g \leqq [\chi + \chi(Y * (A_1 g))] + [Y * (A_2 g)] \equiv h + Kg.$$

Note that $h \in L^{\beta_1}$. We have that

$$g \leqq \sum_{j=1}^{m} K^j h + K^{m+1} g.$$

$K^j h$ is a norm convergent series in L^{β_1} while $K^{m+1} g \to 0$ in L^β. Thus $g \in L^{\beta_1}$. □

Lemma A.1 is closely related to, and in fact implies some results in [8].

References

1. Affleck, I.: Two dimensional disorder in the presence of a uniform magnetic field. J. Phys. C **16**, 5839–5848 (1983)
2. Benilan, Ph., Brezis, H., Crandall, M.: A semilinear equation in $L^1(\mathbb{R}^N)$. Ann. Scuola Norm. Sup. Pisa **2**, 523–555 (1975)

3. Berestycki, H., Gallouet, Th., Kavian, O.: Equations de champs scalaires Euclidiens non linéaires dans le plan. Compt. Rend. Acad. Sci. **297**, 307–310 (1983)
4. Berestycki, H., Gallouet, Th., Kavian, O.: Semilinear elliptic problems in $\mathbb{R}^2$ (in preparation)
5. Berestycki, H., Lions, P.-L.: Existence of stationary states of non-linear scalar field equations. In: Bifurcation phenomena in mathematical physics and related topics. Bardos, C., Bessis, D. (eds.). Proc. NATO ASI, Cargese, 1979, Reidel, 1980
6. Berestycki, H., Lions, P.-L.: Nonlinear scalar field equations. I. Existence of a ground state. **84**, 313–345 (1983). See also II: Existence of infinitely many solutions. Arch. Rat. Mech. Anal. **84**, 347–375 (1983). See also An O.D.E. approach to the existence of positive solutions for semilinear problems in $\mathbb{R}^N$ (with L.A. Peletier). Ind. Univ. Math. J. **30**, 141–157 (1981). See also Une méthode locale pour l'existence de solutions positives de problèmes semilinéaires elliptiques dans $\mathbb{R}^N$. J. Anal. Math. **38**, 144–187 (1980)
7. Berestycki, H., Lions, P.-L.: Existence d'états multiples dans les équations de champs scalaires non linéaires dans le cas de masse nulle. Compt. Rend. Acad. Sci. **297**, **I**, 267–270 (1983)
8. Brezis, H., Kato, T.: Remarks on the Schrödinger operator with singular complex potentials. J. Math. Pures Appl. **58**, 137–151 (1979)
9. Brezis, H., Lieb, E.H.: A relation between pointwise convergence of functions and convergence of functionals. Proc. Am. Math. Soc. **88**, 486–490 (1983)
10. Coleman, S., Glaser, V., Martin, A.: Action minima among solutions to a class of Euclidean scalar field equation. Commun. Math. Phys. **58**, 211–221 (1978)
11. Gilbarg, D., Trudinger, N.S.: Elliptic partial differential equations of second order. Berlin, Heidelberg, New York: Springer 1977
12. Kato, T.: Schrödinger operators with singular potentials. Israel J. Math. **13**, 135–148 (1972)
13. Lieb, E.H.: Some vector field equations. In: Differential equations. Proc. of the Conference Held at the University of Alabama in Birmingham, USA, March 1983, Knowles, I., Lewis, R. (eds.). Math. Studies Series, Vol. 92. Amsterdam: North-Holland 1984
14. Lieb, E.H.: On the lowest eigenvalue of the Laplacian for the intersection of two domains. Invent. Math. **74**, 441–448 (1983)
15. Lions, P.-L.: Principe de concentration-compacité en calcul des variations. Compt. Rend. Acad. Sci. **294**, 261–264 (1982)
16. Lions, P.-L.: The concentration-compactness principle in the calculus of variations: The locally compact case, Parts I and II. Ann. Inst. H. Poincaré, Anal. Non-lin. (submitted)
17. Nirenberg, L.: On elliptic partial differential equations. Ann. Scuola Norm. Sup. Pisa **13**, 115–162 (1959)
18. Pohozaev, S.I.: Eigenfunctions of the equation $\Delta u + \lambda f(u) = 0$. Sov. Math. Dokl. **6**, 1408–1411 (1965)
19. Stampacchia, G.: Equations élliptiques du second ordre à coéfficients discontinus. Montréal: Presses de l'Université de Montréal 1966
20. Strauss, W.A.: Existence of solitary waves in higher dimensions. Commun. Math. Phys. **55**, 149–162 (1977)
21. Keller, C.: Large-time asymptotic behavior of solutions of nonlinear wave equations perturbed from a stationary ground state. Commun. Partial Diff. Equations **8**, 1013–1099 (1983)
22. Strauss, W.A., Vázquez, L.: Existence of localized solutions for certain model field theories. J. Math. Phys. **22**, 1005–1009 (1981)
23. Diaz, J.I.: Nonlinear partial differential equations and free boundaries. London: Pitman (in preparation)

Communicated by A. Jaffe

Received March 30, 1984; in revised form May 18, 1984

Reprinted from JOURNAL OF FUNCTIONAL ANALYSIS
Vol. 62, No. 1, June 1, 1985

Printed in Belgium

Sobolev Inequalities with Remainder Terms

HAÏM BREZIS

*Département de Mathématiques, Université Paris VI,
4, Place Jussieu, 75230 Paris, Cedex 05, France*

AND

ELLIOTT H. LIEB*

*Departments of Mathematics and Physics, Princeton University,
Princeton, New Jersey 08544*

Communicated by the Editors

Received September 14, 1984

The usual Sobolev inequality in $\mathbb{R}^n$, $n \geqslant 3$, asserts that $\|\nabla f\|_2^2 \geqslant S_n \|f\|_{2^*}^2$, with S_n being the sharp constant. This paper is concerned, instead, with functions restricted to bounded domains $\Omega \subset \mathbb{R}^n$. Two kinds of inequalities are established: (i) If $f=0$ on $\partial\Omega$, then $\|\nabla f\|_2^2 \geqslant S_n \|f\|_{2^*}^2 + C(\Omega)\|f\|_{p,w}^2$ with $p = 2^*/2$ and $\|\nabla f\|_2^2 \geqslant S_n \|f\|_{2^*}^2 + D(\Omega)\|\nabla f\|_{q,w}^2$ with $q = n/(n-1)$. (ii) If $f \neq 0$ on $\partial\Omega$, then $\|\nabla f\|_2 + C(\Omega)\|f\|_{q,\partial\Omega} \geqslant S_n^{1/2}\|f\|_{2^*}$ with $q = 2(n-1)/(n-2)$. Some further results and open problems in this area are also presented.

I. INTRODUCTION

The usual Sobolev inequality in $\mathbb{R}^n$, $n \geqslant 3$, for the L^2 norm of the gradient is

$$\|\nabla f\|_2^2 \geqslant S_n \|f\|_{2^*}^2, \qquad 2^* = 2n/(n-2), \tag{1.1}$$

for all functions f with $\nabla f \in L^2$ and with f vanishing at infinity in the weak sense that $\operatorname{meas}\{x \mid |f(x)| > a\} < \infty$ for all $a > 0$ (see [12]). The sharp constant S_n is known to be

$$S_n = \pi n(n-2)[\Gamma(n/2)/\Gamma(n)]^{2/n}. \tag{1.2}$$

* Work partially supported by U.S. National Science Foundation Grant PHY-8116101-A02.

The constant S_n is achieved in (1.1) if and only if

$$f(x) = a[\varepsilon^2 + |x-y|^2]^{(2-n)/2} \tag{1.3}$$

for some $a \in \mathbb{C}$, $\varepsilon \neq 0$, and $y \in \mathbb{R}^n$ [1, 2, 6, 7, 9, 11].

In this paper we consider appropriate modifications of (1.1) when $\mathbb{R}^n$ is replaced by a *bounded* domain $\Omega \subset \mathbb{R}^n$. There are two main problems:

PROBLEM A. If $f = 0$ on $\partial\Omega$, then (1.1) still holds (with L^p norms in Ω, of course), since f can be extended to be zero outside of Ω. In this case (1.1) becomes a strict inequality when $f \neq 0$ (in view of (1.3)). However, S_n is still the sharp constant in (1.1) (since $\|\nabla f\|_2/\|f\|_{2^*}$ is scale invariant). Our goal, in this case, is to give a lower bound to the difference of the two sides in (1.1) for $f \in H_0^1(\Omega)$. In Section II we shall prove the following inequalities (1.4) and (1.6):

$$\|\nabla f\|_2^2 \geqslant S_n \|f\|_{2^*}^2 + C(\Omega) \|f\|_{p,w}^2, \tag{1.4}$$

where $C(\Omega)$ depends on Ω (and n), $p = n/(n-2) = 2^*/2$, and w denotes the weak L^p norm defined by

$$\|f\|_{p,w} = \sup_A |A|^{-1/p'} \int_A |f(x)|\, dx,$$

with A being a set of finite measure $|A|$.

The inequality (1.4) was motivated by the weaker inequality in [3],

$$\|\nabla f\|_2^2 \geqslant S_n \|f\|_{2^*}^2 + C_p(\Omega) \|f\|_p^2, \tag{1.5}$$

which holds for all $p < n/(n-2)$ (with $C_p(\Omega) \to 0$ as $p \to n/(n-2)$). The proof of (1.5) in [3] was very indirect compared to the proof of (1.4) given here. Inequality (1.4) is best possible in the sense that (1.5) cannot hold with $p = n/(n-2)$; this can be shown by taking the f in (1.3), applying a cutoff function to make f vanish on the boundary, and then expanding the integrals (as in [3]) near $\varepsilon = 0$.

An inequality stronger than (1.4), and involving the gradient norm is

$$\|\nabla f\|_2^2 \geqslant S_n \|f\|_{2^*}^2 + D(\Omega) \|\nabla f\|_{q,w}^2, \tag{1.6}$$

with $q = n/(n-1)$. (The reason that (1.6) is stronger than (1.4) is that the Sobolev inequality has an extension to the weak norms, by Young's inequalities in weak L^p spaces.)

Among the open questions concerning (1.4)–(1.6) are the following:

(a) What are the sharp constants in (1.4)–(1.6)? Are they achieved? Except in one case, they are not known, even for a ball. If $n=3$, Ω is a ball of radius R and $p=2$ in (1.5), then $C_2(\Omega)=\pi^2/(4R^2)$; however, this constant is not achieved [3].

(b) What can replace the right side of (1.4)–(1.6) when Ω is unbounded, e.g., a half-space?

(c) Is there a natural way to bound $\|\nabla f\|_2^2 - S_n \|f\|_{2^*}^2$ from below in terms of the "distance" of f from the set of optimal functions (1.3)?

PROBLEM B. If $f \neq 0$ on $\partial\Omega$, then (1.1) does not hold in Ω (simply take $f=1$ in Ω). Let us assume now that Ω is not only bounded but that $\partial\Omega$ (the boundary of Ω) has enough smoothness. Then (1.1) might be expected to hold if suitable boundary integrals are added to the left side. In Section III we shall prove that for $f=\text{constant} \equiv f(\partial\Omega)$ on $\partial\Omega$

$$\|\nabla f\|_2^2 + E(\Omega)\, |f(\partial\Omega)|^2 \geq S_n \|f\|_{2^*}^2. \tag{1.7}$$

On the other hand, if f is not constant on $\partial\Omega$, then the following two inequalities hold.

$$\|\nabla f\|_2^2 + F(\Omega)\, \|f\|_{H^{1/2}(\partial\Omega)}^2 \geq S_n \|f\|_{2^*}^2, \tag{1.8}$$

$$\|\nabla f\|_2 + G(\Omega)\, \|f\|_{q,\partial\Omega} \geq S_n^{1/2} \|f\|_{2^*}, \tag{1.9}$$

with $q=2(n-1)/(n-2)$, which is sharp. (Note the absence of the exponent 2 in (1.9).)

In addition to the obvious analogues of questions (a)–(c) for Problem B, one can also ask whether (1.9) can be improved to

$$\|\nabla f\|_2^2 + H(\Omega)\, \|f\|_{q,\partial\Omega}^2 \geq S_n \|f\|_{2^*}^2. \tag{1.10}$$

We do not know.

If Ω is a ball of radius R, we shall establish that the sharp constant in (1.7) is $E(\Omega)=\sigma_n R^{n-2}/(n-2)$, where σ_n is the surface area of the ball of unit radius in $\mathbb{R}^n$. With this $E(\Omega)$, (1.7) is a strict inequality. Given this fact, one suspects (in view of the solution to Problem A) that some term could be added to the right side of (1.7). However, such a term cannot be any $L^p(\Omega)$ norm of f, as will be shown.

To conclude this Introduction, let us mention two related inequalities. First, if one is willing to replace S_n on the right side of (1.10) by the smaller constant $2^{-2/n}S_n$, then for a ball one can obtain the inequality

$$\int |\nabla f|^2 + I(\Omega)\, \|f\|_{2,\partial\Omega}^2 \geq 2^{-2/n} S_n \|f\|_{2^*}^2. \tag{1.11}$$

This is proved in Section III. Inequalities related to (1.11) were derived by Cherrier [4] for general manifolds.

Second, one can consider the doubly weighted Hardy–Littlewood–Sobolev inequality [7, 10] which in some sense is the dual of (1.1), namely,

$$\left|\iint f(x)\, f(y)\, |x-y|^{-\lambda}\, |x|^{-\alpha}\, |y|^{-\alpha}\, dx\, dy\right| \leqslant P_{\alpha,\lambda,n}\, \|f\|_{p}^{2}, \tag{1.12}$$

with $p' = 2n/(\lambda + 2\alpha)$, $0 < \lambda < n$, $0 \leqslant \alpha < n/p'$. If f is restricted to have support in a bounded domain Ω and if P is (by definition) the sharp constant in $\mathbb{R}^n$, one should expect to be able to add some additional term to the left side of (1.12). When $p = 2$ this is indeed possible, and the additional term is

$$J_n\, |\Omega|^{-\lambda/n} \left\{\int f(x)\, |x|^{-\alpha}\, dx\right\}^2. \tag{1.13}$$

This was proved in [5] for $n = 3$, $\lambda = 2$, $\alpha = \frac{1}{2}$, and Ω being a ball, but the method easily extends (for a ball) to other n, λ. The result (1.13) further extends to general Ω (with the same constant J_n) by using the Riesz rearrangement inequality. On the other hand, when $p \neq 2$, it does not seem to be easy to find the additional term on the left side of (1.12): at least we have not succeeded in doing so. This is an open problem. In particular, in Section III we prove that when $p = \frac{6}{5}$, $n = 3$, $\lambda = 1$, $\alpha = 0$, one cannot even add $\|f\|_1^2$ to the left side of (1.12).

II. Proof of Inequalities (1.4) and (1.6)

Proof of Inequality (1.4). By the rearrangement inequality for the L^2 norm of the gradient we have

$$\|\nabla f^*\|_2 \leqslant \|\nabla f\|_2 \tag{2.1}$$

(see, e.g., [8]); in addition we have

$$\begin{aligned} \|f^*\|_{2^*} &= \|f\|_{2^*}, \\ \|f^*\|_{p,w} &= \|f\|_{p,w}. \end{aligned} \tag{2.2}$$

Here, f^* denotes the symmetric decreasing rearrangement of the function f extended to be zero outside Ω. Therefore, it suffices to consider the case in which Ω is a ball of radius R (chosen to have the same volume as the original domain) and f is symmetric decreasing.

Let $g \in L^{\infty}(\Omega)$ and define u to be the solution of

$$\begin{aligned} \Delta u &= g \quad \text{in} \quad \Omega, \\ u &= 0 \quad \text{on} \quad \partial\Omega. \end{aligned} \tag{2.3}$$

Let

$$\phi(x) = \begin{cases} f(x) + u(x) + \|u\|_{\infty} & \text{in} \quad \Omega, \\ \|u\|_{\infty}(R/|x|)^{n-2} & \text{in} \quad \Omega^c. \end{cases} \tag{2.4}$$

The Sobolev inequality in all of $\mathbb{R}^n$ applied to ϕ yields

$$\int_{\Omega} |\nabla(f+u)|^2 + \|u\|_{\infty}^2 R^{n-2}(n-2)\,\sigma_n \geqslant S_n \|f\|_{2^*}^2 \tag{2.5}$$

since $f \geqslant 0$ and $u + \|u\|_{\infty} \geqslant 0$. Here

$$\sigma_n = 2(\pi)^{n/2}/\Gamma(n/2)$$

is the surface area of the unit ball in $\mathbb{R}^n$. Therefore, we find

$$\int |\nabla f|^2 - 2\int fg + \int |\nabla u|^2 + k\,\|u\|_{\infty}^2 \geqslant S_n \|f\|_{2^*}^2, \tag{2.6}$$

where $k = R^{n-2}(n-2)\,\sigma_n$. Replacing g by λg and u by λu and optimizing with respect to λ we obtain

$$\int |\nabla f|^2 \geqslant S_n \|f\|_{2^*}^2 + \left(\int fg\right)^2 \Big/ \left[\int |\nabla u|^2 + k\,\|u\|_{\infty}^2\right]. \tag{2.7}$$

In inequality (2.7) we can obviously maximize the right side with respect to g. In view of the definition of the weak norm we shall in fact restrict our attention to $g = 1_A$, namely, the characteristic function of some set A in Ω. We shall now establish some simple estimates for all the quantities in (2.7) in which C_n generically denotes constants depending only on n,

$$\int fg = \int_A f, \tag{2.8}$$

$$\int |\nabla u|^2 \leqslant C_n |A|^{1+2/n}, \tag{2.9}$$

$$\|u\|_{\infty} \leqslant C_n |A|^{2/n}. \tag{2.10}$$

Indeed we have, by multiplying (2.3) by u and using Hölder's inequality,

$$\int |\nabla u|^2 = -\int_A u \leqslant \|u\|_{2^*} |A|^{(1/2)+(1/n)}$$
$$\leqslant S_n^{-1/2} \|\nabla u\|_2 |A|^{(1/2)+(1/n)} \tag{2.11}$$

which implies (2.9). Next we have, by comparison with the solution in $\mathbb{R}^n$,

$$\begin{aligned} |u| &\leqslant C_n |x|^{-n+2} * (1_A) \\ &\leqslant C'_n |A|^{2/n} \end{aligned} \tag{2.12}$$

since the function $|x|^{-n+2}$ belongs to $L_w^{n/(n-2)}$. Since $|A| \leqslant |\Omega| = \sigma_n R^n/n$ we obtain

$$\int |\nabla u|^2 + k \|u\|_\infty^2 \leqslant C_n |A|^{4/n} R^{n-2}. \tag{2.13}$$

Hence (1.4) has been proved (for all Ω) with a constant

$$C(\Omega) = C_n |\Omega|^{(2-n)/n}. \tag{2.14}$$

Proof of Inequality (1.6). To a certain extent the previous proof can be imitated except for one important ingredient, namely, the rearrangement technique cannot be used since it is not true that $\|\nabla f\|_{q,w} \leqslant \|\nabla f^*\|_{q,w}$. (However, it is still true that we can replace f by $|f|$ without changing any of the norms in (1.6), and thus we may and still assume that $f \geqslant 0$.) Consequently we have to use a direct approach and the constant $D(\Omega)$ in (1.6) will not depend only on $|\Omega|$; it will in fact depend on the capacity of Ω. It is an open question whether (1.6) holds with $D(\Omega)$ depending only on $|\Omega|$. Our result is that

$$D(\Omega) = C_n/\text{cap}(\Omega). \tag{2.15}$$

We begin as before with (2.3), but (2.4) is replaced by

$$\phi = \begin{cases} f + u + \|u\|_\infty & \text{in } \Omega, \\ \|u\|_\infty v & \text{in } \Omega^c, \end{cases} \tag{2.16}$$

where v is the solution of

$$\begin{aligned} \Delta v &= 0 \quad \text{in } \Omega^c, \\ v &= 1 \quad \text{on } \partial\Omega, \end{aligned} \tag{2.17}$$

with $v \to 0$ at infinity. By definition,

$$\operatorname{cap}(\Omega) = \int |\nabla v|^2. \tag{2.18}$$

Inequality (2.7) still holds but with the constant k replaced by $k = \operatorname{cap}(\Omega)$. Also we note that (2.7) can be written as

$$\int |\nabla f|^2 \geqslant S_n \|f\|_{2^*}^2 + \left(\int \nabla f \cdot \nabla u\right)^2 \Big/ \left[\int |\nabla u|^2 + k \|u\|_\infty^2\right], \tag{2.19}$$

which holds for any $u \in C_0^\infty(\Omega)$. By density, (2.19) still holds for every u in $H_0^1 \cap L^\infty$ (the reason is that for every such u there is a sequence $u_j \in C_0^\infty(\Omega)$ with $u_j \to u$ in H_0^1 and $\|u_j\|_\infty \to \|u\|_\infty$).

We now choose u to be the solution of (2.3) with

$$g = \frac{\partial}{\partial x_i}\left[\left(\operatorname{sgn} \frac{\partial f}{\partial x_i}\right) 1_A\right]. \tag{2.20}$$

This function u is in L^∞ as we now verify. We can write

$$u = w + h,$$

where w satisfies $\Delta w = g$ in all of $\mathbb{R}^n$, namely,

$$w = C_n |x|^{2-n} * g. \tag{2.21}$$

Clearly h is harmonic and $h = -w$ on $\partial\Omega$. Therefore $\|h\|_\infty \leqslant \|w\|_{\infty, \partial\Omega} \leqslant \|w\|_\infty$ and hence $\|u\|_\infty \leqslant 2\|w\|_\infty$. On the other hand,

$$w = C_n \left(\frac{\partial}{\partial x_i} |x|^{2-n}\right) * \left[\left(\operatorname{sgn} \frac{\partial f}{\partial x_i}\right) 1_A\right],$$

and thus

$$|w| \leqslant C_n (n-2) |x|^{1-n} * 1_A. \tag{2.22}$$

Since $|x|^{1-n} \in L_w^{n/(n-1)}$ we obtain

$$\|u\|_\infty \leqslant 2\|w\|_\infty \leqslant C_n' |A|^{1/n}. \tag{2.23}$$

Next, let us estimate $\int |\nabla u|^2$. Multiplying (2.3) by u we have

$$\int |\nabla u|^2 = \int (\operatorname{sgn} \partial f/\partial x_i) 1_A (\partial u/\partial x_i) \leqslant \left[\int |\nabla u|^2\right]^{1/2} |A|^{1/2}$$

and thus

$$\int |\nabla u|^2 \leqslant |A|. \tag{2.24}$$

Finally, since $f=0$ on $\partial\Omega$,

$$\int \nabla f \cdot \nabla u = -\int f\, \Delta u = \int |\partial f/\partial x_i|\, 1_A. \tag{2.25}$$

Using these estimates in (2.19) we find

$$\int |\nabla f|^2 \geqslant S_n \|f\|_{2^*}^2 + C_n \left(\int_A |\partial f/\partial x_i|\right)^2 \Big/ (\text{cap}(\Omega)\, |A|^{2/n}),$$

since $|A|^{1-(2/n)} \leqslant |\Omega|^{1-(2/n)} \leqslant S_n^{-1}\, \text{cap}(\Omega)$ by Sobolev's inequality applied to the function $\tilde{v} = v$ in Ω^c and $\tilde{v} = 1$ in Ω. This completes the proof of (1.6) with the constant given in (2.15).

III. Proofs of (1.7)–(1.9) and Related Matters

Proof of (1.8). Let us define

$$\phi = \begin{cases} f & \text{in } \Omega, \\ w & \text{in } \Omega^c, \end{cases} \tag{3.1}$$

where w is the harmonic function that vanishes at infinity and agrees with f on $\partial\Omega$. Using ϕ in (1.1) we find

$$\int_\Omega |\nabla f|^2 + \int_{\Omega^c} |\nabla w|^2 \geqslant S_n \|f\|_{2^*}^2. \tag{3.2}$$

On the other hand, we have

$$\int_{\Omega^c} |\nabla w|^2 \sim \|f\|_{H^{1/2}(\partial\Omega)}^2. \tag{3.3}$$

This concludes the proof of (1.8).

Proof of (1.7). Now suppose that f is a constant on $\partial\Omega$. We shall first investigate the case that Ω is a ball of radius R centered at zero. In this case $w(x) = f(\partial\Omega)\, R^{n-2}\, |x|^{2-n}$. Inequality (3.2) then yields (1.7) with

$$\begin{aligned} E(\Omega) = \text{cap}(\Omega) &= \sigma_n R^{n-2}/(n-2) \\ &= \frac{n\,|\Omega|}{n-2} \left\{\frac{\sigma_n}{n\,|\Omega|}\right\}^{2/n}. \end{aligned} \tag{3.4}$$

Furthermore, (1.7) is a strict inequality with this $E(\Omega)$ because the function ϕ in (3.1) is not of the form (1.3). Also, $E(\Omega)$ given by (3.4) is the sharp constant in (3.4). To see this we apply (1.7) with $f=f_\varepsilon$ given by (1.3) with $a=1$ and $y=0=$ center of the ball. We have

$$\int_{\mathbb{R}^n} |\nabla f_\varepsilon|^2 = S_n \, \| f_\varepsilon \|^2_{2^*,\mathbb{R}^n}. \tag{3.5}$$

On the other hand, as $\varepsilon \to 0$

$$\begin{aligned} \int_{\mathbb{R}^n} |\nabla f_\varepsilon|^2 &= \int_\Omega |\nabla f_\varepsilon|^2 + \int_{\Omega^c} |\nabla f_\varepsilon|^2 \\ &= \int_\Omega |\nabla f_\varepsilon|^2 + \operatorname{cap}(\Omega) \, | f_\varepsilon(\partial\Omega)|^2 + o(1). \end{aligned} \tag{3.6}$$

Here we have to note that as $\varepsilon \to 0$ for $|x| > R$

$$f_\varepsilon(x) \to |x|^{2-n}$$

in the appropriate topologies. On the other hand,

$$\int_{\mathbb{R}^n} | f_\varepsilon|^{2^*} - \int_\Omega | f_\varepsilon|^{2^*} = \int_{\Omega^c} | f_\varepsilon|^{2^*} \to C.$$

Thus

$$\| f_\varepsilon \|^2_{2^*,\mathbb{R}^n} = \| f_\varepsilon \|^2_{2^*,\Omega} + o(1). \tag{3.7}$$

This proves that $E(\Omega)$ in (1.7) is greater than or equal to $\operatorname{cap}(\Omega)$ when Ω is a ball, and thus that (3.4) is sharp.

The same calculation with f_ε as above shows that if Ω is a ball there is no inequality of the type

$$\int_\Omega |\nabla f|^2 + \operatorname{cap}(\Omega) \, | f(\partial\Omega)|^2 \geqslant S_n \, \| f \|^2_{2^*} + d \, \| f \|^2_1 \tag{3.8}$$

with $d > 0$, because the additional term $\| f_\varepsilon \|_1 = O(1)$ as $\varepsilon \to 0$.

Now we consider a general domain with $f(\partial\Omega) = \text{constant} = C$. We can assume $C \geqslant 0$ and note that we can also assume $f \geqslant C$ in Ω. (This is so because replacing f by $| f - C| + C \geqslant f$ does not decrease the L^{2^*} norm and leaves $\|\nabla f\|_2$ invariant.) Consider the function $g = f - C \geqslant 0$ which vanishes on $\partial\Omega$ and hence can be extended to be zero on Ω^c. Apply to g the rearrangement inequality for the L^2 norm of the gradient, as was done in

Section II. Finally consider $\tilde{f} \equiv g^* + C$ in the ball Ω^* whose volume is $|\Omega|$. Since $\tilde{f}(\partial\Omega^*) = C = f(\partial\Omega)$ we have

$$\int_{\Omega^*} |\nabla\tilde{f}|^2 + E(\Omega^*)\, |f(\partial\Omega)|^2 \geqslant S_n \, \|\tilde{f}\|^2_{2^*,\Omega^*}.$$

As we remarked, $\|\nabla f\|_2 \geqslant \|\nabla\tilde{f}\|_2$. Also since $f \geqslant C$, it is easy to check that $\|f\|_{2^*} = \|\tilde{f}\|_{2^*}$.

The conclusion to be drawn from this exercise is that (1.7) holds for general Ω with $E(\Omega)$ given by (3.4), namely, $\text{cap}(\Omega^*)$. We also note that (1.7), with this $E(\Omega)$, is strict, since it is strict for a ball.

QUESTION. Is $E(\Omega)$ given by (3.4) the sharp constant in general?

Proof of (1.9). Given f in Ω we consider the harmonic function h in Ω which equals f on $\partial\Omega$. We write

$$f = h + u \tag{3.9}$$

with $u = 0$ on $\partial\Omega$ and thus

$$\int |\nabla u|^2 \geqslant S_n \, \|u\|^2_{2^*}. \tag{3.10}$$

On the one hand

$$\int |\nabla u|^2 = \int |\nabla(f-h)|^2 = \int |\nabla f|^2 - \int |\nabla h|^2 \tag{3.11}$$

(note that $\int_\Omega |\nabla h|^2 = \int_{\partial\Omega} h(\partial h/\partial n) = \int_{\partial\Omega} f(\partial h/\partial n) = \int_\Omega \nabla f \, \nabla h$). On the other hand, by the triangle inequality,

$$\|u\|_{2^*} \geqslant \|f\|_{2^*} - \|h\|_{2^*}. \tag{3.12}$$

Inserting (3.11) and (3.12) in (3.10) we obtain

$$\|\nabla f\|_2 + \|h\|_{2^*} \geqslant S_n^{1/2} \, \|f\|_{2^*}. \tag{3.13}$$

Next we claim that

$$\|h\|_{2^*} \leqslant G(\Omega) \, \|f\|_{q,\partial\Omega} \tag{3.14}$$

with $q = 2(n-1)/(n-2)$, which will complete the proof of (1.9). The proof of (3.14) is a standard duality argument. Indeed, let ψ be the solution of

$$\begin{aligned} \Delta\psi &= Y \quad \text{in} \quad \Omega, \\ \psi &= 0 \quad \text{on} \quad \partial\Omega, \end{aligned} \tag{3.15}$$

where Y is some arbitrary function in L^t. We have, by multiplying by h and integrating by parts,

$$\int_{\Omega} hY = \int_{\partial\Omega} f \frac{\partial\psi}{\partial n}. \tag{3.16}$$

However, the L^p regularity theory shows that $\psi \in W^{2,t}$ with $\|\psi\|_{W^{2,t}(\Omega)} \leqslant C\,\|Y\|_t$. In particular, $\|\nabla\psi\|_{W^{1,t}(\Omega)} \leqslant C\,\|Y\|_t$ and, by trace inequalities,

$$\left\| \frac{\partial\psi}{\partial n} \right\|_{r,\partial\Omega} \leqslant C\,\|Y\|_t, \tag{3.17}$$

where

$$\frac{1}{r} = \frac{n-t}{t(n-1)}. \tag{3.18}$$

Therefore, by (3.16) and Hölder's inequality,

$$\left| \int hY \right| \leqslant C\,\|f\|_{q,\partial\Omega}\,\|Y\|_t, \tag{3.19}$$

where $1/r + 1/q = 1$. Since (3.19) holds for all Y we conclude that

$$\|h\|_{t'} \leqslant C\,\|f\|_{q,\partial\Omega},$$

which coincides with (3.14) since $t' = 2^*$ when $q = 2(n-1)/(n-2)$.

Finally, we claim that there is no inequality of the type (1.9) with $q < 2(n-1)/(n-2)$. Indeed, suppose (1.9) holds with some such q. We choose $f = f_\varepsilon$ as in (1.3) with $a = 1$ and $y \in \partial\Omega$. It is obvious that as $\varepsilon \to 0$

$$\int_{\Omega} |\nabla f_\varepsilon|^2 \Big/ \int_{\mathbb{R}^n} |\nabla f_\varepsilon|^2 = 1/2 + o(1),$$

$$\int_{\Omega} |f_\varepsilon|^{2^*} \Big/ \int_{\mathbb{R}^n} |f_\varepsilon|^{2^*} = 1/2 + o(1),$$

while

$$\int_{\mathbb{R}^n} |\nabla f_\varepsilon|^2 = S_n\,\|f_\varepsilon\|^2_{2^*,\mathbb{R}^n} \qquad \text{and} \qquad \|f_\varepsilon\|_{q,\partial\Omega} / \|f_\varepsilon\|_{2^*} = o(1).$$

This contradicts (1.9).

Remark. The last exercise with f_ε given above shows that it is not possible to apply rearrangement techniques when f is not constant on $\partial\Omega$, even if Ω is a ball. It also shows that there is *no* inequality for all $f \in H^1$ of the type

$$\|\nabla f\|_2^2 + C\,\|f\|_{q,\Omega}^2 \geqslant S_n\,\|f\|_{2^*}^2$$

with $q < 2^*$.

Proof of (1.11). Let Ω be a ball of radius R centered at zero. For simplicity, assume $R = 1$. Define

$$g(x) = \begin{cases} f(x), & |x| \leqslant 1, \\ |x|^{2-n} f(x\,|x|^{-2}), & |x| \geqslant 1, \end{cases} \tag{3.20}$$

and apply the usual Sobolev inequality (1.1) to g. We note (by a change of variables) that

$$\begin{aligned} \int_\Omega g^{2^*} &= \int_{\Omega^c} g^{2^*}, \\ \int_\Omega |\nabla g|^2 &= \int_{\Omega^c} |\nabla g|^2 - (n-2)\,\|f\|_{2,\partial\Omega}^2. \end{aligned} \tag{3.21}$$

Inserting (3.21) into (1.1) yields (1.11) with $I(\Omega) = (n-2)/2$.

Remark on the Hardy–Littlewood–Sobolev Inequality

Consider the inequality (in $\mathbb{R}^3$)

$$I(f) \leqslant P\,\|f\|_{6/5}^2, \tag{3.22}$$

with

$$I(f) = \iint f(x)\,f(y)\,|x-y|^{-1}\,dx\,dy \geqslant 0. \tag{3.23}$$

The sharp constant P is known to be [7]

$$P = 4^{5/3}/[3\pi^{1/3}]. \tag{3.24}$$

Let Ω be a ball of radius one centered at zero and assume that $f = 0$ outside Ω. In this case, (3.22) is strict because the only functions that give equality in (3.22) are of the form [7]

$$f_\varepsilon(x) = a[\varepsilon^2 + |x-y|^2]^{-5/2}. \tag{3.25}$$

For $f=0$ outside Ω, we ask whether (3.22) can be improved to

$$C \, \|f\|_1^2 + I(f) \leqslant P \, \|f\|_{6/5}^2. \tag{3.26}$$

Our conclusion is that (3.26) fails for any $C>0$.

Take $f=\bar{f}_\varepsilon \equiv f_\varepsilon 1_\Omega$ with f_ε given by (3.25) and with $y=0$ and with $a=a_\varepsilon$ chosen so that $\|f_\varepsilon\|_{6/5,\mathbb{R}^3}=1$. The function f_ε satisfies the following (Euler) equation on $\mathbb{R}^3$,

$$\frac{1}{|x|} * f_\varepsilon = P f_\varepsilon^{1/5}. \tag{3.27}$$

However, for $|x|<1$

$$\left(\frac{1}{|x|} * \bar{f}_\varepsilon\right)(x) + K_\varepsilon = \left(\frac{1}{|x|} * f_\varepsilon\right)(x), \tag{3.28}$$

where K_ε is a constant bounded above by $D_\varepsilon = \int_{|x|>1} f_\varepsilon$. Multiply (3.27) by $\bar{f}_\varepsilon$ and integrate over Ω. Then

$$I(\bar{f}_\varepsilon) + T_\varepsilon \, \|\bar{f}_\varepsilon\|_1^2 \geqslant I(\bar{f}_\varepsilon) + K_\varepsilon \int \bar{f}_\varepsilon = P \, \|\bar{f}_\varepsilon\|_{6/5}^{6/5} \geqslant P \, \|\bar{f}_\varepsilon\|_{6/5}^2, \tag{3.29}$$

where $T_\varepsilon = D_\varepsilon / \int \bar{f}_\varepsilon$. From (3.29), we see that (3.26) fails if $C > T_\varepsilon$ for any $\varepsilon > 0$. However, it is obvious that $T_\varepsilon \to 0$ as $\varepsilon \to 0$.

References

1. T. Aubin, Problèmes isopérimétriques et espaces de Sobolev, *C. R. Acad. Sci. Paris* **280A** (1975), 279–281; *J. Diff. Geom.* **11** (1976), 573–598.
2. G. A. Bliss, An integral inequality, *J. London Math. Soc.* **5** (1930), 40–46.
3. H. Brezis and L. Nirenberg, Positive solutions of nonlinear elliptic equations involving critical Sobolev exponents, *Comm. Pure Appl. Math.* **36** (1983), 437–477.
4. P. Cherrier, Problèmes de Neumann nonlinéaires sur les variétés Riemanniennes, *J. Funct. Anal.* **57** (1984), 154–206.
5. I. Daubechies and E. Lieb, One-electron relativistic molecules with Coulomb interaction, *Comm. Math. Phys.* **90** (1983), 497–510.
6. B. Gidas, W. M. Ni, and L. Nirenberg, Symmetry of positive solutions of nonlinear elliptic equations in $\mathbb{R}^n$, *in* "Mathematical Analysis and Applications" (L. Nachbin, Ed.), pp. 370–401, Academic Press, New York, 1981.
7. E. Lieb, Sharp constants in the Hardy–Littlewood–Sobolev and related inequalities, *Ann. of Math.* **118** (1983), 349–374.
8. E. Lieb, Existence and uniqueness of the minimizing solution of Choquard's nonlinear equation, *Stud. Appl. Math.* **57** (1977), 93–105.
9. G. Rosen, Minimum value for c in the Sobolev inequality $\|\phi\|_6 \leqslant c \, \|\nabla\phi\|_2$, *SIAM J. Appl. Math.* **21** (1971), 30–32.

10. E. Stein and G. Weiss, Fractional integrals in n-dimensional Euclidean space, *J. Math. Mech.* **7** (1958), 503–514. See also, G. Hardy and J. Littlewood, Some properties of fractional integrals (1), *Math. Z.* **27** (1928), 565–606.
11. G. Talenti, Best constant in Sobolev inequality, *Ann. Mat. Pura Appl.* **110** (1976), 353–372.
12. H. Brezis and E. Lieb, Minimum action solution of some vector field equations, *Comm. Math. Phys.* **96** (1984), 97–113. See Remark 3 on p. 100.

Invent. math. 102, 179–208 (1990)

Inventiones mathematicae

Gaussian kernels have only Gaussian maximizers*

Elliott H. Lieb

Department of Mathematics, Princeton University, Princeton, NJ 08544, USA

Oblatum 19-XII-1989

Abstract. A Gaussian integral kernel $G(x, y)$ on $\mathbf{R}^n \times \mathbf{R}^n$ is the exponential of a quadratic form in x and y; the Fourier transform kernel is an example. The problem addressed here is to find the sharp bound of G as an operator from $L^p(\mathbf{R}^n)$ to $L^q(\mathbf{R}^n)$ and to prove that the $L^p(\mathbf{R}^n)$ functions that saturate the bound are necessarily Gaussians. This is accomplished generally for $1 < p \leqq q < \infty$ and also for $p > q$ in some special cases. Besides greatly extending previous results in this area, the proof technique is also essentially different from earlier ones. A corollary of these results is a fully multidimensional, multilinear generalization of Young's inequality.

I. Introduction

The classic Hausdorff-Young-Titchmarsh [T] inequality for Fourier integrals states that for $1 \leqq p \leqq 2$ the Fourier transform on $L^p(\mathbf{R}^n)$ is a bounded map into $L^{p'}(\mathbf{R}^n)$ with a bound that is at most 1; here $1/p' + 1/p = 1$. In 1961 Babenko [BA] showed that when p' is an even integer greater than 2 and $n = 1$ the bound is in fact less than 1, and he determined its value. This bound is achieved for Gaussian functions and Babenko states, but does not demonstrate explicitly, that Gaussians are the only functions with this property. Babenko's method was to apply analytic function theory to the Euler-Lagrange equation associated with the maximization problem.

The Fourier integral is but one example of a transform given by a Gaussian integral kernel $G(x, y)$, i.e., the exponential of a quadratic plus linear form in x and y. In the Fourier transform case in $\mathbf{R}^n$ the kernel is $G(x, y) = \exp\{-2i(x, y)\}$. Another well known example in $\mathbf{R}^n$ is the purely real operator $\tilde{\mathscr{G}} = \exp\{t\Delta + 2tx \cdot \nabla\}$ on Gauss space (with measure $d\mu = \exp\{-|x|^2\}\,dx$) investigated by Nelson [N1; N2] as an operator from $L^p(\mathbf{R}^n, \mathrm{d}\mu)$ to $L^q(\mathbf{R}^n, \mathrm{d}\mu)$. In

* Work partially supported by U.S. National Science Foundation grant PHY-85-15288-A03

Invent. Math. *102*, 179–208 (1990)

terms of Lebesgue measure, this amounts to considering the kernel

$$G(x, y) = \exp\left\{ -\frac{1}{q}|x|^2 + \frac{1}{p}|y|^2 - \frac{|y - cx|^2}{(1 - c^2)} \right\}$$

from $L^p(\mathbf{R}^n)$ to $L^q(\mathbf{R}^n)$ for $0 \leqq c = e^{-t} < 1$. Nelson defined the operator $\mathcal{G}$ by $(\mathcal{G}f)(x) = \int G(x, y) f(y)\,dy$ and showed that $\mathcal{G}$ is bounded from $L^p(\mathbf{R}^n)$ to $L^q(\mathbf{R}^n)$ when $p \leqq q$ if and only if $(q - 1)c^2 \leqq p - 1$; he also derived the explicit value of the bound—which again is achieved when f is a Gaussian. This is the famous hypercontractivity theorem. [In [N1] Nelson showed that $\mathcal{G}$ is bounded if c is small enough; Glimm [GL] used this fact plus the spectral gap in the generator to show that $\tilde{\mathcal{G}}$ is a contraction on Gauss space for some still smaller c. Finally Nelson [N2] proved the sharp bound as stated above. In 1976 Neveu [NE] and Brascamp and Lieb [BL] found other proofs, and Simon [SI] found a proof for $p = 2$ and $q = 2, 4, 6, 8 \ldots$. Recently, Carlen and Loss [CL] have used their method of competing symmetries to construct another proof of the hypercontractivity theorem.] However, Nelson's method seems incapable of showing that Gaussians are the *only* maximizers; the proof of this fact, as well as a completely different proof (using rearrangement inequalities) of the hypercontractivity theorem was given by Brascamp and Lieb [BL]. The method in [CL] also yields uniqueness. Nelson's original proof used stochastic integrals and Gaussian processes in $\mathbf{R}^n$ (in fact it even extends to infinite dimensions). Segal [S] showed how to use Minkowski's inequality [HLP] to reduce the $\mathbf{R}^n$ case of Nelson's kernel to the $\mathbf{R}^1$ case; he also showed that $\tilde{\mathcal{G}}$ is a contraction on Gauss space for small c. The $\mathbf{R}^1$ case was simplified by Gross [G] who showed the equivalence of hypercontractivity with logarithmic Sobolev inequalities and built up one-dimensional Gauss measure from two-point measures via the central limit theorem. See the survey by Davies et al. [DGS].

In his important 1975 paper, Beckner [B1; B2] used the Nelson-Gross machinery and the Hermite semigroup to settle the question raised by Babenko. By using the tensor product structure of Fourier transforms and an application of Minkowski's inequality related to, but distinct from, Segal's [S], he reduced the $\mathbf{R}^n$ case to the $\mathbf{R}^1$ case. He also showed that for all $1 \leqq p \leqq 2$ the sharp constant in the Hausdorff-Young-Titchmarsh inequality is given by Gaussian functions—as found by Babenko. However, this method also leaves open the question of whether Gaussian functions are the *only* maximizers.

Since then the Nelson-Gross-Beckner method has been extended to other complex (as distinct from purely real or purely imaginary) Gaussian kernels in $\mathbf{R}^n$ (i.e., the complex Mehler kernel) [C; E; J; W]. In this paper the *general* problem in $\mathbf{R}^n$ in the $p \leqq q$ case will be settled by a *completely different method* and, moreover, *the maximizers will be shown to be Gaussian functions*. Some of the $p > q$ cases will be settled as well. Before discussing the earlier results in detail it is necessary to define the problem more completely.

The most general **Gaussian kernel** on $\mathbf{R}^n \times \mathbf{R}^n$ is

$$G(x, y) = \exp\left\{ -(x, Ax) - (y, By) - 2(x, Dy) + 2\left(L, \begin{pmatrix} x \\ y \end{pmatrix}\right)\right\} \tag{1.1}$$

and its action on complex valued, measurable functions $f\colon \mathbf{R}^n \to \mathbf{C}$, is formally given by

$$(\mathcal{G}f)(x) = \int G(x, y) f(y)\,\mathrm{d}y\ . \tag{1.2}$$

In (1.1) A, B and D are (complex) $n \times n$ matrices with A and B being symmetric while L is a vector in $\mathbf{C}^{2n}$. The Fourier transform corresponds to $A = B = 0$, $L = 0$ and $D = iI$, with I denoting the identity.

Notation. If α and β are vectors in $\mathbf{C}^n$ then $(\alpha, \beta) \equiv \sum_{i=1}^n \alpha_i \beta_i$ and *not* $\sum_{i=1}^n \bar{\alpha}_i \beta_i$. Lebesgue integration over $\mathbf{R}^n$ is denoted simply by $\int dx$ whenever the n in question is clear from the context. The $L^p(\mathbf{R}^n)$ norm of a measurable function f will be denoted by $\|f\|_p$, i.e., $\{\int |f(x)|^p dx\}^{1/p}$. The notation

$$\begin{pmatrix} A & D \\ D^T & B \end{pmatrix} = M + iN \tag{1.3}$$

will also be used, where M and N are real, symmetric $2n \times 2n$ matrices. The sole condition imposed on G is that M is positive semidefinite. G is said to be **nondegenerate** if M is positive definite, while G is said to be **degenerate** if M has a zero eigenvalue. The Fourier transform kernel and Nelson's kernel with $(q-1)c^2 = (p-1)$ are examples of degenerate kernels. The operator $\mathcal{G}$ should perhaps be written $\mathcal{G}_G$, but this will not be done since the pairing of $\mathcal{G}$ and G will always be clear from the context.

The linear operator $\mathcal{G}$ associated to G will be studied as an operator from $L^p(\mathbf{R}^n)$ to $L^q(\mathbf{R}^n)$ for $1 < p < \infty$ and $1 < q < \infty$. (The cases p or $q = 1$ or ∞ can also be analyzed by the methods of this paper but they will be omitted since these cases involve extra technical considerations.) When G is nondegenerate the definition of $\mathcal{G}$ in (1.2) makes sense (by Hölder's inequality) but if G is degenerate then (1.2) is meaningless unless f is also in $L^p(\mathbf{R}^n) \cap L^1(R^n)$. Assuming that $\mathcal{G}$, when restricted to $L^p(\mathbf{R}^n) \cap L^1(\mathbf{R}^n)$, is bounded from $L^p(\mathbf{R}^n)$ to $L^q(\mathbf{R}^n)$ then, for any $f \in L^p(\mathbf{R}^n)$, $\mathcal{G}f \in L^q(\mathbf{R}^n)$ is uniquely defined by taking *any* sequence $f_j \in L^p(\mathbf{R}^n) \cap L^1(\mathbf{R}^n)$ that converges to f in $L^p(\mathbf{R}^n)$ and then noting that $\mathcal{G}f = \lim_{j\to\infty} \mathcal{G}f_j$ is well defined since $\mathcal{G}f_j$ is a Cauchy sequence in $L^q(\mathbf{R}^n)$. This definition is well known and is, in fact, the way that the Fourier transform is defined when $1 < p \leqq 2$.

Associated to G and the numbers p and q with $1 \leqq p \leqq \infty$ and $1 \leqq q \leqq \infty$ is the ratio

$$\mathcal{R}_{p\to q}(f) = \frac{\|\mathcal{G}f\|_q}{\|f\|_p} \tag{1.4}$$

for $f \in L^p(\mathbf{R}^n)$, $f \neq 0$ and, in case G is degenerate, $f \in L^1(\mathbf{R}^n)$ as well. The norm of $\mathcal{G}$ from $L^p(\mathbf{R}^n)$ to $L^q(\mathbf{R}^n)$ is defined to be

$$C_{p\to q} = \sup_f \mathcal{R}_{p\to q}(f) \tag{1.5}$$

in which the supremum is over the class of f's just stated. In case $0 \neq f \in L^p(\mathbf{R}^n)$ and $C_{p\to q} < \infty$ and

$$\|\mathcal{G}f\|_q = C_{p\to q}\|f\|_p$$

(using the above definition of $\mathcal{G}f$ as a limit when G is degenerate) then f is said to be a **maximizer** for $\mathcal{G}$ (or for G). If there is any ambiguity about the G under discussion (e.g., in Theorem 3.3) the notation $\mathcal{R}_{p\to q}(G, f)$ and $C_{p\to q}(G)$ will be used.

Functions from $\mathbf{R}^n$ to $\mathbf{C}$ of the form

$$g(x) = \mu \exp\{-(x, Jx) + (l, x)\} \tag{1.6}$$

with $0 \neq \mu \in \mathbf{C}$, $l \in \mathbf{C}^n$ and J a symmetric $n \times n$ matrix with $\mathrm{Re}(J)$ positive *definite* will be called **Gaussian functions.** In case $L = 0$ in (1.1) or $l = 0$ in (1.6) then G (resp. g) will be called a **centered Gaussian kernel** (resp. **function**). If A, B, D and L in (1.1) are real then G is said to be a **real Gaussian kernel**. Likewise, if J and l (but not necessarily μ) in (1.6) are real then g is said to be a **real Gaussian function.**

A preliminary simplification of G can be made. Without loss of generality it can be assumed that A and B are real matrices because the imaginary part of B can be absorbed into f in (1.4) without changing $\|f\|_p$. The imaginary part of A can be omitted without changing $\|\mathscr{G}f\|_q$. For the same reason the vector L can be assumed to be real. Furthermore, when G is nondegenerate then we can also set L (which is now real) equal to zero. The reason is simply that the affine change of variables $\begin{pmatrix} x \\ y \end{pmatrix} \mapsto \begin{pmatrix} x \\ y \end{pmatrix} - V$, with V being the unique solution of the equation $MV = L$ in $\mathbf{R}^{2n}$, eliminates the real linear term from (1.1) and merely changes $C_{p \to q}$ into $C_{p \to q} \exp\{(L, V)\}$. When G is degenerate, L can also be eliminated in the same way provided $MV = L$ has a solution. Because $\mathrm{Rank}(M) < 2n$ in the degenerate case, such a solution conceivably might not exist, but it turns out that a solution does indeed exist whenever $\mathscr{G}$ is bounded. This is the content of Lemma 2.2 below. Therefore, *without loss of generality, the only G's that need to be studied are those for which*

(*i*) A and B are real, symmetric $n \times n$ matrices,
(*ii*) $L = 0$, i.e., G is centered.

These assumptions will be made in the theorems in this paper.

On the other hand, suppose that the supremum of $\mathscr{R}_{p \to q}(f)$ in (1.5) is taken over Gaussian functions *only* (which are automatically in $L^p(\mathbf{R}^n)$ for every p). Then, according to Lemma 2.3 below, only centered Gaussian functions need be considered in (1.5). This is a considerable simplification that is not altogether obvious and it is important in the application of Theorem 4.1 which states that this restricted supremum is all that need be considered.

The results of this paper can be summarized as follows. Three cases are treated. With the assumptions (i) and (ii) above,

(*A*) D is real and $1 < p < \infty$ and $1 < q < \infty$
(*B*) D is imaginary and either $1 < p \leqq 2$ and $1 < q < \infty$ or else $1 < p < \infty$, and $2 \leqq q < \infty$.
(*C*) D is complex and $1 < p \leqq q < \infty$.

If G is *nondegenerate* then $\mathscr{R}_{p \to q}$ has exactly one maximizer and it is a centered Gaussian function. These are Theorems 3.2, 3.3 and 3.4.

If G is *degenerate* then in all cases

$$C_{p \to q} = \sup_g \mathscr{R}_{p \to q}(g) \,, \tag{1.7}$$

where the supremum is over centered Gaussian functions. This is Theorem 4.1. Furthermore, if the supremum in (1.7) is achieved for some Gaussian function then, when $p < q$, *every maximizer is a Gaussian function*—as Theorems 4.5 shows. Theorem 4.3 gives a sufficient condition for the achievement of the maximum in (1.7) in the degenerate case; in Case (A) it is necessary as well. Thus, Case (A) is settled completely: in the degenerate case with $p \geqq q$ there is no maximizer of any kind, while if $p < q$ all maximizers are Gaussian functions.

In general, the question of the existence of a maximizer in the degenerate case is a subtle one. For the Fourier transform (which is both Case (B) and (C)), *every* function in $L^2(\mathbf{R}^n)$ is a maximizer when $p = q = 2$; on the other hand the seemingly harmless modification of the Fourier transform in 4.2(5) below is bounded but has *no maximizer of any kind* when $q = p' \geqq 2$. When $q = p' > 2$ the Fourier transform on $\mathbf{R}^1$ has a three real parameter family of maximizers, $f(y) = \exp\{-Jy^2 + ly\}$ with $J > 0$ and $l \in \mathbf{C}$. When $p < q$ the convolution kernel $G(x, y) = \exp\{-(x-y)^2\}$ on $\mathbf{R}^1$ has a one real parameter family of maximizers, $f(y) = \exp\{-Jy^2 + ly\}$ with $l \in \mathbf{R}$ and $J = \frac{p'}{q'} - 1$; when $p = q$, G is bounded but there is no maximizer (see 4.2 below). There does not seem to be any simple rule. In simple cases (which include all the standard ones in $\mathbf{R}^n$ and all the cases in $\mathbf{R}^1$) the existence of a Gaussian maximizer in (1.7) can be decided by computation. Otherwise, (1.7) reduces to a complicated algebraic problem and precise conditions are not given here. Moreover it is not even proved that the absence of a Gaussian maximizer in (1.7) precludes the existence of a non-Gaussian maximizer—although a conjecture to this effect is made in 4.4.

All these results extend to Gaussian kernels on $\mathbf{R}^m \times \mathbf{R}^n$, in which A is $m \times m$, B is $n \times n$, D is $m \times n$ and $L \in \mathbf{C}^{m+n}$. The proof is given in Sect. V. This generalization, while it is an easy one, does occur in applications, e.g., the entropy bound for coherent states in [L1].

Multilinear Gaussian forms are discussed in Sect. VI and it is proved there that the methods and results of Sects. II–V carry through for real forms. As an application of the real multilinear result in Sect. 6.1, the *fully multidimensional Young inequality* for K functions (which was left unresolved in [BL], p. 162) is proved in 6.2. The method of proof is, of course, quite different from that in [BL]; there, rearrangement inequalities were used and they were not flexible enough to encompass the fully multidimensional case.

The relationship of the results of this paper to earlier results on Gaussian kernels (beyond [BA; N1; N2; B1; B2]) can be summarized as follows. In 1976 Brascamp and Lieb [BL] found the norm for Case (A) in $\mathbf{R}^n$ (Theorem 7) and proved that Gaussian functions are the unique maximizer in $\mathbf{R}^1$ in the degenerate case (Theorem 13); this latter proof easily extends to $\mathbf{R}^n$ and to the nondegenerate case. In fact, by a simple change of variables (see the proof of Theorem 4.3 below) the $\mathbf{R}^n$ Case (A) reduces to a simple tensor product of $\mathbf{R}^1$ kernels. In 1979 Coifman et al. [C] used Beckner's result and an interpolation technique to deduce the norm for the complex Mehler kernel in $\mathbf{R}^1$ for $q = p' \geqq 2$ (which is in Case (C)). In the same year Weissler [W] extended Nelson's and Beckner's results to the complex Mehler kernel in $\mathbf{R}^1$ with the exception of $2 < p \leqq q < 3$ and $\frac{3}{2} < p \leqq q < 2$. In 1988, Epperson [E] found the norm for the following nondegenerate cases in $\mathbf{R}^1$: Case (C), Case (B), the case $p \geqq 2 \geqq q$. He also found the norm for certain $\mathbf{R}^1$ cases $q < p < 2$ and $2 < q < p$ with sufficiently nondegenerate kernels (Theorem 2.10), and for the $\mathbf{R}^1$ degenerate Case (C) if $A > 0$ and $B > 0$ (corresponding to Theorem 4.3 here).

The only complex cases in $\mathbf{R}^n$ that were known prior to Epperson's work were the simple tensor products of $\mathbf{R}^1$ kernels; these could be analyzed for $p \leqq q$ via Minkowski's inequality, as shown by Beckner [B1; B2]. Epperson was able

Invent. Math. *102*, 179–208 (1990)

to handle the nondegenerate Case (C) for which there is an $n \times n$ complex symmetric matrix W with $\|W\| \leqq 1$ such that $A = W(I - W^2)^{-1}W - \frac{1}{q}I$, $B = (I - W^2)^{-1} - \frac{1}{p'}I$ and $D = W(I - W^2)^{-1}$. Here, I is the identity matrix.

It will be seen from the above summary that all the previous cases, except for Epperson's $\mathbf{R}^1$ cases of $p \geqq 2 \geqq q$ and the special $q < p < 2$ and $2 < q < p$ cases, are covered in the cases (A), (B) and (C) treated in this paper. Moreover cases (A), (B) and (C) are resolved here in full $\mathbf{R}^n$ generality (i.e., not only for simple n-fold tensor products of $\mathbf{R}^1$ kernels). The main methodological point of this paper, however, is that all the previous results, except for [BL] and [BA], ultimately rely on the Nelson-Gross machinery which, while it is natural in its original context of quantum field theory and Gauss measures, is conceptually complicated in the context of general Gaussian kernels with Lebesgue measure. The two settings (Gauss measure and Lebesgue measure) for Gaussian kernels are mathematically equivalent, however, and the choice is a matter of taste. Lebesgue measure is used in this paper because it is felt that it is more natural to retain translation invariance (e.g., in the Fourier transform). Prior to Epperson's work all results in the field, except for [BL] and [BA] came from translating Gauss measure bounds for products of complex $\mathbf{R}^1$ Mehler kernels into $\mathbf{R}^n$ results via Beckner's Minkowski lemma. The proofs here use only Minkowski's inequality and simple facts about analytic functions (which appear to be unrelated to Babenko's use of analyticity—the Euler-Lagrange equation is not used).

Basically there is one idea that runs through Theorems 3.1, 3.3 and 4.5, although the technicalities are different in each. The main idea is to study $\mathscr{G} \otimes \mathscr{G}$ from $L^p(\mathbf{R}^{2n})$ to $L^q(\mathbf{R}^{2n})$ and use Minkowski's inequality. By considering the $\mathscr{G} \otimes \mathscr{G}$ maximizer $F(y_1, y_2) = f\left(\frac{y_1 + y_2}{\sqrt{2}}\right) f\left(\frac{y_1 - y_2}{\sqrt{2}}\right)$, where f is a maximizer for $\mathscr{G}$, it is possible to conclude that f must be a Gaussian. It will be noted that some of the proofs are long, and so it may appear at first that their structure is not really very simple. To a large extent the length is due to the fact that proving uniqueness raises technical considerations that would be absent if only inequalities are proved, e.g., it is not sufficient here to prove the inequalities for a dense set of smooth functions.

Apart from the extension to $\mathbf{R}^n$ (which is handled here in a natural way) the main new theorem in this paper is that a maximizer *must be* a Gaussian, and it is unique in the nondegenerate case. In the degenerate case $C_{p \to q}(G)$ is determined by examining only Gaussian functions and, if a Gaussian maximizer exists, every maximizer is a Gaussian. This is Theorem 4.5 and it can be useful as in [L1] and [L2]. Except for the real case [BL], it was previously known that Gaussian functions were *among* the maximizers. The one exception to this rule was pointed out by Beckner (private communication) for the Fourier transform from $L^p(\mathbf{R}^n)$ to $L^{p'}(\mathbf{R}^n)$ with the restriction $p' \geqq 4$. His proof that a maximizer must be a Gaussian function in this case uses a result in [BL]; the proof is

$$\|f\|_p^2 \geqq \mu_r \|f * f\|_r \geqq \mu_r (C_r^B)^n \|\widehat{f * f}\|_{r'} = \mu_r (C_r^B)^n \|(\hat{f})^2\|_{r'} = \mu (C_r^B)^n \|\hat{f}\|_{p'}^2$$

with $r' = p'/2 \geqq 2$, with $(C_r^B)^n$ being the sharp Beckner (or Babenko) constant for the Fourier transform (denoted by $\widehat{\ }$), and with μ_r being the sharp constant in Young's convolution inequality which was derived simultaneously in [B1, B2] and in [BL]. A Gaussian function $f(y) = \exp\{-Jy^2 + ly\}$ with $J > 0$ and $l \in \mathbf{C}$ gives equality above. However, [BL] (Theorem 13) proved that such functions are the *only* ones that give equality in Young's inequality.

It is a pleasure to acknowledge my debt to Eric Carlen. He helped to stimulate my interest in this problem and to understand the literature in the field. He also critically examined the work as it took shape. Thanks are also due to the Institute for Advanced Study for its hospitality during part of this work, and to Michael Loss for valuable discussions.

II. Some basic properties of Gaussians

2.1. Lemma (nondegenerate Gaussian kernels are compact and have maximizers). *Let G be a centered, nondegenerate Gaussian kernel in $\mathbf{R}^n \times \mathbf{R}^n$ as in* (1.1) *with M in* (1.3) *positive definite and $L = 0$. Let $1 < p < \infty$ and $1 < q < \infty$. Then $\mathscr{G}$ in* (1.2) *is a compact operator from $L^p(\mathbf{R}^n)$ to $L^q(\mathbf{R}^n)$ and there is at least one maximizer $f \in L^p(\mathbf{R}^n)$(i.e., $\mathscr{R}_{p\to q}(f) = C_{p\to q}$).*

Every such maximizer $f\colon \mathbf{R}^n \to \mathbf{C}$, has the following three properties, in which α and β are positive constants that depend on G, p and q but not on f.

(*a*) *There is an entire analytic function of order at most 2, $m\colon \mathbf{C}^n \to \mathbf{C}$, such that $f(x) = |m(x)|^{p'} m(x)^{-1}$ for $x \in \mathbf{R}^n$. Here $1/p + 1/p' = 1$. Moreover, for $z \in \mathbf{C}^n$,*

$$|m(z)| \leqq \alpha \| f \|_p^{p-1} \exp\{\beta |z|^2\} .$$

(*b*) *The function $|f|^{2(p-1)}$ from $\mathbf{R}^n$ to $\mathbf{R}$ has an extension to an entire analytic function from $\mathbf{C}^n$ to $\mathbf{C}$ whose order is at most 2. If $g\colon \mathbf{C}^n \to \mathbf{C}$ is this extension then for $z \in \mathbf{C}^n$*

$$|g(z)| \leqq \alpha \| f \|_p^{2(p-1)} \exp\{\beta |z|^2\} .$$

(*c*) *For $x \in \mathbf{R}^n$*

$$|f(x)| \leqq \alpha \| f \|_p \exp\{-\beta(x, x)\} .$$

Finally, if $f_j \in L^p(\mathbf{R}^n)$ for $j = 1, 2, 3, \ldots$ is an L^p bounded maximizing sequence for G(i.e., $\mathscr{R}_{p\to q}(f_j) \to C_{p\to q}$) then there is a function $f \in L^p(\mathbf{R}^n)$ and a subsequence $j(1), j(2), \ldots$ such that $f_{j(k)} \to f$ strongly in $L^p(\mathbf{R}^n)$ as $k \to \infty$. If $f \neq 0$ (i.e., if $\| f_j \|_p \to 0$ as $j \to \infty$) then f is a maximizer.

Proof. For any $f \in L^p(\mathbf{R}^n)$, Hölders inequality can be used to deduce

$$|(\mathscr{G} f)(x)| \leqq T(x) \| f \|_p \tag{1}$$

with $T(x) = \| G(x, \cdot) \|_{p'}$. Simple computation shows that there are positive numbers γ and δ depending only on G and p such that $|T(x)| \leqq \gamma \exp\{-\delta(x, x)\}$. The fact that G is nondegenerate is crucial for this result. The fact that $T \in L^1(\mathbf{R}^n) \cap L^\infty(\mathbf{R}^n)$ shows that $\mathscr{G}$ is bounded from $L^p(\mathbf{R}^n)$ to $L^q(\mathbf{R}^n)$. Now suppose that $f_j \in L^p(\mathbf{R}^n)$ is a sequence that converges weakly in $L^p(\mathbf{R}^n)$ to some $f \in L^p(\mathbf{R}^n)$ as $j \to \infty$. Since, for each $x \in \mathbf{R}^n$, $G(x, \cdot)$ is in $L^{p'}(\mathbf{R}^n)$, it follows that $(\mathscr{G} f_j)(x) \to (\mathscr{G} f)(x)$ as $j \to \infty$ for each $x \in \mathbf{R}^n$. It can be assumed that the f_j and f satisfy $\| f_j \|_p$ and $\| f \|_p \leqq C$ for some $C > 0$ and hence, from (1), the functions $\mathscr{G} f_j$ and $\mathscr{G} f$ are bounded pointwise by the function CT. Since $T \in L^q(\mathbf{R}^n)$, $\| \mathscr{G} f_j - \mathscr{G} f \|_q \to 0$ by dominated convergence. Thus $\mathscr{G}$ takes weakly convergent sequences in $L^p(\mathbf{R}^n)$ into strongly convergent sequences in $L^q(\mathbf{R}^n)$, and so $\mathscr{G}$ is compact.

Now let f_j be a bounded maximizing sequence, i.e., $\mathscr{R}_{p\to q}(f_j) \to C_{p\to q}$ as $j \to \infty$. We can assume $\|f_j\|_p = 1$ for each j. By the Banach-Alaoglu theorem, there is an $f \in L^p(\mathbf{R}^n)$ and a subsequence $j(1), j(2), \ldots$ such that $f_j \to f$ weakly in $L^p(\mathbf{R}^n)$. As is well known, $\|f\|_p \leqq 1$. Then, by the strong convergence proved above

$$C_{p\to q} = \lim_{k\to\infty} \|\mathscr{G} f_{j(k)}\|_q = \|\mathscr{G} f\|_q \leqq C_{p\to q}\|f\|_p \leqq C_{p\to q}\,.$$

This implies that $\|f\|_p = 1$ and that f is a maximizer. Moreover, the fact that $\|f\|_p = 1$ implies (by the uniform convexity of the L^p norm) that $f_{j(k)}$ converges to f strongly in $L^p(\mathbf{R}^n)$. Thus, the first and last assertions of the lemma have been proved.

It remains to prove that a maximizer f satisfies conditions (a), (b) and (c) and it suffices to assume that $\|f\|_p = 1$. There is a function $h \in L^{q'}(\mathbf{R}^n)$ such that $\|h\|_{q'} = 1$ and $C_{p\to q} = \|\mathscr{G} f\|_q = \int h(x)(\mathscr{G} f)(x)\mathrm{d}x$. Let

$$m(y) = \int G(x,y)h(x)\mathrm{d}x = \mathrm{e}^{-(y,By)} \int \mathrm{e}^{-(x,Ax)-2(x,Dy)} h(x)\mathrm{d}x \tag{2}$$

so that, as in the proof of (1) above, $|m(y)| \leqq W(y) \equiv \mu \exp\{-\nu(y,y)\}$ for suitable positive numbers μ and ν which depend only G and q. Hölder's inequality implies that the function $(x,y) \mapsto h(x)G(x,y)f(y)$ is in $L^1(\mathbf{R}^n \times \mathbf{R}^n)$, and Fubini's theorem then implies that $\|\mathscr{G} f\|_q = \int m(y)f(y)\mathrm{d}y$. If $m(y) \equiv |m(y)|\exp\{\mathrm{i}\theta(y)\}$, the optimum choice for f is $f(y) = [|m(y)|/\|m\|_{p'}]^{p'-1}\exp\{-\mathrm{i}\theta(y)\}$, for otherwise $\mathscr{R}_{p\to q}(f)$ can be increased.

The function $m: \mathbf{R}^n \to \mathbf{C}$ has an extension to an entire analytic function on $\mathbf{C}^n$ of order at most 2. This can be seen easily from the representation (2) above and Hölder's inequality; if $y_j = u_j + \mathrm{i}v_j$ for $j = 1, \ldots, n$ and $D = E + \mathrm{i}H$ with u_j, v_j, E and H real then

$$\begin{aligned}|m(y)| &\leqq \exp\{(v,Bv)-(u,Bu)\}\left[\int \exp\{-q'(x,Ax)-2q'(x,Eu)+2q'(x,Hv)\}\mathrm{d}x\right]^{1/q'}\\ &= (\text{const.})\exp\{(v,Bv)-(u,Bu)+(Eu-Hv,A^{-1}(Eu-Hv))\}\,.\end{aligned}$$

Thus $|m(y)| \leqq (\text{const.})\exp\{(\text{const.})[(u,u)+(v,v)]\}$ which implies that the order of m is at most 2. This establishes conclusion (a). Since m is entire, the function $y \mapsto m^*(y) \equiv \overline{m(\bar{y})}$ (with the bar denoting complex conjugate) is also entire, and hence $N(y) \equiv m(y)m^*(y)$ is also entire with order at most 2 and with a pointwise bound that is independent of f. However, when $y \in \mathbf{R}^n$(i.e., $v_j = 0$ for all j) then $N(y) = |m(y)|^2$. Conclusion (b) is then an immediate consequence of the relation between f and m which implies that for $y \in \mathbf{R}^n$, $|f(y)|^{2(p-1)} = \|m\|_{p'}^{-2}|m(y)|^2 = \|m\|_{p'}^{-2}N(y)$; thus $|f|^{2(p-1)}$ has an analytic extension of order at most 2, namely $\|m\|_{p'}^{-2}N$. It only has to be shown that $\|m\|_{p'}^{-2}$ is universally bounded, but this follows from the relation $C_{p\to q} = \|\mathscr{G} f\|_q = \int mf = \|m\|_{p'}$.

Conclusion (c) follows from the fact that when $y \in \mathbf{R}^n$ then $|f(y)| = [|m(y)|/\|m\|_{p'}]^{p'-1} \leqq W(y)^{p'-1}\|m\|_{p'}^{1-p'}$. □

The next two lemmas validate the assertion in Sect. I that linear terms can be eliminated from Gaussians.

2.2. Lemma (elimination of linear terms from Gaussian kernels). *Let G be the (degenerate or nondegenerate) Gaussian kernel given in* (1.1) *with positive definite or semidefinite real quadratic form M in* (1.3) *and with real linear term $L \in \mathbf{R}^{2n}$. Let G_0 denote the Gaussian kernel with no linear term, which is obtained from G by setting*

$L=0$, *i.e.*, $G_0(x,y) = G(x,y)\exp\left\{-2\left(L, \begin{pmatrix} x \\ y \end{pmatrix}\right)\right\}$. *Let* $1 \leqq p \leqq \infty$ *and* $1 \leqq q \leqq \infty$. *Then the following conditions are equivalent.*

(i) $\mathscr{G}_0$ *is bounded from* $L^p(\mathbf{R}^n)$ *to* $L^q(\mathbf{R}^n)$ *and the equation* $MV = L$ *has a solution* $V \in \mathbf{R}^{2n}$.

(ii) $\mathscr{G}$ *is bounded from* $L^p(\mathbf{R}^n)$ *to* $L^q(\mathbf{R}^n)$.

In case these conditions are both satisfied the relation between the norms is

$$C_{p\to q}(G) = C_{p\to q}(G_0)\exp\{(L, V)\}\,.$$

The number (L, V) *is uniquely defined even if the vector* V *is not unique.* $\mathscr{G}$ *has a unique maximizer if and only if* $\mathscr{G}_0$ *has one.*

Proof. (i) $\Rightarrow$ (ii). This was explained in Sect. I. Simply change variables; writing $V = \begin{pmatrix} a \\ b \end{pmatrix}$, let $x \to x + a$ and $y \to y + b$. Then

$$G \to G_0 \exp\left\{(L, V) - 2\mathrm{i}\left(\begin{pmatrix} x \\ y \end{pmatrix}, NV\right) - \mathrm{i}(V, NV)\right\}.$$

The imaginary terms above do not affect the norm. Since M is Hermitian L must be orthogonal to $\mathscr{K} \equiv$ kernel of $M \subset \mathbf{R}^{2n}$, while any two solutions V_1 and V_2 differ by an element of $\mathscr{K}$. Thus, (L, V) is unique. This change of variables also shows that $\mathscr{G}$ has a unique maximizer if and only if $\mathscr{G}_0$ has one.

(ii) $\Rightarrow$ (i). Suppose that $MV = L$ has no solution. Then, since M is Hermitean, L is not orthogonal to $\mathscr{K}$ and thus there is a vector $W = \begin{pmatrix} s \\ t \end{pmatrix} \in \mathscr{K}$ such that the number $P = (W, L)$ is positive. Make the change of variables $x \to x + s$ and $y \to y + t$. Then, since $MW = 0$, G becomes

$$\tilde{G}(x,y) = G(x,y)\exp\left\{-\mathrm{i}(W, NW) - 2\mathrm{i}\left(\begin{pmatrix} x \\ y \end{pmatrix}, NW\right) + 2P\right\}.$$

The change of variables is an isometry so the norm of $\tilde{\mathscr{G}}$ is the same as the norm of $\mathscr{G}$ and, since the imaginary terms are irrelevant, we have $C_{p\to q}(G) = C_{p\to q}(\tilde{G}) = e^{2P}C_{p\to q}(G)$. This is a contradiction since $C_{p\to q}(G) \neq 0$. Thus $MV = L$ has a solution and the same change of variables can be made as before to derive the relation between the norms of $\mathscr{G}$ and $\mathscr{G}_0$. □

2.3. Lemma (elimination of linear terms from maximizers). *Let* G *be a centered Gaussian kernel (degenerate or nondegenerate) and let* $1 < p < \infty$ *and* $1 < q < \infty$. *Assume* $\mathscr{G}: L^p(\mathbf{R}^n) \to L^q(\mathbf{R}^n)$ *is bounded (which is automatically true in the nondegenerate case). If* $g(x) = \exp\{-(x, Jx) + (l, x)\}$ *is a Gaussian function that maximizes* $\mathscr{R}_{p\to q}(g)$ *among all Gaussian functions then* $g_0(x) = \exp\{-(x, Jx)\}$ *is also a maximizer. Moreover, if* $\mathscr{R}_{p\to q}(g)$ *does not have a maximizer among Gaussian functions (which can happen only if* G *is degenerate) then the supremum of* $\mathscr{R}_{p\to q}(g)$ *over Gaussian functions equals the supremum over centered Gaussian functions. Finally, if* G *is nondegenerate then* $g = g_0$, *i.e.*, $l = 0$, *and therefore* g *is centered.*

Proof. Consider the functions $g_\lambda(x) = \exp\{-(x, Jx) + \lambda(l, x)\}$ with λ a real parameter. Clearly $g_\lambda \in L^p(\mathbf{R}^n)$ for all λ and, by a well known property of Gaussian

Invent. Math. *102*, 179–208 (1990)

integrals,

$$\|g_\lambda\|_p = \|g_0\|_p e^{\alpha\lambda^2} \quad \text{and} \quad \|\mathcal{G} g_\lambda\|_q = \|\mathcal{G} g_0\|_q e^{\beta\lambda^2}$$

for some real constants α and β. There are three cases to be considered:

(*i*) $\alpha > \beta$. By setting $\lambda = 0$, $\mathcal{R}_{p\to q}$ is increased, i.e., $\mathcal{R}_{p\to q}(g_0) > \mathcal{R}_{p\to q}(g)$. This means that g is not a maximizer—which is a contradiction.
(*ii*) $\alpha < \beta$. By letting λ tend to infinity we conclude that $\mathcal{R}_{p\to q}$ (and hence also $\mathcal{G}$) is unbounded—which is a contradiction.
(*ii*) $\alpha = \beta$. In this case g_λ is a maximizer for every λ and hence g_0 is a maximizer, as claimed.

These considerations prove all but the last sentence of the lemma.

If G is nondegenerate it is possible to go further. Consider the following sequence of functions with $\lambda = j$, namely $h_j = Z_j g_j$ for $j = 1, 2, 3, \ldots$, where the numbers Z_j are chosen so that $\|h_j\|_p = 1$ for each j. This is a bounded maximizing sequence and, by a trivial modification of the last part of Lemma 2.1 (using the fact that a nonzero $L^p(\mathbf{R}^n)$ weak limit of Gaussian functions is a Gaussian function), there is a nonzero Gaussian function $h \in L^p(\mathbf{R}^n)$ and a subsequence $j(1), j(2), \ldots$ such that $h_{j(k)} \to h$ strongly in $L^p(\mathbf{R}^n)$ as $k \to \infty$. If $l \neq 0$, however, it is easy to check that $h_j \to 0$ weakly in $L^p(\mathbf{R}^n)$ as $j \to \infty$. This contradicts the supposed strong convergence to a nonzero function. □

III. Nondegenerate gaussian kernels

A main ingredient in the following theorems is Minkowski's inequality for integrals. It was exploited by Beckner [B1; B2] to prove that the sharp bound for the tensor product of two operators (e.g., Fourier transforms) is often the product of the individual bounds. In particular, the bound for the Fourier transform from $L^p(\mathbf{R}^n)$ to $L^{p'}(\mathbf{R}^n)$ is $(C_p^B)^n$, where C_p^B is the sharp constant for $\mathbf{R}^1$. A proof of Minkowski's inequality can be found in [HLP]. Of crucial importance here is the sharp form in which the necessary and sufficient condition for equality is specified; this condition was not used before to analyze Gaussian kernels.

3.1. Lemma (Minkowski's inequality). *Let $f: \mathbf{R}^n \times \mathbf{R}^m \to [0, \infty]$ be Lebesgue measurable and let $1 \leqq r < \infty$. Suppose that the measurable function M, defined for almost every $x \in \mathbf{R}^n$ by*

$$M(x) \equiv \int_{\mathbf{R}^m} f(x, y)^r \, dy ,$$

is finite for almost every x and that $M^{1/r} \in L^1(\mathbf{R}^n)$. Then the measurable function

$$N(y) \equiv \int_{\mathbf{R}^n} f(x, y) \, dx$$

is finite for almost every $y \in \mathbf{R}^m$ and

$$\left\{ \int_{\mathbf{R}^m} N^r \right\}^{1/r} \leqq \int_{\mathbf{R}^n} M^{1/r} . \qquad (*)$$

Furthermore, if $r > 1$ and if there is equality in $(*)$ *then there are nonnegative, measurable functions $A \in L^1(\mathbf{R}^n)$ and $B \in L^r(\mathbf{R}^m)$ such that*

$$f(x, y) = A(x)B(y)$$

for almost every $(x, y) \in \mathbf{R}^n \times \mathbf{R}^m$.

Remark. This lemma extends to an arbitrary pair of measure spaces (X, μ) and (Y, ν) in place of $(\mathbf{R}^n, dx)$ and $(\mathbf{R}^m, dy)$ when μ and ν are sigma finite.

As a first application of Minkowski's inequality the uniqueness of maximizers for real, nondegenerate Gaussian kernels for *all p* and *q* will be proved. This is Case (A) of Section I. It is to be noted that the order of integration in Theorem 3.2 is as in [S] and is opposite to that of Theorem 3.4 and opposite to the order in Beckner's lemma. Analyticity considerations play only a subsidiary role in Theorem 3.2 and can be bypassed if desired, but they are important later. Theorem 3.2 was already essentially contained in [BL] Theorems 7 and 13. The following proof is offered because (*i*) it is different from the [BL] approach and (*ii*) it illustrates the techniques of the present paper.

3.2. Theorem (unique Gaussian maximizer for all p and q in the real nondegenerate case). *Let G be a real, nondegenerate, centered Gaussian kernel, i.e., the matrix N in (1.3) is zero. Let $1 < p < \infty$ and $1 < q < \infty$. Then $\mathscr{G}$ has exactly one maximizer, f, (up to a multiplicative constant) from $L^p(\mathbf{R}^n)$ to $L^q(\mathbf{R}^n)$ and f is a real, centered Gaussian, i.e., $f(x) = \exp\{-(x, Jx)\}$ with J being a real, positive definite matrix.*

Proof. Consider the linear operator $\mathscr{G}^{(2)} = \mathscr{G} \otimes \mathscr{G} : L^p(\mathbf{R}^{2n}) \to L^q(\mathbf{R}^{2n})$ given by the Gaussian kernel $G^{(2)}((x_1, x_2), (y_1, y_2)) = G(x_1, y_1)G(x_2, y_2)$ with x_1, x_2, y_1 and y_2 in $\mathbf{R}^n$. The first goal is to prove that $C_{p\to q}(G^{(2)}) = C_{p\to q}(G)^2$. If $F \in L^p(\mathbf{R}^{2n})$ then $(y_1, y_2) \mapsto G^{(2)}((x_1, x_2), (y_1, y_2))\, F(y_1, y_2)$ is in $L^1(\mathbf{R}^{2n})$ for every (x_1, x_2) because $G^{(2)}$ is nondegenerate. Fubini's theorem and Minkowski's inequality yield

$$\|\mathscr{G}^{(2)}F\|_q^q = \int \{ \int | \int (\int G(x_1, y_1)G(x_2, y_2)F(y_1, y_2)\mathrm{d}y_1)\mathrm{d}y_2|^q \mathrm{d}x_1 \}\,\mathrm{d}x_2 \tag{1}$$

$$\leqq \int \{ \int [\int G(x_2, y_2)|K(x_1, y_2)|\mathrm{d}y_2]^q \mathrm{d}x_1 \}\mathrm{d}x_2$$

$$(\text{with } K(x_1, y_2) = \int G(x_1, y_1)F(y_1, y_2)\mathrm{d}y_1) \tag{2}$$

$$\leqq \int \{ \int [\int G(x_2, y_2)^q |K(x_1, y_2)|^q \mathrm{d}x_1]^{1/q} \mathrm{d}y_2\}^q \mathrm{d}x_2 \tag{3}$$

$$\leqq (C_{p\to q}(G))^q \int \{ \int G(x_2, y_2)[\int |F(y_1, y_2)|^p \mathrm{d}y_1]^{1/p} \mathrm{d}y_2\}^q \mathrm{d}x_2 \tag{4}$$

$$\leqq (C_{p\to q}(G))^{2q} \{ \int\int |F(y_1, y_2)|^p \mathrm{d}y_1\,\mathrm{d}y_2\}^{q/p}. \tag{5}$$

(Notes: (2) → (3) is Minkowski's inequality. (3) → (4) uses $C_{p\to q}(G) \geqq \mathscr{R}_{p\to q}(F(\cdot, y_2))$ for each y_2. (4) → (5) uses $C_{p\to q}(G) \geqq \mathscr{R}_{p\to q}((\int |F(y_1, \cdot)|^p \mathrm{d}y_1)^{1/p})$. The fact that $G(x, y) \geqq 0$ is crucial. Here the x_1 integration was done before the x_2 integration; in Theorem 3.4 the x_2 integration will be done first.) Inequalities (1)–(5) establish that $C_{p\to q}(G^{(2)}) \leqq C_{p\to q}(G)^2$. Clearly, by considering F's of the product form $F(y_1, y_2) = h(y_1)h(y_2)$, the reverse inequality is obtained, and so the goal is reached.

Suppose now that $F : \mathbf{R}^{2n} \to \mathbf{C}$ is a maximizer for $G^{(2)}$. Since $G^{(2)}$ is nondegenerate, it has a maximizer by Lemma 2.1. Since $G(x, y) > 0$ for all x and y, it is clear that $F = \lambda|F|$ and $|\lambda| = 1$, for otherwise replacing F by $|F|$ will increase the quotient $\mathscr{R}_{p\to q}$ for $G^{(2)}$. It can be assumed henceforth that $F \geqq 0$. Since F is

Invent. Math. *102*, 179–208 (1990)

a maximizer all the inequalities in (1)–(5) must be equalities. Equality of (2) and (3) implies, by Lemma 3.1, that for almost every x_2 there are measurable functions A_{x_2} and $B_{x_2}: \mathbf{R}^n \to [0, \infty)$ such that

$$G(x_2, y_2)K(x_1, y_2) = A_{x_2}(x_1)B_{x_2}(y_2) \tag{6}$$

for almost every x_1 and y_2. Since $G > 0$, this equation can be divided by $G(x_2, y_2)$ to obtain $K(x_1, y_2) = A_{x_2}(x_1)E_{x_2}(y_2)$ with $E_{x_2}(y) \equiv B_{x_2}(y)/G(x_2, y)$. However, $K(x_1, y_2)$ is independent of x_2 and therefore if any particular value of x_2 is chosen for which (6) holds for almost every x_1 and y_2, and if the functions A and $E: \mathbf{R}^n \to [0, \infty)$ are defined by $A \equiv A_{x_2}$ and $E \equiv E_{x_2}$ for this value of x_2, then

$$K(x_1, y_2) = A(x_1)E(y_2)$$

for almost every x_1 and y_2. If this equation is multiplied by $G(x_2, y_2)$ and integrated over y_2 the result is

$$(\mathscr{G}^{(2)}F)(x_1, x_2) = A(x_1)Z(x_2)$$

for almost every x_1 and x_2 with $Z = \mathscr{G}E$. Since $G^{(2)} > 0$, both A and Z are strictly positive functions.

There is a function $H \in L^{q'}(\mathbf{R}^{2n})$ with $\|H\|_{q'} = 1$, such that $\|\mathscr{G}^{(2)}F\|_q = \int H \cdot \mathscr{G}^{(2)}F$. In fact

$$H(x_1, x_2) = (\text{const.})[(\mathscr{G}^{(2)}F)(x_1, x_2)]^{q-1} = (\text{const.})A(x_1)^{q-1}Z(x_2)^{q-1}\,.$$

The point here is that H is a product function. Then, as in the proof of Lemma 2.1, F satisfies

$$F(y_1, y_2) = (\text{const.})\left\{\iint G(x_1, y_1)G(x_2, y_2)H(x_1, x_2)\mathrm{d}x_1\,\mathrm{d}x_2\right\}^{p'-1} = \alpha(y_1)\beta(y_2) \tag{7}$$

for some positive function α and $\beta: \mathbf{R}^n \to [0, \infty)$. In brief, F *must be* a product function, and this fact is crucial for the next step.

One example of a maximizer is $F(y_1, y_2) = f(y_1)f(y_2)$, where f is an $L^p(\mathbf{R}^n) \to L^q(\mathbf{R}^n)$ maximizer for G (whose existence is guaranteed by Lemma 2.1). For the reason given before about F, we can and do assume that $f(x) \geqq 0$ for all $x \in \mathbf{R}^n$.

A more interesting maximizer is

$$F(y_1, y_2) = f\left(\frac{y_1 - y_2}{\sqrt{2}}\right)f\left(\frac{y_1 + y_2}{\sqrt{2}}\right). \tag{8}$$

Here, the essential property of $O(2)$ rotation invariance of products of centered Gaussians and of Lebesgue measure is being exploited. If θ is any fixed angle and if x'_1, x'_2, y'_1, y'_2 in $\mathbf{R}^n$ are defined by $x'_1 = x_1\cos\theta - x_1\sin\theta$, $x'_2 = x_1\sin\theta + x_2\cos\theta$, $y'_1 = y_1\cos\theta - y_2\sin\theta$, $y'_2 = y_1\sin\theta + y_2\cos\theta$, the $O(2)$ invariance of Lebesgue measure is that $\mathrm{d}x_1\mathrm{d}x_2 = \mathrm{d}x'_1\mathrm{d}x'_2$ and $\mathrm{d}y_1\mathrm{d}y_2 = \mathrm{d}y'_1\mathrm{d}y'_2$. The $O(2)$ invariance of centered Gaussian functions is that $g(x_1)g(x_2) = g(x'_1)g(x'_2)$, while for centered Gaussian kernels $G(x_1, y_1)G(x_2, y_2) = G(x'_1, y'_1)G(x'_2, y'_2)$. With the choice $\theta = \pi/4$, these observations lead to (8). Combining (7) and (8),

$$f\left(\frac{y_1 - y_2}{\sqrt{2}}\right)f\left(\frac{y_1 + y_2}{\sqrt{2}}\right) = \alpha(y_1)\beta(y_2)\,. \tag{9}$$

for almost every y_1 and y_2.

Equation (9) implies that f is a Gaussian. Instead of proving this in full generality for $L^p(\mathbf{R}^n)$ functions, as is done by Carlen [CA], it is easier to simplify the proof here by taking the $2(p-1)^{th}$ power of (9) and by taking advantage of the analyticity result Lemma 2.1(b). Introducing $h = f^{2(p-1)}$, $\gamma = \alpha^{2(p-1)}$ and $\delta = \beta^{2(p-1)}$, it is seen from (9) (by fixing y_2) that γ is analytic; likewise δ is analytic. Thus, (9) holds for *all* y_1 and y_2 because when two analytic functions on $\mathbf{C}^n \times \mathbf{C}^n$ agree almost everywhere on $\mathbf{R}^n \times \mathbf{R}^n$ then they agree everywhere. Furthermore f never vanishes for real y because if $f(Y) = 0$ then, setting $y_1 = y_2 + \sqrt{2}Y$, we would have that $0 = \gamma(y_2 + \sqrt{2}Y)\delta(y_2)$ for all y_2; this is impossible, given that γ and δ are analytic, unless $\gamma \equiv 0$ or $\delta \equiv 0$, which contradicts the assumption that $f \not\equiv 0$. Thus, the logarithms of h, γ and δ are real analytic and

$$\ln\left[h\left(\frac{y_1 - y_2}{\sqrt{2}}\right)\right] + \ln\left[h\left(\frac{y_1 + y_2}{\sqrt{2}}\right)\right] = \ln[\gamma(y_1)] + \ln[\delta(y_2)]\,. \tag{10}$$

If ∂_i denotes the derivative with respect to the ith coordinate, and ∂_i with respect to y_1 and ∂_j with respect to y_2 is taken in (10), then

$$(\partial_i\partial_j \ln h)\left(\frac{y_1 - y_2}{\sqrt{2}}\right) = (\partial_i\partial_j \ln h)\left(\frac{y_1 + y_2}{\sqrt{2}}\right),$$

which implies that the function $\partial_i\partial_j \ln h$ is a constant (call it $4(1-p)J_{ij}$) and therefore $\ln[f(y)] = \dfrac{1}{2(p-1)}\ln[h(y)] = -(y, Jy) + (l, y)$ for some vector l. According to Lemma 2.3, $l = 0$ since G is centered and nondegenerate. This completes the proof that f must be a centered Gaussian.

It remains to prove that f is unique (i.e., the matrix J above is unique). One way would be to compute $\mathscr{R}_{p\to q}(\exp\{-(x, Jx)\})$ for G and then deduce that there is only one optimum J. A very much easier route is to suppose that there are two maximizers f^1 and f^2 with $f^i(y) = \exp\{-(y, J^i y)\}$. Then, for the same reason as before ($O(2)$ symmetry) the function

$$F(y_1, y_2) = f^1\left(\frac{y_1 - y_2}{\sqrt{2}}\right) f^2\left(\frac{y_1 + y_2}{\sqrt{2}}\right) \tag{11}$$

is a maximizer for $\mathscr{G}^{(2)}$. There are two ways in which this implies that $f^1 = f^2$. The first is to use (7), namely F must be a product function, and to note that this product structure is true if and only if $J^1 = J^2$. The second way is to note that since the F in (11) is never zero and, since (3) $\to$ (4) must be an equality, we have that the function $y_1 \mapsto h_{y_2}(y_1) \equiv F(y_1, y_2)$ must be a maximizer for $\mathscr{G}$ for almost every y_2. Although the function h_{y_2} is a Gaussian for each y_2, the Gaussian will have a linear term for each $y_2 \neq 0$ unless $J^1 = J^2$. However, Lemma 2.3 precludes the existence of such a linear term, so $J^1 = J^2$. □

The next theorem concerns Case (B) of Sect. I.

3.3. Theorem (unique Gaussian maximizers in the imaginary, nondegenerate case). *Let G be a centered, nondegenerate Gaussian kernel with a real diagonal part and a purely imaginary off-diagonal part, i.e.,*

$$G(x, y) = \exp\{-(x, Ax) - (y, By) - 2i(x, Dy)\}$$

Invent. Math. *102*, 179–208 (1990)

where A, B and D are real $n \times n$ matrices and A and B are positive definite. Let $1 < p \leqq 2$ and $1 < q < \infty$ or else $1 < p < \infty$ and $2 \leqq q \leqq \infty$. Then, in either case, $\mathscr{G}$ has exactly one maximizer, f, (up to a multiplicative constant) from $L^p(\mathbf{R}^n)$ to $L^q(\mathbf{R}^n)$ and this f is a real, centered Gaussian, i.e., $f(x) = \exp\{-(x, Jx)\}$ with J being a real, positive definite matrix.

Proof. Assume at first that D is nonsingular. Since A and B are positive definite there are nonsingular real matrices U and V so that the change of variables $x \to Ux$ and $y \to Vy$ changes A and B to the identity matrix, I, that is $I = U^TAU = V^TBV$, where T denotes transpose. Then $(x, Dy) \to (x, \tilde{D}y)$ with $\tilde{D} = U^TDV$. The polar decomposition of $\tilde{D}$ is $\tilde{D} = W|\tilde{D}|$, where W is orthogonal and $|\tilde{D}|$ is positive definite (the assumption that D is nonsingular is used here). Then there is an orthogonal matrix Y such that $Y^T|\tilde{D}|Y$ is diagonal and there is a real diagonal matrix Z such that $ZY^T|\tilde{D}|YZ = I$. Now make one more change of variables: $x \to WYZx$ and $y \to YZy$ so that $(x, \tilde{D}y) \to (WYZx,\ W|\tilde{D}|YZy) = (x, y)$ and $(x, x) = (x, Ix) \to (WYZx,\ WYZx) = (x, Z^2x)$ and $(y, y) \to (YZy, YZy) = (y, Z^2y)$. These two changes of variables affect $\mathscr{R}_{p\to q}$ in a trivial way (involving only p and q and the determinants of U, V and Z) and, most importantly, take Gaussian functions into Gaussian functions. In short, it can be assumed without loss of generality that G has the canonical form

$$G(x, y) = \exp\{-(x, Ax) - (y, Ay) - 2i(x, y)\}\,, \tag{1}$$

where A is positive definite and diagonal.

By duality $C_{p\to q}(G) = C_{q'\to p'}(G^T)$ with $G^T(x, y) \equiv G(y, x) = G(x, y)$, so it suffices to consider only the case $1 < p \leqq 2$ and $1 < q < \infty$. It is easily seen that $(\mathscr{G}f)(x) = \exp\{-(x, Ax)\}\hat{h}(x)$ where $\hat{h}$ is the Fourier transform of the function $h(y) = \exp\{-(y, Ay)\}f(y)$. Since $f \in L^p(\mathbf{R}^n)$ it has a Fourier transform $\hat{f}$, and Beckner's theorem (which will also be proved here in Theorem 4.1 and 4.2(1)) states that $\|\hat{f}\|_{p'} \leqq (C_p^B)^n\|f\|_p$, where Beckner's constant C_p^B is the sharp constant for the $p \to p'$ norm of the Fourier transform in $\mathbf{R}^1$. By the convolution formula, $\hat{h}$ satisfies

$$\hat{h}(x) = \mu \int \exp\{-(x - y, A^{-1}(x - y))\}\hat{f}(y)\mathrm{d}y\,,$$

where $\mu > 0$ is a constant which depends only on A. Therefore $(\mathscr{G}f)(x) = \mu(\tilde{\mathscr{G}}\hat{f})(x)$ where $\tilde{G}$ is the real, centered, nondegenerate Gaussian

$$\tilde{G}(x, y) = \exp\{-(x, Ax) - (x - y, A^{-1}(x - y))\}\,. \tag{2}$$

Thus

$$\mathscr{R}_{p\to q}(G, f)\|f\|_p = \mu\mathscr{R}_{p'\to q}(\tilde{G}, \hat{f})\|\hat{f}\|_{p'} \leqq \mu C_{p'\to q}(\tilde{G})\|\hat{f}\|_{p'} \leqq \mu C_{p'\to q}(\tilde{G})(C_p^B)^n\|f\|_p\,, \tag{3}$$

from which it follows that $C_{p\to q}(G) \leqq \mu(C_p^B)^n C_{p'\to q}(\tilde{G})$. However, equality can be achieved in (3) in exactly one way (up to a multiplicative constant). By Theorem 3.2 there is exactly one choice for $\hat{f}$ that will make the first inequality in (3) into an equality. This $\hat{f}$ is a real, centered Gaussian, $\hat{f}(x) = \exp\{-(x, Jx)\}$. Its inverse transform f is also a real, centered Gaussian, i.e., $f(x) = (\text{const.})\exp\{-(x, J^{-1}x)\}$. The second inequality in (3) (Beckner's) is an equality for any *real* Gaussian (in particular, our f), and therefore f is the unique maximizer as asserted in the theorem.

In case D is singular, a change of variables similar to the above replaces the canonical form (1) by

$$G(x, y) = \exp\{-(x, Ax) - (y, Ay) - 2\mathrm{i}(x, Py)\}$$

where $P = \begin{pmatrix} I & 0 \\ 0 & 0 \end{pmatrix}$ is a diagonal projection onto $\mathbf{R}^m$ (with $m < n$ being the rank of D) and $A = \begin{pmatrix} \alpha & 0 \\ 0 & I \end{pmatrix}$ with α positive definite, $m \times m$ and diagonal. Writing $x \in \mathbf{R}^n$ as (x_1, x_2) with $x_1 \in \mathbf{R}^m$ and $x_2 \in \mathbf{R}^{n-m}$, define $g: \mathbf{R}^m \to \mathbf{C}$ by

$$g(y_1) = \int_{\mathbf{R}^{n-m}} \exp\{-(y_2, y_2)\} f(y_1, y_2) \mathrm{d}y_2 , \tag{4}$$

and $G^*: \mathbf{R}^m \times \mathbf{R}^m \to (0, \infty)$ by $G^*(x_1, y_1) = \exp\{-(x_1, \alpha x_1) - (y_1, \alpha y_1) - 2\mathrm{i}(x_1, y_1)\}$. Then, using Fubini's theorem and the same analysis as before with α and m in place of A and n, and with $\tilde{G}^*: \mathbf{R}^m \times \mathbf{R}^m \to (0, \infty)$ as given in (2),

$$\mathscr{R}_{p\to q}(G, f) \| f \|_{L^p(\mathbf{R}^n)} = \nu \mathscr{R}_{p\to q}(G^*, g) \| g \|_{L^p(\mathbf{R}^m)} \leqq \nu\mu C_{p'\to q}(\tilde{G}^*)(C_p^B)^m \| g \|_{L^p(\mathbf{R}^m)} \tag{5}$$

where ν is the $L^q(\mathbf{R}^{n-m})$ norm of the Gaussian function $\exp\{-(x_2, x_2)\}$. As was the case for (3) and the subsequent argument, equality in (5) is uniquely achieved by a real, centered Gaussian

$$g(y_1) = \exp\{-(y_1, Ey_1)\} , \tag{6}$$

where E is a real, positive definite $m \times m$ matrix. By Hölder's or Minkowski's inequality, it follows from (4) that

$$\| g \|_{L^p(\mathbf{R}^m)} \leqq N \| f \|_{L^p(\mathbf{R}^n)} \tag{7}$$

where N is the $L^{p'}(\mathbf{R}^{n-m})$ norm of the Gaussian function $\exp\{-(y_2, y_2)\}$. Equality in (7) is compatible with (6) if and only if $f(y) = f(y_1, y_2) = (\text{const.})\exp\{-y_1, Ey_1) - (p'-1)(y_2, y_2)\}$. The reason is that equality in (7) requires that $f(y_1, y_2) = \lambda(y_1)\exp\{-(p'-1)(y_2, y_2)\}$ for almost every $y_1 \in \mathbf{R}^m$. Then, computing the integral in (4), one finds that $\exp\{-(y_1, Ey_1)\} = g(y_1) = (\text{const.})\,\lambda(y_1)$. □

Finally, Case (C) of Sect. I will be considered.

3.4. Theorem (unique Gaussian maximizer when $p \leqq q$ in the general nondegenerate case). *Let G be a centered, nondegenerate Gaussian kernel and let $1 < p \leqq q < \infty$. Then $\mathscr{G}$ has exactly one maximizer, f, (up to a multiplicative constant) from $L^p(\mathbf{R}^n)$ to $L^q(\mathbf{R}^n)$ and f is a centered Gaussian function.*

Proof. As in the proof of Theorem 3.2, the key is to study the kernel $G^{(2)} = G \otimes G$ by means of Minkowski's inequality. Now, however, the x_2 integration is done first. Thus, for $F \in L^p(\mathbf{R}^{2n})$,

$$\| \mathscr{G}^{(2)} F \|_q^q = \int \{ \int | \int G(x_2, y_2)(\int G(x_1, y_1) F(y_1, y_2) \mathrm{d}y_1) \mathrm{d}y_2 |^q \mathrm{d}x_2 \} \mathrm{d}x_1 \tag{1}$$

$$= \int \{ \int | \int G(x_2, y_2) K(x_1, y_2) \mathrm{d}y_2 |^q \mathrm{d}x_2 \} \mathrm{d}x_1$$

$$\text{(with } K(x_1, y_2) = \int G(x_1, y_1) F(y_1, y_2) \mathrm{d}y_1) \tag{2}$$

$$\leqq (C_{p\to q}(G))^q \int \{ \int | K(x_1, y_2)|^p \mathrm{d}y_2 \}^{q/p} \mathrm{d}x_1 \tag{3}$$

$$\leqq (C_{p\to q}(G))^q \{ \int [\int | K(x_1, y_2)|^q \mathrm{d}x_1]^{p/q} \mathrm{d}y_2 \}^{q/p} \tag{4}$$

$$\leqq (C_{p\to q}(G))^{2q} \{ \iint | F(y_1, y_2)|^p \mathrm{d}y_1 \mathrm{d}y_2 \}^{q/p} . \tag{5}$$

Invent. Math. *102*, 179–208 (1990)

(Notes: (2) → (3) uses $C_{p\to q}(G) \geqq \mathscr{R}_{p\to q}(K(x_1, \cdot))$ for each x_1. (3) → (4) is Minkowski's inequality for the exponent $r = q/p \geqq 1$ and the function $|K(x_1, y_2)|^p$. (4) → (5) uses $C_{p\to q}(G) \geqq \mathscr{R}_{p\to q}(F(\cdot, y_2))$ for each y_2.) This inequality (along with consideration of F's of the form $F(y_1, y_2) = h(y_1)h(y_2)$) shows that

$$C_{p\to q}(G^{(2)}) = C_{p\to q}(G)^2 .$$

Suppose now that $F: \mathbf{R}^{2n} \to \mathbf{C}$ is a maximizer for $\mathscr{G}^{(2)}$. Since $G^{(2)}$ and G are nondegenerate, maximizers exist for each of them by Lemma 2.1. Then all the inequalities in (1)–(5) must be equalities. In particular, inequality (4) → (5) implies that the function $y_1 \mapsto F(y_1, y_2)$ must either be the zero function or it must be a maximizer for $\mathscr{G}$ for almost every $y_2 \in \mathbf{R}^n$. (It is well known that this function is in $L^p(\mathbf{R}^n)$ for almost every y_2.) As in the proof of Theorem 3.2, the $O(2)$ invariance of $G^{(2)}$ implies that the function given by

$$F(y_1, y_2) = f\left(\frac{y_1 - y_2}{\sqrt{2}}\right) f\left(\frac{y_1 + y_2}{\sqrt{2}}\right) \tag{6}$$

is a maximizer for $\mathscr{G}^{(2)}$ when f is a maximizer for $\mathscr{G}$, as will henceforth be assumed. Thus, for almost every z in $\mathbf{R}^n$, the function $g_z(y) = F(y, z)$ is in $L^p(\mathbf{R}^n)$ and either (a) it is a maximizer for $\mathscr{G}$ or (b) g_z is the zero function. The second possibility (b) can be excluded by Lemma 2.1 (b). If $g_z \equiv 0$ then, from (6), $f(w) = 0$ for all w in some set $A \subset \mathbf{R}^n$ of positive Lebesgue measure. But $|f|^{2(p-1)}$ is analytic and this is impossible unless $f \equiv 0$. Thus it can be assumed that g_z is indeed a maximizer for almost every z, i.e., $g_z \not\equiv 0$.

In fact g_z is an $L^p(\mathbf{R}^n)$ maximizer for *every* $z \in \mathbf{R}^n$. To prove this assertion, fix z and let $z_1, z_2, \ldots$ be any sequence in $\mathbf{R}^n$ such that $z_j \to z$ as $j \to \infty$ and such that g_{z_j} is an $L^p(\mathbf{R}^n)$ maximizer for each j. Such a sequence exists because g_z is a maximizer for z's in a dense set. Define $h_j(y) = Z_j g_{z_j}(y)$ where Z_j is chosen so that $\|h_j\|_p = 1$ for each j. By Lemma 2.1, there is a subsequence (still denoted by h_j) and a maximizing function $h \in L^p(\mathbf{R}^n)$ such that $h_j \to h$ strongly as $j \to \infty$. By passing to a further subsequence this convergence can also be assumed to be pointwise almost everywhere. However, translation is a continuous operation in $L^p(\mathbf{R}^n)$ and thus, by passing to a further subsequence, $f((y + z_j)/\sqrt{2})$ converges pointwise to $f((y + z)/\sqrt{2})$ for almost every y. Likewise, by passing to a further subsequence, $f((y - z_j)/\sqrt{2})$ converges pointwise to $f((y - z)/\sqrt{2})$ for almost every y. It follows then that the maximizer h satisfies

$$h(y) = f\left(\frac{y - z}{\sqrt{2}}\right) f\left(\frac{y + z}{\sqrt{2}}\right) \lim_{j\to\infty} Z_j$$

for almost every y. Therefore $\lim_{j\to\infty} Z_j$ exists and g_z is a maximizer for every $z \in \mathbf{R}^n$.

Our first application of this result will be the proof that there is a Gaussian maximizer. Take $z = 0$ so that $f^{(2)}(y) \equiv f(y/\sqrt{2})^2$ is a maximizer. Then apply the same conclusion to $f^{(2)}$ so that $f^{(4)}(y) \equiv f(y/2)^4$ is also a maximizer. Repeating this indefinitely, the sequence of $L^p(\mathbf{R}^n)$ functions given by

$$g_j(y) = N_j f\left(\frac{y}{\sqrt{j}}\right)^j \tag{7}$$

is a sequence of maximizers for $j = 2, 4, 8, 16, \ldots$. The number N_j is chosen in each case so that $\|g_j\|_p = 1$. Using Lemma 2.1 again we infer the existence of a subsequence (still denoted by j) and a maximizer g such that $g_j \to g$ strongly in $L^p(\mathbf{R}^n)$ and pointwise almost everywhere. Our goal will be to prove that g is a Gaussian. This can be inferred from the central limit theorem, but the following argument is more direct and will be needed later for the proof that *every* maximizer is a Gaussian.

The first step is to prove that $f(0) \neq 0$. Recall from Lemma 2.1(b) that $R \equiv |f|^{2(p-1)}$ is analytic. Likewise $S \equiv |g|^{2(p-1)}$ is also analytic and

$$S(y) = \lim_{j \to \infty} N_j^{2(p-1)} R\left(\frac{y}{\sqrt{j}}\right)^j \tag{8}$$

for almost all $y \in \mathbf{R}^n$. Since $S_j(y) \equiv N_j^{2(p-1)} R(y/\sqrt{j})^j$ is the $2(p-1)^{th}$ power of the modulus of a maximizer with unit $L^p(\mathbf{R}^n)$ norm (namely g_j), Lemma 2.1(b) states that the analytic extension of S_j is uniformly bounded on compact subsets of $\mathbf{C}^n$. The almost everywhere convergence in (8) then implies (by Vitali's theorem) that (8) holds for all $y \in \mathbf{C}^n$ and that all partial derivatives with respect to y of the sequence of functions S_j also converge as $j \to \infty$ to the corresponding derivatives of S. However, it is easily seen by Leibniz's rule that if $R(0) = 0$ then *every* derivative of S_j at $y = 0$ converges to zero as $j \to \infty$. This is impossible unless $S(y)$ vanishes identically, which contradicts the fact that $\|g\|_p = 1$.

The second step is to prove that g is a Gaussian. By Lemma 2.1(a), for $y \in \mathbf{R}^n$, $f(y) = |m(y)|^{p'}/m(y)$, where $m\colon \mathbf{C}^n \to \mathbf{C}$ is entire analytic. Since $f(0) \neq 0$, also $m(0) \neq 0$ and hence there is a neighborhood U of $0 \in \mathbf{C}^n$ on which f has an analytic extension and on which f is never zero. [Reason: $m_1(y) \equiv \mathrm{Re}(m(y))$ can be written as a Taylor series for $y \in \mathbf{R}^n$, and so can $m_2(y) \equiv \mathrm{Im}(m(y))$. Consequently m_1 and m_2 extend to entire functions. Then $(m_1^2 + m_2^2)^{p'/2}$ is analytic on U since $m_1(0)^2 + m_2(0)^2 = |m(0)|^2 \neq 0$.] Therefore f has a logarithm, H, which is analytic on U, i.e., $f(y) = f(0)\exp\{H(y)\}$. The function H can be written as $H(y) = (V, y) - (y, Jy) + O(y^3)$ for some $V \in \mathbf{C}^n$ and J a symmetric matrix. For each $y \in \mathbf{R}^n$, the point $y/\sqrt{j}$ lies in U for all sufficiently large j and therefore, by (7),

$$g(y) = \lim_{j \to \infty} N_j f(0)^j \exp\{\sqrt{j}(V, y) - (y, Jy) + O(y^3 j^{-1/2})\}$$

for almost every $y \in \mathbf{R}^n$. The factor $\exp\{O(y^3 j^{-1/2})\}$ converges to 1 as $j \to \infty$ and thus

$$g(y) = \exp\{-(y, Jy)\} \lim_{j \to \infty} N_j f(0)^j \exp\{\sqrt{j}(V, y)\}\,.$$

Clearly this last limit can exist for almost every y if and only if $V = 0$ and $N_j f(0)^j$ has a finite limit (which cannot be zero since $\|g\|_p = 1$). This proves that g must be a Gaussian as claimed (and hence $\mathrm{Re}(J)$ is positive definite) but we also note that the argument also proves the following three statements: Whenever f is a maximizer then (*i*) f is analytic in some complex neighborhood of 0; (*ii*) $f(0) \neq 0$; (*iii*) $(\partial f/\partial y^i)(0) = 0$, for $i = 1, \ldots, n$.

The second assertion of the theorem is that every other maximizer, f, is proportional to the one just found, namely $g(y) = \exp\{-(y, Jy))\}$. Instead of (6) take

$$F(y_1, y_2) = g\left(\frac{y_1 - y_2}{\sqrt{2}}\right) f\left(\frac{y_1 + y_2}{\sqrt{2}}\right)$$

Invent. Math. *102*, 179–208 (1990)

which is obviously also a maximizer for $\mathscr{G}^{(2)}$. By the same reasoning as before, F has the property that $y \mapsto k_z(y) \equiv F(y, z)$ is a maximizer for each fixed $z \in \mathbf{R}^n$. By the three statements just made above, we conclude that k_z is analytic near 0,

$$k_z(0) \neq 0 \quad \text{and} \quad (\partial k_z/\partial y^i)(0) = 0 \, .$$

This is equivalent to the statement that for every $z \in \mathbf{R}^n$, f is analytic near $z/\sqrt{2}$, $f(z/\sqrt{2}) \neq 0$ and

$$(\partial f/\partial y^i)\left(\frac{z}{\sqrt{2}}\right) = [-Jz]_i f\left(\frac{z}{\sqrt{2}}\right),$$

which shows that $f = g$. □

IV. Degenerate Gaussian kernels

In the three cases (A), (B) and (C) of Sect. I, which correspond to Theorems 3.2, 3.3 and 3.4, every nondegenerate Gaussian kernel has a unique maximizer which is a Gaussian function. By taking suitable limits the following formula 4.1(*), which is one of the main results of this paper, can be deduced for the $L^p(\mathbf{R}^n)$ to $L^q(\mathbf{R}^n)$ norm of degenerate kernels. This formula is, of course, trivially true in the nondegenerate case.

4.1. Theorem (the sharp bound for degenerate kernels). *Let G be a centered Gaussian kernel as in (1.1) with $L = 0$ and let p and q satisfy the appropriate conditions given in (A), (B) or (C) of Sect. I, according to the properties of G. Then $\mathscr{G}$ is bounded from $L^p(\mathbf{R}^n)$ to $L^q(\mathbf{R}^n)$ if and only if the following supremum is finite, in which case the supremum is equal to $C_{p\to q}$.*

$$\sup_g \mathscr{R}_{p\to q}(g) = C_{p\to q}\,, \tag{$*$}$$

where the supremum is taken over all centered Gaussian functions, and in Cases (A) and (B) they can be restricted to be real.

Proof. For each $\varepsilon > 0$ let $h_\varepsilon(x) \equiv \exp\{-\varepsilon(x, x)\}$ and define $G_\varepsilon(x, y) \equiv G(x, y)h_\varepsilon(x)h_\varepsilon(y)$, which is nondegenerate. Correspondingly, there is the linear operator $\mathscr{G}_\varepsilon$. For each $f \in L^p(\mathbf{R}^n)$

$$\begin{aligned}\mathscr{R}_{p\to q}(G_\varepsilon, f)\|f\|_p = \|h_\varepsilon \mathscr{G}(h_\varepsilon f_\varepsilon)\|_q &< \|\mathscr{G}(h_\varepsilon f_\varepsilon)\|_q \leqq C_{p\to q}(G)\|h_\varepsilon f_\varepsilon\|_p \\ &< C_{p\to q}(G)\|f\|_p\,.\end{aligned} \tag{1}$$

This proves that $C_{p\to q}(G_\varepsilon) < C_{p\to q}(G)$.

On the other hand, assuming that $C_{p\to q}(G) < \infty$, for each $\delta > 0$ there is an $f^\delta \in L^p(\mathbf{R}^n)$ with $\|f^\delta\|_p = 1$ such that $\|\mathscr{G}f^\delta\|_q > C_{p\to q}(G) - \delta$. Then

$$C_{p\to q}(G_\varepsilon) \geqq \mathscr{R}_{p\to q}(G_\varepsilon, f^\delta) = \|\mathscr{G}_\varepsilon(f^\delta)\|_q = \|h_\varepsilon \mathscr{G}(h_\varepsilon f^\delta)\|_q\,. \tag{2}$$

As $\varepsilon \to 0$, $h_\varepsilon f^\delta \to f^\delta$ strongly in $L^p(\mathbf{R}^n)$, so $\mathscr{G}(h_\varepsilon f^\delta) \to \mathscr{G}(f^\delta)$ strongly in $L^q(\mathbf{R}^n)$. This implies that $h_\varepsilon \mathscr{G}(h_\varepsilon f^\delta) \to \mathscr{G}(f^\delta)$ strongly in $L^q(\mathbf{R}^n)$ as well, and thus, from (2), $\liminf_{\varepsilon\to 0} C_{p\to q}(G_\varepsilon) \geqq C_{p\to q}(G) - \delta$. Since δ is arbitrary, and in view of (1),

$$\lim_{\varepsilon\to 0} C_{p\to q}(G_\varepsilon) = C_{p\to q}(G)\,. \tag{3}$$

A similar argument shows that (3) holds even if $C_{p\to q}(G) = \infty$.

Now let g_ε denote the maximizer for G_ε, which is a centered Gaussian function. Assume $\|g_\varepsilon\|_p = 1$. Then

$$\|h_\varepsilon g_\varepsilon\|_p C_{p\to q}(G) \geqq \|h_\varepsilon g_\varepsilon\|_p \mathscr{R}_{p\to q}(G, h_\varepsilon g_\varepsilon) = \|\mathscr{G}(h_\varepsilon g_\varepsilon)\|_q$$
$$\geqq \|h_\varepsilon \mathscr{G}(h_\varepsilon g_\varepsilon)\|_q = \|\mathscr{G}_\varepsilon(g_\varepsilon)\|_q = C_{p\to q}(G_\varepsilon) . \tag{4}$$

Assuming $\mathscr{G}$ to be bounded, (4) together with (3) and the fact that $\|h_\varepsilon g_\varepsilon\|_p \leqq \|g_\varepsilon\|_p = 1$ implies that $\|h_\varepsilon g_\varepsilon\|_p \to 1$ as $\varepsilon \to 0$. Then

$$C_{p\to q}(G) = \lim_{\varepsilon\to 0} C_{p\to q}(G_\varepsilon) \leqq \lim_{\varepsilon\to 0} \mathscr{R}_{p\to q}(G, h_\varepsilon g_\varepsilon) \leqq C_{p\to q}(G) . \tag{5}$$

This proves the theorem in the bounded case since $h_\varepsilon g_\varepsilon$ is a Gaussian function (which is real in Cases (A) and (B)).

In case $\mathscr{G}$ is unbounded, (4) and (5) imply that

$$\infty = \lim_{\varepsilon\to 0} C_{p\to q}(G_\varepsilon) \leqq \lim_{\varepsilon\to 0} \|h_\varepsilon g_\varepsilon\|_p \mathscr{R}_{p\to q}(G, h_\varepsilon g_\varepsilon) \leqq \lim_{\varepsilon\to 0} \mathscr{R}_{p\to q}(G, h_\varepsilon g_\varepsilon)$$

which proves the theorem since $h_\varepsilon g_\varepsilon$ is a Gaussian function. □

4.2. Remarks and examples. Theorem 4.1(∗) is a formula for the $L^p(\mathbf{R}^n) \to L^q(\mathbf{R}^n)$ norm of $\mathscr{G}$. The same formula is, of course, also valid for nondegenerate kernels, but in that case we are assured that there is precisely one g that achieves the supremum. In the degenerate case a maximizer may not exist—*even if $\mathscr{G}$ is bounded*—as the examples below show. In any event, the evaluation of this formula is, in general, a difficult nonlinear algebraic exercise, although it is simple in many applications.

For example, when $G(x, y) = \exp\{2i(x, y)\}$ (the Fourier transform kernel), it is easy to deduce from 4.1(∗) that G is bounded if and only if $q = p' \geqq 2$, in which case a Gaussian function is a maximizer if and only if it has the form

$$g(x) = \mu \exp\{-(x, Jx) + (l, x)\}$$

with J positive definite, real and symmetric and $l \in \mathbf{C}$. Both J and l are arbitrary. This g is not necessarily centered even though G is. In the degenerate case it is not asserted that every maximizer must be centered when G is centered. The sharp constant is then $C_{p\to p'} = (C_p^B)^n$ with

$$C_p^B = \pi^{1/p'} p^{1/2p} (p')^{-1/2p'} . \tag{1}$$

[Note: The Fourier transform is an example of *both* Cases (B) and (C). While the proof of Theorem 3.3 (Case (B)) required 4.1(∗) and 4.2(1), the proof of Theorem 3.4 (Case (C)) did not. Therefore no circular reasoning is involved because 3.4 ⇒ 4.1(∗) for Case (C) ⇒ 4.2(1) ⇒ 3.3 ⇒ 4.1(∗) for Case (B).]

Another example is the (real convolution operator $G(x, y) = \exp\{-\lambda(x-y, x-y)\}$ which, using Theorem 4.1, turns out to be bounded if and only if $p \leqq q$ (see [BL] Section 4 for more details). There is a maximizing Gaussian function if and only if $p < q$ and it must have the form

$$g(x) = \exp\{-J(x, x) + (l, x)\} \tag{2}$$

Invent. Math. *102*, 179–208 (1990)

with $J = \lambda\left(\frac{p'}{q'} - 1\right)$ and with $l \in \mathbf{R}^n$ arbitrary. Also

$$(C_{p\to q})^{2/n} = \left(\frac{\pi}{\lambda}\right)^{1/p' + 1/q} (p' - q')^{1/p - 1/q} p^{1/p} (q')^{1/p'} q^{-1/q} (p')^{-1/q'} . \tag{3}$$

When $p = q$ the limiting value $C_{p\to q} = \pi/\lambda$ is correct but, since $J = 0$ in this case, there is no Gaussian maximizer. Indeed, there is *no maximizer of any kind* in this case. To prove this, note that $G(x, y) = H(x - y)$ with $H(x) = \exp\{-\lambda(x, x)\}$ and $\int H(x-y) f(y) \mathrm{d}y = \int f(x-y) H(y) \mathrm{d}y$. Then, by Minkowski's inequality,

$$\begin{aligned}\left\{\int \left|\int f(x-y) H(y) \mathrm{d}y\right|^p \mathrm{d}x\right\}^{1/p} &\leqq \int \left\{\int |f(x-y)|^p H(y)^p \mathrm{d}x\right\}^{1/p} \mathrm{d}y \\ &= \|f\|_p \int H(y) \mathrm{d}y = \left(\frac{\pi}{\lambda}\right)^{n/2} \|f\|_p .\end{aligned} \tag{4}$$

Since the condition in Lemma 3.1 for equality is clearly not satisfied, and since $(\pi/\lambda)^{n/2}$ has already been shown to be the sharp bound, a maximizer cannot exist.

A second example of a degenerate G that is bounded but does not have a maximizer is the following modification of the Fourier transform in $\mathbf{R}^1$ with $\lambda > 0$.

$$G_\lambda(x, y) = \exp\{-\lambda y^2 - 2\mathrm{i}xy\} . \tag{5}$$

It is easily verified for all p that $\mathscr{R}_{p\to q}(g)$ is unbounded on *complex* Gaussian functions when $q < 2$. Thus, it can be assumed that $q \geqq 2$, which places us in Case (B) of Sect. I. If $f_J(x) = \exp\{-Jx^2\}$ is an arbitrary Gaussian function, one finds that when $q \geqq 2$ the optimum choice is J real and

$$[\mathscr{R}_{p\to q}(f_J)]^2 = \pi^{1/q + 1/p'} p^{1/p} q^{-1/q} J^{1/p} (\lambda + J)^{-1/q'} . \tag{6}$$

By maximizing this with respect to J one finds that $C_{p\to q}$ is finite whenever $p \geqq q'$ and $C_{p\to q} = \infty$ when $p < q'$. If $p = q'$ there is no J that maximizes the right side of (6) (i.e., $J \to \infty$), although the right side is bounded. Indeed, *there is no maximizer* of any kind when $p = q'$. If there were a maximizer $f \in L^p(\mathbf{R}^1)$ then, by imitating the proof of Theorem 4.1, it is easily seen that $C_{p\to p'}(G_{\mu\lambda}) > C_{p\to p'}(G_\lambda)$ when $0 < \mu < 1$. This contradicts the conclusion of Theorem 4.1 which states that the supremum over J of the right side of (6) correctly gives $C_{p\to p'}(G_\lambda)$ for every λ, but this supremum is obviously independent of λ.

These examples motivate the following theorem.

4.3. Theorem (a condition for Gaussian maximizers). *Let G be a degenerate Gaussian kernel with the property that the $n \times n$ real, symmetric matrices A and B in* (1.1) *are both positive definite. If $1 < p \leqq q < \infty$ then $\mathscr{G}$ is bounded from $L^p(\mathbf{R}^n)$ to $L^q(\mathbf{R}^n)$. If, additionally, $p < q$ then $\mathscr{G}$ has a maximizer which is a Gaussian function.*

If G is also real then obviously A and B must be positive definite if $\mathscr{G}$ is bounded at all. In this real, degenerate case $\mathscr{G}$ is unbounded when $1 < q < p < \infty$ and $\mathscr{G}$ has no maximizer of any kind when $1 < p = q < \infty$.

Proof. It can be assumed that G is centered and, as in the proof of Theorem 3.3, we can use the fact that A and B are positive definite to change variables so that $G(x, y)$ is brought into the canonical form

$$G(x, y) = \exp\{-(x, x) - (y, y) - 2(x, Ey) - 2\mathrm{i}(x, Hy)\} , \tag{1}$$

where E and H are real matrices and E is also diagonal. In the real case $H = 0$. Since $M = \begin{pmatrix} I & E \\ E & I \end{pmatrix}$ must be positive semidefinite, the eigenvalues $e_1, \ldots, e_n$ of E must be in the interval $[-1, 1]$. Since G is degenerate at least one of the e_i's (say e_1) is $+1$ or -1 and, by changing y to $-y$ if necessary, we can assume that $e_1 = -1$. Thus, $G(x, y)$ contains the factor $\exp\{-(x_1 - y_1)^2\}$.

In the real case, $H = 0$, G in (1) is seen to be a tensor product of operators on $\mathbf{R}^1$, i.e., $G(x, y) = G_1(x_1, y_1) \ldots G_n(x_n, y_n)$. If $p > q$ the operator $\mathscr{G}_1$ corresponding to e_1 is unbounded, as shown in 4.2, so $\mathscr{G}$ is unbounded as well. In case $p \leqq q$ the Minkowski inequality argument in the first part of the proof of Theorem 3.2 (applied sequentially to $\mathscr{G}_1, \mathscr{G}_2, \ldots, \mathscr{G}_n$) shows that any maximizer, F, for $\mathscr{G}$ must be of the product form, i.e., $F(y_1, \ldots, y_n) = f_1(y_1) \ldots f_n(y_n)$ and each f_i must be a maximizer for the corresponding $\mathscr{G}_i$. When $p = q$, however, $\mathscr{G}_1$ does not have a maximizer as stated in 4.2 and therefore $\mathscr{G}$ has no maximizer. When $p < q$ we know from 4.2 that each $\mathscr{G}_i$ has a Gaussian maximizer g_i. Since $\prod_i g_i(x_i)$ is a Gaussian function on $\mathbf{R}^n$, the proof for the real case is complete.

For the general case with $p \leqq q$, let $G^0(x, y)$ be the real kernel given by (1) but with H set equal to zero and let $\mathscr{G}^0$ denote the corresponding operator. If $f \in L^p(\mathbf{R}^n) \cap \mathbf{L}^1(\mathbf{R}^n)$ then clearly $\mathscr{R}_{p\to q}(G, f) \leqq \mathscr{R}_{p\to q}(G^0, F)$, where $F \equiv |f|$. Since $\mathscr{G}^0$ is bounded when $p \leqq q$, then $\mathscr{G}$ is also bounded. Referring now to Theorem 4.1 let G_ε be the kernel defined in that proof, i.e., $G_\varepsilon(x, y) = G(x, y)h_\varepsilon(x)h_\varepsilon(y)$ with $h_\varepsilon(x) = \exp\{-\varepsilon(x, x)\}$, and let g_ε denote its unique Gaussian maximizer with $\|g_\varepsilon\|_p = 1$. Let $g_\varepsilon(x) = \mu_\varepsilon \exp\{-(x, J_\varepsilon x) - i(x, K_\varepsilon x)\}$ with J and K real, symmetric and with J_ε positive definite. Define $g_\varepsilon^0(x) = \mu_\varepsilon \exp\{-x, J_\varepsilon x)\}$. Let $\varepsilon \to 0$ through the sequence $\varepsilon = 1/j$ with $j = 1, 2, 3. \ldots$ There is a subsequence of the j's (which we continue to denote by j) such that the eigenvectors of J_ε and K_ε have limits as $j \to \infty$ (because the manifold $O(n)$ is compact). The corresponding eigenvalues of J_ε must be uniformly bounded away from 0 and ∞ since otherwise $\mathscr{R}_{p\to q}(G^0, g_\varepsilon^0)$ will converge to zero, as the following computation shows.

Apart from irrelevant constants, $\|g_\varepsilon^0\|_p = |J_\varepsilon|^{-1/p}$, where $|\cdot|$ denotes the determinant. Also, $\|\mathscr{G}^0(g_\varepsilon^0)\|_q = |J_\varepsilon + I|^{-1}|I - E(J_\varepsilon + I)^{-1}E|^{-1/q}$. Using the fact that $|I - M^T M| = |I - MM^T|$ for any real matrix, M, we have $|I - E(J_\varepsilon + I)^{-1}E| = |I - (J_\varepsilon + I)^{-1/2}E^2(J_\varepsilon + I)^{-1/2}| = |J_\varepsilon + I|^{-1}|J_\varepsilon + I - E^2| \geqq |J_\varepsilon + I|^{-1}|J_\varepsilon|$. Therefore

$$\mathscr{R}_{p\to q}(G^0, g_\varepsilon^0) \leqq |J_\varepsilon|^{1/p - 1/q}|J_\varepsilon + I|^{-1/q'} = \prod_{i=1}^{n} (J_\varepsilon^i)^{1/p - 1/q}(1 + J_\varepsilon^i)^{-1/q'},$$

where the J_ε^i's are the eigenvalues of J_ε. Since $p < q$ the function $t \mapsto t^{1/p - 1/q}(1 + t)^{-1/q'}$ is bounded and goes to zero as $t \to 0$ or $t \to \infty$.

The possibility that $\mathscr{R}_{p\to q}(G^0, g_\varepsilon^0) \to 0$ is not allowed by 4.1(*) and the fact that $\mathscr{R}_{p\to q}(G^0, g_\varepsilon^0) \geqq \mathscr{R}_{p\to q}(G, g_\varepsilon)$. Thus we can pass to a further subsequence such that J_ε has a positive definite limit J as $\varepsilon \to 0$. This implies that μ_ε also has a finite, nonzero limit μ.

The eigenvalues of K_ε must also stay bounded away from infinity for otherwise g_ε would tend weakly to zero in $L^p(\mathbf{R}^n)$ and then the function $\mathscr{G}(g_\varepsilon)$ would tend to zero pointwise. (This is so because the function $y \mapsto G(x, y)\exp\{-\frac{1}{2}(y, Jy)\}$ is in $L^{p'}(\mathbf{R}^n)$ for each x.) But $\mathscr{G}(g_\varepsilon)$ is bounded above pointwise by $\mathscr{G}^0(g_\varepsilon^0)$, and the pointwise convergence to zero would imply by dominated convergence that $\mathscr{G}(g_\varepsilon)$ converges to zero in $L^q(\mathbf{R}^n)$ norm. Thus, by passing to a further subsequence, J_ε and

K_ε have limits J and K. From this it follows that g_ε converges strongly in $L^p(\mathbf{R}^n)$ norm to $g(x) = \mu \exp\{-(x, Jx) - \mathrm{i}(x, Kx)\}$.

The Gaussian function g is the desired maximizer for $\mathscr{G}$. First note that $h_\varepsilon g \to g$ in $L^p(\mathbf{R}^n)$ norm as $\varepsilon \to 0$. Also $g_\varepsilon \to g$, and thus we can write $h_\varepsilon g_\varepsilon = g + \Delta_\varepsilon$ with $\delta_\varepsilon \equiv \|\Delta_\varepsilon\|_p \to 0$ as $\varepsilon \to 0$. Then, since $\mathscr{G}$ is bounded,

$$C_{p\to q}(G) \geqq \mathscr{R}_{p\to q}(G, g) \geqq \mathscr{R}_{p\to q}(G, h_\varepsilon g_\varepsilon) - C_{p\to q}\delta_\varepsilon .$$

Taking the limit $\varepsilon \to 0$,

$$C_{p\to q}(G) \geqq \mathscr{R}_{p\to q}(G, g) \geqq \limsup_{\varepsilon\to 0} \mathscr{R}_{p\to q}(G, h_\varepsilon g_\varepsilon) .$$

But by Eq. (5) of the proof of Theorem 4.1, this latter limit equals $C_{p\to q}(G)$. □

4.4. Remarks and conjectures. Formula 4.1(∗) gives the sharp bound. The question that is incompletely resolved here is whether there is a Gaussian maximizer in the degenerate case or, indeed, any maximizer at all. In the cases of most interest (e.g., Nelson's kernel of Sect. I and the Fourier transform) the existence of a Gaussian maximizer can easily be verified by simple computation. The general case is algebraically complex, although Theorem 4.3 does give a criterion for a Gaussian maximizer and it completely settles the case of real Gaussian kernels. Indeed, as shown in 4.2, a maximizer need not exist even if $\mathscr{G}$ is bounded.

The examples given here lead to the following *conjectures.*

(1) If there is a maximizer for cases (A), (B) or (C) of Sect. I then there is a Gaussian maximizer.
(2) There is a maximizer in these cases if and only if the unique Gaussian maximizer g_ε for the mollified kernel $G_\varepsilon(x, y) = G(x, y)h_\varepsilon(x)h_\varepsilon(y)$ defined in the proof of Theorem 4.1 has a strong limit g in $L^p(\mathbf{R}^n)$ as $\varepsilon \to 0$.

Maximizers need not be unique, as shown in 4.2, but *if there is any Gaussian maximizer for $p < q$ then every maximizer is a Gaussian.* This is Theorem 4.5, and it completely settles the Fourier transform case, for example. (Note that when $p = q = 2$, every function in $L^2(\mathbf{R}^n)$ is a maximizer for the Fourier transform and thus there is at least one case in which there are maximizers that are not Gaussians.)

Theorem 4.5 also completely settles the real Case (A) because, by Theorem 4.3, no maximizer exists in this case when $p \geqq q$ and a Gaussian maximizer does exist when $p < q$.

4.5. Theorem (when $p < q$, a Gaussian maximizer implies all maximizers are Gaussians). *Let $1 < p < q < \infty$ and let G be a degenerate Gaussian kernel. Assume that $\mathscr{G}$ is a bounded operator from $L^p(\mathbf{R}^n)$ to $L^q(\mathbf{R}^n)$ and that g is a Gaussian function that is a maximizer for $\mathscr{G}$. If $f \in L^p(\mathbf{R}^n)$ is another maximizer for $\mathscr{G}$ then f is also a Gaussian (but f is not necessarily proportional to g and f is not necessarily centered even if G is).*

Proof. Step 1. According to Lemmas 2.2 and 2.3 it can be assumed without loss of generality that both G and g are centered. As in the proof of Theorem 3.4, we study the kernel $G^{(2)} = G \otimes G$. For $F \in L^p(\mathbf{R}^{2n}) \cap L^1(\mathbf{R}^{2n})$ the inequalities (1)–(5) there are valid and we conclude that $C_{p\to q}(G^{(2)}) = (C_{p\to q})^2$, where $C_{p\to q} \equiv C_{p\to q}(G)$.

Step 2. If $f \in L^p(\mathbf{R}^n)$ is a maximizer for $\mathscr{G}$ then, using $O(2)$ invariance again,

$$F(y_1, y_2) \equiv f\left(\frac{y_1 + y_2}{\sqrt{2}}\right) g\left(\frac{y_1 - y_2}{\sqrt{2}}\right) \tag{1}$$

is obviously a maximizer for $\mathscr{G}^{(2)}$ if f is also in $L^1(\mathbf{R}^n)$, in which case $F \in L^1(\mathbf{R}^{2n})$. If $f \notin L^1(\mathbf{R}^n)$ consider the mollified function $F_j(y_1, y_2) = F(y_1, y_2) \exp\{-(y_1 + y_2, y_1 + y_2)/j\}$ for $j = 1, 2, \ldots$, which is in $L^1(\mathbf{R}^{2n})$. Clearly $F_j \to F$ strongly in $L^p(\mathbf{R}^{2n})$ as $j \to \infty$. The function $\mathscr{G}^{(2)} F_j$ can be computed as a $dy_1 dy_2$ integral of $G^{(2)} F_j$ and the result (using the $O(2)$ invariance of $G^{(2)}$ and a change of variables) is

$$(\mathscr{G}^{(2)} F_j)(x_1, x_2) = (\mathscr{G} f_j)\left(\frac{x_1 + x_2}{\sqrt{2}}\right) \cdot (\mathscr{G} g)\left(\frac{x_1 - x_2}{\sqrt{2}}\right)$$

with $f_j(y) = f(y) \exp\{-2(y, y)/j\}$. Now the qth power integral of $\mathscr{G}^{(2)} F_j$ can be computed by changing variables again and the result is $\| \mathscr{G}^{(2)} F_j \|_q = \| \mathscr{G} f_j \|_q \| \mathscr{G} g \|_q$. However $\| \mathscr{G} f_j \|_q \to \| \mathscr{G} f \|_q = C_{p \to q} \| f \|_p$ as $j \to \infty$ since $f_j \to f$ in $L^p(\mathbf{R}^n)$ norm, and we conclude that $\| \mathscr{G}^{(2)} F \|_q = \lim_{j \to \infty} \| \mathscr{G}^{(2)} F_j \|_q$ (by definition) $= (C_{p \to q})^2 \| F \|_p$, so that F is indeed a maximizer for $\mathscr{G}^{(2)}$.

Step 3. Since g is a Gaussian, it is obvious that the function $z \mapsto F(z, y)$ is in $L^p(\mathbf{R}^n) \cap L^1(\mathbf{R}^n)$ for each y and therefore that

$$K(x, y) \equiv \int G(x, z) F(z, y) dz \tag{2}$$

is well defined for each x and y in $\mathbf{R}^n$. Since $\mathscr{G}$ is a bounded operator, the function $x \mapsto K(x, y)$ is in $L^q(\mathbf{R}^n)$ for each y. We now assert that the function $y \mapsto K(x, y)$ is in $L^p(\mathbf{R}^n)$ for almost every $x \in \mathbf{R}^n$ and that this function satisfies

$$\{ \int [\int |K(x, y)|^q dx]^{p/q} dy \}^{q/p} = \int \{ \int |K(x, y)|^p dy \}^{q/p} dx , \tag{3}$$

with the understanding that both sides of (3) are finite. Formally, this assertion is a consequence of inequality (3) $\to$ (4) in the proof of Theorem 3.4 and the fact that all the inequalities (1)–(5) must be equalities since $F(y_1, y_2)$ is a maximizer for $\mathscr{G}^{(2)}$. If $F \in L^1(\mathbf{R}^{2n})$ this would be correct, but if $F \notin L^1(\mathbf{R}^{2n})$ a proof is needed.

Set $F_j(y_1, y_2) = F(y_1, y_2) \exp\{-(y_2, y_2)/j\}$ for $j = 1, 2, \ldots$. Clearly $F_j \in L^1(\mathbf{R}^{2n})$ and $F_j \to F$ strongly in $L^p(\mathbf{R}^{2n})$ as $j \to \infty$. (Note that this F_j is not the same one as in step 2.) Let $K_j(x, y)$ be as in (2) with F replaced by F_j, so that $K_j(x, y) = K(x, y) \exp\{-(y, y)/j\}$. The inequalities (1)–(5) in the proof of Theorem 3.4 are then valid with F replaced by F_j. As $j \to \infty$ the left side of these inequalities, namely $\| \mathscr{G}^{(2)} F_j \|_q^q$, converges to $\| \mathscr{G}^{(2)} F \|_q^q = (C_{p \to q})^{2q} \| F \|_p^q \equiv (C_{p \to q})^q Z$ since F is a maximizer. Likewise, the right side, namely $(C_{p \to q})^{2q} \| F_j \|_p^q$ also converges to $(C_{p \to q})^q Z$ since $F_j \to F$. Therefore the numbers

$$B_j \equiv \{ \int [\int |K_j(x, y)|^q dx]^{p/q} dy \}^{q/p} - \int \{ \int |K_j(x, y)|^p dy \}^{q/p} dx \tag{4}$$

(which are nonnegative by Minkowski's inequality) must converge to zero as $j \to \infty$. Moreover, each term in B_j is bounded by $(C_{p \to q})^q \| F_j \|_p^q < Z$, and each term converges to Z as $j \to \infty$ (because of inequalities (1)–(5)). The first term in B_j is

$$A_j \equiv \left\{ \int [\int |K(x, y)|^q dx]^{p/q} \exp\left\{ -\frac{p}{j}(y, y) \right\} dy \right\}^{q/p}$$

and, by the monotone convergence theorem, A_j converges to $A \equiv$ (the left side of (3)). Therefore $A = Z$. The second term in B_j is

$$D_j \equiv \int \left\{ \int |K(x, y)|^p \exp\left\{ -\frac{p}{j}(y, y) \right\} dy \right\}^{q/p} dx .$$

Invent. Math. *102*, 179–208 (1990)

The inner integral (call it $E_j(x)$) converges (by monotone convergence) to $E(x) \equiv \int |K(x, y)|^p dy$. The function E is measurable since it is the monotone limit of measurable functions E_j. Then $\int \{E_j\}^{q/p}$ converges to $\int \{E\}^{q/p}$ by monotone convergence, so D_j converges to the right side of (3). But, as stated above, D_j also converges to Z, so the two sides of (3) are equal and $E(x)$ is finite for almost every x, as asserted.

Step 4. Since $q > p$, the strong form of Minkowski's inequality and the equality in (3) implies the existence of measurable functions α and $\beta : \mathbf{R}^n \to [0, \infty)$ such that

$$|K(x, y)| = \alpha(x)\beta(y) \tag{5}$$

for almost every x and y in $\mathbf{R}^n$. Writing $G(x, y) = \exp\{-(x, Ax) - (y, By) - 2(x, Dy)\}$ as usual (with A and B real, symmetric, positive definite), and writing $g(y) = \exp\{-(y, Jy)\}$ (with J symmetric and $\mathrm{Re}(J)$ positive definite) a simple computation gives

$$\begin{aligned} K(x, y) = \exp\{&-(x, Ax) + (D^T x, (B + J)^{-1} D^T x) \\ &- (y, Jy)\} Q((B + J)y - D^T x) \end{aligned} \tag{6}$$

with $Q : \mathbf{C}^n \to \mathbf{C}$ given by

$$Q(w) = \exp\{-(w, (B + J)^{-1} w)\} \int f\left(\frac{z}{\sqrt{2}}\right) \exp\{-(z, (B + \tfrac{1}{2} J)z) + 2(z, w)\}\, dz\,. \tag{7}$$

Evidently Q is an entire analytic function of order at most 2. Define the function $M : \mathbf{R}^{2n} \to \mathbf{C}$ by $M(x, y) = Q((B + J)y - D^T x)$. Plainly, since Q is entire M has an extension to an entire analytic function from $\mathbf{C}^{2n}$ to $\mathbf{C}$; call this extension N. The function $N^* : \mathbf{C}^{2n} \to \mathbf{C}$ defined by $N^*(x, y) = \overline{N(\bar{x}, \bar{y})}$ for x and $y \in \mathbf{C}^n$ is also entire analytic, and thus $P \equiv NN^*$ is entire analytic as well. It is also true that $P(x, y) = |M(x, y)|^2$ when x and y are in $\mathbf{R}^n$. From (5) and (6),

$$P(x, y) = \gamma(x)\delta(y) \tag{8}$$

for almost every x and y in $\mathbf{R}^n$, and where γ and $\delta : \mathbf{R}^n \to [0, \infty)$ are the measurable functions given by $\gamma(x) = \alpha(x)^2 \exp\{2(x, Ax) - 2\mathrm{Re}((D^T x, (B + J)^{-1} D^T x))\}$ and $\delta(y) = \beta(y)^2 \exp\{2\mathrm{Re}((y, Jy))\}$. If y_0 is a value of y such that $\delta(y_0) \neq 0$ and such that (8) holds for almost every x, we see by substituting this y_0 in (8) that γ has an extension to an entire analytic function. Likewise, δ has an extension. Thus (8) holds for *every* x and y in $\mathbf{C}^n$ (because if two entire functions agree almost everywhere on $\mathbf{R}^n \times \mathbf{R}^n$ then they agree on all of $\mathbf{C}^n \times \mathbf{C}^n$).

Now suppose that $\gamma(x_0) = 0$ for some x_0 in $\mathbf{C}^n$. Then, by (8), $P(x_0, y) = 0$ for every $y \in \mathbf{C}^n$, which implies that for each y either (*i*) $N(x_0, y) = 0$ or (*ii*) $N^*(x_0, y) = 0$. This, in turn, means that for each $y \in \mathbf{C}^n$ either (i) $N(x_0, y) \equiv Q((B + J)y - D^T x_0) = 0$ or (ii) $N(\bar{x}_0, \bar{y}) \equiv Q((B + J)\bar{y} - D^T \bar{x}_0) = 0$. Necessarily, either case (i) holds for all y in some set $S \subset \mathbf{C}^n$ of positive $2n$-dimensional Lebesgue measure $\mathscr{L}^{2n}$ or case (ii) holds in some set S of positive $\mathscr{L}^{2n}$ measure. As y ranges over S both $(B + J)y$ and $(B + J)\bar{y}$ range over sets of positive $\mathscr{L}^{2n}$ measure (because $\mathrm{Re}(B + J)$ is positive definite and therefore $\mathit{Rank}(B + J) = n$). An analytic function that vanishes on a set of positive $\mathscr{L}^{2n}$ measure vanishes identically, and thus Q would vanish identically if $\gamma(x_0) = 0$. This contradicts the fact that $K(x, y)$ is not identically zero. Thus, the assumption that $\gamma(x_0) = 0$ is not possible, and it will be assumed henceforth that $\gamma(x) \neq 0$ for all $x \in \mathbf{C}^n$.

Define the set $A = \{y \in \mathbf{R}^n : \delta(y) \neq 0\} \subset \mathbf{R}^n$. This set A has positive n-dimensional Lebesgue measure $\mathscr{L}^n$, for otherwise $K(x, y) = 0$, $\mathscr{L}^{2n}$ almost everywhere. (In fact $\mathscr{L}^n(\mathbf{R}^n \sim A) = 0$ because δ is analytic and δ does not vanish identically, but this fact is not needed.) For $y \in A$, the function $Z_y : \mathbf{C}^n \to \mathbf{C}$ defined by $Z_y(x) = K(x, y)$ is entire analytic of order at most 2 and never zero (because $\gamma(x)$ is never zero). Then Z_y has the form

$$Z_y(x) = K(x, y) = \exp\{-(x, T_y x) - (R_y, x) + \mu_y\}\,, \tag{9}$$

where T_y is a complex, symmetric matrix, $R_y \in \mathbf{C}^n$ and $\mu_y \in \mathbf{C}$ (all of which depend on y). I thank Eric Carlen for the simple proof of this fact, which is that Z_y, being zero free, has an entire analytic logarithm, i.e., $Z_y = \exp\{H_y\}$. Then, since Z_y has order at most 2, $|H_y(x)|$ is bounded above by (const.) $|x|^2$. By a well known argument using Cauchy's integral formula, H_y must be a polynomial whose order is at most 2, i.e., Z_y has the form stated in (9).

Step 5. As noted in step 2, the function $x \mapsto K(x, y)$ is in $L^q(\mathbf{R}^n)$ for almost every $y \in \mathbf{R}^n$. By (4) $\to$ (5) of Theorem 3.4, the function $z \mapsto F(z, y)$ (which is in $L^p(\mathbf{R}^n)$ for almost every y) must be a maximizer of $\mathscr{R}_{p \to q}$ for almost every y. (Note that $z \mapsto F(z, y)$ cannot be the zero function for any y since g never vanishes.) Thus there is at least one point $y_0 \in \mathbf{R}^n$ such that $\delta(y_0) \neq 0$ and (9) holds and such that $z \mapsto F(z, y_0)$ is a maximizer in $L^p(\mathbf{R}^n)$. Fix this y_0 henceforth and denote the matrix in (9) simply by T. There is then a function $h \in L^{q'}(\mathbf{R}^n)$ with $\|h\|_{q'} = 1$ such that

$$\int h(x) K(x, y_0) \mathrm{d}x = \|K(\cdot, y_0)\|_q = C_{p \to q} \|F(\cdot, y_0)\|_p\,. \tag{10}$$

Since $K(\cdot, y_0) \in L^q(\mathbf{R}^n)$, the matrix T must satisfy $\mathrm{Re}(T)$ is positive definite and therefore $K(\cdot, y_0)$ is a Gaussian. The optimum h satisfies $h(x) = $ (const.) $|K(x, y_0)|^q / K(x, y_0)$ for $x \in \mathbf{R}^n$ and therefore h is also a Gaussian (and hence $h \in L^1(\mathbf{R}^n)$). As remarked in step 3, $F(\cdot, y_0)$ is in $L^1(\mathbf{R}^n)$. Therefore the function $(x, y) \mapsto h(x) G(x, y) F(y, y_0)$ is in $L^1(\mathbf{R}^{2n})$ and Fubini's theorem can be applied to (10). Thus,

$$\int h(x) K(x, y_0) \mathrm{d}x = \int \left\{ \int h(x) G(x, z) \mathrm{d}x \right\} F(z, y_0) \mathrm{d}z\,. \tag{11}$$

Since h is a Gaussian, the inner integral in (11) (call it $k(z)$) is also a Gaussian. Since $F(\cdot, y_0)$ is a maximizer, $F(z, y_0) = (\text{const.}) |k(z)|^{p'} / k(z) \equiv r(z)$ for almost every $z \in \mathbf{R}^n$. Clearly r is a Gaussian and, by (1)

$$f\left(\frac{z + y_0}{\sqrt{2}}\right) g\left(\frac{z - y_0}{\sqrt{2}}\right) = r(z) \tag{12}$$

for almost every $z \in \mathbf{R}^n$. Setting $z = w - y_0$, (12) yields $f(w/\sqrt{2}) = r(w - y_0)/g((w - 2y_0)/\sqrt{2})$, which is a Gaussian (in w) as asserted in the theorem. □

V. Gaussian kernels from $L^p(\mathbf{R}^n)$ to $L^q(\mathbf{R}^m)$

This section consists essentially of a simple remark, but it can be a useful one in applications, e.g., in [L1]. Let G be a Gaussian kernel on $\mathbf{R}^m \times \mathbf{R}^n$ with $m \neq n$, i.e., $G(x, y)$ is given by (1.1) with A $m \times m$ symmetric, B $n \times n$ symmetric, D $m \times n$ and

$L \in C^{m+n}$, and with M in (1.3) a positive semidefinite $(m+n) \times (m+n)$ matrix. Evidently Lemmas 2.1, 2.2 and 2.3 continue to hold in this case, and it can be assumed without loss of generality that A and B are real and $L = 0$. The linear operator $\mathscr{G}$ from $L^p(\mathbf{R}^n)$ to $L^q(\mathbf{R}^m)$ and the norm $C_{p\to q}(G)$ are defined, *mutatis mutandis*, as in Sect. I. The remark is the following.

5.1. Theorem (extension to $m \neq n$). *Let G be a Gaussian kernel on $\mathbf{R}^m \times \mathbf{R}^n$ as defined above. Then all the preceding theorems and lemmas in this paper holds, mutatis mutandis, in this more general case.*

Proof. Suppose $m < n$ and extend G to a Gaussian kernel, $\tilde{G}$, on $\mathbf{R}^n \times \mathbf{R}^n$ by

$$\tilde{G}(x, y) = h(x_1)G(x_2, y)$$

where $x \in \mathbf{R}^n$ is written as (x_1, x_2) with $x_1 \in \mathbf{R}^{n-m}$ and $x_2 \in \mathbf{R}^m$, and where $h(x_1) \equiv \exp\{-(x_1, x_1)\}$. Let $\tilde{\mathscr{G}}$ be the corresponding operator from $L^p(\mathbf{R}^n)$ to $L^q(\mathbf{R}^n)$. Note that $\tilde{G}$ has the same properties as G, i.e., the degeneracy or nondegeneracy of $\tilde{G}$ is the same as that of G; $\tilde{G}$ is in Case (A), (B) or (C) if G is; the $n \times n$ matrix $\begin{pmatrix} I & 0 \\ 0 & A \end{pmatrix}$ is positive definite if and only if A is. Also, $\tilde{\mathscr{G}}$ is unbounded if $\mathscr{G}$ is, and it will be assumed henceforth that $\mathscr{G}$ is bounded.

If $f \in L^p(\mathbf{R}^n)$ then evidently, as functions in $L^q(\mathbf{R}^n)$, $(\tilde{\mathscr{G}} f)(x) = h(x_1)(\mathscr{G} f)(x_2)$ and thus $\|\tilde{\mathscr{G}} f\|_{L^p(\mathbf{R}^n)} = \|h\|_{L^q(\mathbf{R}^{n-m})} \|\mathscr{G} f\|_{L^q(\mathbf{R}^m)}$. This proves that $C_{p\to q}(\tilde{G}) = C_{p\to q}(G)\|h\|_{L^q(\mathbf{R}^{n-m})}$ and that f is a maximizer for $\tilde{\mathscr{G}}$ if and only if f is a maximizer for $\mathscr{G}$. This concludes the $m < n$ case.

If $m > n$ duality can be used: $C_{p\to q}(G) = C_{q'\to p'}(G^T)$ where $G^T(x, y) = G(y, x)$. This changes the $m > n$ case into the $m < n$ case and, since all the theorems in this paper are "duality invariant", the $m > n$ case is proved. □

Remark. Clearly the proof of Theorem 5.1 is such that if other cases with $m = n$ are settled in the future then Theorem 5.1 for $m \neq n$ holds for those cases as well.

VI. Multilinear forms in the real case and Young's inequality

After Sects. I to V were completed, Eric Carlen suggested that the same methods should yield similar results for real multilinear forms. Indeed this is so and the proof is outlined here (the omitted details are merely a repetition of those given before). Some remarks about the complex case will also be made here. Finally, Theorem 6.2 contains an application of the result in Sect. 6.1 for real multilinear forms: The truly multidimensional generalization of Young's inequality, which was surmised in [BL, p. 162], will be proved.

6.1. **Multilinear forms.** For $i = 1, 2, \ldots, K$ let n_i be a positive integer and let x_i denote a point in $\mathbf{R}^{n_i}$. The point $X = (x_1, \ldots, x_K)$ denotes a point in $\mathbf{R}^N$ with $N = \sum_{i=1}^K n_i$. Let $G(X)$ be a "Gaussian kernel", i.e.,

$$G(X) = \exp\left\{-\sum_{i=1}^K \sum_{j=1}^K (x_i, A_{ij}x_j) + 2(L, X)\right\},$$

where A_{ij} is a $n_i \times n_j$ matrix with $A_{ij} = A_{ji}^T$, and where $L \in \mathbf{C}^N$. The $N \times N$ symmetric matrix A is the matrix whose blocks are the A_{ij}'s and G is said to be **nondegenerate** if $M \equiv \mathrm{Re}(A)$ is positive definite. Otherwise $M \geqq 0$ and G is **degenerate**.

Let $P = (p_1, \ldots, p_K)$ satisfy $1 < p_i < \infty$ for each i. The **multilinear form** is

$$\mathscr{G}(f_1, \ldots, f_K) = \int G(x_1, \ldots, x_K) \prod_{i=1}^{K} f_i(x_i) dx_1 \ldots dx_K , \tag{1}$$

where the integration is over $\mathbf{R}^{n_1} \times \mathbf{R}^{n_2} \times \ldots \times \mathbf{R}^{n_K}$ and each $f_i \in L^{p_i}(\mathbf{R}^{n_i})$. The problem is to evaluate

$$C_P = \sup \mathscr{G}(f_1, \ldots, f_K) \tag{2}$$

where the supremum is over f_i's with $\| f_i \|_{p_i} = 1$. As before, if G is degenerate we have to take $f_i \in L^{p_i}(\mathbf{R}^{n_i}) \cap L^1(\mathbf{R}^{n_i})$ and then take limits.

The cases treated in Sects. I to V correspond to $K = 2$ with $p_1 = p$ and $p_2 = q'$. The case $K = 1$ is trivial—by Hölder's inequality.

Lemma 2.1 is easily generalized to the complex, nondegenerate multilinear case; the details are left to the reader. The conclusion of Lemma 2.1 holds for each f_i in a maximizing set $(f_1, \ldots, f_K)$. The conclusions (a), (b) and (c) follow by fixing all the f_j's with $j \neq i$ and then investigating the dependence of $\mathscr{G}(f_1, \ldots, f_K)$ on f_i.

Lemma 2.2 obviously carries through as well; that is A_{ii} can be assumed to be real and G can be assumed to be centered, i.e., $L = 0$. Likewise Lemma 2.3 carries through: When G is centered (i.e., $L = 0$) and when the supremum in (2) is restricted to Gaussian functions f_i, then each f_i can be taken to be centered and, in the nondegenerate case, each f_i *must be* centered.

Let us now turn to the real case, i.e., each A_{ij} is real and $L = 0$. *Theorem* 3.2 *for the nondegenerate case carries through for every choice* of P. The maximizing K-tuple $(f_1, \ldots, f_K)$ is unique (up to multiplicative constants) and each $f_i(x) = \exp\{-(x, J_i x)\}$ with J_i being real and positive definite. To prove that f_1, say, has this property we write (with $q = p_1'$)

$$C_P = \sup \| \tilde{\mathscr{G}}(f_2, \ldots, f_K) \|_q$$

where the supremum is on $f_2, \ldots, f_K$ with $\| f_j \|_{p_j} = 1$ and where

$$\tilde{\mathscr{G}}(f_2, \ldots, f_K)(x_1) = \int G(x_1, \ldots, x_K) \prod_{j=2}^{K} f_j(x_j) dx_2 \ldots dx_K .$$

As before, we replace $\tilde{\mathscr{G}}$ by $\tilde{\mathscr{G}}^{(2)} = \tilde{\mathscr{G}} \otimes \tilde{\mathscr{G}}$ and $f_2(x_2), \ldots, f_K(x_K)$ by $F_2(x_2, y_2), \ldots, F_K(x_K, y_K)$ with $F_j \in L^{p_j}(\mathbf{R}^{2n_j})$. To imitate the inequalities (1)–(5) in Theorem 3.2, define

$$K(x_1, y_2, \ldots, y_K) = \int G(x_1, x_2, \ldots, x_K) \prod_{j=2}^{K} F_j(x_j, y_j) dx_2 \ldots dx_K .$$

Then, proceeding as in (1)–(5) (and with the F_j nonnegative for the same reason as before)

$$\begin{aligned} \| \tilde{\mathscr{G}}^{(2)}(F_2, \ldots, F_K) \|_q^q &= \int [\int G(y_1, \ldots, y_K) K(x_1, y_2, \ldots, y_K) dy_2 \ldots dy_K]^q dx_1 dy_1 \\ &\leqq \int \{ \int G(y_1, \ldots, y_K) [\int K(x_1, y_2, \ldots, y_K)^q dx_1]^{1/q} \\ &\quad \times dy_2 \ldots dy_K \}^q dy_1 \\ &\leqq (C_P)^q \int \{ \int G(y_1, \ldots, y_K) \prod_{j=2}^{K} h_j(y_j) dy_2 \ldots dy_K \}^q dy_1 \\ &\quad (\text{with } h_j(y) = [\int F_j(x, y)^{p_j} dx]^{1/p_j}) \\ &\leqq (C_P)^{2q} \prod_{j=2}^{K} \| F_j \|_{p_j}^q . \end{aligned}$$

Invent. Math. *102*, 179–208 (1990)

As before, Minkowski's inequality implies that

$$K(x_1, y_2, \ldots, y_K) = A(x_1)E(y_2, \ldots, y_K),$$

which then implies that

$$\tilde{\mathscr{G}}^{(2)}(F_2, \ldots, F_K)(x_1, y_1) = A(x_1)Z(y_1).$$

However, $\tilde{\mathscr{G}}^{(2)}(F_2, \ldots, F_K) = F_1^{p_1-1}$, and hence F_1 is a product function. The rest of the proof is identical to the proof of Theorem 3.2.

By taking limits, the analogue of Theorem (4.1) hold in the degenerate case whenever it is known that the nondegenerate case has a Gaussian maximizer K-tuple. *In particular, Theorem* 4.1 *holds in the real case.* The analogue of 4.1(*) is that C_P is given by (2) with the supremum restricted to centered Gaussian functions. Likewise, Theorem 4.3 extends to the multilinear case under the same assumption about the nondegenerate case; the analogous hypothesis is that each A_{ii} is positive definite. *In particular Theorem* 4.3 *holds in the real case.*

These results can be used to derive the sharp constants in the fully multidimensional generalized Young's inequality. Recall that Young's original inequality states that if $f \in L^p(\mathbf{R}^n)$ and $g \in L^q(\mathbf{R}^n)$ then $f * g \in L^r(\mathbf{R}^n)$ with $1/p + 1/q = 1 + 1/r$; here $*$ denotes convolution. The sharp constant in this inequality was derived simultaneously by Beckner [B1, B2] and by Brascamp and Lieb [BL]. Another way to state the inequality is that

$$\int_{\mathbf{R}^n}\int_{\mathbf{R}^n} h(x)f(x-y)g(y)\mathrm{d}x\mathrm{d}y \leqq C\|h\|_{r'}\|f\|_p\|g\|_q \tag{3}$$

with $1/p + 1/q + 1/r' = 2$. The Beckner, Brascamp-Lieb result is that C can be determined by restricting f, g and h to be Gaussian functions. (These, in fact, are the only maximizers, as shown in [BL].)

Young's inequality (3) was generalized in several ways in [BL]. The first way is to allow an arbitrary number of functions $f_1, \ldots, f_K$ instead of merely three as in (3). These are functions from $\mathbf{R}^n$ to $\mathbf{C}$ and $f_j \in L^{p_j}(\mathbf{R}^n)$. This is Theorem 7 of [BL]. The integration is then over $(\mathbf{R}^m)^n$ and the arguments of the f_j's are taken to be $((a_1^j, x_1), \ldots, (a_n^j, x_n)) \in \mathbf{R}^n$, where $a_l^j \in \mathbf{R}^m$ are specified vectors and $x_i \in \mathbf{R}^m$. Unfortunately, this is not a fully mn-dimensional generalization of the $n = 1$ result because $\mathbf{R}^{mn}$ is split unnaturally into $(\mathbf{R}^m)^n$. Following Theorem 7 in [BL] we asked whether the full generalization is possible and Theorem 6.2 below gives it.

A second generalization was the incorporation of a fixed Gaussian function in the integral, as in Theorem 6 of [BL]. Again, the Gaussian in [BL] was completely general when $n = 1$, but not otherwise. In Theorem 6.2 it is completely general.

6.2. Theorem (fully generalized Young's inequality). *Fix $K > 1$, $n_1, \ldots, n_K$ and $p_1, \ldots, p_K > 1$ as before. Let $M \geqq 1$ be an integer and let B_i (for $i = 1, \ldots, K$) be a linear mapping from $\mathbf{R}^M$ to $\mathbf{R}^{n_i}$. For nonnegative functions $f_1, \ldots, f_K$, with $f_i \in L^{p_i}(\mathbf{R}^{n_i})$ consider*

$$I(f_1, \ldots, f_K) = \int_{\mathbf{R}^M} \prod_{i=1}^K f_i(B_i x)\mathrm{d}x. \tag{1}$$

More generally, let $g: \mathbf{R}^M \to \mathbf{R}^+$, $g(x) = \exp\{-(x, Jx)\}$, be a fixed, centered, real Gaussian function and consider

$$I_g(f_1, \ldots, f_K) = \int_{\mathbf{R}^M} \prod_{i=1}^K f_i(B_i x)g(x)\mathrm{d}x. \tag{2}$$

Let

$$C_g \equiv \sup_{f_1,\ldots,f_K} \{I_g(f_1,\ldots,f_K) : \|f_i\|_{p_i} = 1\} \tag{3}$$

and similarly for C *(with* I_g *replaced by* I*). Then*

$$C_g = \sup\{I_g(f_1,\ldots,f_K) : f_1,\ldots,f_K \text{ are real, centered Gaussian functions with } \|f_i\|_{p_i} = 1\}, \tag{4}$$

and similarly for C.

Proof. Suppose the theorem is false and that the right side of (4) (call it D_g) is strictly smaller than C_g. (Alternatively, $D < C$.) Then there are nonnegative summable functions that are *not all* Gaussians, $f_1,\ldots,f_K$ of unit L^{p_i} norm, such that $I_g(f_1,\ldots,f_K) > D_g$ (or $I(f_1,\ldots,f_K) > D$).

Consider the functions $f_i^{(l)} : \mathbf{R}^{n_i} \to \mathbf{R}^+$ given by $f_i^{(l)} \equiv f_i * g_i^{(l)}$ for l a positive integer, where $g_i^{(l)}(x) \equiv (l/\pi)^{n_i/2} \exp\{-l(x,x)\}$ is an $L^1(\mathbf{R}^{n_i})$ normalized Gaussian function. We note that $\|f_i^{(l)}\|_{p_i} \leqq 1$ and that $f_i^{(l)} \to f_i$ in $L^{p_j}(\mathbf{R}^{n_j})$ as $l \to \infty$. By passing to a subsequence (henceforth still denoted by l) we can assume that $f_i^{(l)}(x) \to f_i(x)$ for almost every x in $\mathbf{R}^{n_i}$.

Evidently we can assume that $M \geqq \max\{n_1,\ldots,n_K\}$ and that the rank of B_i is n_i for all i. Otherwise, I or I_g involves knowledge of some f_i only on a hyperplane in $\mathbf{R}^{n_i}$ and this means that I or I_g can be made arbitrarily large (with all f_i's being Gaussian functions) while preserving $\|f_i\|_{p_i} = 1$; the theorem would then be true in this case because both sides of (4) would be infinite. Similarly, the mapping $W \equiv J + \sum_{i=1}^K B_j^* B_j$ (with $*$ denoting adjoint) from $\mathbf{R}^M$ to $\mathbf{R}^M$ is positive definite; otherwise I_g can again be made arbitrarily large with Gaussian f_i's. A similar condition holds for I with $J = 0$. Since B_i is linear and has full rank n_i, the almost everywhere pointwise convergence of $f_i^{(l)}$ to f_i in $\mathbf{R}^{n_i}$ implies that $f_i^{(l)}(B_i x) \to f_i(B_i x)$ for almost every x in $\mathbf{R}^M$.

By Fatou's lemma

$$C'_g \equiv \liminf_{l\to\infty} I_g(f_i^{(l)},\ldots,f_K^{(l)}) \geqq I_g(f_1,\ldots,f_K) > D_g \tag{5}$$

and similarly for C' (with I in place of I_g). By Fubini's theorem, however,

$$I_g(f_1^{(l)},\ldots,f_K^{(l)}) = \int_{\mathbf{R}^N} G_g^{(l)}(y_1,\ldots,y_K) \prod_{i=1}^K f_i(y_i)\mathrm{d}y_1\ldots\mathrm{d}y_K . \tag{6}$$

Here $N = \sum_{i=1}^K n_i$ as in Sect. 6.1, $y_i \in \mathbf{R}^{n_i}$, and $G_g^{(l)}$ is the centered Gaussian kernel

$$G_g^{(l)}(y_1,\ldots,y_K) = \int_{\mathbf{R}^M} \prod_{i=1}^K g_i^{(l)}(B_i x - y_i) g(x)\mathrm{d}x . \tag{7}$$

Similarly, (6) and (7) hold for I in place of I_g by deleting the g. (Note: Because W is positive definite, the integral in (7) is always finite.)

The number C'_g defined in (5) is either finite or infinite. In either case, there is some finite integer k such that $C''_g \equiv I_g(f_1^{(k)},\ldots,f_K^{(k)}) > D_g$. However, by (6) we see that C''_g is a multilinear form as in 6.1 (1). Such a form has the property, as we have seen in Section 6.1, that its supremum over f_i's with $\|f_i\|_{p_i} = 1$ is equal to its supremum over real, centered Gaussian functions. But if we set all the f_i's equal to Gaussian functions we have that $f_i^{(k)}$'s are also Gaussian functions and $\|f_i^{(k)}\|_{p_i} \leqq 1$.

This means that $C_g'' \leqq D_g$, and this is a contradiction. The same proof holds for I in place of I_g. □

References

[BA] Babenko, K.I.: An inequality in the theory of Fourier integrals. Izv. Akad. Nauk SSR Ser. Mat. **25**, 531–542; (1961) English transl. Am. Math. Soc. Transl. (2) **44**, 115–128 (1965)

[B1] Beckner, W.: Inequalities in Fourier analysis. Ann. Math. **102**, 159–182 (1975)

[B2] Beckner, W.: Inequalities in Fourier analysis on $\mathbf{R}^n$. Proc. Natl. Acad. Sci. USA **72**, 638–641 (1975)

[BL] Brascamp, H.J., Lieb, E.H.: Best constants in Young's inequality, its converse, and its generalization to more than three functions. Adv. Math. **20**, 151–173 (1976)

[CA] Carlen, E.: Superadditivity of Fisher's information and logarithmic Sobolev inequalities. J. Funct. Anal. (in press)

[CL] Carlen, E., Loss, M.: Extremals of functionals with competing symmetries. J. Funct. Anal. **88**, 437–456 (1990)

[C] Coifman, R., Cwikel, M., Rochberg, R., Sagher, Y., Weiss, G.: Complex interpolation for families of Banach spaces. Am. Math. Soc. Proc. Symp. Pure Math. **35**, 269–282 (1979)

[DGS] Davies, E.B., Gross, L., Simon, B.: Hypercontractivity: a bibliographic review. Proceedings of the Hoegh-Krohn memorial conference. Albeverio, S. (ed.) Cambridge: Cambridge University Press, 1990

[E] Epperson, Jr., J.B.: The hypercontractive approach to exactly bounding an operator with complex Gaussian kernel. J. Funct. Anal. **87**, 1–30 (1989)

[GL] Glimm, J.: Boson fields with nonlinear self-interaction in two dimensions. Commun. Math. Phys. **8**, 12–25 (1968)

[G] Gross, L.: Logarithmic Sobolev inequalities. Am. J. Math. **97**, 1061–1083 (1975)

[HLP] Hardy, G.H., Littlewood, J.E., Pólya, G.: Inequalities. See Theorem 202 on p. 148. Cambridge: Cambridge University Press 1959

[J] Janson, S.: On hypercontractivity for multipliers on orthogonal polynomials. Ark. Mat. **21**, 97–110 (1983)

[L1] Lieb, E.H.: Proof of an entropy conjecture of Wehrl. Commun. Math. Phys. **62**, 35–41 (1978)

[L2] Lieb, E.H.: Integral bounds for radar ambiguity functions and Wigner distributions. J. Math. Phys. **31**, 594–599 (1990)

[N1] Nelson, E.: A quartic interaction in two dimensions in: Mathematical theory of elementary particles. Goodman, R., Segal, I. (eds.), pp. 69–73. Cambridge: M.I.T. Press 1966

[N2] Nelson, E.: The free Markov field. J. Funct. Anal. **12**, 211–227 (1973)

[NE] Neveu, J.: Sur l'esperance conditionelle par rapport à un mouvement Brownien. Ann. Inst. H. Poincaré Sect. B. (N.S.) **12**, 105–109 (1976)

[S1] Simon, B.: A remark on Nelson's best hypercontractive estimates. Proc. Am. Math. Soc. **55**, 376–378 (1976)

[S] Segal, I.: Construction of non-linear local quantum processes: I. Ann. Math. **92**. 462–481 (1970)

[T] Titchmarsh, E.C.: A contribution to the theory of Fourier transforms. Proc. London Math. Soc. Ser. 2, **23**, 279–289 (1924)

[W] Weissler, F.B.: Two-point inequalities, the Hermite semigroup, and the Gauss-Weierstrass semigroup. J. Funct. Anal. **32**, 102–121 (1979)

J. Math. Phys. *31*, 594–599 (1990)

Integral bounds for radar ambiguity functions and Wigner distributions

Elliott H. Lieb
Departments of Mathematics and Physics, Princeton University, P. O. Box 708, Princeton, New Jersey 08544

(Received 10 November 1989; accepted for publication 22 November 1989)

An upper bound is proved for the L^p norm of Woodward's ambiguity function in radar signal analysis and of the Wigner distribution in quantum mechanics when $p>2$. A lower bound is proved for $1\leqslant p<2$. In addition, a lower bound is proved for the entropy. These bounds set limits to the sharpness of the peaking of the ambiguity function or Wigner distribution. The bounds are best possible and equality is achieved in the L^p bounds if and only if the functions f and g that enter the definition are both Gaussians.

I. INTRODUCTION

The *ambiguity function* introduced by Woodward[1] is important in radar signal analysis. It is a function of two real variables, τ (the time) and ω (with $2\pi\omega$ being the frequency), and is defined as follows in terms of two given functions f and g of one variable:

$$A_{f,g}(\tau,\omega)=\int f\left(t-\frac{1}{2}\tau\right)g^*\left(t+\frac{1}{2}\tau\right)e^{-2\pi i\omega t}\,dt. \tag{1.1}$$

(Our conventions will be that * as a superscript denotes complex conjugate and all integrals are from $-\infty$ to $+\infty$.) Strictly speaking, $A_{f,g}$ is called the *cross-ambiguity function* of f and g while $A_{f,f}$ is the *proper ambiguity function* of f. Usually, one assumes that f and g are square integrable, which guarantees that the integrand of (1.1) is a summable function of t for every τ. The summability can also be guaranteed by Hölder's inequality and the alternative assumption that $f\in L^a$ and $g\in L^b$ (with $1/a+1/b=1$ and $1\leqslant a,b\leqslant\infty$), as in Definition 2 below, and this generalized hypothesis will often be made in this paper.

There is a simple relation between $A_{f,g}$ and $W_{f,g}$, the (*cross*) *Wigner distribution of f and g* used in quantum mechanics and defined by

$$W_{f,g}(\tau,\omega)=\int f\left(\tau+\frac{1}{2}s\right)g^*\left(\tau-\frac{1}{2}s\right)e^{-2\pi i\omega s}\,ds. \tag{1.2}$$

The relation is

$$W_{f,g}(\tau,\omega)=2A_{f,g^-}(-2\tau,2\omega), \tag{1.3}$$

where f^- denotes the function given by

$$f^-(t)\equiv f(-t). \tag{1.4}$$

$W_{f,f}$ is called *the Wigner distribution* (*or density*) *of f*. Because of (1.3) the bounds obtained here for $A_{f,g}$ apply *mutatis mutandis* to $W_{f,g}$.

Ideally, one would like to choose f and g so that $A_{f,g}$ is sharply peaked around some point (τ_0,ω_0) but, as is well known, there are severe limitations to the peaking that can be achieved. These limitations are inherent in the definition (1.1). Let us define, for $p>0$,

$$I_{f,g}(p)=\int\int |A_{f,g}(\tau,\omega)|^p\,d\tau\,d\omega. \tag{1.5}$$

If $A_{f,g}$ were highly peaked then $I_{f,g}(p)$ would be very large for large p and very small for small p. The dividing line is $p=2$ since, by the Parseval's inversion formula, we have the identity

$$I_{f,g}(2)=\int |f(t)|^2dt\int |g(t)|^2\,dt. \tag{1.6}$$

In this paper, limitations on the sharpness of $A_{f,g}$ will be established by proving that $I_{f,g}(p)$ is universally bounded above when $p>2$ (Theorem 1) and universally bounded below when $1\leqslant p<2$ (Theorem 2). For $1<p<2$ and $2<p<\infty$ the bounds will be shown to be saturated if and *only* if f and g are Gaussians. It is remarkable that Gaussians both maximize and minimize $I_{f,g}(p)$, depending on the value of p.

When $p=2$ the identity (1.6) holds for any f and g, so the obvious quantity to consider is the derivative with respect to p of $I_{f,g}(p)$ at $p=2$ under the normalization assumption that the right side of (1.6) is unity. This derivative, multiplied by -2, is the *entropy* given by

$$S_{f,g}=-\int\int |A_{f,g}(\tau,\omega)|^2\ln|A_{f,g}(\tau,\omega)|^2\,d\tau\,d\omega, \tag{1.7}$$

with $0\ln 0\equiv 0$. It will be proved that when the right side of (1.6) is unity the integral in (1.7) is well defined and (Theorem 3)

$$S_{f,g}\geqslant 1. \tag{1.8}$$

This constant is sharp since it is achieved by Gaussians.

To state the theorems precisely it is first necessary to make some definitions.

Definition 1: $f(t)$ is said to be a *Gaussian* if

$$f(t)=\exp[-\alpha t^2+\beta t+\gamma], \tag{1.9}$$

with α, β, and γ being complex numbers and with $\mathrm{Re}(\alpha)>0$; $f(t)$ is a *real Gaussian* if α, β, and γ are real numbers with $\alpha>0$. Two functions f and g are said to be a *matched Gaussian pair* if they are both Gaussians with the same α but with possibly different β's and γ's.

Definition 2: For $0<p<\infty$

$$\|f\|_p\equiv\left\{\int |f(t)|^pdt\right\}^{1/p} \tag{1.10}$$

and for $p=\infty$

$$\|f\|_\infty\equiv\text{ess sup}|f(t)|. \tag{1.11}$$

 0022-2488/90/030594-06$03.00

We say that $f \in L^p$ if and only if the right side of (1.10) or (1.11) is finite.

Definition 3: Let $0<p\leqslant\infty$ and define q by $1/q+1/p=1$, i.e., $q=p/(p-1)$. Note that $\infty\geqslant q\geqslant 1$ if $1\leqslant p\leqslant\infty$, and $0>q>-\infty$ if $0<p<1$. Then C_p is defined to be

$$C_p = p^{1/2p}|q|^{-1/2q}, \tag{1.12}$$

for $p\neq 1$ or ∞ while

$$C_1 = C_\infty = 1. \tag{1.13}$$

Note that $C_2=1$.

Definition 4: Let p and q be as in Definition 3 with $1<p<\infty$ and let a and b satisfy $1/a+1/b=1$ with $1\leqslant a\leqslant\infty, 1\leqslant b\leqslant\infty$. We define $H(p,a,b)\geqslant 0$ by

$$H(p,a,b)^2 = abp^{-2}|p-2|^{2-p} \times |p-a|^{-1+p/a}|p-b|^{-1+p/b}, \tag{1.14}$$

with the convention that $0^0\equiv 1$. When a or $b=p$

$$H(p,a,b)^2 = \frac{1}{p}\left(\frac{p}{p-1}\right)^{p-1}$$

and

$$H(1,1,\infty) = H(1,\infty,1) = 1. \tag{1.15}$$

We also define $K(p,a,b)\geqslant 0$ by

$$K(p,a,b)^2 = p^{-2}2^{2-p}a^{p/a}b^{p/b}$$

and

$$K(1,1,\infty)=\sqrt{2}. \tag{1.16}$$

The following relations (with $1/p+1/q=1$) are noteworthy for $p>1$:

$$H(p,a,b) = C_q^p\{C_{a/q}C_{b/q}/C_{p/q}\}^{p/q}, \tag{1.17}$$

$$H(p,a,b)^{1/p}H(q,a,b)^{1/q} = K(p,a,b)^{1/p}K(q,a,b)^{1/q} = a^{1/a}b^{1/b}p^{-1/p}q^{-1/q}. \tag{1.18}$$

Theorem 1: *Let $p>2$ and assume that f and $g\in L^2$. Then*
(a) $I_{f,g}(p)\leqslant(2/p)\{\|f\|_2\|g\|_2\}^p$. (1.19)
(b) Equality is achieved in (1.19) if and only if f and g are a matched Gaussian pair.
(c) More generally, if $f\in L^a$, $g\in L^b$ with $1/a+1/b=1$ and with $p/(p-1)\leqslant a\leqslant p$ and $p/(p-1)\leqslant b\leqslant p$ then

$$I_{f,g}(p)\leqslant H(p,a,b)\{\|f\|_a\|g\|_b\}^p. \tag{1.20}$$

When both a and $b>p/(p-1)$ equality is achieved in (1.20) if and only if f and g are Gaussians that satisfy

$$\begin{aligned} f(t) &= \exp[-(\alpha m' + iA)t^2+\beta t+\gamma], \\ g(t) &= \exp[-(\alpha n' + iA)t^2+\tilde\beta t+\tilde\gamma], \end{aligned} \tag{1.21}$$

with α, A real, $\alpha>0$, and $\beta,\tilde\beta,\gamma,\tilde\gamma$ complex and with $m'=a(p-1)/(ap-a-p)$ and $n'=b(p-1)/(bp-b-p)$. [Note that $(p-1)/(p-2)<m',n'<\infty$ under the stated conditions.] When a or $b=p/(p-1)$, (1.20) is best possible, but equality is never achieved.

(d) If the additional condition that $g=f$ is imposed (which means that the proper ambiguity function $A_{f,f}$ is being considered) or else that $g=f^-$ (which means that the proper Wigner distribution $W_{f,f}$ is being considered) then (1.20) can be improved. In these cases (and with a and b restricted as before)

$$I_{f,f}(p) \text{ and } I_{f,f}(p)\leqslant K(p,a,b)\{\|f\|_a\|f\|_b\}^p. \tag{1.22}$$

Equality is achieved in (1.22) if and only if f is any Gaussian. Note that a or $b=p/(p-1)$ is allowed here.

Remarks: (1) Even if f and g are Gaussians, it is not possible to have equality in (1.20) for *all a and b simultaneously*, as (1.21) shows.

(2) In view of the symmetry of $A_{f,g}$ between the pair f,g and the Fourier transforms $\hat f,\hat g$ expressed by (2.4) below, Theorem 1 remains true if f and g are replaced by $\hat f$ and $\hat g$ on the right side of (1.19) *et seq.*

In the case that p is an even integer, Theorem 1(a) and (b) [under the additional assumption for (b) that f and g are twice continuously differentiable and never vanish] was proved by Price and Hofstetter[2] by an ingenious application of the Cauchy–Schwarz inequality. They conjectured Theorem 1(a) and (b) for all $p>2$ in their footnote 10. The Price–Hofstetter bounds have found application in the work of Janssen[3] for example.

The next theorem gives reversed inequalities for $p<2$.

Theorem 2: *Assume that $t\mapsto f(t-\frac{1}{2}\tau)g^*(t+\frac{1}{2}\tau)$ is in L^1 for every τ, so that the definition (1.1) of $A_{f,g}(\tau,\omega)$ makes sense. (This L^1 condition can be satisfied, for example, by assuming that $f\in L^\alpha$ and $g\in L^\beta$ for some $1\leqslant\alpha,\beta\leqslant\infty$ with $1/\alpha+1/\beta=1$.) Let $1\leqslant p<2$ and assume that $0<I_{f,g}(p)<\infty$. Then f and $g\in L^r$ for every $p\leqslant r\leqslant q\equiv p(p-1)$. Moreover, for every pair a,b with $p\leqslant a,b\leqslant q$ and with $1/a+1/b=1$ we have that*

$$I_{f,g}(p)\geqslant H(p,a,b)\{\|f\|_a\|g\|_b\}^p. \tag{1.23}$$

In particular,

$$I_{f,g}(p)\geqslant(2/p)\{\|f\|_2\|g\|_2\}^p. \tag{1.24}$$

If $g=f$ or $g=f^-$ [as in Theorem 1(d)] then (1.23) can be improved to

$$I_{f,g}(p)\geqslant K(p,a,b)\{\|f\|_a\|g\|_b\}^p. \tag{1.25}$$

If $1<p<2$ equality is achieved in (1.23) if and only if f and g satisfy (1.21) et seq., but with m' and n' replaced by $|m'|$ and $|n'|$, respectively. Equality in (1.25) occurs if and only if f is any Gaussian. If $p=1$ and $a,b>1$ equality occurs in (1.23) if f and g are given by (1.21), but $\alpha|m'|$ and $\alpha|n'|$ have to be interpreted as αa and αb, respectively (since $|m'|/|n'|\to a/b$ but $m',n'\to 0$ as $p\to 1$).

Remarks: (3) When $p=1$ and $a,b>1$ the Gaussians referred to in the last part of Theorem 2 are, in fact, the only functions for which equality holds in (1.21). A proof can be constructed by using ideas in Ref. 4, but it will not be given here. The uniqueness of Gaussian minimizers for $p=1$ and $a=b=2$ is closely related to and can be inferred from a theorem of Hudson[5] (see also Ref. 6) which says that the only way in which the function $A_{f,g}(\tau,\omega)$ can be a non-negative function of τ and ω is when $f=\lambda g$ for some $\lambda>0$ and f is a Gaussian. (Actually, Hudson does this in the context of the Wigner distribution, but that is immaterial; also he proves the theorem only for $W_{f,f}$ but his method, extends to the general case.) The connection is established by first noting the relation for summable $A_{f,g}$ (which is easy to derive—at least formally)

$$\int A_{f,g}(\tau,\omega)dt\,d\omega = 2\int f(t)g^*(t)dt. \quad (1.26)$$

On the other hand, by Theorem 2(a) with $p = 1$,

$$\int |A_{f,g}(\tau,\omega)|d\tau\,d\omega \geqslant 2\|f\|_2\|g\|_2. \quad (1.27)$$

If $A_{f,g} \geqslant 0$, the left sides of (1.26) and (1.27) are identical, which then requires that $f = \lambda g$ and that equality holds in (1.27). Thus $A_{f,g} \geqslant 0$ is equivalent to equality in (1.24) for $p = 1$.

(4) Theorem 2(c) is striking when $p = a = 1$ and $b = \infty$. Then

$$\int |A_{f,g}(\tau,\omega)|d\tau\,d\omega \geqslant \|f\|_1\|g\|_\infty. \quad (1.28)$$

This says that if f is fixed and $g \to 0$ in all L^p norms except $p = \infty$, then $\int|A|$ does not go to zero. [For example, $g(t) = \exp[-\lambda t^2]$ with $\lambda \to \infty$.] The Fourier transform also has this property [cf. (2.9)] and it is inherited by $A_{f,g}$.

A tempting conjecture is that inequality (1.24), at least, should hold if $0 < p < 1$. Our proof fails in this case because Lemma 1 below requires $p \geqslant 1$.

It is instructive to compare Theorems 1(a) and (1.24) by considering Gaussians $f(t) = \exp(-\alpha t^2)$ and $g(t) = \exp(-\beta t^2)$ with $\mathrm{Re}\,\alpha$ and $\mathrm{Re}\,\beta > 0$. Then one finds

$$I_{f,g}(p)\|f\|_2^{-p}\|g\|_2^{-p} = (2/p)[\mathrm{Re}\,\alpha\,\mathrm{Re}\,\beta\,]^{p/4-1/2}|(\alpha+\beta^*)/2|^{1-p/2}. \quad (1.29)$$

Since $\mathrm{Re}\,\alpha\,\mathrm{Re}\,\beta \leqslant \frac{1}{4}|\alpha+\beta^*|^2$ one sees, for Gaussians, that (1.19) holds for $p \geqslant 2$ and that the reverse inequality holds for *all* $0 < p < 2$, and that equality requires $\alpha = \beta$ in both cases.

Theorem 3: *Assume that f and $g \in L^2$ with $\|f\|_2\|g\|_2 = 1$. Then*

$S_{f,g} \geqslant 1.$

Equality is achieved iff f and g are a matched Gaussian pair.

Remarks: (5) It is possible to show that equality is achieved in Theorem 3 only when f and g are matched Gaussians. The proof is complicated and will not be given; the reader is invited to find a simple proof.

The method of proof of these three theorems follows closely the methods used in Ref. 7 to prove L^p bounds of coherent state transforms. The coherent state transform of f is $A_{f,g}(-\tau,-\omega)\exp(i\pi\omega\tau)$ with g being the fixed Gaussian $g(t) = \tau^{-1/4}\exp(-t^2/2)$. From the mathematical point of view there is, however, a genuinely new development in the present paper, namely the proof that Gaussians uniquely saturate the bounds. This uses Ref. 4.

II. PRELIMINARY LEMMAS

The following convention for the Fourier transform $\hat{f}$ of a function f will be employed:

$$\hat{f}(\omega) = \int f(t)e^{-2\pi i\omega t}dt, \quad (2.1)$$

so that

$$f(t) = \int \hat{f}(\omega)e^{2\pi i\omega t}d\omega \quad (2.2)$$

and Parseval's relation is

$$\|f\|_2 = \|\hat{f}\|_2. \quad (2.3)$$

The equality (1.6) follows from (2.3). Some other important facts about $A_{f,g}$ which follow easily from (2.3), the Cauchy–Schwarz inequality and a change of integration variables are

$$A_{f,g}(\tau,\omega) = A_{\hat{f},\hat{g}}(-\omega,\tau), \quad (2.4)$$

$$A^*_{f,g}(\tau,\omega) = A_{g,f}(-\tau,-\omega), \quad (2.5)$$

$$|A_{f,g}(\tau,\omega)| \leqslant \|f\|_2\|g\|_2. \quad (2.6)$$

More generally, if $f \in L^a$, $g \in L^b$ with $1/a + 1/b = 1$ and $a \geqslant 1, b \geqslant 1$, as in Theorems 1 and 2, Hölder's inequality yields the pointwise bound

$$|A_{f,g}(\tau,\omega)| \leqslant \|f\|_a\|g\|_b. \quad (2.7)$$

Inequality (2.6) is important because it implies that $\ln|A_{f,g}(\tau,\omega)|^2 \leqslant 0$ when $\|f\|_2\|g\|_2 = 1$ and hence $S_{f,g}$ is always well defined by the right side of (1.7) (although it might be $+\infty$).

Three inequalities in Fourier analysis will be needed. The first fact is the sharp constant in the Hausdorff–Young inequality (2.8) proved by Beckner.[8] The criterion for equality is due to Lieb.[4]

Lemma 1: *Let $2 \leqslant p \leqslant \infty$ and $1/q = 1 - 1/p = (p-1)/p$. If $f \in L^q$ then $\hat{f} \in L^p$ and*

$$\|\hat{f}\|_p \leqslant C_q\|f\|_q. \quad (2.8)$$

Conversely, let $1 \leqslant p \leqslant 2$ and assume $f \in L^r$ for some $1 \leqslant r \leqslant 2$, in which case $\hat{f}$ exists by (2.8) (with $q \equiv r$ there.) If $\hat{f} \in L^p$ then $f \in L^q$ with $q = (p-1)/p$ and

$$\|\hat{f}\|_p \geqslant C_q\|f\|_q. \quad (2.9)$$

Equality is achieved in (2.8) when $2 < p < \infty$ and in (2.9) when $1 < p < 2$ if and only if f is any Gaussian with α real and β, γ complex in (1.9).

Proof: Inequality (2.8) is Beckner's result, and the condition for equality when $2 < p < \infty$ is proved in Ref. 4. For (2.9), let $g \equiv \hat{f}$. Since $f \in L^r$, $g \in L^s$ [with $s = r/(r-1) \geqslant 2$]. Therefore, $g \in L^p \cap L^s$ and hence, by convexity, $g \in L^2$. Thus $\hat{g}$ exists and, by the L^2 Fourier inversion formula, $\hat{g} = f^-$. By (2.8), $f^- \in L^q$ and (using $C_qC_p = 1$) $C_q\|f\|_q = C_q\|f^-\|_q \leqslant \|g\|_p = \|\hat{f}\|_p$. Obviously, the condition for equality when $1 < p < 2$ follows from the $2 < p < \infty$ result.

Q.E.D.

Remark: (6) The classical Hausdorff–Young inequality is (2.8) but with C_q replaced by the larger value 1.

The next inequality is the sharp constant in Young's inequality, which was found simultaneously by Beckner[8] and by Brascamp and Lieb.[9] The uniqueness part (b) is due to Brascamp and Lieb.[9] In the following a midline asterisk denotes convolution

$$(f*g)(t) = \int f(t-s)g(s)ds. \quad (2.10)$$

J. Math. Phys. *31*, 594–599 (1990)

Lemma 2: *Let $1/m+1/n=1+1/r$ with $1\leqslant m\leqslant\infty$, $1\leqslant n\leqslant\infty$, $1\leqslant r\leqslant\infty$. Then, when $f\in L^m$ and $g\in L^n$, $f*g\in L^r$ and (a)*

$$\|f*g\|_r\leqslant(C_mC_n/C_r)\|f\|_m\|g\|_n. \tag{2.11}$$

(b) When $m>1$ and $n>1$, equality holds in (2.11) if and only if

$$f(t)=\exp[-\alpha m't^2+\beta t+\gamma],$$
$$g(t)=\exp[-\alpha n't^2+\tilde\beta t+\tilde\gamma], \tag{2.12}$$

with $\alpha>0$ real, $\beta,\gamma,\tilde\beta,\tilde\gamma$ complex but with $\mathrm{Im}(\beta)=\mathrm{Im}(\tilde\beta)$. *Here, $m'=m/(m-1)$ and $n'=n/(n-1)$. If $m=1$ or $n=1$ and $r>1$, (2.11) is at best possible but equality is never achieved. If $m=n=r=1$, equality is achieved when f and g are any pair of non-negative, real valued functions.*

(c) If $g^*=f$ or $g^*=f^-$

$$\|f*g\|_r\leqslant\tfrac12 r^{-1/2r}(2m)^{1/2m}(2n)^{1/2n}\|f\|_m\|f\|_n. \tag{2.13}$$

For all $m\geqslant1$ and $n\geqslant1$ and $r>1$ equality is achieved in (2.13) if and only if f is a Gaussian given by (1.9) with α real and with β real (if $g^*=f$) or β complex (if $g^*=f^-$).

Remarks: (7) The classical inequality of Young is (2.11) but with C_mC_n/C_r replaced by the larger value 1.

(8) Lemma 2(c) was not given in Ref. 9 because it did not occur to us at the time that it might be useful. It is however, a simple consequence of the analysis in Ref. 9.

The third inequality is the converse of Young's inequality. It was first proved by Leindler[10] with 1 in place of C_mC_n/C_r. The sharp form below is due to Brascamp and Lieb.[9]

Lemma 3: *Let f(t) and g(t) be non-negative, real-valued functions that are not identically zero and assume that $f*g\in L^r$. Let $1/m+1/n=1+1/r$ with $0<m\leqslant1$, $0<n\leqslant1$. Note that $0<r\leqslant1$. Then $f\in L^m$ and $g\in L^n$ and (a)*

$$\|f*g\|_r\geqslant(C_mC_n/C_r)\|f\|_m\|g\|_n. \tag{2.14}$$

Equality holds in (2.14) when $m<1$ and $n<1$ if and only if

$$f(t)=\exp[\alpha m't^2+\beta t+\gamma],$$
$$g(t)=\exp[\alpha n't^2+\tilde\beta t+\tilde\gamma], \tag{2.15}$$

with $\alpha>0$ real and $\beta,\gamma,\tilde\beta,\tilde\gamma$ real. Here, $m'=m/(m-1)<0$ and $n'=n/(n-1)<0$.

(b) If $g^*=f$ or $g^*=f^-$ (2.14) can be improved to

$$\|f*g\|_r\geqslant\tfrac12 r^{-1/2r}(2m)^{1/2m}(2n)^{1/2n}\|f\|_m\|f\|_n, \tag{2.16}$$

with equality (for all m and n) if and only if f is a real Gaussian.

Remark: (9) Lemma 3(b) was not given in Ref. 9 but it is a simple consequence of the analysis given there.

The next lemma is an extension of the Cauchy functional equation to quadratics. [One form of Cauchy's equation is $\xi(t-\frac12\tau)\eta(t+\frac12\tau)=\rho(\tau)$ with ξ and η being Lebesque measurable functions; the only solution is $\xi(t)=be^{-At},\eta(t)=ce^{At}$, and $\rho(\tau)=bce^{A\tau}$ for some constants A, b, c.]

Lemma 4: *Let ξ and η be complex valued, Lebesgue measurable functions on* **R** *that satisfy $|\xi(t)|=|\eta(t)|=1$ for all t. Suppose there are real valued functions, μ and ν, on* **R** *(which are not a priori measurable) such that for almost every τ the following holds for almost every t:*

$$\xi(t-\tfrac12\tau)\eta(t+\tfrac12\tau)=\exp[i\mu(\tau)t+i\nu(\tau)]. \tag{2.17}$$

Then there are real constants, A, α, β, γ and δ such that

$$\xi(t)=\exp[iAt^2+i\alpha t+i\gamma],$$
$$\eta(t)=\exp[-iAt^2-i\beta t-i\delta]. \tag{2.18}$$

Proof: Let $\mathscr{B}$ denote the set of τ such that (2.17) holds for almost all t. Let $X(t)=\xi(t)\exp(-t^2)$ and $Y(t)=[1/\eta(t)]\exp(-t^2)$. Using the definition (2.1) of the Fourier transform, it is a simple matter to use the Gaussian bound on $X(t)$ to deduce that $\hat X$ is an entire analytic function of order at most 2, i.e., $|\hat X(\omega)|\leqslant\exp[C+D|\omega|^2]$ for suitable $C,D>0$ and all $\omega\in\mathbf{C}$. (In fact, $|\hat X(\omega)|\leqslant\sqrt\pi\exp[\pi^2(\mathrm{Im}\,\omega)^2]$.) The same is true of $\hat Y(\omega)$. From (2.17), for every $\tau\in\mathscr{B}$ the following holds for almost every t:

$$X(t-\tfrac12\tau)=Y(t+\tfrac12\tau)\exp\{t[i\mu(\tau)+2\tau]+i\nu(\tau)\}. \tag{2.19}$$

Taking Fourier transforms of (2.19) with respect to t we find that

$$\hat X(\omega)\exp(-\pi i\omega\tau)=\hat Y\Big(\omega-\frac{\mu(\tau)}{2\pi}+\frac{i\tau}{\pi}\Big)\times\exp\Big[\pi i\omega\tau-\frac12 i\mu(\tau)\tau+i\nu(\tau)-\tau^2\Big]. \tag{2.20}$$

We claim that $\hat X(\omega)$ has no zeros, for otherwise suppose that $\hat X(\omega_0)=0$. Then $\hat Y(\omega)=0$ whenever ω satisfies

$$\omega=\omega_0-(1/2\pi)\mu(\tau)+(i/\pi)\tau, \tag{2.21}$$

for some $\tau\in\mathscr{B}$. As τ ranges over the uncountable set $\mathscr{B}$, the right side of (2.21) ranges over an uncountable set in the complex plane. [Note that $\mu(\tau)$ is real and $i\tau$ is imaginary so there can be no cancellation in (2.21).] The only entire function with uncountably many zeros is the zero function, so $\hat Y(\omega)\equiv0$. This implies that $Y(t)=0$, which is a contradiction. By reversing the roles of X and Y we find that $\hat Y(\omega)$ has no zeros. Because $\hat X$ and $\hat Y$ are entire analytic and zero free they have analytic logarithms, e.g., $\hat X(\omega)=\exp[\phi(\omega)]$ for some entire analytic function ϕ. Since $\hat X$ has order at most 2, $|\phi(\omega)|\leqslant C|\omega|^2+D$ for suitable $C,D>0$. But then ϕ must be a polynomial of order 2, i.e., $\hat X$ is a Gaussian. The same is true of $\hat Y$. By taking the inverse Fourier transform, we have that X and Y are Gaussians, which, by inspection, proves (2.18). Q.E.D.

III. PROOF OF THEOREM 1

Step 1: Fix $\tau\in\mathbf{R}$. Since $f\in L^a$ and $g\in L^b$ with $1/a+1/b=1$, the function $t\to f(t-\frac12\tau)g(t+\frac12\tau)$ is in L^1. Since $A_{f,g}$ is the Fourier transform of this L^1 function, we can use Lemma 1 with $q=p/(p-1)<2$ in place of p there and obtain

$$\int |A_{f,g}(\tau,\omega)|^p\, d\omega$$

$$\leqslant C_q^p \left\{ \int \left| f\left(t-\frac{1}{2}\tau\right)\right|^q \left| g\left(t+\frac{1}{2}\tau\right)\right|^q dt \right\}^{p/q}. \qquad (3.1)$$

Note that the right-hand integral may be finite or infinite—depending on τ. If it is infinite then (3.1) is trivially true; if it is finite then the use of Lemma 1 is justified. We shall see in step 2 that this integral is finite for almost every τ.

Step 2: The integral on the right side of (3.1) is just the convolution

$$J(\tau) \equiv (|f^-|^q * |g|^q)(\tau). \qquad (3.2)$$

Integrating (3.1) over τ and applying Lemma 2 to $J(\tau)$ with $r = p/q > 1$ and $m = a/q \geqslant 1, n = b/q \geqslant 1$, we have

$$I_{f,g}(p) \leqslant C_q^p \|J\|_r^r$$
$$\leqslant C_q^p (C_m C_n / C_r)^r \{\| |f^-|^q \|_m \| |g|^q \|_n\}^r$$
$$= C_q^p (C_m C_n / C_r)^r \{\|f\|_a \|g\|_b\}^p. \qquad (3.3)$$

The inequalities (1.19) and (1.20) are obtained by using (1.17).

Step 3: It is an elementary exercise to show that Gaussians of the form (1.21) give equality in (3.1) and (3.3), and hence that $H(p,a,b)$ is the sharp constant in (1.19) and (1.20). We want to prove that these Gaussians *uniquely* saturate the bounds. Assume that $m > 1$ and $n > 1$. If there is equality in (1.19) or (1.20) then (3.1) must be an equality for almost every τ and (3.3) must be an equality. By Lemma 1, the following must be true for almost every τ:

$$f(t - \tfrac{1}{2}\tau)\, g^*(t + \tfrac{1}{2}\tau) = D(\tau)\exp[-\sigma(\tau)t^2 + \delta(\tau)t] \qquad (3.4)$$

for almost every t, with $\sigma(\tau) \in \mathbf{R}$ and $D(\tau), \delta(\tau) \in \mathbf{C}$. By Lemma 2, equality in (3.3) requires

$$|f(t)| = \exp[-\alpha m' t^2 + \beta t + \gamma],$$
$$|g(t)| = \exp[-\alpha n' t^2 + \bar\beta t + \bar\gamma], \qquad (3.5)$$

with $m' = m/(m-1)$, $n' = n/(n-1)$, $\alpha > 0$, and $\beta, \gamma, \bar\gamma \in \mathbf{R}$.

Let us define $\xi(t) = f(t)/|f(t)|$ and $\eta(t) = g^*(t)/|g(t)|$, which makes sense since $f(t)$ and $g(t)$ never vanish by (3.5). Then, comparing (3.4) and (3.5), we find that ξ and η satisfy the hypotheses of Lemma 4. The conclusion of Lemma 4, together with (3.5), gives (1.21).

Step 4: When $a \to p/(p-1)$ then $m' \to \infty$ and $n \to r$. By taking limits of Gaussians in (1.21) with $m' \to \infty$ we see that (1.20) is best possible in this case. Equality is never achieved, however. An informal way to see this is to note tht m' must be infinity. A formal proof is to note that (2.11) or (3.3) cannot be an equality when $m = 1$ and $n = r$ [as is stated in Lemma 2(b)] because of the strict convexity of the L^r norm.

Step 5: When $g = f$ or $g = f^-$ we proceed as in steps 1 to 3, making the appropriate changes and using lemma 2(c). From this we infer (1.22) and conclude that f must be a Gaussian in order to have equality. Upon inserting a Gaussian (1.9) for f and g (or g^-) in (1.1), one finds by inspection that equality in (1.22) does not impose any restriction on the Gaussian. Q.E.D.

IV. PROOF OF THEOREM 2

Before proving this theorem, it is perhaps worth noting a proof strategy that works when $a = p$ or $b = p$, but otherwise yields a weaker result. This strategy does not require Lemma 3. From Parseval's relation one has the identity

$$R_{f,g,h,j} \equiv \int A_{f,g}(\tau,\omega) A^*_{h,j}(\tau,\omega)\, d\tau\, d\omega$$

$$= \int f(t) h^*(t) dt \int g^*(t) j(t) dt, \qquad (4.1)$$

for any four functions f,g,h, and j. Let $f = |f| e^{i\phi}$ and $g = |g| e^{i\psi}$ and choose $h(t) = |f(t)|^{a-1} e^{i\phi(t)}$ and $j(t) = |g(t)|^{b-1} e^{i\psi(t)}$. Then

$$R_{f,g,h,j} = \|f\|_a^a \|g\|_b^b. \qquad (4.2)$$

On the other hand, by Hölder's inequality,

$$|R_{f,g,h,j}| \leqslant I_{f,g}(p)^{1/p} I_{h,j}(q)^{1/q}. \qquad (4.3)$$

If $1 < p < 2$, then $q = p/(p-1) > 2$ and we can use Theorem 1(c) for the right-most factor in (4.3):

$$\{I_{h,j}(q)\}^{1/q} \leqslant H(q,b,a)^{1/q} \|f^{a-1}\|_b \|g^{b-1}\|_a. \qquad (4.4)$$

Since $\|f^{a-1}\|_b = \|f\|_a^{a-1}$ and $\|g^{b-1}\|_a = \|g\|_b^{b-1}$, we can combine (4.1)–(4.4) with (1.18) to obtain

$$I_{f,g}(p) \geqslant H(p,a,b) L(p,a,b)^p \{\|f\|_a \|g\|_b\}^p, \qquad (4.5)$$

where

$$L(p,a,b) = p^{1/p} q^{1/q} a^{-1/a} b^{-1/b}. \qquad (4.6)$$

If a or $b = p$ then $L(p,a,b) = 1$ and (4.5) is the desired inequality. Unfortunately, if $p < a$ then $b < q$, $L(p,a,b) < 1$ and (4.5) is too weak.

We now turn to a proof of Theorem 2 which makes use of Lemma 3. When $p > 1$, which is the case we consider first, the proof is virtually the same, *mutatis mutandis*, as for Theorem 1.

Step 1: Using inequality (2.9) (with $r = 1$) we have that (3.1) holds, but with the reversed inequality. Note that the left side of (3.1) is finite for almost every τ since $\int\{\int |A_{f,g}| d\omega\} d\tau < \infty$ by assumption.

Step 2: By (3.2) and Lemma 3, (3.3) holds with the reversed inequality. In particular, $f \in L^a$ and $g \in L^b$. This proves (1.23). Similarly, Lemma 3(b) leads to (1.25). The cases of equality for $1 < p < 2$ are handled in the same way as in step 1 of the proof of Theorem 1.

Finally we turn to the case $p = 1$.

Step 3: Suppose $p = 1 < a \leqslant b < \infty$. Then (1.23) holds for every $p \leqslant a$. As p decreases from a to 1, $H(p,a,b)$ converges to $H(1,a,b)$. On the other hand, $B(\tau,\omega) \equiv |A_{f,g}(\tau,\omega)| / \|f\|_a \|g\|_b \leqslant 1$ by (2.7) so $B(\tau,\omega)^p$ increases monotonically as p decreases. Therefore, by Lebesgue's monotone convergence theorem, $\int B(\tau,\omega)^p d\tau\, d\omega$ converges to $\int B(\tau,\omega) d\tau\, d\omega$ as $p \downarrow 1$ and this, together with (1.23) for $p > 1$, establishes (1.23) for $p = 1$. A similar proof holds for (1.25).

Step 4: Suppose $p = a = b = 1$. For each $a,b > 1$ such that $1/a + 1/b = 1$ inequality (1.23) holds by step 3. As $a \downarrow 1$ and $b \uparrow \infty$ we have that $H(1,a,b) \to H(1,1,\infty)$. Also, it is a

J. Math. Phys. *31*, 594–599 (1990)

standard fact that $\|f\|_a \to \|f\|_1$ and $\|g\|_b \to \|g\|_\infty$. A similar proof works for Eq. (1.25). Q.E.D.

V. PROOF OF THEOREM 3

It is assumed that f and $g \in L^2$ and $\|f\|_2 \|g\|_2 = 1$. By (1.6), $I_{f,g}(2) = 1$ and, by (2.6), $|A_{f,g}(\tau,\omega)| \leqslant 1$ for all τ and ω. Let $p > 2$ whence, by Theorem 1, $I_{f,g}(p) \leqslant 2/p$. If we define, for $\epsilon > 0$,

$$K(\epsilon) \equiv \epsilon^{-1}\{I_{f,g}(2) - I_{f,g}(2+2\epsilon)\}, \tag{5.1}$$

we have that

$$K(\epsilon) \geqslant (1+\epsilon)^{-1}. \tag{5.2}$$

Assume now that $S_{f,g}$ defined by (1.7), is finite; otherwise the inequality (1.8) is trivial. (Note that $|A_{f,g}| \leqslant 1$ implies that $0 \leqslant S_{f,g} \leqslant \infty$.) We claim that

$$\lim_{\epsilon \downarrow 0} K(\epsilon) = S_{f,g}, \tag{5.3}$$

which, in view of (5.2), proves the inequality. Since $|A_{f,g}| \leqslant 1$ we have, for each τ and ω, that

$$0 \leqslant \epsilon^{-1}|A_{f,g}|^2(1 - |A_{f,g}|^{2\epsilon}) \leqslant -|A_{f,g}|^2 \ln|A_{f,g}|^2. \tag{5.4}$$

(The last inequality is simply $1 + \epsilon \ln X \leqslant X^\epsilon$ for all $X > 0$.) Now $K(\epsilon)$ is just the integral of the middle function in (5.4) (which is non-negative), and we see that this function is uniformly dominated by an integrable function. Furthermore, as $\epsilon \downarrow 0$ the middle function in (5.4) converges pointwise to the right-hand function. Equation (5.3) then follows by Lebesgue's dominated convergence theorem. Q.E.D.

ACKNOWLEDGMENTS

I thank E. Carlen, I. Daubechies, P. Flandrin, and A. Grossman for helpful discussions and A. J. E. M. Janssen for a helpful correspondence and for encouraging me to write this paper. In fact, the results in this paper have already been quoted and used by Janssen.[11]

This work was partially supported by U. S. National Science Foundation Grant PHY 85-15288-A03.

[1] P. M. Woodward, *Probability and Information Theory with Applications to Radar* (McGraw-Hill, New York, 1953), p. 120.

[2] R. Price and E. M. Hofstetter, "Bounds on the volume and height distributions of the ambiguity function," IEEE Trans. Inf. Theory **IT-11**, 207–214 (1965).

[3] A. J. E. M. Janssen, "Positivity properties of phase-plane distribution functions," J. Math. Phys. **25**, 2240–2252 (1984).

[4] E. H. Lieb, "Gaussian kernels have only Gaussian maximizers," to be published in Invent. Math. (1990).

[5] R. L. Hudson, "When is the Wigner quasi-probabilty density non-negative?," Rep. Math. Phys. **6**, 249–252 (1974).

[6] A. J. E. M. Janssen, "Bilinear phase-plane distribution functions and positivity", J. Math. Phys. **26**, 1986–1994 (1985).

[7] E. H. Lieb, "Proof of an entropy conjecture of Wehrl," Commun. Math. Phys. **62**, 35–41 (1978).

[8] W. Beckner, "Inequalities in Fourier analysis," Ann. Math. **102**, 159–182 (1975).

[9] H. J. Brascamp and E. H. Lieb, "Best constants in Young's inequality, its converse, and its generalization to more than three functions," Adv. Math. **20**, 151–173 (1976).

[10] L. Leindler, "On a certain converse of Hölder's inequality. II," Acta Math. Szeged. **33**, 217–223 (1972).

[11] A. J. E. M. Janssen, "Wigner weight functions and Weyl symbols of non-negative definite linear operators," Philips J. Res. **6**, 7–42 (1989).

Part VII

Inequalities Related to Harmonic Maps

C. R. Acad. Sc. Paris, t. 303, Série I, n° 5, 1986 **207**

CALCUL DES VARIATIONS. — *Estimations d'énergie pour des applications de* $\mathbf{R}^3$ *à valeurs dans* S^2. Note de **Haïm Brezis, Jean-Michel Coron** et **Elliott H. Lieb,** présentée par Jean Leray.

On résout deux problèmes concernant des applications φ avec des singularités ponctuelles d'un domaine $\Omega \subset \mathbf{R}^3$, à valeurs dans S^2. Le premier est de déterminer le minimum de l'énergie de φ lorsque la position et le degré topologique des singularités est prescrit. Dans le second problème Ω est la boule unité et $\varphi = g$ est donné sur $\partial\Omega$. On montre que $g(x/|x|)$ minimise l'énergie si et seulement si $g = \mathrm{Cte}$ ou bien $g(x) = \pm R\, x$ et R est une rotation.

CALCULUS OF VARIATIONS. — Energy estimates for $\mathbf{R}^3 \to S^2$ mappings.

Two problems concerning maps φ with point singularities from a domain $\Omega \subset \mathbf{R}^3$ to S^2 are solved. The first is to determine the minimum energy of φ when the location and topological degree of the singularities are prescribed. In the second problem Ω is the unit ball and $\varphi = g$ is given on $\partial\Omega$; we show that the only cases in which $g(x/|x|)$ minimizes the energy is $g = $ const. or $g(x) = \pm R\, x$ with R a rotation.

On considère divers problèmes liés à des estimations d'énergie pour des applications φ de $\mathbf{R}^3$ dans S^2 qui sont discontinues en des points isolés.

1. SINGULARITÉS PRESCRITES. — On fixe des points $a_1, a_2, \ldots, a_N$ dans $\mathbf{R}^3$ et des entiers $d_1, d_2, \ldots, d_N \in \mathbf{Z}$ avec $d_i \neq 0$ pour tout i. On introduit la classe d'applications $\varphi : \mathbf{R}^3 \to S^2$ définie par :

$$\mathscr{E} = \left\{ \varphi \in C\left(\mathbf{R}^3 \setminus \bigcup_{i=1}^{N} \{a_i\};\ S^2\right) \middle| \int_{\mathbf{R}^3} |\nabla \varphi|^2 < \infty \text{ et } \deg(\varphi, a_i) = d_i \right\}.$$

Ici, $\nabla\varphi$ est entendu au sens de $\mathscr{D}'(\mathbf{R}^3)$ et $\deg(\varphi, a_i)$ désigne le degré topologique de φ restreint à une sphère centrée en a_i et de rayon r assez petit $(r < r_i = \operatorname{Inf}_{j \neq i} |a_i - a_j|)$. On vérifie aisément que $\mathscr{E}$ est non vide si et seulement si

$$\sum_{i=1}^{N} d_i = 0 \tag{1}$$

et on fera cette hypothèse dans la suite.

On s'intéresse à l'énergie minimale de déformation

$$E = \operatorname*{Inf}_{\varphi \in \mathscr{E}} \int_{\mathbf{R}^3} |\nabla \varphi|^2. \tag{2}$$

Cette quantité, qui a l'homogénéité d'une longueur, dépend très explicitement de la position des points a_i et des degrés d_i. Afin d'exprimer cette dépendance on introduit la notion de connexion minimale.

On dit que a_i est un *point positif* (resp. *négatif*) si $d_i > 0$ (resp. $d_i < 0$). Soit

$$Q = \sum_{d_i > 0} d_i = -\sum_{d_i < 0} d_i$$

la somme des degrés positifs. On fait la liste des points positifs en répétant chaque point d_i fois. On désigne cette liste par $p_1, p_2, \ldots, p_Q$. On procède de la même manière avec les points négatifs en répétant chacun d'eux $|d_i|$ fois. On désigne cette liste par $n_1, n_2, \ldots, n_Q$. On pose

$$L = \operatorname*{Min}_{\sigma} \sum_{i=1}^{Q} |p_i - n_{\sigma(i)}| \tag{3}$$

où le minimum est pris sur le groupe des permutations σ de l'ensemble $\{1, 2, \ldots, Q\}$. Une *connexion minimale* est la réunion des segments $C = \bigcup_{i=1}^{Q} [p_i, n_{\sigma(i)}]$ où σ est l'une des permutations qui réalise le minimum dans (3). Bien entendu, il peut exister plusieurs connexions minimales.

On désigne par δ_C la mesure de Hausdorff de la connexion minimale C, c'est-à-dire $\delta_C = \sum_{i=1}^{Q} \delta_{I_i}$ où $I_i = [p_i, n_{\sigma(i)}]$ et δ_I est la mesure de Hausdorff uniforme sur le segment I.

Théorème 1. — *On a*

$$E = 8\pi L. \tag{4}$$

De plus, l'infimum en (2) *n'est pas atteint; si* (φ_n) *est une suite minimisante pour* (2), *alors il existe une sous-suite* (φ_{n_k}) *et une connexion minimale C telles que* φ_{n_k} *converge vers une constante p.p. et* $|\nabla \varphi_{n_k}|^2$ *converge au sens des mesures vers* $8\pi\delta_C$.

Insistons sur le fait que, même s'il existe plusieurs connexions minimales, alors $|\nabla \varphi_{n_k}|^2$ se concentre sur *une seule* connexion minimale (et non pas sur une réunion de connexions minimales). Ceci n'est pas le cas pour le problème de minimisation en D [*voir* (7)].

Principe de la démonstration de (4). — On procède en deux étapes. Pour l'estimation supérieure $E \leqq 8\pi L$, on considère d'abord le cas d'un *dipole*, c'est-à-dire, un point positif p de degré $+1$, et un point négatif n de degré -1 séparés par une distance L. Étant donné $\varepsilon > 0$, on construit explicitement une application $\varphi_\varepsilon \in \mathscr{E}$ telle que $\int |\nabla \varphi_\varepsilon|^2 \leqq 8\pi L + \varepsilon$, avec φ_ε constante en dehors d'un voisinage d'ordre ε du segment $[p, n]$. Dans le cas général on prouve que $E \leqq 8\pi L$ en recollant des dipoles.

L'estimation inférieure, $E \geqq 8\pi L$, est plus difficile. A cet effet, on introduit un concept très utile. A toute application $\varphi \in \mathscr{E}$ on associe le champ de vecteurs D, de composantes

$$D = (\varphi \cdot \varphi_y \wedge \varphi_z, \varphi \cdot \varphi_z \wedge \varphi_x, \varphi \cdot \varphi_x \wedge \varphi_y) \quad (\text{où } \varphi_x = \partial\varphi/\partial x, \ldots).$$

On montre que

$$\operatorname{div} D = 4\pi \sum_{i=1}^{N} d_i \delta_{a_i} = 4\pi \left(\sum_{i=1}^{Q} \delta_{p_i} - \sum_{i=1}^{Q} \delta_{n_i} \right) \equiv 4\pi\rho. \tag{5}$$

Comme, d'autre part, on a

$$|2D| \leqq |\nabla \varphi|^2 \tag{6}$$

il vient

$$E \geqq 8\pi \operatorname*{Inf}_{\operatorname{div} D = \rho} \int |D|. \tag{7}$$

Un argument de dualité conduit à l'égalité

$$\operatorname*{Inf}_{\operatorname{div} D = \rho} \int |D| = \operatorname*{Max}_{\zeta \in K} \int \zeta \, d\rho$$

où $K = \{\zeta : \mathbf{R}^3 \to \mathbf{R} \mid \|\zeta\|_{\text{Lip}} \leqq 1\}$ et $\|\zeta\|_{\text{Lip}} = \operatorname*{Sup}_{x \neq y} |\zeta(x) - \zeta(x)| / |x - y|$.

C. R. Acad. Sc. Paris, t. 303, Série I, n° 5, 1986

On prouve enfin que $\operatorname*{Max}_{\zeta \in K} \int \zeta \, d\rho = L$ à l'aide d'un théorème de Kantorovich (*voir* [1] et [2]) et du théorème de Birkhoff sur les matrices doublement stochastiques (*voir* par exemple [3]).

Remarque 1. — La relation (4) s'étend à des situations plus générales. Considérons, par exemple, un ouvert Ω de $\mathbf{R}^3$ contenant les points a_i et soit

$$\mathscr{E}_1 = \left\{ \varphi \in C\left(\Omega \setminus \bigcup_{i=1}^{N} \{a_i\}; S^2\right) \middle| \int_\Omega |\nabla \varphi|^2 < \infty \text{ et } \deg(\varphi, a_i) = d_i \right\},$$

$$E_1 = \operatorname*{Inf}_{\varphi \in \mathscr{E}_1} \int_\Omega |\nabla \varphi|^2.$$

Alors on a $E_1 = 8\pi L_1$ où $L_1 = \operatorname*{Min}_{\sigma} \sum_{i=1}^{Q} D(p_i, n_{\sigma(i)})$ et

$$D(p, n) = \operatorname{Min}\{|p-n|, \operatorname{dist}(p, \partial\Omega) + \operatorname{dist}(n, \partial\Omega)\}.$$

Dans le même ordre d'idées on peut considérer

$$\mathscr{E}_2 = \{\varphi \in \mathscr{E}_1 \mid \varphi \text{ est constante sur } \partial\Omega\}$$

et

$$E_2 = \operatorname*{Inf}_{\varphi \in \mathscr{E}_2} \int_\Omega |\nabla \varphi|^2.$$

Alors on a $E_2 = 8\pi L_2$ où $L_2 = \operatorname*{Min}_{\sigma} \sum_{i=1}^{Q} d_\Omega(p_i, n_{\sigma(i)})$ et $d_\Omega(p, n)$ désigne la distance géodésique dans Ω entre p et n.

Remarque 2. — On peut englober les cas précédents dans une situation encore plus générale où l'on remplace les points a_i par des « trous » H_i (compacts disjoints de Ω). Pour définir $\deg(\varphi, H_i)$ on procède de manière similaire au cas d'un point. La conclusion est encore que $E = 8\pi L$ où L fait intervenir une distance appropriée entre les trous. Ici, on ne fait plus l'hypothèse $d_i \neq 0$ et les trous de degré zéro peuvent jouer un rôle dans le calcul de la distance entre les trous.

2. Minimisation avec condition aux limites. — Soit Ω un ouvert borné de $\mathbf{R}^3$ et soit $g : \partial\Omega \to S^2$ une donnée au bord. On s'intéresse au problème

$$\text{(8)} \qquad E(g) \equiv \operatorname{Min}\left\{ \int_\Omega |\nabla \varphi|^2 \,\middle|\, \varphi \in H^1(\Omega; S^2) \text{ et } \varphi = g \text{ sur } \partial\Omega \right\}.$$

Il est clair que le minimum en (8) est atteint et on sait, d'après un résultat de Schoen et Uhlenbeck [4], que si φ réalise le minimum, alors φ admet au plus un nombre fini de points de discontinuité.

Nos résultats principaux sont les suivants :

Théorème 2. — *On suppose que* $\Omega = \{x \in \mathbf{R}^3 \mid |x| < 1\}$ *et que* $g(x) = x$. *Alors* $\psi(x) \equiv x/|x|$ *réalise le minimum dans* (8). *En fait*, $\psi(x)$ *est l'unique minimum dans* (8).

Théorème 3. — *On suppose que* $\Omega = \{x \in \mathbf{R}^3 \mid |x| < 1\}$ *et que* g *est quelconque. Alors* $\psi(x) = g(x/|x|)$ *ne réalise pas le minimum dans* (8), *excepté si* $\pm g$ *est une rotation ou une constante.*

Revenant au cas d'un domaine Ω général et d'une donnée g arbitraire, il résulte des théorèmes 2 et 3, de [4] et [5], le

COROLLAIRE 4. — *On suppose que* φ *réalise le minimum dans* (8), *alors le degré de* φ *en chaque point singulier* x_0 *est* ± 1 *et*

$$\varphi(x) \simeq \pm \mathrm{R}(x-x_0)/|x-x_0| \quad \textit{quand}\; x \to x_0,$$

où R *est une rotation.*

Principe de la démonstration du théorème 3. — Il est clair que si $\psi(x)=g(x/|x|)$ réalise le minimum dans (8), alors nécessairement g est une application harmonique. Si le degré de g est ± 1 et que g n'est pas une isométrie alors on peut diminuer l'énergie en « déplaçant la singularité vers le centre de masse » de $|\nabla g|^2$, i. e. $\int_{\partial\Omega} |\nabla g|^2 \sigma \, d\sigma \neq 0$. Si le degré de g est différent de 0, ± 1, alors on peut diminuer l'énergie en éclatant la singularité en plusieurs points.

Remarque 3. — La motivation originale de ce travail est liée à des questions qui apparaissent dans l'étude des cristaux liquides (*voir* [6], [7], [8]). (Dans ce cas, il faut remplacer S^2 par $\mathbf{R}P^2$ ce qui se fait facilement, *voir* [9]). Le corollaire 4 explique le fait que seules les singularités de degré ± 1 sont observées expérimentalement (*voir* par exemple [10]) (dans un travail antérieur, Hardt-Kinderlehrer et Lin [11] avaient établi que le degré des singularités est majoré par une constante universelle).

Nous remercions J. Ericksen et D. Kinderlehrer qui ont attiré notre attention sur ces questions.
Le détail des démonstrations paraîtra dans [9].

Reçue le 12 mai 1986.

RÉFÉRENCES BIBLIOGRAPHIQUES

[1] L. V. KANTOROVICH, *Dokl. Akad. Nauk S.S.S.R.*, 37, n° 7-8, 1942, p. 227-229.
[2] S. T. RACHEV, *Theory of Probability and its Appl.*, 29, 1985, p. 647-676.
[3] H. MINC, Permanents, *Encyclopedia of Math. and Appl.*, 6, Addison-Wesley, Reading, Mass, 1978.
[4] R. SCHOEN et K. UHLENBECK, *J. Diff. Geom.*, 17, 1982, p. 307-335 et 18, 1983, p. 253-268.
[5] L. SIMON, *Annals of Math.*, 118, 1983, p. 525-571.
[6] P. G. DE GENNES, *The physics of liquid crystals*, Clarendon Press, Oxford, 1974.
[7] M. KLEMAN, *Points, lignes, parois*, Les Éditions de Physique, Orsay, 1977.
[8] J. ERICKSEN, in *Advances in liquid crystals*, 2, G. BROWN éd., Acad. Press, New York, 1976.
[9] H. BREZIS, J.-M. CORON et F. LIEB, Harmonic maps with defects (à paraître).
[10] W. BRINKMAN et P. CLADIS, *Physics Today*, 35, 1982, p. 48-54.
[11] R. HARDT, D. KINDERLEHRER et F. H. LIN, en préparation.

H. B. : *Université Paris-VI*, 4, *place Jussieu*, 75252 *Paris Cedex* 05;
J.-M. C. : *École Polytechnique*, 91128 *Palaiseau Cedex;*
E. L. : *I.H.E.S.*, 91440 *Bures-sur-Yvette* et *Princeton University.*

BULLETIN (New Series) OF THE
AMERICAN MATHEMATICAL SOCIETY
Volume 17, Number 2, October 1987

SINGULARITIES OF ENERGY-MINIMIZING MAPS FROM THE BALL TO THE SPHERE

FREDERICK J. ALMGREN, JR. AND ELLIOTT H. LIEB

We study maps φ from the unit ball $\mathbf{B}$ in $\mathbf{R}^3$ to the unit sphere $\mathbf{S}^2$ in $\mathbf{R}^3$ which minimize Dirichlet's energy integral

$$\mathcal{E}(\varphi) = \int_{\mathbf{B}} |\nabla\varphi|^2 dV.$$

If such a φ minimized Dirichlet's integral among mappings into $\mathbf{R}^3$ rather than being constrained to lie in $\mathbf{S}^2$ it would then be a classical smooth harmonic function. A minimizing constrained φ, however, sometimes has isolated point discontinuities. We here announce several new estimates on the number and arrangement of such singular points [**AL**]. The φ's we consider have well defined values ψ on the boundary $\partial\mathbf{B}$ of $\mathbf{B}$, and the boundary Dirichlet's energy integral is

$$\partial\mathcal{E}(\psi) = \int_{\partial\mathbf{B}} |\nabla_T\psi|^2 dA,$$

where $\nabla_T\psi$ denotes the tangential gradient. In our theorems and examples below each ψ has finite energy. One of our principal results is

MAIN THEOREM. *Suppose φ minimizes Dirichlet's integral among all functions mapping $\mathbf{B}$ to $\mathbf{S}^2$ and having boundary value function ψ on $\partial\mathbf{B}$. Then the number of points of discontinuity of φ is bounded by a constant times $\partial\mathcal{E}(\psi)$.*

This linear law is noteworthy because examples illustrate linear growth of the number of singularities with $\partial\mathcal{E}(\psi)$ while other examples show that the number of singularities cannot be bounded by $\mathcal{E}(\varphi)$. This shows that the number and location of singularities cannot be inferred from simple energy comparisons alone. The subtlety of this estimate is further illustrated by

EXAMPLES. *There are boundary value functions ψ for which the minimizing φ's are unique and have an arbitrarily large number of singular points stacked arbitrarily high near the boundary—like bubbles in a pan of water that is almost ready to boil. The number of stacks is also arbitrarily large.*

Such examples show the necessity of an analysis containing several different length scales in proving the principal result above—the length scale of a singular point is its distance to the boundary.

Received by the editors April 20, 1987.
1980 *Mathematics Subject Classification* (1985 *Revision*). Primary 58E20; Secondary 58E30, 82A50.

One might expect that if ψ mapped $\partial\mathbf{B}$ to cover only small area in $\mathbf{S}^2$ then there could not be too many singular points of φ in $\mathbf{B}$. Indeed, prior to our work, all examples of boundary values ψ with many singularities also had boundary mapping area proportional to $\partial\mathcal{E}(\psi)$. Such a relationship turns out not to hold in general and, as another of our principal results, we show

EXAMPLES. *For any preassigned number N, there is a smooth boundary value mapping ψ of $\partial\mathbf{B}$ to $\mathbf{S}^2$ with the following properties*: (i) *the image of ψ in $\mathbf{S}^2$ consists of a single smooth curve Γ near the equator (ψ thus has zero mapping area), and* (ii) *any minimizing φ has at least N singularities.*

One key ingredient of these examples is the existence of two different parametrizations of Γ from the boundary ∂D of the unit disk D such that the least energy extension of the first parametrization maps D to cover the north pole of $\mathbf{S}^2$ while the least energy extension of the second parametrization maps D to cover the south pole. This then leads directly to an example in which $\mathbf{B}$ is replaced by a large solid torus with cross-section D and the boundary parametrizations alternate as one goes along the torus. We effectively embed such a torus in $\mathbf{B}$ using the conformal equivalence between the disk and the half-plane.

Another natural question one might ask is whether minimizers respect boundary value symmetries (if any), as is true for classical harmonic functions. This is not the case as we illustrate by

EXAMPLES. *There are boundary value functions ψ which are symmetric about the midplane of $\mathbf{B}$ but for which any minimizer cannot possess such a symmetry (nor can its set of discontinuities).*

The basic existence and regularity (interior and boundary) theorems for φ's and ψ's as above appear in papers of R. Schoen and K. Uhlenbeck [**SU1**, **SU2**]. It is their work which guarantees that the interior discontinuities for φ's are isolated. The uniqueness of tangential approximations at such points of discontinuity follows from the work of L. Simon [**S**]. Following initial estimates by R. Hardt, D. Kinderlehrer, and M. Luskin [**HKL**], H. Brezis, J.-M. Coron, and Lieb showed that the only possible tangential approximation to a minimizing φ at any singular point is the function $x/|x|$ composed with an orthogonal mapping of $\mathbf{S}^2$ [**BCL**]. Hardt and F. H. Lin showed in [**HL**] how to construct boundary values ψ which would guarantee many singularities in a minimizing φ. Except for this, little was known about the number and location of singularities in a minimizer when the present work began.

Much of the basic analysis in the literature mentioned above has been based ultimately on compactness arguments, i.e. failure of a desired estimate for all constants leads to an impossible situation. Such compactness arguments are central to the present work as well; they lead fairly directly to the following important estimate (Hardt and Lin have informed us of their independent discovery of this fact).

THEOREM. *The distance between any two singularities p and q in a minimizing φ is at least a fixed constant multiple of the distance from p to $\partial\mathbf{B}$.*

Another compactness argument which combines the theorem above with the boundary regularity theory enables us to conclude that the existence of a singularity at distance δ from $\partial\mathbf{B}$ implies that the boundary function ψ must have nearby Dirichlet integral at scales comparable to δ independent of boundary energy distribution at much larger or much smaller scales. A combinatorial analysis on a Cayley tree based on these differing length scales permits us to sum these different energies in proving our main theorem.

As one might suspect our main theorem remains true (with appropriate constants) if $\mathbf{B}$ is replaced by considerably more general domains in $\mathbf{R}^3$, while the second theorem holds with the same constant.

One of the original motivations for studying mappings to $\mathbf{S}^2$ (or to $\mathbf{RP}^2$) was the mathematical analysis of liquid crystal configurations—in this context one usually regards φ as a unit vectorfield in $\mathbf{B}$. Because we base our analysis on compactness arguments we can also readily conclude that a unit vectorfield φ which minimizes any nematic liquid crystal energy integral sufficiently close to Dirichlet's integral must have at most isolated point discontinuities and the number of these discontinuities is dominated by boundary energy.

References

[**AL**] F. J. Almgren, Jr. and E. H. Lieb, *Singularities of energy-minimizing maps from the ball to the sphere: examples, counterexamples, and bounds*, in preparation.

[**BCL**] H. Brezis, J.-M. Coron and E. H. Lieb, *Harmonic maps with defects*, Comm. Math. Physics **107** (1986), 649–705.

[**HKL**] R. Hardt, D. Kinderlehrer and M. Luskin, *Remarks about the mathematical theory of liquid crystals*, IMA Preprint #276, October 1986.

[**HL**] R. Hardt and F. H. Lin, *A remark on* $\mathbf{H}^1$ *mappings*, Manuscripta Math. **56** (1986), 1–10.

[**SU1**] R. Schoen and K. Uhlenbeck, *A regularity theory for harmonic maps*, J. Differential Geom. **17** (1982), 307–335.

[**SU2**] ——, *Boundary regularity and the Dirichlet problem of harmonic maps*, J. Differential Geom. **18** (1983), 253–268.

[**S**] L. Simon, *Asymptotics for a class of non-linear evolution equations with applications to geometric problems*, Ann. of Math. (2) **118** (1983), 525–571.

Department of Mathematics, Princeton University, Princeton, New Jersey 08544

With F. Almgren and W. Browder in Springer Lecture Notes in Math. *1306*, 1–22 (1988)

CO-AREA, LIQUID CRYSTALS, AND MINIMAL SURFACES[1]

F. Almgren, W. Browder, and E. H. Lieb
Department of Mathematics, Princeton University
Princeton, New Jersey 08544, USA

Abstract. Oriented n area minimizing surfaces (integral currents) in $\mathcal{M}^{m+n}$ can be approximated by level sets (slices) of nearly m-energy minimizing mappings $\mathcal{M}^{m+n} \to \mathbf{S}^m$ with essential but controlled discontinuities. This gives new perspective on multiplicity, regularity, and computation questions in least area surface theory.

In this paper we introduce a collection of ideas showing relations between co-area, liquid crystals, area minimizing surfaces, and energy minimizing mappings. We state various theorems and sketch several proofs. A full treatment of these ideas is deferred to another paper.

Problems inspired by liquid crystal geometries.[2] Suppose Ω is a region in 3 dimensional space $\mathbf{R}^3$ and f maps Ω to the unit 2 dimensional sphere $\mathbf{S}^2$ in $\mathbf{R}^3$. Such an f is a unit vectorfield in Ω to which we can associate an 'energy'

$$\mathcal{E}(f) = \left(\frac{1}{8\pi}\right) \int_\Omega |Df|^2 \, d\mathcal{L}^3;$$

here Df is the differential of f and $|Df|^2$ is the square of its Euclidean norm—in terms of coordinates,

$$|Df(x)| = \sum_{k=1}^{3}\sum_{i=1}^{3} \left(\frac{\partial f^k}{\partial x_i}(x)\right)^2$$

for each x. The factor $1/8\pi$ which equals 1 divided by twice the area of $\mathbf{S}^2$ is a useful normalizing constant. It is straightforward to show the existence of f's of least energy for given boundary values (in an appropriate function space).

Such boundary value problems have been associated with liquid crystals.[3] In this context, a "liquid crystal" in a container Ω is a fluid containing long rod like molecules whose directions are specified by a unit vectorfield. These molecules have a preferred alignment relative to each other—in the present case the preferred alignment is parallel. If we imagine the molecule orientations along

[1] This research was supported in part by grants from the National Science Foundation

[2] The research which led to the present paper began as an investigation of a possible equality between infimums of m-energy and the n area of area minimizing n dimensional area minimizing manifolds in $\mathbf{R}^{m+n}$ suggested in section VIII(C) of the paper, *Harmonic maps with defects* [**BCL**] by H. Brezis, J-M. Coron, and E. Lieb. Although the specific estimates suggested there do not hold (by virtue of counterexamples [**MF**][**W1**][**YL**]) their general thrust does manifest itself in the results of the present paper.

[3] See, for example, the discussion by R. Hardt, D. Kinderlehrer, and M. Luskin in [**HKL**].

$\partial\Omega$ to be fixed (perhaps by suitably etching container walls) then interior parallel alignment may not be possible. In one model the system is assumed to have 'free energy' given by our function $\mathcal{E}$ and the crystal geometry studied is that which minimizies this free energy.

If Ω is the unit ball and $f(x) = x$ for $|x| = 1$, then there is no continuous extension of these boundary values to the interior; indeed the unique least energy f is given by setting $f(x) = x/|x|$ for each x. It turns out that this singularity is representative, and the general theorem is that *least energy f's exist and are smooth except at isolated points p of discontinuity where 'tangential structure' is $\pm x/|x|$ (up to a rotation), e.g. f has local degree equal to ± 1* [SU] [BCL VII].

As a further step towards an understanding of the geometry of of energy minimizing f's one might seek estimates on the number of points of discontinuity which such an f can have—e.g. if the boundary values are not to wild must the number of points of discontinuity be not too big?[4] An alternative problem to this is to seek a lower bound on the energy when the points of discontinuity are prescribed together with the local degrees of the mapping being sought. This question has a surprisingly simple answer as follows.

THEOREM. *Suppose $p_1, \dots, p_N$ are points in $\mathbf{R}^3$ and $d_1, \dots, d_N \in \mathbf{Z}$ are the prescribed degrees with $\sum_{i=1}^N d_i = 0$. Let $\inf \mathcal{E}$ denote the infimum of the energies of (say, smooth) mappings from $\mathbf{R}^3 \sim \{p_1, \dots, p_N\}$ to $\mathbf{S}^2$ which map to the 'south pole' outside some bounded region in $\mathbf{R}^3$ and which, for each i, map small spheres around p_i to $\mathbf{S}^2$ with degree d_i. Then $\inf \mathcal{E}$ equals the least mass $\mathbf{M}(T)$ of integral 1 currents T in $\mathbf{R}^3$ with*

$$\partial T = \sum_{i=1}^{N} d_i [\![p_i]\!].$$

This fact (stated in slightly different language) is one of the central results of [BCL]. We would like to sketch a proof in two parts: first by showing that $\inf \mathcal{E} \le \inf \mathbf{M}$ (with the obvious meanings) and then by showing that $\inf \mathbf{M} \le \inf \mathcal{E}$. The proof of the first part follows [BCL] while the second part is new. It is in this second part that the coarea formula makes its appearance.

Proof that $\inf \mathcal{E} \le \inf \mathbf{M}$. The first inequality is proved by construction as illustrated in Figure 1. We there represent that case in which N equals 2 and p_1 and p_2 are distinct points with $d_1 = -1$ and $d_2 = +1$. We choose and fix a smooth curve C connecting these two points and orient C by a smoothly varying unit tangent vector field ς which points away from p_1 and towards p_2. The associated 1 dimensional integral current is $T = \mathbf{t}(C, 1, \varsigma)$ and its mass $\mathbf{M}(T)$ is the length of C since the density specified is everywhere equal to 1.[5] We now choose (somewhat arbitrarily)

[4] As it turns out, away from the boundary of Ω, the number of these points is bounded a *priori* independent of boundary values.

[5] Formally, a 1 current such as T is a linear functional on smooth differential 1 forms in $\mathbf{R}^3$. If φ is such a 1 form then

$$T(\varphi) = \int_{x \in C} \langle \varsigma(x), \varphi(x) \rangle \, d\mathcal{H}^1 x.$$

To each point p in $\mathbf{R}^3$ is associated the 0 dimensional current $[\![p]\!]$ which maps the smooth function ψ to the number $\psi(p)$. See Appendix A.4.

3

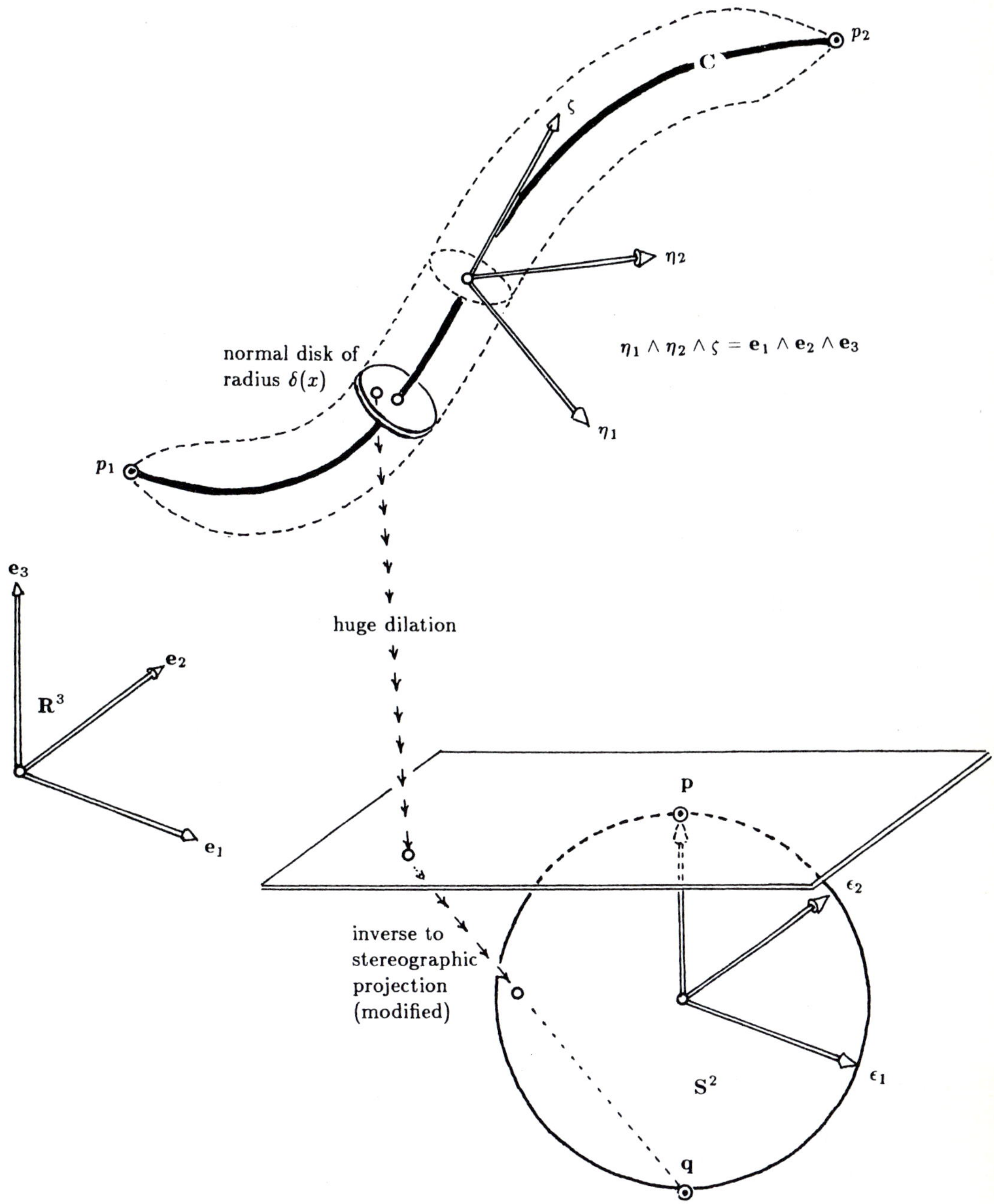

Figure 1. Construction of a mapping f (indicated by dashed arrows) from $\mathbf{R}^3$ to $\mathbf{S}^2$ having energy $\mathcal{E}(f)$ not much greater than the length of the curve C connecting the points p_1 and p_2. Small disks normal to C map by f to cover $\mathbf{S}^2$ once in a nearly conformal way. This implies that small spheres around p_1 map to $\mathbf{S}^2$ with degree -1 while small spheres around p_2 map with degree $+1$. The **1** current $\mathbf{t}(C, 1, \varsigma)$ is the slice $\langle \mathbf{E}^3, f, \mathbf{p}\rangle$ of the Euclidean 3 current $\mathbf{E}^3$ by the mapping f and the 'north pole' $\mathbf{p}$ of $\mathbf{S}^2$.

and fix two smoothly varying unit normal vector fields η_1 and η_2 along C which are perpendicular to each other and for which, at each point x of C, the 3-vector $\eta_1(x) \wedge \eta_2(x) \wedge \varsigma(x)$ equals the orienting 3-vector $\mathbf{e}_1 \wedge \mathbf{e}_2 \wedge \mathbf{e}_3$ for $\mathbf{R}^3$. These two vector fields are a 'framing' of the normal bundle of C.

We then construct a mapping γ of $\mathbf{R}^2$ onto the unit 2 sphere $\mathbf{S}^2$ which is a slight modification of the inverse to stereographic projection. To construct such γ we fix a huge radius R in $\mathbf{R}^2$ and require: (i) if $|y| \leq R$ then $\gamma(y)$ is that point in $\mathbf{S}^2$ which maps to y under stereographic projection $\mathbf{S}^2 \to \mathbf{R}^2$ from the south pole $\mathbf{q}$ of $\mathbf{S}^2$; (ii) if $|y| \geq 2R$ then $\gamma(y) = \mathbf{q}$; (iii) for $R < |y| < 2R$, $\gamma(y)$ is suitably interpolated. See Appendix A.2.

Next we choose some smoothly varying (and very small) radius function δ on C which vanishes only at the endpoints p_1 and p_2.

Finally, as our mapping f from $\mathbf{R}^3$ to $\mathbf{S}^2$ with which to estimate $\mathcal{E}(f)$ we specify the following. If p in $\mathbf{R}^3$ can be written $p = x + s\eta_1(x) + t\eta_2(x)$ for some x in C and some s and t with $s^2 + t^2 \leq \delta(x)^2$, then

$$f(p) = \gamma\left(\frac{2Rs}{\delta(x)}, \frac{2Rt}{\delta(x)}\right).$$

Otherwise, $f(p) = \mathbf{q}$. We leave it as an exercise to the reader to use the fact that γ is conformal for $|y| < R$ to check that $\mathcal{E}(f)$ very nearly equals $\mathbf{M}(T)$; see Appendix A.2. The remainder of the proof that $\inf \mathcal{E} \leq \inf \mathbf{M}$ is also left to the reader.

Proof that $\inf \mathbf{M} \leq \inf \mathcal{E}$. Suppose that f does map $\mathbf{R}^3$ to $\mathbf{S}^2$, has degree d_i at each p_i, and maps to the south pole outside some bounded region. From dimensional considerations one would expect that for most points w in $\mathbf{S}^2$ the inverse image $f^{-1}\{w\}$ would be a collection of curves connecting the various points $p_1, \ldots, p_N$. H. Federer's *coarea formula* is what enables one to quantify this idea; see Appendix A.5. This formula asserts

$$\int_{w \in \mathbf{S}^2} \mathcal{H}^1(f^{-1}\{w\})\, d\mathcal{H}^2 w = \int_{x \in \mathbf{R}^3} J_2 f(x)\, d\mathcal{L}^3 x;$$

here $\mathcal{H}^1$ and $\mathcal{H}^2$ are Hausdorff's 1 and 2 dimensional measures in $\mathbf{R}^3$ and $\mathcal{L}^3$ is Lebesgue's 3 dimensional measure for $\mathbf{R}^3$. Also $J_2 f(x)$ here denotes the 2 dimensional Jacobian of f at x and a key observation (as noted in [**BCL**]) is that $J_2 f(x)$ is always less than or equal to half of $|Df(x)|^2$ with equality only if the differential mapping $Df(x)\colon \mathbf{R}^3 \to \operatorname{Tan}(\mathbf{S}^2, f(x))$ is maximally conformal; see Appendix A.1.3. Also central to the present analysis is the manner in which the curves $f^{-1}\{w\}$ connect the various points $p_1, \ldots, p_N$ and how they relate to the prescribed degrees $d_1, \ldots, d_N$. This connectivity is naturally measured by the current structure of these $f^{-1}\{w\}$'s which comes from the slicing theory for currents; see Appendix A.5. To set this up we regard $\mathbf{R}^3$ as the Euclidean current $\mathbf{E}^3$ (oriented by the 3 vector $\mathbf{e}_1 \wedge \mathbf{e}_2 \wedge \mathbf{e}_3$). The *slice* of $\mathbf{E}^3$ by the map f at the point w in $\mathbf{S}^2$ is the current

$$\langle \mathbf{E}^3, f, w \rangle = \mathbf{t}(f^{-1}\{w\}, 1, \varsigma);$$

the meanings here are the same as for the current T discussed above. A check of orientations and degrees shows that

$$\partial\langle \mathbf{E}^3, f, w\rangle = \sum_{i=1}^{N} k_i [\![p_i]\!];$$

compare with our construction of η_1 and η_2 above. It follows immediately that

$$\begin{aligned}
4\pi \inf \mathbf{M}(T) &= \mathcal{H}^2(\mathbf{S}^2) \inf \mathbf{M}(T) \\
&\le \int_{w \in \mathbf{S}^2} \mathbf{M}(\langle \mathbf{E}^3, f, w\rangle)\, d\mathcal{H}^2 w \\
&= \int_{\mathbf{R}^3} J_2 f \, d\mathcal{L}^3 \\
&= \left(\frac{1}{2}\right) \int_{\mathbf{R}^3} |Df|^2 \, d\mathcal{L}^3.
\end{aligned}$$

This finishes the proof that $\inf \mathbf{M} \le \inf \mathcal{E}$.

First Generalization. Since the methods used in the proofs of the two inequalities are quite general one might correctly suspect that considerable generalization is possible. Suppose, for example, we fix $B = \{p_1, \dots, p_N\}$ as a general boundary set and let $\mathcal{F}_0$ be the family of those mappings f of $\mathbf{R}^3$ to $\mathbf{S}^2$ which are locally Lipschitzian except possibly on B, which map to the southpole outside some bounded region, and which have finite energy. Since deformations of mappings in $\mathcal{F}_0$ do not alter discrete combinatorial structures we are led to study properties of homotopy classes $\Pi(\mathcal{F}_0)$ of mappings in $\mathcal{F}_0$—it is most useful here if our homotopies $[0,1] \times \mathbf{R}^3 \to \mathbf{S}^2$ are permitted to have isolated point discontinuties; see Appendix A.3.

Our conditions about mapping degrees above generalize to requirements about degrees $\mathbf{d}(f, S)$ of f on general integral 2 dimensional cycles S in $\mathbf{R}^3 \sim B$. It turns out that such a degree $\mathbf{d}(f, S)$ depends only on the homotopy class of f and on the homology class of S.

It also turns out that the relative homology classes of the slices $\langle \mathbf{E}^3, f, w\rangle$ depend only on the homotopy class $[f]$ of f. We denote this homology class by $s[f]$.

The *Kronecker index* is a pairing between 2 dimensional cycles S in $\mathbf{R}^3 \sim B$ and 1 currents T having boundary in B. In general the Kronecker index $\mathbf{k}(S, T)$ is the sum over points of intersection of S and T of an index of relative orientations; see Appendix A.6

These various ideas are related in the following theorem.

THEOREM. *The diagram below is commutative. Furthermore, s is an isomorphism, and d and k are injections.*

$$\begin{array}{ccc}
 & & \mathbf{H}_1(\mathbf{R}^3, B; \mathbf{Z}) \\
 & \nearrow s & \\
\Pi(\mathcal{F}_0) & & \downarrow k \\
 & \searrow d & \\
 & & \mathrm{Hom}(\mathbf{H}_2(\mathbf{R}^3 \sim B, \mathbf{Z}), \mathbf{Z})
\end{array}$$

Here

$s[f] = \text{“}[f^{-1}\{w\}]\text{”} = [\langle \mathbf{E}^3, f, w\rangle] =$ the integral homology class of the 1 current slice;

$d[f][S] = \mathbf{d}(f, S) =$ the degree of f on the 2 cycle S;

$k[T][S] = \mathbf{k}(S, T) =$ the Kronecker Index of the 2 cycle S and the 1 current T.

Our relations between energy minimization and area minimization become the following.

THEOREM. *Suppose that P is an integral 1 current in $\mathbf{R}^3$ with the support of ∂P in B. Suppose also that $T^{\mathbf{Z}}$ has least mass among all integral 1 currents which are homologous to P over the integers $\mathbf{Z}$ and that $T^{\mathbf{R}}$ has least mass among all integral 1 currents which are homologous to P over the real numbers $\mathbf{R}$. Then*

$$\mathbf{M}(T^{\mathbf{Z}}) = \inf\{\mathcal{E}(f) : s[f] = [P]\}$$

and

$$\mathbf{M}(T^{\mathbf{R}}) = \inf\{\mathcal{E}(f) : d[f] = k[P]\}.$$

Moreover, $\mathbf{M}(T^{\mathbf{Z}}) = \mathbf{M}(T^{\mathbf{R}})$ (because of our special situation).

Further generalizations. The essential ingredients of the analyses above remain, for example, if $\mathbf{R}^3$ is replaced by a general $m+n$ dimensional manifold $\mathcal{M}$ (without boundary) which is smooth, compact, and oriented (or $\mathcal{M} = \mathbf{R}^{m+n}$), and B is replaced by a sufficiently nice (possibly empty) compact subset of $\mathcal{M}$ of dimension $n-1$. To study n dimensional integral currents in $\mathcal{M}$ having boundary in B we consider mappings f of $\mathcal{M}$ to a sphere of the complementary dimension m. The spaces $\mathcal{F}$ and $\mathcal{F}_0$ of such mappings and the homotopy classes $\Pi(\mathcal{F})$ are specified in sections A.3.1 and A.3.2 of the Appendix. Some discontinuities are essential.[6] It seems worthwhile to consider three different energies $\mathcal{E}_1$, $\mathcal{E}_2$, and $\mathcal{E}_3$ for mappings in $\mathcal{F}_0$. $\mathcal{E}_1$ is a normalization of the usual 'n energy' of mappings, $\mathcal{E}_3$ is a normalized Jacobian integral associated with the coarea formula, and $\mathcal{E}_2$ is an intermediate energy; see Appendix A.3.2. As indicated above, mapping degrees and the Kronecker index have general meanings which are set forth in sections A.6 and A.7 of the Appendix. These various ideas are related as the following theorem shows.

THEOREM. *The diagram of mappings below is well defined and is commutative. In particular, the images of d and k and j in $\mathbf{Hom}(\mathbf{H}_m(\mathcal{M} \sim B, \mathbf{Z}), \mathbf{Z})$ are the same. Furthermore, s is an*

[6] Suppose $m = 2$ and $n = 5$ and $\mathcal{M} = \mathbf{R}^7$, and B is a smoothly embedded copy of 2 dimensional complex projective space $\mathbf{CP}(2)$. Then there are no continuous mappings f from the complement of B to $\mathbf{S}^2$ such that small 2 spheres S which link B once map to $\mathbf{S}^2$ with degree one. Any f satisfying such a linking condition for general position S's near B must have interior discontinuities of dimension at least 3.

7

isomorphism.

$$\begin{array}{ccccc} & & \mathbf{H}_n(\mathcal{M},B;\mathbf{Z}) & \xrightarrow{c} & \mathbf{H}_n(\mathcal{M},B;\mathbf{R}) \\ & \nearrow s & & \searrow c & \uparrow i \\ \Pi(\mathcal{F}) & & \downarrow k & & c[\mathbf{H}_n(\mathcal{M},B;\mathbf{Z})] \\ & \searrow d & & \swarrow j & \\ & & \mathbf{Hom}(\mathbf{H}_m(\mathcal{M}\sim B,\mathbf{Z}),\mathbf{Z}) & & \end{array}$$

Here

$s[f] =$ "$[f^{-1}\{p\}]$" $= [\langle [\![\mathcal{M}]\!], f, p\rangle] =$ the integral homology class of the n current slice;

$d[f][S] = \mathbf{d}(f,S) =$ the degree of f on the m cycle S;

$k[T][S] = \mathbf{k}(S,T) =$ the Kronecker index of the m cycle S and the n current T;[7]

c is induced by the coefficient inclusion $\mathbf{Z} \to \mathbf{R}$;

i is the inclusion; and

j is defined by commutivity.

We defer proof of this theorem to our fuller treatment of this subject. The natural setting and generality of such relationships are still under investigation.

The relations between energy minimization and area minimization then become the following.

MAIN THEOREM. *Suppose P is an integral current in $\mathcal{M}$ with the support of ∂P contained in B so that the integral homology class $[P]$ of P belongs to $\mathbf{H}_n(\mathcal{M},B;\mathbf{Z})$. Let $T^{\mathbf{Z}}$ be an integral current of least mass among all integral currents belonging to the same integral homology class as P in $\mathbf{H}_n(\mathcal{M},B,\mathbf{Z})$, and let $T^{\mathbf{R}}$ be an integral current of least mass among all integral currents belonging to the same real homology class as P in $\mathbf{H}_n(\mathcal{M},B,\mathbf{R})$. Then*

$$\mathbf{M}(T^{\mathbf{Z}}) = \inf\{\mathcal{E}_1(f): s[f] = [P]\} = \inf\{\mathcal{E}_2(f): s[f] = [P]\} = \inf\{\mathcal{E}_3(f): s[f] = [P]\}$$

and

$$\mathbf{M}(T^{\mathbf{R}}) = \inf\{\mathcal{E}_1(f): d[f] = k[P]\} = \inf\{\mathcal{E}_2(f): d[f] = k[P]\} = \inf\{\mathcal{E}_3(f): d[f] = k[P]\}.$$

[7] Suppose $m = 2$ and $n = 1$ and $\mathcal{M}$ is a 3 dimensional real projective space $\mathbf{RP}(3)$ and $T = \mathbf{t}(\mathcal{N}, 1, \varsigma)$; here $\mathcal{N}$ is a 1 dimensional real projective space $\mathbf{RP}(1)$ sitting in $\mathbf{RP}(3)$ in the usual way and ς is some orientation function. Since T is not a boundary while $2T$ is, we conclude that the homology class

$$[T] \in \mathbf{H}_1(\mathcal{M}, \emptyset;\ \mathbf{Z}) = \mathbf{Z}_2$$

is not the 0 class although $\mathbf{k}(S,T) = 0$ for each 2 cycle S in $\mathcal{M}$. In particular, the mapping k is generally not an injection.

In general, of course, $\mathbf{M}(T^{\mathbf{R}}) < \mathbf{M}(T^{\mathbf{Z}})$. Although we again defer complete proofs to our fuller treatment of this subject, it does seem useful to sketch some of the main ideas.

Proof of the inequality "$\inf \mathcal{E} \leq \inf \mathbf{M}$". The proof here is again by construction. We will indicate the main ingredients in a special case. Suppose, say, $\mathcal{M} = \mathbf{R}^{m+n}$, B is polyhedral, and T is an integral n current which is mass minimizing subject to some appropriate constraints as in the Main Theorem above. We will construct a mapping $f: \mathbf{R}^{m+n} \to \mathbf{S}^m$ in the relevant homotopy class such that $\mathcal{E}_1(f), \mathcal{E}_2(f)$, and $\mathcal{E}_3(f)$ are nearly equal and are not much bigger that $\mathbf{M}(T)$. By virtue of the Strong Approximation Theorem for integral currents [FH1 4.2.20] we can modify T slightly to become simplicial with only a slight increase in mass.

Suppose then that we can express

$$T = \sum_{\alpha=1}^{M} \mathbf{t}(\Delta_\alpha^n, z_\alpha, \varsigma_\alpha)$$

as a 'simplicial' integral current (with the obvious interpretation). For each $k = 0, \ldots, n$ we denote by K_k the collection of closed k simplexes which occur as k dimensional faces of n simplexes among the Δ_α^n's. We then choose numbers $0 < \delta_n << \delta_{n-1} << \delta_{n-2} << \ldots << \delta_0 << 1$ and define sets $N_0, N_1, \ldots, N_n$ in $\mathbf{R}^{m+n}$ by setting

$$N_0 = \{x: \operatorname{dist}(x, \cup K_0) < \delta_0\}$$

and, for each $k = 1, \ldots, n$ set

$$N_k = \{x: \operatorname{dist}(x, \cup K_k) < \delta_k\} \sim (N_{k-1} \cup N_{k-2} \cup \ldots \cup N_0).$$

We assume that $\delta_0, \ldots, \delta_n$ have been chosen so that the distinct components of each N_k correspond to distinct k simplexes in K_k.

We now define mappings $f_{n+1}, f_n, \ldots, f_0 = f$ as follows.

First, the mapping $f_{n+1}: \mathbf{R}^{m+n} \sim (N_n \cup \ldots \cup N_0) \to \mathbf{S}^m$ is defined by setting $f_{n+1}(x) = \mathbf{q}$ for each x.

Second, the mapping $f_n: \mathbf{R}^{m+n} \sim (N_{n-1} \cup \ldots \cup N_0) \to \mathbf{S}^m$ is constructed geometrically in virtually the same manner as the mapping g in the example A.8 in the Appendix. Details are left to the reader.

Third, the mapping $f_{n-1}: \mathbf{R}^{m+n} \sim (N_{n-2} \cup \ldots \cup N_0) \to \mathbf{S}^m$ is constructed geometrically in a manner virtually identical with the construction of the mapping $f_{\delta,r}$ of example A.8 of the Appendix (with δ, r replaced by $\delta_n/2, \delta_{n-1}$ respectively there). The mapping f_{n-1} is Lipschitz across parts of $n-1$ simplexes which do not lie in B and is discontinuous on those $n-1$ simplexes which contain part of ∂T.

Assuming $f_{n+1}, f_n, \ldots, f_{k+1}$ have been constructed we define

$$f_k: \mathbf{R}^{m+n} \sim (N_{k-1} \cup \ldots \cup N_0) \to \mathbf{S}^m$$

as follows. Each point v in $N_k \sim (N_{k-1} \cup \ldots \cup N_0)$ can be written uniquely in the form $v = v_0 + (v - v_0)$ where v_0 is the unique closest point in $\cup K_k$ to v and $|v - v_0| < \delta_k$. If $v \neq v_0$ we note that

$$v_1 = v_0 + \delta_k \left(\frac{v - v_0}{|v - v_0|} \right) \in \mathrm{dmn}(f_{k+1})$$

and we set $f_k(v) = f_{k+1}(v_1)$. A direct extension of the estimates used for the example A.8 of the Appendix shows that the energies $\mathcal{E}_1(f), \mathcal{E}_2(f)$, and $\mathcal{E}_3(f)$ very nearly equal $\mathbf{M}(T)$.

Proof of the inequality "$\inf \mathbf{M} \leq \inf \mathcal{E}$". The argument here is a direct extension of the corresponding argument given above and is left to the reader.

Remarks.

(1) One of the main reasons for analyzing relations between the energy of mappings and the area of currents is that it provides a way to study n dimensional area minimizing *integral* currents (whose geometry is not specified ahead of time) by studying functions and integrals over the given ambient manifold. This seems the first such scheme which works in general codimensions. For *real* currents, however, differential forms play a role roughly analogous to that of our function spaces $\mathcal{F}_0$; in this regard see, for example, the paper of H. Federer, *Real flat chains, cochains, and variational problems* [**F2** 4.10(4), 4.11(2)]. Incidentally, in the language of [**F2** 5.12, page 400], examples show that the equation in question there is not always true under the alternative hypotheses of [**F2** 5.10].

(2) Suppose C consists of smooth simple closed curves in $\mathbf{R}^3$ oriented by ς. Suppose also for positive integers ν we have reasonable mappings f_ν from the complement of C in $\mathbf{R}^3$ to the circle $\mathbf{S}^1$ with the property that small circles which link C once are mapped to $\mathbf{S}^1$ by f_ν with degree ν. Because of the dimensions we have

$$\mathcal{E}_1(f_\nu) = \mathcal{E}_2(f_\nu) = \mathcal{E}_3(f_\nu) = \left(\frac{1}{2\pi}\right) \int |Df_\nu| \, d\mathcal{L}^3.$$

If f_ν is nearly $\mathcal{E}_1$ energy minimizing then for most w's in $\mathbf{S}^1$ the slice

$$T_\nu(w) = \langle \mathbf{E}^3 \, , \, f_\nu \, , \, w \rangle \in \mathbf{I}_2(\mathbf{R}^3)$$

will be defined with $\partial T_\nu(w) = \mathbf{t}(C \, , \nu \, , \varsigma)$ and will be nearly mass minimizing. H. Parks, in his memoir, *Explicit determination of area minimizing hypersurfaces, II* [**PH**], used a similar energy for mappings to the real numbers $\mathbf{R}$ (instead of to $\mathbf{S}^1$) and was able to exhibit an algorithm for finding area minimizing surfaces. The technique used by Parks requires that C be extreme, i.e. that it lie on the boundary of its convex hull. The analysis of our paper on the other hand applies to any collection of curves which, for example, may be knotted or linked in any way. One of our hopes is to develop a method of computation analogous to that of Parks.

(3) Suppose that C and the mappings f_ν have the same meaning as in (2) above. If θ denotes the usual (multiple-valued radian) angle function on $\mathbf{S}^1$ then $d\theta$ as a well defined closed 1-form

whose pullbacks $f_\nu^\sharp d\theta$ give closed 1 forms on the complement of C in $\mathbf{R}^3$ with $|f_\nu^\sharp d\theta| = |Df_\nu|$. For fixed x_0 in the complement of C we define functions g_ν mapping the complement of C to $\mathbf{S}^1$ by requiring that

$$\theta \circ g_\nu(x) = \theta \circ f_\nu(x_0) + \int_\gamma f_\nu^\sharp d\theta \quad (\text{mod } 2\pi)$$

for each x (with the obvious meanings); here $\gamma(x)$ denotes any oriented path in the complement of C starting at x_0 and ending at x. It is immediate to check that $g_\nu = f_\nu$ for each ν. If we write $\nu = \lambda \cdot \mu$ for some λ and μ and define $h_\lambda(x)$ in $\mathbf{S}^1$ by requiring

$$\theta \circ h_\lambda(x) = \int_\gamma \left(\frac{1}{\mu}\right) f_\nu^\sharp d\theta \quad (\text{mod } 2\pi)$$

for γ as above. The mapping h_λ maps small circles with the same degrees as does f_λ. Taking $\mu = \nu$ we readily conclude, for example, that

$$\inf\{\mathbf{M}(T) : \partial T = \mathbf{t}(C, \nu, \varsigma)\} = \nu \cdot \inf\{\mathbf{M}(T) : \partial T = \mathbf{t}(C, 1, \varsigma)\}$$

for each ν. This estimate implies that integral and real mass minimizing 2 currents having boundary $\mathbf{t}(C, 1, \varsigma)$ have the same masses [**F2** 5.8]; although this has been known for some time, the present proof by factoring mappings seems new and simpler. This fact (and our proof) extend to $n-1$ dimensional boundaries in general manifolds $\mathcal{M}$ of dimension $n+1$ with, for example, the property that each 1 cycle is a boundary. There are counterexamples to such equalities in higher codimensions given first by L. C. Young [**YL**] and later by F. Morgan [**MF**] and B. White [**W1**]. How *badly* such an equality can fail remains an important open question. It is not even known, for example, if the number

$$\inf\{\mathbf{M}(S)/\mathbf{M}(T) : S, T \in \mathbf{I}_2(\mathbf{R}^4, \mathbf{R}^4) \text{ are mass minimizing with } 0 \neq \partial S = 2\partial T\}$$

is positive; note, however, the isoperimetric inequality [**A1** 2.6].

(4) Suppose $\mathcal{M}$ is a complex submanifold of some complex projective space $\mathbf{CP}(n)$ (or, more generally, $\mathcal{M}$ is a Kähler manifold). Then any complex analytic (meromorphic) function f from $\mathcal{M}$ to the Riemann Sphere $\mathbf{CP}(1) = \mathbf{S}^2$ has integral current slices which are absolutely mass minimizing in their integral homology classes [**F1** 5.4.19]. Such f's are thus necessarily maximally conformal and minimize each of the energies $\mathcal{E}_1$, $\mathcal{E}_2$, and $\mathcal{E}_3$ among functions in the same homotopy classes.

(5) In the context of this paper, if the mass minimizing current T being sought happens to be unique then most slices of nearly minimizing mappings will be close to that current. In a sense this describes the asymptotic behavior of a sequence $\{f_k\}_k$ of mappings in $\mathcal{F}_0$ converging towards energy minimization; in particular, the real currents

$$\left\{\left(\frac{1}{(m+1)\alpha(m+1)}\right) [\![\mathcal{M}]\!] \llcorner f_k^\sharp \sigma^\bullet \right\}_k$$

must converge to T as $k \to \infty$. If $m = 2$ then the energy $\mathcal{E}_1$ is Dirichlet's integral which is widely studied in the general theory of harmonic mappings between manifolds pioneered by J. Eells and J. Sampson.

In any codimension m each n dimensional mass minimizing integral current is a *regular* minimal submanifold except possibly on a singular set of dimension not exceeding $n - 2$ as shown by F. Almgren in [**A2**]. It is not yet clear to what extent the present new setup will provide new tools for study of the regularity and singularity properties of mass minimizing integral currents. This could be one of its most important potential uses.

APPENDIX

When not otherwise specified we follow the general terminology of pages 669-671 of H. Federer's treatise, *Geometric Measure Theory* [**F1**] or the newer standardized terminology of the 1984 AMS Summer Research Institute in Geometric Measure Theory and the Calculus of Variations as summarized in pages 124-130 of F. Almgren's paper, *Deformations and multiple-valued functions* [**A1**].

A.1 Terminology.

A.1.1 We fix positive integers m and n and suppose that $\mathcal{M}$ is an $m+n$ dimensional submanifold (without boundary) of $\mathbf{R}^N$ (some N) which is smooth, compact, and oriented by the continuous unit $(m+n)$-vectorfield $\xi: \mathbf{M} \to \wedge_{m+n}\mathbf{R}^N$; alternatively $\mathcal{M} = \mathbf{R}^{m+n}$ with standard orthonormal basis vectors $\mathbf{e}_1, \ldots, \mathbf{e}_{m+n}$ and orienting $(m+n)$-vector $\mathbf{e}_1 \wedge \ldots \wedge \mathbf{e}_{m+n}$. We also suppose that B is a finite (possibly empty) union of various (curvilinear) $n-1$ simplexes $\Delta_1, \Delta_2, \ldots, \Delta_J$ associated with some smooth triangulation of $\mathcal{M}$.

A.1.2 We denote by $\mathbf{S}^m$ the unit sphere in $\mathbf{R} \times \mathbf{R}^m = \mathbf{R}^{1+m}$ with its usual orientation given by the unit m-vectorfield $\sigma: \mathbf{S}^m \to \wedge_m \mathbf{R}^{1+m}$; in particular, for each $w \in \mathbf{S}^m \subset \mathbf{R}^{1+m} = \wedge_1 \mathbf{R}^{1+m}$, $\sigma(w) = *w$. It is convenient to let $z, y_1, \ldots, y_m$ denote the usual orthonormal coordinates for $\mathbf{R} \times \mathbf{R}^m$ and also let $\mathbf{p}, \varepsilon_1, \ldots, \varepsilon_m$ be the associated orthonormal basis vectors. In particular, $\sigma(\mathbf{p}) = *\mathbf{p} = \varepsilon_1 \wedge \ldots \wedge \varepsilon_m$. We regard $\mathbf{p}$ as the 'north pole' of $\mathbf{S}^m$. The 'south pole' is $\mathbf{q} = -\mathbf{p}$. We denote by σ^* the differential m form (the 'volume form') on $\mathbf{S}^m$ dual to σ.

A.1.3 If L is a linear mapping $\mathbf{R}^{m+n} \to \mathbf{R}^m$ then the polar decomposition theorem guarantees the existence of orthonormal coordinates for $\mathbf{R}^{m+n}$ and $\mathbf{R}^m$ with respect to which L has the matrix representation

$$L = \begin{pmatrix} \lambda_1 & 0 & \cdots & 0 & 0 & \cdots & 0 \\ 0 & \lambda_2 & \cdots & 0 & 0 & \cdots & 0 \\ \vdots & \vdots & \ddots & 0 & \vdots & \cdots & \vdots \\ 0 & 0 & \cdots & \lambda_m & 0 & \cdots & 0 \end{pmatrix}$$

with $\lambda_1 \geq \lambda_2 \geq \ldots \geq \lambda_m \geq 0$. In these coordinates we can express the Euclidean norm $|L|$ of L as

$$|L| = (\lambda_1^2 + \lambda_2^2 + \ldots + \lambda_m^2)^{\frac{1}{2}},$$

express the mapping norm $\|L\|$ of L as

$$\|L\| = \lambda_1,$$

and express the mapping norm $\| \wedge_m L\|$ of the linear mapping $\wedge_m L$ of m-vectors induced by L as

$$\| \wedge_m \| = \lambda_1 \cdot \lambda_2 \cdots \lambda_m.$$

Whenever $\lambda_1 \geq \lambda_2 \geq \cdots \geq \lambda_m \geq 0$ we have

$$\lambda_1 \cdot \lambda_2 \cdots \lambda_m \leq \frac{1}{m^{\frac{m}{2}}} (\lambda_1^2 + \lambda_2^2 + \ldots + \lambda_m^2)^{\frac{m}{2}} \leq \lambda_1^m \leq (\lambda_1^2 + \lambda_2^2 + \ldots + \lambda_m^2)^{\frac{m}{2}}.$$

The first two inequalities are equalities if and only if $\lambda_1 = \lambda_2 = \ldots = \lambda_m$. The right hand inequalitiy is an equality if and only if $\lambda_2 = \lambda_3 = \ldots = \lambda_m = 0$.

If f is a mapping and $L = Df(a)$ is the differential of f at a, then $|Df(a)|^2$ is of value of *Dirichlet's integrand* of f at a, and

$$J_m f(a) = \| \wedge_m Df(a)\|$$

is the m *dimensional Jacobian* of f at a.

A.2 Modified Stereographic Projection. Stereographic projection of $\mathbf{S}^m$ onto $\mathbf{R}^m$ from the south pole $\mathbf{q}$ maps $(z,y) \in \mathbf{S}^m \sim \{\mathbf{q}\}$ to $2y/(1+z) \in \mathbf{R}^m$ while the inverse mapping $\gamma_0: \mathbf{R}^m \to \mathbf{S}^m$ sends $y \in \mathbf{R}^m$ to

$$\gamma_0(y) = \left(\frac{4 - |y|^2}{4 + |y|^2}, \quad \frac{4y}{4 + |y|^2} \right) \in \mathbf{S}^m \sim \{\mathbf{q}\}.$$

γ_0 is an orientation preserving conformal diffeomorphism between $\mathbf{R}^m$ and $\mathbf{S}^m \sim \{\mathbf{q}\}$ as is readily checked.

For convenience we let $\theta: \mathbf{S}^m \to [0, \pi]$ denote angular distance in radians (equivalently, geodesic distance in $\mathbf{S}^m$) to $\mathbf{p}$. General level sets of θ are thus $m-1$ spheres of constant latitude while $\theta(\mathbf{p}) = 0$ and $\theta(\mathbf{q}) = \pi$. Also for $(z,y) \in \mathbf{S}^m$ we have $z = \cos\theta(z,y)$ and $|y| = \sin\theta(z,y)$. Latitude lines on $\mathbf{S}^m$ are level sets of the function ω which maps $(z,y) \in \mathbf{S}^m \sim \{\mathbf{p},\mathbf{q}\}$ to

$$\omega(z,y) = \frac{y}{|y|} \in \mathbf{S}^{m-1} \subset \mathbf{R}^m.$$

Certain mappings derived from γ_0 are important in our constructions. If $0 < \delta << 1/2$ is a given very small number we fix $0 < r = r(\delta) << R < \infty$ by requiring that R be the radius of the sphere in $\mathbf{R}^m$ which γ_0 maps to the latitude sphere $\theta = \pi - \delta$ near $\mathbf{q}$ in $\mathbf{S}^m$ and that rR/δ be the radius of the sphere in $\mathbf{R}^m$ which γ_0 maps to the latitude sphere $\theta = \delta$ near $\mathbf{p}$.

We now modify γ_0 to obtain a mapping $\gamma = \gamma_\delta = \gamma_{\delta,1}$ which maps $\mathbf{R}^m$ onto all of $\mathbf{S}^m$ and which maps points y in $\mathbf{R}^m$ with norm less that r^2 to $\mathbf{p}$, maps points y in $\mathbf{R}^m$ with norm greater that 2δ to $\mathbf{q}$, maps points y in $\mathbf{R}^m$ with norm between r and δ to $\gamma_0(Ry/\delta)$ and suitable interpolates in the two remaining annular regions. More precisely, we set

$$\gamma(y) = \begin{cases} \mathbf{p} & \text{if } 0 \le |y| \le r^2 \\ \left(\cos\left(\delta\left(\frac{|y|-r^2}{r-r^2}\right)\right), \sin\left(\delta\left(\frac{|y|-r^2}{r-r^2}\right)\right)\frac{y}{|y|}\right) & \text{if } r^2 \le |y| \le r \\ \gamma_0(\frac{Ry}{\delta}) & \text{if } r \le |y| \le \delta \\ \left(\cos\left(\pi + |y| - 2\delta\right), \sin\left(\pi + |y| - 2\delta\right)\frac{y}{|y|}\right) & \text{if } \delta \le |y| \le 2\delta \\ \mathbf{q} & \text{if } 2\delta \le |y| < \infty. \end{cases}$$

In the region $0 \le |y| \le r$ we estimate that the Lipschitz constant of γ does not exceed $\delta/(r - r^2)$ which is less that $2\delta/r$ since $r < 1/2$. Hence

$$\int_{|y|\le r} |D\gamma|^m \, d\mathcal{L}^m \le m^{\frac{1}{2}} \left(\frac{2\delta}{r}\right)^m \alpha(m) r^m = 2^m m^{\frac{1}{2}} \alpha(m) \delta^m$$

which is small if δ is small.

Similarly, in the region $\delta \le |y| \le 2\delta$ we estimate that the local Lipschitz constants do not exceed 1. Hence

$$\int_{\delta\le|y|\le 2\delta} |D\gamma|^m \, d\mathcal{L}^m < m^{\frac{1}{2}} \alpha(m)(2\delta)^m = 2^m m^{\frac{1}{2}} \alpha(m) \delta^m$$

which is small is δ is small.

Finally we note that, in the region, $r < |y| < \delta$ the mapping γ is conformal so that

$$\int_{r\le|y|\le\delta} J_m\gamma \, d\mathcal{L}^m = \int_{r\le|y|\le\delta} \|D\gamma\|^m \, d\mathcal{L}^m = \frac{1}{m^{\frac{m}{2}}} \int_{r\le|y|\le\delta} |D\gamma|^m \, d\mathcal{L}^m = \mathcal{H}^m(\mathbf{S}^m \cap \theta^{-1}[\delta, \pi - \delta]).$$

Our mapping $\gamma_{\delta,1}$ from $\mathbf{R}^m$ to $\mathbf{S}^m$ preserves orientations and covers once. It is useful to have mappings $\gamma_{\delta,\nu}$ with similar conformal properties but covering ν times. To do this we fix a ratio $\rho = \left(r(\delta)^2/\delta\right)$ and let $\tau(z,y) = (-z, -y_1, y_2, \dots, y_m)$ for $(z,y) \in \mathbf{S}^m$; the map τ thus interchanges the north and south poles of $\mathbf{S}^m$ while preserving orientation. We then define

$$\gamma_{\delta,\nu}(y) = \begin{cases} \gamma_\delta(y) & \text{if } \rho\delta \le |y| < \infty \\ \tau^k \circ \gamma_\delta(y/\rho^k) & \text{if } k \in \{1, \dots, \nu-2\} \text{ and } \rho^{k+1}\delta \le |y| \le \rho^k\delta \\ \tau^{\nu-1} \circ \gamma_\delta(y/\rho^{\nu-1}) & \text{if } 0 \le |y| \le \rho^{\nu-1}\delta. \end{cases}$$

A.3. Mappings and homotopies from $\mathcal{M}$ to $\mathbf{S}^m$ with contolled discontinuities.

A.3.1 Whenever $f: \mathcal{M} \to \mathbf{S}^m$ we denote by

$$C_f$$

the closure of the set of points of discontinuity of f. We then let

$$\mathcal{F}$$

be the collection of all functions $f : \mathcal{M} \to \mathbf{S}^m$ such that the closure of $C_f \sim B$ (recall A.1.1) has dimension not exceeding $n-2$. In case m equals 1 we require that $C_f \subset B$ for functions f in $\mathcal{F}$. Also, if $\mathcal{M}$ is $\mathbf{R}^{m+n}$ we require that $f(x) = \mathbf{q}$ whenever $|x|$ is sufficiently large.

Similarly, whenever $h: [0,1] \times \mathcal{M} \to \mathbf{S}^m$ we denote by

$$C_h$$

the closure of the set of discontinuities of h. We then say that f and g in $\mathcal{F}$ are s-homotopic provided there is a function $h: [0,1] \times \mathcal{M} \to \mathbf{S}^m$ such that $h(0,\cdot) = f$ and $h(1,\cdot) = g$ and also

$$C_h \sim ((\{0\} \times C_f) \cup (\{1\} \times C_g) \cup ([0,1] \times B))$$

lies in $(0,1) \times \mathcal{M}$ and has dimension not exceeding $n-1$ (in case $\mathcal{M}$ is $\mathbf{R}^{m+n}$ we additionally require that $h(t,x) = \mathbf{q}$ for all t when $|x|$ is sufficiently large); such a function h is called an s-homotopy between f and g. We then denote by

$$\Pi(\mathcal{F})$$

the s-homotopy equivalence classes of $\mathcal{F}$.

A.3.2 We denote by

$$\mathcal{F}_0$$

those functions f in $\mathcal{F}$ for which $f|(\mathcal{M} \sim C_f)$ is locally Lipschitz and then associate to each such f three energies $\mathcal{E}_1(f)$, $\mathcal{E}_2(f)$, and $\mathcal{E}_3(f)$ given by setting

$$\mathcal{E}_1(f) = \frac{1}{m^{m/2}(m+1)\alpha(m+1)} \int_{\mathcal{M}} |Df|^m \, d\mathcal{H}^{m+n},$$

$$\mathcal{E}_2(f) = \frac{1}{(m+1)\alpha(m+1)} \int_{\mathcal{M}} \|Df\|^m \, d\mathcal{H}^{m+n},$$

$$\mathcal{E}_3(f) = \frac{1}{(m+1)\alpha(m+1)} \int_{\mathcal{M}} J_m f \, d\mathcal{H}^{m+n}.$$

For some analyses (beyond the scope of this present paper) it is important to recognize that

$$J_m f(x) = |\langle \sigma^*(f(x)), \wedge^m Df(x) \rangle|.$$

We also call the reader's attention to the paper *Homotopy classes in Sobolev spaces and the existence of energy minimizing mappings* [**W2**] by B. White in which p energy minimization is studied in homotopy classes of mappings which are not necessarily continuous.

A.3.3 A basic fact is the following

PROPOSITION.

(1) Each s-homotopy class in $\Pi(\mathcal{F})$ contains a representative f which belongs to $\mathcal{F}_0$ and for which each of the energies $\mathcal{E}_1(f)$, $\mathcal{E}_2(f)$, and $\mathcal{E}_3(f)$ is finite.

(2) Suppose f and g belong to $\mathcal{F}_0$ and are representatives of the same s-homotopy class in $\Pi(\mathcal{F})$. Suppose also that $\mathcal{E}_1(f)$ and $\mathcal{E}_1(g)$ are both finite. Then there is an s-homotopy h between f and g such that $h|([0,1] \times \mathcal{M} \sim C_h)$ is locally Lipschitz and

$$\int_{[0,1]\times\mathcal{M}} |Dh|^m \, d\mathcal{H}^{m+n+1} < \infty.$$

A.4 Currents. A general k (dimensional) current T is a continuous linear functional on an appropriate space of smooth differential k forms in $\mathbf{R}^N$. The boundary of a k current T is the $k-1$ current ∂T which maps a smooth differential $k-1$ form ω to the number $\partial T(\omega) = T(d\omega)$— Stokes's theorem becomes a definition. In this paper we are concerned with currents of the form $T = \mathbf{t}(\Sigma, \theta, \varsigma)$. In writing such an expression we mean that $\mathbf{set}(T) = \Sigma$ is a (bounded) $\mathcal{H}^k$ measurable and $(\mathcal{H}^k, k)$ rectifiable subset of $\mathcal{M}$, and that the *density function* $\theta\colon \Sigma \to \mathbf{R}^+$ is $\mathcal{H}^k \llcorner \Sigma$ summable, and that the *orientation* ς is an $\mathcal{H}^k \llcorner \Sigma$ measurable function whose simple unit k vector values are compatible with the tangent plane structure of Σ. Such a k current T maps a differential k form φ to the number

$$T(\varphi) = \int_{x\in\Sigma} \langle \varsigma(x), \varphi(x) \rangle \, \theta(x) \, d\mathcal{H}^k x.$$

Associated with $\mathcal{M}$ itself is the $m+n$ current

$$[\![\mathcal{M}]\!] = \mathbf{t}(\mathcal{M}, 1, \xi);$$

if $\mathcal{M} = \mathbf{R}^{m+n}$ a standard notation is

$$\mathbf{E}^{m+n} = \mathbf{t}(\mathbf{R}^{m+n}, 1, \xi)$$

with $\xi(x) = \mathbf{e}_1 \wedge \ldots \wedge \mathbf{e}_{m+n}$ for each x.

The area of a current $T = \mathbf{t}(\Sigma, \theta, \varsigma)$ weighted with its density gives its *mass*,

$$\mathbf{M}(T) = \int_\Sigma \theta \, d\mathcal{H}^k = \sup\{T(\varphi) \colon \|\varphi\| \le 1\}.$$

The theorems of this paper relate to minimization of this mass rather than, say, the k areas of the underlying set Σ (which is called the *size* of T and is denoted $\mathbf{S}(T)$). The measure $\|T\|$ associated with mass is thus $\mathcal{H}^k \llcorner \Sigma \wedge \theta$ so that $\mathbf{M}(T) = \|T\|(\mathcal{M}) = \|T\|(\mathbf{R}^N)$.

A general fact about such a current $T = \mathbf{t}(\Sigma, \theta, \varsigma)$ is that its general current boundary ignores closed sets of zero k-1 measure, e.g. if $U \subset \mathbf{R}^N$ is open and the support of ∂T inside U has zero $\mathcal{H}^{k-1}$ measure, then $\partial T(\omega) = 0$ for each ω supported in U [**F1** 4.1.20].

Suppose that $T = \mathbf{t}(\Sigma, \theta, \varsigma)$ is an n current such that the support of ∂T lies in B. Because of our special assumptions about B in **A.1.1** we can use [**F1** 4.1.31] together with our preceding remark to infer for each $k = 1, \ldots J$ the existence of nonnegative real numbers r_k and continuous orientation functions ς_k on Δ_k such that

$$\partial T = \sum_{k=1}^{J} \mathbf{t}(\Delta_k, r_k, \varsigma_k).$$

For general (possibly empty) subsets A and C of $\mathcal{M}$ with $C \subset A$ we denote by $\mathbf{R}_k(A, C)$ the vector space of those k currents $T = \mathbf{t}(\Sigma, \theta, \varsigma)$ with the closure of Σ contained in A such that $\partial T = \mathbf{t}(\Sigma', \theta', \varsigma')$ for some $\Sigma', \theta', \varsigma'$ with the closure of Σ' contained in C. We further let $\mathbf{I}_k(A, C)$ denote the subgroup of those currents $T = \mathbf{t}(\Sigma, \theta, \varsigma)$ in $\mathbf{R}_k(A, C)$ such that θ assumes only positive integer values. It follows from [**F1** 4.2.16(2)] that $\partial T \in \mathbf{I}_{k-1}(C, \emptyset)$ whenever $T \in \mathbf{I}_k(A, C)$.

When convenient we will denote by spt$T$ the support of a current T.

A.5 The coarea formula and slices of currents. A key ingredient of the present paper is slicing the current $[\![\mathcal{M}]\!]$ by mappings $f : \mathcal{M} \to \mathbf{S}^m$ belonging to $\mathcal{F}_0$ and use of the *coarea formula* to estimate the masses of these slices in terms of the energy $\mathcal{E}_3(f)$. As a consequence of [**F1** 3.2.22, 4.3.8, 4.3.11] we infer that for $\mathcal{H}^m$ almost every $w \in \mathbf{S}^m$ the slice

$$\langle [\![\mathcal{M}]\!] , f , w \rangle = \mathbf{t}(f^{-1}\{w\} , 1 , \varsigma)$$

is well defined as an n dimensional current. Here, for $\mathcal{H}^n$ almost every $x \in f^{-1}\{w\}$, if $\eta(x)$ is that simple unit m vector associated with the m plane $\ker Df(x)^{\perp}$ in $\mathrm{Tan}(\mathcal{M}, x)$ for which

$$\langle \eta(x) , \wedge_m Df(x) \rangle \bullet \sigma(w) > 0$$

then we specify $\varsigma(x)$ to be that simple unit n vector associated with $\ker Df(x)$ in $\mathrm{Tan}(\mathcal{M}, x)$ for which $\xi(x) = \eta(x) \wedge \varsigma(x)$; we have used the symbol $\bullet$ to denote the inner product in $\wedge_m \mathbf{R}^{m+1}$.

We further infer from the coarea formula [**F1** 3.2.22] that

$$(m+1)\alpha(m+1)\mathcal{E}_3(f) = \int_{w \in \mathbf{S}^m} \mathbf{M}(\langle [\![\mathcal{M}]\!] , f , w \rangle) \, d\mathcal{H}^m w.$$

Since $\partial[\![\mathcal{M}]\!] = 0$ we readily infer from [**F1** 4.3.1] together with A.3.2 and A.4 above that for $\mathcal{H}^m$ almost every $w \in \mathbf{S}^m$, $\partial\langle [\![\mathcal{M}]\!] , f , w \rangle$ belongs to $\mathbf{I}_{m-1}(B, \emptyset)$.

A.6 Kronecker indices of integral currents. Whenever $S \in \mathbf{I}_m(\mathcal{M}, \mathcal{M})$ and $T \in \mathbf{I}_n(\mathcal{M}, \mathcal{M})$ with

$$\emptyset = \mathrm{spt}\partial S \cap \mathrm{spt}T = \mathrm{spt}S \cap \mathrm{spt}\partial T,$$

there is naturally defined the *Kronecker index* of S and T in $\mathcal{M}$, denoted

$$\mathbf{k}(S,T) = \mathbf{k}(S,T;\mathcal{M}) \in \mathbf{Z}.$$

which is a direct extension of the definitions in [**F1** 4.3.20]. For 'sufficiently regular' such currents

$$S = \mathbf{t}(\Sigma_1, \theta_1, \varsigma_1) \qquad \text{and} \qquad T = \mathbf{t}(\Sigma_2, \theta_2, \varsigma_2)$$

in 'general position', we can write

$$\mathbf{k}(S,T) = \sum_{x\in\Sigma_1\cap\Sigma_2} \theta_1(x)\cdot\theta_2(x)\cdot \operatorname{sign}(\varsigma_1(x)\wedge\varsigma_2(x)\bullet\xi(x)).$$

Among the important facts about the Kronecker index is its ability to characterize real homology classes. We have the following.

PROPOSITION. *Suppose $T_1, T_2 \in \mathbf{I}_n(\mathcal{M}, B)$ with $\partial T_1 = \partial T_2$ and*

$$\mathbf{k}(S,T_1) = \mathbf{k}(S,T_2)$$

for each $S \in \mathbf{I}_m(\mathcal{M}, \emptyset)$ for which both Kronecker indices are defined. Then there is $Q \in \mathbf{R}_{n+1}(\mathcal{M},\mathcal{M})$ such that $\partial Q = T_1 - T_2$.

Proof. In view of [**F1** 4.4.1] it is sufficient to verify the assertion in the context of Lipschitz singular chains of algebraic topology. Moreover it is sufficient to check than an n cycle T in $\mathcal{M}$ is a boundary in case its general position intersections with m cycles S in $\mathcal{M}$ all have Kronecker index zero. This is well known.

A.7 Degrees of mappings of currents. Suppose $f \in \mathcal{F}_0$ and

$$S = \mathbf{t}(\Sigma\,,\theta\,,\varsigma) \in \mathbf{I}_m(\mathcal{M} \sim C_f, \emptyset).$$

Then the m current $f_\sharp S$ in $\mathbf{S}^m$ is naturally defined in accordance with [**F1** 4.1.14, 4.1.15] with $\partial f_\sharp S = 0$ since $\partial S = 0$. We then infer from [**F1** 4.1.31] the existence of an integer $\mathbf{d}(f,S)$ such that

$$f_\sharp S = \mathbf{t}(\mathbf{S}^m\,, \mathbf{d}(f,S)\,, \sigma).$$

We call $\mathbf{d}(f,S)$ the *degree* of f on S. If f and S are 'sufficiently regular' then, for $\mathcal{H}^m$ almost every $w \in \mathbf{S}^m$,

$$\mathbf{d}(f,S) = \sum_{x\in\Sigma\cap f^{-1}\{w\}} \theta(x)\operatorname{sign}\left(\langle\varsigma(x), \wedge_m Df(x)\rangle \bullet \sigma(w)\right).$$

Basic properties of degrees are the following.

PROPOSITION.

(1) The degree $\mathbf{d}(f,S)$ depends only on the real homology class of S in $\mathcal{M} \sim B$. More precisely, if $f \in \mathcal{F}_0$, and $S_1, S_2 \in \mathbf{I}_m(\mathcal{M} \sim C_f, \emptyset)$, and $Q \in \mathbf{R}_{m+1}(\mathcal{M}\sim B, \mathcal{M}\sim B)$ with $\partial Q = S_1 - S_2$, then $\mathbf{d}(f,S_1) = \mathbf{d}(f,S_2)$.

(2) The degree $\mathbf{d}(f,S)$ *depends only on the s-homotopy class of f. More precisely, if* $f,g \in \mathcal{F}_0$ *are s-homotopic and* $S \in \mathbf{I}_m(\mathcal{M} \sim (C_f \cup C_g \cup B), \emptyset)$*, then* $\mathbf{d}(f,S) = \mathbf{d}(g,S)$.

A.8 An example showing relations between integral current slices and boundaries, Kronecker indices, and mapping degrees. Suppose, as illustrated in Figure 2, the following.

(a) $\mathcal{M} = \mathbf{R}^{m+n}$ with its usual orthonormal basis, and

$$U = \mathbf{U}^{m+1}(0,1) \times \mathbf{U}^{n-1}(0,1),$$

is an open set, and

$$\Delta = \{0\} \times \mathbf{U}^{n-1}(0,1)$$

is an $n-1$ disk with orientation function

$$\beta : \Delta \to \{\mathbf{e}_{m+2} \wedge \ldots \wedge \mathbf{e}_{m+n}\}.$$

(b) K and $z_1, \ldots, z_K$ are positive integers and $\epsilon_1, \ldots, \epsilon_K \in \{-1,+1\}$.

(c) For each k the vectors

$$p(k), \eta_1(k), \ldots, \eta_m(k) \in \mathbf{S}^m \times \{0\} \subset \mathbf{R}^{m+1} \times \mathbf{R}^{n-1}$$

are an orthonormal family such that

$$\eta_1(k) \wedge \ldots \wedge \eta_m(k) \wedge p(k) = \mathbf{e}_1 \wedge \ldots \wedge \mathbf{e}_{m+1}$$

and also $p(1), \ldots, p(K)$ are distinct.

(d) For each k we let Π_k denote the n plane spanned by $p(k)$ and $\{0\} \times \mathbf{R}^{n-1}$ and define the n half disk

$$\Delta_k = \Pi_k \cap U \cap \{x : x \bullet p(k) \le 0\}$$

with orientation function

$$\varsigma : \Delta_k \to \{\epsilon_k\, p(k) \wedge \mathbf{e}_{m+2} \wedge \ldots \wedge \mathbf{e}_{m+n}\}.$$

(e) $0 < \delta << r \le s << 1$ are very small numbers and

$$N = U \cap \{x : \operatorname{dist}(x, \Delta) < r\} \qquad \text{and} \qquad N_k = (U \sim N) \cap \{x : \operatorname{dist}(x, \Delta_k < 2\delta\}$$

for each k; we assume that δ is small enough so that the sets $N_1, \ldots, N_k$ are positive distances apart.

(e) We denote by Σ the small m sphere

$$\Sigma = \partial \mathbf{B}^{m+1}(0,s) \times \{0\}$$

with the standard continuous orientation function $\tau : \Sigma \to \wedge_m \mathbf{R}^{m+n}$ determined by requiring

$$x \wedge \tau(x) = s \cdot \mathbf{e}_1 \wedge \ldots \wedge \mathbf{e}_{m+1}$$

19

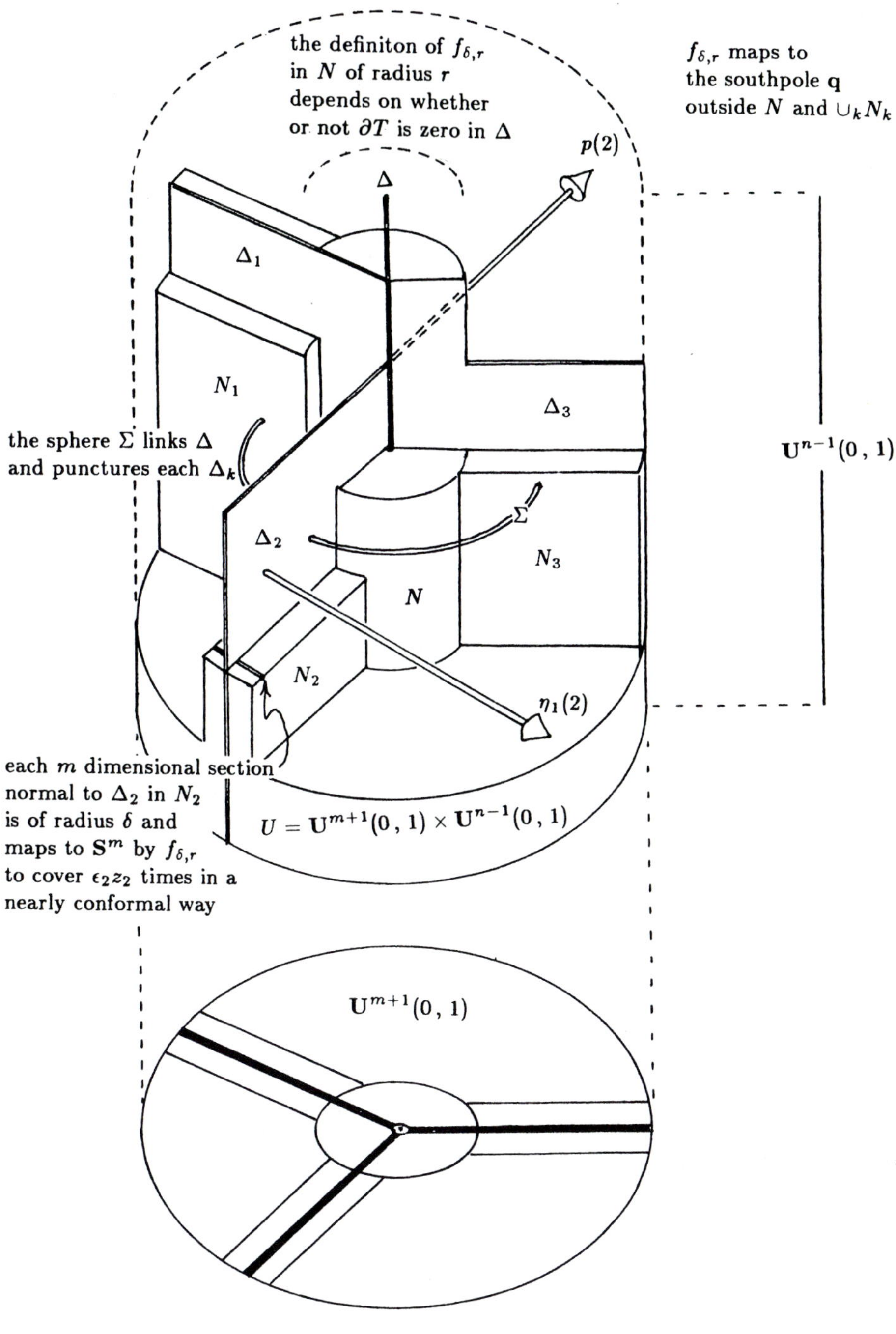

Figure 2. Relations between integral current slices and boundaries, Kronecker indices, and mapping degrees are illustrated by example in Appendix A.8.

for each x in Σ; it follows that

$$\tau(-s \cdot p(k)) = (-1)^{m+1}\eta_1(k) \wedge \ldots \wedge \eta_m(k)$$

for each k. Here $\cdot$ denotes scalar multiplication of a vector. The m sphere Σ 'links' the $n-1$ disk Δ in U while 'puncturing' each Δ_k at the point $-s \cdot p(k)$.

We then set

$$T = \sum_{k=1}^{K} \mathbf{t}(\Delta_k\,,\, z_k\,,\, \epsilon_k \cdot \varsigma_k) \qquad \text{and} \qquad S = \mathbf{t}(\Sigma\,,\,1\,,\,\tau)$$

and estimate

(1) The boundary of T inside U is given by

$$\partial T \llcorner U = \sum_{k=1}^{K} \mathbf{t}(\Delta\,,\, z_k\,,\, \epsilon_k \cdot \beta)$$

[**F1** 4.1.8] so that $\partial T \llcorner U = 0$ if and only if $\sum_{k=1}^{K} \epsilon_k z_k = 0$.

(2) The Kronecker index of S and T is given by

$$\begin{aligned}
\mathbf{k}(S,T) &= \sum_{k=1}^{K} z_k \cdot \tau(-s \cdot p(k)) \wedge \varsigma(-s \cdot p(k)) \bullet \mathbf{e}_1 \wedge \ldots \wedge \mathbf{e}_{m+n} \\
&= \sum_{k=1}^{K} z_k(-1)^{m+1}\eta_1(k) \wedge \ldots \eta_m(k) \wedge \epsilon_k \cdot p(k) \wedge \mathbf{e}_{m+1} \wedge \ldots \wedge \mathbf{e}_{m+n} \bullet \mathbf{e}_1 \wedge \ldots \wedge \mathbf{e}_{m+n} \\
&= (-1)^{m+1} \sum_{k=1}^{K} \epsilon_k z_k
\end{aligned}$$

so that $\mathbf{k}(S,T) = 0$ if and only if $\partial T \llcorner U = 0$.

We now assume $r = s$ and will construct a mapping $g: U \sim N \to \mathbf{S}^m$. We first set $g(x) = \mathbf{q}$ (the southpole) if x lies outside both N and all the N_k's. Each point in each N_k can be written uniquely in the form

$$x + y_1\eta_1(k) + \ldots + y_m\eta_m(k)$$

where x is the unique closest point in Δ_k and $y \in \mathbf{B}^{m+1}(0, 2\delta)$; for each such point we set

$$g(x + y_1\eta_1(k) + \ldots y_m\eta_m(k)) = \gamma_{\delta,z_k}(\epsilon_k \cdot y_1, y_2, \cdots, y_m).$$

Since $r \le s < 1$ our function g is defined on Σ and there is a well defined mapping degree $\mathbf{d}(g, S)$ (with the obvious meaning). Since each γ_{δ,z_k} is orientation preserving (and δ is very small) the orientation of g on Σ near $p(k)$ is determined by ϵ_k and by the inner product

$$\eta_1(k) \wedge \ldots \wedge \eta_m(k) \bullet \tau(-s \cdot p(k)),$$

and we compute

(3) The degree of g on S is given by

$$\begin{aligned}\mathbf{d}(g,S) &= \sum_{k=1}^{K} z_k \epsilon_k \cdot \eta_1(k) \wedge \ldots \wedge \eta_m(k) \bullet \tau(-s \cdot p(k)) \\ &= (-1)^{m+1} \sum_{k=1}^{K} \epsilon_k z_k\end{aligned}$$

so that $\mathbf{d}(g,S) = 0$ if and only if $\partial T \llcorner U = 0$.

The extension of g to a mapping $f = f_{\delta,r}$ on all of U depends on which of two cases occurs.

Case 1. If $\mathbf{d}(g,S) = 0$ we infer from Hurewicz's theorem the existence of a Lipschitz mapping $h: \mathbf{B}^{m+1}(0,r) \to \mathbf{S}^m$ such that

$$h(w) = \begin{cases} g(w,0) & \text{if } |w| = r \\ \mathbf{q} & \text{if } |w| \le r/2. \end{cases}$$

We then define our mapping $f: U \to \mathbf{S}^m$ by setting

$$f(x) = \begin{cases} g(x) & \text{if } x \notin N \\ h(x_1, \ldots, x_{m+1}) & \text{if } x \in N. \end{cases}$$

Case 2. If $\mathbf{d}(g,S) \neq 0$ we define a discontinuous mapping $h: \mathbf{B}^{m+1} \to \mathbf{S}^m$ by setting

$$h(w) = g\left(\frac{rw}{|w|}, 0\right)$$

for each w and, as above, define $f: U \to \mathbf{S}^m$ by setting

$$f(x) = \begin{cases} g(x) & \text{if } x \notin N \\ h(x_1, \ldots, x_{m+1}) & \text{if } x \in N. \end{cases}$$

With the obvious interpretation of $\mathcal{E}_1, \mathcal{E}_2$, and $\mathcal{E}_3$ for function on U, each of these energies of mappings $f_{\delta,r}$ nearly equals the mass of T when δ and r are small (and reasonable choices are made for h in Case 1). More precisely, we have.

$$\lim_{\delta,r\downarrow 0} \mathcal{E}_1(f_{\delta,r}) = \lim_{\delta,r\downarrow 0} \mathcal{E}_2(f_{\delta,r}) = \lim_{\delta,r\downarrow 0} \mathcal{E}_3(f_{\delta,r}) = \mathbf{M}(T) = \sum_{k=1}^{K} z_k \mathcal{H}^n(\Delta_k).$$

It is also straight forward to check that for $\mathcal{H}^m$ almost every $w \in \mathbf{S}^m$ the slice

$$T_w = \langle \mathbf{E}^{m+n} \llcorner U \,,\, f_{\delta,r} \,,\, w \rangle$$

exists with

$$\partial T_w \llcorner U = \partial T \llcorner U,$$

and also if a sequence of δ's and r's converging to 0 is fixed then, for $\mathcal{H}^m$ almost every w in $\mathbf{S}^m$,

$$\lim_{\delta,r\downarrow 0} \langle \mathbf{E}^{m+n} \llcorner U\,,\, f_{\delta,r}\,,\, w\rangle = T.$$

REFERENCES

[A1] F. Almgren, *Deformations and multiple-valued functions*, Geometric Measure Theory and the Calculus of Variations, Proc. Symposia in Pure Math. **44** (1986), 29-130.

[A2] _________, **Q** *valued functions minimizing Dirichlet's integral and the regularity of area minimizing rectifiable currents up to codimension two*, preprint.

[BCL] H. Brezis, J-M Coron, E. Lieb, *Harmonic maps with defects*, Comm. Math. Physics, 1987; see also C. R. Acad. Sc. Paris **303** (1986), 207-210.

[F1] H. Federer, *Geometric Measure Theory*, Springer-Verlag, 1969, XIV + 676 pp.

[F2] _________, *Real flat chains, cochains and variational problems*, Indiana U. Math. J. **24** (1974), 351–407.

[HKL] R. Hardt, D. Kinderlehrer, and M. Luskin, *Remarks about the mathematical theory of liquid crystals*, Institute for Mathematics and its Applications, preprint, 1986.

[MF] F. Morgan, *Area-minimizing currents bounded by higher multiples of curves*, Rend. Circ. Matem. Palermo, (II) **33** (1984), 37–46.

[PH] H. Parks, *Explicit determination of area minimizing hypersurfaces, II*, Mem. Amer. Math. Soc. **60**, March 1986, iv + 90 pp.

[SU] R. Schoen and K. Uhlenbeck, *A regularity theory for harmonic maps*, J. Diff. Geom. **60** (1982), 307–335.

[W1] B. White, *The least area bounded by multiples of a curve*, Proc. Amer. Math. Soc. **90** (1984), 230-232.

[W2] _______, *Homotopy classes in Sobolev spaces and the existence of energy minimizing maps*, preprint.

[YL] L. C. Young, *Some extremal questions for simplicial complexes V. The relative area of a Klein bottle*, Rend. Circ. Matem. Palermo, (II), **12** (1963), 257–274.

With F. Almgren in *Symposia Mathematica, vol. XXX*, 103–118 (1989)

COUNTING SINGULARITIES IN LIQUID CRYSTALS

FREDERICK J. ALMGREN JR. - ELLIOTT H. LIEB

Abstract. *Energy minimizing harmonic maps from the ball to the sphere arise in the study of liquid crystal geometries and in the classical nonlinear sigma model. We linearly dominate the number of points of discontinuity of such a map by the energy of its boundary value function. Our bound is optimal (modulo the best constant) and is the first bound of its kind. We also show that the locations and numbers of singular points of minimizing maps is often counterintuitive; in particular, boundary symmetries need not be respected.*

1. INTRODUCTION

This note is an introduction to and summary of discoveries we have made about the singular behaviour of

- A mathematical model of some liquid crystal geometries
- Dirichlet energy minimizing harmonic maps from regions in $\mathbf{R}^3$ to S^2
- Energy minimizing configurations of a classical nonlinear sigma model ($R^3 \to S^2$).

These phenomena are different facets of a common mathematical analysis set forth in detail in our paper [AL]. There we study vector fields φ of unit length defined in a reasonable region Ω in $\mathbf{R}^3$. In coordinates we can thus write for each $x = (x^1, x^2, x^3)$ in Ω,

$$(1) \qquad \varphi(x) = (\varphi^1(x), \varphi^2(x), \varphi^3(x)) \qquad \text{with} \qquad \sum_{i=1}^{3} \varphi^i(x)^2 = 1.$$

Since our target S^2 is 2-dimensional we could, in principle, describe φ using two functions instead of our three constrained functions. It is easier, however, to work with three functions and a constraint.

The φ's important for us have distribution first derivatives which are square summable. (Caution: the space of such φ's satisfying (1) is not the completion of any space of smooth mappings $\Omega \to S^2$.) The gradients of such φ's are defined for almost every x with norms represented by the formula

$$|\nabla\varphi(x)|^2 = \sum_{i=1}^{3}\sum_{\alpha=1}^{3}\left(\frac{\partial\varphi^i(x)}{\partial x^\alpha}\right)^2$$

which gives the value of **Dirichlet's integrand** at x. The integral of this integrand is **Dirichlet's energy integral** of φ,

$$\mathcal{E}(\varphi) = \int_\Omega |\nabla\varphi|^2 \,\mathrm{d}V,$$

with $\mathrm{d}V = \mathrm{d}x^1\mathrm{d}x^2\mathrm{d}x^3$. Critical points of this energy integral $\mathcal{E}$ are by definition **harmonic functions** and satisfy the associated **Euler-Lagrange partial differential equations**

$$-\Delta\varphi^i(x) = \varphi^i(x)|\nabla\varphi(x)|^2 \qquad (i = 1,2,3).$$

These equations state that a critical φ has vanishing Laplacian in directions in which it is unconstrained. Such an energy functional and associated partial differential equations appear in the physics literature under the rubric of **the nonlinear sigma model**.

Somewhat more generally, reasonable maps $\varphi : \mathcal{M} \to \mathcal{N}$ between Riemannian manifolds $\mathcal{M}$ and $\mathcal{N}$ (often submanifolds of Euclidean vector spaces) have a Dirichlet's energy integral

$$\mathcal{E}_{\mathcal{M}\mathcal{N}}(\varphi) = \int_{\mathcal{M}} |D\varphi|^2 \,\mathrm{d}V_{\mathcal{M}}$$

of which ours is a special case. Alternatively, one can write

$$\mathcal{E}_{\mathcal{M}\mathcal{N}}(\varphi) = \int_{\mathcal{M}} g_{ij}(\varphi(x))G^{\alpha\beta}(x)\left(\frac{\partial\varphi^i}{\partial x^\alpha}(x)\right)\left(\frac{\partial\varphi^j}{\partial x^\beta}(x)\right)\mathrm{d}V_{\mathcal{M}}x$$

where g is the metric on N, G is the metric on M and $\mathrm{d}V_{\mathcal{M}}x = (\det G(x))^{1/2}$ $\mathrm{d}x$. Extremal mappings for such energies are also called **harmonic mappings**. Such mappings often are not continuous and there in an extensive mathematical theory about them.

The φ's mapping Ω to S^2 which are important for us also have well defined **boundary functions** $\psi : \partial\Omega \to S^2$ having **boundary energy**

$$\partial\mathcal{E}(\psi) = \int_{\partial\Omega} |\nabla_T \psi|^2 \mathrm{d}\, A$$

which is finite; here $\nabla_T \psi$ is the tangential gradient of ψ and $\mathrm{d}\, A$ is surface area measure. Associated with such a ψ is the number

$$E(\psi) = \inf\ \{\mathcal{E}(\varphi) : \varphi \text{ has boundary value function } \psi\}.$$

We call φ an **energy minimizing map** for boundary value function ψ if and only if

$$\mathcal{E}(\varphi) = E(\psi).$$

If Ω is any reasonable bounded domain and ψ is any boundary value function of finite energy then there will always be at least one minimizer φ having ψ as boundary values (a compactness argument). Sometimes, however, there can be more than one minimizer. This is one of the fascinations of this simple nonlinear problem; if the target S^2 were replaced by $\mathbb{R}^3$ (i.e. our constaint were removed) then the Euler-Lagrange partial differential equations are (the unconstrained) linear partial differential equations of Laplace, $\Delta\varphi^i = 0$ $(i = 1, 2, 3)$, for which uniqueness is well known.

If our domain Ω is all of $\mathbb{R}^3$ there is no boundary value function ψ, of course. We then say that $\varphi : \mathbb{R}^3 \to S^2$ is a minimizer provided φ cannot be modified on a compact set K to decrease energy in a larger bounded open set containing K.

Liquid crystals

The connection of our energy minimizing φ's with liquid crystals requires explanation. We imagine that Ω is a container containing a liquid crystal. At points x in Ω the liquid determines a **directrix** $n(x)$ lying in real projective space $\mathbb{R}\mathbf{P}^2$. Since $\mathbb{R}\mathbf{P}^2$ is obtained from S^2 by identifying antipodal points, this means intuitively that $n(x)$ is a unit vector like our $\varphi(x)$ except that its head is indistinguishable from its tail. For the liquid crystals with which we are concerned, the energy of n is defined analogously to our $\mathcal{E}$, e.g. zero energy corresponds to parallel alignment. Like our minimizing φ's (as we shall see), any minimizing n will be continuous except at isolated points. This means, in particular, that any minimizing n can locally be lifted to become a minimizing φ having the same energy; this lifting is global in case Ω is simply connected. (see [BCL], p. 686 for details). Thus, for simply connected Ω's, our original problem is equivalent to the liquid crystal problem. In any case, whether or not Ω is simply connected, our estimates on the number of singular points hold for these liquid crystal minimizers. Line singularities do not occur in our model because they would have infinite Dirichlet energy. They do occur in nature, but to model them one, effectively, has to fatten the line and treat it separately (much as in the liquid helium problem). A further complication for liquid crystals is that there are other, more appropriate, integrands which are quadratic in $\nabla\varphi$ and respect rotational symmetry. The general nematic liquid crystal integrand, for example, is of the form

$$K_1(\operatorname{div}\varphi)^2 + K_2(\varphi\cdot\operatorname{curl}\varphi)^2 + K_3(\varphi\wedge\operatorname{curl}\varphi)^2.$$

Our Dirichlet's energy integrand corresponds (except for a fixed boundary term) to setting $K_1 = K_2 = K_3 = 1$ (see [BCL], p. 653). Our methods give information about such liquid crystal geometries (by a compactness argument) only when K_1, K_2 and K_3 are nearly equal.

2. BASIC FACTS ABOUT MINIMIZERS

(A) Existence and regularity of minimizers

As we mentioned above, whenever we have a reasonable domain Ω and boundary function ψ of finite energy, there will always exist a minimizer φ having boundary values ψ. Such a result is included among the general analysis of Dirichlet's integral minimizing mappings between manifolds by R. Schoen and K. Uhlenbeck in their basic papers [SU1] [SU2]. They further showed that a minimizing φ in our context is a real analytic mapping except at isolated points of discontinuity (which are our singularities). Finally, they concluded that a minimizing φ assumes its boundary values smoothly when both $\partial\Omega$ and ψ are comparably smooth.

(B) Monotonicity of energy and tangential approximations

One of the basic technical properties of energy minimizing mappings is usually called **monotonicity**. Whenever φ is a minimizer in $\Omega, y \in \Omega$, and $0 < r < s < R$ so that the ball $B_R(y)$ also lies within Ω, then

$$\frac{1}{r}\int_{B_r(y)} |\nabla\varphi|^2 \mathrm{d}V \leq \frac{1}{s}\int_{B_s(y)} |\nabla\varphi|^2 \mathrm{d}V.$$

For a proof, see [SU1]. (The absence of a corresponding monotonicity estimate is the main reason our analysis of liquid crystals is restricted to the $K_1 = K_2 = K_3$ case). The monotonicity estimate leads fairly directly to the existence of certain tangential approximations to φ at each interior y. A major and deep development occured in a paper of L. Simon [S] which for our problem guarantees the existence of a **unique tangential approximating mapping**. At regular points this approximating mapping is constant. For a singular point y of φ in Ω, Simon's result gives a **unique harmonic mapping** $f : S^2 \to S^2$ such that

$$\varphi(y + t\omega) \to f(\omega) \qquad \text{as} \qquad t \to 0+$$

uniformly for all ω's in S^2 (see [AL]), i.e.

$$\varphi(x) \approx f\left(\frac{x - y}{|x - y|}\right)$$

for x's near y. The correspondence here is in several strong senses (see [AL]). In general, if $f : S^2 \to S^2$ and $F : \mathbf{R}^3 \to S^2$ is defined by setting

$$F(x) = f\left(\frac{x}{|x|}\right)$$

for each $x \neq 0$ then f is harmonic if and only if F is.

EXAMPLE. $f\left(\frac{x}{|x|}\right) = \frac{x}{|x|}$, i.e. f is the identity; see Figure 1.

(C) Harmonic mappings between spheres and mapping degrees

Any continuous mapping $S^2 \to S^2$ has a well defined topological degree measuring the number of times the first sphere covers the second, taking into account the orientations. Since the boundary functions ψ under consideration map S^2 to S^2 and have finite energy, they also have a well defined degree given by the Jacobian integral

$$\deg(\psi) = \frac{1}{4\pi}\int_{\partial\Omega} J(\psi)\,\mathrm{d}A;$$

here $J(\psi)$ is the Jacobian (determinant) function of ψ whose sign is positive or negative at a point depending on whether $D\psi$ preserves or reverses orientations at that point. For continuous ψ's of finite energy these two notions of degree coincide.

All possible harmonic mappings from S^2 to S^2 have been classified for some time. In complex coordinates (resulting from stereographic projection of the S^2's onto $\mathbb{C}$) they are all of the form

$$f(z) = \frac{P(z)}{Q(z)} \qquad \text{or} \qquad f(z) = \frac{P(\bar{z})}{Q(\bar{z})}$$

corresponding to various complex polynomial functions P and Q which are relatively prime. The degree of these f's can be checked to be

$$\deg(f) = \begin{cases} \max(\deg(P), \deg(Q)) & \text{first case;} \\ -\max(\deg(P), \deg(Q)) & \text{second case.} \end{cases}$$

For these harmonic maps $f : S^2 \to S^2$ we also set $F(x) = f\left(\frac{x}{|x|}\right)$ as above and compute for each $0 < R < \infty$ that

$$\int_{|x|<R} |\nabla F|^2 \mathrm{d}V = 8\pi R |\deg(f)|,$$

i.e. the energy does not depend on P and Q except via the degree.

(D) Tangential approximations to minimizers

Suppose $y \in \Omega$ is a singular point of a minimizer φ and the tangential approximation is of the form $F(x) = f\left(\frac{x}{|x|}\right)$ corresponding to one of the harmonic f's given in (C) above. By the **degree of the singular point** y we mean the mapping degree of the associated f. **Which of the possible f's actually occur?** This question was answered by H. Brezis, J-M. Coron, and E. Lieb in their paper [BCL]. The *only* f's that occur are rotations $\mathcal{R}$ and reflections of the f in the above example, i.e.

$$(2) \qquad f(\omega) = \pm\mathcal{R}(\omega), \quad (\omega \in S^2) \qquad \text{with} \quad \deg(f) = \pm 1;$$

see Figure 1. This class does not even include all harmonic maps of degree ± 1. The proof proceeds by a construction of comparison functions. If $|\deg(f)| > 1$ then the energy of F can be decreased by splitting the singularity at the origin into two nearby singularities of lower degree. If $|\deg(f)| = 1$ and $f \neq \pm\mathcal{R}$ then the energy of F can be decreased by moving the singular point slightly.

The paper [BCL] also answered a question that in some sense is complementary to the minimization question we have been studying here. Suppose $y_1, \ldots y_n$ are *fixed* points in Ω and $\mathrm{d}_1, \ldots, \mathrm{d}_n$ are *fixed* degrees associated to these points (not necessarily ± 1).

What is the infimum of energies $\mathcal{E}(\varphi)$ among all φ's which are continuous except at y_i's and map small spheres around each y_i with degree d_i?

The boundary function ψ is not fixed. This infimum is *not* achieved in general. The answer is shown in the Figure 2. Think of each singularity as a source or sink of flux and draw lines to carry the flux between singularities, or between a singularity and the boundary. Then

Fig. 1. Here are shown representations of unit vector fields

$$F(x) = \left(\frac{x}{|x|}\right) \quad \textbf{and} \quad G(x) = \mathcal{R}\left(\frac{x}{|x|}\right)$$

in which $\mathcal{R}$ is a counterclockwise rotation through $45°$. Such arrays minimize Dirichlet's integral energy and are also observed as stable liquid crystal geometries [K].

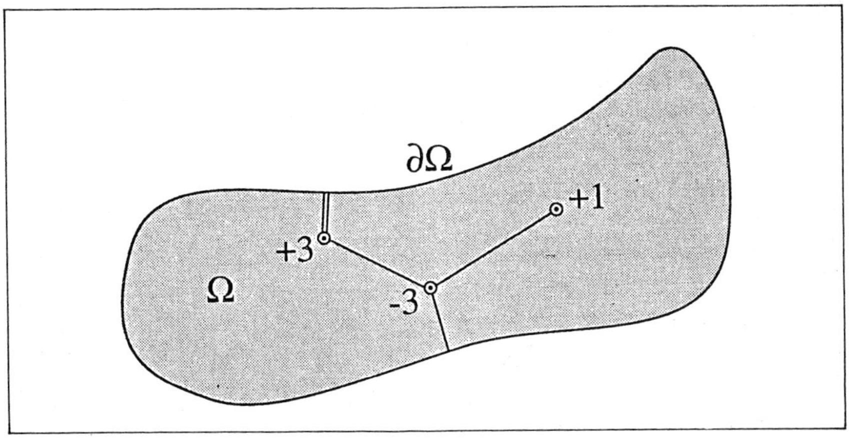

Fig. 2. A region Ω is pictured here containing three prescribed singular points whose degrees $(+3, -3, +1)$ are also prescribed. The least energy of unit vector fields having this singular behavior is the least total mass of oriented line segments connecting these singular points (as currents) either to each other or to the boundary. Such a least length array is illustrated.

$$\inf \mathcal{E}(\varphi) = 8\pi \min \left\{ \sum \text{lengths of lines} \right\}$$

where the minimum is over all ways of constructing the lines. A different proof of this result was later given by F. Almgren, W. Browder, and E. Lieb [ABL] using H. Federer's co-area formula in the context of currents. This is like quark confinement: a plus and minus quark have an energy proportional to their separation.

From this result with *specified* singularities one is tempted to surmise that, in our original minimization problem, *potential* singularities would tend to annihilate each other (if of opposite degrees) or move to $\partial\Omega$. The number of singularities that will occur will be only that required by topology, i.e.

$$\sum_{\text{singularities}} \deg(\text{singularity}) = \deg(\psi) = \frac{1}{4\pi} \int_{\partial\Omega} J(\psi) \, \mathrm{d}A.$$

This surmise is very wrong, as we shall see later in Example 3, and misled us for a long time. Arbitrarly many singularities (of mixed signs) can occur, even if the Jacobian $J(\psi)$ vanishes identically.

(E) Boundary regularity and hot spots

Our main estimates require an extension of the boundary regularity results indicated above in (A). These theorems take several pages merely to state precisely, but the essence of the matter is the following. Assume that $\partial\Omega$ is smooth and take a small patch $P \subset \partial\Omega$ which is roughly a 2-dimensional disk of radius R. One consequence of the boundary regularity theory mentioned in (A) is the following. There is a fixed $\epsilon > 0$, independent of R, with the property that whenever the boundary function ψ satisfies

$$\int_P |\nabla_T \psi|^2 \mathrm{d}A < \epsilon$$

then every minimizer φ is free of singularities in the region

$$K = \left\{ x : x \in \Omega, \operatorname{dist}(x, P) > \frac{1}{2} R\epsilon, \operatorname{dist}\left(x, P_{\frac{1}{2}}\right) < 2R\epsilon \right\},$$

here $P_{\frac{1}{2}}$ is the concentric disk of radius $\frac{1}{2}R$. Note that ϵ is dimensionless. Our **hot spot boundary regularity** theorem (proved in [AL]) asserts the existence of a fixed number $0 < \delta << \epsilon$ such that whenever $P' \subset P$ is a smaller subpatch of radius δR and

$$\int_{P \sim P'} |\nabla_T \psi|^2 \mathrm{d}A < \epsilon$$

then φ is also free of singularities in the region K above. In other words arbitrarly large boundary energy in a very small disk P' cannot by itself induce singularities far away.

3. COUNTING SINGULARITIES

The principal question motivating our work in [AL] is this:
How many singular points $N(\psi)$ is at possible for a minimizing φ to have?
The following possibilities seem plausible at the outset:

$N(\psi) \leq CE(\psi)$	FALSE;	
$N(\psi) \leq C \int_{\partial\Omega} \lvert J(\psi)\rvert \mathrm{d}A$	FALSE;	
$N(\psi) \leq C\partial\mathcal{E}(\psi)$	TRUE	«**The Linear Law**».

here C is a constant, possibly depending on Ω.

The first possibility is false by counterexample – see below. The second possibility was suggested by the work in [BCL] and misled us for some time (had it been true it would have led to a beautiful geometric theory). In fact it is quite false as Example 1 below shows; in particular, $N(\psi)$ can be large while $J(\psi)$ vanishes identically.

Our main result, **The Linear Law**, is optimal (modulo the value of $C = C_\Omega$, of which we have no knowledge since our proof is by contradiction based on compactness arguments). It is, to our knowledge, the first bound of its kind.

The following example given by R. Hardt and F. H. Lin in [HL1] shows that $N(\psi)$ can indeed be proportional to $\partial\mathcal{E}(\psi)$. Choose N well separated small disks in $\partial\Omega$. Our ψ is constructed to wrap each disk D around the target sphere once (essentially by the inverse function to stereographic projection while preserving or reserving orientation as one chooses); each ∂D is mapped to the north pole. The complement of these disks in $\partial\Omega$ is mapped by ψ also to the north pole. Then $\partial\mathcal{E}(\psi) \approx CN$; the constant C is independent of the size of the disks since surface energy is scale invariant. Clearly the orientations of ψ on the disks can be arranged so that the total mapping degree of ψ is either zero or one. It is not hard to prove directly that any minimizing φ having ψ as boundary value function must have at least one singularity close to each tiny disk – otherwise $\mathcal{E}(\varphi)$ would be too large. Thus

$$N(\psi) \geq N \approx C^{-1}\partial\mathcal{E}(\psi).$$

Our first main new result (proved independently by Hardt and Lin in [HL2]) is that singularities cannot be very close if they are well inside Ω.

THEOREM 1. *There is a universal constant C (independent of Ω and ψ) such that whenever y and z in Ω are singular points of a minimizer φ then* $\operatorname{dist}(y,z) \geq C \operatorname{dist}(y,\partial\Omega)$.

The idea of the proof is the following. Fix y and suppose the contrary. Then there will be a sequence of minimizing $\varphi^{(j)}$ with singular points at $z^{(j)}$ and at

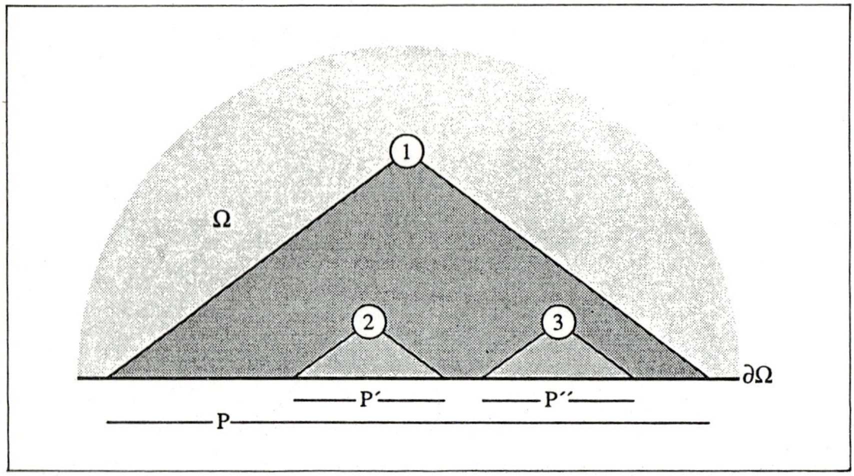

Fig. 3. Pictured here are the «cones of influence» in Ω of three singular points. The presence of singular points $1,2,3$ implies the presence of boundary energy in disks P, P', P'' in $\partial\Omega$. The problem is that these disks are not disjoint so that the total boundary energy is not a simple sum. Nesting of such cones induces a Cayley tree graph in which a combinatorial anaysis overcomes this difficulty.

y such that $z^{(j)} \to y$ as $j \to \infty$. A compactness argument (contradicting the negation) and monotonicity (A) shows that the energy of φ in small balls of radius R about y is *uniformly* greater than $8\pi R$. The limit of a subsequence of the minimizers $\varphi^{(j)}$ is a minimizer which thus can have at worst a singularity of degree ± 1 at y (by equation (2) above). The tangential approximation theorem implies that the energy of the limit φ must be very close to $8\pi R$ for a small R's. This leads to a contradiction because of the continuity of Dirichlet's integral when minimizers converge.

A consequence of Theorem 1 together with equation (2) above is the following.

THEOREM 2. **(Complete classification of energy minimizing maps from $\mathbf{R}^3$ to S^2.)** *Suppose $\varphi : \mathbf{R}^3 \to S^2$ is a minimizer. Then, either φ is a constant mapping or $\varphi = \pm\mathcal{R}\left(\frac{x-y}{|x-y|}\right)$ for some y and $\mathcal{R}$.*

Theorem 1 says that if there are many singularities they have to pile up near $\partial\Omega$. This leads to a difficult geometric-combinatorial problem on different scales proportional to δ^k, where δ is given in (E) above and $k = 1,2,,\ldots$ We attempt to illustrate this in Figure 3. Referring to the ϵ and δ of (E) consider the points 1, 2, and 3 in Ω at distances $R\epsilon$, $R\epsilon\delta$, and $R\epsilon\delta$ above a boundary patch P of radius R and two boundary patches P' and P'' of radii $R\delta$ inside P. The hot spot boundary regularity theorem gives us the following lower bounds for the energy of ψ in P if we consider the various possibilities of having singularities at positions 1, 2, or 3:

Positions occupied	Local boundary energy
(1 alone) or (2 alone) or (3 alone)	ϵ
(1 and 2) or (1 and 3) or (2 and 3)	2ϵ
(1 and 2 and 3)	2ϵ

The source of all our difficulties is that we cannot infer an energy 3ϵ if there are singularities at all three points.

If $S^{(k)}$ denotes the strip $\{x : x \in \Omega, \operatorname{dist}(x, \partial\Omega) \le \epsilon\delta^k\}$, we can effectively decompose each $S^{(k)}$ into cones of height $\epsilon\delta^k$ and base radius δ^k. We then have a Cayley tree whose vertices represent these cones (i.e. a vertex of order $k+1$ is connected to a vertex of order k in the tree if the smaller cone is inside the larger one). A vertex is occupied if its cone has a singularity near the apex; otherwise it is unoccupied. Each occupied vertex gets an energy ϵ if and only if no more than one higher order vertex to which it is pathwise connected is occupied.

The actual details of decomposing each $S^{(k)}$ into cones so that due account is taken of overlaps (and all the other problems that will occur to the reader) involves a complicated covering and counting lemma. The final result is The Linear Law for $N(\psi)$ in terms of $\partial\mathcal{E}(\psi)$, as stated at the beginning of this section.

4. THREE EXAMPLES OF COUNTERINTUITIVE BEHAVIOR

EXAMPLE 1. **Zero Mapping Area.** It is easy to prove for any Ω that if ψ takes values only in some closed hemisphere of S^2 then φ has no singularities. We, however, are able to construct a single curve Γ in S^2 which is a slight perturbation of the equator and, for each N, a smooth boundary value function $\psi^N : \partial\Omega \to S^2$ **having its image equal to** Γ such than any minimizer φ^N having boundary values ψ^N must have at least N singular points. In the example of [AL], Ω is taken to be a ball, but the details of Ω are not important. The Jacobian $J(\psi^N)$ of each ψ^N vanishes identically since its image is one dimensional.

The idea behind the construction appears in the following preliminary problem. Consider reasonable mappings $\varphi : D^2 \to S^2$ from the unit disk D^2 in the plane having two dimensional Dirichlet's integral denoted by $\mathcal{E}_2(\varphi)$. Suppose $\Gamma \subset S^2$ is a smooth embedding of a circle parametrized by a map $P : \partial D^2 \to \Gamma$. The functions φ from D^2 to S^2 having boundary values P can be separated into two homological classes: the +class, in which, heuristically, φ «covers the top of S^2 one more time than it covers the bottom» and, the – class in which φ «covers the bottom one more time than it covers the top»; see Figure 4.

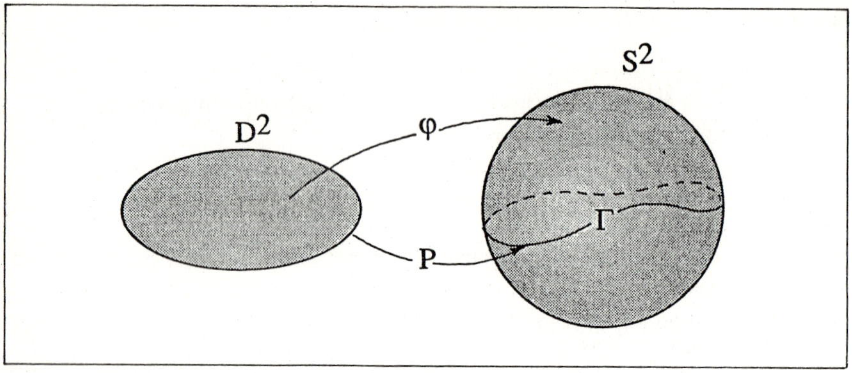

Fig. 4. Illustrated here is one of two homologically distinct classes of mappings $\varphi : D^2 \to S^2$ corresponding to a given boundary parametrization $P : \partial D^2 \to \Gamma$ (the curve Γ is a perturbation of the equator). A «+ function» is one which «covers the northern hemisphere». For some Γ's, the homology type preferred by a least energy mapping can change if the parametrization P is changed. This phenomenon leads ultimately to construction of least energy mappings from the ball to the sphere having many interior singularities but for which the boundary mapping of the sphere to the sphere has zero mapping area (its entire image lies within the curve Γ).

Consider the two numbers

$$E^{\pm}(P) = \inf\{\mathcal{E}_2(\varphi) : \varphi = P \text{ on } \partial D^2 \text{ and } \varphi \in \pm \text{ class}\}.$$

In general $E^+(P)$ will not be the same as $E^-(P)$.

We construct a single Γ having two different (homotopic) parametrizations P^+ and P^- such that

$$E^+(P^+) < E^-(P^+) - \epsilon \text{ and } E^-(P^-) < E^+(P^-) - \epsilon$$

for some $\epsilon > 0$. In other words if the parametrization of Γ changes from P^+ to P^- any *absolute* minimizer φ changes from lying in the + class to lying in the – class.

The next step is to let Ω be a very long solid tube T of radius 1 and length $N(L+1)$. (Actually, T is bent into a torus so that we can ignore the two ends.) As boundary function ψ we alternately paste P^- and P^+ on sections of length L (i.e. each cross-sectional disk has P^- or P^+ on its boundary). In the transitional regions of length 1 we smoothly interpolate between P^- and P^+ (which can be done since they are homotopic). In the transition regions ψ continues to take values only in Γ. See Figure 5.

If L is large enough (depending only on ϵ), it is believable (and we prove it) that φ must be mostly a – function on the P^- disks and it must be mostly a + function

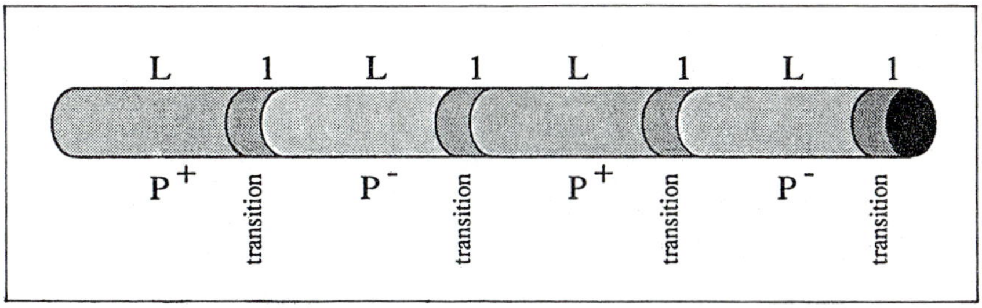

Fig. 5. Illustrated here is a boundary value function $\psi : \partial\Omega \to S^2$ for a long tube domain Ω. The image of ψ is a smooth curve Γ in S^2. On crossectional circles of $\partial\Omega$ the boundary values alternate between intervals of P^+ mappings and intervals of P^- separated by transition intervals. Least energy maps $\varphi : \Omega \to S^2$ with such boundary values map most crossections in P^+ regions to cover the northern hemisphere and map most crossections in P^- regions to cover the southern hemisphere. The minimizer φ therefore has at least one singular point near each transition region.

on the P^+ disks, for otherwise $\mathcal{E}(\varphi)$ would be unnecessarily large. But when φ switches from being a $-$ function to being a $+$ function φ must have a singularity for topological reasons. Thus, φ will have at least N singularities altogether.

The drawback to this example is that the domain T depends on N. To achieve the same result for a fixed domain Ω = unit ball, we first cut the surface ∂T longitudinally (i.e. perpendicular to the disks) and flatten it (key estimates here come from the conformal equivalence of the disk and the upper half plane and the fact that Dirichlet's integral in two dimensions is invariant under conformal reparametrizations of domains).This yields a strip of width 2π and length $N(L+1)$. We also rotate P^+ if necessary so that P^+ and P^- have the same value $\gamma \in \Gamma$ along the cut. Next we shrink the strip to width $(2\pi)^2/N(L+1)$ and length 2π. Finally we paste this strip (which is very narrow since N is large) along the equator of Ω and let $\psi : \partial\Omega \to S^2$ be the old ψ in the strip and let $\psi(x) = \gamma$ for $x \in \partial\Omega$ but $x \notin$ the strip. A somewhat nerve wracking argument shows, as expected, that any minimizer $\varphi : \Omega \to S^2$ must have at least N singularities close to the equator of Ω.

EXAMPLE 2. **Symmetry Breaking** When φ takes values in $\mathbb{R}^3$ instead of S^2, any geometric symmetry of Ω and ψ is inherited by the minimizing φ. The reason is simply that minimizers are unique in the linear case ($\Delta\varphi = 0$). When, as in our case, φ takes values in S^2, the symmetry of Ω and φ can be broken by φ; obviously there must then be several minimizers.

Let Ω be the unit ball in $\mathbb{R}^3$ and let $\psi : \partial\Omega \to S^2$ be the distortion of the identity map illustrated in Figure 6. In small caps N (resp. S) on $\partial\Omega, \psi$ covers

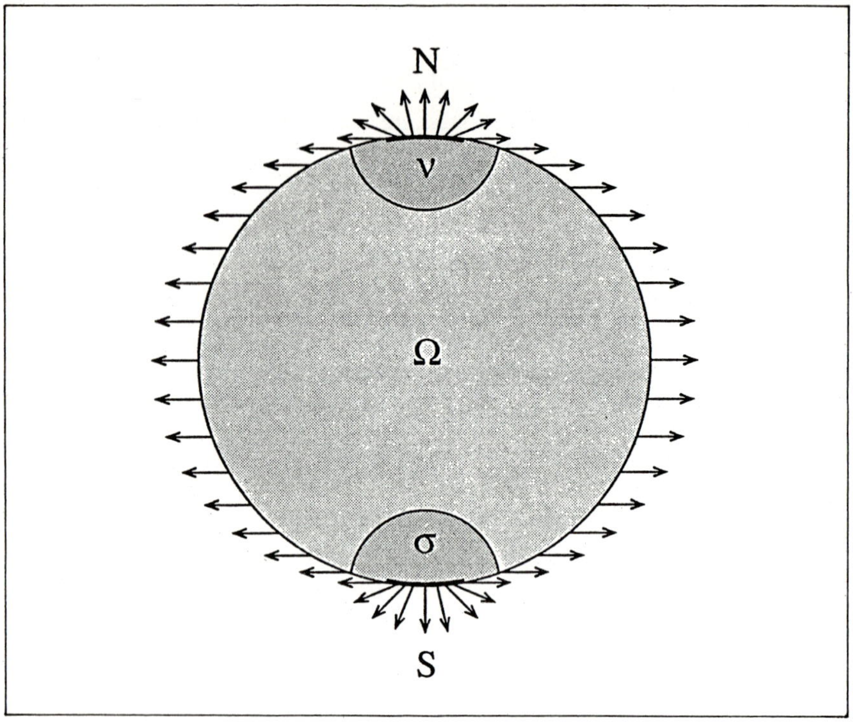

Fig. 6. Here our domain Ω is the unit ball so that $\partial\Omega$ is the unit sphere. Pictured schematically is a special boundary value function $\psi : \partial\Omega \to S^2$ having a mirror image symmetry through the equatorial plane. A small cap N around the north pole maps to cover the entire northern hemisphere of S^2 while a small cap S around the south pole covers the entire southern hemisphere. The sphere less these two caps maps entirely to the equator. Longitude is preserved in each of these regions. No minimizing $\varphi : \Omega \to S^2$ having boundary values ψ can possess such a symmetry since the (necessarily odd) number of singular points must be contained within one of the regions ν and σ near the poles.

the northern (resp. southern) hemisphere of S^2. The two maps are mirror images of each other. On the rest of $\partial\Omega$ between N and S, ψ takes values in the equator of S^2 in the obvious way, i.e. $\psi(x, y, z) = (x^2 + y^2)^{-1/2}(x, y, 0)$.

THEOREM 3. *Any minimizer φ can have singularities only in small shaded regions in Ω, labelled ν and σ, near the caps N and S.*

Since $\deg(\psi) = 1$, this result implies that φ does not inherit the mirror image symmetry through the equatorial disk possessed by ψ. (Our function φ necessarily has an odd number of singularities, and if φ were symmetric, it would necessarily have one on the equatorial disk in Ω.)

The proof of Theorem 3 has two parts. First we show that when N and S are small φ has no singularities in a concentric ball Ω' of radius $1-\epsilon$ for some small ϵ. This is done by a variational (or comparison) argument. Second, we show that there are no singularities in $\{x : 1 \geq |x| > 1-\epsilon \text{ and } \operatorname{dist}(x, \sigma \cap \nu) > \epsilon\}$ by using the boundary regularity (E).

EXAMPLE 3. **Boiling Water** The [BCL] result mentioned in (D) above suggests that + and − singularities tend to annihilate each other. On the other hand, the hot spot boundary regularity mentioned in (E) above suggests that behavior at different length scales (as measured by the distance to $\partial\Omega$) is independent so that + and − singularities *could* coexist provided their distances to $\partial\Omega$ were very different. There would appear to be a conflict here and one of our results is that of the two points of view just mentioned the second one is correct. We have proved the following.

THEOREM 4. *Let Ω be the unit ball and let $p_1, \ldots, p_M$ be any distinct points in $\partial\Omega$. Also let $N_1, \ldots, N_M$ be any positive integers and for each $i = 1, \ldots, M$ let A_i be any sequence of length N_i consisting of +1 's and −1 's. Finally, let $\epsilon > 0$. Then there is a smooth $\psi : \partial\Omega \to S^2$ such that*

(i) $\partial\mathcal{E}(\psi) \leq \epsilon + 8\pi \sum_{i=1}^{M} N_i$.

(ii) The minimizer φ is unique.

(iii) For each $i = 1, \ldots, M$ there are at least N_i singularities stacked nearly vertically above p_i (like bubbles in a pan of water that is about to boil), and these have the specified sequence of degrees given by A_i.

REFERENCES

[ABL] F. ALMGREN, W. BROWDER and E. LIEB: Co-area, liquid crystals and minimal surfaces. In: *Partial Differential Equations*, ed. S. S. Chern, Springer Lecture Notes in Math., **1306**, 1-12 (1988).

[AL] F. ALMGREN and E. LIEB: Singularites of energy minimizing maps from the ball to the sphere: examples counterexamples and bounds. *Ann. of Math.*, **128**, 483-530 (1988). See also: Singularities of energy minimizing maps from the ball to the sphere, *Bull. Amer. Math. Soc.*, **17**, 304-306 (1987).

[BCL] H. BREZIS, J-M. CORON and E. LIEB: Harmonic maps with defects. *Commun. Math. Phys.* **107**, 649-705 (1986).

[HL1] R. HARDT and F. H. LIN: A remark on H^1 mappings. *Manuscripta Math.*, **56**, 1-10 (1986).

[HL2] R. HARDT and F. H. LIN: Stability of singularities of minimizing harmonic maps. *J. Diff. Geom.*, **29**, 113-123 (1989).

[K] M. KLÉMAN: Points, lignes, parois dans les fluides anisotropes et les solides cristalline. *Les Éditiones de Physique (Orsay)*, **I**, 36-37.

[S] L. SIMON: Asymptotics for a class of nonlinear evolution equations with applications to geometric problems. *Ann. of Math.* **118**, 525-571 (1983).

[SU1] R. SCHOEN and K. UHLENBECK: A regularity theory for harmonic maps. *J. Diff. Geom.*, **17**, 307-335 (1982).

[SU2] R. SCHOEN and K. UHLENBECK: Boundary regularity and the Dirichlet problem of harmonic maps. *J. Diff. Geom.*, **18**, 253-268 (1983).

Mathematical Research Letters **1**, 701–715 (1994)

SYMMETRY OF THE GINZBURG LANDAU MINIMIZER IN A DISC

ELLIOTT H. LIEB AND MICHAEL LOSS

ABSTRACT. The Ginzburg-Landau energy minimization problem for a vector field on a two dimensional disc is analyzed. This is the simplest nontrivial example of a *vector field* minimization problem and the goal is to show that the energy minimizer has the full geometric symmetry of the problem. The standard methods that are useful for similar problems involving real valued *functions* cannot be applied to this situation. Our main result is that the minimizer in the class of symmetric fields is stable, i.e., the eigenvalues of the second variation operator are all nonnegative.

I. Introduction

There are many energy minimization problems having a geometric symmetry and for which one can show that the energy minimizer has the same symmetry as the problem itself. Typically this is done by using a rearrangement inequality of some sort. However, and this is the important point, rearrangement inequalities work (if they work at all) only when the variable is a function and not something more complicated like a vector field.

There are several important problems in which the variable is one or more vector or tensor fields and for which the minimizer is believed to be symmetric. Examples include the full multi-field Ginzburg-Landau problem for a superconductor in a magnetic field, the 't Hooft-Polyakov monopole and the Skyrme model (see [LE2] for a review). They are all unresolved. In this paper we analyze the simplest possible nontrivial example of a vector field energy minimization problem—the Ginzburg-Landau problem for a complex scalar field in a disc. It has exercised many authors (see, e.g., [JT], [BBH] and references therein) but no one has been able to show that the obvious symmetric vector field minimizes the energy (except in the

Received October 5, 1994.

Work of E. Lieb partially supported by NSF grant PHY 90-19433 A03.

Work of M. Loss partially supported by NSF grant DMS 92-07703.

With M. Loss in Math. Res. Lett. *1*, 701–715 (1994)

weak coupling regime where convexity holds). In fact, it has not even been shown that the symmetric solution is stable under perturbations, and it is the purpose of this paper to prove just that. We do so by using a mixture of rearrangement inequalities on different components of the vector field and, while our methods are highly specialized to this problem, we believe that it is one of the few examples in which light can be shed on the symmetry of an energy minimizing vector field.

As an illustration of the problem in which the variable, ψ, is a function, one could mention the following: Let $\mathbf{B}_n$ denote the closed unit ball centered at $0 \in \mathbf{R}^n$ and let ψ denote a real valued function on $\mathbf{B}_n$ that vanishes on $\partial\mathbf{B}_n$, the boundary of $\mathbf{B}_n$, and whose gradient is square integrable. Then we set

$$\mathcal{F}(\psi) = \int_{\mathbf{B}_n} |(\nabla\psi)(x)|^2 \mathrm{d}x + \int_{\mathbf{B}_n} [1 - \psi(x)^2]^2 \mathrm{d}x \tag{1.1}$$

and seek to minimize $\mathcal{F}(\psi)$. It is well known that there is a minimizer and that it is spherically symmetric, i.e., $\psi(x) = \psi(y)$ if $|x| = |y|$. The minimizer thus retains the symmetry of the problem. Indeed, more is true: ψ is symmetric decreasing, i.e., $\psi(x) \geq \psi(y)$ if $|x| \leq |y|$. While there are other methods to prove the symmetry, one of the simplest is to do so by using rearrangement inequalities to show that ψ is symmetric decreasing.

The first step in this process is to observe that replacing ψ by $|\psi|$ does not change $|\nabla\psi|^2$ and hence does not change the energy $\mathcal{F}(\psi)$. The second step is to replace $|\psi|$ by the equimeasurable function ψ^* which is defined to be the symmetric decreasing rearrangement of $|\psi|$. Certainly ψ^* satisfies the boundary conditions. The equimeasurability of ψ^* and $|\psi|$ guarantees that $\int[1-\psi^2]^2 = \int[1-\psi^{*2}]^2$. The important inequality concerns the kinetic energy, or Dirichlet integral. It is

$$\int_{\mathbf{B}_n} |\nabla|\psi||^2 \geq \int_{\mathbf{B}_n} |\nabla\psi^*|^2. \tag{1.2}$$

This shows that among the energy minimizers there is at least one that is symmetric decreasing.

We now turn to the Ginzburg-Landau problem in the disc $\mathbf{D} = \mathbf{B}_2$ in $\mathbf{R}^2$, which looks deceptively similar to the above problem. For one thing the variable is now a real vector field $\psi(x) = (f(x), g(x))$ instead of a single function. It is customary to introduce the complex valued function $\phi(x) = f(x) + ig(x)$. The energy functional is

$$\begin{aligned}\mathcal{E}(\psi) &= \int_{\mathbf{D}} \{(\nabla f(x))^2 + (\nabla g(x))^2 + J(f(x)^2 + g(x)^2)\}\mathrm{d}x \\ &= \int_{\mathbf{D}} \{|\nabla\phi|^2 + J(|\phi|^2)\}.\end{aligned} \tag{1.3}$$

Usually, $J : \mathbf{R}^+ \to \mathbf{R}^+$ is taken to be the function $J(t) = \lambda(1-t)^2$ with $\lambda > 0$. For our purposes we can generalize this to J satisfying certain conditions, which we assume henceforth:

(i) $J(0) = \lambda > 0, \quad J(1) = 0, \quad J(t) \geq 0$ if $t > 1$,
(ii) $J(t)$ is monotone decreasing and convex on the interval $[0, 1]$,
(iii) J is twice differentiable on $[0, 1]$.

The gradients of ψ are assumed to be square integrable and the condition on ψ on the boundary of $\mathbf{D}$ is

$$\psi(x) = x = (x_1, x_2) = (\cos\theta, \sin\theta). \tag{1.4}$$

We denote the class of $H^1(\mathbf{D})$ functions satisfying (1.4) by $\mathcal{C}$. *The problem is to minimize $\mathcal{E}(\psi)$ subject to $\psi \in \mathcal{C}$.*

For this problem it is a standard fact that a minimizer exists and satisfies the Euler-Lagrange pair of equations

$$-\Delta\psi + J'(\psi^2)\psi = 0 \tag{1.5}$$

with $\psi^2 = f^2 + g^2$. The obvious conjecture about a minimizer ψ is that it is a "hedgehog", i.e., for some nonnegative function f defined on $[0, 1]$ with $f(1) = 1$

$$\psi(x) = f(r)(\cos\theta, \sin\theta) \tag{1.6}$$

where $r := \sqrt{x_1^2 + x_2^2}$. There is always a function ψ_0 that minimizes the energy in the class of vector fields of type (1.6), and it satisfies (1.5). The problem is to show that this ψ_0 is a global minimizer. In terms of $f(r)$, (1.5) reads

$$-f'' - \frac{1}{r}f' + \frac{1}{r^2}f + J'(f^2)f = 0 \tag{1.7}$$

with $f(0) = 0$ and $f(1) = 1$. The solution to this problem is unique [HH]. It is not hard to see that f is monotone increasing, but this fact is not needed in this paper.

Although we cannot prove the full hedgehog conjecture, we are able to verify that the hedgehog is stable, that is to say that all the eigenvalues of the self-adjoint second variation operator H, defined by the quadratic form,

$$\tfrac{1}{2}\frac{\mathrm{d}^2}{\mathrm{d}\varepsilon^2}\mathcal{E}(\psi_0 + \varepsilon v)\Big|_{\varepsilon=0} = (v, Hv), \tag{1.8}$$

are nonnegative. Specifically H is given by

$$Hv = -\Delta v + J'(\psi_0^2)v + 2J''(\psi_0^2)\langle\psi_0, v\rangle\psi_0 \tag{1.9}$$

for vector fields v that vanish on $\partial\mathbf{D}$. Here $\langle a, b\rangle$ is the inner product on $\mathbf{R}^2$.

We believe that all the eigenvalues of H are strictly positive but we cannot show this. If they are, then we can reach the following conclusion: For small λ the hedgehog is certainly the global minimizer because $\psi \to \mathcal{E}(\psi)$ is strictly convex and hence the global minimizer is unique. If the hedgehog ceases to be the minimizer for large λ then the non-hedgehog minimizer cannot be close to the hedgehog. In other words, a simple bifurcation away from the hedgehog cannot occur.

II. Statements of theorems and lemmas

The following three theorems will be proved in the next section in the order 2, 3, 1. Theorem 1 is our main result. Theorem 3 will require three lemmas which we list here. Lemmas 1 and 2 on rearrangements are well known.

The proof of Theorem 2 uses some simple facts about convexity. This theorem holds for the analogous Ginzburg-Landau problem in $\mathbf{R}^n$ for any n, not just for $n = 2$. Theorem 1 is a Corollary of Theorems 2 and 3.

Theorem 1 (Weak stability of the symmetric minimizer). *The eigenvalues of H in (1.8, 1.9) (with Dirichlet boundary conditions) are all nonnegative. The complex eigenfunctions of H can all be chosen to have the following form*

$$v(r,\theta) = e^{im\theta}\begin{pmatrix} a(r)e^{i\theta} + b(r)e^{-i\theta} \\ -ia(r)e^{i\theta} + ib(r)e^{-i\theta}\end{pmatrix}, \tag{2.1}$$

for suitable real functions $a = a_m$ and $b = b_m$ and with $m = 0, \pm 1, \pm 2, \ldots$. Clearly, $\overline{v}$, the complex conjugate, is also an eigenvector with the same eigenvalue as v. The lowest eigenvalue of H belongs either to $m = 0$ or to $m = \pm 1$.

Remark: Both cases, $m = 0$ or $m = 1$, can occur—depending on J. When $J = 0$, $m = 1$ is optimal with $a(r) = 0$. The lowest eigenfunction of $-\Delta$ is well known to be nodeless. When J is very large the best choice is $m = 0$ with $b(r) \simeq -a(r)$ because $a = -b$ makes $\langle\psi_0, v\rangle$ vanish.

Theorem 2 (Partial convexity of the energy functional $\mathcal{E}(\psi)$). *Suppose $\psi = (\psi_1, \ldots, \psi_n)$ is a real vector field in $H^1(\mathbf{B}_n)$ that satisfies $\psi(x) =$*

x on the boundary of $\mathbf{B}_n$. Suppose that $\psi(x)^2 = \sum_{i=1}^{n} \psi_i(x)^2 \leq 1$ for all x and suppose that each component ψ_i satisfies

$$\int_{\mathbf{S}^{n-1}} \psi_i(r\omega)\mathrm{d}\omega = 0 \tag{2.2}$$

for all r. Define the vector field $\widetilde{\psi}(x)$ by

$$\widetilde{\psi}(r\omega) = h(r)\omega, \qquad \omega \in \mathbf{S}^{n-1} \tag{2.3}$$

where h is the spherical average of $|\psi|^2$, i.e.,

$$h(r) = \left[\frac{1}{|\mathbf{S}^{n-1}|}\int_{\mathbf{S}^{n-1}}\left(\sum_i \psi_i(r\omega)^2\right)\mathrm{d}\omega\right]^{1/2} \tag{2.4}$$

and $|\mathbf{S}^{n-1}| = \int_{\mathbf{S}^{n-1}} \mathrm{d}\omega$. Then (with $\mathcal{E}(\psi)$ given by the obvious generalization to $\mathbf{B}_n$ of (1.3)),

$$\mathcal{E}(\widetilde{\psi}) \leq \mathcal{E}(\psi). \tag{2.5}$$

If we assume that $h(r) > 0$ for all $r > 0$ then equality occurs in (2.5) only if $\psi = \widetilde{\psi}$.

Theorem 3 (Rearrangements of special vector fields). *Suppose that ψ is a vector field in $\mathcal{C}$ and suppose that there exists some fixed vector $\omega_0 \in \mathbf{S}^1$ such that*

$$\psi(t\omega_0) = h(t)\omega_0 \tag{2.6}$$

for all $t \in [-1, 1]$. Then there is a vector field $\widehat{\psi} \in \mathcal{C}$ satisfying (2.6) and, additionally,

$$\text{(i)} \quad \widehat{\psi}(x) = -\widehat{\psi}(-x) \text{ for all } x \in \mathbf{D}, \tag{2.7}$$

$$\text{(ii)} \quad \mathcal{E}(\widehat{\psi}) \leq \mathcal{E}(\psi). \tag{2.8}$$

Remark: The following might help to clarify the relation between Theorems 2 and 3. Write a minimizing $\psi \in \mathcal{C}$ in complex form as

$$\phi(r,\theta) = \sum_{k=-\infty}^{\infty} c_k(r)e^{ik\theta}, \tag{2.9}$$

with $c_k(1) = 0$ if $k \neq 1$ and $c_1(1) = 1$. If $c_0(r) \equiv 0$ then Theorem 2 applies and we learn that the hedgehog is the minimizer, i.e., $c_k(r) \equiv 0$ for $k \neq 1$.

Next suppose that we take a ϕ in the form (2.9) in which only at most two of the c_k's are not identically zero, say c_1 and c_m with $m \neq 1$. Then we claim that we can choose the two c's to be real functions without raising the energy. Having done this, Theorems 2 and 3 apply and we again learn that the energy minimizing choice in this restricted category has $c_m \equiv 0$ for $m \neq 1$. The proof of this assertion is the following. We write $c_j(r) = p_j(r)\exp[i\alpha_j(r)]$ with $p_j \geq 0$ and α_j real. Then

$$|\phi(r,\theta)|^2 = p_1(r)^2 + p_m(r)^2 + 2p_1(r)p_m(r)\cos[(m-1)\theta + \alpha_m(r) - \alpha_1(r)],$$

and we observe two things: If we replace α_1 and α_m by zero then

(i) the gradient term in $\mathcal{E}$ can only decrease;

(ii) the J term does not change because by a trivial shift of θ, the θ integral does not depend on $\alpha_m(r) - \alpha_1(r)$. (The convexity of J plays no role here.)

The lemmas about symmetric decreasing rearrangements that we shall need are the following. The first was basically proved by Chiti [CG] and then by Crandall-Tartar [CT]. For some generalizations see [AL], 2.2 and 2.3.

Lemma 1. *Let f and g be nonnegative functions on $\mathbf{R}^n$ and let $J : \mathbf{R} \to \mathbf{R}^+$ be a convex function with $J(0) = 0$. Then*

$$\int_{\mathbf{R}^n} J(f^*(x) - g^*(x))\mathrm{d}x \leq \int_{\mathbf{R}^n} J(f(x) - g(x))\mathrm{d}x \tag{2.10}$$

where f^* and g^* are the symmetric decreasing rearrangements of f and g.

Lemma 2 (Rearrangements and gradient norms). *For $u \in H_0^1([-a,a])$ define $u^* = |u|^*$. Then $u^* \in H_0^1([-a,a])$ and*

$$\int \left(\frac{\mathrm{d}u^*}{\mathrm{d}x}\right)^2 \leq \int \left(\frac{\mathrm{d}u}{\mathrm{d}x}\right)^2. \tag{2.11}$$

Lemma 3 (Cutting argument). *Let $\psi = (f,g) \in \mathcal{C}$ and assume in addition that $g(x_1, 0) = 0$ for $x_1 \in [-1,1]$. Then there exists $\widetilde{\psi} = (\widetilde{f}, \widetilde{g}) \in \mathcal{C}$ such that for all $x = (x_1, x_2)$ in $\mathbf{D}$*

(i) *$\widetilde{g}(x_1,x_2) \geq x_2$ for $x_2 \geq 0$ and $\widetilde{g}(x_1,x_2) \leq x_2$ for $x_2 \leq 0$,*

(ii) *$|\widetilde{\psi}(x)| \leq 1$ for all $x \in \mathbf{D}$ and hence $\widetilde{f}(x_1,x_2)^2 \leq 1 - x_2^2$,*

(iii) *$\mathcal{E}(\widetilde{\psi}) \leq \mathcal{E}(\psi)$.*

III. Proofs

3A. Proof of Theorem 2. Since $\sum_{i=1}^{n} \psi_1(x)^2 \le 1$, and since $t \mapsto J(t)$ is convex we have, by Jensen's inequality, that

$$\frac{1}{|\mathbf{S}^{n-1}|}\int_{\mathbf{S}^{n-1}} J(|\psi(r\omega)|^2)\mathrm{d}\omega \ge J(h(r)^2) \tag{3A.1}$$

and hence

$$\int_{\mathbf{B}_n} J(|\psi(x)|^2)\mathrm{d}x \ge \int_{\mathbf{B}_n} J(h(r)^2)\mathrm{d}x = \int_{\mathbf{B}_n} J(|\widehat{\psi}(x)|^2)\mathrm{d}x. \tag{3A.2}$$

To estimate the kinetic energy we expand each component, ψ_j, into normalized spherical harmonics, Y_{lm}, with coefficients $c_j^{lm}(r)$.

$$\psi_j(r\omega) = \sum_{l=1}^{\infty}\sum_m c_j^{lm}(r)Y_{lm}(\omega). \tag{3A.3}$$

Here l denotes the irreducible representation of $SO(n)$, while m is a multi-index that labels the rows. The reason $l=0$ is absent is that

$$\int_{\mathbf{S}^{n-1}} \psi_j(r\omega)\mathrm{d}\omega = 0 \text{ for every } 0 < r \le 1.$$

Note that $h(r)^2 = \sum_{j=1}^{n}\sum_{l=1}^{\infty}\sum_m (c_j^{lm}(r))^2/|\mathbf{S}^{n-1}|$. It is well known that

$$\int_{\mathbf{B}_n} |\nabla\psi_j|^2 = \sum_{l=1}^{\infty}\sum_m \int_0^1 \left\{(\mathrm{d}c_j^{lm}/\mathrm{d}r)^2 + \frac{l(l+n-2)}{r^2}(c_j^{lm})^2\right\} r^{n-1}\mathrm{d}r. \tag{3A.4}$$

Since $h\frac{\mathrm{d}h}{\mathrm{d}r} = \sum_{j=1}^{n}\sum_{l=1}^{\infty}\sum_m \left(c_j^{lm}\frac{\mathrm{d}c_j^{lm}}{\mathrm{d}r}\right)\Big/|\mathbf{S}^{n-1}|$ we get, by Schwarz's inequality, that

$$\left|\frac{\mathrm{d}h}{\mathrm{d}r}\right|^2 \le \sum_{j=1}^{n}\sum_{l=1}^{\infty}\sum_m \left(\frac{\mathrm{d}c_j^{lm}}{\mathrm{d}r}\right)^2 \Big/ |\mathbf{S}^{n-1}|. \tag{3A.5}$$

Obviously,

$$\sum_{l=1}^{\infty}\sum_m \int_0^1 l(l+n-2)(c_j^{lm})^2 r^{n-3}\mathrm{d}r \ge \sum_{l=1}^{\infty}\sum_m \int_0^1 (n-1)(c_j^{lm})^2 r^{n-3}\mathrm{d}r \tag{3A.6}$$

with equality only if $c_j^{lm} \equiv 0$ for all $l \geq 2$. In that case we can write

$$\psi_j(x) = \sqrt{n}\sum_{k=1}^{n} \mathrm{d}_j^k(r)\frac{x_k}{r} \tag{3A.7}$$

and $h(r)^2 = \sum_{j,k=1}^{n} d_j^k(r)^2$. In general, by summing over j, we find that (3A.8)

$$\int_{\mathbf{B}_n} |\nabla\psi|^2 \geq |\mathbf{S}^{n-1}| \int_0^1 \{(\mathrm{d}h(r)/\mathrm{d}r)^2 + \frac{n-1}{r^2}h(r)^2\}r^{n-1}\mathrm{d}r = \int_{\mathbf{B}_n} |\nabla\widetilde{\psi}|^2,$$

with equality only if (3A.7) is satisfied.

In short, (2.5) has been proved and we know that equality requires (3A.7). Our final task is to show that equality in (2.5) also requires $d_j^k(r) = h(r)\delta_{k,j}/\sqrt{n}$ when $h(r) > 0$ for all $r > 0$.

Inequality (3A.5) was obtained by using Schwarz's inequality. In order to have equality we must have that

$$\frac{\mathrm{d}}{\mathrm{d}r}d_j^k(r) = \lambda(r)d_j^k(r) \tag{3A.9}$$

for some function $\lambda(r)$ not depending on j and k. By multiplying (3A.9) on both sides by $d_j^k(r)$ and summing over k and j we have $h(r)h'(r) = \lambda(r)h(r)^2$. Since $h(r) > 0$ for all $r > 0$ we have that

$$\lambda(r) = h'(r)/h(r). \tag{3A.10}$$

This function is integrable away from the origin and hence (3A.9) yields

$$d_j^k(r) = \mu(r)d_j^k(1). \tag{3A.11}$$

with $\mu(r) = \exp\left\{-\int_r^1 \lambda(s)\mathrm{d}s\right\}$. By assumption, $d_j^k(1) = n^{-1/2}\delta_{j,k}$, and this yields the desired conclusion with $\mu(r) = h(r)$. □

3B. Proof of Theorem 3. Without loss of generality, we can assume $\omega_0 = (1{,}0)$. Our hypothesis is that if $\psi(x) = (f(x), g(x))$ then $g(x_1, 0) = 0$ for $-1 \leq x_1 \leq 1$. By Lemma 3 (cutting argument) we can assume two important facts about our f and g:

(i) $f(x_1,x_2)^2 \leq 1 - x_2^2$;
(ii) $g(x_1,x_2) \geq x_2$ if $x_2 > 0$ and $g(x_1,x_2) \leq x_2$ if $x_2 \leq 0$.

We can also assume that $f^2 + g^2 \leq 1$.

The first step is to define $\widetilde{g}$ in the following way. For each, fixed x_2 we replace the function $x_1 \mapsto g(x_1, x_2)$ by its (one-dimensional) symmetric decreasing rearrangement $g^*(x_1, x_2)$ if $x_2 \geq 0$ and we replace it by $-|g(x_1, x_2)|^*$ if $x_2 \leq 0$. By (ii) above, $\widetilde{g}$ satisfies the correct boundary condition, (i.e., $g(x_1, x_2) = x_2$ on $\partial\mathbf{D}$), and $\widetilde{g}$ also satisfies (ii) above.

The next step, the definition of $\widetilde{f}(x_1, x_2)$, is a bit more complicated. First, let $f^\#$ be the symmetric *increasing* rearrangement of $|f|$. [Another way to say this is that $1 - f^\# = (1 - |f|)^*$]. Once again, the rearrangement is done on each line $x_2 =$ constant. We note that $f(x_1, x_2)$ is continuous in x_1 for a.e. x_2 (because it is an $H^1(\mathbf{R})$ function for a.e. x_2) and has antisymmetric boundary values at $x_1 = \pm(1 - x_2^2)^{1/2}$. Therefore $f^\#$ is a continuous function of x_1 (indeed, it is an $H^1(\mathbf{R})$ function by Lemma 2) and $f^\#(0, x_2) = 0$. By (i) above, $f^\# \leq (1 - x_2^2)^{1/2}$. Moreover, $f^\# = (1 - x_2^2)^{1/2}$ on $\partial\mathbf{D}$ since $x_1 \mapsto |f(x_1, x_2)|$ is continuous and $|f|$ satisfies the same boundary condition. Now define

$$\widetilde{f}(x_1, x_2) = \begin{cases} f^\#(x_1, x_2) & \text{if } x_1 > 0 \\ -f^\#(x_1, x_2) & \text{if } x_1 < 0 \end{cases} \tag{3B.1}$$

which satisfies the correct conditions on $\partial\mathbf{D}$. We also note that $|\partial\widetilde{f}/\partial x_1| = |\partial f^\#/\partial x_1|$ and $|\partial\widetilde{f}/\partial x_2| = |\partial f^\#/\partial x_2|$.

Our task is to show that these rearrangements decrease both terms in the functional $\mathcal{E}$. We turn to the gradient norms first. By Lemma 2 we have that $\int_\mathbf{D}(\partial\widetilde{g}/\partial x_1)^2$ does not increase and the same is true for $\int_\mathbf{D}(\partial\widetilde{f}/\partial x_1)^2$. We next show that $\int_\mathbf{D}(\partial\widetilde{g}/\partial x_2)^2$ does not increase either. (The argument for $\widetilde{f}$ is essentially identical.) There are several ways to prove this, and one way is the following. An easy approximation argument shows that

$$\int_\mathbf{D}(\partial g/\partial x_2)^2 = \lim_{\delta\to 0} \delta^{-2} \int_\mathbf{D} [g(x_1, x_2 + \delta) - g(x_1, x_2)]^2 dx_1 dx_2 \tag{3B.2}$$

(Here, $g(x_1, x_2)$ has to be extended to be x_2 outside $\mathbf{D}$.) The result we want—that replacement of g by $\widetilde{g}$ does not increase the two sides of (3B.2)—follows from a trivial modification of Lemma 1.

To summarize, the vector field $\widetilde{\psi}$ is in C and its gradient norms are not bigger than those of ψ.

The penultimate step is to prove that $|\widetilde{\psi}(x)|^2 \leq 1$ for all $x \in \mathbf{D}$. Let $K(t) = [\max(0, t)]^2$. Then

$$\int_\mathbf{D} K(g^2 - (1 - f^2)) = 0 \tag{3B.3}$$

since $|\psi|^2 = f^2 + g^2 \leq 1$. By Lemma 1, however,

$$\int_{\mathbf{D}} K(g^2 - (1-f^2)) \geq \int_{\mathbf{D}} K(\widetilde{g}^2 - (1-\widetilde{f}^2)) \tag{3B.4}$$

since $(g^2)^* = \widetilde{g}^2$ for each line, $x_2 =$ constant, and, similarly, $(1-f^2)^* = (1-\widetilde{f}^2)$ since $f^2 \leq 1$ and $\widetilde{f}^2 \leq 1$. If $\widetilde{g}^2 + \widetilde{f}^2 > 1$ on a set of positive measure, the right side of (3B.4) would be positive, but this is precluded by (3B.3).

Finally, we turn to the J term in $\mathcal{E}$. We can define $L(t) = J(1-t)$ for $0 \leq t \leq 1$. (The definition of $L(t)$ for $t < 0$ or $t > 1$ is not needed since $0 \leq t \leq 1$ in our application.) Then, by Lemma 1 and the same reasoning as for K above,

$$\int_{\mathbf{D}} L((1-f^2) - g^2) \geq \int_{\mathbf{D}} L((1-\widetilde{f}^2) - \widetilde{g}^2),$$

which is the same as $\int J(|\psi|^2) \geq \int J(|\widetilde{\psi}|^2)$.

Thus far we have constructed a $\widetilde{\psi}$ with $\mathcal{E}(\widetilde{\psi}) \leq \mathcal{E}(\psi)$ and with $\widetilde{f}(x_1,x_2) = -\widetilde{f}(-x_1,x_2)$ and $\widetilde{g}(x_1,x_2) = \widetilde{g}(-x_1,x_2)$. The final step is to use this $\widetilde{\psi}$ to construct a $\widehat{\psi}$ satisfying $\widehat{\psi}(x) = -\widehat{\psi}(-x)$ and $\mathcal{E}(\widehat{\psi}) \leq \mathcal{E}(\widetilde{\psi})$. Let $\mathbf{D}_+$ denote the upper hemidisc $\{(x_1,x_2) : x_2 \geq 0\} \cap \mathbf{D}$ and $\mathbf{D}_-$ the lower hemidisc. Let $(\widetilde{f}_\pm, \widetilde{g}_\pm)$ denote $\widetilde{\psi}$ restricted to $\mathbf{D}_+$ and $\mathbf{D}_-$. Consider the following two vector fields.

$$\psi_1 = \begin{cases} (\widetilde{f}_+(x_1,x_2), \widetilde{g}_+(x_1,x_2)) & \text{in } \mathbf{D}_+ \\ (\widetilde{f}_+(x_1,-x_2), -\widetilde{g}_+(x_1,-x_2)) & \text{in } \mathbf{D}_- \end{cases}$$

$$\psi_2 = \begin{cases} (\widetilde{f}_-(x_1,-x_2), -\widetilde{g}_-(x_1,-x_2)) & \text{in } \mathbf{D}_+ \\ (\widetilde{f}_-(x_1,x_2), \widetilde{g}_-(x_1,x_2)) & \text{in } \mathbf{D}_- \end{cases}$$

Clearly $\psi_{1,2}(x) = -\psi_{1,2}(-x)$. Also, ψ_1 and ψ_2 are in C because $\widetilde{g}(x_1,0) = 0$. Moreover,

$$\mathcal{E}(\psi_1) + \mathcal{E}(\psi_2) = 2\mathcal{E}(\widetilde{\psi}).$$

Therefore, ψ_1 or ψ_2 is a vector field satisfying the conclusion of Theorem 3. □

3C. Proof of Theorem 1. The basic fact, which we shall prove later, is that the real eigenfunction of H can be chosen to have at least one of the following symmetry properties for all $x \in D$.

$$\begin{aligned} &(a) \quad v(x) = -v(-x) \\ &(b) \quad v(x) = Pv(P^{-1}x) \end{aligned} \tag{3C.1}$$

where

$$P = \begin{pmatrix} 1 & 0 \\ 0 & -1 \end{pmatrix}$$

is the reflection about the x_1-axis in $\mathbf{R}^2$. This is not to say that every eigenfunction has one of these properties, but we do assert that each *eigenvalue* of H has at least one eigenfunction of type (a) or (b). Since we are interested only in the eigenvalues of H, we may assume (a) or (b).

Now consider $\psi_\varepsilon := \psi_0 + \varepsilon v \in \mathcal{C}$, with v as in (3C.1). In case (a), $\psi_\varepsilon(x) = -\psi_\varepsilon(-x)$ for all ε. In case (b), ψ_ε satisfies hypothesis (2.6) of Theorem 3, with $\omega_0 = (1, 0)$. By Theorem 2, in case (a) there is a $\widetilde{\psi}_\varepsilon$ with $\mathcal{E}(\widetilde{\psi}_\varepsilon) \le \mathcal{E}(\psi_\varepsilon)$ and $\widetilde{\psi}_\varepsilon$ is a hedgehog (1.6). Thus,

$$\mathcal{E}(\psi_\varepsilon) \ge \mathcal{E}(\widetilde{\psi}_\varepsilon) \ge \mathcal{E}(\psi_0) \tag{3C.2}$$

since ψ_0 is the energy minimizer among all hedgehogs. In case (b) we first have to use Theorem 3 to obtain an intermediate $\widehat{\psi}$ that satisfies the hypothesis of Theorem 2. Again, (3C.2) holds. Since

$$\mathcal{E}(\psi_\varepsilon) = \mathcal{E}(\psi_0) + \varepsilon^2 \gamma \int_{\mathbf{D}} v^2 + o(\varepsilon^2),$$

where γ is the eigenvalue of H belonging to v, we see from (3C.2) that $\gamma \ge 0$.

There are two ways to derive the symmetry (3C.1) and we shall give both. The first is a fairly general argument and the second involves a detailed study of the eigenfunctions leading to (2.1).

General argument: Let P denote reflection about some axis through the origin. For any eigenfunction, w, its reflection, $(P^*w)(x) := Pw(P^{-1}x)$, is also an eigenfunction with the same eigenvalue. If $v(x) = w(x) + (P^*w)(x)$ is not identically zero for *some* P then v is an eigenfunction satisfying (3C.1) (b). If v vanishes identically for *all* reflections P we claim that w must be of type (3C.1)(a). To see this recall that any rotation $\mathcal{R}$ is the product of two reflections and hence $\mathcal{R}w(\mathcal{R}^{-1}x) = w(x)$, i.e., w is rotationally symmetric. It is easy to see that w must then satisfy $w(x) = k(r)(-x_2, x_1)$ for some function k, and hence w satisfies (3C.1)(a).

Details of eigenfunctions: Let $\mathcal{R}_\alpha$ be the rotation through the angle α and let U_α be its representation given by

$$(U_\alpha v)(x) = \mathcal{R}_\alpha v(\mathcal{R}_{-\alpha} x). \tag{3C.3}$$

U_α is a strongly continuous one-parameter subgroup of the unitary group of $L^2(\mathbf{D}; \mathbf{C}^2)$ and it commutes with H. Its infinitesimal generator is

$$L = i\frac{\partial}{\partial\theta} + i\begin{pmatrix} 0 & 1 \\ -1 & 0 \end{pmatrix}. \tag{3C.4}$$

By standard arguments we can choose the eigenfunctions of H to be eigenfunctions of L. By solving $Lv = \nu v$ we find that $v(\theta)$ must be of the form (2.1). ν must be an integer since $v(0) = v(2\pi)$. Furthermore, a glance at the eigenvalue equation reveals that a and b can be taken to be real.

Next we verify that the lowest eigenvalue belongs to $m = 0$ or $m = \pm 1$. Suppose, on the contrary, that the lowest one belongs to $M > 1$ (m and $-m$ are the same by complex conjugation). Then define a comparison function by $\widehat{v} := e^{-i\theta}v$. Obviously, the J-term is unchanged. The only term that changes in the gradient norm is the replacement of $I := \int_{\mathbf{D}} \left|\frac{\partial}{\partial\theta}v\right|^2 r^{-2}$ by $\widehat{I} := \int_{\mathbf{D}} \left|\frac{\partial}{\partial\theta}e^{-i\theta}v\right|^2 r^{-2}$. One easily computes that

$$I = 2\int_{\mathbf{D}} \{(M+1)^2 a(r)^2 + (M-1)^2 b(r)^2\} r^{-2}$$

and hence $\widehat{I} < I$ if $M > 1$.

Now consider (2.1) with $m = 0$. This vector v satisfies $v(x) = -v(-x)$ (because changing x to $-x$ amounts to changing θ to $\theta + \pi$). Thus, all $m = 0$ eigenfunctions satisfy (3C.1)(a). Indeed all even-m eigenfunctions have this property.

When $m = 1$ we claim that (3C.1)(b) holds—thus completing our proof. Take the real part of (2.1), which is $v(r,\theta) = (a(r)\cos 2\theta + b(r), a(r)\sin 2\theta)$; this satisfies (3C.1)(b) with P being reflection about the x_1-axis. □

3D. Remarks on Lemma 1. The Chiti, Crandall-Tartar theorem requires the convex function J to be even. It is usually stated as

$$J : \mathbf{R}^+ \to \mathbf{R}^+, \quad J(0) = 0$$

and with (2.10) replaced by

$$\int_{\mathbf{R}^n} J(|f^* - g^*|) \le \int_{\mathbf{R}^n} J(|f - g|). \tag{2.10a}$$

It is a simple matter to derive (2.10) from (2.10a) but we are unaware of (2.10) in the literature. Evidently, it suffices to prove (2.10) for the extremal convex functions, i.e., J of the type $J(t) = \lambda(t-a)_+$ or $J(t) = \lambda(t+a)_-$ with $t_\pm := \max\{\pm t, 0\}$ and $a, \lambda \ge 0$. Consider $\lambda(t-a)_+$. By replacing f by $(f-a)_+$ we may as well take $a = 0$ and $\lambda = 2$. Then $2t_+ = |t| + t$. We can assume $f - g \in L^1(\mathbf{R})$, for otherwise the right side of (2.10) is $+\infty$. It is easy to see that $f^* - g^* \in L^1(\mathbf{R})$ as well. Then, since $2t_+ = |t| + t$, (2.10) reads

$$\int \{|f^* - g^*| - |f - g|\} \ge \int \{f - g + g^* - f^*\}.$$

Indeed, the left side is nonnegative by (2.10a) and the right side is zero since f, f^* and g, g^* are equimeasurable.

3E. Remarks on Lemma 2. (2.11) also holds for functions on $\mathbf{R}^n$, but we need only the $\mathbf{R}^1$ version. For some historical remarks about this inequality see [AL] 1.1, 2.6 and 2.7. There are many proofs of it but the simplest (in our view) is in [LE1], Lemma 5. The method of [LE1], Lemma 5 also proves a generalization that would suffice for our needs in the proof of Theorem 3 above. The generalization is that if $f : \mathbf{R}^{n+m} \to \mathbf{R}^+$ and if $f^* : \mathbf{R}^{n+m} \to \mathbf{R}^+$ is the symmetric decreasing rearrangement with respect to the first n variables only, then

$$\int_{\mathbf{R}^{n+m}} |\nabla f^*|^2 \leq \int_{\mathbf{R}^{n+m}} |\nabla f|^2 .$$

3F. Proof of Lemma 3. We shall construct $\widetilde{\psi}$ by a sequence of steps.

Step 1.
$$\psi \mapsto T_1\psi = \begin{cases} \psi & \text{if } |\psi| \leq 1 \\ \psi/|\psi| & \text{if } |\psi| \geq 1 \end{cases}$$

[with $|\psi| = (f^2+g^2)^{1/2}$]. An easy exercise shows that $\|\nabla T_1\psi\|_2 \leq \|\nabla\psi\|_2$. Furthermore, $\int J(|\psi|^2) \geq \int J(|T_1\psi|^2)$ because $J(t) \geq 0$ when $t > 1$ while $J(1) = 0$. Therefore, without loss of generality we can assume that $|\psi(x)| \leq 1$ for all x.

Step 2.
$$\psi \to T_2\psi = (f,h) \quad \text{with } h(x_1,x_2) = \begin{cases} \max\{x_2, g(x_1,x_2)\} & \text{if } x_2 \geq 0 \\ \min\{x_2, g(x_1,x_2)\} & \text{if } x_2 \leq 0. \end{cases}$$

Obviously $|T_2\psi(x)| \geq |\psi(x)|$ for all x. The condition $g(x_1,0) = 0$ guarantees that $T_2\psi \in \mathcal{C}$.

Step 3.
$$\psi \mapsto T_3\psi = T_1T_2\psi.$$

If we write $T_1T_2\psi = (\widehat{f},\widehat{g}) = \widehat{\psi}$ we can easily verify the following for all x (using $|\psi| \leq 1$)

$$(a) \qquad 1 \geq |T_3\psi(x)| \geq |\psi(x)|$$

$$(b) \qquad |\widehat{f}(x)| \leq |f(x)| \ \text{ and } \ sgn\widehat{f}(x) = sgn f(x)$$

$$(c) \qquad |\widehat{g}(x)| \geq |g(x)| \ \text{ and } \ sgn\,\widehat{g}(x) = sgn\,x_2$$

$$(d) \qquad \widehat{g}(x)^2 - g(x)^2 \geq \frac{1-x_2^2}{1+x_2^2}\,[x_2^2 - g(x)^2]_+ .$$

(a) is obvious because T_2 does not decrease $|\psi|$ and T_1 only cuts off $|T_2\psi|$ at 1; but $|\psi| \leq 1$ everywhere. (b) is also obvious because T_2 leaves f

invariant and T_1 can only decrease $|f|$. (c) follows from the facts that T_2 increases $|g|$, the map $t \mapsto t/(f^2+t)$ is monotone increasing for $t > 0$, and $g^2 \le g^2/(f^2+g^2)$ since $f^2+g^2 \le 1$. Indeed, (d) gives a more quantitative estimate. To prove (d) we recall that $T_2\psi =: (f,h)$. If $f^2+h^2 \le 1$ and $x_2^2 \ge g^2$ then $|\widehat{g}| = |h| = |x_2|$ and (d) is certainly true. If $f^2+h^2 > 1$ and $x_2^2 \ge g^2$ then

$$\begin{aligned}\widehat{g}^2 - g^2 &= \frac{x_2^2}{f^2+x_2^2} - g^2 \\ &\ge \frac{x_2^2}{1-g^2+x_2^2} - g^2 \\ &= \frac{1-g^2}{1-g^2+x_2^2}\,[x_2^2-g^2] \\ &\ge \frac{1-x_2^2}{1+x_2^2}\,[x_2^2-g^2].\end{aligned}$$

We claim that $\mathcal{E}(T_3\psi) \le \mathcal{E}(\psi)$. As far as the gradient term is concerned, T_2 replaces g by the harmonic function x_2 on the set where $|g| \le |x_2|$. This certainly lowers the gradient term. The J term does not increase by property (a) above, since $J(t)$ is decreasing for $0 \le t \le 1$.

Now we iterate T_3 and denote $(f_n, g_n) = \psi_n := T_3^n\psi$. By (b) and (c) f_n and g_n are bounded monotone sequences and converge pointwise to limit functions $\widetilde{f}$ and $\widetilde{g}$. Since $\mathcal{E}(\psi)$ is weakly lower semicontinuous we have that $\mathcal{E}(\widetilde{\psi}) \le \mathcal{E}(\psi)$, where $\widetilde{\psi} = (\widetilde{f}, \widetilde{g})$. It is clear that $\widetilde{\psi}$ satisfies the correct boundary conditions and hence is in $\mathcal{C}$.

The only thing left to check is that $\widetilde{g}(x)^2 \ge x_2^2$. If we define $a_n(x) = [x_2^2 - g_n(x)^2]_+$ property (d) can be rewritten as $a_{n+1}(x) \le a_n(x)(2x_2^2/(1+x_2^2))$, which shows that $a_n(x)$ converges to zero pointwise for all $x \in \mathbf{D}$. □

Acknowledgements

We thank László Erdős for many valuable discussions.

References

[AL] F.J. Almgren, Jr. and E.H. Lieb, *Symmetric decreasing rearrangement is sometimes continuous*, J. Amer. Math. Soc. **2** (1989), 683–773.

[BBH] F. Bethuel, H. Brezis and F. Hélein, *Ginzburg-Landau Vortices*, Birkhäuser, 1994.

[CG] G. Chiti, *Rearrangements of functions and convergence in Orlicz spaces*, Appl. Anal. **9** (1979), 23–27.

[CT] M.G. Crandall and L. Tartar, *Some relations between nonexpansive and order preserving mappings*, Proc. Amer. Math. Soc. **78** (1980), 358–390.

[HH] R.M. Hervé and M. Hervé, *Etude qualitative des solutions réeles de l'équation differentielle ...* (to appear).

[JT] A. Jaffe and C. Taubes, *Vortices and Monopoles*, Birkhäuser, 1980.

[LE1] E.H. Lieb, *Existence and uniqueness of the minimizing solution of Choquard's non-linear equation*, Stud. Appl. Math. **57** (1977), 93–105.

[LE2] ______, *Remarks on the Skyrme Model*, Proc. Amer. Math. Soc., Symposia in Pure Math. **54** (1993), 379–384, (Proceedings of Summer Research Institute on Differential Geometry at UCLA, July 8–28, 1990).

Department of Mathematics,Princeton University, P.O. Box 708, Princeton, NJ 08544-0708
E-mail address: lieb@math.princeton.edu

School of Mathematics, Georgia Institute of Technology, Atlanta, GA 30332-0160
E-mail address: loss@math.gatech.edu

Publications of Elliott H. Lieb

1. Second Order Radiative Corrections to the Magnetic Moment of a Bound Electron, Phil. Mag. Vol. **46**, 311–316 (1955).
2. A Non-Perturbation Method for Non-Linear Field Theories, Proc. Roy. Soc. **241A**, 339–363 (1957).
3. (with K. Yamazaki) Ground State Energy and Effective Mass of the Polaron, Phys. Rev. **111**, 728–733 (1958).
4. (with H. Koppe) Mathematical Analysis of a Simple Model Related to the Stripping Reaction, Phys. Rev. **116**, 367–371 (1959).
5. Hard Sphere Bose Gas – An Exact Momentum Space Formulation, Proc. U.S. Nat. Acad. Sci. **46**, 1000–1002 (1960).
6. Operator Formalism in Statistical Mechanics, J. Math. Phys. **2**, 341–343 (1961).
7. (with D.C. Mattis) Exact Wave Functions in Superconductivity, J. Math. Phys. **2**, 602–609 (1961).
8. (with T.D. Schultz and D.C. Mattis) Two Soluble Models of an Antiferromagnetic Chain, Annals of Phys. (N.Y.) **16**, 407–466 (1961).

† 9. (with D.C. Mattis) Theory of Ferromagnetism and the Ordering of Electronic Energy Levels, Phys. Rev. **125**, 164–172 (1962).

† 10. (with D.C. Mattis) Ordering Energy Levels of Interacting Spin Systems, J. Math. Phys. **3**, 749–751 (1962).

11. New Method in the Theory of Imperfect Gases and Liquids, J. Math. Phys. **4**, 671–678 (1963).
12. (with W. Liniger) Exact Analysis of an Interacting Bose Gas. I. The General Solution and the Ground State, Phys. Rev. **130**, 1605–1616 (1963).
13. Exact Analysis of an Interacting Bose Gas. II. The Excitation Spectrum, Phys. Rev. **130**, 1616–1624 (1963).
14. Simplified Approach to the Ground State Energy of an Imperfect Bose Gas, Phys. Rev. **130**, 2518–2528 (1963).
15. (with A. Sakakura) Simplified Approach to the Ground State Energy of an Imperfect Bose Gase. II. The Charged Bose Gas at High Density, Phys. Rev. **133**, A899–A906 (1964).
16. (with W. Liniger) Simplified Approach to the Ground State Energy of an Imperfect Bose Gas. III. Application to the One-Dimensional Model, Phys. Rev. **134**, A312–A315 (1964).
17. (with T.D. Schultz and D.C. Mattis) Two-Dimensional Ising Model as a Soluble Problem of Many Fermions, Rev. Mod. Phys. **36**, 856–871 (1964).

† means the paper appears in this volume.

18. The Bose Fluid, *Lectures in Theoretical Physics, Vol. VIIC,* (Boulder summer school), University of Colorado Press, 175–224 (1965).
19. (with D.C. Mattis) Exact Solution of a Many-Fermion System and its Associated Boson Field, J. Math. Phys. **6**, 304–312 (1965).
20. (with S.Y. Larsen, J.E. Kilpatrick and H.F. Jordan) Suppression at High Temperature of Effects Due to Statistics in the Second Virial Coefficient of a Real Gas, Phys. Rev. **140**, A129–A130 (1965).
21. (with D.C. Mattis) Book *Mathematical Physics in One Dimension*, Academic Press, New York (1966).
† 22. Proofs of Some Conjectures on Permanents, J. of Math. and Mech. **16**, 127–139 (1966).
23. Quantum Mechanical Extension of the Lebowitz-Penrose Theorem on the van der Waals Theory, J. Math. Phys. **7**, 1016–1024 (1966).
24. (with D.C. Mattis) Theory of Paramagnetic Impurities in Semiconductors, J. Math. Phys. **7**, 2045–2052 (1966).
25. (with T. Burke and J.L. Lebowitz) Phase Transition in a Model Quantum System: Quantum Corrections to the Location of the Critical Point, Phys. Rev. **149**, 118–122 (1966).
26. Some Comments on the One-Dimensional Many-Body Problem, unpublished Proceedings of Eastern Theoretical Physics Conference, New York (1966).
27. Calculation of Exchange Second Virial Coefficient of a Hard Sphere Gas by Path Integrals, J. Math. Phys. **8**, 43–52 (1967).
28. (with Z. Rieder and J.L. Lebowitz) Properties of a Harmonic Crystal in a Stationary Nonequilibrium State, J. Math. Phys. **8**, 1073–1078 (1967).
29. Exact Solution of the Problem of the Entropy of Two-Dimensional Ice, Phys. Rev. Lett. **18**, 692–694 (1967).
30. Exact Solution of the F Model of an Antiferroelectric, Phys. Rev. Lett. **18**, 1046–1048 (1967).
31. Exact Solution of the Two-Dimensional Slater KDP Model of a Ferroelectric, Phys. Rev. Lett. **19**, 108–110 (1967).
32. The Residual Entropy of Square Ice, Phys. Rev. **162**, 162–172 (1967).
33. Ice, Ferro- and Antiferroelectrics, in *Methods and Problems in Theoretical Physics, in honour of R.E. Peierls*, Proceedings of the 1967 Birmingham conference, North-Holland, 21–28 (1970).
34. Exactly Soluble Models, in *Mathematical Methods in Solid State and Superfluid Theory*, Proceedings of the 1967 Scottish Universities' Summer School of Physics, Oliver and Boyd, Edinburgh 286–306 (1969).
35. The Solution of the Dimer Problems by the Transfer Matrix Method, J. Math. Phys. **8**, 2339–2341 (1967).
36. (with M. Flicker) Delta Function Fermi Gas with Two Spin Deviates, Phys. Rev. **161**, 179–188 (1967).
† 37. Concavity Properties and a Generating Function for Stirling Numbers, J. Combinatorial Theory **5**, 203–206 (1968).
38. A Theorem on Pfaffians, J. Combinatorial Theory **5**, 313–319 (1968).

39. (with F.Y. Wu) Absence of Mott Transition in an Exact Solution of the Short-Range One-Band Model in One Dimension, Phys. Rev. Lett. *20*, 1445–1448 (1968).
40. Two Dimensional Ferroelectric Models, J. Phys. Soc. (Japan) **26** (supplement), 94–95 (1969).
41. (with W.A. Beyer) Clusters on a Thin Quadratic Lattice, Studies in Appl. Math. **48**, 77–90 (1969).
42. (with C.J. Thompson) Phase Transition in Zero Dimensions: A Remark on the Spherical Model, J. Math. Phys. **10**, 1403–1406 (1969).
43. (with J.L. Lebowitz) The Existence of Thermodynamics for Real Matter with Coulomb Forces, Phys. Rev. Lett. **22**, 631–634 (1969).
44. Two Dimensional Ice and Ferroelectric Models, in *Lectures in Theoretical Physics, XI D*, (Boulder summer school) Gordon and Breach, 329–354 (1969).
45. Survey of the One Dimensional Many Body Problem and Two Dimensional Ferroelectric Models, in *Contemporary Physics: Trieste Symposium 1968*, International Atomic Energy Agency, Vienna, vol. 1, 163–176 (1969).
46. Models, in *Phase Transitions*, Proceedings of the 14th Solvay Chemistry Conference, May 1969, Interscience, 45–56 (1971).
† 47. (with H. Araki) Entropy Inequalities, Commun. Math. Phys. **18**, 160–170 (1970).
48. (with O.J. Heilmann) Violation of the Non-Crossing Rule: The Hubbard Hamiltonian for Benzene, Trans. N.Y. Acad. Sci. **33**, 116–149 (1970). Also in Annals N.Y. Acad. Sci. **172**, 583–617 (1971). (Awarded the 1970 Boris Pregel award for research in chemical physics.)
49. (with O.J. Heilmann) Monomers and Dimers, Phys. Rev. Lett. **24**, 1412–1414 (1970).
50. Book Review of "Statistical Mechanics" by David Ruelle, Bull. Amer. Math. Soc. **76**, 683–687 (1970).
51. (with J.L. Lebowitz) Thermodynamic Limit for Coulomb Systems, in *Systèmes a un Nombre Infini de Degrés de Liberté*, Colloques Internationaux de Centre National de la Recherche Scientifique **181**, 155–162 (1970).
52. (with D.B. Abraham, T. Oguchi and T. Yamamoto) On the Anomolous Specific Heat of Sodium Trihydrogen Selenite, Progr. Theor. Phys. (Kyoto) **44**, 1114–1115 (1970).
53. (with D.B. Abraham) Anomalous Specific Heat of Sodium Trihydrogen Selenite – An Associated Combinatorial Problem, J. Chem. Phys. **54**, 1446–1450 (1971).
54. (with O.J. Heilmann, D. Kleitman and S. Sherman) Some Positive Definite Functions on Sets and Their Application to the Ising Model, Discrete Math. **1**, 19–27 (1971).
55. (with Th. Niemeijer and G. Vertogen) Models in Statistical Mechanics, in *Statistical Mechanics and Quantum Field Theory*, Proceedings of 1970

Ecole d'Eté de Physique Théorique (Les Houches), Gordon and Breach, 281–326 (1971).

56. (with H.N.V. Temperley) Relations between the 'Percolation' and 'Colouring' Problem and Other Graph-Theoretical Problems Associated with Regular Planar Lattices: Some Exact Results for the 'Percolation' Problem, Proc. Roy. Soc. **A322**, 251–280 (1971).
57. (with M. de Llano) Some Exact Results in the Hartree-Fock Theory of a Many-Fermion System at High Densities, Phys. Letts. **37B**, 47–49 (1971).
58. (with J.L. Lebowitz) The Constitution of Matter: Existence of Thermodynamics for Systems Composed of Electrons and Nuclei, Adv. in Math. **9**, 316–398 (1972).
59. (with F.Y. Wu) Two Dimensional Ferroelectric Models, in *Phase Transitions and Critical Phenomena*, C. Domb and M. Green eds., vol. 1, Academic Press 331–490 (1972).
60. (with D. Ruelle) A Property of Zeros of the Partition Function for Ising Spin Systems, J. Math. Phys. **13**, 781–784 (1972).
61. (with O.J. Heilmann) Theory of Monomer-Dimer Systems, Commun. Math. Phys. **25**, 190–232 (1972). Errata **27**, 166 (1972).
62. (with M.L. Glasser and D.B. Abraham) Analytic Properties of the Free Energy for the "Ice" Models, J. Math. Phys. **13**, 887–900 (1972).
63. (with D.W. Robinson) The Finite Group Velocity of Quantum Spin Systems, Commun. Math. Phys. **28**, 251–257 (1972).
64. (with J.L. Lebowitz) Phase Transition in a Continuum Classical System with Finite Interactions, Phys. Lett. **39A**, 98–100 (1972).
65. (with J.L. Lebowitz) Lectures on the Thermodynamic Limit for Coulomb Systems, in *Statistical Mechanics and Mathematical Problems*, Battelle 1971 Recontres, Springer Lecture Notes in Physics **20**, 136–161 (1973).
66. (with J.L. Lebowitz) Lectures on the Thermodynamic Limit for Coulomb Systems, in *Lectures in Theoretical Physics XIV B*, (Boulder summer school), Colorado Associated University Press, 423–460 (1973).

† 67. Convex Trace Functions and the Wigner-Yanase-Dyson Conjecture, Adv. in Math. **11**, 267–288 (1973).

† 68. (with M.B. Ruskai) A Fundamental Property of Quantum Mechanical Entropy, Phys. Rev. Lett. **30**, 434–436 (1973).

† 69. (with M.B. Ruskai) Proof of the Strong Subadditivity of Quantum-Mechanical Entropy, J. Math. Phys. **14**, 1938–1941 (1973).

70. (with K. Hepp) On the Superradiant Phase Transition for Molecules in a Quantized Radiation Field: The Dicke Maser Model, Annals of Phys. (N.Y.) **76**, 360–404 (1973).
71. (with K. Hepp) Phase Transition in Reservoir Driven Open Systems with Applications to Lasers and Superconductors, Helv. Phys. Acta **46**, 573–602 (1973).
72. (with K. Hepp) The Equilibrium Statistical Mechanics of Matter Interacting with the Quantized Radiation Field, Phys. Rev. **A8**, 2517–2525 (1973).
73. (with K. Hepp) Constructive Macroscopic Quantum Electrodynamics, in *Constructive Quantum Field Theory*, Proceedings of the 1973 Erice Sum-

mer School, G. Velo and A. Wightman, eds., Springer Lecture Notes in Physics **25**, 298–316 (1973).

† 74. The Classical Limit of Quantum Spin Systems, Commun. Math. Phys. **31**, 327–340 (1973).

75. (with B. Simon) Thomas-Fermi Theory Revisited, Phys. Rev. Lett. **31**, 681–683 (1973).

† 76. (with M.B. Ruskai) Some Operator Inequalities of the Schwarz Type, Adv. in Math. **12**, 269–273 (1974).

77. Exactly Soluble Models in Statistical Mechanics, lecture given at the 1973 I.U.P.A.P. van der Waals Centennial Conference on Statistical Mechanics, Physica **73**, 226–236 (1974).

78. (with B. Simon) On Solutions to the Hartree-Fock Problem for Atoms and Molecules, J. Chem. Physics **61**, 735–736 (1974).

79. Thomas-Fermi and Hartree-Fock Theory, lecture at 1974 International Congress of Mathematicians, Vancouver. Proceedings, Vol. 2, 383–386 (1975).

† 80. Some Convexity and Subadditivity Properties of Entropy, Bull. Amer. Math. Soc. **81**, 1–13 (1975).

† 81. (with H.J. Brascamp and J.M. Luttinger) A General Rearrangement Inequality for Multiple Integrals, Jour. Funct. Anal. **17**, 227–237 (1975).

† 82. (with H.J. Brascamp) Some Inequalities for Gaussian Measures and the Long-Range Order of the One-Dimensional Plasma, lecture at Conference on Functional Integration, Cumberland Lodge, England. *Functional Integration and its Applications*, A.M. Arthurs ed., Clarendon Press, 1–14 (1975).

83. (with K. Hepp) The Laser: A Reversible Quantum Dynamical System with Irreversible Classical Macroscopic Motion, in *Dynamical Systems*, Battelle 1974 Rencontres, Springer Lecture Notes in Physics **38**, 178–208 (1975). Also appears in *Melting, Localization and Chaos*, Proc. 9th Midwest Solid State Theory Symposium, 1981, R. Kalia and P. Vashishta eds., North-Holland, 153–177 (1982).

84. (with P. Hertel and W. Thirring) Lower Bound to the Energy of Complex Atoms, J. Chem. Phys. **62**, 3355–3356 (1975).

85. (with W. Thirring) Bound for the Kinetic Energy of Fermions which Proves the Stability of Matter, Phys. Rev. Lett. **35**, 687–689 (1975). Errata **35**, 1116 (1975).

86. (with H.J. Brascamp and J.L. Lebowitz) The Statistical Mechanics of Anharmonic Lattices, in the proceedings of the 40th session of the International Statistics Institute, Warsaw, **9**, 1–11 (1975).

† 87. (with H.J. Brascamp) Best Constants in Young's Inequality, Its Converse and Its Generalization to More Than Three Functions, Adv. in Math. **20**, 151–172 (1976).

† 88. (with H.J. Brascamp) On Extensions of the Brunn-Minkowski and Prékopa-Leindler Theorems, Including Inequalities for Log Concave Functions and with an Application to the Diffusion Equation, J. Funct. Anal. **22**, 366–389 (1976).

89. (with J.F. Barnes and H.J. Brascamp) Lower Bounds for the Ground State Energy of the Schroedinger Equation Using the Sharp Form of Young's Inequality, in *Studies in Mathematical Physics*, Lieb, Simon, Wightman eds., Princeton Press, 83–90 (1976).

† 90. Inequalities for Some Operator and Matrix Functions, Adv. in Math. **20**, 174–178 (1976).

91. (with H. Narnhofer) The Thermodynamic Limit for Jellium, J. Stat. Phys. **12**, 291–310 (1975). Errata J. Stat. Phys. **14**, 465 (1976).

92. The Stability of Matter, Rev. Mod. Phys. **48**, 553–569 (1976).

93. Bounds on the Eigenvalues of the Laplace and Schroedinger Operators, Bull. Amer. Math. Soc. **82**, 751–753 (1976).

94. (with F.J. Dyson and B. Simon) Phase Transitions in the Quantum Heisenberg Model, Phys. Rev. Lett. **37**, 120–123 (1976). (See no. 104.)

† 95. (with W. Thirring) Inequalities for the Moments of the Eigenvalues of the Schrödinger Hamiltonian and Their Relation to Sobolev Inequalities, in *Studies in Mathematical Physics*, E. Lieb, B. Simon, A. Wightman eds., Princeton University Press, 269–303 (1976).

96. (with B. Simon and A. Wightman) Book *Studies in Mathematical Physics: Essays in Honor of Valentine Bargmann*, Princeton University Press (1976).

97. (with B. Simon) Thomas-Fermi Theory of Atoms, Molecules and Solids, Adv. in Math. **23**, 22–116 (1977).

98. (with O. Lanford and J. Lebowitz) Time Evolution of Infinite Anharmonic Oscillators, J. Stat. Phys. **16**, 453–461 (1977).

99. The Stability of Matter, Proceedings of the Conference on the Fiftieth Anniversary of the Schroedinger equation, Acta Physica Austriaca Suppl. XVII, 181–207 (1977).

† 100. Existence and Uniqueness of the Minimizing Solution of Choquard's Non-Linear Equation, Studies in Appl. Math. **57**, 93–105 (1977).

101. (with J. Fröhlich) Existence of Phase Transitions for Anisotropic Heisenberg Models, Phys. Rev. Lett. **38**, 440–442 (1977).

102. (with B. Simon) The Hartree-Fock Theory for Coulomb Systems, Commun. Math. Phys. **53**, 185–194 (1977).

103. (with W. Thirring) A Lower Bound for Level Spacings, Annals of Phys. (N.Y.) **103**, 88–96 (1977).

104. (with F. Dyson and B. Simon) Phase Transitions in Quantum Spin Systems with Isotropic and Non-Isotropic Interactions, J. Stat. Phys. *18*, 335–383 (1978).

105. Many Particle Coulomb Systems, lectures given at the 1976 session on statistical mechanics of the International Mathematics Summer Center (C.I.M.E.). In *Statistical Mechanics*, C.I.M.E. 1 Ciclo 1976, G. Gallavotti, ed., Liguore Editore, Naples, 101–166 (1978).

106. (with R. Benguria) Many-Body Atomic Potentials in Thomas-Fermi Theory, Annals of Phys. (N.Y.) **110**, 34–45 (1978).

107. (with R. Benguria) The Positivity of the Pressure in Thomas-Fermi Theory, Commun. Math. Phys. **63**, 193–218 (1978). Errata **71**, 94 (1980).

108. (with M. de Llano) Solitons and the Delta Function Fermion Gas in Hartree-Fock Theory, J. Math. Phys. **19**, 860–868 (1978).
109. (with J. Fröhlich) Phase Transitions in Anisotropic Lattice Spin Systems, Commun. Math. Phys. **60**, 233–267 (1978).
110. (with J. Fröhlich, R. Israel and B. Simon) Phase Transitions and Reflection Positivity. I. General Theory and Long Range Lattice Models, Commun. Math. Phys. **62**, 1–34 (1978). (See no. 124.)
† 111. (with M. Aizenman and E.B. Davies) Positive Linear Maps Which are Order Bounded on C* Subalgebras, Adv. in Math. **28**, 84–86 (1978).
† 112. (with M. Aizenman) On Semi-Classical Bounds for Eigenvalues of Schrödinger Operators, Phys. Lett. **66A**, 427–429 (1978).
113. New Proofs of Long Range Order, in *Proceedings of the International Conference on Mathematical Problems in Theoretical Physics* (June 1977), Springer Lecture Notes in Physics, **80**, 59–67 (1978).
† 114. Proof of an Entropy Conjecture of Wehrl, Commun. Math. Phys. **62**, 35–41 (1978).
115. (with B. Simon) Monotonicity of the Electronic Contribution to the Born-Oppenheimer Energy, J. Phys. B. **11**, L537–L542 (1978).
116. (with O. Heilmann) Lattice Models for Liquid Crystals, J. Stat. Phys. **20**, 679–693 (1979).
117. (with H. Brezis) Long Range Atomic Potentials in Thomas-Fermi Theory, Commun. Math. Phys. **65**, 231–246 (1979).
118. The $N^{5/3}$ Law for Bosons, Phys. Lett. **70A**, 71–73 (1979).
119. A Lower Bound for Coulomb Energies, Phys. Lett. **70A**, 444–446 (1979).
120. Why Matter is Stable, Kagaku **49**, 301–307 and 385–388 (1979). (In Japanese).
121. The Number of Bound States of One-Body Schrödinger Operators and the Weyl Problem, Symposium of the Research Inst. of Math. Sci., Kyoto University, (1979).
122. Some Open Problems About Coulomb Systems, in *Proceedings of the Lausanne 1979 Conference of the International Association of Mathematical Physics*, Springer Lecture Notes in Physics, **116**, 91–102 (1980).
† 123. The Number of Bound States of One-Body Schrödinger Operators and the Weyl Problem, *Proceedings of the Amer. Math. Soc. Symposia in Pure Math.*, **36**, 241–252 (1980).
124. (with J. Fröhlich, R. Israel and B. Simon) Phase Transitions and Reflection Positivity. II. Lattice Systems with Short-Range and Coulomb Interactions. J. Stat. Phys. **22**, 297–347 (1980). (See no. 110.)
125. Why Matter is Stable, Chinese Jour. Phys. **17**, 49–62 (1980). (English version of no. 120).
† 126. A Refinement of Simon's Correlation Inequality, Commun. Math. Phys. **77**, 127–135 (1980).
127. (with B. Simon) Pointwise Bounds on Eigenfunctions and Wave Packets in N-Body Quantum Systems. VI. Asymptotics in the Two-Cluster Region, Adv. in Appl. Math. **1**, 324–343 (1980).

128. The Uncertainty Principle, article in *Encyclopedia of Physics*, R. Lerner and G. Trigg eds., Addison Wesley, 1078–1079 (1981).

† 129. (with S. Oxford) An Improved Lower Bound on the Indirect Coulomb Energy, Int. J. Quant. Chem. **19**, 427–439 (1981).

130. (with R. Benguria and H. Brezis) The Thomas-Fermi-von Weizsäcker Theory of Atoms and Molecules, Commun. Math. Phys. **79**, 167–180 (1981).

131. (with M. Aizenman) The Third Law of Thermodynamics and the Degeneracy of the Ground State for Lattice Systems, J. Stat. Phys. *24*, 279–297 (1981).

132. (with J. Bricmont, J. Fontaine, J. Lebowitz and T. Spencer) Lattice Systems with a Continuous Symmetry III. Low Temperature Asymptotic Expansion for the Plane Rotator Model, Commun. Math. Phys. **78**, 545–566 (1981).

133. (with A. Sokal) A General Lee-Yang Theorem for One-Component and Multi-component Ferromagnets, Commun. Math. Phys. **80**, 153–179 (1981).

134. Variational Principle for Many-Fermion Systems, Phys. Rev. Lett. **46**, 457–459 (1981). Errata **47**, 69 (1981).

135. Thomas-Fermi and Related Theories of Atoms and Molecules, in *Rigorous Atomic and Molecular Physics*, G. Velo and A. Wightman, eds., Plenum Press 213–308 (1981).

136. Thomas-Fermi and Related Theories of Atoms and Molecules, Rev. Mod. Phys. *53*, 603–641 (1981). Errata **54**, 311 (1982). (Revised version of no. 135.)

137. Statistical Theories of Large Atoms and Molecules, in *Proceedings of the 1981 Oaxtepec conference on Recent Progress in Many-Body Theories*, Springer Lecture Notes in Physics, **142**, 336–343 (1982).

138. Statistical Theories of Large Atoms and Molecules, Comments Atomic and Mol. Phys. **11**, 147–155 (1982).

139. Analysis of the Thomas-Fermi-von Weizsäcker Equation for an Infinite Atom without Electron Repulsion, Commun. Math. Phys. **85**, 15–25 (1982).

140. (with D.A. Liberman) Numerical Calculation of the Thomas-Fermi-von Weizsäcker Function for an Infinite Atom without Electron Repulsion, Los Alamos National Laboratory Report, LA-9186-MS (1982).

141. Monotonicity of the Molecular Electronic Energy in the Nuclear Coordinates, J. Phys. B.: At. Mol. Phys. **15**, L63–L66 (1982).

142. Comment on "Approach to Equilibrium of a Boltzmann Equation Solution", Phys. Rev. Lett. **48**, 1057 (1982).

143. Density Functionals for Coulomb Systems, in *Physics as Natural Philosophy: Essays in honor of Laszlo Tisza on his 75th Birthday*, A. Shimony and H. Feshbach eds., M.I.T. Press, 111–149 (1982).

† 144. An L^p Bound for the Riesz and Bessel Potentials of Orthonormal Functions, J. Funct. Anal. **51**, 159–165 (1983).

† 145. (with H. Brezis) A Relation Between Pointwise Convergence of Functions and Convergence of Functionals, Proc. Amer. Math. Soc. **88**, 486–490 (1983).

146. (with R. Benguria) A Proof of the Stability of Highly Negative Ions in the Absence of the Pauli Principle, Phys. Rev. Lett. *50*, 1771–1774 (1983).

† 147. Sharp Constants in the Hardy-Littlewood-Sobolev and Related Inequalities, Annals of Math. **118**, 349–374 (1983).

† 148. Density Functionals for Coulomb Systems (a revised version of no. 143), Int. Jour. Quant. Chem. **24**, 243–277 (1983). An expanded version appears in *Density Functional Methods in Physics*, R. Dreizler and J. da Providencia eds., Plenum Nato ASI Series **123**, 31–80 (1985).

149. The Significance of the Schrödinger Equation for Atoms, Molecules and Stars, lecture given at the Schrödinger Symposium, Dublin Institute of Advanced Studies, October 1983, unpublished Proceedings.

150. (with I. Daubechies) One Electron Relativistic Molecules with Coulomb Interaction, Commun. Math. Phys. **90**, 497–510 (1983).

151. (with I. Daubechies) Relativistic Molecules with Coulomb Interaction, in *Differential Equations, Proc. of the Conference held at the University of Alabama in Birmingham, 1983*, I. Knowles and R. Lewis eds., Math. Studies Series, **92**, 143–148 North-Holland (1984).

152. Some Vector Field Equations, in *Differential Equations, Proc. of the Conference held at the University of Alabama in Birmingham, 1983*, I. Knowles and R. Lewis eds., Math. Studies Series **92**, 403–412 North-Holland (1984).

† 153. On the Lowest Eigenvalue of the Laplacian for the Intersection of Two Domains, Inventiones Math. **74**, 441–448 (1983).

154. (with J. Chayes and L. Chayes) The Inverse Problem in Classical Statistical Mechanics, Commun. Math. Phys. **93**, 57–121 (1984).

† 155. On Characteristic Exponents in Turbulence, Commun. Math. Phys. **92**, 473–480 (1984).

156. Atomic and Molecular Negative Ions, Phys. Rev. Lett. **52**, 315–317 (1984).

157. Bound on the Maximum Negative Ionization of Atoms and Molecules, Phys. Rev. **29A**, 3018–3028 (1984).

158. (with W. Thirring) Gravitational Collapse in Quantum Mechanics with Relativistic Kinetic Energy, Annals of Phys. (N.Y.) **155**, 494–512 (1984).

159. (with I.M. Sigal, B. Simon and W. Thirring) Asymptotic Neutrality of Large-Z Ions, Phys. Rev. Lett. **52**, 994–996 (1984). (See no. 185.)

160. (with R. Benguria) The Most Negative Ion in the Thomas-Fermi-von Weizsäcker Theory of Atoms and Molecules, J. Phys. B: At. Mol. Phys. **18**, 1045–1059 (1985).

† 161. (with H. Brezis) Minimum Action Solutions of Some Vector Field Equations, Commun. Math. Phys. **96**, 97–113 (1984).

† 162. (with H. Brezis) Sobolev Inequalities with Remainder Terms, J. Funct. Anal. **62**, 73–86 (1985).

† 163. Baryon Mass Inequalities in Quark Models, Phys. Rev. Lett. **54**, 1987–1990 (1985).

164. (with J. Fröhlich and M. Loss) Stability of Coulomb Systems with Magnetic Fields I. The One-Electron Atom, Commun. Math. Phys. *104*, 251–270 (1986).

165. (with M. Loss) Stability of Coulomb Systems with Magnetic Fields II. The Many-Electron Atom and the One-Electron Molecule, Commun. Math. Phys. **104**, 271–282 (1986).
166. (with W. Thirring) Universal Nature of van der Waals Forces for Coulomb Systems, Phys. Rev. A **34**, 40–46 (1986).
167. Some Ginzburg-Landau Type Vector-Field Equations, in *Nonlinear systems of Partial Differential Equations in Applied Mathematics*, B. Nicolaenko, D. Holm and J. Hyman eds., Amer. Math. Soc. Lectures in Appl. Math. **23**, Part 2, 105–107 (1986).
168. (with I. Affleck) A Proof of Part of Haldane's Conjecture on Spin Chains, Lett. Math. Phys. **12**, 57–69 (1986).
† 169. (with H. Brezis and J-M. Coron) Estimations d'Energie pour des Applications de $\mathbf{R}^3$ a Valeurs dans S^2, C.R. Acad. Sci. Paris **303** Ser. 1, 207–210 (1986).
170. (with H. Brezis and J-M. Coron) Harmonic Maps with Defects, Commun. Math. Phys. **107**, 649–705 (1986).
171. Some Fundamental Properties of the Ground States of Atoms and Molecules, in *Fundamental Aspects of Quantum Theory*, V. Gorini and A. Frigerio eds., Nato ASI Series B, Vol. 144, 209–214, Plenum Press (1986).
172. (with T. Kennedy) A Model for Crystallization: A Variation on the Hubbard Model, in *Statistical Mechanics and Field Theory: Mathematical Aspects*, Springer Lecture Notes in Physics **257**, 1–9 (1986).
173. (with T. Kennedy) An Itinerant Electron Model with Crystalline or Magnetic Long Range Order, Physica **138A**, 320–358 (1986).
174. (with T. Kennedy) A Model for Crystallization: A Variation on the Hubbard Model, Physica **140A**, 240–250 (1986) (Proceedings of IUPAP Statphys 16, Boston).
175. (with T. Kennedy) Proof of the Peierls Instability in One Dimension, Phys. Rev. Lett. **59**, 1309–1312 (1987).
176. (with I. Affleck, T. Kennedy and H. Tasaki) Rigorous Results on Valence-Bond Ground States in Antiferromagnets, Phys. Rev. Lett. *59*, 799–802 (1987).
177. (with H.-T. Yau) The Chandrasekhar Theory of Stellar Collapse as the Limit of Quantum Mechanics, Commun. Math. Phys. **112**, 147–174 (1987).
178. (with H.-T. Yau) A Rigorous Examination of the Chandrasekhar Theory of Stellar Collapse, Astrophys. Jour. **323**, 140–144 (1987).
† 179. (with F. Almgren) Singularities of Energy Minimizing Maps from the Ball to the Sphere, Bull. Amer. Math. Soc. **17**, 304–306 (1987). (See no. 190.)
180. Bounds on Schrödinger Operators and Generalized Sobolev Type Inequalities, Proceedings of the International Conference on Inequalities, University of Birmingham, England, 1987, Marcel Dekker Lecture Notes in Pure and Appl. Math., W.N. Everitt ed., volume 129, pages 123–133 (1991).
181. (with I. Affleck, T. Kennedy and H. Tasaki) Valence Bond Ground States in Isotropic Quantum Antiferromagnets, Commun. Math. Phys. **115**, 477–528 (1988).

182. (with T. Kennedy and H. Tasaki) A Two Dimensional Isotropic Quantum Antiferromagnet with Unique Disordered Ground State, J. Stat. Phys. **53**, 383–416 (1988).
183. (with T. Kennedy and S. Shastry) Existence of Néel Order in Some Spin 1/2 Heisenberg Antiferromagnets, J. Stat. Phys. **53**, 1019–1030 (1988).
184. (with T. Kennedy and S. Shastry) The XY Model has Long-Range Order for all Spins and all Dimensions Greater than One, Phys. Rev. Lett. **61**, 2582–2584 (1988).
185. (with I.M. Sigal, B. Simon and W. Thirring) Approximate Neutrality of Large-Z Ions, Commun. Math. Phys. **116**, 635–644 (1988). (See no. 159.)
186. (with H.-T. Yau) The Stability and Instability of Relativistic Matter, Commun. Math. Phys. **118**, 177–213 (1988).
187. (with H.-T. Yau) Many-Body Stability Implies a Bound on the Fine Structure Constant, Phys. Rev. Lett. **61**, 1695–1697 (1988).
188. (with J. Conlon and H.-T. Yau) The $N^{7/5}$ Law for Charged Bosons, Commun. Math. Phys. **116**, 417–448 (1988).
† 189. (with F. Almgren and W. Browder) Co-area, Liquid Crystals, and Minimal Surfaces, in *Partial Differential Equations*, S.S. Chern ed., Springer Lecture Notes in Math. **1306**, 1–22 (1988).
190. (with F. Almgren) Singularities of Energy Minimizing Maps from the Ball to the Sphere: Examples, Counterexamples and Bounds, Ann. of Math. **128**, 483–530 (1988).
† 191. (with F. Almgren) Counting Singularities in Liquid Crystals, in *IXth International Congress on Mathematical Physics*, B. Simon, A. Truman, I.M. Davies eds., Hilger, 396–409 (1989). This also appears in: *Symposia Mathematica, vol. XXX,* Ist. Naz. Alta Matem. Francesco Severi Roma, 103–118, Academic Press (1989); *Variational Methods,* H. Berestycki, J-M. Coron, I. Ekeland eds., Birkhäuser, 17–36 (1990); How many singularities can there be in an energy minimizing map from the ball to the sphere?, in *Ideas and Methods in Mathematical Analysis, Stochastics, and Applications,* S. Albeverio, J.E. Fenstad, H. Holden, T. Lindstrom eds., Cambridge Univ. Press, vol. 1, 394–408 (1992).
† 192. (with F. Almgren) Symmetric Decreasing Rearrangement can be Discontinuous, Bull. Amer. Math. Soc. **20**, 177–180 (1989).
† 193. (with F. Almgren) Symmetric Decreasing Rearrangement is Sometimes Continuous, Jour. Amer. Math. Soc. **2**, 683–773 (1989). A summary of this work (using 'rectifiable currents') appears as The (Non)continuity of Symmetric Decreasing Rearrangement in *Symposia Mathematica, vol. XXX,* Ist. Naz. Alta Matem. Francesco Severi Roma, 89–102, Academic Press (1989) and in *Variational Methods,* H. Berestycki, J-M. Coron, I. Ekeland eds., Birkhäuser, 3–16 (1990).
† 194. Two Theorems on the Hubbard Model, Phys. Rev. Lett. **62**, 1201–1204 (1989). Errata **62**, 1927 (1989).
195. (with J. Conlon and H.-T. Yau) The Coulomb gas at Low Temperature and Low Density, Commun. Math. Phys. **125**, 153–180 (1989).

† 196. Gaussian Kernels have only Gaussian Maximizers, Invent. Math. **102**, 179–208 (1990).
† 197. Kinetic Energy Bounds and their Application to the Stability of Matter, in *Schrödinger Operators*, Proceedings Sønderborg Denmark 1988, H. Holden and A. Jensen eds., Springer Lecture Notes in Physics **345**, 371–382 (1989). Expanded version of no. 180.
198. The Stability of Matter: From Atoms to Stars, 1989 Gibbs Lecture, Bull. Amer. Math. Soc. **22**, 1–49 (1990).
† 199. Integral Bounds for Radar Ambiguity Functions and Wigner Distributions, J. Math. Phys. **31**, 594–599 (1990).
200. On the Spectral Radius of the Product of Matrix Exponentials, Linear Alg. and Appl.**141**, 271–273 (1990).
† 201. (with M. Aizenman) Magnetic Properties of Some Itinerant-Electron Systems at $T > 0$, Phys. Rev. Lett. **65**, 1470–1473 (1990).
202. (with H. Siedentop) Convexity and Concavity of Eigenvalue Sums, J. Stat. Phys. **63**, 811–816 (1991).
† 203. (with J.P. Solovej) Quantum Coherent Operators: A Generalization of Coherent States, Lett. Math. Phys. **22**, 145–154 (1991).
204. The Flux-Phase Problem on Planar Lattices, Helv. Phys. Acta **65**, 247–255 (1992). Proceedings of the conference "Physics in Two Dimensions", Neuchatel, August 1991.
205. Atome in starken Magnetfeldern, Physikalische Blätter *48*, 549–552 (1992). Translation by H. Siedentop of the Max-Planck medal lecture (1 April 1992) "Atoms in strong magnetic fields".
206. Absence of Ferromagnetism for One-Dimensional Itinerant Electrons, in *Probabilistic Methods in Mathematical Physics*, Proceedings of the International Workshop Siena, May 1991, F. Guerra, M. Loffredo and C. Marchioro eds., World Scientific pp. 290–294 (1992). A shorter version appears in *Rigorous Results in Quantum Dynamics*, J. Dittrich and P. Exner eds., World Scientific, pp. 243–245 (1991).
207. (with J.P. Solovej and J. Yngvason) Heavy Atoms in the Magnetic Field of a Neutron Star, Phys. Rev. Lett. **69**, 749–752 (1992).
208. (with J.P. Solovej) Atoms in the Magnetic Field of a Neutron Star, in *Differential Equations with Applications to Mathematical Physics*, W.F. Ames, J.V. Herod and E.M. Harrell II eds., Academic Press, pages 221–237 (1993). Also in *Spectral Theory and Scattering Theory and Applications*, K. Yajima, ed., Advanced Studies in Pure Math. **23**, 259–274, Math. Soc. of Japan, Kinokuniya (1994). This is a summary of nos. 215, 216. Earlier summaries also appear in: (a) *Méthodes Semi-Classiques, Colloque internatinal (Nantes 1991)*, Asterisque *210*, 237–246 (1991); (b) *Some New Trends on Fluid Dynamics and Theoretical Physics*, C.C. Lin and N. Hu eds., 149–157, Peking University Press (1993); (c) *Proceedings of the International Symposium on Advanced Topics of Quantum Physics, Shanxi*, J.Q. Lang, M.L. Wang, S.N. Qiao and D.C. Su eds., 5–13, Science Press, Beijing (1993).

209. (with M. Loss and R. McCann) Uniform Density Theorem for the Hubbard Model, J. Math. Phys. **34**, 891–898 (1993).

210. Remarks on the Skyrme Model, in *Proceedings of the Amer. Math. Soc. Symposia in Pure Math. 54*, part 2, 379–384 (1993). (Proceedings of Summer Research Institute on Differential Geometry at UCLA, July 8–28, 1990.)

† 211. (with E. Carlen) Optimal Hypercontractivity for Fermi Fields and Related Noncommutative Integration Inequalities, Commun. Math. Phys. **155**, 27–46 (1993).

212. (with E. Carlen) Optimal Two-Uniform Convexity and Fermion Hypercontractivity, in *Quantum and Non-Commutative Analysis*, Proceedings of June, 1992 Kyoto Conference, H. Araki et.al. eds., Kluwer (1993), pp. 93–111. (Condensed version of no. 211.)

213. (with M. Loss) Fluxes, Laplacians and Kasteleyn's Theorem, Duke Math. Journal **71**, 337–363 (1993).

214. (with V. Bach, R. Lewis and H. Siedentop) On the Number of Bound States of a Bosonic N-Particle Coulomb System, Zeits. f. Math. **214**, 441–460 (1993).

215. (with J.P. Solovej and J. Yngvason) Asymptotics of Heavy Atoms in High Magnetic Fields: I. Lowest Landau Band Region, Commun. Pure Appl. Math. **47**, 513–591 (1994).

216. (with J.P. Solovej and J. Yngvason) Asymptotics of Heavy Atoms in High Magnetic Fields: II. Semiclassical Regions, Commun. Math. Phys. **161**, 77–124 (1994).

217. (with V. Bach, M. Loss and J.P. Solovej) There are No Unfilled Shells in Unrestricted Hartree-Fock Theory, Phys. Rev. Lett. **72**, 2981–2983 (1994).

† 218. (with K. Ball and E. Carlen) Sharp Uniform Convexity and Smoothness Inequalities for Trace Norms, Invent. Math. **115**, 463–482 (1994).

† 219. Coherent States as a Tool for Obtaining Rigorous Bounds, *Proceedings of the Symposium on Coherent States, past, present and future*, Oak Ridge, D.H. Feng, J. Klauder and M.R. Strayer eds., World Scientific (1994), pages 267–278.

220. The Hubbard model – Some Rigorous Results and Open Problems, in Proceedings of 1993 conference in honor of G.F. Dell'Antonio, *Advances in Dynamical Systems and Quantum Physics*, S. Albeverio et al. eds., pp. 173–193, World Scientific (1995). A revised version appears in Proceedings of 1993 NATO ASI *The Hubbard Model*, D. Baeriswyl et al. eds., pp. 1–19, Plenum Press (1995). A further revision appears in *Proceedings of the XIth International Congress of Mathematical Physics*, Paris, 1994, D. Iagolnitzer ed., pp. 392–412, International Press (1995).

221. (with V. Bach and J.P. Solovej) Generalized Hartree-Fock Theory of the Hubbard Model, J. Stat. Phys. **76**, 3–90 (1994).

222. The Flux Phase of the Half-Filled Band, Phys. Rev. Lett. **73**, 2158–2161 (1994).

† 223. (with M. Loss) Symmetry of the Ginzburg-Landau Minimizer in a Disc, Math. Res. Lett. **1**, 701–715 (1994).

224. (with J.P. Solovej and J. Yngvason) Quantum Dots, in *Proceedings of the Conference on Partial Differential Equations and Mathematical Physics*, University of Alabama, Birmingham, 1994, I. Knowles, ed., International Press (1995), pages 157–172.
225. (with J.P. Solovej and J. Yngvason) Ground States of Large Quantum Dots in Magnetic Fields, Phys. Rev. B **51**, 10646–10665 (1995).
226. (with J. Freericks) The Ground State of a General Electron-Phonon Hamiltonian is a Spin Singlet, Phys. Rev. B **51**, 2812–2821 (1995).
227. (with B. Nachtergaele) The Stability of the Peierls Instability for Ring Shaped Molecules, Phys. Rev. B **51**, 4777–4791 (1995).
228. (with B. Nachtergaele) Dimerization in Ring-Shaped Molecules: The Stability of the Peierls Instability in *Proceedings of the XIth International Congress of Mathematical Physics*, Paris, 1994, D. Iagolnitzer ed., pp. 423–431, International Press (1995).
229. (with B. Nachtergaele) Bond Alternation in Ring-Shaped Molecules: The Stability of the Peierls Instability. In Proceedings of the conference *The Chemical Bond*, Copenhagen 1994, Int. J. Quant. Chem. **58**, 699–706 (1996).
230. Fluxes and Dimers in the Hubbard Model, in Proceedings of the International Congress of Mathematicians, Zürich, 1994, S.D. Chatterji ed., vol. 2, pp. 1279–1280, Birkhäuser (1995).
231. (with M. Loss and J. P. Solovej) Stability of Matter in Magnetic Fields, Phys. Rev. Lett. **75**, 985–989 (1995).
232. (with O.J. Heilmann) Electron Density near the Nucleus of a large Atom, Phys. Rev A **52**, 3628–3643 (1995).
233. (with A. Iantchenko and H. Siedentop) Proof of a Conjecture about Atomic and Molecular Cores Related to Scott's Correction, J. reine u. ang. Math. **472**, 177–195 (1996).
234. (with L. Thomas) Exact Ground State Energy of the Strong-Coupling Polaron, Commun. Math. Phys. **183**, 511–519 (1997). Errata **188**, 499–500 (1997).
† 235. (with L. Cafarelli and D. Jerison) On the Case of Equality in the Brunn-Minkowski Inequality for Capacity, Adv. in Math. **117**, 193–207 (1996).
236. (with M. Loss and H. Siedentop) Stability of Relativistic Matter via Thomas-Fermi Theory, Helv. Phys. Acta **69**, 974–984 (1996).
237. Some of the Early History of Exactly Soluble Models, in *Proceedings of the 1996 Northeastern University conference on Exactly Soluble Models,* Int. Jour. Mod. Phys. B **11**, 3–10 (1997).
238. (with H. Siedentop and J.P. Solovej) Stability and Instability of Relativistic Electrons in Magnetic Fields, J. Stat. Phys. **89**, 37–59 (1997).
239. (with H. Siedentop and J-P. Solovej) Stability of Relativistic Matter with Magnetic Fields, Phys. Rev. Lett. **79**, 1785–1788 (1997).
240. Stability of Matter in Magnetic Fields, in *Proceedings of the Conference on Unconventional Quantum Liquids, Evora, Portugal, 1996* Zeits. f. Phys. B **933**, 271–274 (1997).

241. Birmingham in the Good Old Days, in *Proceedings of the Conference on Unconventional Quantum Liquids, Evora, Portugal, 1996* Zeits. f. Phys. B **933**, 125–126 (1997).
242. (with M. Loss) book *Analysis*, American Mathematical Society (1997).
243. Doing Math with Fred, in *In Memoriam Frederick J. Almgren Jr., 1937–1997*, Experimental Math. **6**, 2–3 (1997).
244. (with J.P. Solovej and J. Yngvason) Asymptotics of Natural and Artificial Atoms in Strong Magnetic Fields, in *The Stability of Matter: From Atoms to Stars, Selecta of E. H. Lieb*, W. Thirring ed., second edition, Springer Verlag, pp. 145–167 (1997). This is a summary of nos. 207, 208, 215, 216, 224, 225.
245. Stability and Instability of Relativistic Electrons in Classical Electromagnetic Fields, in Proceedings of Conference on Partial Differential Eqations and Mathematical Physics, Georgia Inst. of Tech., March, 1997, Amer. Math. Soc. Contemporary Math. series, E. Carlen, E. Harrell, M. Loss eds., **217**, 99–108 (1998).
246. (with J. Yngvason) Ground State Energy of the Low Density Bose Gas, Phys. Rev. Lett. **80**, 2504–2507 (1998). arXiv math-ph/9712138, mp_arc 97-631.
247. (with J. Yngvason) A guide to Entropy and the Second Law of Thermodynamics, Notices of the Amer. Math. Soc. **45**, 571–581 (1998). arXiv math-ph/9805005, mp_arc 98-339. http://www.ams.org/notices/199805/lieb.pdf. See no. 266. This paper received the American Mathematical Society 2002 Levi Conant prize for "the best expository paper published in either the Notices of the AMS or the Bulletin of the AMS in the preceding five years".
† 248. (with D. Hundertmark and L.E. Thomas) A Sharp Bound for an Eigenvalue Moment of the One-Dimensional Schroedinger Operator, Adv. Theor. Math. Phys. **2**, 719–731 (1998). arXiv math-ph/9806012, mp_arc 98-753.
† 249. (with E. Carlen) A Minkowski Type Trace Inequality and Strong Subadditivity of Quantum Entropy, in Amer. Math. Soc. Transl. (2), **189**, 59–69 (1999).
250. (with J. Yngvason) The Physics and Mathematics of the Second Law of Thermodynamics, Physics Reports **310**, 1–96 (1999). arXiv cond-mat/9708200, mp_arc 97-457.
251. Some Problems in Statistical Mechanics that I would like to see Solved, 1998 IUPAP Boltzmann prize lecture, Physica A **263**, 491–499 (1999).
252. (with P. Schupp) Ground State Properties of a Fully Frustrated Quantum Spin System, Phys. Rev. Lett. **83**, 5362–5365 (1999). arXiv math-ph/9908019, mp_arc 99-304.
253. (with P. Schupp) Singlets and Reflection Symmetric Spin Systems, Physica A **279**, 378–385 (2000). arXiv math-ph/9910037, mp_arc 99-404.
254. (with R. Seiringer and J.Yngvason) Bosons in a Trap: A Rigorous Derivation of the Gross-Pitaevskii Energy Functional, Phys. Rev A **61**, 043602-1 – 043602-13 (2000). arXiv math-ph/9908027, mp_arc 99-312.
255. (with J. Yngvason) The Ground State Energy of a Dilute Bose Gas, in *Differential Equations and Mathematical Physics, University of Alabama,*

Birmingham, 1999, R. Weikard and G. Weinstein, eds., 295–306, Internat. Press (2000). arXiv math-ph/9910033, mp_arc 99-401.

256. (with M. Loss) Self-Energy of Electrons in Non-perturbative QED, in *Differential Equations and Mathematical Physics, University of Alabama, Birmingham, 1999*, R. Weikard and G. Weinstein, eds. 279–293, Amer. Math. Soc./Internat. Press (2000). arXiv math-ph/9908020, mp_arc 99-305.
257. (with R. Seiringer and J. Yngvason) The Ground State Energy and Density of Interacting Bosons in a Trap, in *Quantum Theory and Symmetries, Goslar, 1999*, H.-D. Doebner, V.K. Dobrev, J.-D. Hennig and W. Luecke, eds., pp. 101–110, World Scientific (2000).
arXiv math-ph/9911026, mp_arc 99-439.
258. (with J. Yngvason) The Ground State Energy of a Dilute Two-dimensional Bose Gas, J. Stat. Phys. **103**, 509–526 (2001). arXiv math-ph/0002014, mp_arc 00-63.
259. (with J. Yngvason) A Fresh Look at Entropy and the Second Law of Thermodynamics, Physics Today **53**, 32–37 (April 2000). arXiv math-ph/0003028, mp_arc 00-123. See also **53**, 11–14, 106 (October 2000).
260. Lieb-Thirring Inequalities, in *Encyclopaedia of Mathematics, Supplement vol. 2,* pp. 311–313, Kluwer (2000). arXiv math-ph/0003039, mp_arc 00-132.
261. Thomas-Fermi Theory, in *Encyclopaedia of Mathematics, Supplement vol. 2,* pp. 455–457, Kluwer (2000). arXiv math-ph/0003040, mp_arc 00-131.
262. (with H. Siedentop) Renormalization of the Regularized Relativistic Electron-Positron Field, Commun. Math. Phys. **213**, 673–684 (2000).
arXiv math-ph/0003001 mp_arc 00-98.
263. (with R. Seiringer and J. Yngvason) A Rigorous Derivation of the Gross-Pitaevskii Energy Functional for a Two-dimensional Bose Gas, Commun. Math. Phys. **224**, 17–31 (2001). arXiv cond-mat/0005026, mp_arc 00-203.
264. (with M. Griesemer and M. Loss) Ground States in Non-relativistic Quantum Electrodynamics, Invent. Math. **145**, 557–595 (2001). arXiv math-ph/0007014, mp_arc 00-313.
265. (with J.P. Solovej) Ground State Energy of the One-Component Charged Bose Gas, Commun. Math. Phys. **217**, 127–163 (2001). Errata **225**, 219–221 (2002). arXiv cond-mat/0007425, mp_arc 00-303.
266. (with J. Yngvason) The Mathematics of the Second Law of Thermodynamics, in *Visions in Mathematics, Towards 2000*, A. Alon, J. Bourgain, A. Connes, M. Gromov and V. Milman, eds., GAFA 2000, no. 1, Birkhäuser, p. 334–358 (2000). See no. 247. mp_arc 00-332.
267. The Bose Gas: A Subtle Many-Body Problem, in *Proceedings of the XIII International Congress on Mathematical Physics, London*, A. Fokas, et al. eds. International Press, pp. 91–111, 2001. arXiv math-ph/0009009, mp_arc 00-351.
268. (with J. Freericks and D. Ueltschi) Segregation in the Falicov-Kimball Model, Commun. Math. Phys. (in press). arXiv math-ph/0107003, mp_arc 01-243.

269. (with G.K. Pedersen) Convex Multivariable Trace Functions, Reviews in Math. Phys. (in press). arXiv math.OA/0107062.
270. (with J. Freericks and D. Ueltschi) Phase Separation due to Quantum Mechanical Correlations, Phys. Rev. Lett. **88**, #106401 (2002). arXiv cond-mat/0110251.
271. (with M. Loss) Stability of a Model of Relativistic Quantum Electrodynamics, Commun. Math. Phys. (in press). arXiv math-ph/0109002, mp_arc 01-315.
272. (with M. Loss) A Bound on Binding Energies and Mass Renormalization in Models of Quantum Electrodynamics, J. Stat. Phys. (in press). arXiv math-ph/0110027.
273. (with R. Seiringer) Proof of Bose-Einstein Condensation for Dilute Trapped Gases. Phys. Rev. Lett. **88**, #170409 (2002). arXiv math-ph/0112032, mp_arc 02-115.
274. (with M. Loss) Stability of Matter in Relativistic Quantum Mechanics, in *Proceedings of QMath8, Taxco*, Amer. Math. Soc. Contemporary Math series (in press).
275. (with J. Yngvason) The Mathematical Structure of the Second Law of Thermodynamics, in *Contemporary Developments in Mathematics 2001*, International Press (in press).
276. (with R. Seiringer, J.P. Solovej and J. Yngvason) The Ground State of the Bose Gas, in *Contemporary Developments in Mathematics 2001*, International Press (in press).

Druck: Strauss Offsetdruck, Mörlenbach
Verarbeitung: Schäffer, Grünstadt